Screen Menus

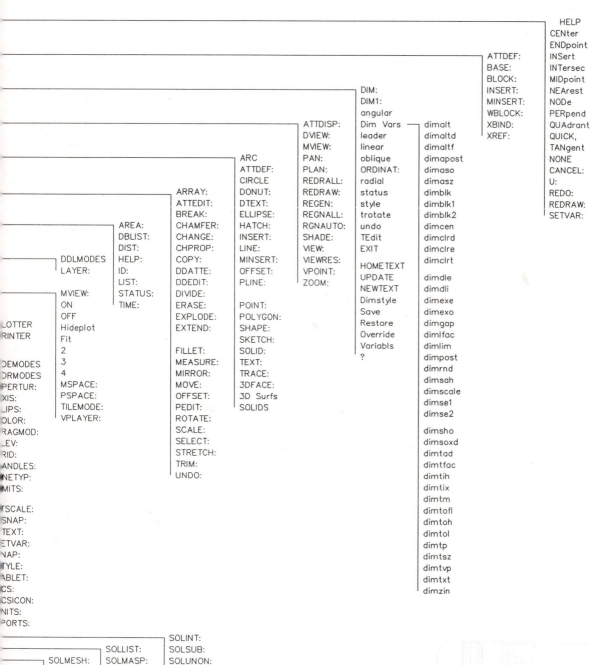

											HELP

HELP
CENter
ENDpoint
INSert
INTersec
MIDpoint
NEArest
NODe
PERpend
QUAdrant
QUICK,
TANgent
NONE
CANCEL:
U:
REDO:
REDRAW:
SETVAR:

ATTDEF:
BASE:
BLOCK:
INSERT:
MINSERT:
WBLOCK:
XBIND:
XREF:

DIM:
DIM1:
angular
Dim Vars
leader
linear
oblique
ORDINAT:
radial
status
style
trotate
undo
TEdit
EXIT

HOMETEXT
UPDATE
NEWTEXT
Dimstyle
Save
Restore
Override
Variabls
?

dimalt
dimaltd
dimaltf
dimapost
dimaso
dimasz
dimblk
dimblk1
dimblk2
dimcen
dimclrd
dimclre
dimclrt

dimdle
dimdli
dimexe
dimexo
dimgap
dimlfac
dimlim
dimpost
dimrnd
dimsah
dimscale
dimse1
dimse2

dimsho
dimsoxd
dimtad
dimtfac
dimtih
dimtix
dimtm
dimtofl
dimtoh
dimtol
dimtp
dimtsz
dimtvp
dimtxt
dimzin

ATTDISP:
DVIEW:
MVIEW:
PAN:
PLAN:
REDRALL:
REDRAW:
REGEN:
REGNALL:
RGNAUTO:
SHADE:
VIEW:
VIEWRES:
VPOINT:
ZOOM:

ARC
ATTDEF:
CIRCLE
DONUT:
DTEXT:
ELLIPSE:
HATCH:
INSERT:
LINE:
MINSERT:
OFFSET:
PLINE:

POINT:
POLYGON:
SHAPE:
SKETCH:
SOLID:
TEXT:
TRACE:
3DFACE:
3D Surfs
SOLIDS

ARRAY:
ATTEDIT:
BREAK:
CHAMFER:
CHANGE:
CHPROP:
COPY:
DDATTE:
DDEDIT:
DIVIDE:
ERASE:
EXPLODE:
EXTEND:

FILLET:
MEASURE:
MIRROR:
MOVE:
OFFSET:
PEDIT:
ROTATE:
SCALE:
SELECT:
STRETCH:
TRIM:
UNDO:

AREA:
DBLIST:
DIST:
HELP:
ID:
LIST:
STATUS:
TIME:

DDLMODES
LAYER:

MVIEW:
ON
OFF
Hideplot
Fit
2
3
4
MSPACE:
PSPACE:
TILEMODE:
VPLAYER:

LOTTER
RINTER

DEMODES
RMODES
PERTUR:
XIS:
LIPS:
OLOR:
RAGMOD:
EV:
RID:
ANDLES:
NETYP:
MITS:

TSCALE:
SNAP:
TEXT:
ETVAR:
NAP:
TYLE:
ABLET:
CS:
CSICON:
NITS:
PORTS:

SOLINT:
SOLSUB:
SOLUNON:
SOLCHAM:
SOLFILL:
SOLCHP:
SOLMOVE:
SOLSEP:

SOLLIST:
SOLMASP:
SOLAREA:

SOLMESH:
SOLWIRE:
SOLFEAT:
SOLPROF:
SOLSECT:

MAT:
VAR:
UCS:
IN:
OUT:
PURG:
OAD

Third Edition

GRAPHICS FOR ENGINEERS

AutoCAD® RELEASE 11

Third Edition

GRAPHICS FOR ENGINEERS

AutoCAD® Release 11

JAMES H. EARLE

Texas A&M University

ADDISON-WESLEY PUBLISHING COMPANY

Reading, Massachusetts • Menlo Park, California • New York
Don Mills, Ontario • Wokingham, England • Amsterdam • Bonn • Sydney
Singapore • Tokyo • Madrid • San Juan • Milan • Paris

Faith Sherlock: Sponsoring Editor
Katherine Harutunian: Associate Editor
Karen Myer: Production Supervisor
Patsy DuMoulin: Production Administrator
Joe Vetere: Technical Art Consultant
Lorraine Hodsdon: Layout Artist
Roy Logan: Manufacturing Supervisor
James H. Earle: Illustrator
Sharon Elwell/Smizer Design: Cover Designer

Many of the designations used by manufacturers and sellers to distinguish their products are claimed as trademarks. Where those designations appear in this book, and Addison-Wesley was aware of a trademark claim, the designations have been printed in initial caps or all caps.

The programs and applications presented in this book have been included for their instructional value. They have been tested with care, but are not guaranteed for any particular purpose. The publisher does not offer any warranties or representations, nor does it accept any liabilities with respect to the programs or applications.

Library of Congress Cataloging-in-Publication Data

Earle, James H.
    Graphics for engineers : AutoCAD release 11 / James H. Earle. —
3rd ed.
        p.   cm.
    Includes index.
    ISBN 0-201-56999-X
    1. Engineering graphics.   2. Mechanical drawing.   3. AutoCAD
(Computer program)   I. Title
T353.E3   1992
604.2'4—dc20   91-22890
                                                        CIP

Reprinted with corrections April, 1992

3 4 5 6 7 8 9 10–HA–95949392

Dedicated to Donald L. Earle
An outstanding welterweight

Preface

The third edition of *Graphics for Engineers* is suitable for a one-semester course, or more. The major additions are chapters on AutoCAD's Release 11 and solid modeling using AutoCAD's AME extension.

Objectives

The objective of this book is to support a course in which the student learns:

- ANSI standards and techniques of preparing engineering drawings.
- How to solve three-dimensional problems by descriptive geometry.
- How to use drafting instruments to prepare drawings.
- How to use computer graphics to prepare engineering drawings.
- How to use graphics as a medium of innovative design.

Above all, this textbook is designed to **help students expand their creative talents and communicate their ideas in an effective manner.**

Format

Graphics for Engineers provides self-instruction examples that enable the student to work independently.

Many examples are presented in a step-by-step sequence to illustrate how problems are solved. A second color is used to emphasize sequential points of problem solution. The problems at the ends of chapters are graduated from simple to difficult to offer a range of assignments.

Computer Graphics

Chapters 3, 23, and 24 are devoted to computer graphics. Chapter 3 gives a general overview of hardware and software. Chapter 23 covers the use of AutoCAD Release 11 for the 286 computer, and Chapter 24 gives an introduction to three-dimensional modeling for the 386 computer using the AME extension with AutoCAD.

AutoCAD applications have been integrated throughout the other chapters to give a dual approach (by pencil and computer) to the solution of most types of problems.

AutoCAD software was selected as the software to feature because it is the most widely sold computer graphics software for the microcomputer. Therefore, students are most likely to encounter AutoCAD in industry.

All computer graphics principles have been presented in illustrations that use a two-step, three-step, or four-step format. Within each step are prompts from AutoCAD as the user would see them on the screen and the responses the user should type into the system. The prompts and responses are in a different typeface to distinguish them from the main text. With this format, the student will be able to progress on his own.

Institutions not equipped to cover computer graphics, or AutoCAD, will find the coverage in this book will provide a general introduction to computer graphics. *Graphics for Engineers* has been designed for use with or without computer graphics: **Learning graphics is the major theme.** Therefore, the student will benefit from reading about computer graphics, even if he does not use computer graphics techniques.

Depth of Coverage

Some material in this book may not be covered in the course for which it is used due to the lack of time available or the type of emphasis placed on the course by the teacher. In some cases, more coverage of certain topics may be desired, but the content of this book is sufficient for a solid, contemporary course in engineering graphics.

The student can easily understand that drawings are used to communicate ideas from which to build projects. Perhaps more importantly, engineering drawings are legal contracts that are the basis for the expenditure of millions of dollars. Enormous saving can be realized from precise, clearly prepared drawings. Conversely, poorly done drawings can result in the waste of time and money with unaffordable results. Also, the competency of the engineer and the evaluation of his professional performance on a project will be supported or refuted by the working drawings that he approved.

Since the course in which this book is used may be the only formal course in the preparation of drawings as legal documents that a student will encounter in his or her career, this textbook should be retained for future reference. Those topics and specialty areas that did not receive full coverage in class could be one that the engineer may need to review in practice.

An Aid to Learning

This third edition of *Graphics for Engineers* has been revised to keep abreast of the needs of education and industry. Above all, the book has been made **as teachable as possible.** That is, the examples, illus-trations, applications, format, texts, and problems have been modified or newly done to assist the teacher in transmitting these principles to the student.

Revision Features

Some of the major revisions features of this third edition of *Graphic for Engineers* are:

- The inclusion of more than one hundred new illustrations.
- The revision of many illustrations.
- The inclusion of new working drawing problems in Chapter 17.
- The revision of Chapter 23 to cover AutoCAD Release 11.
- The inclusion of Chapter 24 to give an introduction to the AME AutoCAD Extension for solid modeling.

A Teaching System

This textbook, used in combination with the supplements list below, comprises a **complete teaching system** for achieving the maximum in efficiency and effectiveness:

Textbook problems

This text contains over 500 problems that offer a range of assignments to aid the student in grasping the principles covered in each chapter.

Textbook problem solutions

Solutions to textbook problems are being developed as teaching aids for the instructor.

Laboratory problems

Eighteen problem books and teachers' guides (with outlines, problem solutions, tests, and test solutions) are available for use with this textbook, and other titles will be introduced in the future. Fourteen of the manuals have computer graphics on the backs of the problem sheets, which allows solution of problems by both computer and pencil. A listing of these books and their source is given in the rear endpapers.

Visual aids

Sixteen modules of *SoftVisuals* are available on disks from which multicolored overhead transparencies can be plotted on transparency film for classroom presentations. Transparency selections can be made from over 500 *SoftVisuals* keyed to this textbook that can be plotted with AutoCAD 2.52 or later versions.

Acknowledgments

We are grateful for the assistance of many who have influenced the development of this volume. Many industries have furnished photographs, drawings, and applications that have been acknowledged in the corresponding legends. The Engineering Design Graphics staff of Texas A&M University have been helpful in making suggestions for the revision of this book. Professor Tom Pollock provided valuable information on various metals for Chapter 20. Professor Leendert Kersten of the University of Nebraska, Lincoln, kindly provided his descriptive geometry computer programs for inclusion, and his cooperation is appreciated.

We are indebted to Mary Ann Zadfar, Josef Woodman, and Joseph Oakey of Autodesk, Inc., for their assistance with AutoCAD. We appreciate the assistance and cooperation of Karen Kershaw of MEGACAD, Inc. After our association with these individuals and their companies, it is understandable why they are leaders in their respective fields.

We are appreciative of the many institutions that have thought enough of our publications to adopt them for classroom use. It is an honor for one's work to be accepted by his colleagues. We are hopeful that this textbook will fill the needs of engineering and technology programs. As always, comments and suggestions for improvement and revision of this book will be appreciated.

College Station, Texas   Jim Earle

Brief Contents

Introduction to Engineering and Technology 1 **1**

The Design Process 15 **2**

The Computer in Design and Drafting 41 **3**

Lettering 48 **4**

Drawing Instruments 57 **5**

Geometric Construction 77 **6**

Multiview Sketching 106 **7**

Multiview Drawing with Instruments 122 **8**

Auxiliary Views 152 **9**

Sections 170 **10**

Screws, Fasteners, and Springs 189 **11**

Gears and Gams 220 **12**

Materials and Processes 236 **13**

Dimensioning 249 **14**

15 Tolerances 277

16 Welding 312

17 Working Drawings 322

18 Reproduction Methods and Drawing Shortcuts 363

19 Pictorials 370

20 Descriptive Geometry 404

21 Civil Engineering Applications 455

22 Graphs 472

23 AutoCAD Computer Graphics 491

24 Introduction to Solid Modeling 593

Appendixes A-1

Index I-1

Contents

1 Introduction to Engineering and Technology 1

1.1 Introduction 1
1.2 Engineering graphics 1
1.3 The technological team 2
1.4 Engineering fields 5
1.5 Technologists and technicians 12
1.6 Drafters 12
Problems 13

2 The Design Process 15

2.1 Introduction 15
2.2 Hunting seat—problem identification 18
2.3 Hunting seat—preliminary ideas 20
2.4 Hunting seat—refinement 21
2.5 Hunting seat—analysis 22
2.6 Hunting seat—decision 25
2.7 Hunting seat—implementation 27
2.8 Introduction to design problems 29

3 The Computer in Design and Drafting 41

3.1 Introduction 41
3.2 Computer-aided design 41
3.3 Applications of computer graphics 42
3.4 CAD/CAM 43
3.5 Hardware systems 43
3.6 CAD software for the microcomputer 47

4 Lettering 48

4.1 Lettering 48
4.2 Tools of lettering 48
4.3 Gothic lettering 49
4.4 Guidelines 49
4.5 Vertical letters 50
4.6 Inclined letters 51
4.7 Spacing numerals and letters 52
4.8 Mechanical lettering 53
4.9 Lettering by computer 54
Problems 56

5 Drawing Instruments 57

5.1 Introduction 57
5.2 Pencil 57
5.3 Papers and drafting media 58
5.4 T-square and board 59
5.5 Drafting machines 59
5.6 Alphabet of lines 60
5.7 Horizontal lines 60
5.8 Vertical lines 61
5.9 Drafting triangles 61
5.10 Protractor 62
5.11 Parallel lines 62
5.12 Perpendicular lines 62
5.13 Irregular curves 63
5.14 Erasing 63
5.15 Scales 63
5.16 Metric scales 68
5.17 The instrument set 70
5.18 Ink drawing 72
5.19 Solutions of problems 75
Problems 75

6 Geometric Construction 77

6.1 Introduction 77
6.2 Angles 77
6.3 Triangles 77
6.4 Quadrilaterals 78
6.5 Polygons 78
6.6 Elements of circles 78
6.7 Geometric solids 79
6.8 Constructing triangles 80
6.9 Constructing polygons 80
6.10 Hexagons 81
6.11 Octagons 81
6.12 Pentagons 81
6.13 Bisecting lines and angles 81
6.14 Revolution of figures 83
6.15 Enlargement and reduction of figures 83

6.16 Division of lines 83
6.17 Arcs through three points 84
6.18 Parallel lines 85
6.19 Points of tangency 85
6.20 Line tangent to an arc 85
6.21 Arc tangent to a line from a point 85
6.22 Arc tangent to two lines 87
6.23 Arc tangent to an arc and a line 89
6.24 Arc tangent to two arcs 89
6.25 Ogee curves 92
6.26 Curve of arcs 92
6.27 Rectifying arcs 92
6.28 Conic sections 93
6.29 Ellipses 93
6.30 Parabolas 95
6.31 Hyperbolas 96
6.32 Spirals 96
6.33 Helixes 96
6.34 Involutes 98
Problems 98

7 Multiview Sketching 106

7.1 The purpose of sketching 106
7.2 Shape description 106
7.3 Six-view drawings 107
7.4 Sketching techniques 107
7.5 Three-view sketch 108
7.6 Circular features 112
7.7 Isometric sketching 115
Problems 118

8 Multiview Drawing with Instruments 122

8.1 Introduction 122
8.2 Orthographic projection 122

8.3 Alphabet of lines 124
8.4 Six-view drawings 125
8.5 Three-view drawings 126
8.6 Arrangement of views 127
8.7 Selection of views 127
8.8 Line techniques 129
8.9 Point numbering 129
8.10 Line and planes 130
8.11 Alternate arrangement of views 130
8.12 Laying out three-view drawings 130
8.13 Two-view drawings 133
8.14 One-view drawings 134
8.15 Incomplete and removed views 134
8.16 Curve plotting 135
8.17 Partial views 135
8.18 Conventional revolutions 136
8.19 Intersections 136
8.20 Fillets and rounds 138
8.21 Left-hand and right-hand views 141
8.22 First-angle projection 142
Problems 142

9 Auxiliary Views 152

9.1 Introduction 152
9.2 Folding-line approach 152
9.3 Auxiliaries projected from the top view 153
9.4 Auxiliaries from the top view— folding-line method 155
9.5 Auxiliaries from the top view— reference-plane method 155
9.6 Auxiliaries from the front view— folding-line method 156
9.7 Auxiliaries from the front view— reference-plane method 159
9.8 Auxiliaries from the profile view— folding-line method 159
9.9 Auxiliaries from the profile— reference-plane method 159
9.10 Auxiliaries of curved shapes 159
9.11 Partial views 160

9.12 Auxiliary sections 160
9.13 Secondary auxiliary views 161
9.14 Elliptical features 163
Problems 163

10 Sections 170

10.1 Introduction 170
10.2 Sectioning symbols 171
10.3 Sectioning assemblies 173
10.4 Full sections 173
10.5 Parts not section-lined 174
10.6 Ribs in section 175
10.7 Half sections 176
10.8 Partial views 176
10.9 Offset sections 178
10.10 Revolved sections 178
10.11 Removed sections 179
10.12 Broken-out sections 181
10.13 Phantom (ghost) sections 181
10.14 Conventional breaks 181
10.15 Conventional revolutions 182
10.16 Auxiliary sections 184
Problems 185

11 Screws, Fasteners, and Springs 189

11.1 Threaded fasteners 189
11.2 Definitions of thread terminology 189
11.3 Thread specifications (English system) 190
11.4 Using thread tables 192
11.5 Metric thread specifications (ISO) 193

11.6 Thread representation 196

11.7 Detailed UN/UNR threads 196

11.8 Detailed square threads 197

11.9 Detailed Acme threads 199

11.10 Schematic threads 199

11.11 Simplified threads 200

11.12 Drawing small threads 200

11.13 Nuts and bolts 201

11.14 Drawing square bolt heads 203

11.15 Drawing hexagon bolt heads 203

11.16 Drawing nuts 205

11.17 Drawing nuts and bolts in combination 206

11.18 Cap screws 206

11.19 Machine screws 206

11.20 Set screws 207

11.21 Miscellaneous screws 208

11.22 Wood screws 209

11.23 Tapping a hole 210

11.24 Washers, lock washers, and pins 210

11.25 Pipe threads 211

11.26 Keys 212

11.27 Rivets 212

11.28 Springs 214

11.29 Drawing springs 215

Problems 216

12 Gears and Cams 220

12.1 Introduction to gears 220

12.2 Spur gear terminology 220

12.3 Tooth forms 222

12.4 Gear ratios 222

12.5 Spur gear calculations 223

12.6 Drawing spur gears 224

12.7 Bevel gear terminology 224

12.8 Bevel gear calculations 225

12.9 Drawing bevel gears 226

12.10 Worm gears 227

12.11 Worm gear calculations 229

12.12 Drawing worm gears 229

12.13 Introduction to cams 230

12.14 Cam motion 230

12.15 Construction of a plate cam 231

12.16 Construction of a cam with an offset follower 234

Problems 234

13 Materials and Processes 236

13.1 Introduction 236

13.2 Iron 236

13.3 Steel 237

13.4 Copper 238

13.5 Aluminum 238

13.6 Magnesium 239

13.7 Properties of materials 239

13.8 Heat treatment of metals 239

13.9 Castings 240

13.10 Forgings 241

13.11 Stamping 243

13.12 Plastics and miscellaneous materials 244

13.13 Machining operations 244

13.14 Surface finishing 248

14 Dimensioning 249

14.1 Introduction 249

14.2 Dimensioning terminology 249

14.3 Units of measurement 250

14.4 English/metric conversions 251

14.5 Dual dimensioning 251

14.6 Metric designation 252

14.7 Aligned and unidirectional numbers 252

14.8   Placement of dimensions   253
14.9   Dimensioning in limited spaces   255
14.10   Dimensioning symbology   256
14.11   Computer dimensioning   256
14.12   Dimensioning prisms   257
14.13   Dimensioning angles   259
14.14   Dimensioning cylinders   259
14.15   Measuring cylindrical parts   260
14.16   Cylindrical holes   260
14.17   Pyramids, cones, and spheres   262
14.18   Leaders   262
14.19   Dimensioning arcs   263
14.20   Fillets and rounds and TYP   263
14.21   Curved surfaces   264
14.22   Symmetrical objects   265
14.23   Finished surfaces   265
14.24   Location dimensions   266
14.25   Location of holes   266
14.26   Objects with rounded ends   268
14.27   Machined holes   269
14.28   Chamfers   271
14.29   Keyseats   271
14.30   Knurling   271
14.31   Necks and undercuts   272
14.32   Tapers   273
14.33   Dimensioning sections   273
14.34   Miscellaneous notes   274
Problems   274

15   Tolerances   277

15.1   Introduction   277
15.2   Tolerance dimensions   277
15.3   Mating parts   279
15.4   Terminology of tolerancing   280
15.5   Basic hole system   281
15.6   Basic shaft system   281
15.7   Metric limits and fits   281
15.8   Preferred sizes and fits   283
15.9   Example problems—
metric system   285

15.10   Preferred metric fits—
nonpreferred sizes   287
15.11   Standard fits—English units   287
15.12   Chain dimensions   288
15.13   Origin selection   288
15.14   Conical tapers   289
15.15   Tolerance notes   289
15.16   General tolerances—metric   289
15.17   Geometric tolerances   291
15.18   Symbology of
geometric tolerances   292
15.19   Limits of size   292
15.20   Three rules of tolerances   293
15.21   Three-datum plane concept   293
15.22   Cylindrical datum features   294
15.23   Datum features at RFS   295
15.24   Datum targets   296
15.25   Tolerances of location   297
15.26   Tolerances of form   300
15.27   Tolerances of profile   300
15.28   Tolerances of orientation   301
15.29   Tolerances of runout   302
15.30   Surface texture   303
Problems   307

16   Welding   312

16.1   Introduction   312
16.2   Weld joints   314
16.3   Welding symbols   314
16.4   Types of welds   315
16.5   Application of symbols   315
16.6   Groove welds   316
16.7   Surface contoured welds   317
16.8   Seam welds   318
16.9   Built-up welds   319
16.10   Welding standards   319
16.11   Brazing   319
16.12   Soft soldering   319

17 **Working Drawings** 322

17.1 Introduction 322
17.2 Working drawings—inch system 322
17.3 Working drawings—metric system 325
17.4 Working drawings—dual dimensions 327
17.5 Laying out a working drawing 328
17.6 Title blocks and parts lists 330
17.7 Scale specification 331
17.8 Tolerances 331
17.9 Part names and numbers 331
17.10 Checking a drawing 332
17.11 Drafter's log 333
17.12 Assembly drawings 333
17.13 Freehand working drawings 334
17.14 Castings and forged parts 336
17.15 Sheet metal drawings 337
Problems 338

18 **Reproduction Methods and Drawing Shortcuts** 363

18.1 Introduction 363
18.2 Reproduction of working drawings 363
18.3 Folding the drawing 365
18.4 Overlay drafting techniques 366
18.5 Paste-on photos 366
18.6 Stick-on materials 367
18.7 Photo drafting 368

19 **Pictorials** 370

19.1 Introduction 370
19.2 Types of pictorials 370

19.3 Oblique pictorials 370
19.4 Oblique drawings 371
19.5 Constructing obliques 371
19.6 Angles in oblique 372
19.7 Cylinders in oblique 373
19.8 Circles in oblique 374
19.9 Curves in oblique 375
19.10 Oblique sketching 376
19.11 Dimensioned obliques 376
19.12 Isometric pictorials 376
19.13 Angles in isometric 378
19.14 Circles in isometric 379
19.15 Cylinders in isometric 382
19.16 Partial circular features 382
19.17 Measuring angles 383
19.18 Curves in isometric 384
19.19 Ellipses on nonisometric planes 386
19.20 Machine parts in isometric 386
19.21 Isometric sections 388
19.22 Dimensioned isometrics 388
19.23 Fillets and rounds 388
19.24 Isometric assemblies 388
19.25 Axonometric pictorials 390
19.26 Perspective pictorials 390
19.27 One-point perspectives 391
19.28 Two-point perspectives 393
19.29 Axonometric pictorials by computer 395
19.30 Perspectives by computer 398
Problems 402

20 **Descriptive Geometry** 404

20.1 Introduction 404
20.2 Techniques of labeling points, lines, and planes 404
20.3 Descriptive geometry by computer 405
20.4 Orthographic projection of a point 406
20.5 Lines 407

20.6   Location of a point on a line   409

20.7   Intersecting and nonintersecting lines   409

20.8   Visibility of crossing lines   409

20.9   Visibility of a line and a plane   409

20.10 Planes   410

20.11 Primary auxiliary view of a line   411

20.12 True length by analytical geometry   413

20.13 The true-length diagram   413

20.14 Slope of a line   414

20.15 Compass bearing of a line   415

20.16 Edge view of a plane   416

20.17 Dihedral angles   417

20.18 Piercing points by auxiliary views   417

20.19 Perpendicular to a plane   419

20.20 Intersections by auxiliary view   419

20.21 Slope of a plane   419

20.22 Successive auxiliary views   420

20.23 Point view of a line   420

20.24 Angle between planes   421

20.25 True size of a plane   422

20.26 Shortest distance from a point to a line   423

20.27 Shortest distance between skewed lines—line method   424

20.28 Angular distance to a line   425

20.29 Angle between a line and a plane— plane method   426

20.30 Intersections and developments   427

20.31 Intersections of lines and planes   428

20.32 Intersections between prisms   430

20.33 Intersection of a plane and cylinder   430

20.34 Intersections between cylinders and prisms   433

20.35 Intersections between two cylinders   434

20.36 Intersections between planes and cones   435

20.37 Intersections between cones and prisms   435

20.38 Intersections between pyramids and prisms   437

20.39 Principles of developments   438

20.40 Development of prisms   439

20.41 Development of oblique prisms   440

20.42 Development of cylinders   441

20.43 Development of oblique cylinders   442

20.44 Development of pyramids   443

20.45 Development of cones   444

20.46 Development of transition pieces   445

20.47 Solution of descriptive geometry problems   447

Problems   447

21   **Civil Engineering Applications**   455

21.1   Introduction   455

21.2   Plot plans   455

21.3   Contour maps and profiles   458

21.4   Profiles   460

21.5   Plan profiles   461

21.6   Cut and fill   462

21.7   Design of a dam   463

21.8   Strike and dip   464

21.9   Distances from a point to a plane   466

21.10 Outcrop   466

Problems   467

22   **Graphs**   472

22.1   Introduction   472

22.2   Size proportions of graphs   473

22.3 Pie graphs   473

22.4 Bar graphs   473

22.5 Linear coordinate graphs   475

22.6 Logarithmic coordinate graphs   481

22.7 Semilogarithmic coordinate graphs   482

22.8 Polar graphs   484

22.9 Schematics   485

Problems   486

23 **AutoCAD Computer Graphics**   491

23.1   Introduction   491

23.2   Starting up   492

23.3   Experimenting   492

23.4   Introduction to plotting   492

23.5   Shutting down   493

23.6   Drawing layers   494

23.7   Setting screen parameters   496

23.8   Utility commands   500

23.9   Custom-designed lines   501

23.10 Making a drawing—lines   502

23.11 Selection of entities   504

23.12 Erasing and breaking lines   505

23.13 UNDO command   506

23.14 TRACE command   507

23.15 POINT command   507

23.16 Drawing circles   508

23.17 Tangent options of the CIRCLE command   508

23.18 Drawing arcs   509

23.19 FILLET command   509

23.20 CHAMFER command   510

23.21 POLYGON command   511

23.22 Enlarging, reducing, and panning drawings   511

23.23 CHANGE command   512

23.24 CHPROP command   513

23.25 POLYLINE (PLINE) command   514

23.26 PEDIT command   515

23.27 HATCH command   517

23.28 Text and numerals   518

23.29 The STYLE command   520

23.30 Moving and copying drawings   521

23.31 Mirroring drawings   522

23.32 Mirrored text (MIRRTEXT)   522

23.33 OSNAP (object snap)   522

23.34 ARRAY command   524

23.35 DONUT command   524

23.36 SCALE command   524

23.37 STRETCH command   525

23.38 ROTATE command   526

23.39 TRIM command   526

23.40 EXTEND command   527

23.41 DIVIDE command   527

23.42 MEASURE command   528

23.43 OFFSET command   528

23.44 BLOCKS   529

23.45 External references   530

23.46 Transparent commands   531

23.47 VIEW command   531

23.48 Inquiry commands   532

23.49 Dimensioning principles   533

23.50 Dimensioning variables—introduction   534

23.51 Ordinate Dimensions   536

23.52 Dimensioning arcs and circles   537

23.53 Dimensioning angles   538

23.54 Dimensioning variables   539

23.55 Associative dimensioning   541

23.56 Special arrowheads   543

23.57 Toleranced dimensions   544

23.58 Oblique pictorials   544

23.59 Isometric pictorials   544

23.60 ELLIPSE command   546

23.61 Introduction to 3D extrusions   547

23.62 Fundamentals of 3D drawing   549

23.63 The coordinate systems   550

23.64 The DVIEW command   553

23.65 Basic 3D forms   556

23.66 3D polygon meshes   558

23.67 The RULESURF command   559

23.68 The TABSURF command   559

23.69 The REVSURF command   560

23.70 The EDGESURF command   560

23.71 The PFACE command   561

23.72 LINE, PLINE, and
3DPOLY commands   561

23.73 3DFACE command   562

23.74 XYZ filters   564

23.75 New drawing in 3D   565

23.76 Object with an inclined surface
567

23.77 Model space and paper space   568

23.78 Drawing with Tilemode = 1   570

23.79 Drawing with Tilemode = 0   572

23.80 Drawing the meshes   575

23.81 Plotting a drawing   575

23.82 Attributes   578

23.83 Attribute extract (ATTEXT)   580

23.84 Grid rotation   581

23.85 Digitizing with the tablet   581

23.86 SKETCH command   582

23.87 Slide shows   582

23.88 SETVAR command   584

Problems   584

24.3 Primitives: cone (SOLCONE)   594

24.4 Primtives: cylinder (SOLCYL)   594

24.5 Primitives: sphere
(SOLSPHERE)   594

24.6 Primitives: torus (SOLTORUS)   595

24.7 Primitives: wedge
(SOLWEDGE)   595

24.8 Extrusions (SOLEXT)   595

24.9 Solid revolution (SOLREV)   596

24.10 Solidify command (SOLIDIFY)   596

24.11 Subtracting solids (SOLSUB)   596

24.12 Adding solids (SOLUNION)   596

24.13 Separating solids (SOLSEP)   596

24.14 Chamfer (SOLCHAM)   597

24.15 Fillet (SOLFILL)   598

24.16 Change solid (SOLCHP)   598

24,17 Solid move (SOLMOVE)   598

24.18 Sections (SOLHPAT and
SOLSECT)   598

24.19 Solid inquiry commands   599

24.20 Solid representations   599

Appendixes   A-1

Index   I-1

24 Introduction to Solid Modeling   593

24.1 Introduction   593

24.2 Primitives: box (SOLBOX)   593

Introduction to Engineering and Technology

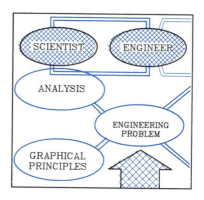

1

1.1
Introduction

This book introduces engineering design concepts and applies engineering graphics to the design process. We give examples that have an engineering problem at the core and that require organization, analysis, problem-solving graphical principles, communication, and skill (Fig. 1.1).

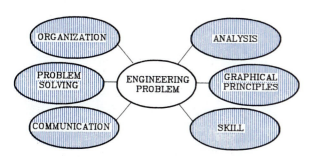

FIGURE 1.1 Problems in this text require a total engineering approach with the engineering problem as the central theme.

Creativity and imagination are essential to the engineering profession. Albert Einstein said, "Imagination is more important than knowledge, for knowledge is limited, whereas imagination embraces the entire world . . . stimulating progress, or, giving birth to evolution" (Fig. 1.2).

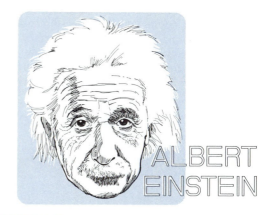

FIGURE 1.2 "Imagination is more important than knowledge."

1.2
Engineering graphics

Engineering graphics is the total field of graphical problem solving and includes two areas of specialization: descriptive geometry and working drawings. Other areas that can be used for a wide variety of applications are nomography, graphical mathematics, empirical equations, technical illustration, vector analysis, data analysis, and computer graphics. Graphics is one of the designer's main methods of thinking, solving problems, and communicating ideas.

Descriptive Geometry

Gaspard Monge (1746–1818) is the "father of descriptive geometry" (Fig. 1.3). While a military student in France, young Monge used this graphical method to solve design problems related to fortifications and battlements. For not solving problems of this type by the usual (long and tedious) mathematical process, he was scolded by his headmaster. Only after long explanations and comparisons of both methods' solutions was he able to convince the faculty that his graphical method solved problems in considerably less time. Descriptive geometry was such an improvement on the mathematical method that it was kept a military secret for 15 years before it was allowed to be taught as part of the technical curriculum. During Napoleon's reign, Monge became a scientific and mathematical aide to the emperor.

Descriptive geometry is the projection of three-dimensional figures onto a two-dimensional plane of paper in a manner that allows geometric manipulations to determine lengths, angles, shapes, and other descriptive information about the figures.

FIGURE 1.3 Gaspard Monge, the "father of descriptive geometry."

1.3

The technological team

So rapidly has the scope of technology broadened that it has become necessary for professional responsibilities to be performed by people with specialized training. Technology has become a team effort involving the scientist, engineer, technologist, technician, and craftsman (Fig. 1.4).

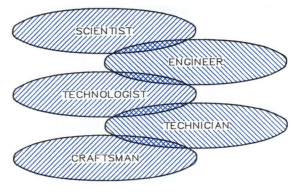

THE TECHNOLOGICAL TEAM

FIGURE 1.4 The technological team.

Scientist

The scientist is a researcher who seeks to establish new theories and principles through experimentation and testing (Figs. 1.5 and 1.6). Since primarily interested in isolating significant relationships, scientists often show little concern for applying specific principles. Scientific discoveries are the basis for the development of practical applications that may not exist until years after discovery.

FIGURE 1.5 Scientists conduct research to establish fundamental relationships. This chemist is trying to determine the best combination of ingredients to yield a better tire. (Courtesy of Uniroyal, Inc.)

FIGURE 1.6 This geologist is studying seismological charts to determine the likelihood of petroleum deposits. (Courtesy of Texas Eastern; photo by Bob Thigpen.)

Engineer

Engineers are trained in science, mathematics, and industrial processes to prepare them to apply the principles discovered by scientists (Fig. 1.7). They are concerned with converting raw materials and power sources into needed products and services. Creatively applying these principles to new products or systems is the design process, the engineer's most unique function. Generally, the engineer uses known principles and available resources to achieve a practical end at a reasonable cost.

FIGURE 1.7 These engineers are discussing geological data and area surveys to arrive at the best placement of exploratory oil wells. (Courtesy of Texas Eastern; photo by Bob Thigpen.)

Technologist

Technologists are usually graduates of four-year programs in engineering technology where they are trained in science, mathematics, and industrial processes. Whereas the engineer is responsible for research and design, the technologist is concerned with the application of engineering principles to planning, detail design, and production (Fig. 1.8).

Technologists apply their knowledge of engineering principles, manufacturing, and testing to assist in the implementation of projects and products. They also provide support and act as a liaison to the consumer once the products are in service.

FIGURE 1.8 Engineering technologists combine their knowledge of production techniques with the design talents of the engineers to produce a product. (Courtesy of Omark Industries, Inc.)

Technician

Technicians assist the engineer and technologist at a level below the technologist's. In general, they work as liaisons between the technologist and the craftsman (Fig. 1.9). Their work may vary from conducting routine laboratory experiments to supervising craftsmen in manufacturing or construction. Usually the technician is required to have two years of technical training beyond high school.

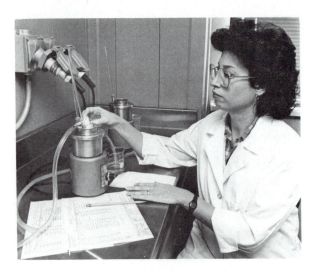

FIGURE 1.9 This engineering technician performs laboratory tests as a member of the technological team. (Courtesy of Texas Eastern; photo by Bob Thigpen.)

Craftsman

Craftsmen are responsible for implementing the engineering design by producing it according to the engineer's specifications. Craftsmen may be machinists who fabricate the product's components or electricians

who assemble electrical components. The craftsman's ability to produce a part according to design specifications is as necessary as the engineer's ability to design the part. Craftsmen include electricians, welders, machinists, fabricators, drafters, and members of many other occupational groups (Fig. 1.10).

Designer

The designer may be an engineer, an inventor, or a person who has special talents for devising creative solutions. Designers often do not have an engineering background especially in newer technologies where there may be little precedent for the work involved. Thomas A. Edison (Fig. 1.11) had little formal education, but he had an exceptional ability to design and implement some of the world's most useful inventions.

FIGURE 1.11 Thomas A. Edison had essentially no formal education, but he gave the world some of its most creative designs.

FIGURE 1.10 This craftsman is a welder skilled in joining metal parts according to prescribed specifications. (Courtesy of Texas Eastern: *TE Today*; photo by Bob Thigpen.)

Stylist

Concerned with the outward appearance of a product rather than the development of a functional design (Fig. 1.12), stylists may design an automobile body or the configuration of an electric iron. An automobile stylist considers the functional requirements of the body, driver vision, enclosure of passengers, space for a power unit, and so forth. But the stylist is not involved with the design of internal details of the product such as the engine or steerage linkage. The stylist must have a high degree of aesthetic awareness and a feel for the consumer's acceptance of designs.

FIGURE 1.12 **The stylist is more concerned with the outward appearance of a product than the functional aspects of its design. (Courtesy of Ford Motor Corp.)**

Aerospace engineers may also specialize in a particular product, such as conventional-powered planes, jet-powered military aircraft, rockets, satellites, or manned space capsules.

Aerospace engineers can be divided into two major areas: research engineering and design engineering. The research engineer investigates known principles in search of new ideas and concepts. The design engineer translates these new concepts into workable applications for improving the state of the art. This approach has elevated the field of aerospace engineering from the Wright brothers' first flight at Kittyhawk, North Carolina, to the penetration of outer space.

> The professional society for aerospace engineers is the American Institute of Aeronautics and Astronautics (AIAA).

1.4
Engineering fields

Recent changes in engineering include the emergence of the technologist and technician and the growing number of women pursuing engineering careers. Indeed, more than 15% of today's freshman engineering students are women.

Aerospace engineering

Aerospace engineering deals with all aspects (all speeds and altitudes) of flight. Aerospace engineering assignments range from complex vehicles traveling 350 million miles to Mars to hovering aircraft used in deep-sea exploration. In the space exploration branch of this profession, aerospace engineers work on all types of aircraft and spacecraft—state-of-the-art missiles and rockets, as well as conventional propeller-driven and jet-powered planes (Fig. 1.13).

Second only to the auto industry in sales, the aerospace industry contributes immeasurably to national defense and the economy. Specialized areas include (1) aerodynamics, (2) structural design, (3) instrumentation, (4) propulsion systems, (5) materials, (6) reliability testing, and (7) production methods.

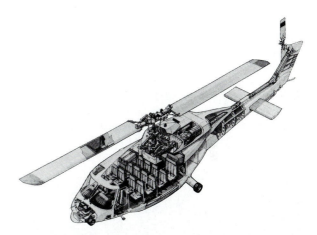

FIGURE 1.13 **The helicopter is a product of the skill and ability of the aerospace engineer. (Courtesy of Bell Helicopter Textron.)**

Agricultural engineering

Agricultural engineers are trained to serve the world's largest industry, agriculture. Agricultural engineering problems deal with the production, processing, and handling of food and fiber.

MECHANICAL POWER The agricultural engineer who works with manufacturers of farm equipment is concerned with gasoline and diesel engine equipment such as pumps, irrigation machinery, and tractors. Machinery must be designed for the electrical curing of hay, milk processing and pasteurizing, fruit processing, and heating environments in which to raise livestock and poultry. Farm machinery designed by agricultural engineers has been largely responsible for the increased productivity in agriculture.

FARM STRUCTURES The construction of barns, shelters, silos, granaries, processing centers, and other agricultural buildings requires specialists in agricultural engineering. Engineers must understand heating, ventilation, and chemical changes that might affect the storage of crops.

ELECTRICAL POWER Agricultural engineers design electrical systems and select equipment that will provide efficient operation to meet the requirements of a situation. They may serve as consultants or designers for manufacturers or processors of agricultural products.

SOIL AND WATER CONTROL The agricultural engineer is responsible for devising systems for improving drainage and irrigation systems, resurfacing fields, and constructing water reservoirs (Fig. 1.14). These activities may be performed in association with the U.S. Department of Agriculture, the U.S. Department of the Interior, state agricultural universities, consulting engineering firms, or irrigation companies.

FIGURE 1.14 Agricultural engineers build canals such as this one to conserve water and irrigate lands that would not otherwise be productive. (Courtesy of the U.S. Bureau of Reclamation; photo by J. F. Santa.)

Most agricultural engineers are employed in private industry especially by manufacturers of heavy farm equipment and specialized lines of field, barnyard, and household equipment; electrical service companies; and distributors of farm equipment and supplies. Although few agricultural engineers live on farms, it is helpful if they have a clear understanding of agricultural problems, farming, crops, animals, and farmers themselves. Thus agricultural engineering has helped increase the farmer's efficiency: Today, a farmer produces enough food for 80 people, whereas one hundred years ago, a farmer was able to feed only 4 people.

> The professional society for agricultural engineers is the American Society for Agricultural Engineers.

Chemical engineering

Chemical engineering involves the design and selection of equipment that facilitates the processing and manufacturing of large quantities of chemicals. Chemical engineers design unit operations, including fluid transportation through ducts or pipelines, solid material transportation through pipes or conveyors, heat transfer from one fluid or substance to another through plate or tube walls, absorption of gases by bubbling them through liquids, evaporation of liquids to increase concentration of solutions, distillation under carefully controlled temperatures to separate mixed liquids, and many other similar chemical processes. Chemical engineers may employ chemical reactions of raw products such as oxidation, hydrogenation, reduction, chlorination, nitration, sulfonation, pyrolysis, and polymerization (Fig. 1.15).

> Process control and instrumentation are important specialities in chemical engineering. With the measurement of quality and quantity by instrumentation, process control is fully automatic.

Chemical engineers develop and process chemicals such as acids, alkalies, salts, coal-tar products, dyes, synthetic chemicals, plastics, insecticides, and fungicides for industrial and domestic uses. They help develop drugs and medicine, cosmetics, explosives, ceramics, cements, paints, petroleum products, lubricants, synthetic fibers, rubber, and detergents. They

FIGURE 1.15 The efforts of scientists, chemical engineers, and others are realized in the construction and operation of a refinery that produces vital products for our economy. (Courtesy of Houston Oil and Refining Co.)

also design equipment for food preparation and canning plants.

Approximately 80% of chemical engineers work in the manufacturing industries, primarily the chemical industry. The other 20% or so work for government agencies, independent research institutes, and as independent consulting engineers. New fields requiring chemical engineers are nuclear sciences, rocket fuels, and environmental pollution.

The professional society for chemical engineers is the American Institute of Chemical Engineers (AIChE).

Civil engineering

Civil engineering, the oldest branch of engineering, is closely related to practically all of our daily activities. The buildings we live and work in, the transportation we use, the water we drink, and the drainage and sewage systems we rely on are all the results of civil engineering. Civil engineers design and supervise the construction of roads, harbors, airfields, tunnels, bridges, water supply and sewage systems, and many other types of structures.

CONSTRUCTION ENGINEERS manage the resources, workers, finances, and materials needed for construction projects, which vary from erecting skyscrapers to moving concrete and earth.

CITY PLANNERS develop plans for the future growth of cities and the systems related to their operation. Street planning, zoning, and industrial site development are problems that city planners solve.

STRUCTURAL ENGINEERS design and supervise the erection of buildings, dams, powerhouses, stadiums, bridges, and other structures.

HYDRAULIC ENGINEERS work with the behavior of water from its conservation to its transportation. They design wells, canals, dams, pipelines, drainage systems, and other methods of controlling and using water and petroleum products.

TRANSPORTATION ENGINEERS develop and improve railroads and airlines in all phases of their operations. Railroads are built, modified, and maintained under the supervision of transportation engineers. Design and construction of airport runways, control towers, passenger and freight stations, and aircraft hangars are also done by transportation engineers.

HIGHWAY ENGINEERS develop the complex network of highways and interchanges for moving automobile traffic. These systems require the design of tunnels, culverts, and traffic control systems (Fig. 1.16).

SANITARY ENGINEERS help maintain public health through the purification of water and control of

FIGURE 1.16 This photograph shows several civil engineering projects, including highways, waterways, bridges, and structures. (Courtesy of Los Angeles Dept. of Highways.)

water pollution and sewage. These systems involve the design of pipelines, treatment plants, dams, and so on.

Many civil engineers find positions in administration and municipal management. Most civil engineers are associated with federal, state, and local government agencies and the construction industry. Many work as consulting engineers for architectural firms and independent consulting engineering firms. The remainder work for public utilities, railroads, educational institutions, steel industries, and other manufacturing industries.

> The professional society for civil engineering is the American Society of Civil Engineers (ASCE). Founded in 1852, it is the oldest engineering society in the United States.

Electrical engineering

Electrical engineers are concerned with the use and distribution of electrical energy. The two main divisions of electrical engineering are (1) power, which deals with the control of large amounts of energy used by cities and large industries, and (2) electronics, which deals with the small amounts of power used for communications and automated operations that have become part of our everyday lives. These two divisions

FIGURE 1.17 Electrical engineers work with engineers from other disciplines to design power plants such as this one. (Courtesy of Kaiser Engineers.)

of electrical engineering have many areas of specialization.

POWER GENERATION poses many electrical engineering problems from the development of transmission equipment to the design of generators for producing electricity (Fig. 1.17).

POWER APPLICATIONS are quite numerous in homes, where toasters, washers, dryers, vacuum cleaners, and lights are used. Of total energy consumption, only about one quarter is used in the home. About half of all energy is used by industry for metal refining, heating, motor drives, welding, machinery controls, chemical processes, plating, electrolysis, and so forth.

ILLUMINATION is required in nearly every area of modern life. Improving the efficiency and economy of illumination systems is a challenging area for the electrical engineer.

COMPUTERS are a gigantic industry and the domain of electrical engineers. Used with industrial electronics, computers have changed industry's manufacturing and production processes, resulting in greater precision and fewer employees.

COMMUNICATIONS is devoted to the improvement of radio, telephone, telegraph, and television systems, the nerve centers of most industrial operations.

INSTRUMENTATION is the study of systems of electronic instruments used in industrial processes. Extensive use has been made of the cathode-ray tube and the electronic amplifier in industry and atomic power reactors. Increasingly, instrumentation is applied to medicine for diagnosis and therapy.

MILITARY ELECTRONICS is used in practically all areas of military weapons and tactical systems from the walkie-talkie to the distant radar networks for detecting enemy aircraft. Remote-controlled electronic systems are used for navigation and interception of guided missiles.

More electrical engineers are employed than any other type of engineer. The increasing need for electrical equipment, automation, and computerized systems is expected to contribute to the growth of this field.

The professional society for electrical engineers is the Institute of Electrical and Electronic Engineers (IEEE). Founded in 1884, the IEEE is the world's largest technical society.

Industrial engineering

Industrial engineering, one of the newer engineering professions, is defined by the National Professional Society of Industrial Engineers as follows:

> Industrial engineering is concerned with the design, improvement, and installation of integrated systems of men, materials, and equipment. It draws upon specialized knowledge and skill in the mathematical, physical, and social sciences together with the principles and methods of engineering analysis and design to specify, predict, and evaluate the results to be obtained from such systems.

Industrial engineering differs from other branches of engineering in that it is more closely related to people and their performance and working conditions (Fig. 1.18). Often the industrial engineer is a manager of people, machines, materials, methods, money, and the markets involved.

The industrial engineer may be responsible for plant layout, the development of plant processes, or the determination of operating standards that will improve the efficiency of a plant operation. They also design and supervise systems for improved safety and production.

Specific areas of industrial engineering include management, plant design and engineering, electronic data processing, systems analysis and design, control of production and quality, performance standards and measurements, and research. Industrial engineers are also increasingly involved in implementing automated production systems.

People-oriented areas include the development of wage incentive systems, job evaluation, work measurement, and the design of environmental systems. Industrial engineers are often involved in management–labor agreements that affect the operation and production of an industry.

More than two thirds of all industrial engineers are employed in manufacturing industries. Others work for insurance companies, construction and mining firms, public utilities, large businesses, and governmental agencies.

The professional society for industrial engineers is the American Institute of Industrial Engineers (AIIE), which was organized in 1948.

Mechanical engineering

The mechanical engineer's major areas of specialization are power generation, transportation, aeronautics, marine vessels, manufacturing, power services, and atomic energy.

POWER GENERATION requires that prime movers (machines that convert natural energy into work) be developed to power electrical generators that will produce electrical energy. Mechanical engineers design and supervise the operation of steam engines, turbines, internal combustion engines, and other prime movers (Fig. 1.19).

TRANSPORTATION including trucks, buses, automobiles, locomotives, marine vessels, and aircraft are designed and manufactured by mechanical engineers.

AERONAUTICS requires mechanical engineers to develop aircraft engines as well as aircraft controls and environmental systems.

FIGURE 1.18 Industrial engineers lay out and design complex industrial facilities to provide efficient production. This plant produces and packages computers. (Courtesy of Texas Instruments.)

FIGURE 1.19 Mechanical engineers designed and supervised the building of this 1.7 liter 4-cylinder automobile engine. (Courtesy of Chrysler Corp.)

MARINE VESSELS powered by steam, diesel, or gas-generated engines are designed by mechanical engineers, as are power services throughout the vessel such as light, water, refrigeration, and ventilation.

MANUFACTURING requires mechanical engineers to design new products and factories to build them in. Economy of manufacturing and uniform quality of products are major functions of manufacturing engineers. The professional society for manufacturing engineers is the Society of Manufacturing Engineers (SME).

POWER SERVICES include the movement of liquids and gases through pipelines, refrigeration systems, elevators, and escalators. These mechanical engineers must have a knowledge of pumps, ventilation equipment, fans, and compressors.

ATOMIC ENERGY needs mechanical engineers for the development and handling of protective equipment and materials. The mechanical engineer plays an important role in the construction of nuclear reactors.

The professional society for mechanical engineers is the American Society of Mechanical Engineers (ASME).

Mining and metallurgical engineering

Mining engineers are responsible for extracting minerals from the earth and preparing them for use by manufacturing industries. Working with geologists to locate ore deposits, which are exploited through the construction of tunnels and underground operations, mining engineers must have an understanding of safety, ventilation, water supply, and communications. Two main areas of metallurgical engineering are (1) extractive metallurgy, the extraction of metal from raw ores to form pure metals, and (2) physical metallurgy, the development of new products and alloys.

Many metallurgical engineers work on the development of machinery for electrical equipment and in the aircraft and aircraft parts industries. The development of new lightweight, high-strength materials for space flight vehicles, jet aircraft, missiles, and satellites will increase the need for metallurgical engineers (Fig. 1.20). Mining engineers who work at mining sites are usually employed near small, out-of-the-way communities, whereas those in research and consulting often work in metropolitan areas.

The professional society for mining and metallurgical engineers is the American Institute of Mining, Metallurgical, and Petroleum Engineering (AIME).

Nuclear engineering

The earliest work in the nuclear field has been for military and defense applications. However, nuclear

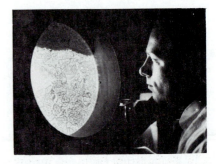

FIGURE 1.20 A metallograph shows the structure of an alloy that may be used in the construction of a refinery unit. Materials are specially developed for specific applications by metallurgical engineers. (Courtesy of Exxon Corp.)

power for domestic needs is being developed for the medical profession and other areas.

Peaceful applications of nuclear engineering are divided into two major areas: radiation and nuclear power reactors. Radiation is the propagation of energy through matter or space in the form of waves. In atomic physics, radiation includes fast-moving particles (alpha and beta rays, free neutrons, and so on), gamma rays, and x rays. Nuclear science is closely allied with botany, chemistry, medicine, and biology.

FIGURE 1.21 This pulsing reactor, operated by a nuclear engineer, tests the effect of extremely high radiation on delicate equipment. (Courtesy of General Dynamics Corp.)

The production of nuclear power in the form of mechanical or electrical power is a major peaceful use of nuclear energy (Fig. 1.21). For the production of electrical power, nuclear energy is used as the fuel for producing steam that will drive a turbine generator in the conventional manner.

Although nuclear engineering degrees are offered at the bachelor's level, advanced degrees are recommended for this area. Most of the nuclear engineer's training centers on the design, construction, and op-

eration of nuclear reactors. Other areas of study include the processing of nuclear fuels, thermonuclear engineering, and the use of various nuclear by-products.

> The professional society for nuclear engineers is the American Nuclear Society.

Petroleum engineering

The recovery of petroleum and gases is the primary concern of petroleum engineers, but they also develop methods for the transportation and separation of various products. Moreover, they are responsible for the improvement of drilling equipment and its economy of operation (Fig. 1.22). In exploring for petroleum, the petroleum engineer is assisted by the geologist and by instruments like the airborne magnetometer, which indicates uplifts on the earth's subsurfaces that could hold oil or gas.

Oil well drilling is supervised by the petroleum engineer, who also develops the equipment to most efficiently remove the oil. When oil is found, the petroleum engineer must design piping systems to remove and transport the oil to its next point of processing. Processing itself is a joint project with chemical engineers.

FIGURE 1.22 The petroleum engineer supervises the operation of offshore platforms used in the exploration for oil. (Courtesy of Texas Eastern; photo by Bob Thigpen.)

The Society of Petroleum Engineers is a branch of AIME, which includes mining and metallurgical engineers and geologists.

1.5
Technologists and technicians

As technology has become more complex, two new members of the technological team have emerged: the technologist and the technician. The primary mission of both is to assist the engineer at a technical level below that of the engineer and above that of the craftsman.

THE TECHNOLOGIST works under the supervision of an engineer but has considerable responsibility, thereby freeing the engineer for more advanced applications. Most technologists have a four-year college background in a specialty area of engineering technology that enables them to perform semiprofessional jobs with a high degree of skill. Their interest in the practical aspects of engineering qualifies them to offer advice to the engineer about production specifications and on-the-site procedures when new projects are being designed.

THE TECHNICIAN performs tasks less technical than those performed by the technologist. Most technicians have graduated from a two-year technical pro-

gram that enables them to be repairers, inspectors, production specialists, or surveyors (Fig. 1.23). They may work under the supervision of an engineer, but ideally they are supervised by a technologist. In turn, the technician coordinates the activities between the technologist and the craftsman (Fig. 1.24). These levels of responsibility ensure that the proper skills and qualifications are available throughout the chain of command.

FIGURE 1.24 Technicians monitor quality control in the assembly and testing of microcomputers. (Courtesy of Tandy Corporation.)

FIGURE 1.23 These technicians are making a cartographic survey. Data from the tellurometer are being recorded in a notebook. (Courtesy of the U.S. Forest Service.)

1.6
Drafters

Today's drafters carry a great responsibility in assisting the engineer and designer. The experienced drafter may be involved in preparing complex drawings, selecting materials, detailing designs, and writing specifications.

DESIGN AND CONSTRUCTION DRAWINGS are made by drafters to explain how to fabricate, build, or erect a project or product in fields such as aerospace engineering, architecture, machine design, mechanical engineering, and electrical and electronic engineering.

TECHNICAL ILLUSTRATION is a type of graphics that is usually prepared as three-dimensional pictorials to illustrate a project or product as realistically as possible. Technical illustration is the most artistic area of engineering design graphics.

MAPS, GEOLOGICAL SECTIONS, AND HIGHWAY PLATS are used for locating property lines, physical features, strata, right-of-way, building sites, bridges, dams, mines, utility lines, and so forth. Drawings of this type are usually prepared as permanent ink drawings.

There are three levels of certification for drafters: drafters, design drafters, and engineering designers.

DRAFTERS are graduates of a two-year, post-high-school curriculum in engineering design graphics.

DESIGN DRAFTERS complete two-year programs in an approved junior college or technical institute.

ENGINEERING DESIGNERS are drafters who have completed a four-year college course in engineering design graphics. Graduates of these programs can become certified as technologists.

Computerized systems are being adopted by industry to improve the drafter's productivity. Computer graphics systems do not lessen the need for drafters or

FIGURE 1.25 These drafters are working at a computer using AutoCAD software to generate technical drawings. (Courtesy of Autodesk, Inc.)

engineers to know graphical principles; rather, they offer a different medium of expression (Fig. 1.25). In Chapter 10, we discuss computer graphics systems and their uses in engineering graphics. Further demonstration of the use of the computer in engineering graphics problems can be found throughout the text.

Problems

1. Write a report that outlines the specific duties and relationships between the scientist, engineer, craftsman, designer, and stylist in an engineering field of your choice. For example, explain this relationship for an engineering team involved in an aspect of civil engineering. Your report should be supported by factual information obtained from interviews, brochures, or library references.

2. Write a report that investigates the employment opportunities, job requirements, professional challenges, and activities of your chosen branch of engineering or technology. Illustrate this report with charts and graphs where possible for easy interpretation. Compare your personal abilities and interests with those required by the profession.

3. Arrange a personal interview with a practicing engineer, technologist, or technician in the field of your interest. Discuss with him or her the general duties and responsibilities of the position to gain a better understanding of this field. Summarize your interview in a written report.

4. Write to the professional society of your field of study for information about this area. Prepare a notebook of these materials for easy reference. Include in the notebook a list of books that would provide career information for the engineering student.

Addresses of professional societies

Publications and information from these societies were used in preparing this chapter.

American Ceramic Society
65 Ceramic Drive, Columbus, Ohio 43214

The American Institute of Aeronautics and Astronautics
1290 Avenue of the Americas, New York, N.Y. 10019

American Institute of Chemical Engineers
345 East 47th Street, New York, N.Y. 10017

American Institute for Design and Drafting
3119 Price Road, Bartlesville, Okla. 74003

The American Institute of Industrial Engineers
345 East 47th Street, New York, N.Y. 10017

American Institute of Mining, Metallurgical, and Petroleum Engineering
345 East 47th Street, New York, N.Y. 10017

American Nuclear Society
244A East Ogden Avenue, Hinsdale, Ill. 60521

American Society of Agricultural Engineers
2950 Niles Road, St. Joseph, Mich. 49085

American Society of Civil Engineers
345 East 47th Street, New York, N.Y. 10017

American Society for Engineering Education
11 DuPont Circle, Suite 200, Washington, D.C. 20036

American Society of Mechanical Engineers
345 East 47th Street, New York, N.Y. 10017

The Institute of Electrical and Electronic Engineers
345 East 47th Street, New York, N.Y. 10017

National Society of Professional Engineers
2029 K Street, N.W., Washington, D.C. 20006

Society of Petroleum Engineers (AIME)
6300 North Central Expressway, Dallas, Texas 75206

Society of Women Engineers
United Engineering Center, Room 305,
345 East 47th Street, New York, N.Y. 10017

The Design Process

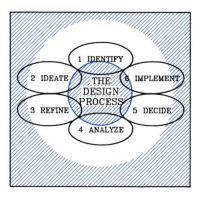

2.1
Introduction

Design, the responsibility that most distinguishes the engineer from the scientist and technician, is the act of devising an original solution to a problem by a combination of principles, resources, and products.

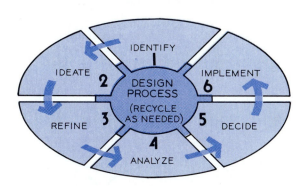

FIGURE 2.1 The steps of the design process. Each step can be recycled when needed.

In this book, we emphasize a six-step design process: (1) problem identification, (2) preliminary ideas, (3) refinement, (4) analysis, (5) decision, and (6) implementation (Fig. 2.1). Designers work sequentially from step to step, but they may recycle to previous steps as they progress.

Problem identification

Most engineering problems are not clearly defined at the outset, so they must be identified before an attempt is made to solve them (Fig. 2.2). A prominent concern in our society is air pollution, but before this problem can be solved, we must identify what air pollution is and what causes it. Is pollution caused by automobiles, factories, atmospheric conditions that harbor impurities, or geographic features that contain impure atmospheres? When you enter a street intersection where traffic is unusually congested, do you identify the reasons for the congestion? Are there too many cars? Are the signals poorly synchronized? Are there visual obstructions?

Problem identification requires a good deal more study than just a simple statement like "solve air pol-

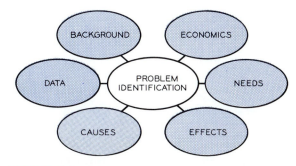

FIGURE 2.2 Problem identification requires accumulating as much information about the problem as possible before attempting a solution.

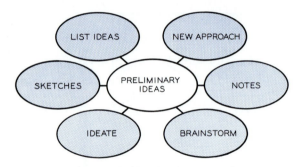

FIGURE 2.3 Preliminary ideas are developed after the identification process has been completed. All ideas should be listed and sketched to give the designer a broad selection to work from.

lution.'' You will need to gather data of several types: field data, opinion surveys, historical records, personal observations, experimental data, and physical measurements and characteristics (Fig. 2.2).

Preliminary ideas

The second step is to accumulate as many ideas for solving the problem as possible (Fig. 2.3). Preliminary ideas should be broad enough to allow for unique solutions that could revolutionize present methods. Many rough sketches of preliminary ideas should be made and retained to generate original ideas and stimulate the design process. Ideas and comments should be noted on the sketches.

Figure 2.4 shows the relationship between creativity and accumulating information during the de-

sign process. You have no limitations on generating preliminary ideas; be as creative as possible, as wild as you like. During the later stages of the design process, the need for creativity diminishes, and the need for information increases.

The Xerox Conference Copier (Fig. 2.5) is an example of an advanced design idea to convert drawings and notes on an easel board to paper copies. Anything drawn on the board can be immediately duplicated on paper by the press of a button.

FIGURE 2.5 The Xerox Conference Copier is a unique design that duplicates anything drawn on the easel board into multiple paper copies. (Courtesy of the Xerox Corporation.)

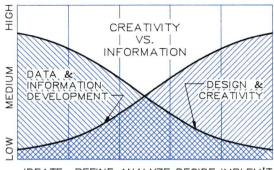

FIGURE 2.4 Engineering creativity is highest during the initial stages of the design process, whereas data and information development increase during the final stages.

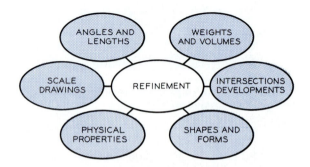

FIGURE 2.6 Refinement begins with the construction of scale drawings of the better preliminary ideas. Descriptive geometry and graphical methods are used to find necessary geometric characteristics.

Refinement

Next, several of the better preliminary ideas are selected for refinement to determine their true merits. Rough sketches are converted to scale drawings that will permit space analysis, critical measurements, and the calculation of areas and volumes affecting the design (Fig. 2.6). Consideration is given to spatial relationships, angles between planes, lengths of structural members, intersections of surfaces and planes.

Designing the structural system of the mall shown in Fig. 2.7 required the application of descriptive geometry. The configuration of the supports were drawn to scale, and the lengths of the members were determined, as were the angles of bend. The connectors had to be designed using descriptive geometry for the members to join precisely.

Analysis

The step of the design process where engineering and scientific principles are most often used is analysis (Fig. 2.8)—the evaluation of the best designs to deter-

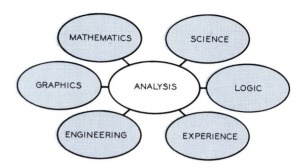

FIGURE 2.8 In the analysis phase of the design process, all available technological methods, from science to graphics, are used to evaluate the refined designs.

mine the comparative merits of each with respect to cost, strength, function, and market appeal. Graphical methods of analysis are means of checking a solution. Data that are difficult to mathematically interpret can be graphically analyzed. Models constructed at reduced scales are also a valuable analytical tool. They help to establish relationships of moving parts and outward appearances and to evaluate other design characteristics.

Decision

At this stage, a decision must be made. A single design must be selected as the solution of the design problem (Fig. 2.9). Often, the final design is a compromise that offers many of the best features of several designs. The decision may be made by the designer alone, or it may be made by several associates. Regardless of who decides, graphics is a primary means of presenting the proposed designs for a decision. The outstanding as-

FIGURE 2.7 The refinement of this structural system of a mall required using descriptive geometry and other graphical methods. (Courtesy of Kaiser Engineers.)

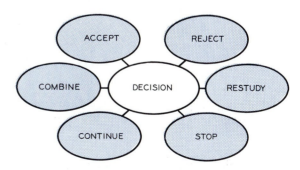

FIGURE 2.9 Decision is the selection of the best design or design features to be implemented.

pects of each design usually lend themselves to graphical presentations that compare manufacturing costs, weights, operational characteristics, and other data that would be considered before arriving at the final decision.

Implementation

The final design concept must be presented in a workable form. Working drawings and specifications are usually used as the instruments for fabrication of a product, whether a small piece of hardware or a huge bridge (Fig. 2.10). Workers must have detailed instructions for the manufacture of each part, measured to a thousandth of an inch to ensure its proper manufacture. Working drawings must be explicit enough to provide a legal contractual basis for the contractor's bid on the job.

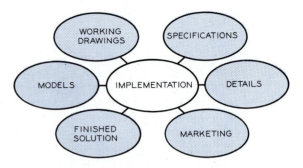

FIGURE 2.10 During implementation, the final step of the design process, drawings and specifications are prepared from which the final product can be constructed.

2.2
Hunting seat—problem identification

We use the hunting seat example to illustrate the problem identification step of the design process.

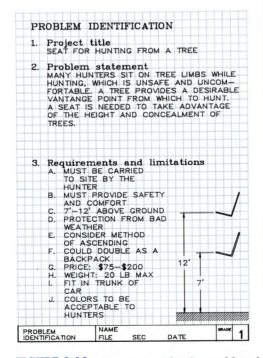

FIGURE 2.11 A worksheet for the problem identification step for the hunting seat design.

Hunting seat

Many hunters hunt from trees to obtain a better vantage point. Design a seat that provides the hunter with comfort and safety while hunting from a tree and that meets the requirements of economy and the limitations of hunting.

WORKSHEET COMPLETION The worksheet in Fig. 2.11 is typical of the information the designer needs to understand the background of the problem.

TITLE AND PROBLEM STATEMENT The title of the project is recorded along with a brief problem statement.

REQUIREMENTS AND LIMITATIONS The requirements and limitations are listed along with any sketches that would aid in understanding the problem. You might have to list some requirements as questions for the time being.

After further investigation, you should list the limits—for example, must cost between $60 and $100. It is better to give a range of prices or weights rather than try to be exact. Catalogues offering similar products are one source of information on sales prices, weights, and sizes.

NEEDED INFORMATION How many hunters are there? How many hunt from trees or elevated blinds? This and similar information is available from your state game office.

What is the average income of the hunter? How much do they spend on their hobby per year? Sporting-goods dealers could help you by sharing their experiences, and perhaps they could direct you to other sources for answers to these questions.

MARKET CONSIDERATIONS The designer must think about cost control even in the problem identification step (Figs. 2.11 and 2.12).

Figure 2.13 shows how an item is priced from wholesale to retail. Percentages vary by product: There is less profit in retail food sales than in furniture sales. If an item retails for $50, the production cost cannot be more than about $20 to maintain the necessary margins.

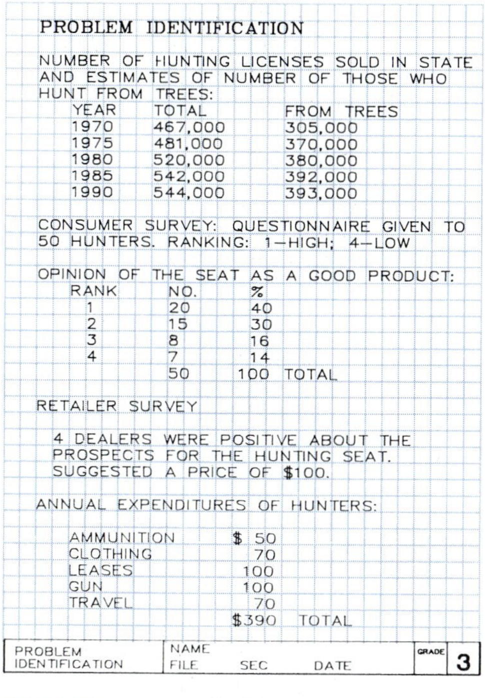

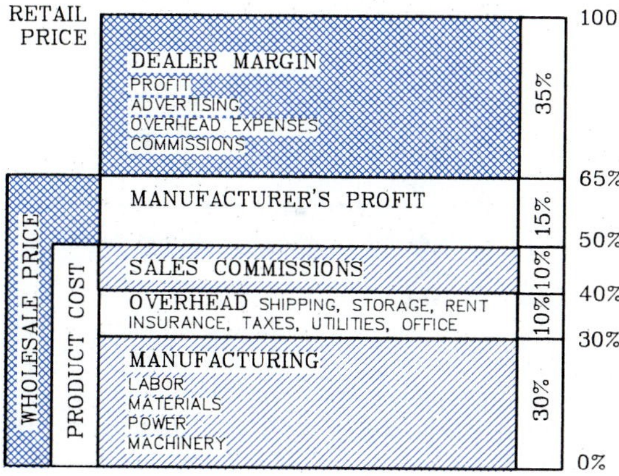

FIGURE 2.13 A model showing the breakdown of expenses and costs involved in arriving at the retail price of a product.

FIGURE 2.12 A worksheet for collecting data.

Another method of collecting information is surveying hunters about the merits of introducing a hunting seat to the market (Fig. 2.12). The data show that 20 out of 50 people gave the market protential of the seat a high ranking.

GRAPHS Data are easier to interpret if presented graphically. For example, Fig. 2.14 compares the number of hunters with those who hunt from trees.

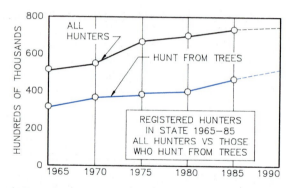

FIGURE 2.14 The survey data plotted in this graph describes the population trends of potential customers for the hunting seat.

FIGURE 2.15 A worksheet listing the brainstorming ideas recorded by a design team.

Thorough problem identification includes graphs, sketches, and schematics that improve the communication of the findings and the conclusions of the designer.

The problem indentification of this problem is not yet complete. By following the method in this example, you should be able to incorporate your own innovations to arrive at a more thorough problem identification.

2.3
Hunting seat—preliminary ideas

BRAINSTORMING IDEAS are gathered from a brainstorming session with classmates. All ideas are listed on a worksheet (Fig. 2.15). Remember, wild ideas are encouraged. The better ideas are then selected and their features described on the worksheet (Fig. 2.16). You may list more features than would be possible to include in a single design; be sure no ideas are forgotten or lost at this stage.

SKETCHES OF PRELIMINARY IDEAS are drawn on worksheets, using rapid freehand techniques. Orthographic views and pictorial methods are both used. Thoughts or questions that come to mind during this sketching should be noted on the drawings. Lettering and sketching techniques need not be highly detailed or precisely executed.

In Fig. 2.17, ideas have been adapted from various types of chairs, lawn chairs in particular. Each

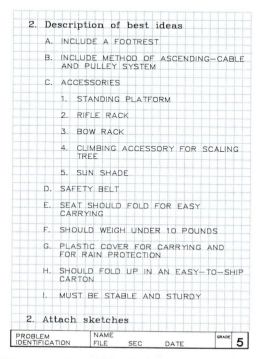

FIGURE 2.16 A description of the better ideas selected from the original brainstorming ideas.

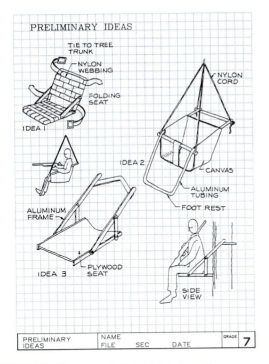

PRELIMINARY IDEAS

TIE TO TREE TRUNK

NYLON WEBBING

FOLDING SEAT

IDEA I

NYLON CORD

IDEA 2

CANVAS

ALUMINUM TUBING

FOOT REST

ALUMINUM FRAME

PLYWOOD SEAT

IDEA 3

SIDE VIEW

| PRELIMINARY IDEAS | NAME
FILE   SEC   DATE | GRADE **7** |

FIGURE 2.17 **A worksheet for presenting preliminary ideas on the development of a hunting seat.**

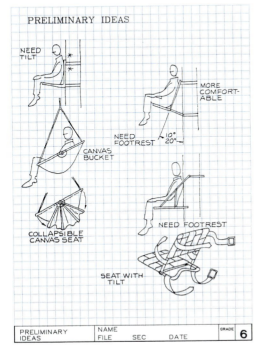

PRELIMINARY IDEAS

NEED TILT

MORE COMFORTABLE

NEED FOOTREST   10°
20°

CANVAS BUCKET

COLLAPSIBLE CANVAS SEAT

NEED FOOTREST

SEAT WITH TILT

| PRELIMINARY IDEAS | NAME
FILE   SEC   DATE | GRADE **6** |

FIGURE 2.18 **Additional preliminary ideas illustrate design concepts for the hunting seat problem.**

idea is numbered for identification. Another worksheet (Fig. 2.18) shows a fourth idea, and idea #2 is modified to include a footrest and suggests a method of guying the seat while suspended.

Many other ideas of this type need to be developed and sketched before leaving the preliminary ideas step of the design process.

2.4
Hunting seat—refinement

REFINEMENT   The features to be incorporated into the design are listed on a worksheet (Fig. 2.19).

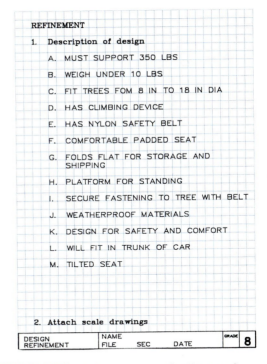

REFINEMENT

1. Description of design

    A. MUST SUPPORT 350 LBS

    B. WEIGH UNDER 10 LBS

    C. FIT TREES FOM 8 IN TO 18 IN DIA

    D. HAS CLIMBING DEVICE

    E. HAS NYLON SAFETY BELT

    F. COMFORTABLE PADDED SEAT

    G. FOLDS FLAT FOR STORAGE AND SHIPPING

    H. PLATFORM FOR STANDING

    I. SECURE FASTENING TO TREE WITH BELT

    J. WEATHERPROOF MATERIALS

    K. DESIGN FOR SAFETY AND COMFORT

    L. WILL FIT IN TRUNK OF CAR

    M. TILTED SEAT

2. Attach scale drawings

| DESIGN REFINEMENT | NAME
FILE   SEC   DATE | GRADE **8** |

FIGURE 2.19 **A design's specifications and desirable features are listed on a worksheet of this type. In this example, the hunting seat is refined.**

Idea #2 is refined in Fig. 2.20, where a scale drawing of the seat is shown orthographically. Tubular parts, like the separator bars, are blocked in to expedite the drawing process; also, some hidden lines are omitted.

It is important that refinement drawings be made to scale to give an accurate proportion of the design

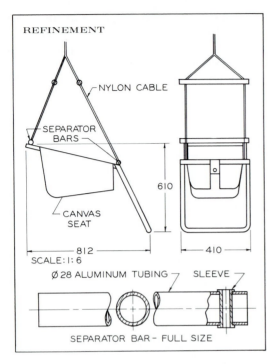

FIGURE 2.20 A refinement drawing of idea #2 for a hunting seat. Only general dimensions are given on the scale drawing.

and to serve as a basis for finding angles, lengths, shapes, and other geometric specifications. Only overall dimensions are given on the drawing, but specific details are shown for the separator bar to explain an idea for a sleeve to protect the nylon cord from being cut. Another design concept is refined in Fig. 2.21. Once again, only the major dimensions are given on these scaled refinement drawings made with instruments. These worksheets do not represent a complete refinement of the design; they are merely examples of the type of drawings required in this step of the design process.

2.5
Hunting seat—analysis

To illustrate a method of analyzing a product design, we return to the hunting seat problem.

HUNTING SEAT   Many hunters hunt from trees to obtain a better vantage point. Design a seat that provides the hunter with comfort and safety while hunting from a tree and that meets the requirements of economy and the limitations of hunting.

Analysis

The worksheets in Fig. 2.22 list the major areas of analysis. Additional worksheets should be used to elaborate on each of these areas as required.

FIGURE 2.21 Another design concept for a hunting seat is shown as a refinement drawing.

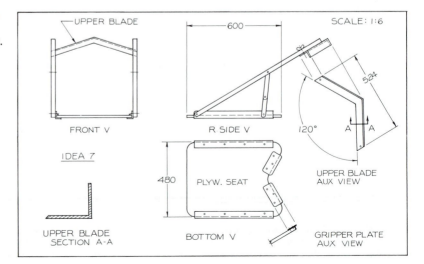

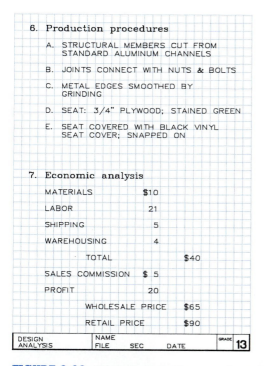

ANALYSIS
1. Function

    A. PROVIDES A METHOD OF CLIMBING TREES

    B. PROVIDES COMFORTABLE SEATING

    C. PROVIDES DECK FOR STANDING

2. Human engineering

    A. SAFETY BELT INCLUDED

    B. FOOTREST FOR COMFORT

    C. PORTABLE: 10–15 LBS IN WEIGHT

    D. SHOULDER STRAPS FOR CARRYING

    E. 360 DEGREES OF VISION

3. Market & consumer acceptance

    A. POTENTIAL MARKET
       1. STATE: 40,000
       2. NATION: 1,800,000

    B. CHEAPER THAN DEER STAND BY 100%–500%

    C. AFFORDABLE AT $100

    D. ADVERTISE IN HUNTING MAGAZINES

    E. RETAIL THROUGH SPORTING GOODS STORES AND DIRECT MAIL

DESIGN ANALYSIS	NAME FILE SEC DATE	GRADE **11**

4. Physical description

    A. PLATFORM 19" X 24", STAINED

    B. FITS TREES OF 6"–18" DIAMETERS

    C. STRAPS FOR CARRYING ON BACK

    D. HAND–CLIMBER DEVICE INCLUDED

    E. SEAT FOLDS FLAT TO 20" X 36"

    F. WEIGHT–10 LBS

    G. NYLON SAFETY BELT

    H. VINYL PADDED SEAT

5. Strength

    A. SUPPORTS 300 LBS

    B. SAFTEY BELT SUPPORTS 350 LBS

    C. SHOULDER STRAPS SUPPORT 350 LBS

    D. HAND CLIMBER SUPPORTS 400 LBS

    E. FOOT REST SUPPORTS 200 LBS WITHOUT BENDING

    F. TREE TRUNK STRAP SUPPORTS 400 LBS

DESIGN ANALYSIS	NAME FILE SEC DATE	GRADE **12**

6. Production procedures

    A. STRUCTURAL MEMBERS CUT FROM STANDARD ALUMINUM CHANNELS

    B. JOINTS CONNECT WITH NUTS & BOLTS

    C. METAL EDGES SMOOTHED BY GRINDING

    D. SEAT: 3/4" PLYWOOD; STAINED GREEN

    E. SEAT COVERED WITH BLACK VINYL SEAT COVER; SNAPPED ON

7. Economic analysis

MATERIALS	$10	
LABOR	21	
SHIPPING	5	
WAREHOUSING	4	
TOTAL		$40
SALES COMMISSION	$ 5	
PROFIT	20	
WHOLESALE PRICE		$65
RETAIL PRICE		$90

DESIGN ANALYSIS	NAME FILE SEC DATE	GRADE **13**

FIGURE 2.22   **Worksheet used to analyze the various features of a hunting seat design.**

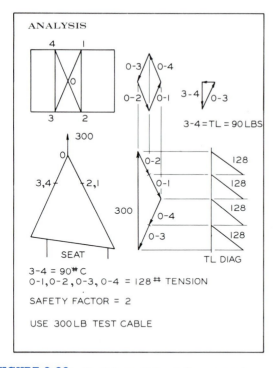

ANALYSIS

4   1

O

3   2

0-3   0-4

0-2   0-1

3-4   0-3

3-4 = TL = 90 LBS

300

O

3,4   2,1

SEAT

0-2   128

0-1   128

300

0-4   128

0-3   128

TL DIAG

3-4 = 90# C
0-1, 0-2, 0-3, 0-4 = 128# TENSION

SAFETY FACTOR = 2

USE 300 LB TEST CABLE

FIGURE 2.23 On this worksheet, the support system of a proposed hunting seat is analyzed with graphical vectors.

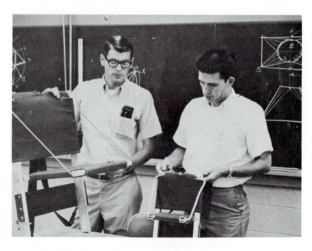

FIGURE 2.24 Full-size and half-size models were built by Keith Sherman and Larry Oakes to aid them in analyzing their design.

The strength of the hunting seat's support system of one design is determined by using graphical vectors (Fig. 2.23). Once the loads in each support cable are found, the proper size cable can be selected. For further analysis, models are constructed at a reduced scale and at full size (Fig. 2.24).

A commercial version of the seat is illustrated in Fig. 2.25, where it is tested to measure its functional features, including the method of using it to climb a

FIGURE 2.25 **A.** The hunting seat used as a platform for standing. **B.** The hunting seat used for sitting.

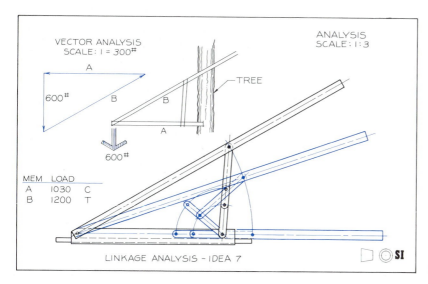

FIGURE 2.26 The drawing is used to analyze the linkage system of the hunting seat, and a vector diagram is used to determine the forces in the members when the seat is loaded to maximum.

tree. An analysis drawing of the hunting seat (Fig. 2.26) illustrates the operation of the linkage system that permits the seat to collapse into a single plane for carrying ease. The forces in the members are also found graphically by using vector analysis. Figure 2.27 shows the overall features and physical properties of the Baker Tree Stand. These features are helpful to a consumer in making a purchase.

2.6 ────────────────────────────
Hunting seat—decision

Oral and technical reports present the basis for deciding whether a recommended design should be implemented.

ACCEPTANCE A design may be accepted in its entirety, which is a compliment to the designer's research and problem-solving abilities.

REJECTION A design recommendation may be rejected in its entirety. Changes in the economic climate or moves by competitors may make the design unprofitable.

COMPROMISE If a design is not approved in its entirety, a compromise might be suggested in one or more areas; for example, the initial production run

BAKER TREE STAND FEATURES:
1. Platform 19″ x 24″ (456 sq. in.) stained.
2. Back pack wt. 10 lbs.
3. Tested to hold 560 lbs.
4. Fits trees 5″ to 18″ diameter.
5. Riveted assembly folds flat for carrying.
6. Hand Climber fits inside frame.
7. Back packs with Strap Assembly.
8. Safety Belt with Extension and a Tie Down (for safety strap) included.

FIGURE 2.27 A summary of the features and physical properties of the Baker Tree Stand. (Courtesy of Baker Manufacturing Co.)

might be increased or decreased, or several features might be modified to make a design more attractive.

Decision—hunting seat

The design of a hunting seat is used to illustrate the decision step of the design process. We restate the problem below.

HUNTING SEAT Many hunters hunt from trees to obtain a better vantage point. Design a seat that provides the hunter with comfort and safety while hunting from a tree and that meets the requirements of economy and the limitations of hunting.

DECISION CHART The chart shown in Fig. 2.28 can be used to compare the available designs. Each idea is listed and given a number for identification.

Next, maximum values for the various factors of analysis are assigned so that the total of all factors is ten points. These values will vary from product to product, so use your best judgment to determine

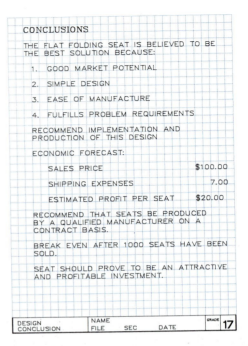

FIGURE 2.29 The decision is summarized on this worksheet to give the designer's conclusion and recommendation concerning the next step, implementation.

them. You can now evaluate each factor of the competing designs.

The vertical columns of numbers are added to determine the design with the highest total, the best design, perhaps. But your instincts may disagree with your numerical analysis. If so, you should have enough faith in your judgment not to be restricted by your numerical analysis. Decision will always remain the most subjective part of the design process.

CONCLUSION Once a decision has been made, it should be clearly stated along with the reasons for the design's acceptance (Fig. 2.29). Additional information may be given such as number to be produced initially, selling price per unit, expected profit per unit, expected sales during the first year, number that must be sold to break even, and its most marketable features.

It is possible that you would recommend a design not be implemented. The design process should not then be considered a failure; a negative decision could save an investor from large losses.

DECISION
1. Decision for evaluation

DESIGN 1	FOLDING SEAT
DESIGN 2	CANVAS SEAT
DESIGN 3	PLATFORM SEAT
DESIGN 4	
DESIGN 5	
DESIGN 6	

MAX	FACTORS	1	2	3	4
3.0	FUNCTION	2.0	2.3	2.5	
2.0	HUMAN FACTORS	1.6	1.4	1.7	
0.5	MARKET ANALYSIS	0.4	0.4	0.4	
1.0	STRENGTH	1.0	1.0	1.0	
0.5	PRODUCTION	0.3	0.2	0.4	
1.0	COST	0.7	0.6	0.8	
1.5	PROFITABILITY	1.1	1.0	1.3	
0.5	APPEARANCE	0.3	0.4	0.4	
10	TOTALS	7.4	7.3	8.5	

DESIGN DECISION | NAME FILE SEC DATE | GRADE **16**

FIGURE 2.28 A worksheet with a decision table used to evaluate the developed design alternatives.

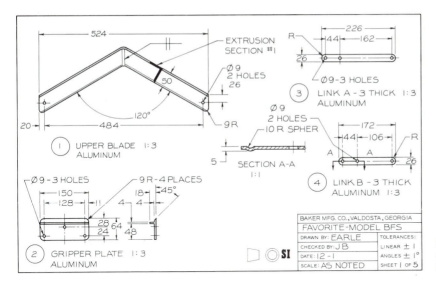

FIGURE 2.30 A working drawing sheet of parts of a hunting seat design. Sheet 1 of 5.

2.7

Hunting seat— implementation

The design of a hunting seat illustrates the implementation step of the design process. We restate the problem.

HUNTING SEAT Many hunters hunt from trees to obtain a better vantage point. Design a seat that pro-

vides the hunter with comfort and safety while hunting from a tree and that meets the requirements of economy and the limitations of hunting.

Four working drawing sheets (Figs. 2.30–2.33) have been prepared to present the details of the hunting seat design. The fifth sheet (Fig. 2.34) is an assembly drawing that illustrates how the parts are assembled once they have been made. (This particular design was developed and patented and is marketed by Baker Manufacturing Company, Valdosta, Georgia. It is the Baker Favorite Seat, Patent No. 3460649.)

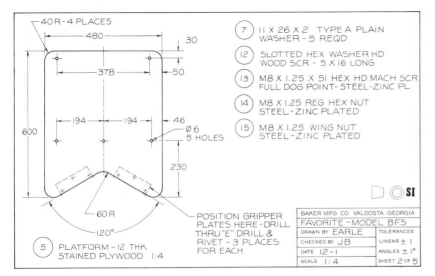

FIGURE 2.31 A working drawing sheet of parts of a hunting seat design. Sheet 2 of 5.

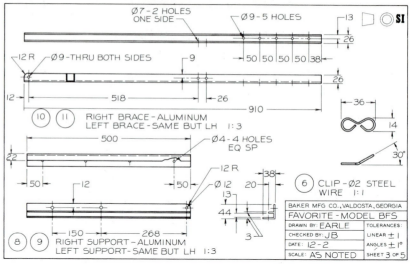

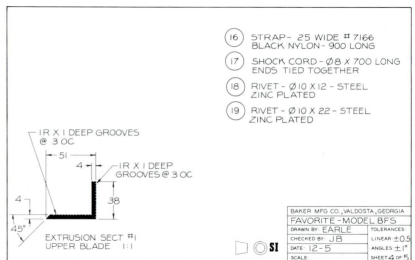

FIGURE 2.32 A working drawing sheet of parts of a hunting seat design. Sheet 3 of 5.

FIGURE 2.33 A working drawing sheet of parts of a hunting seat design. Sheet 4 of 5.

FIGURE 2.34 An assembly drawing that demonstrates how the parts of the hunting seat design are assembled. Sheet 5 of 5.

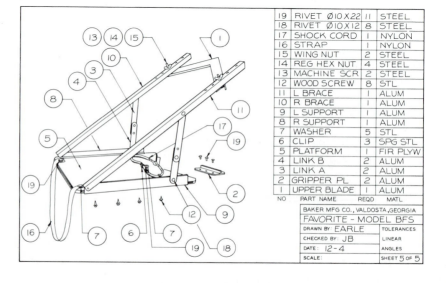

All parts have been dimensioned in millimeters, which are metric units. Standard parts purchased from suppliers are not drawn but are itemized on the drawing, given parts numbers, and listed in the parts list on the assembly drawing. Figure 2.34 is an assembly drawing, which is a pictorial with the different parts identified by balloons attached to leaders. Each part is listed in the parts list by number with general information to describe it.

PACKAGING The Baker Favorite Seat is packaged in a corrugated cardboard box and weighs approximately 10 pounds (Fig. 2.35). The box has been customized to advertise the product and so that shipping personnel can easily recognize each model (Fig. 2.36).

STORAGE The periods before hunting seasons will require more inventory than other times of the year. An inventory of seats waiting to be sold adds to the cost of overhead, that is, interest payments, warehouse rent, warehouse personnel, and loading equipment.

SHIPPING Shipping costs for all types of carriers (rail, motor freight, air delivery, mail services) must be evaluated. The shipping cost for a single Baker Favorite Seat with its accessories is $5–$10 depending on the distance shipped. The cost per unit is reduced by about 50% when shipped in bundles of ten to the same destination.

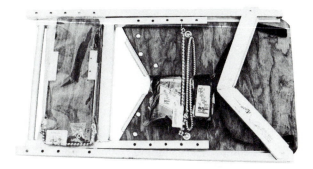

FIGURE 2.36 The hunting seat is folded into a flat position for ease of packaging before shipment.

ACCESSORIES Examples of accessories are fold-down seats, hand climbers, and add-on seats (Fig. 2.37). Accessories provide for the special needs of the hunters, increase the marketability of the product, and increase sales volume.

The retail price of the Baker Seat is about $90, five or six times more than the cost of the materials and labor to manufature it. Retailers are given approximately a 40% margin, and distributors earn about 10%. The remainder of the overhead includes advertising costs and other expenses already mentioned. All expenses must be absorbed by the consumer who ultimately buys the product.

2.8
Introduction to design problems

The following design problems are of two types, short design problems and product design problems. The short problems can be solved by a single person or team within an hour or two.

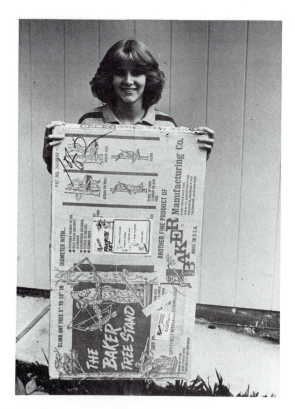

FIGURE 2.35 The Baker Favorite Seat is shipped in a corrugated cardboard box to the retailer or consumer.

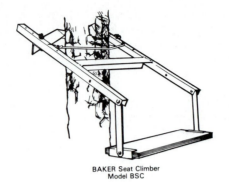

A Conversion Kit, Model CK, will convert Hand Climbers to Seat Climbers.

BAKER Seat Climber
Model BSC

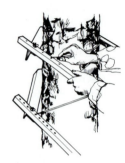

FIGURE 2.37 Examples of accesories designed to accompany the hunting seat. (Courtesy of Baker Manufacturing Co.)

An accessory to the Seat Climber is the Padded Pouch — Model PP.

Secure Seat Climber and Tree Stand with tie down for added safty while hunting.

The product design problems are more comprehensive and require more investigation and background research to fully understand the problem, market needs, and other design factors. These problems are better suited to solution by a team rather than by a single person.

The solution to these problems should follow the steps of the hunting seat example in this chapter.

Short design problems

The following short problems can be completed in less than two periods:

1. **Lamp bracket.** Design a simple bracket to attach a desk lamp to a vertical wall for reading in bed. The lamp should be easily removable so that it can be used as a conventional desk lamp.

2. **Towel bar.** Design a towel bar for a kitchen or bathroom. Determine optimum size, and consider styling, ease of use, and method of attachment.

3. **Pipe aligner for welding.** The initial problem of joining pipes with a butt weld is the alignment of the pipes in the desired position. Design a device with which to align pipes for on-the-job welding. For ease of operation, a hand-held device would be desirable. Assume the pipes will vary in diameter from 2 to 4 in.

4. (Fig. 2.38) **Film reel design.** The film reel used on projectors is often difficult to thread because of the limited working space. The figure shows a typical 12-in-diameter movie reel. Redesign this type of reel to allow more space for threading.

5. **Sideview mirror.** In most cars, rearview mirrors are attached to the side of the automobile to im-

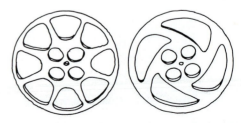

FIGURE 2.38 Problem 4. A typical movie projector reel.

prove the driver's view of the road. Design a side-view mirror that is an improvement over those you are familiar with. Consider the aerodynamics of your design, protection from inclement weather, and other factors that would affect the function of the mirror.

6. **Railing-post mount.** An ornamental iron railing is to be attached to a wooden porch surface supported by several 1-in^2 tubular posts. Design the mounting piece that will attach the posts to the surface. Figure 2.39 shows a typical railing.

   A second attachment is needed to assemble the railing with the support post at each end. If two screw holes are to be used, design the part that can secure the two perpendicular members.

7. **Nail feeder.** Workers lose time in covering a roof with shingles if they have to fumble for nails. Design a device that can be attached to a worker's chest and that will hold nails in such a way that they will be fed in a lined-up position ready for driving.

8. **Cupboard door closer.** Kitchen cupboard and cabinet doors are usually not self-closing and thus are safety hazards and unsightly. Design a device that will close doors left partly open. It would be advantageous to provide a means for disengaging the closer when desired.

9. **Paint-can holder.** Paint cans are designed with a simple wire bail that, when held, makes it difficult to get a paint brush in the can. Wire bails are also painful to hold for any length of time. Design a holding device that can be easily attached and removed from a gallon-size paint can

FIGURE 2.40 Problem 12. A movie projector. (Courtesy of Eastman Kodak.)

($6\frac{1}{2} \times 7\frac{1}{2}$ in.). Consider human factors such as comfort, grip, balance, and function.

10. **Self-closing, self-opening hinge.** Interior doors of residences tend to remain partly open. The edge of the door that is ajar in the middle of the hall can be a hazard. Design a hinge that will hold the door in a completely open position.

11. **Tape cartridge storage unit.** Tape players are often used in automobiles. Design a storage unit that will hold several of these cartridges in an orderly fashion so that the driver can select and insert them with minimum motion and distraction. Determine the best location for this unit and the method of attachment to an automobile.

12. (Fig. 2.40) **Slide projector elevator.** Most commercial slide and movie projectors have adjustment feet to raise the projector to the proper position for casting an image on a screen. Study a slide projector to determine the specific needs and limitations of an adjustment. Design a device to serve this purpose as part of the original design or as an accessory that could be used on existing projectors.

13. **Book holder for reading in bed.** As a student, you may often desire to read while lying in bed. Design a holder that can be used for supporting a book in the desired position.

14. **Table leg design.** Do-it-yourselfers build a variety of tables using slab doors, plywood, and commercially available legs. Table tops come in several sizes, but table heights are fairly standard.

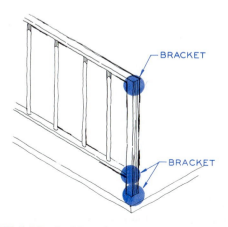

FIGURE 2.39 Problem 6. A porch railing system.

Determine what the standard table heights are, and design a family of legs that can be attached to table tops with screws. Indicate the method of manufacture, size, cost and method of attachment.

15. **Canoe mounting system.** Canoes and light boats are often transported on the top of automobiles on a luggage rack or similar attachment. Design an accessory that will enable a single person to remove and load a boat on top of an automobile. This attachment should accommodate aluminum boats from 14 to 17 ft long and weighing from 100 to 200 lb. Give specifications for a method of securing the boat after it is on top of the automobile.

16. **Toothbrush holder.** Design a toothbrush holder than can be attached to a bathroom wall and that can hold a drinking cup and two toothbrushes.

17. **Napkin holder.** Design a device that will hold 25 paper napkins on a dining room table for easy access.

18. **Book holder.** Design a holder that will support your textbook on your drawing table in a position that will make it easier to read and more accessible.

19. **Clothes hook.** Design a clothes hook that can be attached to a closet door for hanging clothes. It should be easy to manufacture and simple to use.

20. **Automobile ashtray.** Design an ashtray that can be attached to the dashboard of any automobile. It should be easy to attach and to remove.

21. **Pencil holder.** Design a holder that can be attached to the interior of the car for holding pencils or pens.

22. **Door stop.** Design a door stop that can be attached to a vertical wall or floor to prevent the door knob from bumping the wall.

23. **Teaching aid.** Design an apparatus that can be used by a teacher to illustrate the principles of orthographic projection. Investigate the market potential of such a product.

24. **Cup dispenser.** Design a paper-cup dispenser that can be attached to a vertical wall. This dispenser should hold a series of cups 2 in. in diameter that measure 6 in. tall when stacked together.

25. **Drawer handle.** Design a handle that would be satisfactory for a standard file cabinet drawer.

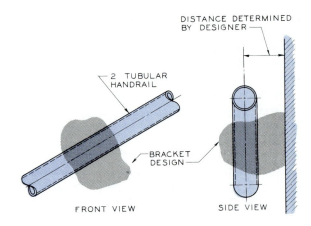

FIGURE 2.41   **Problem 27. A handrail bracket.**

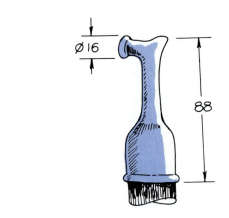

FIGURE 2.42   **Problem 28. A latchpole hanger.**

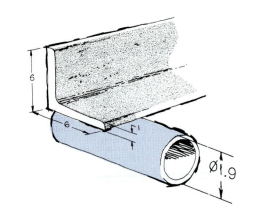

FIGURE 2.43   **Problem 29. A pipe clamp.**

26. **Paper dispenser.** Design a dispenser that will hold a 6-×-24-in. roll of wrapping paper. The paper will be used on a table top for wrapping packages.

27. (Fig. 2.41) **Handrail bracket.** Design a bracket that will support a tubular handrail to be used on a staircase. Consider the weight that the handrail must support.

28. (Fig. 2.42) **Latchpole hanger.** Design a hanger that can be used to support a latchpole from a vertical wall. It should be easy to install and use.

29. (Fig. 2.43) **Pipe clamp.** A pipe with a 4-in. diameter must be supported by angles spaced 8 ft apart. Design a clamp that will support the pipe without drilling holes in the angles.

30. **TV yoke.** Design a yoke that will support a TV set from the ceiling of a classroom and that will permit it to be adjusted at the best position for viewing.

31. **Flagpole socket.** Design a flagpole socket that is to be attached to a vertical wall. Determine the best angle of inclination for the flagpole.

32. **Crutches.** Design a portable crutch that could be used by a person with a temporary leg injury.

33. **Cup holder.** Design a holder that will support a soft-drink can or bottle in the interior of an automobile.

34. **Gate hinge.** Design a hinge that could be attached to a 3-in-diameter tubular post to support a 3-ft-wide gate.

35. (Fig. 2.44) **Safety lock.** Design a safety lock that

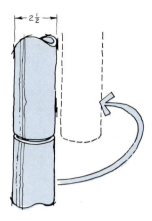

FIGURE 2.45 Problem 36. A tubular hinge.

will hold a high-voltage power switch in either the off or on positions to prevent an accident.

36. (Fig. 2.45) **Tubular hinge.** Design a hinge that can be used to hinge 2.5-inch OD high-strength aluminum pipe in the manner shown in this figure. A hinge of this type is needed for portable scaffolding.

37. (Fig. 2.46) **Miter jig.** Design a jig that can be used for assembling wooden frames at 90° angles. The stock for the frames is to be rectangular in cross sections that vary from 0.75 × 1.5 in. to 1.60 × 3.60 in. Outside dimensions vary from 10 to 24 in.

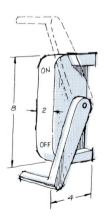

FIGURE 2.44 Problem 35. A safety lock.

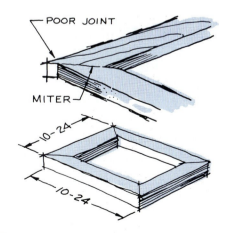

FIGURE 2.46 Problem 37. A miter jig.

38. (Fig. 2.47) **Base hardware.** Design the hardware needed at the points indicated for a standard volleyball net. The 7-ft pipes are supported by crossing 2 × 4-in. boards. Design the hardware needed at points *a*, *b*, and *c*.

39. (Fig. 2.48) **Conduit connector hanger.** Design a support that will attach to a ¾-in. conduit that will support a channel used as an adjustable raceway for electrical wiring. Your design should permit ease of adjustment.

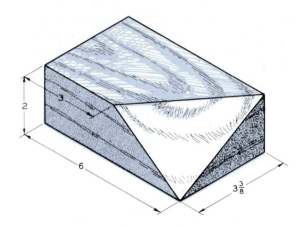

FIGURE 2.49 Problem 40. A fixture design.

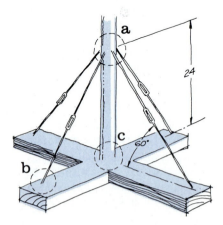

FIGURE 2.47 Problem 38. Base hardware for a vollyball net.

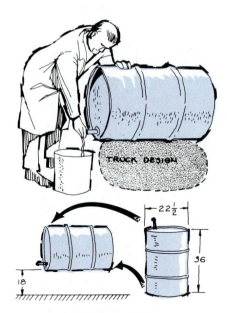

FIGURE 2.50 Problem 41. A drum truck.

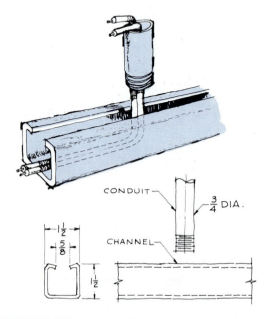

FIGURE 2.48 Problem 39. A conduit connector hanger.

40. (Fig. 2.49) **Fixture design.** Design a fixture that will permit a small-scale manufacturer to saw the corner of the block as shown in this figure.

41. (Fig. 2.50) **Drum truck.** Design a truck that can be used for moving a 55-gal drum of turpentine (7.28 lb per gallon). The drum will be kept in a horizontal position, but it would be advantageous to incorporate a feature into the truck permitting the drum to be set upright.

Product design problems

A product design involves problems developing a device that will perform a specific function and be mass-produced and sold to a broad market.

42. **Hunting blind.** Hunters of geese and ducks must remain concealed while hunting. Design a portable hunting blind to house two hunters. This blind should be completely portable so that it can be carried in separate sections by each of the hunters. Specify its details and how it is to be assembled and used.

43. **Convertible drafting table.** Design a drafting table that can enclose a drafting machine, thereby concealing it and protecting it when not needed.

44. (Fig. 2.51) **Writing table for a folding chair.** Design a writing-tablet arm for a folding chair that could be used in an emergency or when a class needs more seating. To allow easy storage, the arm must fold with the chair.

45. (Fig. 2.52) **Firewood caddy.** Design a cart that can be used for carrying firewood to a fireplace from the outdoors. This cart, loaded with wood, should be easy to handle when climbing steps. Include other features that would make the caddy attractive as a marketable product.

46. (Fig. 2.53) **Sensor-retaining device.** The instrumentation Department of the Naval Oceanographic Office uses underwater sensors to learn more about the ocean. These sensors, which each weigh 75 lb, are submerged on cables from a boat on the surface. The winch used to retrieve the sensor frequently overruns (continues pulling when it has been retrieved), causing the cable to break and the sensor to be lost. Design a safety device that will retain the sensor if the cable is broken when the sensor reaches a pulley.

47. **Flexible trailer hitch.** In combat zones, vehicles must tow trailers where terrain may be uneven and hazardous. Design a trailer hitch that will provide for the most extreme conditions possible. Study the problem requirements and limitations to identify the parameters your design must function within.

48. **Workers' stilts—human engineering.** Workers who apply gypsum board and other types of wallboards to the interiors of buildings must work on scaffolds or wear some type of stilts to be able to reach the ceiling to nail the 4- × -8-ft boards

FIGURE 2.51 Problem 44. Folding chair with a writing tablet attached.

FIGURE 2.52 Problem 45. A firewood caddy for hauling firewood.

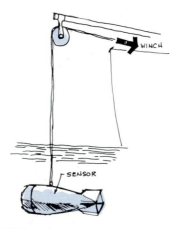

FIGURE 2.53 Problem 46. An underwater sensor.

into position. Design stilts that will provide workers with access to an 8-ft ceiling while permitting them to nail ceiling panels with comfort.

49. **Pole-vault uprights.** Many pole-vaulters are exceeding the 18-ft height in track meets, which introduces a problem for the officials of this event. The pole-vault uprights must be adjusted for each pole-vaulter by moving them forward or backward plus or minus eighteen in. Also, the crossbar must be replaced with great difficulty at these heights by using forked sticks and ladders. Develop a more efficient set of uprights that can be easily repositioned and that will allow the crossbar to be replaced with greater ease.

50. **Sportsman's chair.** Analyze the need for a sportsman's chair that could be used for camping, for fishing from a bank or boat, at sporting events, and for as many other purposes as you can think of. The need is not for a special-purpose chair but for a chair suitable for a variety of uses to fully justify it as a marketable item.

51. **Portable toilet.** Design a portable toilet unit for the camper and outdoorsperson. The unit should be highly portable, with consideration given to the method of waste disposal. Evaluate the market potential for this product.

52. **Child carrier for a bicycle.** Design a seat that can be used to carry a small child as a passenger on a bicycle. Assume the bicycle will be ridden by an older youth or an adult. Determine the age of the child who would probably be carried as a passenger.

53. **Lawn-sprinkler control.** Design a sprinkler that can be used to water irregularly shaped yards while giving a uniform coverage. This sprinkler should be adjustable so that it can be adapted, within its range, to yards of any shape. Also consider a method of cutting the water off at certain sprinkler positions to prevent the watering of patios or other areas that are to remain dry.

54. **Power lawn-fertilizer attachment.** The rotary-power lawn mower emits a force through its outlet caused by the air pressure from the rotating blades. This force might be used to distribute fertilizer while the lawn is being mowed. Design an attachment for a power mower that could spread fertilizer while the mower is performing its usual cutting operation.

55. **Car and window washer.** Design an attachment for the typical garden hose that would apply water and agitation (for optimum action) to the surface being cleaned. Consider other applications of the force exerted by water pressure in the performance of yard and household chores.

56. **Projector cabinet.** Design a cabinet that could serve as an end table or some other function while also housing a slide projector ready for use at any time. The cabinet might also serve as storage for slide trays. It should have electrical power for the projector. Evaluate the market potential for a multipurpose cabinet of this type.

57. **Heavy-appliance mover.** Design a device that can be used for moving large appliances, such as stoves, refrigerators, and washers, about the house. This product would not be used often—only for rearranging, cleaning, and servicing the appliances.

58. **Car jack.** The average car jack does not attach itself adequately to the automobile's frame or bumper, introducing a severe safety problem. Design a jack that would be an improvement over existing jacks and possibly employ a different method of applying a lifting force to a car. Consider the various types of terrain on which the device must serve.

59. **Map holder.** The driver of an automobile traveling alone in an unfamiliar part of the country must frequently refer to a map. Design a system that will give the driver a ready view of the map in a convenient location in the car. Also provide a means of lighting the map during night driving that will not distract the driver.

60. **Bicycle-for-two adapter.** Design the parts and assembly required to convert two bicycles into one bicycle-built-for-two (tandem). Work from an existing bicycle, and consider how each rider can equally share in the pedaling. Determine the cost of your assembly and its method of attachment to the average bicycle.

61. **Automobile unsticker.** Design a kit to be carried in the car trunk in a minimum of space containing the items required to "unstick" a car when no other help is available. This kit can contain one or several items. Investigate the need for such a kit and the main factors that lead to the loss of traction.

62. Stump remover. Assume a number of tree stumps must be removed from the ground to clear land for construction. The stumps are dead with partially deteriorated root systems and require a force of approximately two thousand pounds to remove them. Design an apparatus that could be attached to a car bumper that could be used to remove the stumps by either pushing or pulling.

63. Gate opener. An annoyance to farmers and ranchers is the necessity of opening and closing gates when driving from one fenced area to the next. A gate that could be opened and closed without getting out of the vehicle would be desirable. Design a manually operated gate that would appeal to this market.

64. Paint mixer. Design a product that could be used by the paint store or paint contractor to quickly mix paint in the store and on the job. Determine the standard-size paint cans your mixer will be designed for.

65. Mounting for an outboard motor on a canoe. Unlike a square-end boat, the pointed-end canoe does not provide a suitable surface for attaching an outboard motor. Design an attachment that will adapt an outboard motor to a canoe.

66. Automobile coffee maker. Adequate heat is available in the automobile's power system to prepare coffee in minimum time. Design an attachment as an integral system of an automobile that will serve coffee from the dashboard area. Consider the type of coffee to be used (instant or regular), the method of changing or adding water, the spigot system, and similar details.

67. Baby seat (cantilever). Design a child's chair that can be attached to a standard table top and that will support the child at the required height. The chair should be designed to ensure that the child cannot crawl out or detach it from the table top. A possible solution could be a design that would cantilever from the table top, using the child's body as a means of applying the force necessary to grip the table top. The design would be further improved if the chair were collapsible or suitable for other purposes.

68. Miniature-TV support. Miniature television sets for close viewing are available with a 5- × -5-in. screen. An attachment is needed that would support sets ranging in size from 6 × 6 in. to 7 × 7 in. for viewing from a bed. Determine the placement of the set for best viewing results. Provide adjustments on the support that will be used to position the set properly.

69. Panel applicator. A worker who applies 4-×-8-ft gypsum board or paneling must be assisted by a helper who holds the panel in position while it is being nailed to the ceiling. Design a device that could be used in this capacity; it should be collapsible, economical, and versatile. The average ceiling height is 8 ft, but provide adjustments that would adapt the device to lower or higher ceilings.

70. Backpack. Design a backpack that can be used by the outdoorsperson who must carry supplies while hiking. Most of your design effort should be devoted to adapting the backpack to the human body for maximum comfort and the leverage for carrying a load over a long time. Can other uses be made of your design?

71. Automobile controls. Design driving controls that can be easily attached to the standard automobile that will permit an injured person to drive a car without the use of his or her legs. This device should be easy to operate and provide maximum safety.

72. Bathing apparatus. Design an apparatus that would help a wheelchair-bound person who does not have use of his or her legs to get in and out of a bathtub without help from others.

73. Adjustable TV base. Design a TV base to support full-size TV sets that would allow maximum adjustment: up and down, rotation about a vertical axis and a horizontal axis. Design the base to be as versatile as possible.

74. Trailer jack. Design a trailer jack that can be used to repair flat tires that may occur on a boat trailer. It may be possible to build in jack devices as permanent features of the trailer.

75. Projector cabinet. Design a portable cabinet, which could be left permanently in a classroom, that would house a slide projector and movie projector. The cabinet should provide both convenience and security from vandalism and theft.

76. (Fig. 2.54) Boat loader. Design a rack and a system whereby a single person could load a boat weighing 110 lb on top of a car for transporting

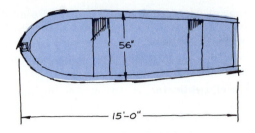

FIGURE 2.54 Problem 76. Boat specifications.

home owner. The mixer is seldom used, so it should be sufficiently simple and economical to justify its being built.

80. **Boat trailer.** Design a trailer that supports the boat under the trailer rather than on top of it. Develop this design so that a boat can be launched in water more shallow than is required now.

81. **Washing machine.** Design a manually operated washing machine that can be used by people in less-developed countries that may not have power. This could be considered an "undesign" of a powered washing machine.

82. (Fig. 2.56) **Pickup truck hoist.** Design a tailgate that can be attached to the tailgate of a pickup truck and that can be used for raising and lowering loads from the ground to the floor of the truck. Design it to be operated without a motor.

from site to site. Use the boat specifications shown in the figure.

77. (Fig. 2.55) **Door opener.** Design a method whereby a trucker at a loading dock could open the warehouse door without having to get out of the truck. The doors are dimensioned in the figure, and the dock extends 8 ft from the doors.

78. **Projector eraser.** Design a device that would erase grease pencil markings from the acetate roll of an overhead projector as the acetate is cranked past the stage of the projector.

79. **Cement mixer.** Design a portable and simple cement mixer that can be operated manually by a

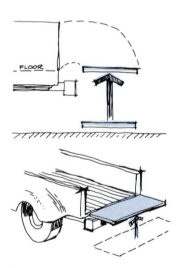

FIGURE 2.56 Problem 82. A pickup truck hoist.

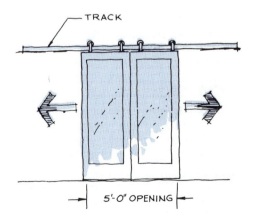

FIGURE 2.55 Problem 77. Door dimensions.

83. (Fig. 2.57) **Writing tablet for the handicapped.** Design a writing tablet that will permit a person to write from a prone position using a series of mirrors. This problem involves applying human engineering considerations.

84. (Fig. 2.58) **Portable seat.** Design a portable seat that can be used in as many applications as possible, such as at football games, in fishing boats, and at campsites. It should be easy to carry and collapsible.

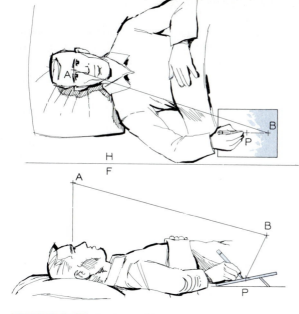

FIGURE 2.57 Problem 83. A writing tablet for the handicapped.

FIGURE 2.58 Problem 84. A portable seat.

FIGURE 2.59 Problem 85. A portable display booth.

85. (Fig. 2.59) **Display booth.** Design a portable display booth that can be used behind or with an 8-ft table. Your design should consist of panels that stand behind the table on which your company's name and product advertising can be integrated. It must be collapsible so that it can be easily carried by one person as airplane luggage.

86. (Fig. 2.60) **Exerciser.** Design a simple device that can be used specifically for aiding in doing push-ups and sit-ups. It should sell for around $20 and store easily in the minimum of space.

87. (Fig. 2.61) **Patio grill.** Design a portable charcoal grill for cooking on the patio. Consider how it would be cleaned, stored, and used. Study the competing products on the market to develop a marketable grill.

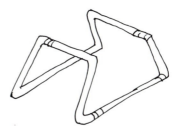

FIGURE 2.60 Problem 86. A push-up and sit-up exerciser.

FIGURE 2.61 Problem 87. A patio grill.

FIGURE 2.62 Problem 88. A shop bench.

88. (Fig. 2.62) **Shop bench.** Design a shop bench that will be marketable, serve a need, be collapsible for ease of storage, and be adjustable. Could it also be designed to serve as a step stool? Design it to accommodate other accessories such as vises, anvils, and electrical appliances.

89. (Fig. 2.63) **Chimney cover.** Design a chimney cover that can be closed from inside the house for repelling rain and saving lost heat and cooling. Evaluate your design's market potential, and analyze its manufacturing cost.

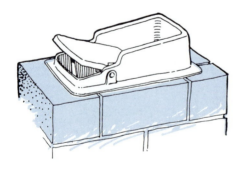

FIGURE 2.63 Problem 89. A chimney cover.

90. (Fig. 2.64) **Punching bag platform.** Design a punching bag platform for supporting a speed bag. Your design should offer an easy means of adjusting the platform's height. It should be portable and easy to ship in a flat box.

91. (Fig. 2.65) **Shopping caddy.** Design a portable caddy that a shopper can use for carrying parcels. It should be as lightweight as possible.

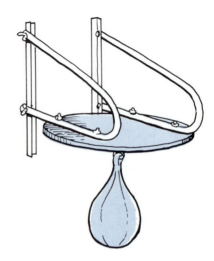

FIGURE 2.64 Problem 90. A punching bag platform.

FIGURE 2.65 Problem 91. A shopping caddy.

The Computer in Design and Drafting

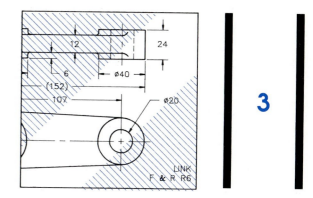

3

3.1
Introduction

The use of computers in engineering and related fields is widespread, and great growth is still expected. It is increasingly important for students of engineering and technology to become familiar with the nature and prospects of computer technologies. We devote this chapter to computer usage in drafting and design. To familiarize you with how computer graphics can help apply the principles of this text, we emphasize techniques of computer-aided design drafting (CADD).

3.2
Computer-aided design

Computer-aided design (CAD) is the process of solving design problems with the aid of computers. This includes computer generation and modification of graphic images on a video display, printing these images as hard copy on a plotter or printer, analysis of design data, and electronic storage and retrieval of design information. Many CAD systems perform these functions in an integrated fashion, which can increase the designer's productivity manyfold.

Computer-aided design drafting (CADD), a subset of CAD, is the computer-assisted generation of working drawings and other engineering documents. The CADD user generates graphics by interactive communication with the computer. Graphics are displayed

on a video display and can be converted into hard copy with a plotter or printer.

Most engineers agree that the computer does not change the nature of the design process and that it is simply a tool to improve efficiency and productivity. The designer and the CAD system should be viewed as a design team: The designer provides knowledge, creativity, and control, whereas the computer is able to generate accurate, easily modifiable graphics; to perform complex design analysis at great speeds; and to store and recall design information. Occasionally, the computer can augment or replace many of the engineer's other tools, but it is important to remember that it does not change the fundamental role of the designer.

Advantages of using the computer in design and drafting

Depending on the nature of the problem and the sophistication of the computer system, computer use can afford the designer or drafter several potential advantages.

1. **Easier creation and correction of working drawings.** In a CAD system, working drawings can be created using function commands and digitizing methods (assigning numerical coordinates). Complicated changes and corrections are made using a few keystrokes.

2. **Easier visualization of drawings.** In many systems, different views of the same part can be dis-

41

42

played quickly and easily. In systems with three-dimensional capabilities, a part can even be rotated on the CRT screen.

3. **Drawings can be stored and easily referenced for modification.** Modified designs can be made from one original in far less time than it would take using a manual approach. Design databases (libraries of designs) can be created in some systems. These databases can also store standard parts and symbols for easy recall. Many systems are configured so that information in the databases can be easily accessed by others in an organization, such as management or production personnel.

4. **Quick and convenient solution of computational design analysis problems.** Because the computer offers a tremendous advantage in ease of design analysis, the designer can rigorously analyze each design, thus speeding up the design refinement stage.

5. **Simulation and testing of designs.** Some systems enable the engineer to simulate the operation of a design and to perform tests and analyses in which the part is subjected to a variety of conditions or stresses. This capacity may improve or replace the process of building models and prototypes.

6. **Increased accuracy.** The accuracy of the computer lessens the chance for error. Many CAD systems are capable of detecting errors and will inform the user when data or designs are incorrect.

3.3
Applications of computer graphics

Computer graphics is almost limitless in its applications to engineering and technical fields. Almost all graphical solutions that can be done with a pencil can be done with the computer, and often they can be done more productively. Applications can vary from three-dimensional modeling and finite element analysis to two-dimensional drawings and mathematical calculations.

Automobile bodies are designed at the computer (Fig. 3.1). As part of this process, Fig. 3.2 shows a scanning device that records measurements of the life-size tape drawing of a new vehicle. Once recorded and stored in the computer, the data can be manipulated

FIGURE 3.1 The Ford Taurus and Mercury Sable received more computer-aided design and engineering than any other car in the company's history. (Courtesy of Ford Motor Company.)

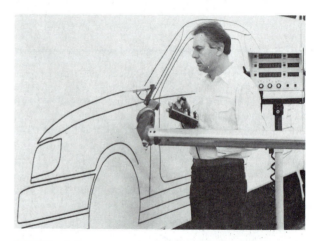

FIGURE 3.2 A scanning device is used to record measurements from a life-size tape drawing of a new vehicle. Once stored, this data can be used for other design functions. (Courtesy of Ford Motor Company.)

to form the three-dimensional shape of the body style on the computer monitor from which three-dimensional clay models are milled (Fig. 3.3).

Advanced applications, once the domain of large computer systems, can now be done on microcomputers. Figure 3.4 illustrates how a portion of the earth's surface was modeled and shown three dimensionally by using a microcomputer. Using simple keyboard commands, this site plan can be viewed from an infinite number of positions to obtain desired vantage points.

In Fig. 3.5, a perspective drawing of an urban area was plotted using microcomputer software. This

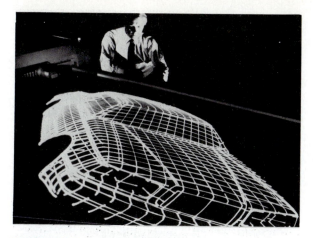

FIGURE 3.3   Once design data is stored in the computer's memory, it can be manipulated to obtain variations in body design. (Courtesy of Ford Motor Company.)

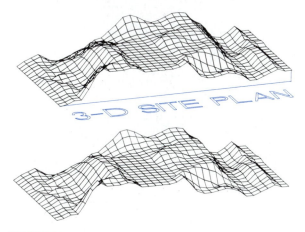

FIGURE 3.4   A three-dimensional representation of a site plan. The lower version is slightly different from the upper one since hidden lines on the surface have been hidden by the computer. (Courtesy of LANDCADD Inc.)

program lets the viewer be positioned at any height and location in the site. The viewer can "walk through" the city by successively changing positions.

A microcomputer was used to produce the layer of the electronic circuit board shown in Fig. 3.6. Circuit board drawings are drawn as much as five times their actual size and then photographically reduced.

3.4
CAD/CAM

An important application of CAD lies in the field of manufacturing. **Computer-aided design/computer-aided manufacturing** (CAD/CAM) describes a system that can design a part or product, devise the essential production steps, and electronically communicate this information to manufacturing equipment like robots (Fig. 3.7). A CAD/CAM system offers many potential advantages over traditional manufacturing systems, including less design effort through the use of CAD and CAD databases, more efficient material use, reduced lead time, greater accuracy, and improved inventory functioning.

In **computer-integrated manufacturing** (CIM) a computer or system of computers coordinates all stages of manufacturing, which enables manufacturers to custom design products efficiently and economically.

3.5
Hardware systems

Computer-aided design systems have three major components: the designer, hardware, and software. **Hardware** is the physical components of a computer

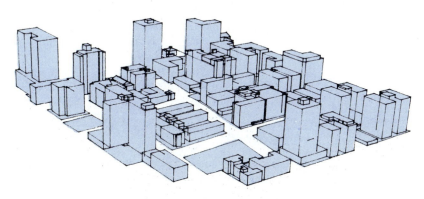

FIGURE 3.5   An urban area has been plotted as a perspective drawing with hidden lines removed using MegaCADD's Design Board 3D. (Courtesy of MegaCADD, Inc.)

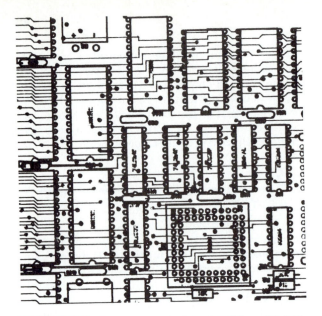

FIGURE 3.6 This is one of nine circuit board trace layers designed on the computer using Auto-Board System II.

FIGURE 3.7 This automatic welding system for Chrysler LeBaron GTS and Dodge Lancer car bodies uses computer-controlled robots for consistent welds of all components in the unitized body structure. This assembly plant also features energy-efficient electric robot welders that require less maintenance and provide a high degree of accuracy. (Courtesy of Chrysler Corp.)

system, and **software** is the programmer's instructions to the computer. The hardware of a computer graphics system includes the computer, terminal, input devices (digitizers, light pens), and output devices (plotters, printers) (Fig. 3.8).

Computer

Computers receive input from the user, execute the instructions contained in the input, and then produce some form of output. A sequence of instructions called a **program** controls the computer's activities. The part of the computer that follows the program's instructions is the **central processing unit** (CPU).

FIGURE 3.8 A Hewlett-Packard Vectra with a computer, keyboard, monitor, and mouse.

Computers are often classified by size. **Mainframes,** the largest computers, are big, fast, powerful, and expensive. **Minicomputers,** smaller and less costly than mainframes, are used by many small businesses. **Microcomputers,** the smallest computers, are widely used for both personal and business applications. Rapid advances in hardware technology and software capability, along with continually falling prices, have brought microcomputers into wide use in engineering graphics applications in industry, government, and education.

Terminal

The **terminal** allows the user to communicate with the computer. It typically consists of a keyboard, a

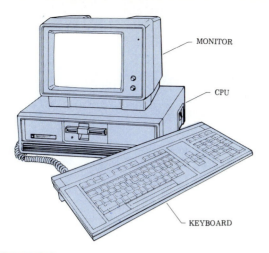

FIGURE 3.9 **The basic components of a computer system: CPU, keyboard, and monitor.**

cathode-ray tube, and the interconnections between these devices and the computer (Fig. 3.9).

KEYBOARD The keyboard allows the user to communicate with the computer through a set of alphanumeric and function keys. Keyboards generally resemble typewriters but include many other function keys, some of which may be user-defined.

CATHODE-RAY TUBE (CRT) A CRT is a video display device consisting of a tube with a phosphor-coated screen. An electron gun throws a beam that sweeps out rows, called **raster lines,** on the screen.

Each raster line consists of a number of dots called **pixels.** Images are generated on the screen by turning pixels on and off. Raster-scanned CRTs refresh the picture display many times per second. One measure of the quality of the pictures that can be produced on a screen is **resolution,** the number of pixels per inch that can be drawn on the screen. Higher-quality graphics can be drawn on higher-resolution screens.

Raster-scan technology has largely replaced the **vector-refreshed tube** used in early video displays. In this type of display, each line in the picture is continuously redrawn by the computer. Because the display produced a collection of lines instead of a set of individual pixels, vector-display pictures are less realistic than raster-scanned displays.

Input devices

DIGITIZER The digitizer is a graphics input device that can communicate information in a picture to the computer for display, storage, or modification (Fig. 3.10). The user lays a drawing or sketch on a digitizer board and scans it, thereby converting the picture to a digital or computerized form based on the *xy* coordinates of individual points. Since a digitizer can also input symbols, a fully labeled drawing can be represented on a CRT screen and stored for later use.

LIGHT PEN Digitized pictures on a screen can be created or modified point by point using a light pen. The user places this device in contact with the CRT screen, thereby telling the computer the position of the pen on the screen (Fig. 3.11).

FIGURE 3.10 **A digitizer board is made up of a set of coordinates that correspond to points on the CRT screen. (Courtesy of GTCO Corporation.)**

FIGURE 3.11 **A light pen allows the user to modify drawings by touching the CRT screen with it. (Courtesy of Ford Motor Company.)**

MOUSE A mouse is a device that can be rolled on a table top to move a cursor (a mark on the screen that indicates location) around a CRT screen. The mouse can be used to activate commands or change information on the screen (Fig. 3.12).

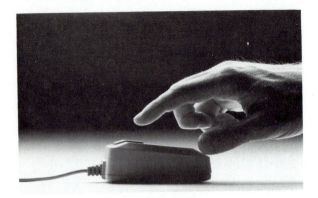

FIGURE 3.12 A mouse can move the cursor around a CRT screen. The user activates commands or makes changes by pressing the button(s). (Courtesy of Apollo Computer, Inc.)

FIGURE 3.13 A joy stick can be used to "steer" a cursor around a CRT screen. (Courtesy of Apple Computer, Inc.)

JOY STICK A joy stick allows the user to "steer" a cursor around the screen by tilting a lever in appropriate directions (Fig. 3.13).

Output devices

PLOTTER The plotter is a machine directed by the computer to make a drawing. Two types of plotters are

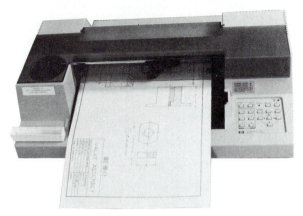

FIGURE 3.14 A Hewlett-Packard 7475 plotter for making Size A and Size B drawings.

the **flatbed** plotter and the **drum,** or **roll-feed,** plotter. In the flatbed plotter, paper is attached to the bed, and the pen is moved about the paper in raised or lowered position to complete the drawing (Fig. 3.14). In the drum plotter, a special type of paper is held on a spool and rolled over a rotating drum (Fig. 3.15). As

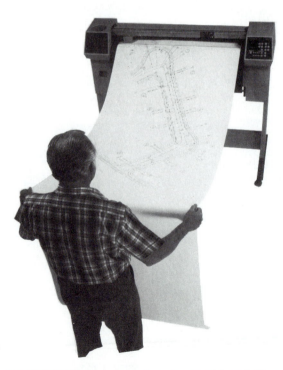

FIGURE 3.15 A roll-feed plotter used for plotting the output. (Courtesy of Hewlett-Packard.)

the drum rotates, the pen suspended above the paper moves left or right along the drum.

PRINTER The printer is a device operated by the computer that makes images on paper. An **impact printer,** which works like a typewriter, forms characters by forcing typefaces to impact with an inked ribbon and paper. **Dot-matrix printers,** a popular, inexpensive type of impact printer, have printheads composed of a rectangle of pins, each of which can be raised or lowered to form a character. These patterns of pins are forced against a ribbon to make dotted characters on paper. Dot-matrix printers are useful for graphics because the patterns of dots can be made to correspond to lighted pixels on a CRT screen. However, the output of dot-matrix printers is of lower quality than some other printers, so they have limited usefulness in graphics applications.

FIGURE 3.16 The PlotMaster thermal transfer plotter/printer produces fast, high-resolution color hard copy for computer-aided drawings. (Courtesy of CalComp.)

Nonimpact printers form characters from a distance by using ink sprays, laser beams, photography, or heat. One example, the **ink-jet printer,** uses electrical fields to direct jets of ink to appropriate spots on the paper (Fig. 3.16). Like the dot-matrix printer, the ink-jet printer forms a pattern of dots, thus limiting its accuracy or resolution. One feature of this type of printer is that multicolor graphics can be generated by using multiple jet nozzles.

As printer technology makes rapid strides in cost reduction and output quality, it will be found increasingly useful for graphics applications. Still, the need for strict accuracy makes the use of plotters more attractive for many applications.

3.6
CAD software for the microcomputer

The availability of CAD software for the microcomputer continues to increase, and each upgrade is more powerful than the previous version. Before long, almost all computer graphics will be done on the microcomputer (Fig. 3.17).

The major advantages of CAD microcomputer software packages are cost and ease of upgrading when improvements in the software are made.

Among the more popular computer graphics software packages for the microcomputer are AutoCAD, VersaCAD, MegaCADD, DynaPerspective, and Cadkey. Most of these packages require a computer with at least a 20 Mb hard disk and 640 Kb of RAM. Figure 3.17 shows a typical IBM-compatible graphics system where a mouse is used to draw on the color monitor.

FIGURE 3.17 A typical computer-graphics workstation comprised of a Hewlett-Packard Vectra with a color monitor, mouse, and AutoCAD software.

Lettering

4

4.1
Lettering

All drawings are supplemented with notes, dimensions, and specifications that must be lettered. The ability to construct legible freehand letters is an important skill to develop since it affects the usage and interpretation of a drawing.

4.2
Tools of lettering

The best pencils for lettering on most surfaces are in the H–HB grade range, with an F pencil being the most commonly used grade. Some papers and films are coarser than others and may require a harder pencil lead. To give the desired line width, the point of the pencil should be slightly rounded (Fig. 4.1); a needle point will break off when pressure is applied.

When lettering, the pencil should be revolved slightly between your fingers as the strokes are being made so that the lead will wear down gradually and evenly. Bear down firmly to make letters black and

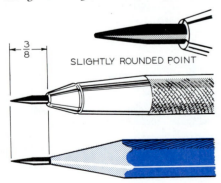

FIGURE 4.1 Good lettering begins with a properly sharpened pencil point. The point should be slightly rounded, not a needle point. The F pencil is usually the best grade for lettering.

FIGURE 4.2 When lettering a drawing, use a protective sheet under your hand to prevent smudges. Your lettering will be best when you are working from a comfortable position; you may wish to turn your paper for the most natural strokes.

bright for good reproduction. To prevent smudging your drawing while lettering, place a sheet of paper under your hand to protect the drawing (Fig. 4.2).

4.3
Gothic lettering

The type of lettering recommended for engineering drawings is **single-stroke Gothic lettering,** so called because the letters are made with a series of single strokes, and the letter form is a variation of Gothic lettering.

> Two general categories of Gothic lettering are **vertical** and **inclined** (Fig. 4.3). Although equally acceptable, they both should not be used on the same drawing.

4.4
Guidelines

> The most important rule of lettering is **use guidelines at all times.**

This applies whether you are lettering a paragraph or a single letter or numeral. Figure 4.4 shows the method of constructing and using guidelines. Use a sharp pencil in the 3H–5H grade range, and draw

FIGURE 4.3 Two types of Gothic lettering recommended by engineering standards are vertical and inclined lettering.

these lines lightly, just dark enough for them to be seen.

Most lettering is done with the capital letters $\frac{1}{8}$ in. (3 mm) high. The spacing between lines of lettering should be no closer than half the height of the capital letters, $\frac{1}{16}$ in. in this case.

Lettering guides

The two instruments used most often for drawing guidelines are the **Braddock-Rowe lettering triangle** and the **Ames lettering instrument.**

The Braddock-Rowe triangle is pierced with sets of holes for spacing guidelines (Fig. 4.5B). The numbers under each set of holes represent thirty-seconds of an inch. For example, the numeral 4 represents $\frac{4}{32}$ in. or $\frac{1}{8}$ in. for making uppercase (capital) letters. Some triangles are marked for metric lettering in millimeters. In Fig. 4.5, intermediate holes are provided for guidelines for lowercase letters, which are not as tall as the capital letters.

With a horizontal straightedge held firmly in position, place the Braddock-Rowe triangle against its edge. A sharp 4H pencil is placed in one hole of the desired set of holes to contact the drawing surface, and

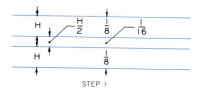

FIGURE 4.4 Lettering guidelines.

Step 1 Letter heights, *H*, are laid off, and light guidelines are drawn with a 4H pencil. The spacing between the lines should be no closer than *H*/2.

Step 2 Vertical guidelines are drawn as light, thin lines. These are randomly spaced to serve as visual guides for lettering.

Step 3 The letters are drawn with single strokes using a medium-grade pencil, H–HB. The guidelines need not be erased since they are drawn lightly.

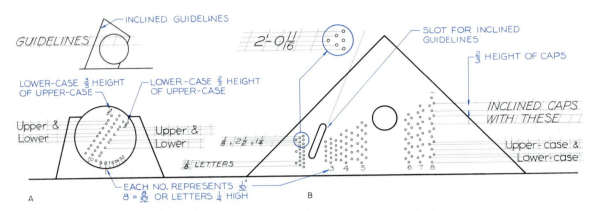

FIGURE 4.5 **A.** The Ames lettering guide can be used to draw guidelines for vertical or inclined uppercase and lowercase letters. The dial is set to the desired number of thirty-seconds of an inch for the height of uppercase letters. **B.** The Braddock-Rowe triangle can be used as a 45° triangle and as an instrument for constructing guidelines. The numbers designating the guidelines represent thirty-seconds of an inch.

the pencil point is guided across the paper to draw the guideline while the triangle slides against the straight-edge. This is repeated as the pencil point is moved successively to each hole until the desired number of guidelines are drawn. An oblique slot for drawing guidelines for inclined lettering is cut in the triangle. Slanting guidelines are spaced randomly by eye.

The Ames lettering guide (Fig. 4.5A) is a similar device with a circular dial for selecting the proper spacing of guidelines. Again, the numbers around the dial represent thirty-seconds of an inch. The number 8 represents $\frac{8}{32}$ in. or guidelines for drawing capital letters that are $\frac{1}{4}$ in. tall.

4.5
Vertical letters

Vertical capital letters

Figure 4.6 shows the capital letters for the single-stroke Gothic alphabet. Each letter is drawn inside a square box of guidelines to help you learn their correct proportions. Some letters require the full area of the box, some require less space, and a few require more space. Each straight-line stroke should be drawn as a single stroke; for example, the letter A is drawn with three single strokes. Letters composed of curves can best be drawn in segments; the letter O can be drawn by joining two semicircles to form the full circle.

FIGURE 4.6 The uppercase letters used in single-stroke Gothic lettering. Each letter is drawn inside a square to help you learn their correct proportions.

Memorize the shape of each letter given in this alphabet. Small wiggles in your strokes will not detract from your lettering if the letter forms are correct. Figure 4.7 shows examples of poor lettering.

Vertical lowercase letters

Figure 4.8 shows an alphabet of lowercase letters. Lowercase letters are either two thirds or three fifths as tall as the uppercase letters they are used with. Both of these ratios are labeled on the Ames guide, but only the two-thirds ratio is available on the Braddock-Rowe triangle.

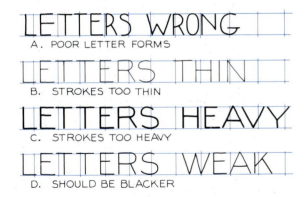

FIGURE 4.7 **There are many ways to letter poorly. A few of them, and the reasons the lettering is inferior, are shown here.**

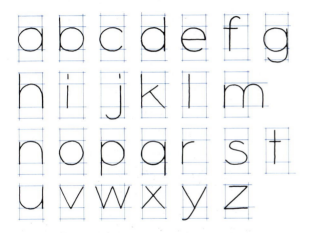

FIGURE 4.8 **The lowercase alphabet used in single-stroke Gothic lettering. The body of each letter is drawn inside a square to help you learn the proportions.**

FIGURE 4.9 **Uppercase and lowercase letters are sometimes used together. The ratio of the lowercase letters to the uppercase letters will be either two thirds or three fifths. The Ames guide has both, and the Braddock-Rowe triangle has only the three-fifths ratio.**

Some lowercase letters have **ascenders** that extend above the body of the letter, such as the letter b; some have **descenders** that extend below the body, such as the letter y. The ascenders are the same length as the descenders.

The guidelines in Fig. 4.8 form perfect squares about the body of each letter to illustrate the proportions. Several of these letters have bodies that are perfect circles that touch all sides of the squares. In Fig. 4.9, capital and lowercase letters are used together.

Vertical numerals

Figure 4.10 shows vertical numerals for single-stroke Gothic lettering. Each number is enclosed in a square box of guidelines. Each number is also made the same height as the capital letters being used, usually $\frac{1}{8}$ in. high. The numeral, 0 (zero) is an oval, whereas the letter O is a perfect circle in vertical lettering.

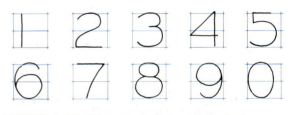

FIGURE 4.10 **The numerals for single-stroke Gothic lettering. Each numeral is drawn inside a square to help you learn their proportions.**

4.6
Inclined letters

Inclined uppercase letters

Inclined uppercase letters (capitals) have the same heights and proportions as vertical letters; the only difference is their 68° inclination (Fig. 4.11). Inclined guidelines should be drawn using the Braddock-Rowe triangle or the Ames guide.

Inclined lowercase letters

Inclined lowercase letters are drawn in the same manner as vertical lowercase letters (Fig. 4.12). Ovals (ellipses) are used instead of the circles used in vertical

FIGURE 4.11 The inclined uppercase alphabet for single-stroke Gothic lettering.

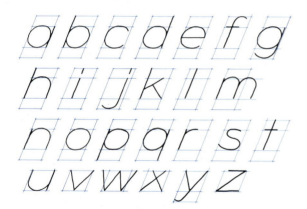

FIGURE 4.12 The inclined lowercase alphabet for single-stroke Gothic lettering. The body of each letter is drawn inside a rhombus to help you learn the proportions.

FIGURE 4.13 The inclined numerals for single-stroke Gothic lettering. Each number is drawn inside a rhombus to help you learn the proportions.

lettering. The angle of inclination is 68°, the same as is used for uppercase letters.

Inclined numerals

Figure 4.13 shows the inclined numerals that should be used with inclined lettering, and Fig. 4.14 shows the use of inclined letters and numbers in combination. The guidelines in Fig. 4.14 were constructed using the Braddock-Rowe triangle.

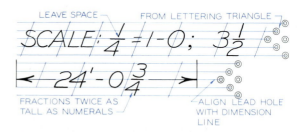

FIGURE 4.14 Inclined common fractions are twice as tall as single numerals. Inch marks are omitted when numerals are used to show dimensions.

4.7 ─────────────
Spacing numerals and letters

Common fractions are twice as tall as single numerals (Fig. 4.15). A separate set of holes for common fractions is given on the Braddock-Rowe triangle and on the Ames guide. These are equally spaced $\frac{1}{16}$ in. apart with the centerline being used for the fraction's crossbar.

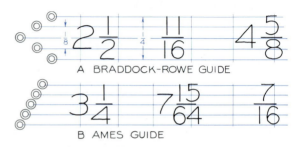

FIGURE 4.15 Common fractions are twice as tall as single numerals. Guidelines for these can be drawn by using the Ames guide or the Braddock-Rowe triangle.

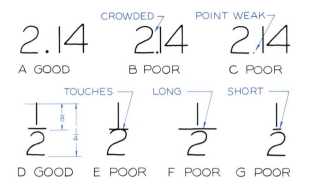

FIGURE 4.16 Examples of poorly spaced numerals that result in inferior lettering.

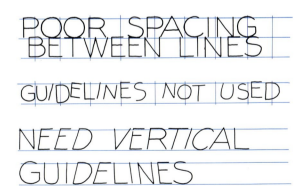

FIGURE 4.17 Proper spacing of letters is necessary for good lettering and appearance. The areas between letters should be approximately equal.

FIGURE 4.18 Always leave space between lines of lettering. After constructing guidelines, use them. Use vertical guidelines to improve the angle of your vertical strokes.

When numbers are used with decimals, space should be provided for the decimal point (Fig. 4.16). Figure 4.16D shows the correct method of drawing common fractions, and Figs. 4.16E–G show several errors that are often encountered.

When letters are grouped together to spell words, the area between the letters should be approximately equal for the most pleasing result (Fig. 4.17). Figure 4.18 shows the incorrect use of guidelines and other violations of good lettering practice.

4.8
Mechanical lettering

Drawings and illustrations to be reproduced by a printing process are usually drawn in India ink. Several mechanical aids for ink lettering are available.

The Rapidograph lettering template (Fig. 4.19) can be placed against a fixed straightedge for aligning the letters. You move the template from position to position while drawing each letter (with a lettering pen) through the raised portion of the template where the holes form the letters.

FIGURE 4.19 A typical India ink fountain pen and template that can be used for mechanical lettering. (Courtesy of Koh-I-Noor Rapidograph, Inc.)

Another system of mechanical lettering uses a grooved template along with a scriber that follows the grooves and inks the letters on the drawing surface (Fig. 4.20). A standard India ink technical pen can be unscrewed from its barrel and attached to the scriber (Fig. 4.20). Many templates of varying styles of lettering and symbols are available for this system of mechanical lettering.

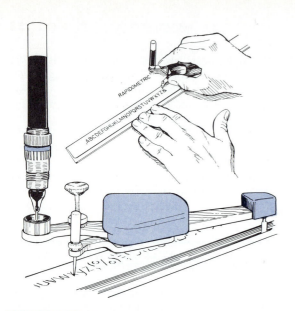

FIGURE 4.20 Templates and scribers of this type are available for mechanical lettering.

4.9
Lettering by Computer

Although AutoCAD offers a wide range of lettering (text) fonts, the ROMANS font shown in Fig. 4.21 is the one that best approximates the single-stroke Gothic text recommended for engineering and technical drawings. An example of a multiple-stroke font, ROMANT, is shown in Fig. 4.22. You will notice that several strokes are used to form the letters.

DTEXT (dynamic text) is the command that is

```
ABCDEFGHIJK
KLMNOPQRST
UVWXYZ&%01
23456789abc
defghijklmnop
rstuvwxyz
```

FIGURE 4.21 Text drawn with the ROMANS font is best for technical and engineering drawings.

```
ABCDEFG
HIJKLMNO
PQRSTUVW
XYZ&%
```
A

```
ABCDEFG
HIJKLMNO
PQRSTUVW
XYZ&%
```
B

FIGURE 4.22 **A. Vertical uppercase letters drawn with the** ROMANT **font. B. Inclined uppercase letters drawn with the** ROMANT **font. The inclination angle is measured from the vertical.**

most often used to apply text to a drawing since it displays the text across the screen as it is typed. The TEXT command can be used, but the text is not displayed until after it has been typed:

```
Command: DTEXT
Justify/Style/<Start point>: J
(CR)
Align/Fit/Center/Middle/Right/
TL/TC/TR/ML/MC/MR/BL/BC/BR:
```

The abbreviations represent the insertion points for a string of text. Figure 4.23 illustrates the meaning of these abbreviations. Examples of different insertion points for text are illustrated in Fig. 4.24.

By using the STYLE command, a single font, ROMANT, can be modified to yield text in any of the formats shown in Fig. 4.25.

```
Command: STYLE (CR)
```

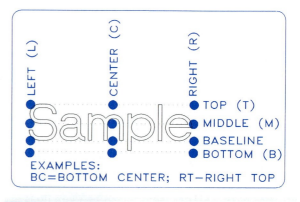

FIGURE 4.23 **Under the** JUSTIFY **option of the** DTEXT **command are several alternatives for inserting text into a drawing. For example, BC means bottom center of a line of text.**

54

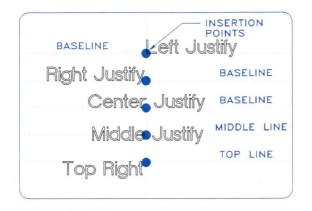

Examples of text inserted in a drawing using various options under DTEXT/JUSTIFY.

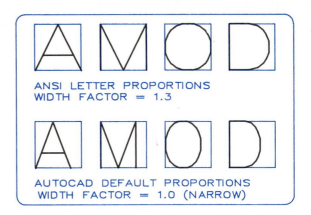

By setting the width factor of the ROMANS font to 1.30, the text will match the ANSI standards for engineering lettering.

```
Text style name (or ?)
<Standard>: NEWFONT (CR)
Font file <TXT>: ROMANT
Height <0>: .125
Width factor <1>: 1.3
Obliquing angle <0>: (CR)
Backwards? <N>: (CR)
Upside-down? <N>: (CR)
Vertical? <N>: (CR)
NEWFONT is now the current
style.
```

Each of these prompts can be responded to in a variety of ways to yield types of text. The prompt that asks for font file must be answered with a name of one of

AutoCAD's fonts as shown in Fig. 23.67. Notice that the width factor was set to 1.30 to better match the recommended letter proportions as shown in Fig. 4.26.

When text is inserted using DTEXT, a box appears on the screen to represent the height of a letter (Fig. 4.27). By using the select key of the mouse, this box can be moved to any position on the screen and a new starting point selected. By pressing the enter button (carriage return), the box will automatically reposition itself for the next line of type. The text STYLE remains in use until the STYLE option of the DTEXT command is selected, and a new style is specified.

NARROW TEXT — 0.8 WIDE ROMANT
REGULAR WIDTH — 1.00 WIDE
WIDE TEXT — 1.30 WIDE
VERY WIDE — 1.8 WIDE
INCLINED TEXT — 22° INCLINED
BACKWARDS — BACKWARDS
UPSIDE DOWN — UPSIDE DOWN

Styles of lettering can be created to your specifications by changing the variables of the STYLE command.

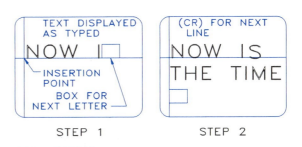

STEP 1   STEP 2

DTEXT command.

Step 1 Command: DTEXT Select insertion point: **(Cursor-select.)**
Height <.18>: .125
Angle of rotation <0>: 0
Text: NOW IS **(CR)**

Step 2 **(Box moves to start of next line.)**
Text: THE TIME **(CR)** **(Box moves to start of next line. (CR to save the text.)**

Problems

Lettering problems are to be presented on Size AV (8½ × 11-inch) paper, plain or grid, using the format shown in Fig. 4.28.

1. Practice lettering the vertical uppercase alphabet shown in Fig. 4.28. Construct each letter three times: three A's, three B's, and so on. Use a medium-weight pencil—H, F, or HB.

2. Practice lettering vertical numerals and the lowercase alphabet as shown in Fig. 4.29. Construct each letter and numeral two times: two 1's, two 2's, and so on. Use a medium-weight pencil—H, F, or HB.

3. Practice lettering the inclined uppercase alphabet shown in Fig. 4.11. Construct each letter three times. Use a medium-weight pencil—H, F, or HB.

4. Practice lettering the vertical numerals and the lowercase alphabet shown in Figs. 4.8 and 4.10.

Construct each letter two times. Use a medium-weight pencil—H, F, or HB.

5. Construct guidelines for ⅛-in. capital letters starting ¼ in. from the top border. Each guideline should end ½ in. from the left and right borders. Using these guidelines, letter the first paragraph of the text of this chapter. Use all vertical capitals. Spacing between the lines should be ⅛ in.

6. Repeat Problem 5, but use all inclined capital letters. Use inclined guidelines to help you uniformly slant your letters.

7. Repeat Problem 5, but use vertical capitals and lowercase letters in combination. Capitalize only those words that are capitalized in the text.

8. Repeat Problem 5, but use inclined capitals and lowercase letters in combination. Capitalize only those words that are capitalized in the text.

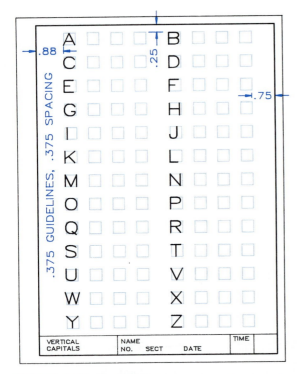

FIGURE 4.28   Problem 1. Construct each vertical uppercase letter three times.

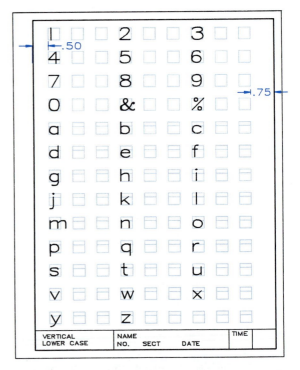

FIGURE 4.29   Problem 2. Construct each vertical numeral and lowercase letter two times.

Drawing Instruments

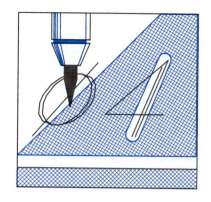

5

5.1
Introduction

Preparing technical drawings is possible only by being knowledgeable of and skillful in the use of drafting instruments. Your skill and productivity will increase as you become more familiar with using the available tools. Persons with little artistic ability can produce professional technical drawings when they learn to use drawing instruments properly.

5.2
Pencil

Pencils may be the conventional wood pencil or the lead holder, which is a mechanical pencil (Fig. 5.1). Both types are identified by numbers and letters at their ends. Sharpen the end of the pencil opposite these markings so that you do not sharpen away the identity of the grade of lead.

FIGURE 5.1 The mechanical pencil (lead holder) or the wood pencil can be used for mechanical drawing. The ends of these pencils are labeled to indicate the grade of the pencil lead.

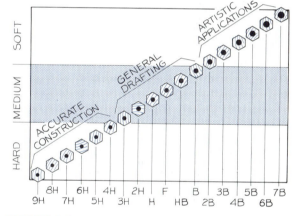

FIGURE 5.2 The hardest pencil lead is 9H, and the softest is 7B. Note the diameters of the hard leads are smaller than the soft leads.

Pencil grades range from the hardest, 9H, to the softest, 7B (Fig. 5.2). The pencils in the medium-grade range, 3H–3B, are most often used for drafting work.

It is important that your pencil be properly sharpened. This can be done with a small knife or a drafter's pencil sharpener, which removes the wood and leaves approximately ⅜ inch of lead exposed (Fig. 5.3). The point can then be sharpened to a conical point with a sandpaper pad by stroking the sandpaper with the pencil point while revolving the pencil (Fig. 5.4). Excess graphite is wiped from the point with a cloth or tissue.

A pencil pointer used by professional drafters (Fig. 5.5) can be used to sharpen wood and mechani-

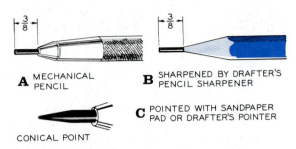

A MECHANICAL PENCIL

B SHARPENED BY DRAFTER'S PENCIL SHARPENER

C POINTED WITH SANDPAPER PAD OR DRAFTER'S POINTER

CONICAL POINT

FIGURE 5.3 **The drafting pencil should be sharpened to a tapered conical point (not a needle point) with a sandpaper pad or other type of sharpener.**

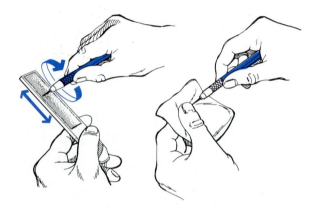

FIGURE 5.4 **The drafting pencil is revolved about its axis as you stroke the sandpaper pad to form a conical point. The graphite is wiped from the sharpened point with a tissue or cloth.**

FIGURE 5.5 **The professional drafter often uses a pencil pointer of this type to sharpen pencils.**

cal pencils. Simply insert the pencil in the hole and revolve it to sharpen the lead. Other types of small hand-held point sharpeners are also available.

5.3
Papers and drafting media

SIZES The surface a drawing is made on must be carefully selected to yield the best results for a given application. Sheet sizes are specified by letters such as Size A, Size B, and so forth. These sizes, which are listed below, are multiples of either the standard $8\frac{1}{2}$-$\times$-11-inch sheet or the 9-$\times$-12-inch sheet.

Size A	$8\frac{1}{2}'' \times 11''$	$9'' \times 12''$
Size B	$11'' \times 17''$	$12'' \times 18''$
Size C	$17'' \times 22''$	$18'' \times 24''$
Size D	$22'' \times 34''$	$24'' \times 36''$
Size E	$34'' \times 44''$	$36'' \times 48''$

DETAIL PAPER When drawings are not to be reproduced by the diazo process (which is a blue-line print), an opaque paper, called **detail paper,** can be used as the drawing surface. The higher the rag content (cotton additive) of the paper, the better its quality and durability. Preliminary layouts can be drawn on detail paper and then traced onto the final tracing surface.

TRACING PAPER A thin, translucent paper used for making detail drawings is **tracing paper** or **tracing vellum.** These papers permit the passage of light through them so that drawings can be reproduced by the diazo, or blue-line, process. The tracing papers that yield the best reproductions are most translucent. Vellum is tracing paper that has been chemically treated to improve its translucency, but it does not retain its original quality as long as do high-quality, untreated tracing papers.

TRACING CLOTH **Tracing cloth** is a permanent drafting medium for both ink and pencil drawings. It is made of cotton fabric that has been covered with a compound of starch to provide a tough, erasable drafting surface that yields excellent blue-line reproductions. More stable than paper, tracing cloth does not change its shape with variations in temperature and humidity as much as does

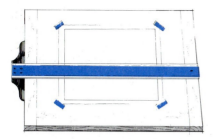

FIGURE 5.6 The T-square and drafting board are the basic tools used by the student drafter. The drawing paper is taped to the board with drafting tape.

5.4
T-square and board

The T-square and drafting board are the basic equipment used by the beginning drafter (Fig. 5.6). With its head in contact with the edge of the drawing board, the T-square can be moved for drawing parallel horizontal lines. Drawing paper should be attached to the board parallel to the blade of the T-square (Fig. 5.7). The drafting board is made of basswood, which is lightweight but strong. Standard sizes of boards are 12 × 14 inches, 15 × 20 inches, and 21 × 26 inches.

tracing paper. Erasures can be repeatedly made on tracing cloth without damaging the surface, which is especially important when drawing with ink.

POLYESTER FILM An excellent drafting surface is polyester film, which is available under several trade names such as Mylar. It is more transparent, stable, and tough than paper or cloth. It is also waterproof and difficult to tear.

Mylar film is used for both pencil and ink drawings. Some films specify that a plastic-lead pencil be used, whereas others adapt well to standard lead pencils. Ink will not wash off with water and will not erase with a dry eraser; erasures can be made with a dampened hand-held eraser.

FIGURE 5.8 The professional who uses a drafting station may work in an environment similar to the one shown here. (Courtesy of Martin Instrument Co.)

FIGURE 5.7 The drafting machine is often used instead of the T-square and drafting board. (Courtesy of Keuffel & Esser Co., Morristown, N.J.)

5.5
Drafting machines

Although the T-square is used in industry and the classroom, most professional drafters prefer the **mechanical drafting machine** (Fig. 5.7). This machine is attached to the table top and has fingertip controls for drawing lines at any angle.

Figure 5.8 shows a modern, fully equipped drafting station. Today, many offices are equipped with computer graphics stations to supplement manual equipment and techniques (Fig. 5.9).

FIGURE 5.9 More and more professional work stations are being equipped with computer graphics equipment. (Courtesy of T&W Systems.)

5.6
Alphabet of lines

The type of line produced by a pencil depends on the hardness of the lead, drawing surface, and technique of the drafter. Figure 5.10 shows examples of the standard lines, or **alphabet of lines**, and the recommended pencils for drawing the lines. These pencil grades may vary greatly with the drawing surface being used. Guidelines are very light lines (just dark enough to be seen) used to aid in lettering and laying out a drawing; a 4H pencil is recommended for drawing most guidelines.

5.7
Horizontal lines

A horizontal line is drawn using the upper edge of your horizontal straightedge and, for the right-handed person, drawing the line from left to right (Fig. 5.11). Your pencil should be held in a vertical plane to make a 60° angle with the drawing surface. As horizontal lines are drawn, the pencil should be rotated about its axis to allow its point to wear evenly (Fig. 5.12). If necessary, lines can be darkened by drawing over them one or more times. For the best line, a small

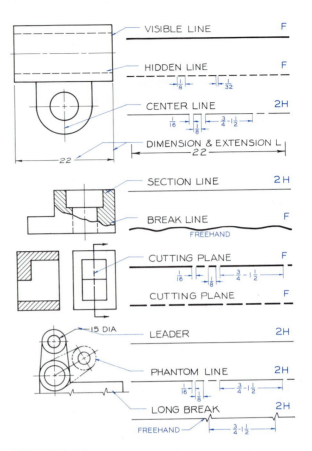

FIGURE 5.10 The alphabet of lines varies in width. The full-size lines are shown in the right column along with the recommended pencil grades for drawing them.

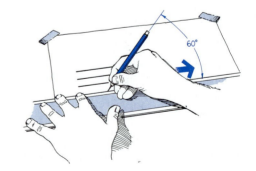

FIGURE 5.11 Horizontal lines are drawn with a pencil held in a plane perpendicular to the paper and at 60° to the surface. These lines are drawn left to right along the upper edge of the T-square.

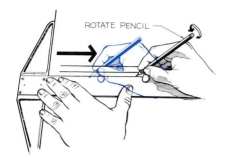

FIGURE 5.12 As the horizontal lines are drawn, the pencil should be rotated about its axis so that the point will wear down evenly.

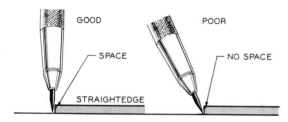

FIGURE 5.13 The pencil point should be held in a vertical plane and inclined 60° to leave a space between the point and the straightedge.

space should be left between the straightedge and the pencil point (Fig. 5.13).

5.8
Vertical lines

A triangle is used with a straightedge for drawing vertical lines. While the straightedge is held firmly with one hand, the triangle can be positioned where needed and the vertical lines drawn (Fig. 5.14). Vertical lines are drawn upward along the left side of the triangle while holding the pencil in a vertical plane at 60° to the drawing surface.

5.9
Drafting triangles

The two most often used triangles are the 45° triangle and the 30°–60° triangle. The 30°–60° triangle is spec-

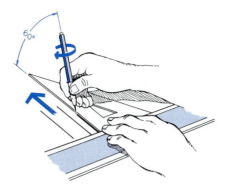

FIGURE 5.14 Vertical lines are drawn along the left side of a triangle in an upward direction with the pencil held in a vertical plane at 60° to the surface.

ified by the longer of the two sides adjacent to the 90° angle (Fig. 5.15). Standard sizes of 30°–60° triangles range, in 2-inch intervals, from 4 to 24 inches.

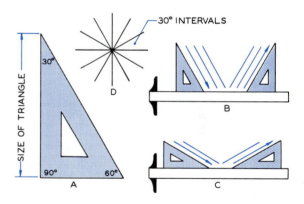

FIGURE 5.15 The 30°–60° triangle can be used to construct lines spaced at 30° intervals throughout 360°.

The 45° triangle is specified by the length of the sides adjacent to the 90° angle. These range from 4 to 24 inches at 2-inch intervals, but the 6- and 10-inch sizes are adequate for most classroom applications. Figure 5.16 shows the various angles that can be drawn with this triangle. By using the 45° and 30°–60° triangles in combination, angles can be drawn at 15° intervals throughout 360° (Fig. 5.17).

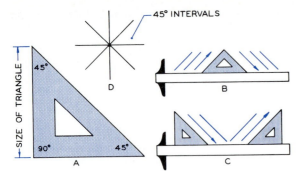

FIGURE 5.16 The 45° triangle can be used to draw lines at 45° intervals throughout 360°.

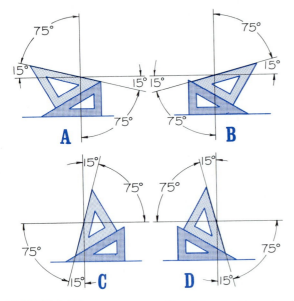

FIGURE 5.17 By using the 30°–60° triangle in combination with the 45° triangle, angles can be drawn at 15° intervals.

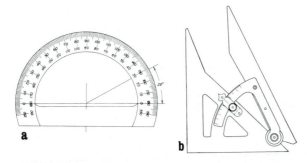

FIGURE 5.18 The semicircular protractor can be used to measure angles. The adjustable triangle can be used as a drawing edge and to measure angles.

5.10
Protractor

When lines must be drawn or measured at multiples of other than 15°, a **protractor** is used (Fig. 5.18). Protractors are available as semicircles (180°) or circles (360°). An adjustable triangle serves as both a protractor and a drawing edge (Fig. 5.18B).

5.11
Parallel lines

A series of lines can be drawn parallel to a given line by using a triangle and straightedge (Fig. 5.19). The 45° triangle is placed parallel to a given line and is held in contact with the straightedge (which may be another triangle). By holding the straightedge in one position, the triangle can be moved to various positions for drawing a series of parallel lines.

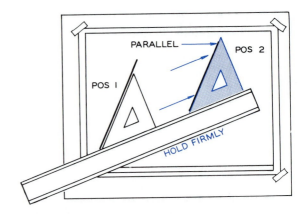

FIGURE 5.19 A straightedge and a 45° triangle can be used to draw a series of parallel lines. The straightedge is held firmly in position, and the triangle is moved from position 1 to position 2.

5.12
Perpendicular lines

Perpendicular lines can be constructed by using either of the standard triangles. A 30°–60° triangle is used with a straightedge or another triangle to draw line 3–4 perpendicular to line 1–2 (Fig. 5.20). One edge of

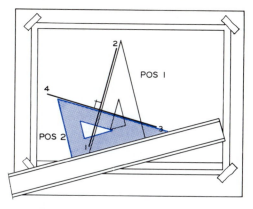

FIGURE 5.20 A 30°–60° triangle and a straightedge can be used to construct a line perpendicular to line 1–2. The triangle is aligned with line 1–2 and is then rotated to position 2 to construct line 3–4.

the triangle is placed parallel to line 1–2 in position 1 with the straightedge in contact with the triangle. By holding the straightedge in place, the triangle is rotated and moved to position 2 to draw the perpendicular line.

5.13
Irregular curves

Curves that are not arcs must be drawn with an **irregular curve.** These plastic curves come in a variety of sizes and shapes, but the one shown in Fig. 5.21 is typical. In this figure, the irregular curve connects a

series of points to form a smooth curve. The **flexible spline** is an instrument used for drawing long, irregular curves (Fig. 5.22). The spline is held in position by weights while the curve is drawn.

5.14
Erasing

Erasing should be done with the softest eraser that will serve the purpose. For example, ink erasers should not be used to erase pencil lines because ink erasers are coarse and may damage the surface of the paper. When working in small areas, an **erasing shield** can prevent your accidentally erasing adjacent lines (Fig. 5.23). All erasing should be followed by removing the "crumbs" with a dusting brush; do not use your hands for this, or you may smudge your drawing. A cordless model electric eraser (Fig. 5.24) uses several grades of erasers for erasing ink and pencil lines.

5.15
Scales

All engineering drawings require the use of scales to measure lengths, sizes, and so forth. Scales may be flat or triangular and are made of wood, plastic, or metal. Figure 5.25 shows triangular engineers' and architects' scales. Most scales are either 6 or 12 inches long. In

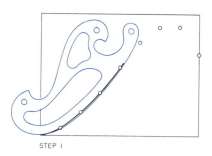 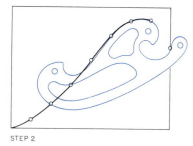 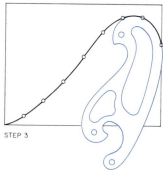

FIGURE 5.21 Use of the irregular curve.

Step 1 The irregular curve is positioned to pass through as many points as possible, and a portion of the curve is drawn.

Step 2 The irregular curve is positioned for drawing another portion of the connecting curve.

Step 3 The last portion is drawn to complete the smooth curve. Most irregular curves must be drawn in separate steps.

FIGURE 5.22 A flexible spline can be used for drawing large irregular curves. The spline is held in position by weights.

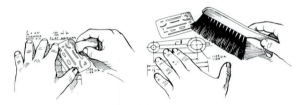

FIGURE 5.23 The erasing shield is used for erasing in tight spots. The dusting brush is used to remove the erased material. Brushing with the palm of your hand will smear the drawing.

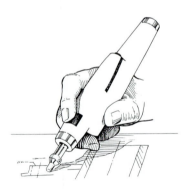

FIGURE 5.24 This cordless electric eraser is typical of those used by professional drafters.

this section, we cover the architects', engineers', mechanical engineers', and metric scale.

Architects' scale

The architects' scale is used to dimension and scale features encountered by the architect such as cabinets, plumbing, and electrical layouts. Most indoor measurements are made in feet and inches with an archi-

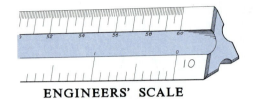

ENGINEERS' SCALE

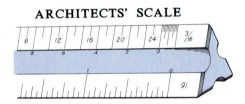

ARCHITECTS' SCALE

FIGURE 5.25 The architects' scale measures in feet and inches, whereas the engineers' scale measures in decimal units.

tects' scale. Figure 5.26 shows the basic form for indicating the scale being used. This form should be used in the title block or in some other prominent location on the drawing. Since the dimensions made with the architects' scale are in feet and inches, it is necessary to convert all dimensions to decimal equivalents (all feet or all inches) before the simplest arithmetic can be performed.

ARCHITECTS' SCALE

BASIC FORM $\text{SCALE:} \dfrac{X}{X} = 1'\text{-}0$ FROM END OF SCALE

TYPICAL SCALES

SCALE: FULL SIZE (USE 16-SCALE)

SCALE: HALF SIZE (USE 16-SCALE)

SCALE: 3 = 1'-0   *SCALE: $1\frac{1}{2}$ = 1'-0*

SCALE: $1\frac{1}{2}$ = 1'-0   *SCALE: $\frac{3}{4}$ = 1'-0*

SCALE: $\frac{1}{2}$ = 1'-0   *SCALE: $\frac{3}{8}$ = 1'-0*

SCALE: $\frac{3}{16}$ = 1'-0   *SCALE: $\frac{1}{8}$ = 1'-0*

SCALE: $\frac{3}{32}$ = 1'-0

FIGURE 5.26 The basic form for indicating the scale on the architects' scale and the variety of scales available.

SCALE: FULL SIZE The 16 scale is used for measuring full-size lines (Fig. 5.27A). An inch on the 16 scale is divided into sixteenths to match the ruler used by the carpenter. This example is measured to be $3\frac{1}{8}''$. When the measurement is less than 1 ft, a zero may precede the inch measurements; note that the inch marks are omitted.

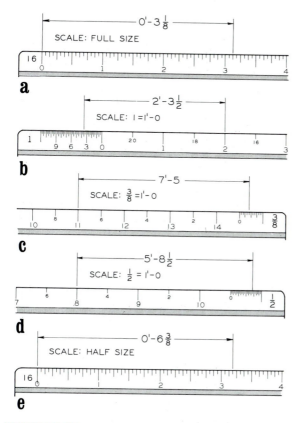

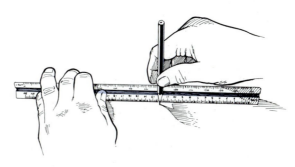

FIGURE 5.28 When marking off measurements along a scale, hold your pencil vertically for the most accurate measurement.

SCALE: $\frac{1}{2}$ = 1′−0 A line is measured to be $5'-8\frac{1}{2}$ in Fig. 5.27D.

SCALE: HALF SIZE The 16 scale is used to measure or draw a line that is half size. This is sometimes specified as Scale: 6 = 12 (inch marks omitted). The line in Fig. 5.27E is measured to be $0'-6\frac{3}{8}$.

A couple of pointers: When marking measurements, hold your pencil vertically for the greatest accuracy (Fig. 5.28). And when indicating dimensions in feet and inches, they should be in the form shown in Fig. 5.29. (Notice the fractions are twice as tall as the whole numerals.)

Engineers' scale

The engineers' scale is a decimal scale on which each division is a multiple of ten units. Because it is used for making drawings of outdoor engineering projects—streets, structures, land measurements, and other large topographical dimensions—it is sometimes called the civil engineers' scale.

Since the measurements are in decimal form, it is easy to perform arithmetic operations; there is no

FIGURE 5.27 Examples of lines measured using an architects' scale.

SCALE: 1 = 1′−0 In Fig. 5.27B, a line is measured to its nearest whole foot (2 ft in this case), and the remainder is measured in inches at the end of the scale ($3\frac{1}{2}$ in.) for a total of $2'-3\frac{1}{2}$. At the end of each architects' scale, a foot is divided into inches for measuring dimensions less than a foot. The scale $1'' = 1'-0$ is the same as saying 1 in. is equal to 12 in., or a $\frac{1}{12}$ size.

SCALE: $\frac{3}{8}$ = 1′−0 When this scale is used, $\frac{3}{8}$ in. represents 12 in. on a drawing. Figure 5.27C is measured to be $7'-5$.

OMIT INCH MARKS   ZERO HERE   ZERO OPTIONAL

FIGURE 5.29 Inch marks are omitted according to current standards, but foot marks are shown. When the inch measurement is less than a whole inch, a leading zero is used. When representing feet, a zero is optional if the measurement is less than a foot.

ENGINEERS' SCALES

BASIC FORM SCALE: 1 = XX   *FROM END OF ENGR. SCALE*

EXAMPLE SCALES

10	SCALE: 1=10';	SCALE: 1 = 1,000'
20	SCALE: 1=200';	SCALE: 1 = 20 LB
30	SCALE: 1= 0.3;	SCALE: 1 = 3,000'
40	SCALE: 1=4';	SCALE: 1 = 40'
50	SCALE: 1=50';	SCALE: 1 = 500'
60	SCALE: 1=6;	SCALE: 1 = 0.6'

FIGURE 5.30 The basic form for indicating the scale on the engineers' scale and the variety of scales available.

need to convert from one unit to another, as there is when using the architects' scale. Figure 5.30 shows the form for specifying scales on the engineers' scale, for example, Scale: 1 = 10'. Each end of the scale is labeled 10, 20, 30 (and so on), which indicates the number of units per inch on the scale. Many combinations may be obtained by moving the decimal places of a given scale, as Fig. 5.30 shows.

10 SCALE In Fig. 5.31A, the 10 scale is used to measure a line drawn at the scale of 1 = 10'. The line is 32.0 ft long.

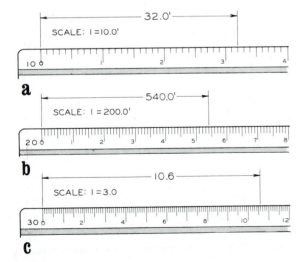

FIGURE 5.31 Examples of lines measured with the engineers' scale.

20 SCALE In Fig. 5.31B, the 20 scale is used to measure a line drawn at a scale of 1 = 200.' The line is 540.0 ft long.

30 SCALE In Fig. 5.31C, a line of 10.6 (inch marks omitted) is measured using the scale of 1 = 3.

Figure 5.32 shows the format for indicating measurements in feet and inches.

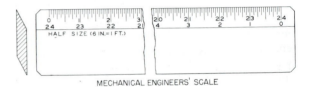

FIGURE 5.32 When using English units (inches), decimal fractions do not have leading zeros, and inch marks are omitted. Be sure to provide adequate space for decimal points between the numbers. Foot marks are shown.

FIGURE 5.33 The mechanical engineers' scales are used for measuring small parts at scales of half size, quarter size, and one-eighth size. These units are in inches with common fractions.

Mechanical engineers' scale

The mechanical engineers' scale is used to draw small parts (Fig. 5.33) in inches using common fractions. These scales are available in ratios of half size, one-quarter size, and one-eighth size. For example, on the half-size scale, 1 inch represents 2 inches, and on the quarter-size scale, 1 inch represents 4 inches.

English system

The English system (Imperial system) of measurement has been used in the United States, Britain, and Canada since these countries were established. The English system is based on arbitrary units of the inch, foot, cubit, yard, and mile (Fig. 5.34). Because there is no common relationship between these units, the system is cumbersome to use when simple arithmetic is performed; for example, finding the area of a rec-

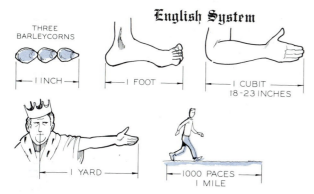

FIGURE 5.34 The units of the English system were based on arbitrary dimensions.

tangle that is 25 inches by 6¾ yards is a complex problem.

Metric system—SI units

The metric system was proposed by France in the fifteenth century. In 1793, the French National Assembly agreed that the meter (m) would be one ten-millionth of the meridian quadrant of the earth (Fig. 5.35). Fractions of the meter were expressed as decimal fractions. Debate continued until an international commission officially adopted the metric system in 1875. Since a slight error in the first measurement of the meter was found, the meter was later established

as equal to 1,650,763.73 wavelengths of the orange-red light given off by krypton-86 (Fig. 5.35).

The international organization charged with the establishment and promotion of the metric system is the **International Standards Organization** (ISO). The system they have endorsed is called **Système International d'Unités** (International System of Units) and is abbreviated SI. Figure 5.36 shows the basic SI units with their abbreviations. It is important that lowercase and uppercase abbreviations be used as shown.

SI UNITS			DERIVED UNITS		
LENGTH	METER	m	AREA	SQ METER	m²
MASS	KILOGRAM	kg	VOLUME	CU METER	m³
TIME	SECOND	s	DENSITY	KILOGRAM/CU MET	kg/m³
ELECTRICAL			PRESSURE	NEWTON/SQ MET	N/m²
CURRENT	AMPERE	A			
TEMPERATURE	KELVIN	K			
LUMINOUS					
INTENSITY	CANDELA	cd			

FIGURE 5.36 The basic SI units and their abbreviations. The derived units have come into common usage.

To make them easier to use, several practical units of measurement have been derived from basic SI units (Fig. 5.37). Note that degrees Celsius (centigrade) is recommended over Kelvin, the official temperature measurement. When using Kelvin, the freezing and boiling temperatures are 273.15 K and 373.15 K, respectively.

PARAMETER	PRACTICAL UNITS		SI EQUIVALENT
TEMPERATURE	DEGREES CELSIUS	°C	0°C = 273.15 K
LIQUID VOLUME	LITER	l	l = dm³
PRESSURE	BAR	BAR	BAR = 0.1 MPa
MASS WEIGHT	METRIC TON	t	t = 10³ kg
LAND MEASURE	HECTARE	ha	ha = 10⁴ m²
PLANE ANGLE	DEGREE	°	1° = π/180 RAD

FIGURE 5.37 These practical metric units are a few of those that are widely used because they are easier to deal with than the official SI units.

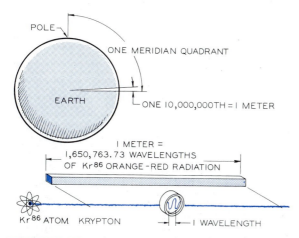

FIGURE 5.35 Originally based on the dimensions of the earth, the meter was later based on the wavelength of krypton-86. A meter is 39.37 inches.

Many SI units have prefixes to indicate placement of the decimal. Figure 5.38 shows the more common of these. Figure 5.39 compares several English and SI units.

VALUE		PREFIX	SYMBOL
1 000 000	$= 10^6$	$=$ MEGA	M
1 000	$= 10^3$	$=$ KILO	k
100	$= 10^2$	$=$ HECTO	h
10	$= 10^1$	$=$ DEKA	da
1	$= 10^0$		
0.1	$= 10^{-1}$	$=$ DECI	d
0.01	$= 10^{-2}$	$=$ CENTI	c
0.001	$= 10^{-3}$	$=$ MILLI	m
0.000 001	$= 10^{-6}$	$=$ MICRO	u

FIGURE 5.38 The prefixes and abbreviations used to indicate the decimal placement for SI measurements.

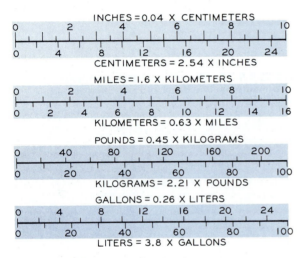

FIGURE 5.39 A comparison of metric units with those used in the English system of measurement.

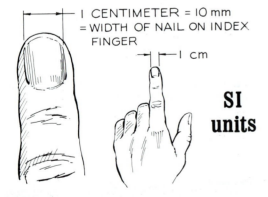

FIGURE 5.40 The width of the nail on your index finger is approximately equal to 1 centimeter, or 10 millimeters.

METRIC SCALES FROM END OF SCALE

BASIC FORM SCALE: 1:2

TYPICAL SCALES

SCALE: 1:1 (1mm=1mm; 1cm=1cm; ETC)

SCALE: 1:20 (1mm=20mm; 1mm=2cm)

SCALE: 1:300 (1mm=300mm; 1mm=0.3m)

OTHERS: 1:125; 1:250; 1:500

FIGURE 5.41 The basic form for indicating the scale on the metric scale and the variety of scales available.

5.16
Metric scales

> The basic unit of measurement on an engineering drawing is the **millimeter** (mm), which is one-thousandth of a meter, or one-tenth of a centimeter.

These units are understood unless otherwise specified on a drawing. The width of the fingernail of your index finger can serve as a convenient gauge to approximate the dimension of one centimeter, or ten millimeters (Fig. 5.40). Figure 5.41 shows the form for indicating metric scales.

> Decimal fractions are unnecessary on drawings dimensioned in millimeters; thus the dimensions are usually rounded off to whole numbers except for those measurements dimensioned with specified tolerances.

For metric units less than 1, a zero is placed in front of the decimal. In the English system, the zero is omitted from inch measurements (Fig. 5.42).

SCALE 1:1 The full-size metric scale (Fig. 5.43) shows the relationship between the metric units of the dekameter, centimeter, millimeter, and micrometer. There are 10 dekameters in a meter, 100 centi-

---

FIGURE 5.42 When decimal fractions are shown in metric units, a zero precedes the decimal. Be sure to allow adequate space for the decimal point when numbers with decimals are lettered.

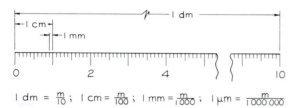

$$1\,dm = \frac{m}{10}; \quad 1\,cm = \frac{m}{100}; \quad 1\,mm = \frac{m}{1000}; \quad 1\,\mu m = \frac{m}{1\,000\,000}$$

FIGURE 5.43 The dekameter is one tenth of a meter; the centimeter is one hundredth of a meter; a millimeter is one thousandth of a meter; and a micrometer is one millionth of a meter.

meters in a meter, 1000 millimeters in a meter, and 1,000,000 micrometers in a meter. A line of 59 mm is measured in Fig. 5.44A.

SCALE 1:2 This scale is used when 1 mm represents 2 mm, 20 mm, 200 mm, and so forth. The line in Fig. 5.44B is 106 mm long.

SCALE 1:3 A line of 165 mm is measured in Fig. 5.44C, where 1 mm represents 3 mm.

Other scales

Many other metric (SI) scales are used: 1:250, 1:400, 1:500, and so on. The scale ratios mean one unit represents the number of units to the right of the colon. For example, 1:20 means 1 millimeter equals 20 mm, or 1 centimeter equals 20 cm, or 1 meter equals 20 m.

Metric symbols

When drawings are made in metric units, this can be noted in the title block or elsewhere using the SI symbol (Fig. 5.45), which indicates Système International. The two views of the partial cone are used to denote whether the orthographic views were drawn in the

U.S. system (third-angle projection) or the European system (first-angle projection).

Scale conversion

Appendix 2 gives tables for converting inches to millimeters, but this conversion can be performed by multiplying decimal inches by 25.4 to obtain millimeters.

An architect's scale must be multiplied by 12 to convert it to an approximate metric scale. For example, Scale: $\frac{1}{8}$ = 1'–0 is the same as $\frac{1}{8}$ in. = 12 in. or 1

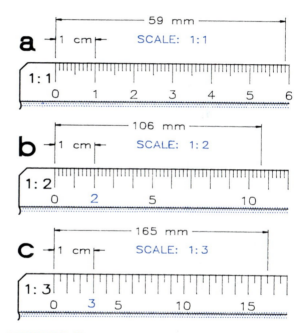

FIGURE 5.44 Examples of lines measured with metric scales.

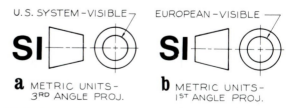

FIGURE 5.45 The large SI indicates that the measurements are in metric units. The partial cones indicate that the views are arranged using the third-angle projection (the U.S. system) or the first-angle projection (the European system).

OMIT COMMAS AND GROUP INTO THREES
1 000 000 000   NOT   1,000,000,000

USE RAISED DOT FOR MULTIPLICATION
N•M   OR   NM

INDICATE DIVISION BY EITHER
kg/m   OR   kg m^{-1}

USE ZERO PRECEDING DECIMALS
0.72 mm   NOT   .72 mm

INDICATE SI SCALES AS
SCALE: 1:2 SI

FIGURE 5.46   General rules to be used with the SI system.

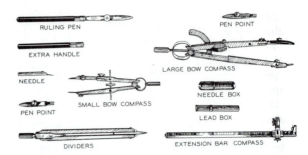

FIGURE 5.47   The parts of a set of drafting instruments.

in. = 96 in. This scale closely approximates the metric scale of 1:100. Many of the scales used in the metric system cannot be converted to exact English scales, but the metric scale of 1:60 converts exactly to the scale of 1 = 5'–0.

Expression of metric units

Figure 5.46 gives the general rules for expressing SI units. Commas are not used between sets of zeros; instead, a space is left between them.

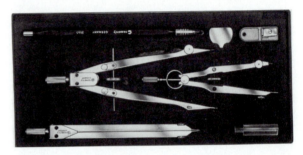

FIGURE 5.48   Instruments usually come as a cased set. (Courtesy of Gramercy Guild.)

5.17
The instrument set

Figure 5.47 shows a basic set of drawing instruments. Although these can be bought separately, they are available as sets in cases similar to the one shown in Fig. 5.48.

Compass

The **compass** is used to draw circles and arcs in ink and in pencil (Fig. 5.49). To obtain good results with the compass, its pencil point must be sharpened on its outside with a sandpaper board (Fig. 5.50). A bevel cut of this type gives the best all-around point for drawing a circle. When the compass point is set in the drawing surface, it should be inserted just enough for a firm set, not to the shoulder of the point. When the table top has a hard covering, several sheets of paper

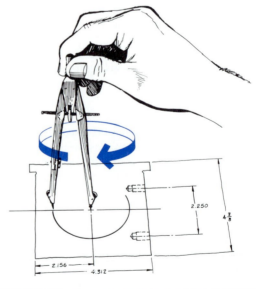

FIGURE 5.49   The compass is used for drawing circles.

should be placed under the drawing to provide a seat for the compass point.

Bow compasses are provided in some sets (Fig. 5.51) for drawing small circles. For large circles, bars are provided to extend the range of the large bow compass. A beam compass can be used (Fig. 5.52) for still larger circles. Small circles and ellipses can be effectively drawn with a circle template that is aligned with the perpendicular centerlines of the circle. The circle or ellipse is drawn with a pencil to match the other lines of the drawing (Fig. 5.53).

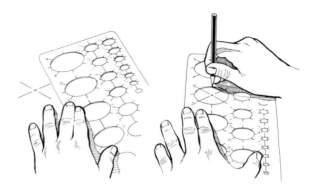

FIGURE 5.53 Circle templates can be used for drawing circles without the use of a compass. The circle or ellipse template is aligned with the centerlines.

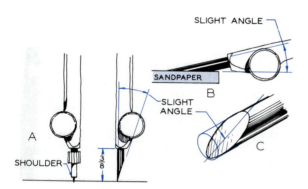

FIGURE 5.50 **A.** The pencil point should be about the same length as the compass point. **B.** The compass lead should be sharpened from the outside on a sandpaper pad. **C.** The lead should be sharpened to a wedge point.

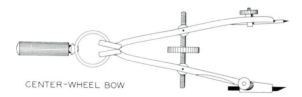

FIGURE 5.51 A small bow compass for drawing circles of about one-inch radius.

FIGURE 5.52 Large circles can be drawn with the beam compass. Ink attachments are also available.

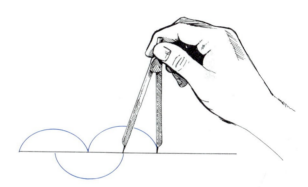

FIGURE 5.54 Dividers are used to step off measurements.

Dividers

The **dividers** look much like a compass but are used for laying off and transferring dimensions onto a drawing. For example, equal divisions can be stepped off rapidly along a line (Fig. 5.54). As each measurement is made, a slight impression is made in the drawing surface with the dividers' points.

Dividers can be used to transfer dimensions from a scale to a drawing (Fig. 5.55) or to divide a line into a number of equal parts. Small bow dividers can be used for transferring smaller dimensions, such as the spacing between the guidelines for lettering (Fig. 5.56).

Proportional dividers

Dimensions can be transferred from one scale to another by using a special type of dividers, the **propor-**

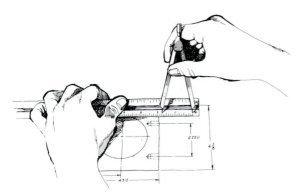

FIGURE 5.55 Dividers are also used to transfer dimensions from a scale to a drawing.

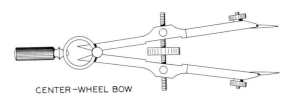

CENTER−WHEEL BOW

FIGURE 5.56 A type of a bow divider for transferring small dimensions, such as the spacing between guidelines for lettering.

FIGURE 5.57 A proportional divider can be used for making measurements that are proportional to other dimensions.

FIGURE 5.58 The pen is inked between the nibs with the spout on the ink bottle cap.

tional dividers. The central pivot point can be moved to vary the ratio of the spacing at one end of the dividers to the ratio at the other end (Fig. 5.57).

5.18
Ink drawing

Unlike pencil drawings, ink drawings remain dark and distinct. Ink lines also reproduce better than pencil.

Materials for ink drawing

A good grade of tracing paper can be used for ink drawings, but erasing errors may result in holes in the paper and the loss of your time. Therefore, tracing film or tracing cloth should be used, which will withstand many erasings and corrections.

When using drafting film, the drawing should be made on the matte surface according to the manufacturer's directions. A cleaning solution is available that can be used to prepare the surface for ink and to remove spots that might not properly take the ink. Tracing cloths need to be prepared for inking by applying a coating of powder or pounce to absorb oily spots that will otherwise repel an ink line. India ink—a dense, black carbon ink that is much thicker and faster drying than fountain pen ink—is used for engineering drawings.

To prevent clogging, ink should be removed from instruments before drying.

Ruling pen

The ruling pen should be inked with the spout on the cap of the ink bottle (Fig. 5.58). Experiment with your pen to learn the proper amount of ink to apply to the nibs. When drawing horizontal lines, the ruling pen is held in the same position as a pencil (Fig. 5.59), maintaining a space between the nibs and straightedge. An extra margin of safety can be obtained by placing a triangle or template under the straightedge.

An alternative pen that can be used is the technical ink fountain pen (Fig. 5.60). These pens come in sets with pen points of various sizes that are used for the alphabet of lines (Fig. 5.61). Lines drawn with this type of pen dry faster than those drawn with ruling pens because the ink is applied in a thinner layer.

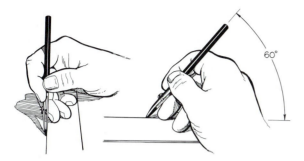

FIGURE 5.59 The ruling pen is held in a vertical plane at 60° to the drawing surface.

FIGURE 5.60 An India ink technical pen that can be used for making ink drawings.

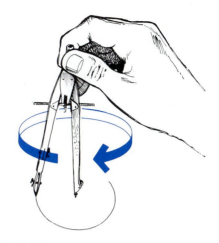

FIGURE 5.62 An attachment can be used with a large bow compass for drawing circles and arcs in ink.

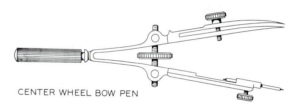

CENTER WHEEL BOW PEN

FIGURE 5.63 A small bow compass for drawing arcs in ink up to a radius of about one inch.

6 x 0
5 x 0
4 x 0
3 x 0
00
0
1
2
2½
3
4
6
7
8
9
10
12
14

*Approximate only. (Line widths will vary, depending on type of surface, type of ink, speed at which line is drawn, etc.)

FIGURE 5.61 This chart shows the variety of technical pens available for drawing lines of graduated widths. (Courtesy of Koh-I-Noor Rapidograph, Inc.)

Inking compass

The inking compass is usually the same compass used for the circles drawn by pencil, with the inking attachment inserted in place of the pencil attachment. The circle can be drawn with one continuous line, as Fig. 5.62 shows. Figure 5.63 shows a compass for drawing smaller circles. The spring bow compass can be used to draw pencil and ink circles about one-eighth inch in radius. Larger circles can be drawn using the extension bar with the large compass (Fig. 5.64). Special compasses are available for drawing circles with a technical fountain pen (Fig. 5.65). These pens screw into an adapter that fits the compass.

Order of inking

When a drawing is to be inked, begin by locating the centerlines and tangent points of the arcs. This construction should be laid out in pencil before inking. When a drawing is composed mostly of straight lines,

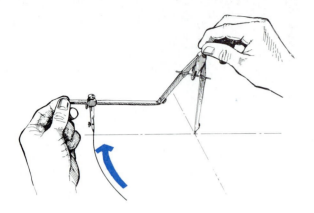

FIGURE 5.64 An extension bar can be used with a bow compass for drawing large arcs.

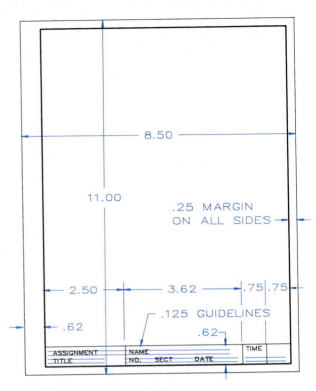

FIGURE 5.66 The format and title strip for a Size A sheet (8½″ × 11″) suggested for solving the problems at the end of each chapter. When the sheet is in the vertical format, it will be called a Size AV.

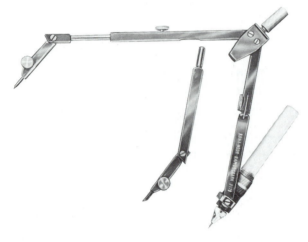

FIGURE 5.65 Special compasses with adapters are available for using technical ink pens for drawing arcs. (Courtesy of Koh-I-Noor Rapidograph, Inc.)

begin at the top and draw all the horizontal lines as you progress from the top of the sheet to the bottom. After allowing the last horizontal line to dry, ink the vertical lines by beginning with the far left (if you are right-handed) and moving across the drawing to the right, away from the wet lines.

Templates

Various templates are available for drawing nuts and bolts, circles and ellipses, architectural symbols, and many other applications. Templates work best when

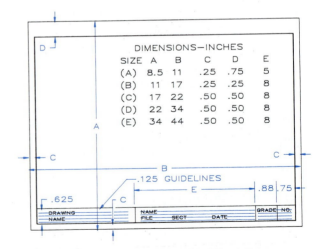

FIGURE 5.67 The format for Size AH (an 11″ × 8½″ sheet in a horizontal position) and the sizes of other sheets. The dimensions under columns A through E give the various layouts.

DIMENSIONS—INCHES

SIZE	A	B	C	D	E
(A)	8.5	11	.25	.75	5
(B)	11	17	.25	.25	8
(C)	17	22	.50	.50	8
(D)	22	34	.50	.50	8
(E)	34	44	.50	.50	8

used with technical fountain pens rather than the traditional ruling pen.

5.19
Solutions of problems

The following formats are suggested for the layout of problem sheets. Most problems will be drawn on $8\frac{1}{2}$- × -11-inch sheets, as Fig. 5.66 shows. A title strip is suggested in this figure, with a border as shown. Guidelines should be drawn very lightly to be only faintly visible. The $8\frac{1}{2}$- × -11-inch sheet in the vertical format is called Size AV throughout the rest of this textbook. When this sheet is in the horizontal format, as Fig. 5.67 shows, it will be called Size AH.

Figure 5.67 shows the standard sizes of sheets, from Size A through Size E. Figure 5.68 shows an alternative title strip for Sizes B, C, D, and E. Guidelines should always be used for lettering title strips.

A smaller title block and parts list is given above, which are placed in the lower right-hand corner of the sheet against the borders. When both are used on the same drawing, the parts list is placed directly above and in contact with the title block or title strip.

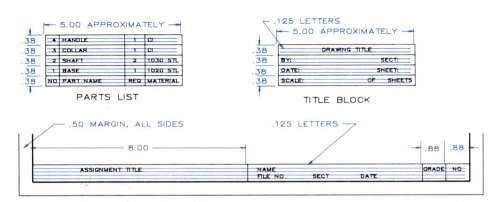

FIGURE 5.68 A title strip that can be used on sheet Sizes B, C, D, and E instead of the one given in Fig. 5.67.

Problems

Construct the problems (Figs. 5.69–5.71) on Size AH ($8\frac{1}{2}$- × -11-inch) paper, plain or with a printed grid, using the format shown in Fig. 5.66. Use pencil or ink as assigned by your instructor. Two problems can be drawn per sheet using the scale of each square equals .125 in., or 3 mm. One double-size problem can be drawn per sheet.

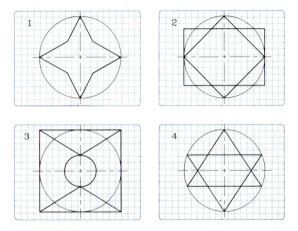

FIGURE 5.69 Problems 1–4.

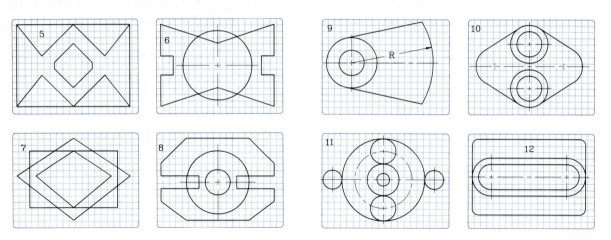

FIGURE 5.70 Problems 5–8.

FIGURE 5.71 Problems 9–12.

Geometric Construction

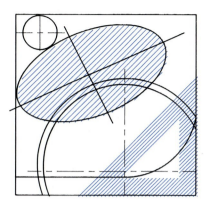

6.1
Introduction

Many graphical problems can be solved only by using geometry and geometric construction. Mathematics was an outgrowth of graphical construction, so the two areas are closely related. The proofs of many principles of plane geometry and trigonometry may be developed by using graphics. Moreover, graphical methods can be applied to algebra and arithmetic, and virtually all problems of analytical geometry can be solved graphically.

6.2
Angles

A fundamental requirement of geometric construction is the construction of lines that join at specified angles with each other. Figure 6.1 gives the definitions of various angles.

The unit of angular measurement is the degree, and a circle has 360 degrees. A degree (°) can be divided into 60 parts called minutes ('), and a minute can be divided into 60 parts called seconds ("). An angle of 15°32'14" is an angle of 15 degrees, 32 minutes, and 14 seconds.

6.3
Triangles

The **triangle** is a three-sided polygon (or figure) that is named according to its shape. The four types of triangles are the **scalene, isosceles, equilateral,** and **right triangle** (Fig. 6.2). The sum of the angles inside a triangle is always 180°.

FIGURE 6.1 **Standard types of angles and their definitions.**

77

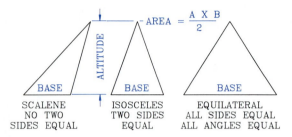

FIGURE 6.2 **Types of triangles and their definitions.**

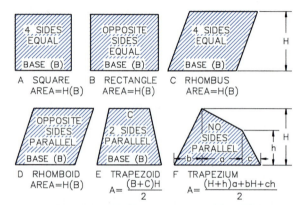

FIGURE 6.3 **Types of quadrilaterals (four-sided plane figures).**

6.4
Quadrilaterals

A **quadrilateral** is a four-sided figure of any shape. The sum of the angles inside a quadrilateral is 360°. Figure 6.3 shows the various types of quadrilaterals and the equations for the areas of these figures.

6.5
Polygons

A **polygon** is a multisided plane figure of any number of sides. (The triangle is a three-sided polygon, and the quadrilateral is a four-sided polygon.) If the sides of the polygon are equal in length, the polygon is a **regular polygon.** Figure 6.4 shows four types of regular polygons. Note that a regular polygon can be inscribed in a circle and that all the corner points will lie on the circle.

Other regular polygons not pictured are the **heptagon** (7 sides), the **nonagon** (9 sides), the **decagon** (10 sides), and the **dodecagon** (12 sides). The sums of the angles inside any polygon can be found by the equation

$$S = (n - 2) \times 180°,$$

where n equals the number of sides of the polygon.

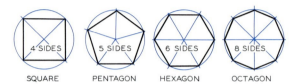

FIGURE 6.4 **Regular polygons inscribed in circles.**

6.6
Elements of circles

A circle can be divided into a number of parts, each of which has its own special name (Fig. 6.5). These terms, and others that deal with elements of circles, are used throughout this book.

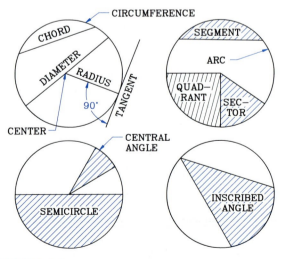

FIGURE 6.5 **Definitions of the elements of a circle.**

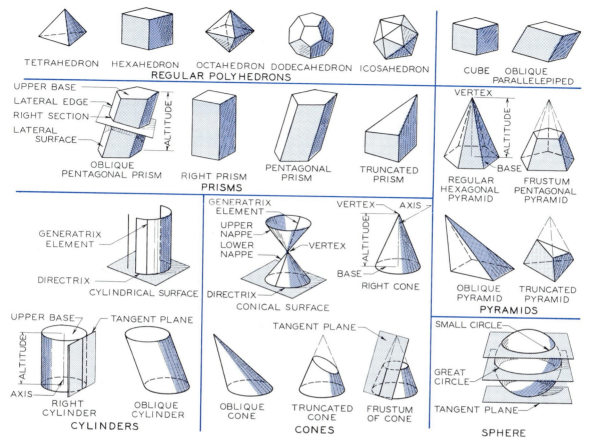

FIGURE 6.6 Types of geometric solids and their elements and definitions.

6.7
Geometric solids

Figure 6.6 shows the various types of solid geometric shapes, along with their names and definitions.

POLYHEDRA A multisided solid formed by intersecting planes is called a **polyhedron.** If the faces of a polyhedron are regular polygons, it is a **regular polyhedron.** The five regular polyhedra are the **tetrahedron** (4 sides), the **hexahedron** (6 sides), the **octahedron** (8 sides), the **dodecahedron** (12 sides), and the **icosahedron** (20 sides).

PRISMS A **prism** is a solid that has two parallel bases that are equal in shape. The bases are connected by sides that are parallelograms. The line from the

center of one base to the center of the other is the **axis.** An axis that is perpendicular to the bases is an **altitude,** and the prism is a **right prism.** If the axis is not perpendicular to the base, the prism is an **oblique prism.** A prism that has been cut off to form a base that is not parallel to the other is called a **truncated prism.** A **parallelepiped** is a prism with a base that is either a rectangle or a parallelogram.

PYRAMIDS A **pyramid** is a solid with a polygon as a base and triangular faces that converge at a point called the **vertex.** The line from the vertex to the center of the base is the **axis.** If the axis is perpendicular to the base, it is the **altitude** of the pyramid, and the pyramid is a **right pyramid.** If the axis is not perpendicular to the base, the pyramid is an **oblique pyramid.** A truncated pyramid is called a **frustum** of a pyramid.

CYLINDERS A **cylinder** is formed by a line or element (called a **generatrix**) that moves about the circle while remaining parallel to its axis. The axis of a cylinder connects the centers of each end of a cylinder. If the axis is perpendicular to the bases, it is the **altitude** of a **right cylinder.** If the axis does not make a 90° angle with the base, the cylinder is an **oblique cylinder.**

CONES A **cone** is also formed by a generatrix, one end of which moves about the circular base while the other end remains at a fixed point called the **vertex.** The line from the center of the base to the vertex is the **axis.** If the axis is perpendicular to the base, it is called the **altitude,** and the cone is a **right cone.** A truncated cone is called a **frustum** of a cone.

SPHERES A **sphere** is generated by the plane of a circle that is revolved about one of its diameters to form a solid. The ends of an axis through the center of the sphere are called **poles.**

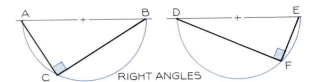

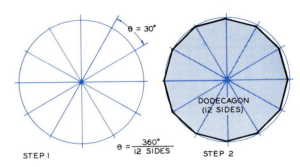

FIGURE 6.8 **Any angle inscribed in a semicircle will be a right angle.**

FIGURE 6.9 **The regular polygon.**

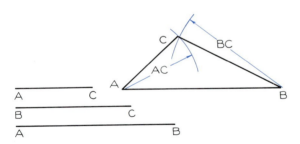

FIGURE 6.7 **When three sides are given, a triangle can be drawn with a compass.**

6.9
Constructing polygons

A regular polygon (having equal sides) can be inscribed in a circle or circumscribed about a circle. When inscribed, all the corner points will lie along the circle (Fig. 6.9). For example, a 12-sided polygon is constructed by dividing the circle into 12 sectors and connecting the points to form the polygon.

6.8
Constructing triangles

When three sides of a triangle are given, the triangle can be constructed by using a compass, as Fig. 6.7 shows. Only one triangle can be found when the sides are given by this method, called **triangulation.** A right triangle can be constructed by inscribing it in a semicircle, as Fig. 6.8 shows. Any triangle inscribed in a semicircle will always be a right triangle.

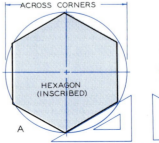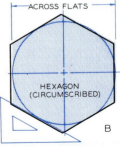

FIGURE 6.10 **A circle can be inscribed or circumscribed to form a hexagon by using a 30°–60° triangle.**

6.10 Hexagons

The **hexagon,** a 6-sided regular polygon, can be inscribed and circumscribed (Fig. 6.10). Hexagons are drawn with 30°–60° triangles either inside or outside the circles. Note that the circle represents the distance from corner to corner when inscribed, and from flat to flat when circumscribed.

6.11 Octagons

The **octagon,** an 8-sided regular polygon, can be inscribed in, or circumscribed about, a circle (Fig. 6.11) by using a 45° triangle. A second method inscribes the octagon inside a square (Fig. 6.12).

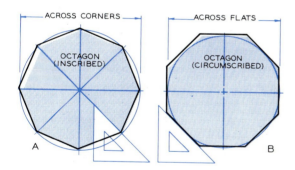

FIGURE 6.11 A circle can be inscribed or circumscribed to form an octagon with a 45° triangle.

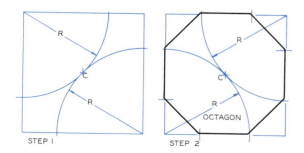

FIGURE 6.12 Octagon in a square.

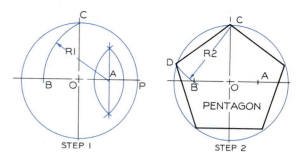

FIGURE 6.13 The pentagon.

Step 1 Bisect radius *OP* to locate point *A*. With *A* as the center and *AC* as the radius *R1*, locate point *B* on the diameter.

Step 2 With point *C* as the center and *BC* as the radius *R2*, locate point *D* on the arc. Line *CD* is the chord that can be used to locate the other corners of the pentagon.

6.12 Pentagons

The **pentagon,** a 5-sided regular polygon, can be inscribed in, or circumscribed about, a circle. Figure 6.13 shows another method of constructing a pentagon. This construction is performed with a compass and straightedge.

COMPUTER METHOD Polygons can be drawn using AutoCAD's POLYGON option under the DRAW command. One option allows you to give the number of sides, select the center, select the radius, and indicate whether the polygon is to be inscribed in, or circumscribed about, the circle (Fig. 6.14).

The second option allows you to select the endpoints of one edge of the polygon, and it will be drawn with the specified number of sides.

6.13 Bisecting lines and angles

Figure 6.15 shows two methods of finding the midpoint, or perpendicular bisector, of a line. The first method, which can be used to find the midpoint of an

SIDES=7
LOCATE CENTER

STEP 1

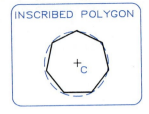

INSCRIBED POLYGON

STEP 2

FIGURE 6.14

Step 1 Command: <u>POLYGON</u> **(CR)**
Number of sides: <u>7</u> **(CR)**
Edge<Center of polygon>: **(Select center C.)**
Inscribed in circle/Circumscribed
about circle (I/C): <u>I</u> **(CR)**

Step 2 Radius of circle: **(Select radius with cursor.) (The polygon is inscribed inside the imaginary circle.)**

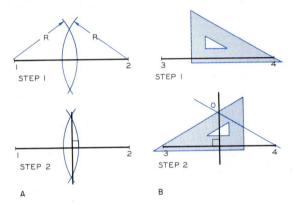

FIGURE 6.15 Bisecting a line.

A line can be bisected by using a compass and any radius or a standard triangle and a straightedge.

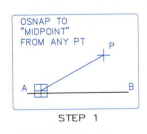

OSNAP TO
"MIDPOINT"
FROM ANY PT

STEP 1

MIDPOINT OF
LINE AB

STEP 2

FIGURE 6.16   Midpoint by computer.

Step 1 Find the midpoint of *AB* in the following manner:
Command: <u>LINE</u> **(CR)** From point: <u>P</u> **(Locate *P* anywhere.)**
To point: <u>OSNAP</u> **(Select** MIDPOINT **mode.) (Select any point on line** *AB*.**)**

Step 2 The line from point *P* will be drawn to the midpoint of line *AB*.

---

arc or a straight line, uses a compass to construct a perpendicular to a line. The second method uses a standard triangle.

> **COMPUTER METHODS**   The midpoint of a line can be found by computer (Fig. 6.16) by using the MIDPOINT mode of OSNAP and drawing a line from any point, *P*, to the line. The line will automatically snap to the given line's midpoint. The angle in Fig. 6.17 can be bisected with a compass by drawing three arcs.

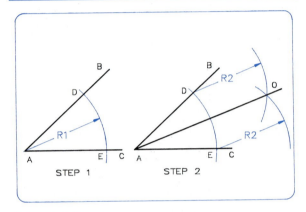

STEP 1   STEP 2

FIGURE 6.17   Bisecting an angle.

Step 1 Swing an arc of any radius to locate points *D* and *E*.

Step 2 Draw two equal arcs from *D* and *E* to locate point 0. Line *A*0 is the bisector of the angle.

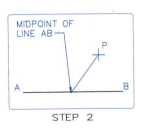

DRAW ARC
LOCATE P
ANYWHERE

STEP 1

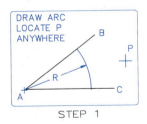

OSNAP FROM P
TO "MIDPOINT"
OF ARC

STEP 2

FIGURE 6.18   Bisecting an angle by computer.

Step 1 **Using the** ARC **command, any radius, and center** *A*, **draw an arc that** SNAPs **to lines** *AC* **and** *AB*.

Step 2 Command: <u>LINE</u> **(CR)** From point: **(Select** *P* **anywhere.)**
To point: (OSNAP) Midpoint of **(Select arc.)**
(The line from *P* **is drawn to the midpoint of the arc, which locates the bisector.)**

COMPUTER METHOD A second computer method is to draw an arc at any position between the two given lines using their point of intersection as the center (Fig. 6.18). Using the DRAW command and the MIDPOINT mode of OSNAP, a line is drawn from any point to the arc, which will be the arc's midpoint. The bisector can then be drawn from vertex *A* to this midpoint.

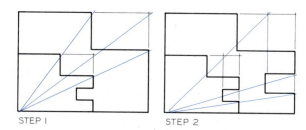

STEP I   STEP 2

FIGURE 6.20   Enlargement of a figure.

Step 1   A proportional enlargement is made by using a series of diagonals drawn through a single point, the lower left-hand corner in this case.

Step 2   Additional diagonals are drawn to locate the other features of the object. This process can be reversed for reducing an object.

6.14
Revolution of figures

Figure 6.19 demonstrates rotating a triangle about one of its points. Where the triangle is rotated about point 1 of line 1–4, point 4 is rotated to its desired position using a compass. Points 2 and 3 are found by triangulation by drawing arcs with radii 4–2 and 4–3 from 4' to complete the rotated view.

larger rectangle is drawn proportional to the small one. The upper right-hand notch is located using the same technique. This method can also be used to reduce a larger drawing.

6.16
Division of lines

It is often necessary to divide a line into several equal parts when a convenient scale is not available for this purpose. For example, suppose a 6-inch line must be divided into seven equal parts. The method shown in Fig. 6.21 is an efficient way to solve this problem.

An application of this principle is used for locating lines on a graph that are equally spaced (Fig.

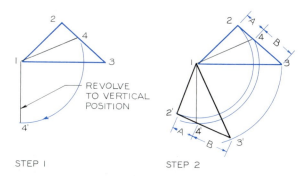

STEP I   STEP 2

FIGURE 6.19   Rotation of a figure.

Step 1   A plane figure can be rotated about any point. Line 1–4 is rotated about point 1 to its desired position with a compass.

Step 2   Points 2' and 3' are located by measuring distances *A* and *B* from 4.

6.15
Enlargement and reduction of figures

In Fig. 6.20, the small figure is enlarged by using a series of radial lines from the lower left-hand corner. The smaller figure is completed as a rectangle, and the

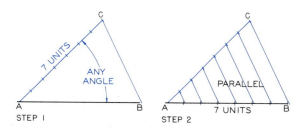

STEP I   STEP 2

FIGURE 6.21   Division of line.

Step 1   Line *AB* is divided into seven equal divisions by constructing a line through *A* and dividing it into seven known units with your dividers. Point *C* is connected to point *B*.

Step 2   A series of lines are drawn parallel to *CB* to locate the divisions along line *AB*.

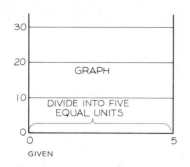

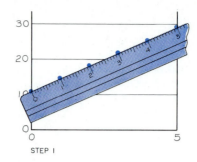

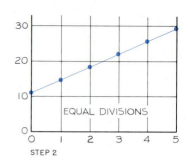

FIGURE 6.22 Division of a space.

Step 1 To divide a graph into five equal divisions along the *x* axis, a scale with five units of measurement that approximate the horizontal distance is laid across the graph. Align the 0 and 5 markings with the lines.

Step 2 Construct vertical lines through the points found in Step 1. This method can also be used to calibrate the divisions along the *y* axis.

6.22). A scale with the desired number of units (0 to 5) is laid across from left to right on the graph. Vertical lines are drawn through these points to divide the graph into five equal divisions.

6.17
Arcs through three points

An arc can be drawn through any three points by connecting the points with two lines (Fig. 6.23). Perpendicular bisectors are found for each line to locate the center at point *C*. The radius is drawn, and the lines *AB* and *BD* become chords of the circle.

This system can be reversed to find the center of a given circle or arc. Draw two chords that intersect at

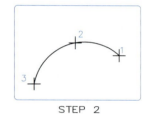

FIGURE 6.24 Three-point arc by computer.

Step 1 Command: ARC (CR)
Center/<Start point>: (Locate pt. 1)
Center/End <Second point>: (Locate pt. 2)

Step 2 Endpoint: (Locate pt. 3)

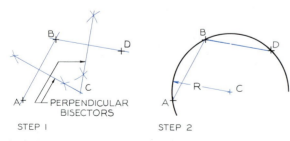

FIGURE 6.23 Arc through three points.

Step 1 Connect the points with two lines and find their perpendicular bisectors. The bisectors will intersect at the center, *C*.

Step 2 Using the center, *C*, and the distance to the points as the radius, construct the arc through the points.

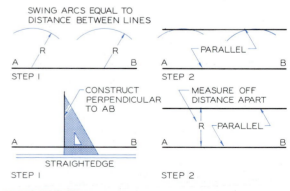

FIGURE 6.25 Construction of parallel lines.

Either of the above methods can be used for constructing one line parallel to another. The first method uses a compass and straightedge; the other method uses a triangle and T-square.

84

a point on the circumference and bisect them. The perpendicular bisectors will intersect at the center of the circle.

> **COMPUTER METHOD** Using the `ARC` command and the `3-point` option, a counterclockwise arc can be drawn through any three points selected on the screen (Fig. 6.24).

6.18
Parallel lines

A line can be drawn parallel to another by using either of the methods shown in Fig. 6.25. The first method uses a compass to draw two arcs to locate a parallel line that is the desired distance away. The second method requires constructing a perpendicular from a given line and measuring the distance, *R*, to locate the parallel, which is drawn with a straightedge.

6.19
Points of tangency

A point of tangency is the theoretical point where a straight line joins an arc or where two arcs join. In Fig. 6.26, a line is tangent to an arc. The point of tangency

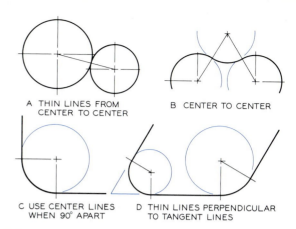

A THIN LINES FROM CENTER TO CENTER

B CENTER TO CENTER

C USE CENTER LINES WHEN 90° APART

D THIN LINES PERPENDICULAR TO TANGENT LINES

FIGURE 6.27 Thin lines that extend beyond the curves from the centers are used to mark points of tangency. These lines should always be shown in this manner.

is located by constructing a thin perpendicular to the line from the center of the arc. Figure 6.27 shows the conventional methods of marking points of tangency.

6.20
Line tangent to an arc

Figure 6.28 shows the method of finding the exact point of tangency between a line and an arc. Point *A* and the arc are given. Point *A* is connected to the center in Step 1, *AC* is bisected in Step 2, and *T* is located in Step 3. The point of tangency could also have been found by using a standard triangle, as Fig. 6.29 shows.

> **COMPUTER METHOD** A line can be drawn from a point tangent to an arc by using the `TANGENT` option of the `OSNAP` command (Fig. 6.30). When prompted for the second point on the line, the snap target is placed over the arc, and the tangent is drawn.

6.21
Arc tangent to a line from a point

If an arc is to be constructed tangent to line *CD* at *T* (Fig. 6.31) and pass through point *P*, a perpendicular bisector of *TP* is drawn. A perpendicular to *CD* is

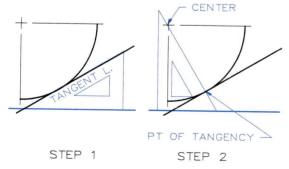

CENTER

TANGENT L.

PT OF TANGENCY

STEP 1   STEP 2

FIGURE 6.26 Locating a tangent point.

Step 1 Align your triangle with the tangent line while holding it firmly against a straightedge.

Step 2 Hold the straightedge in position, rotate the triangle, and construct a line through the center that is perpendicular to the line.

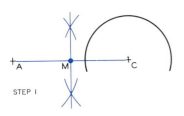

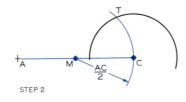

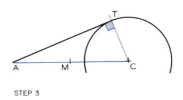

STEP 1 STEP 2 STEP 3

FIGURE 6.28 Line from a point tangent to an arc.

Step 1 Connect point *A* with center *C*. Locate point *M* by bisecting *AC*.

Step 2 Using point *M* as the center and *MC* as the radius, locate point *T* on the arc.

Step 3 Draw the line from *A* to *T* that is tangent to the arc of point *T*.

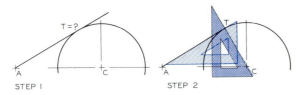

STEP 1 STEP 2

FIGURE 6.29 Line from a point tangent to an arc.

Step 1 A line can be drawn from point *A* tangent to the arc by eye.

Step 2 By rotating your triangle, the point of tangency can be located at the 90° angle with the line that passes through the center.

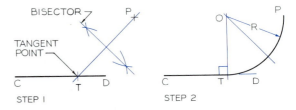

STEP 1 STEP 2

FIGURE 6.31 Arc through two points.

Step 1 If an arc must be tangent to a given line at *T* and pass through *P*, find the perpendicular bisector of the line *TP*.

Step 2 Construct a perpendicular to the line at *T* to intersect the bisector. The arc is drawn from center *O* with radius *OT*.

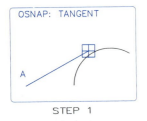

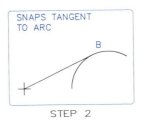

STEP 1 STEP 2

FIGURE 6.30 Line tangent to an arc by computer.

Step 1 Command: <u>LINE</u> **(CR)** From point: <u>A</u> To point: **(**OSNAP—Tangent mode**)** Tangent to:

Step 2 Select point on the arc, and line *AB* is drawn tangent to the arc.

drawn at *T* to locate the center at 0. A similar problem in Fig. 6.32 requires you to draw an arc of a given radius that will be tangent to line *AB* and pass through point *P*. In this case, the point of tangency on the line is not known until the problem has been solved.

COMPUTER METHOD An arc can be drawn tangent to a line from the end of a line by selecting the ARC option and CONTINUE immediately after the line has been drawn (Fig. 6.33). The arc will begin at the end of the line, and the other end can be DRAGged to the location desired.

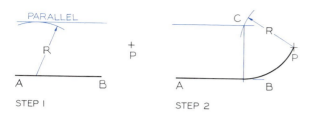

STEP 1   STEP 2

FIGURE 6.32   Arc tangent to a line and a point.

Step 1   When an arc of a given radius is to be drawn tangent to a line and through a point P, draw a line parallel to AB and R from it.

Step 2   Draw an arc from P with radius R to locate the center at C. The arc is drawn with radius R and center C.

6.22
Arc tangent to two lines

An arc of a given radius can be constructed tangent to two nonparallel lines if the radius is given. This method is shown in Fig. 6.34, where two lines form an acute angle. The same steps are used to find an arc tangent to two lines that form an obtuse angle (Fig. 6.35). In both cases, the points of tangency are located with thin lines drawn from the centers through the points of tangency.

A different technique can be used to find an arc of a given radius that is tangent to perpendicular lines (Fig. 6.36). This method will work only for perpendicular lines.

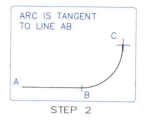

STEP 1   STEP 2

FIGURE 6.33   Arc tangent to a line by computer.

Step 1   Command LINE (CR)
From point: (select pt. A)
To point: (select pt. B)
To point: (CR)
Command: ARC (CR)
Center/<Start point >: (CR)
End point: (select desired end point)

Step 2   Move cursor to point C to locate the end of the curve that is tangent to AB at B.

COMPUTER METHOD   An arc can be drawn tangent to two nonparallel lines by using the FIL-LET command (Fig. 6.37). You may assign the radius length and select a point on each line; the arc will be drawn, and the lines will be automatically trimmed. If tangent points need to be shown, locate the center of the arc by using the CENTER option of the DIM: command. Lines can be OSNAPped perpendicular to AB and CD from the center, C.

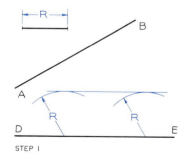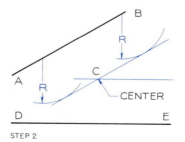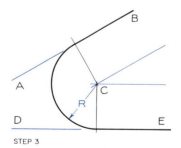

STEP 1   STEP 2   STEP 3

FIGURE 6.34   Arc tangent to two lines—acute angle.

Step 1   Construct a line parallel to DE with the radius of the specified arc R.

Step 2   Draw a second construction line parallel to AB to locate the center C.

Step 3   Thin lines are drawn from C perpendicular to AB and DE to locate the points of tangency. The tangent arc is drawn using the center C.

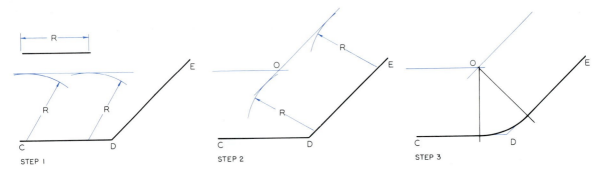

STEP 1   STEP 2   STEP 3

FIGURE 6.35   Arc tangent to two lines—obtuse angle.

Step 1   Using the specified radius R, construct a line parallel to CD.

Step 2   Construct a line parallel to DE that is distance R from it to locate center O.

Step 3   Construct thin lines from center O perpendicular to lines CD and DE to locate the points of tangency. Draw the arc using radius R and center O.

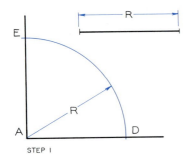
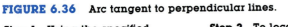
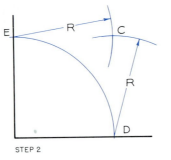

STEP 1   STEP 2   STEP 3

FIGURE 6.36   Arc tangent to perpendicular lines.

Step 1   Using the specified radius, R, locate points D and E by using center A.

Step 2   To locate point C, swing two arcs using the radius, R, that was used in Step 1.

Step 3   Locate the tangent points with lines from C. Draw the arc with radius R.

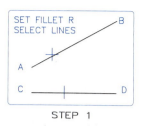

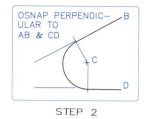

STEP 1   STEP 2

FIGURE 6.37 Arc tangent to two lines by computer.

Step 1 Command: FILLET **(CR)**
Polyline/Radius/<Select two objects>: R **(CR) (To define radius.)**
Enter fillet radius <0.0000>: .75 **(CR)**
Command: **(CR)**
FILLET Polyline/Radius/<Select two objects>: **(Select Pts. AB and CD.)**

Step 2 Command: LINE **(CR)** From point: OSNAP-CENTER **(CR)** center of **(Select point on the arc.)** To point: PERPEND **(CR)** perpend to **(Select point on AB, and a perpendicular is drawn to the point of tangency. Extend the line beyond AB. Locate the tangent point on CD in the same manner.)**

6.23
Arc tangent to an arc and a line

Figure 6.38 shows the steps for constructing an arc tangent to an arc and a line. A variation of this principle of construction is shown in Fig. 6.39, where the arc is drawn parallel to an arc and line with the arc in a reverse position.

6.24
Arc tangent to two arcs

A third arc is drawn tangent to two given arcs in Fig. 6.40. Thin lines are drawn from the centers to locate the points of tangency. This tangent arc is concave from the top. A convex arc can be drawn tangent to the given arcs if the radius of the arc is greater than the radius of either of the given arcs (Fig. 6.41).

A variation of this problem is shown in Fig. 6.42, where an arc of a given radius is drawn tangent to the top of one arc and the bottom of the other. A similar problem is shown in Fig. 6.43, where an arc is drawn tangent to a circle and a larger arc.

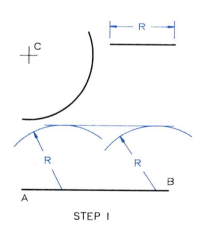

STEP 1

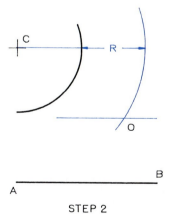

STEP 2

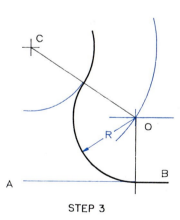

STEP 3

FIGURE 6.38 Arc tangent to an arc and a line.

Step 1 Construct a line parallel to AB that is R from it. Use thin construction lines.

Step 2 Add radius R to the extended radius from point C. Use this lengthened radius to locate point O.

Step 3 Lines OC and OT are drawn to locate the tangency points. The arc with radius R and center O is drawn.

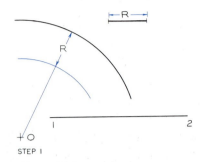

STEP 1

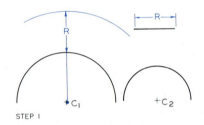

STEP 2

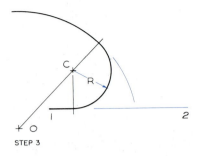

STEP 3

FIGURE 6.39   Arc tangent to an arc and a line.

Step 1   The specified radius, *R*, is subtracted from the radius through the arc's center at *O*. A concentric arc is drawn with the shortened radius.

Step 2   A line parallel to 1–2 is drawn a distance of *R* from it to locate the center, point *C*.

Step 3   The tangent points are located with lines from *O* through *C* and through *C* perpendicular to 1–2. Draw the tangent arc with radius *R*.

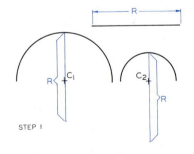

STEP 1

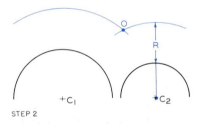

STEP 2

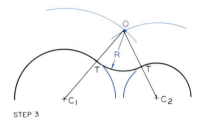

STEP 3

FIGURE 6.40   Arc tangent to two arcs—concave arc.

Step 1   The radius of one circle is extended, and the radius *R* is added to it. The extended radius is used for drawing a concentric arc.

Step 2   The radius of the other circle is extended, and the radius *R* is added to it. The extended radius is used to construct an arc and to locate point *O*, the center.

Step 3   The centers are connected with point *O* to locate the points of tangency. The arc is drawn tangent to the two arcs with radius *R*.

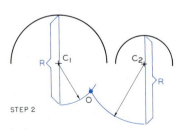

STEP 1

STEP 2

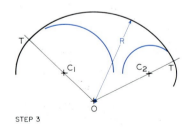

STEP 3

FIGURE 6.41   Arc tangent to two arcs—convex arc.

Step 1   The radius of each arc is extended from the arc past its center, and the specified radius *R* is laid off from the arcs along these lines.

Step 2   The distance from each center to the ends of the extended radii are used for drawing two arcs to locate the center *O*.

Step 3   Thin lines from *O* through centers C_1 and C_2 locate the points of tangency. The arc is drawn using point *O* as the center.

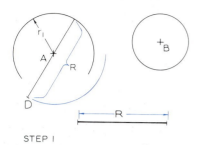 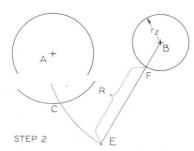 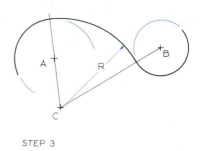

STEP 1 STEP 2 STEP 3

FIGURE 6.42 Arc tangent to two circles.

Step 1 The specified radius R is laid off from the arc along the extended radius to locate point D. Radius AD is used to construct a concentric arc.

Step 2 The radius through center B is extended, and the radius R is added to it from point F. Radius BE is used to locate the center C.

Step 3 The tangent arc is drawn with center C and radius R. The points of tangency are located with thin lines from C through the given centers.

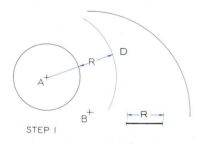 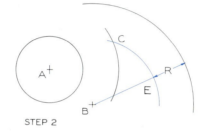 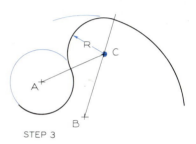

STEP 1 STEP 2 STEP 3

FIGURE 6.43 Arc tangent to two arcs.

Step 1 Radius R is added to the radius from center A. Radius AD is used to draw a concentric arc with center A.

Step 2 Radius R is subtracted from the radius through B. Radius BE is used to construct an arc to locate center C.

Step 3 The points of tangency are located with thin lines BC and AC. The tangent arc is drawn with center C.

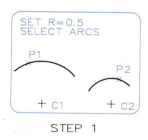 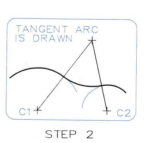

STEP 1 STEP 2

FIGURE 6.44 Arc tangent to two arcs.

Step 1 Command: <u>FILLET</u> (CR)
Polyline/Radius/<Select two lines>: <u>R</u> (CR)
Enter fillet radius <current>: <u>.5</u> (CR)

Step 2 Command: <u>FILLET</u> (CR)
Polyline/Radius/<Select two objects>: <u>P1</u>, <u>P2</u> (Select points on each arc.)
(The tangent arc is drawn, and the given arcs are trimmed. Locate tangent points by drawing lines from center to center.)

COMPUTER METHOD An arc is automatically drawn tangent to two arcs by using the FIL-LET command (Fig. 6.44). After entering the FILLET command, give the desired radius when prompted. You will be returned to the COMMAND mode. Press the carriage return (CR) to return to the FILLET command.

When prompted, select the two arcs with your cursor; the tangent arc will be drawn and the given arcs trimmed at the points of tangency. Locate the tangent points by drawing lines from the center to C1 and C2.

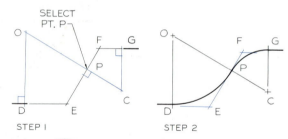

FIGURE 6.45 An ogee curve.

Step 1 To draw an ogee curve between two parallel lines, draw line *EF* at any angle. Locate a point of your choosing along *EF*, *P* in this case. Find the tangent points by making *FG* equal to *FP* and *DE* equal to *EP*. Draw perpendiculars at *G* and *D* to intersect the perpendicular bisector at *O* and *C*.

Step 2 Using radii *CP* and *OP*, at centers *O* and *C*, draw the two tangent arcs to complete the ogee curve.

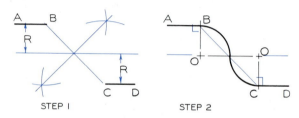

FIGURE 6.46 An ogee curve.

Step 1 To draw an ogee curve formed by two equal arcs passing through points *B* and *C*, draw a line between the points. Bisect the line *BC*, and draw a line parallel to *AB* and *CD* to find the radius, *R*.

Step 2 Construct perpendiculars at *B* and *C* to locate the centers at both points *O*. Draw the arcs to complete the ogee curve.

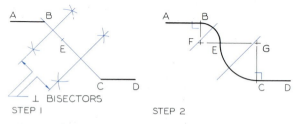

FIGURE 6.47 The unequal ogee curve.

Step 1 Two parallel lines are to be connected by an ogee curve that passes through *B* and *C*. Draw line *BC*, and select point *E* on the line. Bisect *BE* and *EC*.

Step 2 Construct perpendiculars at *B* and *C* to intersect the bisectors to locate centers *F* and *G*. Locate the points of tangency, and draw the ogee curve using radii *FB* and *GC*.

6.25
Ogee curves

The **ogee curve** is a double curve formed by tangent arcs. By constructing two arcs tangent to three intersecting lines, the ogee curve in Fig. 6.45 was found.

An ogee curve can be drawn between two parallel lines (Fig. 6.46) from points *B* to *C* by geometric construction. Figure 6.47 shows an alternative method of drawing an ogee curve that passes through points *B*, *E*, and *C*.

6.26
Curve of arcs

An irregular curve formed with tangent arcs can be constructed as shown in Fig. 6.48. The radii of the arcs are selected to give the desired curve by moving from one set of points to the next.

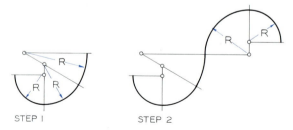

FIGURE 6.48 A curve of arcs.

Step 1 A series of arcs can be joined to form a smooth curve. Begin with the small arc, extend the radius through its center, and draw the second and then the third.

Step 2 The curve can be reversed by extending the radius in the opposite direction and repeating the same process.

6.27
Rectifying arcs

An arc is rectified when its true length is laid out along a straight line. Figure 6.49 shows a method of rectifying an arc. Another method of rectifying an arc uses the mathematical equation for finding the circumfer-

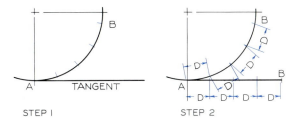

STEP I   STEP 2

FIGURE 6.49   **To rectify an arc.**

Step 1   **Construct a line tangent to the arc, and divide the arc into a series of equal divisions from _A_ to _B_.**

Step 2   **The chordal distances, _D_, along the arc are laid out along the straight line until point _B_ is located.**

ence of the circle. Since a circle has 360°, the arc of a 30° sector is one twelfth of the full circumference (360 ÷ 30 = 12). Therefore, if the circumference is 12 inches, the 30° arc equals 1 inch.

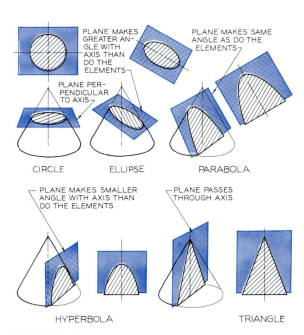

FIGURE 6.50   **The conic sections are formed by passing cutting planes at various angles through right cones. The conic sections are the circle, ellipse, parabola, hyperbola, and triangle.**

6.28
Conic sections

Conic sections are plane figures that can be scribed graphically as well as mathematically. They are formed by passing imaginary cutting planes through a right cone (Fig. 6.50).

6.29
Ellipses

The **ellipse** is a conic section formed by passing a plane through a right cone at an angle (Fig. 6.50). The ellipse is mathematically defined as the path of a point that moves in such a way that the sum of the distances from two focal points is a constant. The largest diameter of an ellipse is always the true length and is called the **major diameter.** The shortest diameter is perpendicular to the major diameter and is called the **minor diameter.**

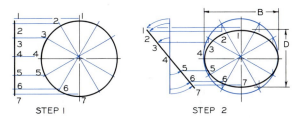

STEP I   STEP 2

FIGURE 6.51   **An ellipse by revolution.**

Step 1   **When the edge view of a circle is perpendicular to the projectors between its adjacent view, the view will be a true circle. Mark equally spaced points along the arc, and project them to the edge.**

Step 2   **Revolve the edge of the circle to the desired position, and project the points to the circular view, which will now appear as an ellipse. The points are projected vertically downward to new positions.**

The construction of an ellipse is found by revolving the edge view of a circle, as Fig. 6.51 shows. This ellipse could have been drawn using the ellipse template shown in Fig. 6.52. The angle between the line of sight and the edge view of the circle (or the one closest to this size) is the angle of the ellipse template that should be used. Ellipse templates are available in

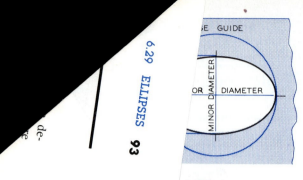

SE GUIDE

MINOR DIAMETER

OR DIAMETER

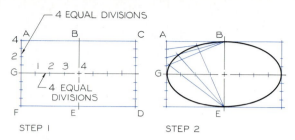

4 EQUAL DIVISIONS

STEP 1

STEP 2

FIGURE 6.54 Ellipse—parallelogram method.

Step 1 An ellipse can be drawn inside a rectangle or parallelogram by dividing the horizontal centerline into the same number of equal divisions as the shorter sides, *AF* and *CD*.

Step 2 The construction of the curve in one quadrant is shown by using sets of rays from *E* and *B* to plot the points.

...mplate.

When the eage ... le is revolved so that the line of sight between the two views is not perpendicular to the edge view, the circle will appear as an ellipse. The major diameter remains constant, but the minor diameter will vary. The angle between the line of sight and the edge view of the circle is the angle of the ellipse template.

intervals of 5° and in variations in size of the major diameter of about ⅛ inch (Fig. 6.53).

The ellipse can also be constructed inside a rectangle or parallelogram (Fig. 6.54), where a series of points is plotted to form an elliptical curve. Two circles can be used for constructing an ellipse by making the

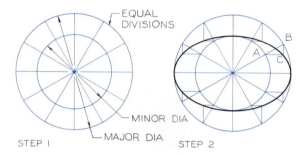

EQUAL DIVISIONS

MINOR DIA

MAJOR DIA

STEP 1   STEP 2

FIGURE 6.55 Ellipse—circle method.

Step 1 Two concentric circles are drawn with the large one equal to the major diameter and the small one equal to the minor diameter. Divide them into equal sectors.

Step 2 Plot points on the ellipse by projecting downward from the large curve to intersect horizontal construction lines drawn from the intersections on the small circle.

diameter of the large circle equal to the major diameter and the diameter of the small circle equal to the minor diameter (Fig. 6.55).

COMPUTER METHOD Using the ELLIPSE command, you can draw an ellipse by selecting the endpoints of the major diameter. The third point defines the length of minor radius; the ellipse is then drawn (Fig. 6.56).

Another option of the ELLIPSE command allows you to select the center, the end of the major radius, and the end of the minor radius; the ellipse is then drawn. Isometric ellipses can also be drawn to isometric pictorial drawings.

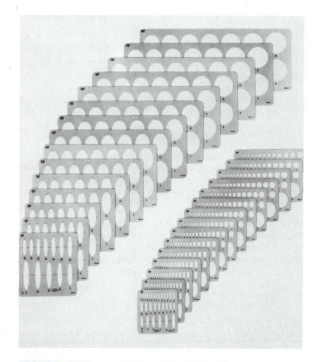

FIGURE 6.53 Ellipse templates are calibrated at 5° intervals from 15° to 60°. (Courtesy of Timely Products, Inc.)

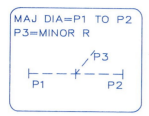 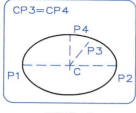

STEP 1   STEP 2

FIGURE 6.56   Ellipse—computer method.

Step 1   Command: <u>ELLIPSE</u> **(CR)**
`<Axis endpoint 1>/Center:` **(Select P1.)**
`Axis endpoint 2:` **(Select P2.)**

Step 2   `<Other axis distance>/Rotation:`
(Select minor radius, P3, in any direction.)
(The ellipse is drawn. The option, `Rotation`, could have been used to specify the orientation of the major diameter.)

The **conjugate diameters** of an ellipse are diameters parallel to the tangents at the ends of each, as Fig. 6.57 shows. A single ellipse has an infinite number of sets of conjugate diameters.

When the ellipse and a pair of conjugate diameters are given, the major and minor diameters of the ellipse can be found by using the method illustrated in Fig. 6.58. When the conjugate diameters are not given, they can be constructed as shown in Fig. 6.57, and the major and minor diameters can be found. The major and minor diameters are necessary for using the ellipse template.

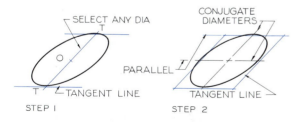

STEP 1   STEP 2

FIGURE 6.57   Conjugate diameters.

Step 1   A conjugate diameter is parallel to the tangents at the ends of another conjugate diameter. A diameter is selected, and parallel tangents are drawn.

Step 2   A conjugate diameter is drawn parallel to the horizontal tangents, and the inclined tangents are drawn parallel to the conjugate diameter found in Step 1.

The mathematical equation of an ellipse is

$$\frac{x^2}{a^2} + \frac{y^2}{b^2} = 1, \quad \text{where } a, b \neq 0.$$

6.30
Parabolas

The **parabola** is mathematically defined as a plane curve, each point of which is equidistant from a straight line (called a **directrix**) and its focal point. The parabola is a conic section formed when the cutting plane makes the same angle with the base of a cone as do the elements of the cone.

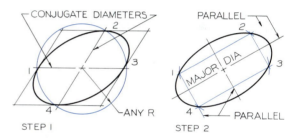

STEP 1   STEP 2

FIGURE 6.58   Finding the axes of an ellipse.

Step 1   When the conjugate diameters of an ellipse are given, you can find the major and minor diameters of the ellipse by drawing a circle of any radius from the intersection of the diameters.

Step 2   The circle cuts four points along the ellipse. The points are connected to form a rectangle. The major and minor diameters are parallel to the sides of the rectangle.

Figure 6.59 shows the construction of a parabola using its mathematical definition. A parabolic curve can also be constructed geometrically by dividing the two perpendicular lines into the same number of divisions (Fig. 6.60). The parabola is drawn through the plotted points with an irregular curve. Figure 6.61 shows a third method of construction using parallelograms.

The mathematical equation of the parabola is

$$y = ax^2 + bx + c, \quad \text{where } a \neq 0.$$

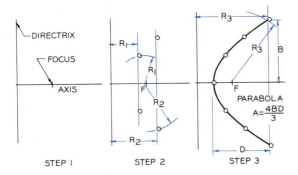

FIGURE 6.59 Parabola—mathematical method.

Step 1 Draw an axis perpendicular to a line (a directrix). Choose a point for the focus, *F*.

Step 2 Locate points by using a series of selected radii to plot points on the curve. For example, draw a line parallel to the directrix and R_2 from it. Swing R_2 from *F* to intersect the line and plot the point.

Step 3 Continue the process with a series of arcs of varying radii until an adequate number of points have been found to complete the curve.

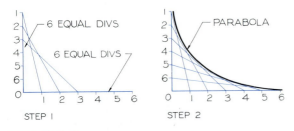

FIGURE 6.60 Parabola—tangent method.

Step 1 Construct two lines at a convenient angle, and divide each of them into the same number of divisions. Connect the points with a series of diagonals.

Step 2 When finished, construct the parabolic curve to be tangent to the diagonals.

6.31
Hyperbolas

The **hyperbola** is a two-part conic section. Mathematically, it is defined as the path of a point that moves in such a way that the difference of its distances

from two focal points is a constant. Figure 6.62 shows the construction of a hyperbola using this definition.

Figure 6.63 shows a second method of construction. Two perpendicular lines are drawn through point *B* as asymptotes. The hyperbolic curve becomes more nearly parallel and closer to the asymptotes as the hyperbola is extended, but the curve never merges with the asymptotes.

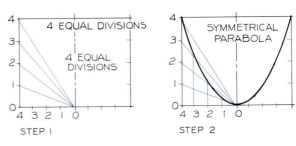

FIGURE 6.61 Parabola—parallelogram method.

Step 1 Construct a rectangle or parallelogram to contain the parabola, and locate its axis parallel to the sides through *O*. Divide the sides into equal divisions. Connect the divisions with point *O*.

Step 2 Construct lines parallel to the sides (vertical in this case) to locate the points along the rays from *O*. Draw the parabola.

6.32
Spirals

The **spiral** is a coil that begins at a point and becomes larger as it travels around the origin. A spiral lies in a single plane. Figure 6.64 shows the steps for constructing a spiral.

6.33
Helixes

The **helix** is a curve that coils around a cylinder or cone at a constant angle of inclination. Examples of helixes are corkscrews or threads on a screw. Figure 6.65 shows a helix constructed about a cylinder, and Fig. 6.66 shows a helix constructed about a cone.

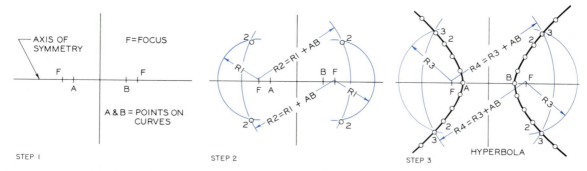

FIGURE 6.62 The hyperbola.

Step 1 A perpendicular is drawn through the axis of symmetry, and focal points *F* are located equidistant from it on both sides. Points on the curve, *A* and *B*, are located equidistant from the perpendicular at a location of your choice but between the focal points.

Step 2 Radius *R*1 is selected to draw arcs using focal points *F* as the centers. *R*1 is added to *AB* (the distance between the nearest points on the hyperbolas) to find *R*2. Radius *R*2 is used to draw arcs using the focal points as centers. The intersections of *R*1 and *R*2 establish points 2 on the hyperbola.

Step 3 Other radii are selected and added to distance *AB* to locate additional points in the same manner as described in Step 2. A smooth curve is drawn through the points to form the hyperbolic curves.

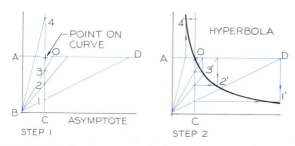

FIGURE 6.63 The equilateral hyperbola.

Step 1 Two perpendiculars are drawn through *B*, and any point *O* on the curve is located. Horizontal and vertical lines are drawn through *O*. Line *CO* is divided into equal divisions, and rays from *B* are drawn through them to the horizontal line.

Step 2 Horizontal construction lines are drawn from the divisions along line *OC*, and lines from *AD* are projected vertically to locate points 1′ through 4′ on the curve.

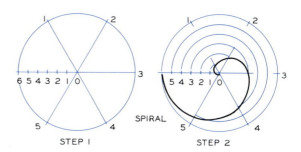

FIGURE 6.64 The spiral.

Step 1 Draw a circle and divide it into equal parts. The radius is divided into the same numer of equal parts (six in this example).

Step 2 By beginning on the inside, draw arc 0–1 intersect radius 0–1. Then swing arc 0–2 to rad 0–2, and continue until the last point is rea which lies on the original circle.

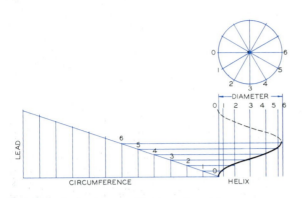

FIGURE 6.65 The helix.

Divide the top view of the cylinder into equal divisions, and project them to the front view. Lay out the circumference and the height of the cylinder, which is the **lead**. Divide the circumference into the same number of equal parts by taking the measurements from the top view. Project the points along the inclined rise to their respective elements to find the helix.

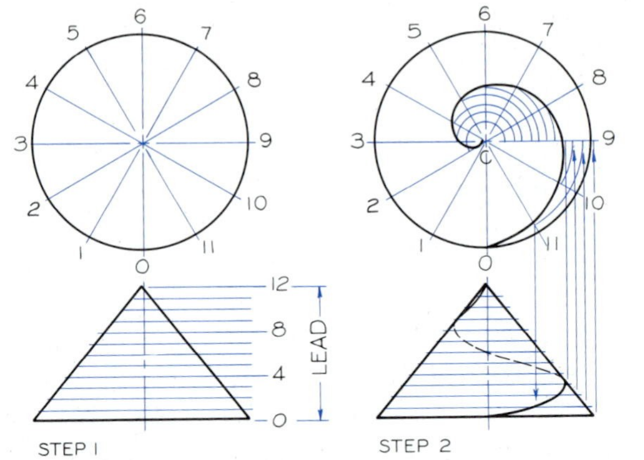

FIGURE 6.66 A conical helix.

Step 1 Divide the cone's base into equal parts. Pass a series of horizontal cutting planes through the front view of the cone. Use the same number as the divisions on the base (12 in this case).

Step 2 Project all the divisions along the front view of the cone to line *C* 9, and draw a series of arcs from center *C* to their respective radii in the top view to plot the points. Project the points to their respective cutting planes in the front view.

6.34

Involutes

The **involute** is the path of the end of a line as it unwinds from a line or plane. In Fig. 6.67, an involute is formed by unwind̶i̶n̶g̶ ̶̶̶e from a rectangle. Successively d̶i̶f̶f̶ ̶̶̶̶̶e equal in length to the ̶̶̶̶̶̶̶̶̶̶hvolute.

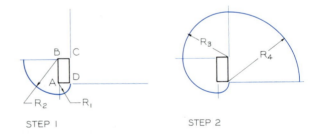

STEP 1 STEP 2

FIGURE 6.67 The involute.

Step 1 Side *AD* is used as a radius for drawing arc *AD*. *AB* is added to *AD* to form radius *R*2. Draw a second arc using center *B*.

Step 2 The two remaining sides are used to unwind the involute back to its point of origin.

on Size AV paper ̶̶̶̶3, where Problems ̶̶̶̶id is equal to 0.20 ̶̶̶̶̶ 10 scale to lay

out the problems. By equating each grid to 5 mm, you can use your full-size metric scale to lay out and solve the problems.

Show your construction and mark all points of

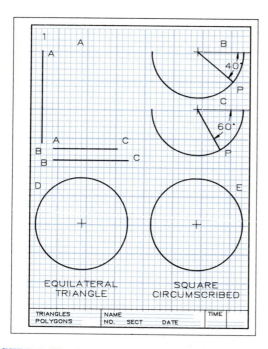

FIGURE 6.68 Problems 1A–1E. Basic constructions.

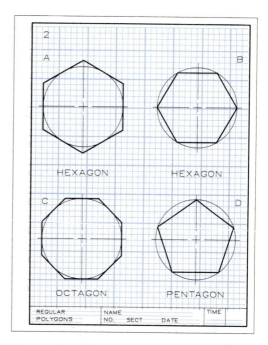

FIGURE 6.69 Problems 2A–2D. Construction of regular polygons.

tangency, as discussed in the chapter.

1. A. Draw triangle *ABC* using the given sides.

B–C. Inscribe an angle in the semicircles with the vertexes at point *P*.

D. Inscribe a three-sided regular polygon inside the circle.

E. Circumscribe a four-sided regular polygon about the circle.

2. A. Circumscribe a hexagon about the circle.

B. Inscribe a hexagon in the circle.

C. Circumscribe an octagon about the circle.

D. Construct a pentagon inside the circle using the compass method.

3. A. Bisect the lines.

B. Bisect the angles.

C. Rotate the triangle 60° in a clockwise direction about point *A*.

D. Enlarge the given shape to the size indicated by the diagonal.

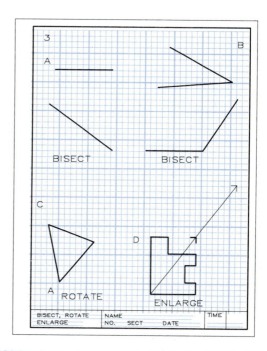

FIGURE 6.70 Problems 3A–3D. Basic constructions.

4. A. Divide *AB* into seven equal parts. Draw the construction line through *A* for your construction.

B. Divide the space between the two vertical lines into four equal divisions. Draw three vertical lines at these divisions that are equal in length to the given lines.

C. Construct an arc with radius *R* that is tangent to the line at *J* and passes through point *P*.

D. Construct an arc with radius *R* that is tangent to the line and passes through *P*.

5. A. Construct a line from *P* that is tangent to the semicircle. Locate the points of tangency. Use the compass method.

B–D. Construct arcs with the given radii tangent to the lines.

6. A–D. Construct arcs that are tangent to the arcs or lines. The radii are given for each problem.

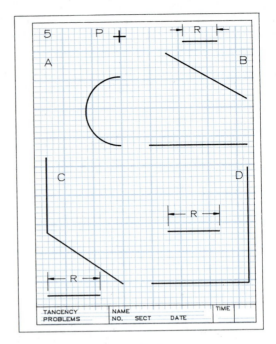

FIGURE 6.72 Problems 5A–5D. Tangency construction.

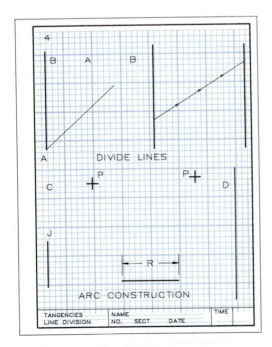

FIGURE 6.71 Problems 4A–4D. Tangency construction.

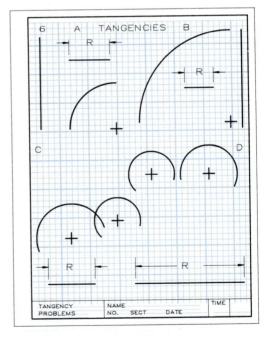

FIGURE 6.73 Problems 6A–6D. Tangency construction.

7. **A–D.** Construct ogee curves that connect the ends of the given lines and pass through points *P* where given.

8. **A–B.** Using the given radii, connect the given arcs with a tangent arc as indicated in the sketches.

9. **A–B.** Rectify the arc along the given line by dividing the circumference into equal divisions and laying them off with your dividers.

   C. Construct an ellipse inside the rectangular layout.

   D. Construct an ellipse inside the large circle. The small circle represents the minor diameter.

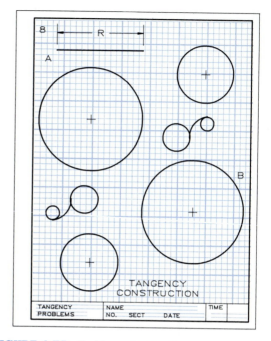

FIGURE 6.75 Problems 8A–8B. Tangency construction.

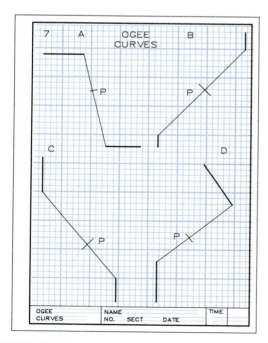

FIGURE 6.74 Problems 7A–7D. Ogee curve construction.

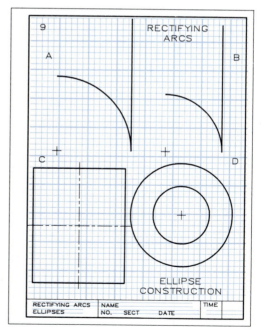

FIGURE 6.76 Problems 9A–9D. Rectifying an arc, ellipse construction.

10. **A.** Construct an ellipse inside the circle when the edge view has been rotated 45° as shown.

   B. Using the focal point *F* and the directrix, plot and draw the parabola formed by these elements.

11. **A.** Using the focal point *F*, points *A* and *B* on the curve, and the axis of symmetry, construct the hyperbolic curve.

   B. Construct a hyperbola that passes through 0. The perpendicular lines are asymptotes.

   C. Construct a spiral by using the four divisions marked along the radius.

12. **A–B.** Construct helixes that have a rise equal to the heights of the cylinder and cone. Show construction and the curve in all views.

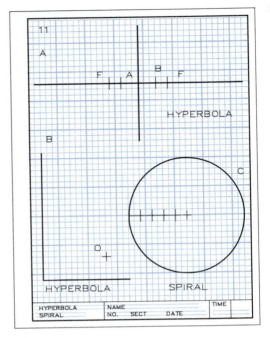

FIGURE 6.78 Problems 11A–11C. Hyperbola and spiral construction.

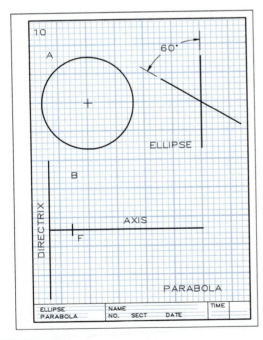

FIGURE 6.77 Problems 10A–10B. Ellipse and parabola construction.

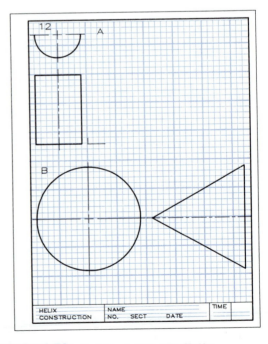

FIGURE 6.79 Problems 12A–12B. Helix construction.

13–23. (Figs. 6.80–6.90) Construct these problems on Size A sheets, one problem per sheet. Select the proper scale that will best fit the problem to the sheet. Mark all points of tangency and strive for good line quality.

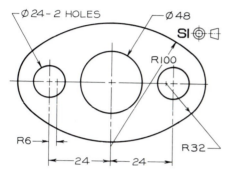

FIGURE 6.80 Problem 13. Gasket.

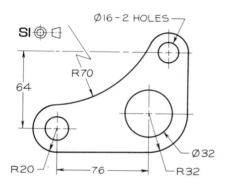

FIGURE 6.81 Problem 14. Lever crank.

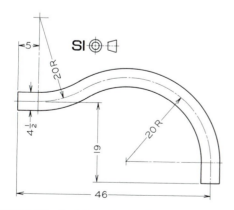

FIGURE 6.82 Problem 15. Road tangency.

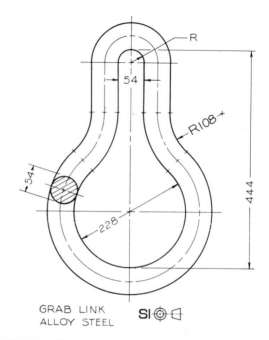

GRAB LINK
ALLOY STEEL

FIGURE 6.83 Problem 16. Grab link.

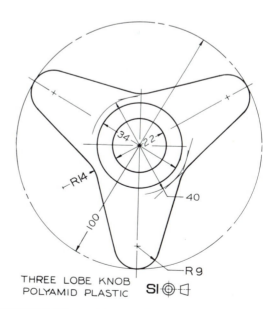

THREE LOBE KNOB
POLYAMID PLASTIC

FIGURE 6.84 Problem 17. Three-lobe knob.

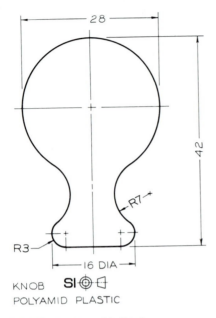

KNOB SI ⊕ ⊟
POLYAMID PLASTIC

FIGURE 6.85 Problem 18. Knob.

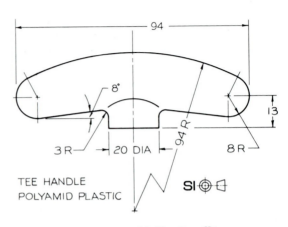

TEE HANDLE
POLYAMID PLASTIC SI ⊕ ⊟

FIGURE 6.87 Problem 20. Tee handle.

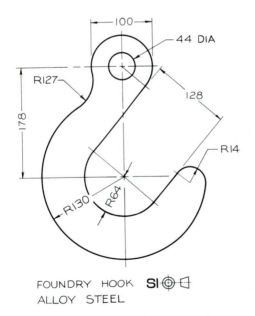

FOUNDRY HOOK SI ⊕ ⊟
ALLOY STEEL

FIGURE 6.86 Problem 19. Foundry hook.

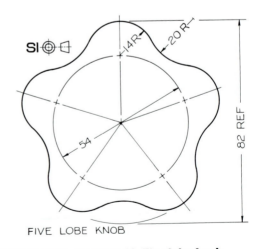

FIVE LOBE KNOB

FIGURE 6.88 Problem 21. Five-lobe knob.

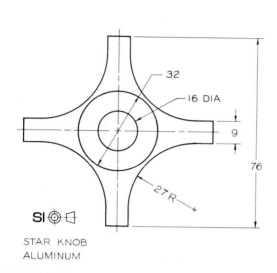

STAR KNOB
ALUMINUM

FIGURE 6.89 Problem 22. Star knob.

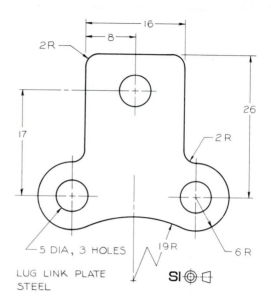

LUG LINK PLATE
STEEL

FIGURE 6.90 Problem 23. Lug link plate.

Multiview Sketching

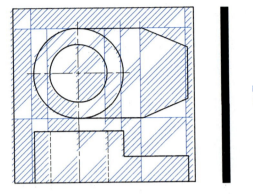

7.1
The purpose of sketching

Sketching is as much a thinking process as it is a communication technique. Designers develop their ideas by making many sketches and revising them before finally arriving at the desired solution.

Sketching should be a rapid method of drawing. Designers who develop their sketching skills can assign drafting work to assistants who can then prepare the finished drawings from their sketches. If sketches are not sufficiently clear to communicate the ideas to someone else, the designer likely has not thought out the solution well enough. The ability to communicate by any means is a great asset, and sketching is one of the more powerful techniques.

7.2
Shape description

Figure 7.1 shows a pictorial of an object with the top, front, and right-side views of the object. Each view is two dimensional.

This system is called **orthographic,** or **multiview projection.** In multiview projection, it is im-

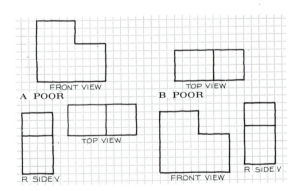

FIGURE 7.2 Poor arrangement of views.

A. These views are sketched incorrectly; the views are scrambled.

B. These views are nearly correct, but they do not project from view to view.

FIGURE 7.1 The three views—top, front, and right side—describe the object by multiview projection.

portant that the views be located as shown in Fig. 7.1. The top view is placed over the front view since both views share the dimension of width. The side view is placed to the right of the front view where these views share the dimension of height. The distance between the views can vary, but the views must be positioned so that they project from each other as shown here.

Figure 7.2 shows several examples of poorly arranged views. Although these individual views are correct, they are hard to interpret because the views are not placed in their standard positions.

Figure 7.3 shows three views of a coffee cup using the principles of multiview projection. To simplify the views, hidden lines have been omitted.

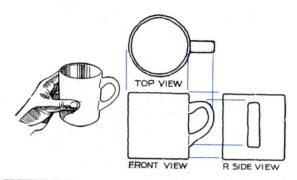

FIGURE 7.3   **Three orthographic views of a cup.**

7.3
Six-view drawings

Six principal views may be found for any object by using the rules of orthographic, or multiview, projection. The directions of sight for the six orthographic views are shown in Fig. 7.4, where the views are drawn in their standard positions. The width dimension is common to the top, front, and bottom views. Height is common to the right-side, front, left-side, and rear views.

Seldom will an object be so complex as to require six orthographic views, but if six views are needed, they should be arranged as shown in this figure.

7.4
Sketching techniques

Sketching means **freehand drawing** without the use of instruments or straightedges. Medium-weight pencils, such as H, F, or HB grades, are the best pencils for sketching. Figure 7.5 shows the standard lines used in multiview drawing and their respective line weights.

By sharpening the pencil point to match the desired line width, you will be able to use the same grade of pencil for all lines (Fig. 7.6). A line drawn

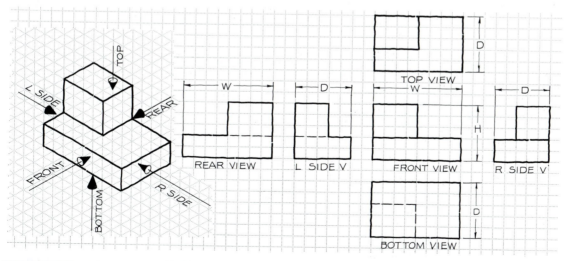

FIGURE 7.4   **Six principal views can be sketched by looking at the object in the directions indicated by the lines of sight. Note how the dimensions are placed on the views. Height (H) is shared by all four of the horizontally positioned views.**

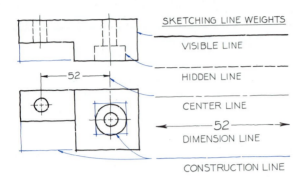

FIGURE 7.5 These lines are examples of those that you should sketch with an F or an HB pencil when drawing views of an object. Some lines are thinner than others, and all except construction lines are black.

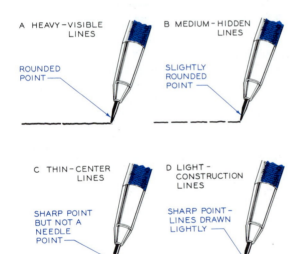

FIGURE 7.6 The alphabet of lines that are sketched freehand are all made with the same pencil grade (F or HB). The variation in the lines is achieved by varying the sharpness of the pencil point.

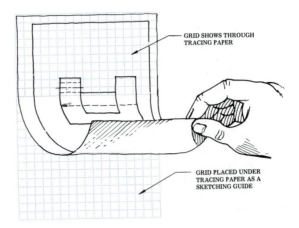

FIGURE 7.7 To aid you in freehand sketching, a grid can be placed under a sheet of tracing paper.

freehand should have a freehand appearance; no attempt should be made to give the line the appearance of one drawn by instruments. However, sketching technique can be improved by using a printed grid on sketching paper or by overlaying a printed grid with translucent tracing paper (Fig. 7.7) so that the grid can be seen through the paper.

When a freehand sketch is made, some lines will be vertical and others horizontal or angular. If you do not tape your drawing to the table top, you will be able to position the sheet for the most comfortable strokes, which are (for the right-handed drafter) from left to right (Fig. 7.8). Finally, for the best effect, the lines sketched to form the various views should intersect as indicated in Fig. 7.9.

7.5
Three-view sketch

Figure 7.10 shows three views of an object with height, width, and depth dimensions given. Each view is labeled. Figure 7.11 shows the steps of drawing three orthographic views of a similar part on a printed grid. The most commonly used combination of views are the front, top, and right-side views, as shown in this figure. First, the overall dimensions of the object are sketched. Then the slanted surface is drawn in the top view and projected to the other views. Lastly, the lines are darkened; the views labeled; and the overall dimensions of height, width, and depth applied.

When surfaces are slanted, they will not appear true shape in the principal views of orthographic projection (Fig. 7.12). Surfaces that do not appear true size either are **foreshortened** or appear as **edges.** In Fig. 7.12C, two planes of the object are slanted; thus

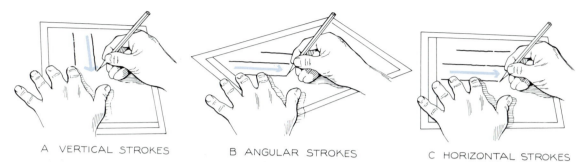

FIGURE 7.8 Freehand sketching techniques.

A VERTICAL STROKES   B ANGULAR STROKES   C HORIZONTAL STROKES

A. Vertical lines should be sketched in a downward direction.

B. Angular strokes can be sketched left to right if you rotate your sheet slightly.

C. Horizontal strokes are made best in a left-to-right direction. Always sketch from a comfortable position, and turn your paper if necessary.

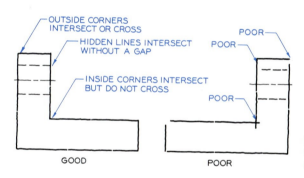

OUTSIDE CORNERS INTERSECT OR CROSS

HIDDEN LINES INTERSECT WITHOUT A GAP

INSIDE CORNERS INTERSECT BUT DO NOT CROSS

POOR

POOR

POOR

GOOD

POOR

FIGURE 7.9 For good sketches, follow these examples of good technique. Compare the good drawing with the poor one.

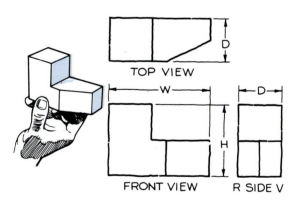

TOP VIEW

W

D

H

FRONT VIEW   R SIDE V   D

FIGURE 7.10 The standard arrangement of drawing three views of an object with its dimensions and labels.

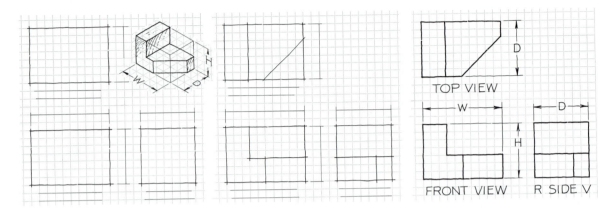

FIGURE 7.11   Three-view sketching.

Step 1   Block in the views by using the overall dimensions. Allow proper spacing for labeling and dimensioning the views.

Step 2   Remove the notches, and project from view to view.

Step 3   Check your layout for correctness; then darken the lines, and complete the labels and dimensions.

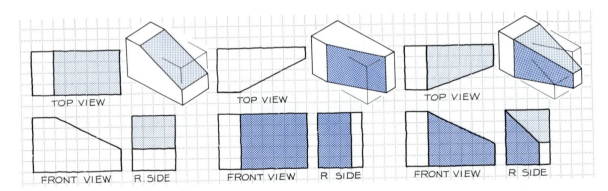

FIGURE 7.12   Views of planes.

A.   The plane appears as an edge in the front view, and it is foreshortened in the top and side views.

B.   The plane is an edge in the top view and foreshortened in the front and side views.

C.   These two planes appear foreshortened in the right-side view. Each appears as an edge in the top and front views.

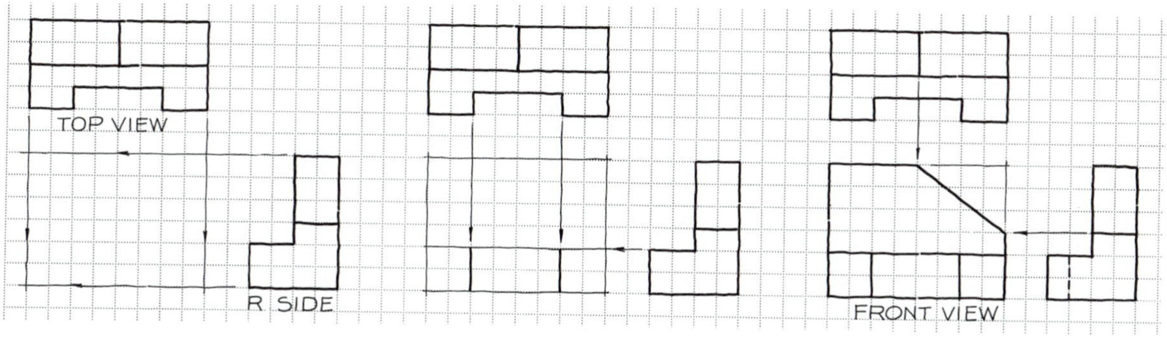

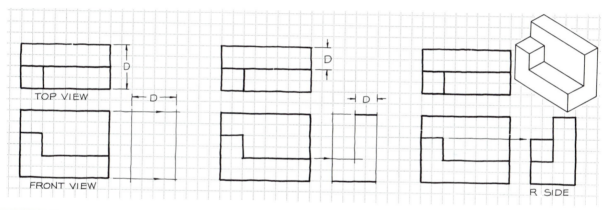

FIGURE 7.13 Missing front views.

Step 1 When two views are given and the third is required, begin by projecting the overall dimensions from the top and right-side views.

Step 2 The various features of the object are sketched using construction lines.

Step 3 The features are completed, the views checked, and the lines darkened to the proper line quality.

FIGURE 7.14 Missing side view.

Step 1 To find the right-side view when the top and front views are given, block in the view with the overall dimensions.

Step 2 Develop the features of the view by analyzing the views together. Use light construction lines.

Step 3 Check the views for correctness, and darken the lines to their proper line weight.

both appear foreshortened in the right-side view.

A good exercise for analyzing the given views is to find the missing view when two views are given (Fig. 7.13). The right-side view is found in Fig. 7.14, where the top and front views are given. The right-side view has the depth in common with the top view and the height in common with the front view. Knowing this enables us to block in the side view in

Step 1. The side view is developed in Step 2 and completed in Step 3.

Another exercise is to complete the views when some or all of them have missing lines (Fig. 7.15). A pictorial is provided to help you analyze the given views.

Figure 7.16 shows three views of an aircraft bracket, which contains several curved corners.

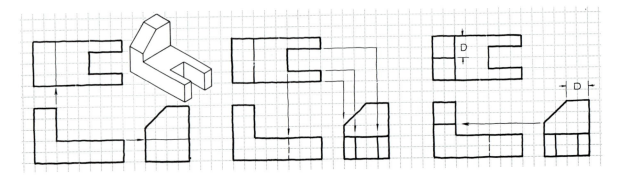

FIGURE 7.15 Missing lines.

Step 1 Lines may be missing in all views in this type of problem. The first missing line is found by projecting the edges of the planes from the front view.

Step 2 The notch in the top view is projected to the front and side views. The line in the front view is hidden.

Step 3 The line formed by the beveled surface is found in the front view by projecting from the side view.

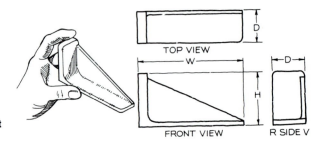

FIGURE 7.16 Three orthographic views of an aircraft part.

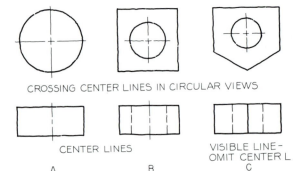

CROSSING CENTER LINES IN CIRCULAR VIEWS

CENTER LINES

A   B

VISIBLE LINE – OMIT CENTER L

C

FIGURE 7.17 Centerlines are used to indicate the centers of circles and the axes of cylinders. They are drawn as very thin lines. When they coincide with visible or hidden lines, centerlines are omitted.

7.6

Circular features

Centerlines indicate the features are true circles or cylinders (Figure 7.17). In circular views, centerlines cross to indicate the center of the circle. Centerlines consist of short dashes, which should cross in the circular views, spaced at 1-inch intervals along the line. If a centerline coincides with an object line—visible or hidden—the centerline should be omitted since the object lines are more important (Fig. 7.17C).

The application of centerlines is shown in Fig. 7.18, where they indicate whether or not circles and arcs are concentric (share the same centers). The centerline should extend beyond the arc by about $\frac{1}{8}$ inch. Figure 7.19 shows the correct manner of applying centerlines. The circular view clearly indicates that the cylinders are concentric since each shares the same centerlines.

Sketching circles

Circles can be sketched by using light guidelines along with centerlines (Fig. 7.20). Since it is difficult to draw a freehand circle in one continuous line, short arcs are drawn using the guidelines. If you fail to become rea-

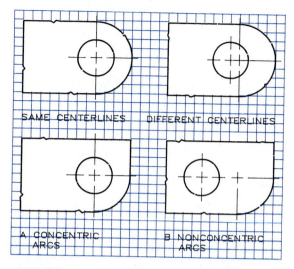

FIGURE 7.18 The centerline should extend beyond the last arc that has the same center. When the arcs are not concentric, separate centerlines should be drawn.

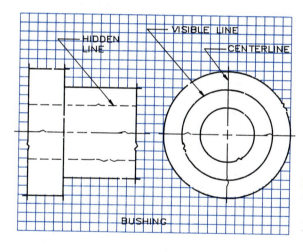

FIGURE 7.19 Here you can see the application of centerlines of concentric cylinders and the relative weight of hidden lines, visible lines, and centerlines.

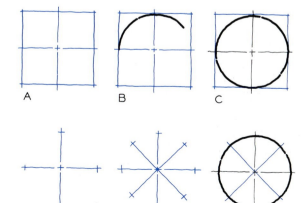

FIGURE 7.20 Circles can be sketched using either of the construction methods shown here. The use of guidelines is essential to freehand sketching of circles and arcs.

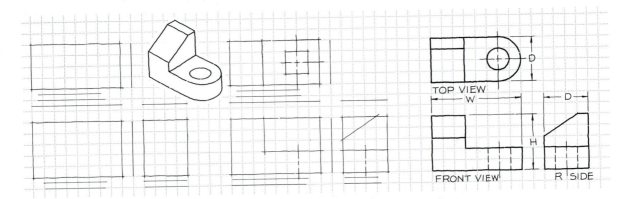

FIGURE 7.21 Circular features in orthographic views.

Step 1 To draw orthographic views of the object shown, begin by blocking in the overall dimensions. Leave room for the labels and dimensions.

Step 2 Construct the centerlines and the squares about the centerlines in which the circles will be drawn. Show the slanted surface in the side view.

Step 3 Sketch the arcs, and darken the final lines of the views. Label the views, and show the dimensions of W, D, and H.

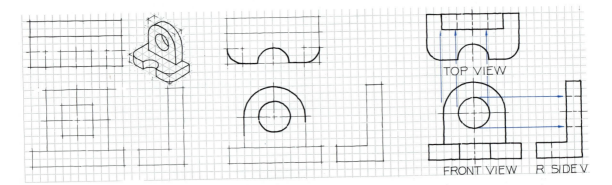

FIGURE 7.22 Orthographic views with multiple circular features.

Step 1 When sketching orthographic views with circular features, you should begin by sketching the centerlines and guidelines.

Step 2 Using the guidelines, sketch the circular features. These can be darkened as they are drawn if they will be final lines.

Step 3 The outlines of the circular features are found by projecting from the views found in Step 2. All final lines are darkened.

sonably skilled at sketching circles, use a circle template or compass to lightly draw the circle or arc; then darken the line freehand to match the other lines of your sketch.

Figure 7.21 shows the construction of three orthographic views with circular features. The circular features are located with centerlines and guidelines in Step 2; then they are sketched in and darkened in Step 3.

A similar example is given in Fig. 7.22, where the object consists of circular features and arcs. The circles should be drawn first so that their corresponding rectangular views (such as the hidden hole in the top view) can be found by projecting from the circular view.

7.7
Isometric sketching

Another type of three-dimensional pictorial is the **isometric drawing,** which may be drawn on a specially printed grid composed of a series of lines making 60° angles with one another (Fig. 7.23). The squares in the orthographic views can be laid off along the isometric grid, as Step 1 shows. The notch is located in the same manner in Steps 2 and 3 to complete the isometric pictorial.

Angles cannot be measured with a protractor in isometric pictorials; they must be drawn by measuring coordinates along the three axes of the printed grid. In Fig. 7.24, the ends of the angular slope are located by measuring the direction of width and height. When an object has two sloping planes that intersect (Fig. 7.25), it is necessary to draw the sloping planes one at a time to find point B. The line from A to B is the line of intersection between the two planes.

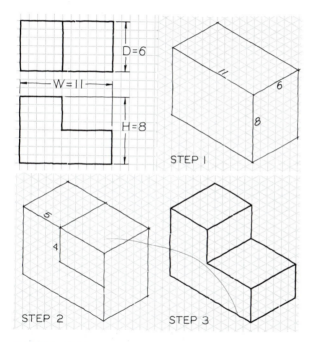

FIGURE 7.23 Isometric pictorial sketching.

Step 1 When orthographic views of a part are given, an isometric pictorial can be sketched on a printed isometric grid. Construct a box using the overall given dimensions.

Step 2 The notch can be located by measuring over five squares as shown in the orthographic views. The notch is measured four squares downward.

Step 3 The pictorial is completed, and the lines are darkened. This is a three-dimensional pictorial, whereas the orthographic views are two dimensional views.

Circles in isometric pictorials

Circles, which will appear as ellipses in isometric pictorials, can be sketched by locating their centerlines as shown in Step 1 of Fig. 7.26. The center must be equidistant from the top, bottom, and end of the front view.

Figure 7.27 shows two methods of constructing elliptical views of circles.

Figure 7.28 shows the technique of sketching ellipses in isometric to represent a cylinder. This same technique is used to draw an object with a semicircular end in Fig. 7.29.

Figure 7.30A shows a part composed of several cylindrical forms. These three views are used as the basis for an isometric pictorial shown in the steps of construction. When an isometric grid is not used, the axes of the isometric sketch are positioned 120° apart. In other words, the height dimension is vertical, and the width and depth dimensions make 30° angles with the horizontal direction on your paper.

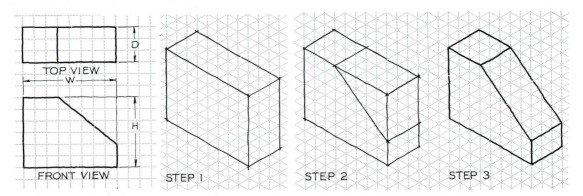

FIGURE 7.24 Angles in isometric pictorials.

Step 1 Begin by drawing a box using the overall dimensions given in the orthographic views. Count the squares, and transfer them to the isometric grid.

Step 2 Angles cannot be measured with a protractor. Angles will be either larger or smaller than their true measurements. Find each end of the angle by measuring along the axes.

Step 3 The ends of the angles are connected. Dimensions can be measured only in directions parallel to the three axes.

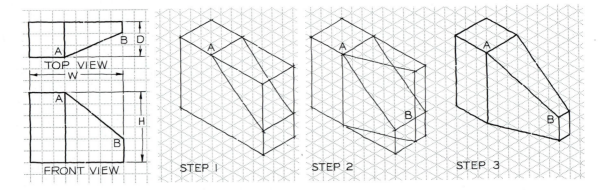

FIGURE 7.25 Double angles in isometric pictorials.

Step 1 When part of an object has a double angle, begin by constructing the overall box and finding one of the angles.

Step 2 Find the second angle that locates point *B*. Point *A* will connect to point *B* to give the intersection line.

Step 3 The final lines are darkened. Line *AB* is the line of intersection between the two sloping planes.

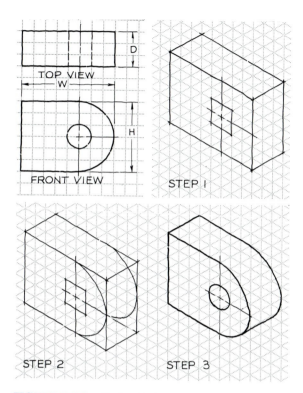

FIGURE 7.26 Circles in isometric pictorials.

Step 1 Construct a box using the overall orthographic dimensions. Draw the centerlines and a square (rhombus) of guidelines around the circular hole.

Step 2 Draw the pictorial views of the arcs tangent to the boxes formed by the guidelines. These arcs will appear elliptical rather than circular.

Step 3 Construct the hole, and darken the lines. Hidden lines are normally omitted in pictorial sketches.

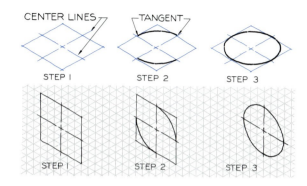

FIGURE 7.27 Sketching ellipses.

Two methods of sketching ellipses are shown: one without a grid and one with a grid. When a grid is not used, the centerlines are drawn at 30° to the horizontal (for a horizontal circle). A rhombus is drawn about them, and the ellipse is sketched inside the guidelines. When there is a grid, the lines of the grid become the guidelines.

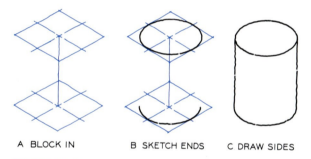

FIGURE 7.28 Sketching circular features.

Several examples of sketching circles and cylinders. In all cases, guidelines are used to aid in proportional sketching.

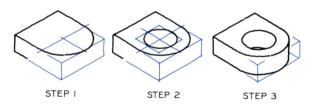

FIGURE 7.29 To make an isometric sketch of a cylinder, the ends are blocked in with rhombuses that are connected by the axis of the cylinder. Ellipses are sketched tangent to the rhombuses, and the sides of the cylinders are drawn.

118

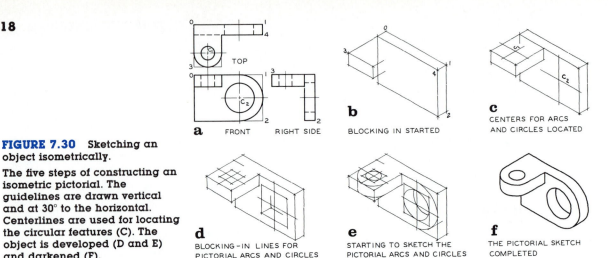

a FRONT   RIGHT SIDE

b BLOCKING IN STARTED

c CENTERS FOR ARCS AND CIRCLES LOCATED

FIGURE 7.30 Sketching an object isometrically.

The five steps of constructing an isometric pictorial. The guidelines are drawn vertical and at 30° to the horizontal. Centerlines are used for locating the circular features (C). The object is developed (D and E) and darkened (F).

d BLOCKING-IN LINES FOR PICTORIAL ARCS AND CIRCLES

e STARTING TO SKETCH THE PICTORIAL ARCS AND CIRCLES

f THE PICTORIAL SKETCH COMPLETED

Problems

These sketching problems should be drawn on Size A (8½-×-11-inch) paper, with or without a printed grid. A typical format for this size sheet is shown in Fig. 7.31, where a 0.20″ grid is given. (This grid can be converted to an approximate metric grid by equating each square to 5 mm.) All sketches and lettering should be neatly executed by applying the principles covered in this chapter. Figures 7.32 and 7.33 contain the problems and instructions.

FIGURE 7.31 The layout of a Size A sheet for sketching problems.

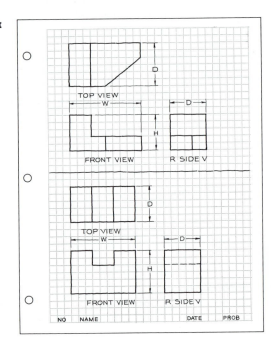

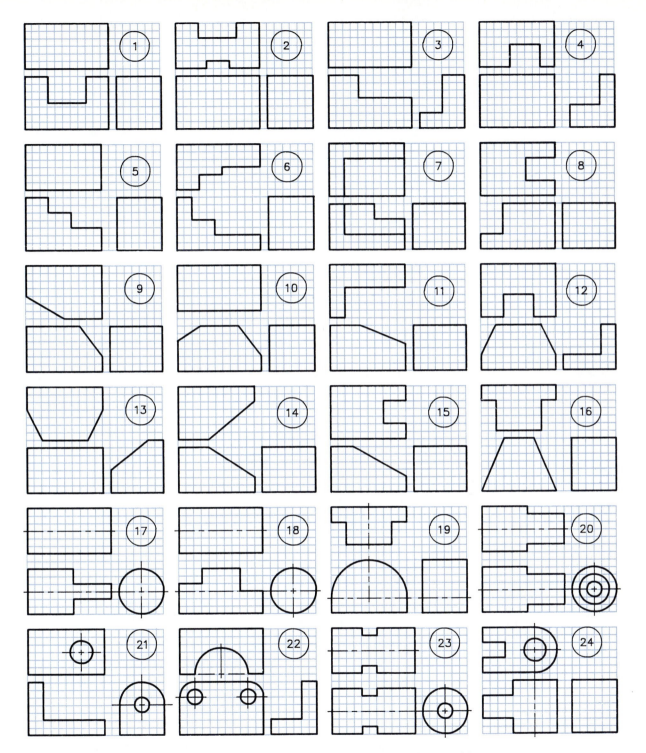

FIGURE 7.32 On Size A paper, sketch top, front, and right-side views of the problems assigned. Two problems can be drawn on each sheet. Give the overall dimensions of *W*, *D*, and *H*, and label each view.

FIGURE 7.33 Multiview problems and isometric sketching.

On Size A paper, sketch the top, front, and right-side views of the problems assigned. Supply the lines that may be missing from all views. Then sketch isometric pictorials of the object assigned, two per sheet.

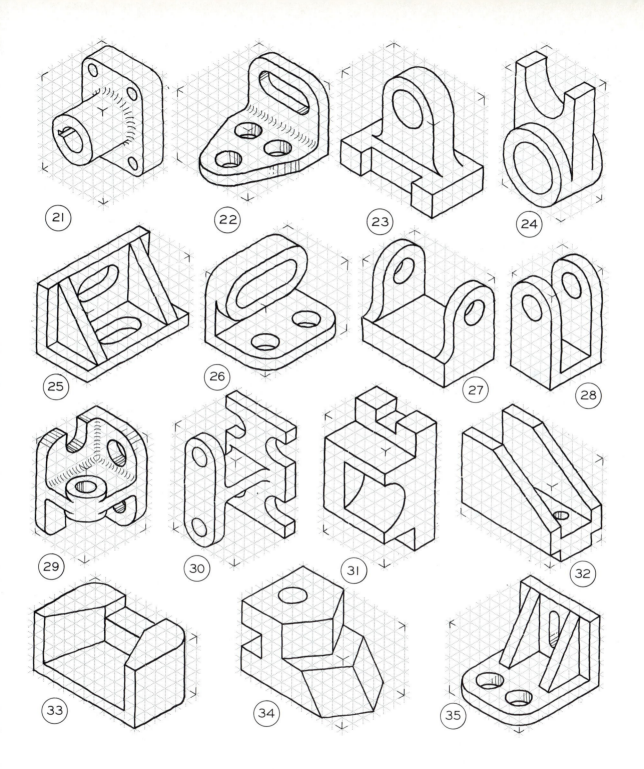

21

22

23

24

25

26

27

28

29

30

31

32

33

34

35

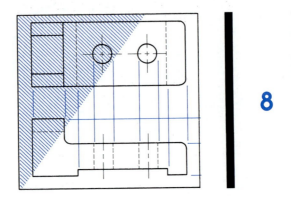

Multiview Drawing with Instruments

8

8.1

Introduction

Multiview drawing, the system of representing three-dimensional objects by separate views arranged in a standard manner, is readily understood by the technical community. Because multiview drawings are usually executed with instruments and drafting aids, they are often called **mechanical drawings.** They are called **working,** or **detail drawings,** when dimen-

sions and notes are added to complete the specifications of the parts that have been drawn.

8.2

Orthographic projection

The artist is likely to represent objects impressionistically, but the drafter must represent them precisely. The method of preparing a precise, detailed, clearly

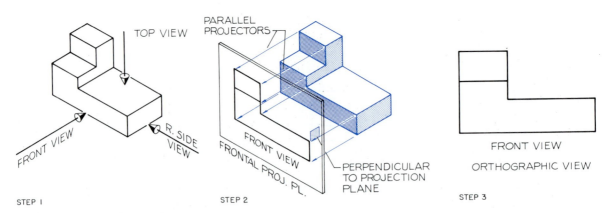

FIGURE 8.1 Orthographic projection.

Step 1 Three mutually perpendicular lines of sight are drawn to obtain three views of the object.

Step 2 The frontal plane is a vertical plane on which the front view is projected with parallel projectors perpendicular to the frontal plane.

Step 3 The resulting view is the front view of the object. This is a two-dimensional orthographic view.

understood drawing is **orthographic projection,** or **multiview drawing.**

> By **orthographic projection,** the views of an object are projected perpendicularly onto projection planes with parallel projectors (Fig. 8.1).

The front view is projected onto a vertical frontal plane with parallel projectors. The resulting front view is two dimensional since it has no depth and lies in a single plane described by two dimensions, width and height.

The top view of the same object is projected onto a horizontal projection plane perpendicular to the frontal projection plane in Fig. 8.2A. The right-side view is projected onto a vertical profile plane perpendicular to both the horizontal and frontal planes (Fig. 8.2B).

Imagine that the same object has been enclosed in a glass box showing the frontal, horizontal, and profile projection planes (Fig. 8.3). While in the glass

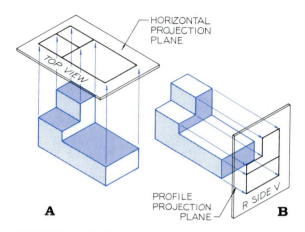

FIGURE 8.2 **A.** The top view is projected onto a horizontal projection plane. **B.** The right-side view is projected onto a vertical profile plane perpendicular to the horizontal and frontal planes.

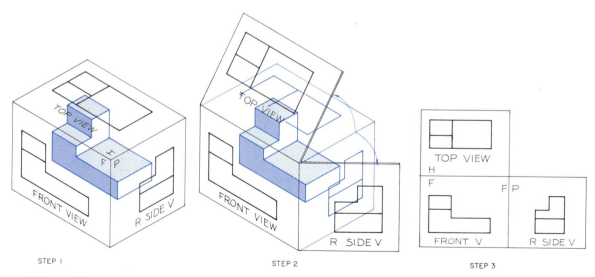

FIGURE 8.3 Glass-box theory.

Step 1 Imagine that the object has been placed inside a box formed by the horizontal, frontal, and profile planes onto which the top, front, and right-side views have been projected.

Step 2 The three projection planes are then opened into the plane of the drawing surface.

Step 3 The three views are positioned with the top view over the front view and the right-side view to the right. The planes are labeled *H, F,* and *P* at the fold lines.

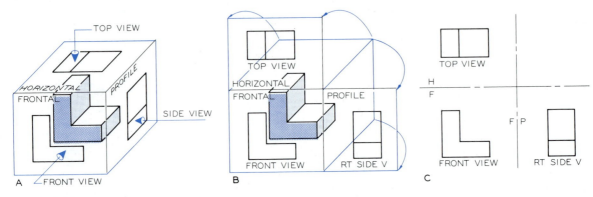

FIGURE 8.4 Principal projection planes.

A. The three principal projection planes of orthographic projection can be thought of as planes of a glass box.

B. The views of an object are projected onto the projection planes, which are opened into the plane of the drawing surface.

C. The outlines of the planes are omitted. The fold lines are drawn and labeled.

box, the object's views are projected onto the projection planes. Then the box is opened into the plane of the drawing surface to give the standard positions for the three orthographic views.

A similar example of this principle is shown by the object in the projection box in Fig. 8.4. The three views are positioned in the same manner: the top view over the front view and the right-side view to the right of the front view.

> The three principal projection planes of orthographic projection are the **horizontal** (H), **frontal** (F), and **profile** (P) planes.

Any view projected onto one of these principal planes is a **principal view.** The dimensions of an object used to show its three-dimensional form are height (H), width (W), and depth (D).

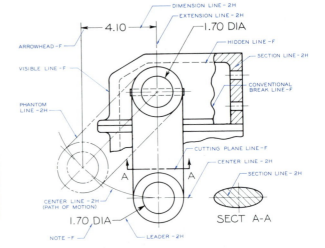

FIGURE 8.5 The line weights and suggested pencil grades recommended for orthographic views.

8.3
Alphabet of lines

Using proper line weights will greatly improve a drawing's readability and appearance. All lines should be drawn dark and dense, as if drawn with ink. Only by their width should the lines vary—except, that is, for

guidelines and construction lines, which are drawn very lightly for lettering and laying out drawings.

The lines of an orthographic view are labeled, along with the suggested pencil grades for drawing them, in Fig. 8.5. The lengths of dashes in hidden lines and centerlines are drawn longer as the size of a drawing increases. Figure 8.6 gives additional specifications for these lines.

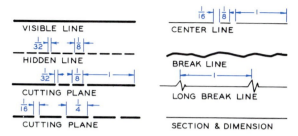

FIGURE 8.6 A comparison of the line weights for orthographic views. These dimensions will vary for different sizes of drawings and should be approximated by eye.

COMPUTER LINES Smaller computer graphics plotters vary the widths of lines on a drawing by multiple-pen accessories that hold pens of different point widths, usually 0.7 mm and 0.3 mm wide. Different pen widths or colors are specified with the LINETYPE command. AutoCAD provides a command, LTSCALE, that varies dash and space lengths in dashed lines (Fig. 8.7).

A comparison of lines drawn with LTSCALEs of 0.4 and 0.2 and pen widths of 0.7 mm and 0.3 mm is given. Once the drawing has been completed, the line scales are changed with the LTSCALE command by changing one number (from 0.4 to 0.2 in this example), and the views are automatically redrawn to show the revised lines (Fig. 8.8).

Step 1 These lines were drawn using AutoCAD's standard LINETYPEs and two pens. The LTSCALE factor was 0.4, which affects the spacing between the dashes and the lengths of the dashes.

Step 2 By changing the LTSCALE factor from 0.4 to 0.2, the dashes and the spaces between them are reduced in size.

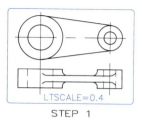

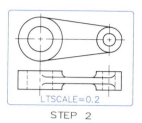

FIGURE 8.8 Line weights by computer.

Step 1 This two-view drawing was made using an LTSCALE of 0.4, which resulted in the omission of dashes in centerlines and hidden lines with dashes that were too long.

Step 2 By activating the LTSCALE and giving it a factor of 0.2, the views are redrawn, and centerlines and hidden lines appear with the appropriate dashes.

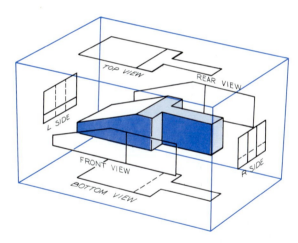

FIGURE 8.9 Six principal views of an object can be drawn in orthographic projection. You can imagine that the object is in a glass box with the views projected onto the six planes.

8.4
Six-view drawings

If you visualize an object placed inside a glass box, you will see there are two horizontal planes, two frontal planes, and two profile planes (Fig. 8.9).

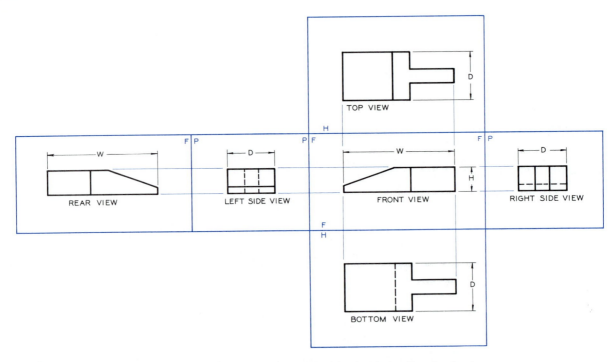

FIGURE 8.10 Once the box is completely opened into a single plane, the six views are positioned to describe the object. The outlines of the planes are usually omitted. They are shown here to help you relate this figure to Fig. 8.9.

Therefore, the maximum number of principal views that can be used to represent an object is six.

The top and bottom views are projected onto horizontal planes, the front and rear views onto frontal planes, and the right- and left-side views onto profile planes.

To draw the views on a sheet of paper, imagine the glass box is opened up into the plane of the drawing paper; the views will then appear as shown in Fig. 8.10. The top view is placed over, and the bottom view under, the front view; the right-side view is to the right of, and the left-side view to the left of, the front view; the rear view is to the left of the left-side view.

Height, width, and depth are three dimensions of an object necessary to give its size. The standard arrangement of the six views allows some of the views to share dimensions by projection. For example, the height dimension, which is shown only once between the front and right-side views, applies to the four horizontally arranged views. Furthermore, the width dimension is placed between the top and front views, but it also applies to the bottom view.

Projectors align the views both horizontally and vertically about the front view in Fig. 8.10. Each side of the fold lines of the glass box is labeled H, F, or P (horizontal, frontal, or profile) to identify the projection planes on a given side of the fold lines.

8.5
Three-view drawings

Because three views are usually adequate to describe an object, the most commonly used orthographic arrangement is the **three-view drawing**, consisting of front, top, and right-side views. The object used in the previous example is shown placed in a glass box in Fig. 8.11, which is opened onto the plane of the draw-

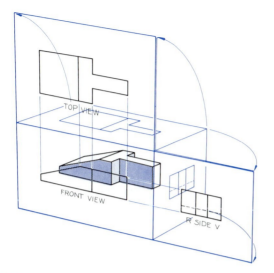

FIGURE 8.11 Three-view drawings are commonly used for describing small machine parts. The glass box is used to illustrate how the views are projected onto their projection planes.

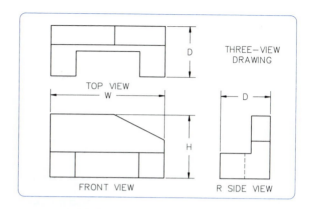

FIGURE 8.13 This three-view orthographic drawing was made by AutoCAD and an A-B size plotter using two pen sizes (0.7 mm and 0.3 mm). The same principles of orthographic projection are used whether drawings are made manually or by computer.

ing surface. The resulting three-view arrangement is shown in Fig. 8.12, where the views are labeled and dimensioned.

COMPUTER VIEWS The three-view drawing of a part (Fig. 8.13) was drawn by using two pen widths, 0.7 mm and 0.3 mm, and the DRAW and DIMension commands of AutoCAD.

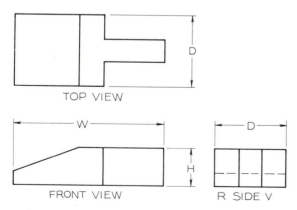

FIGURE 8.12 The resulting three-view drawing of the object from Fig. 8.11.

8.6
Arrangement of views

Figure 8.14A shows the standard positions for a three-view drawing: The top and side views are projected directly from the front view. The views are properly labeled and dimensioned. Views that are arranged in a nonstandard sequence (Fig 8.14B) and that do not project from view to view (Fig. 8.14C) are incorrect. Figure 8.15 emphasizes these rules of arrangement and dimensioning.

8.7
Selection of views

When drawing an object by orthographic projection, you should select the views with the fewest hidden lines. In Fig. 8.16A, the right-side view is preferred to the left-side view because it has fewer hidden lines.

Although the three-view arrangement of top, front, and right-side views is the most commonly used, the top, front, and left-side views is equally acceptable

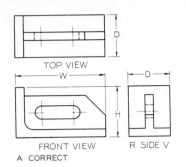

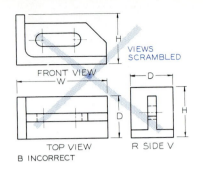

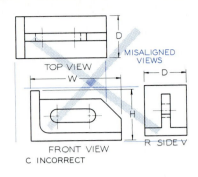

FIGURE 8.14 Positioning orthographic views.

A. This is a correct arrangement of views, labels, and dimensions. The views project from each other in proper alignment.

B. These views are scrambled into unconventional positions, making it hard to interpret them. Dimensions are needlessly repeated.

C. These views are misaligned, so they do not project from one to the other. This is an incorrect arrangement.

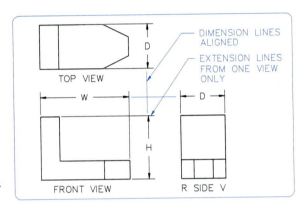

FIGURE 8.15 Dimension and extension lines used in three-view orthographic projection should be aligned as shown in this computer drawing. Notice that extension lines are drawn from only one view when dimensions are placed between two views.

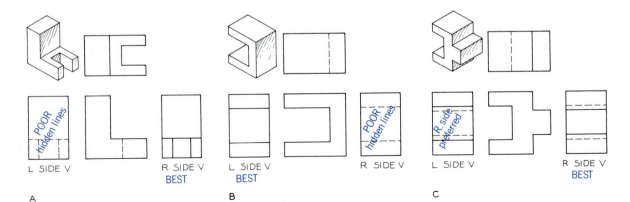

FIGURE 8.16 Selection of views.

A. In orthographic projection, you should select the sequence of views with the fewest hidden lines.

B. The left-side view has fewer hidden lines; therefore, this view is selected over the right-side view.

C. When both views have an equal number of hidden lines, the right-side view is traditionally selected.

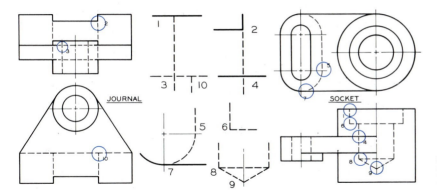

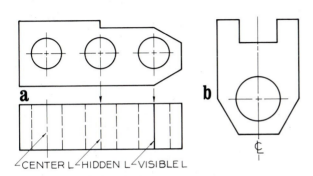

FIGURE 8.17 When lines intersect in orthographic projection, they should intersect as shown here.

(Fig. 8.16B) if the left-side view has fewer hidden lines than the right-side view.

The most descriptive view is usually selected as the front view. Some objects have standard views that are regarded as the front view, top view, and so forth. A chair, for example, has front and top views that are recognized as such by everyone; therefore, a chair's accepted front view should be used as the orthographic front view.

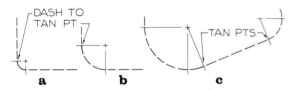

FIGURE 8.18 Hidden lines in orthographic projection that are composed of curves should be drawn in this manner.

FIGURE 8.19 **A.** The order of importance (precedence) of lines is visible lines, hidden lines, and centerlines. **B.** The symbol made of the letters *C* and *L* is used to label a centerline when needed on symmetrical parts.

8.8
Line techniques

As drawings become more complex, you will encounter more instances of lines overlapping and intersecting in ways similar to those shown in Fig. 8.17. This illustration shows the techniques of handling most types of intersecting lines. Figure 8.18 shows the methods of drawing hidden lines composed of lines and arcs.

You should become familiar with the order of precedence (priority) of lines (Fig. 8.19): The most important line is the visible object line; it is shown regardless of any other line lying behind it. Of next importance is the hidden object line, which takes precedence over the centerline.

> **CENTERLINES BY COMPUTER** Figure 8.20 shows a method of ensuring that centerline dashes cross properly when drawn by computer. Centerlines must be drawn in connected segments (Fig. 8.20B).

8.9
Point numbering

The method of numbering points and lines of an object will be helpful to you in constructing orthographic views. For example, Fig. 8.21 shows an object that has been numbered to aid in the construction of the miss-

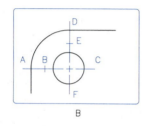

A

B

FIGURE 8.20 Centerlines by computer.

A. When centerlines for a rounded corner are drawn from 1 to 2 and from 3 to 4, the center dashes do not cross at the center.

B. Centerline dashes can be made to cross at the center by drawing a line from A to B to C, where the BC segment extends equally beyond the center on both sides. Line DF is drawn from D to E to F in the same manner.

ing front view when the top and side views are given. By projecting selected points from the top and side views, the front view of the object can be found.

8.10
Line and planes

An orthographic view of a line can appear **true length, foreshortened,** or as a **point** (Fig. 8.22). When a line appears true length, it must be parallel to

the reference line in the previous view. A plane in orthographic projection can appear **true size, foreshortened,** or as an **edge** (Fig. 8.22).

8.11
Alternate arrangement of views

Although the right-side view is usually placed to the right of the front view (Fig. 8.23), the side view can be projected from the top view (Fig. 8.24); this is advisable when the object has a much larger depth than height.

8.12
Laying out three-view drawings

The depth dimension applies to both the top and side views, but these views are usually positioned where this dimension does not project between them (Fig. 8.25). The depth dimension can be graphically transferred by using a 45° line, an arc, or a pair of dividers.

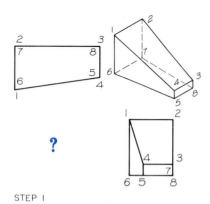

STEP I

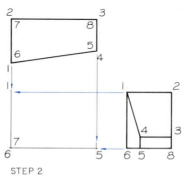

STEP 2

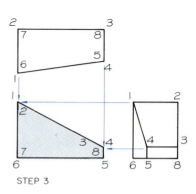

STEP 3

FIGURE 8.21 Point numbering.

Step 1 When a missing orthographic view is to be drawn, it is helpful to number the points in the given views.

Step 2 Points 1, 5, 6, and 7 are found by projecting from the given views of these points.

Step 3 The plotted points are connected to form the missing front view.

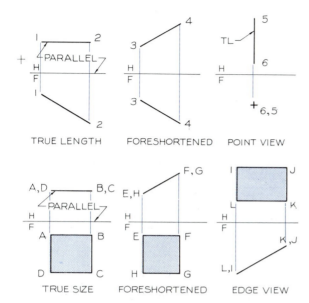

TRUE LENGTH   FORESHORTENED   POINT VIEW

TRUE SIZE   FORESHORTENED   EDGE VIEW

FIGURE 8.22   Lines and planes.

A line will appear in orthographic projection as true length, foreshortened, or a point. A plane in orthographic projection will appear as true size, foreshortened, or an edge.

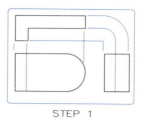
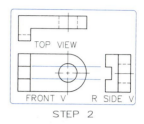

STEP 1   STEP 2

FIGURE 8.23   Three views by computer.

Step 1   The three views are blocked in. Notice that the depth dimension in the top view is the same as the depth in the right-side view.

Step 2   The circular hole and its centerlines are drawn in all views. The notch is drawn in the side view and projected to the top and front views.

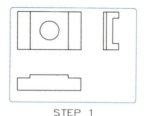

STEP 1   STEP 2

FIGURE 8.24   Alternate-position side view.

Step 1   The right-side view of this part is projected from the top view to save space since the depth dimension is large.

Step 2   The views are completed, and dimensions and labels are added.

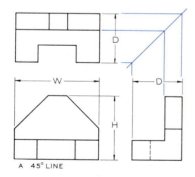
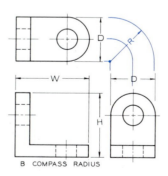
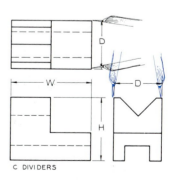

A 45° LINE

B COMPASS RADIUS

C DIVIDERS

FIGURE 8.25   Transferring depth dimensions.

A. The depth dimension can be projected from the top view to the right-side view by constructing a 45° line positioned as shown.

B. The depth dimension can be projected from the top view to the side view by using a compass and center point.

C. The depth dimension can be transferred from the top view to the side view by using dividers—the most desirable method.

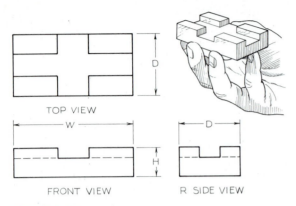

FIGURE 8.26 A three-view drawing of an object.

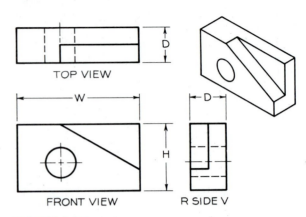

FIGURE 8.29 A three-view drawing of an object.

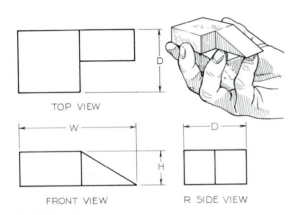

FIGURE 8.27 A three-view drawing of an object.

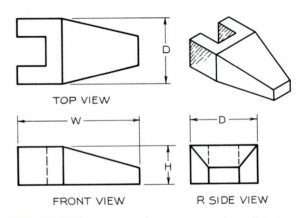

FIGURE 8.30 A three-view drawing of an object.

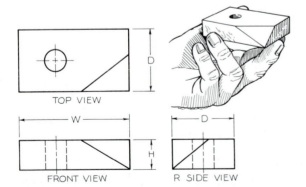

FIGURE 8.28 A three-view drawing of an object.

Figures 8.26–8.30 show examples of three-view orthographic drawings of objects.

THREE-VIEW DRAWINGS BY COMPUTER
Figure 8.31 illustrates the steps for constructing three views of an object by computer methods. The overall outlines of the views are drawn with the LINE command, and the views are labeled with the TEXT command in Step 1. Other lines are added in Step 2 by projecting from view to view. In Step 3, the remaining visible and hidden lines are drawn.

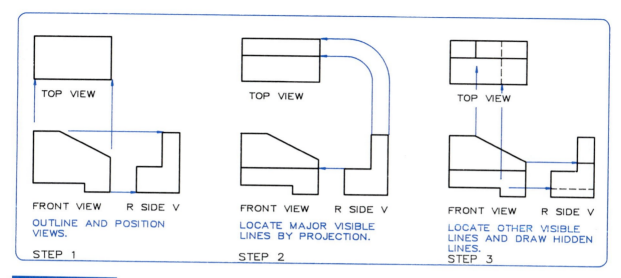

TOP VIEW

FRONT VIEW   R SIDE V

OUTLINE AND POSITION
VIEWS.

STEP 1

TOP VIEW

FRONT VIEW   R SIDE V

LOCATE MAJOR VISIBLE
LINES BY PROJECTION.

STEP 2

TOP VIEW

FRONT VIEW   R SIDE V

LOCATE OTHER VISIBLE
LINES AND DRAW HIDDEN
LINES.
STEP 3

FIGURE 8.31   Three views by computer.

Step 1   Using the LINE command, draw the outlines of the
views and label them.

Step 2   By orthographic projection, draw other visible lines.

Step 3   Draw the remaining visible and hidden lines. Change
LAYERs before drawing hidden
lines.

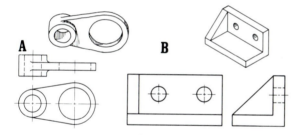

FIGURE 8.32   These objects can be adequately
described with two orthographic views.

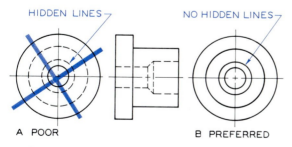

HIDDEN LINES

A POOR

NO HIDDEN LINES

B PREFERRED

FIGURE 8.33   Cylindrical objects can be depicted
with two views. Always select views with the fewest
hidden lines.

8.13
Two-view drawings

It is good economy of time and space to use only the
views necessary to depict an object. Figure 8.32 shows
objects that require only two views. Cylindrical objects
need only two views, as Fig. 8.33 shows. Because it is
preferable to select the views with the fewest hidden
lines, the right-side view is the better view in this example.

TWO-VIEW DRAWINGS BY COMPUTER
Figure 8.34 shows the steps in making a two-view
drawing of an object. The ARC and FILLET
commands are used to draw the semicircular feature and the rounded corners in Step 2.

The MIRROR command can be used to reduce
construction by drawing only half (or one quarter)
of a view and then mirroring the drawing to give
the other symmetrical half (Fig. 8.35).

The object in Fig. 8.36, which is composed of
arcs and tangent lines, is constructed as a half top
view, which is mirrored along line *AB*. Fillets are
drawn in Step 2 with the FILLET command.

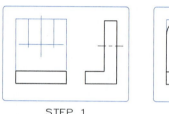
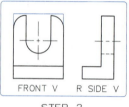

STEP 1   STEP 2

FIGURE 8.34 A two-view drawing by computer.

Step 1 The front and side views are drawn using the overall dimensions.

Step 2 The FILLET command is used to draw the corners in the front view, and the ARC command is used to draw the semicircular arc.

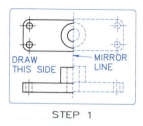
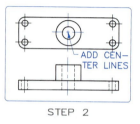

STEP 1   STEP 2

FIGURE 8.35 Mirroring by computer.

Step 1 To save drawing time, the top and front views are drawn as half views, and MIRRORed about the centerline.

Step 2 Centerlines that coincide with the MIRROR line should be drawn after the mirroring to prevent the centerline from being drawn twice.

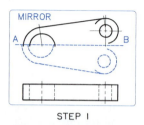
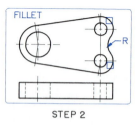

STEP 1   STEP 2

FIGURE 8.36 Circular features by computer.

Step 1 Draw the top view as a half view and MIRROR it about *AB*. Locate tangent points by using the PERPendicular option of the OSNAP command.

Step 2 The fillet is found with the FILLET command. You must select the radius and points on the arcs that the fillet is to be tangent to.

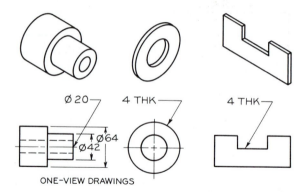

ONE-VIEW DRAWINGS

FIGURE 8.37 Objects that are cylindrical or of uniform thickness can be described with only one orthographic view and supplementary notes.

8.14
One-view drawings

Cylindrical parts and those with a uniform thickness can be described in one view. In both cases, notes explain the missing feature or dimension (Fig. 8.37).

8.15
Incomplete and removed views

The right- and left-side views of the part in Fig. 8.38 would be hard to interpret if all hidden lines were shown as specified by the rules of orthographic projection. Therefore, it is best to omit lines that confuse a clear understanding of the views.

Often it is difficult to show a feature because of its location. Standard views can be confusing when lines overlap other features. The view indicated by the

FIGURE 8.38 Unnecessary and confusing hidden lines are omitted in the side views to improve their clarity.

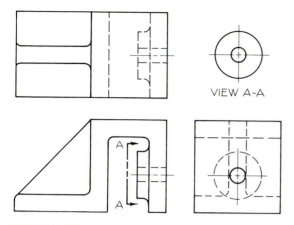

FIGURE 8.39 A removed view, indicated by the directional arrows, can be used to draw views in new hard-to-see locations.

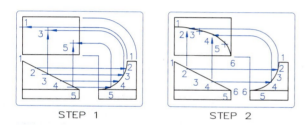

STEP 1   STEP 2

FIGURE 8.40 Plotting curved lines by computer.

Step 1 A series of points is found in the front and side views by orthographic projection. Points 1, 3, and 5 are projected to the top view.

Step 2 Points 2 and 4 are projected to the top view, and a PLINE curve is drawn to FIT the points to complete the top view.

directional arrows in Fig. 8.39 is more clearly shown when removed to an isolated position.

8.16
Curve plotting

An irregular curve can be drawn by following the rules of orthographic projection, as Fig. 8.40 shows. Plotting begins by locating points along the curve in two given views. These points are projected to the top

view where each point is located, and the points are then connected by a smooth curve. In Fig. 8.41, an ellipse is plotted in the top view by projecting from the front and side views. You will find it helpful to number points on curves that are being located by projection.

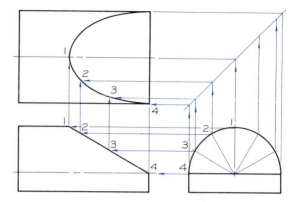

FIGURE 8.41 The ellipse in the top view was found by numbering points in the front and side views and then projecting them to the top view.

8.17
Partial views

A partial view can be used to save time and space when the parts are symmetrical or cylindrical. By omitting the rear of the circular top view in Fig. 8.42, space can be saved without sacrificing clarity. Or, to

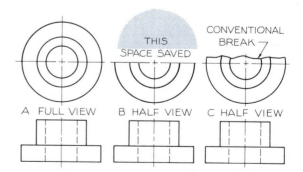

A FULL VIEW   B HALF VIEW   C HALF VIEW

FIGURE 8.42 To save space and drawing time, the top view of a cylindrical part can be drawn as a partial view using either of these methods.

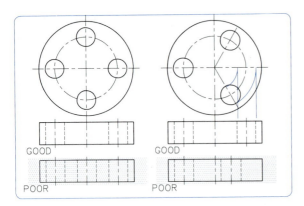

FIGURE 8.43 Revolving holes.

Left: A true projection of equally spaced holes gives a misleading impression that the center hole passes through the center of the plate.

Right: A conventional view is used to show the true radial distances of the holes from the center by revolution. The third hole is omitted.

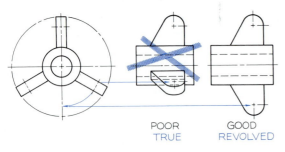

FIGURE 8.44 Symmetrically positioned external features, such as these lugs, are revolved to their true-size positions for the best views.

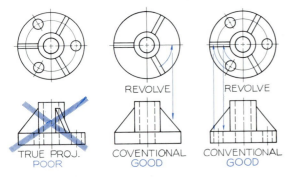

FIGURE 8.45 The conventional methods of revolving holes and ribs in combination for improved clarity.

make it more apparent that a portion of the view has been omitted, a break may be used.

8.18
Conventional revolutions

The readability of an orthographic view may be improved if the rules of projection are violated.

> Established violations of rules that are customarily made for the sake of clarity are called **conventional practices.**

When holes are symmetrically spaced in a circular plate, as shown in Fig. 8.43, it is conventional practice to show them at their true radial distance from the center of the plate. This requires an imagined revolution of the holes in the top view. This principle of revolution also applies to symmetrically positioned features, such as the three lugs on the outside of the part in Fig. 8.44. The conventional view is better than the true orthographic projection. Figure 8.45 shows the conventional and desired method of drawing holes and ribs in combination.

Another conventional practice is illustrated in Fig. 8.46, where an inclined feature is revolved to a horizontal position in the front, so that it can be drawn as true size in the top view. The revolution of the part is not drawn since it is an imagined revolution.

Figure 8.47 shows other parts whose views are improved by revolution. It is desirable to show the top-view features at 45°, so they will not coincide with the centerlines. The front views are drawn by imagining the features have been revolved. A closely related type of conventional view is the true-size developed view where a bent piece of material is drawn as if it were flattened out (Fig. 8.48).

8.19
Intersections

In orthographic projection, the intersection between planes results in a line that describes the object. In Fig. 8.49, examples of views are shown where lines may or may not be required.

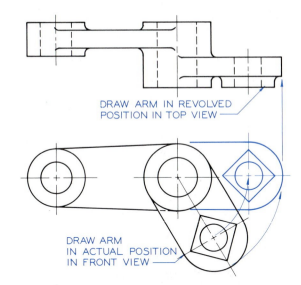

DRAW ARM IN REVOLVED
POSITION IN TOP VIEW

DRAW ARM
IN ACTUAL POSITION
IN FRONT VIEW

FIGURE 8.46 The arm in the front view is imagined to be revolved so that its true length can be drawn in the top view. This is an accepted conventional practice.

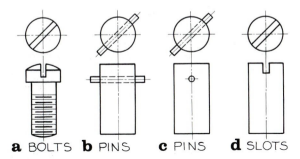

a BOLTS **b** PINS **c** PINS **d** SLOTS

FIGURE 8.47 Parts like these are drawn at 45° angles in the top views, but the front views are drawn to show the details as revolved views.

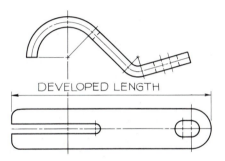

DEVELOPED LENGTH

FIGURE 8.48 Objects that have been shaped by bending thin stock can be shown as true-size developed views.

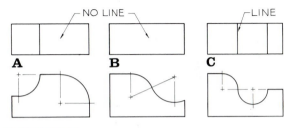

NO LINE — LINE

A **B** **C**

FIGURE 8.49 Object lines are drawn only where there are sharp intersections or where arcs are tangent at their centerlines, as in part C.

Figure 8.50 shows the standard types of intersections between cylinders. Figures 8.50A and B are conventional intersections, which means they are approximations drawn for ease of construction. Figure 8.50C shows a true intersection between cylinders of equal diameters. Figures 8.51 and 8.52 show similar intersections, and Fig. 8.53 shows the types of conventional intersections formed by holes in cylinders.

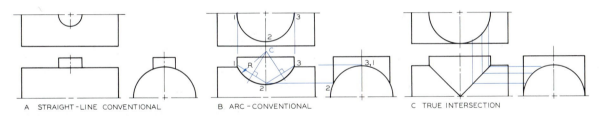

A STRAIGHT-LINE CONVENTIONAL   B ARC-CONVENTIONAL   C TRUE INTERSECTION

FIGURE 8.50 The conventional methods of showing intersections between cylinders. Except for part C, these are approximations.

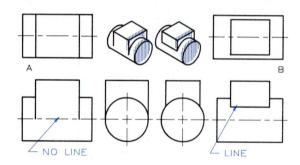

FIGURE 8.51 Conventional intersections between prisms and cylindrical shapes.

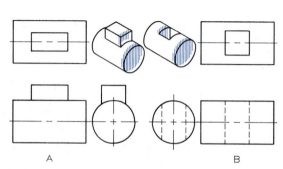

FIGURE 8.52 Conventional intersections between prisms and cylindrical shapes.

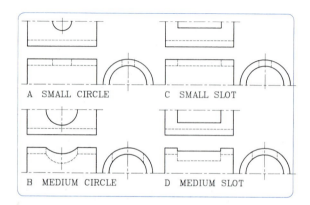

A SMALL CIRCLE   C SMALL SLOT

B MEDIUM CIRCLE   D MEDIUM SLOT

FIGURE 8.53 Conventional intersections between cylinders and holes piercing them.

8.20
Fillets and rounds

Fillets and **rounds** are rounded corners used on castings, such as the body of the Collet Index Fixture shown in Fig. 8.54. A fillet is an inside rounding, and a round is an external rounding. The radii of fillets and rounds may be many sizes, but they are usually about $\frac{1}{4}$ inch. Fillets and rounds are used on castings for added strength and improved appearance.

FIGURE 8.54 The edges of this Collet Index Fixture are rounded to form fillets and rounds. The surface of the casting is rough except where it has been machined. (Courtesy of Hardinge Brothers Inc.)

A casting will have square corners only when its surface has been finished, which is the process of machining away part of the surface to a smooth finish (Fig. 8.55B).

Finished surfaces are indicated by placing a **finish mark** (V) on the edge views of the finished surfaces whether the edges are visible or hidden. Figure 8.56 shows alternative finish marks.

Note (in Fig. 8.55C) that a **boss** is a raised cylindrical feature that is thickened to receive a shaft or to be threaded, and that the curve formed by a fillet at a point of tangency is a **runout.** Figure 8.57 shows the techniques of showing fillets and rounds on orthographic views.

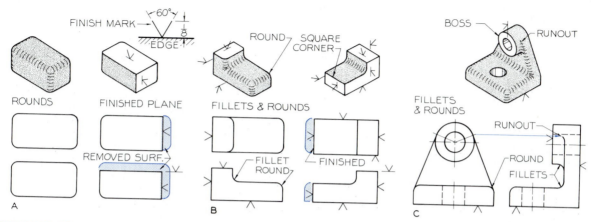

FIGURE 8.55 Fillets and rounds.

A. When a surface has been finished by machining, rounds are removed, and the corners are squared. A finish mark is indicated by a V placed on the edge view of the finished surfaces.

B. A fillet is a rounded inside corner. The rounds are removed when the outside surfaces are finished. The fillets can be seen only in the front view.

C. The views of an object with fillets and rounds must be drawn in a way that calls attention to them.

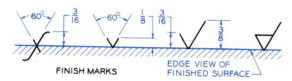

FIGURE 8.56 Alternative finish marks are applied to the edge views of the finished surface, whether hidden or visible.

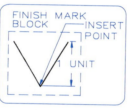

STEP 1

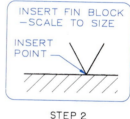

STEP 2

FIGURE 8.58 A finish mark by computer.

Step 1 Draw a finish mark that is 1 inch high. Make a BLOCK of it named FIN. Select the insertion point to be the point of the vee.

Step 2 INSERT the block, FIN, by selecting the point on the edge view of the finished surface. Scale the block to 0.125, and its height will be ⅛ inch high.

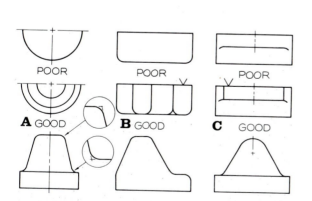

FIGURE 8.57 Examples of conventionally drawn fillets and rounds.

A computer-drawn finish mark can be saved as a BLOCK that can be used repeatedly and rapidly (Fig. 8.58). If the finish mark is drawn 1 inch high, it will be easy to scale it when INSERTed. For example, a scale factor of 0.125 will reduce it to ⅛ inch. By saving the BLOCK as a WBLOCK, it can be used on different drawing files, not just the one it was created on.

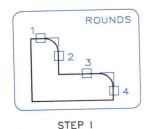
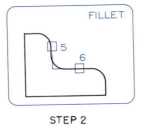

STEP I   STEP 2

FIGURE 8.59 Fillets and rounds by computer.

Step 1 Command: FILLET (CR)
Polyline Radius〈Select two objects〉: R (CR)
Enter fillet radius 〈0.000〉: 0.50 (CR)
Command: (CR) FILLET Polyline Radius〈Select two objects〉: 1 and 2
(Repeat and select 3 and 4 to draw round.)

Step 2 Press Return, and select points 5 and 6 on inside lines to construct a fillet. The straight lines that have the fillets and rounds applied to them are trimmed to their proper lengths.

Fillets and rounds can be drawn by computer as shown in Fig. 8.59. The object is first drawn with square intersections, and the FILLET command is used to round the corners for both fillets and rounds.

Figure 8.60 shows a comparison of intersections and runouts of parts with and without fillets and rounds. Large runouts are constructed as an eighth of a circle with a compass, as Fig. 8.61 shows. Small run-

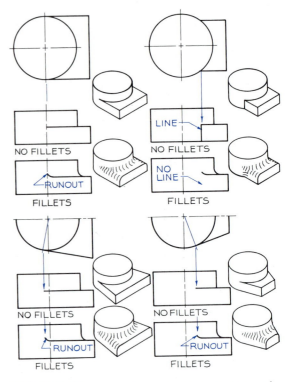

FIGURE 8.60 Intersections between features of objects. Intersections with fillets have runouts.

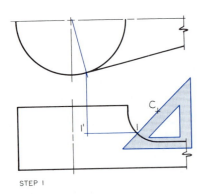
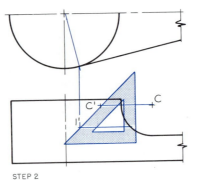
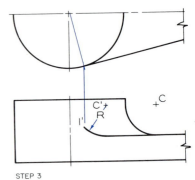

FIGURE 8.61 Plotting runouts.

Step 1 Find the point of tangency in the top view, and project it to the front view. A 45° triangle is used to find point 1, which is projected to point 1'.

Step 2 A 45° triangle is used to locate point C', which is on the horizontal projector from the center of the fillet, C.

Step 3 The radius of the fillet is used to draw the runout with C' as the center. The runout arc is equal to one eighth of a circle.

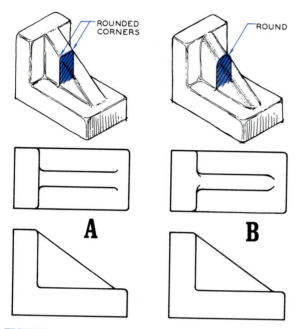

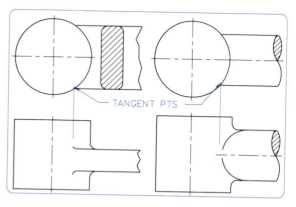

FIGURE 8.64 Conventional runouts of different cross-sectional shapes.

FIGURE 8.62 Runouts are shown for differently shaped ribs. Part A has fillets, and part B has rounded edges.

outs can be drawn with a circle template. Runouts on orthographic views will reveal much about the details of an object. For example, the runout in the top view of Fig. 8.62A tells us the rib has rounded corners, whereas the top view of Fig. 8.62B tells us the rib is completely round. Figures 8.63 and 8.64 illustrate methods of showing other types of intersections.

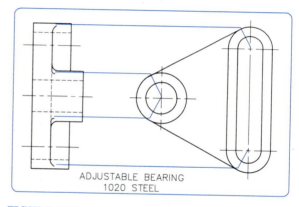

FIGURE 8.65 A typical application of runouts on a part with fillets and rounds.

> **COMPUTER METHOD** Figure 8.65 shows a drawing of a part with runouts at its tangent points. The runouts are plotted with the ARC command after the tangent points have been projected from the right-side view.

8.21
Left-hand and right-hand views

Two parts are often required that are "mirror images" of each other (Fig. 8.66). The drafter can reduce drawing time by drawing views of only one of the parts and

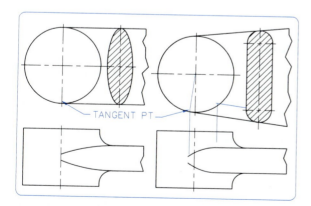

FIGURE 8.63 Conventional runouts of different cross-sectional shapes.

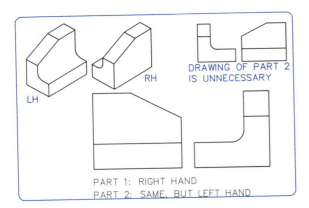

FIGURE 8.66 When left- and right-hand mirror parts are needed, only one view is drawn and labeled. The other view need not be drawn but should be indicated by a note.

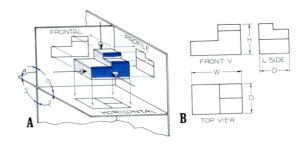

FIGURE 8.67 First-angle projection.

A. The first angle of projection is used by many of the countries that use the metric system. You imagine that the object is placed above the horizontal plane and in front of the frontal plane.

B. The views are drawn in this location, which is different from the third-angle of projection used in the United States.

labeling these views. A note can be added to indicate that the other matching part has the same dimensions.

8.22
First-angle projection

The examples in this chapter are presented as third-angle projections, where the top view is placed over the front view, and the right-side view is placed to the right of the front view. This method is used extensively in the United States, Britain, and Canada. Most of the rest of the industrial world uses **first-angle** projection.

The first-angle system is illustrated in Fig. 8.67, where an object is placed above the horizontal plane and in front of the frontal plane. When these projection planes are opened onto the surface of the drawing paper, the front view is projected over the top view, and the left-side view is placed to the right of the front view.

It is important that the angle of projection be indicated on a drawing to aid in the interpretation of the views. This is done by placing a truncated cone in or near the title block (Fig. 8.68). When metric units of measurement are used, the cone and SI symbol are placed together on the drawing.

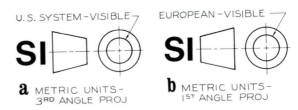

FIGURE 8.68 The angle of projection used to prepare a set of drawings is indicated by a truncated cone, which is placed in or near the title block of a drawing.

Problems

The following problems are to be drawn as orthographic views on Size A or Size B paper, as assigned by your instructor.

1–11. (Figs. 8.69–8.79) Draw the given views using the dimensions provided, and then construct the missing top, front, or right-side views. Use Size A sheets, and draw one or two problems per sheet.

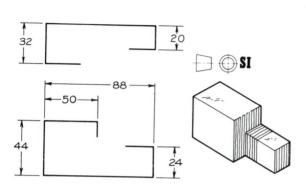

FIGURE 8.69 Problem 1: Guide block.

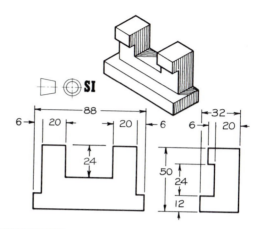

FIGURE 8.72 Problem 4: Lock catch.

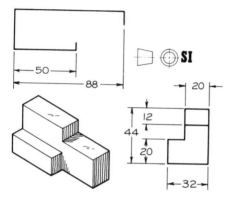

FIGURE 8.70 Problem 2: Double step.

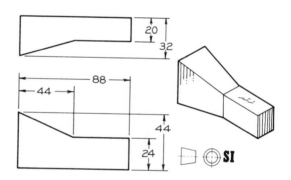

FIGURE 8.73 Problem 5: Two-way adjuster.

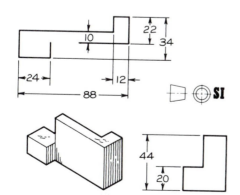

FIGURE 8.71 Problem 3: Adjustable stop.

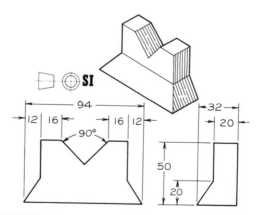

FIGURE 8.74 Problem 6: 90° Vee block.

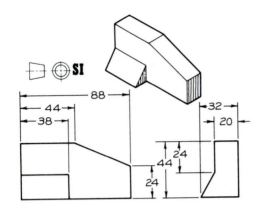

FIGURE 8.75 Problem 7: Filler.

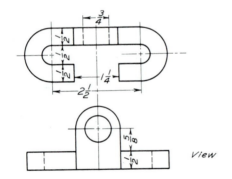

FIGURE 8.76 Problem 8: Slide stop.

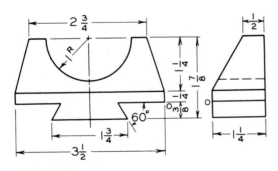

FIGURE 8.77 Problem 9: Shaft support.

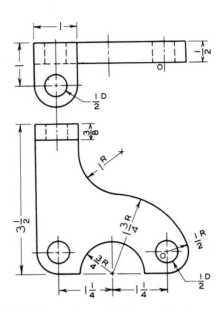

FIGURE 8.78 Problem 10: Support brace.

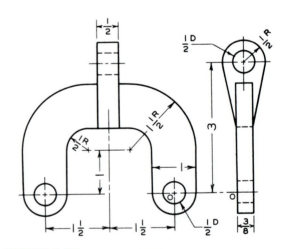

FIGURE 8.79 Problem 11: Yoke.

12–48. (Figs. 8.80–8.115) Construct the necessary orthographic views to describe the objects in these figures. Draw the views on Size A or Size B sheets. Label the views and show the overall dimensions of *W*, *D*, and *H*.

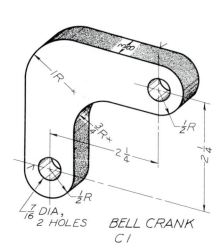

FIGURE 8.80 Problem 12: Bell crank.

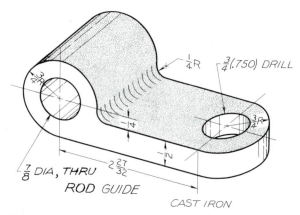

FIGURE 8.82 Problem 14: Rod guide.

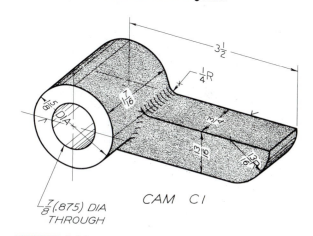

FIGURE 8.83 Problem 15: Cam.

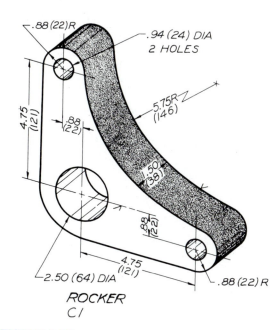

FIGURE 8.81 Problem 13: Rocker.

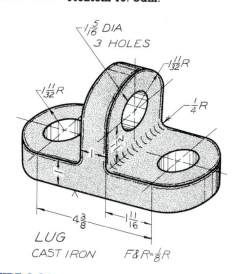

FIGURE 8.84 Problem 16: Lug.

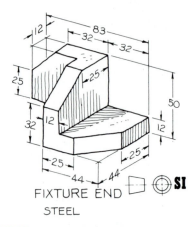

FIXTURE END
STEEL

FIGURE 8.85 Problem 17: Fixture end.

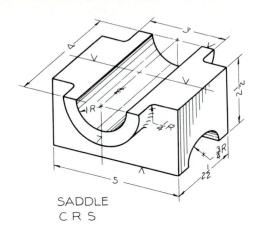

SADDLE
C R S

FIGURE 8.88 Problem 20: Saddle.

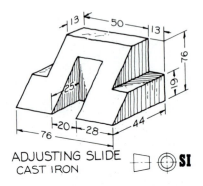

ADJUSTING SLIDE
CAST IRON

FIGURE 8.86 Problem 18: Adjusting slide.

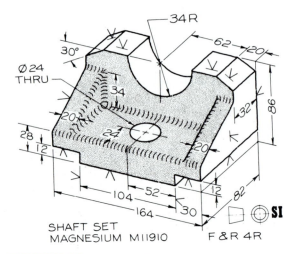

SHAFT SET
MAGNESIUM M11910

F & R 4R

FIGURE 8.89 Problem 21: Shaft set.

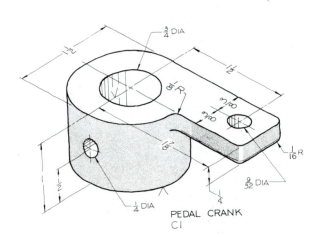

PEDAL CRANK
CI

FIGURE 8.87 Problem 19: Pedal crank.

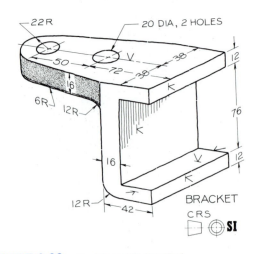

BRACKET
C R S

FIGURE 8.90 Problem 22: Bracket.

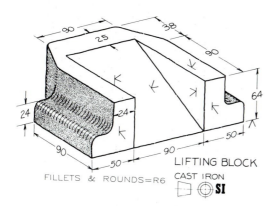

FIGURE 8.91 Problem 23: Lifting block.

LIFTING BLOCK
CAST IRON

FILLETS & ROUNDS=R6

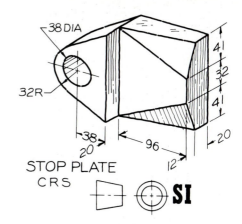

STOP PLATE
CRS

FIGURE 8.94 Problem 26: Stop plate.

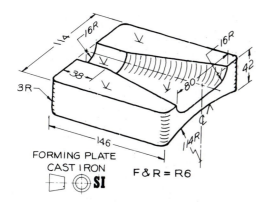

FORMING PLATE
CAST IRON
F & R = R6

FIGURE 8.92 Problem 24: Forming plate.

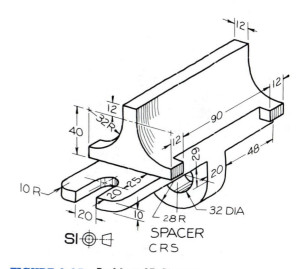

SPACER
CRS

FIGURE 8.95 Problem 27: Spacer.

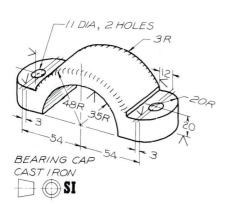

11 DIA, 2 HOLES

BEARING CAP
CAST IRON

FIGURE 8.93 Problem 25: Bearing cap.

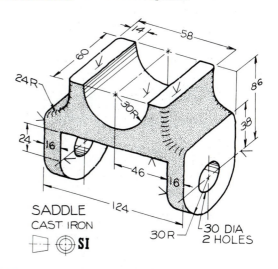

SADDLE
CAST IRON

30 DIA
2 HOLES

FIGURE 8.96 Problem 28: Saddle.

147

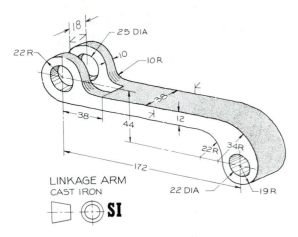

FIGURE 8.97 Problem 29: Linkage arm.

LINKAGE ARM
CAST IRON

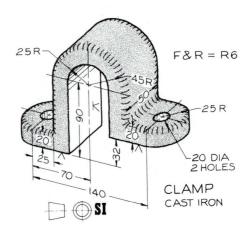

CLAMP
CAST IRON

FIGURE 8.100 Problem 32: Clamp.

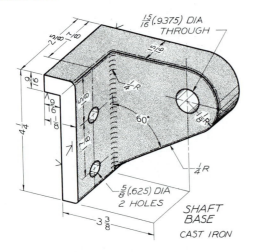

SHAFT
BASE
CAST IRON

FIGURE 8.98 Problem 30: Shaft base.

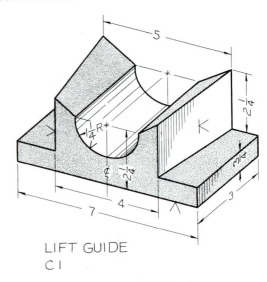

LIFT GUIDE
C I

FIGURE 8.101 Problem 33: Lift guide.

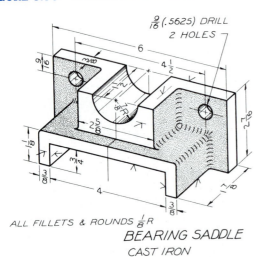

ALL FILLETS & ROUNDS $\frac{1}{8}$R
BEARING SADDLE
CAST IRON

FIGURE 8.99 Problem 31: Bearing saddle.

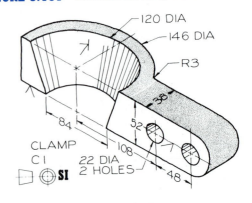

CLAMP
C I

FIGURE 8.102 Problem 34: Clamp.

148

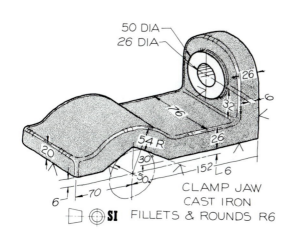

FIGURE 8.103   Problem 35: Clamp jaw.

50 DIA
26 DIA
26
6
32
76
20
54 R
26
30
30
152
6
6
70
CLAMP JAW
CAST IRON
SI   FILLETS & ROUNDS R6

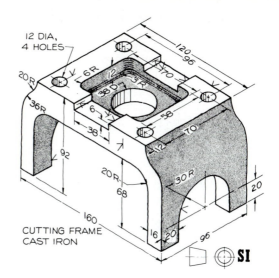

FIGURE 8.105   Problem 37: Cutting frame.

12 DIA,
4 HOLES
120
96
20 R
6 R
12
70
36 R
38 D
3 R
6
58
38
92
70
20 R
68
12
30 R
20
160
16
20
96
CUTTING FRAME
CAST IRON
SI

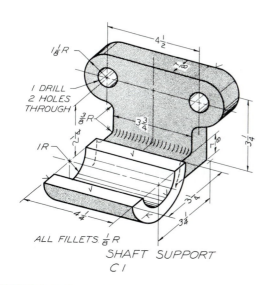

FIGURE 8.104   Problem 36: Shaft support.

$1\frac{1}{8}$ R
$4\frac{1}{2}$
$\frac{3}{10}$
1 DRILL
2 HOLES
THROUGH
$\frac{3}{8}$ R
$3\frac{3}{4}$
$1\frac{1}{4}$
$3\frac{1}{4}$
1 R
$\frac{7}{8}$
$4\frac{1}{4}$
$3\frac{1}{4}$
$3\frac{1}{4}$
ALL FILLETS $\frac{1}{8}$ R
SHAFT SUPPORT
C I

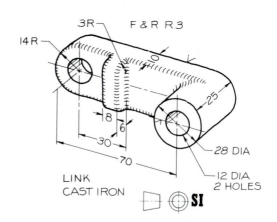

FIGURE 8.106   Problem 38: Link.

3 R
F & R R3
14 R
10
8
25
6
30
28 DIA
70
12 DIA
2 HOLES
LINK
CAST IRON
SI

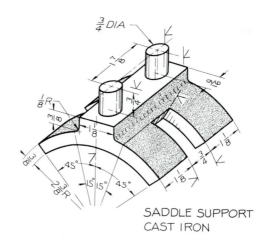

FIGURE 8.107   Problem 39: Saddle support.

$\frac{3}{4}$ DIA
10
$\frac{9}{16}$
$\frac{1}{8}$ R
$\frac{3}{8}$
$\frac{3}{4}$
$\frac{1}{8}$
$\frac{1}{8}$
$\frac{3}{4}$
$\frac{1}{8}$
$\frac{3}{8}$
45°
$2\frac{3}{8}$ R
15° 15°
45°
SADDLE SUPPORT
CAST IRON

149

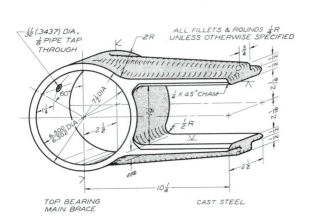

FIGURE 8.108 Problem 40: Top bearing.

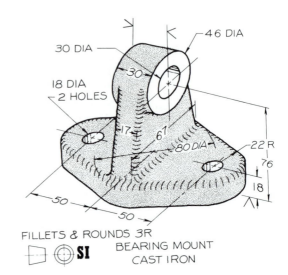

FIGURE 8.110 Problem 42: Bearing mount.

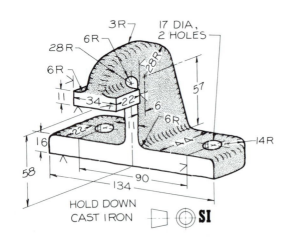

FIGURE 8.109 Problem 41: Hold down.

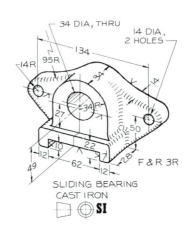

FIGURE 8.111 Problem 43: Sliding bearing.

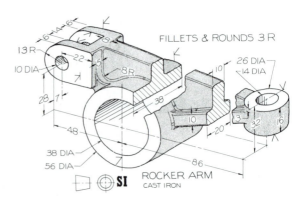

FIGURE 8.112 Problem 44: Rocker arm.

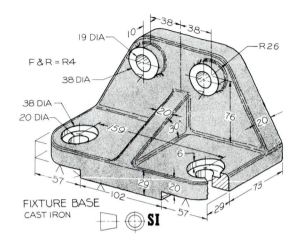

FIGURE 8.113 Problem 45: Fixture base.

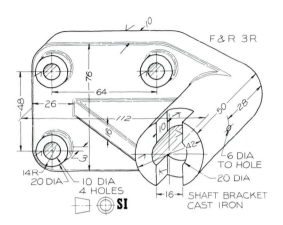

FIGURE 8.115 Problem 47: Shaft bracket.

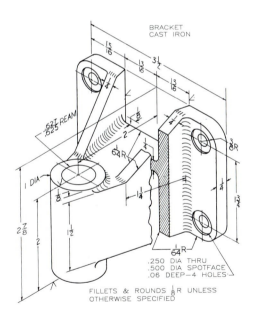

FIGURE 8.114 Problem 46: Swivel attachment.

Auxiliary Views

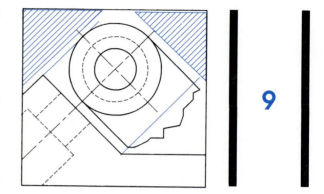

9

9.1
Introduction

A plane not parallel to one of the principal projection planes is a nonprincipal plane that will **not** appear true size in a principal view; it can be found true size only on an **auxiliary plane** parallel to the nonprincipal plane. This nonprincipal view is called an **auxiliary view.**

An inclined surface of an object (Fig. 9.1) does not appear true size in the top view because it is not parallel to the horizontal projection plane. However, the inclined surface will appear true size if an auxiliary view is projected perpendicularly from the edge view of the plane in the front view.

9.2
Folding-line approach

The three principal orthographic planes are the **frontal** (F), **horizontal** (H), and **profile** (P) planes.

> A primary auxiliary plane is perpendicular to one of the principal planes but oblique to the other two, and a **primary auxiliary view** is projected from a **primary orthographic view.**

Auxiliary planes can be thought of as planes that fold from principal planes, as Fig. 9.2 shows. The

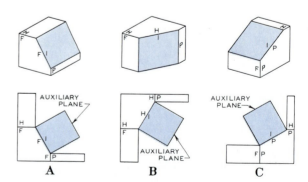

FIGURE 9.1 When a surface appears as an inclined edge in a principal view, it can be found true size by an auxiliary view. In part a, the top view is foreshortened, but this plane is true size in part b, the auxiliary view.

FIGURE 9.2 A primary auxiliary plane can be folded from the frontal, horizontal, or profile planes. The fold lines are labeled F-1, H-1, and P-1.

152

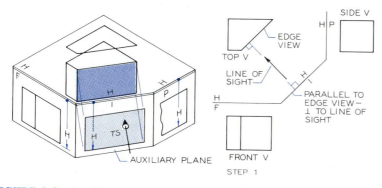

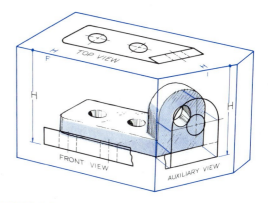

FIGURE 9.3 Auxiliary view from the top—folding-line method.

Given An object with an inclined surface.

Required Find the inclined surface true size by an auxiliary view.

Step 1 Construct a line of sight perpendicular to the edge view of the inclined surface. Draw the H-1 fold line parallel to the edge, and draw the H-F fold line between the top and front views.

Step 2 Project the four corners of the edge parallel to the line of sight. Locate the corners by measuring perpendicularly from the horizontal plane with height (*H*) dimensions.

plane in Fig. 9.2A folds from the frontal plane to make a 90° angle with it. The fold line between the two planes is labeled F-1, where F is an abbreviation for frontal, and 1 represents **first,** or **primary,** auxiliary plane. Figures 9.2B and 9.2C illustrate the positions for auxiliary planes that fold from the horizontal and profile planes.

9.3
Auxiliaries projected from the top view

The inclined plane in Fig. 9.3 is an edge in the top view, and it is perpendicular to the horizontal plane. An auxiliary plane can be drawn parallel to the inclined surface, and the view projected onto it will be a true-size view of it.

> A surface must appear as an **edge** in a principal view before it can be found as true size in a primary auxiliary view.

Fold line H-1 is drawn parallel to the edge view of the inclined plane in Step 1. The line of sight is drawn perpendicular to the edge view. Each corner of the in-

FIGURE 9.4 A pictorial showing the relationship of the auxiliary projection plane is used to find the true-size view of the inclined surface.

clined plane is projected perpendicularly to the auxiliary plane and is located by transferring the dimension of height (*H*) from the side or front views.

A similar example is the object shown in the glass box in Fig. 9.4. Since the inclined surface in the top view appears as an edge, it can be found true size in a primary auxiliary view. The height (*H*) is transferred from the front view to the auxiliary view since both views are measured from the same horizontal plane. The auxiliary plane is rotated about the H-1 fold line into the plane of the top view in Fig. 9.5.

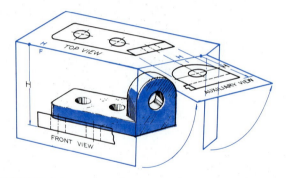

FIGURE 9.5 The auxiliary plane is opened into the plane of the top view by revolving it about the H-1 fold line.

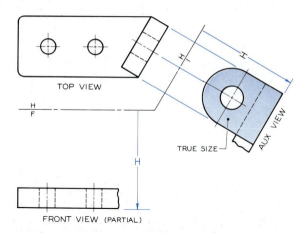

FIGURE 9.6 When the object is drawn on a sheet of paper, it is laid out in this manner. The front view is drawn as a partial view since the omitted part is shown true size in the auxiliary view.

When drawn on a sheet of paper, the drawing of this object would appear as shown in Fig. 9.6. The front view is shown as a partial view because the omitted portion would have been hard to draw, and it would not have been true size. The auxiliary view shows the true-size view of the inclined surface.

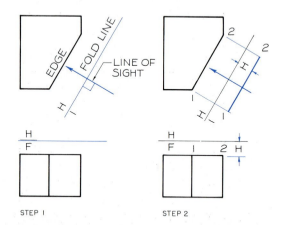

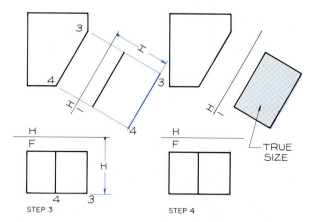

FIGURE 9.7 Construction of an auxiliary view.

Step 1 The line of sight is drawn perpendicular to the edge view of the inclined surface. The H-1 fold line is drawn parallel to the edge view of the inclined surface. An H-F fold line is drawn between the given views.

Step 2 Points 1 and 2 are found by transferring the height (*H*) dimensions from the front view to the auxiliary view.

Step 3 Points 3 and 4 are found in the same manner using the dimensions of height (*H*).

Step 4 The corner points are connected to complete the true-size view of the inclined plane.

9.4

Auxiliaries from the top view—folding-line method

Figure 9.7 shows the steps of constructing an auxiliary view projected from the top view to find the true-size view of the inclined surface. Since the inclined surface appears as an edge in the top view, it can be found true size in a primary auxiliary view. The line of sight is drawn perpendicular to the edge view, and the fold line is drawn parallel to the edge. Height *(H)* is transferred from the front view.

The fold line, drawn thin but black, is labeled H-1. It is also helpful to number or letter points on the views. The projectors are construction lines, and they should be drawn with a hard pencil (3H–4H) just dark enough to be seen.

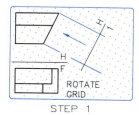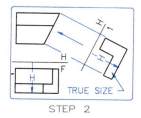

FIGURE 9.8 **Auxiliary view by computer.**

Step 1 To perpendicularly project an auxiliary from the edge of the inclined surface in the top view, the grid is rotated, using the SNAP command and RO-TATE option. The H-1 reference line is drawn parallel to the edge, and projectors are drawn perpendicular to the H-1 line.

Step 2 The auxiliary view is found by transferring height dimensions measured in the front view and transferring them perpendicularly from the H-1 line. The auxiliary view is the true-size view of the inclined surface.

COMPUTER METHOD Using computer graphics, the true-size view of a plane is found (Fig. 9.8) by projecting from the top view, where the inclined plane appears as an edge. With AutoCAD's RO-TATE option of the SNAP command, the grid is rotated so that it is parallel to the edge view of the plane in the top view (Step 1). The true-size auxiliary view is found by projecting perpendicularly from the top view of the plane and locating the auxiliary view with height dimensions transferred from the front view.

The reference plane shown as a horizontal edge in the front view is a **horizontal reference plane** (HRP). The HRP will appear as an edge view of the inclined surface from which the auxiliary is projected. The auxiliary view will lie between the HRP and the top view.

9.5

Auxiliaries from the top view—reference-line method

A similar method of locating an auxiliary view uses a reference plane instead of the fold line (Fig. 9.9). Instead of placing a fold line between the top and front views, a reference plane is passed through the bottom of the front view. The height *(H)* dimensions are measured upward from the reference plane instead of downward from a fold line.

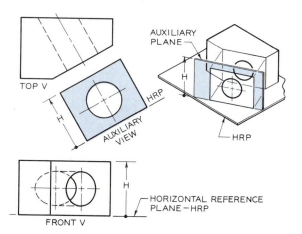

FIGURE 9.9 A horizontal reference plane can be used instead of the folding-line technique to construct an auxiliary view. Instead of placing the reference plane between the top and auxiliary views, it is placed outside the auxiliary view.

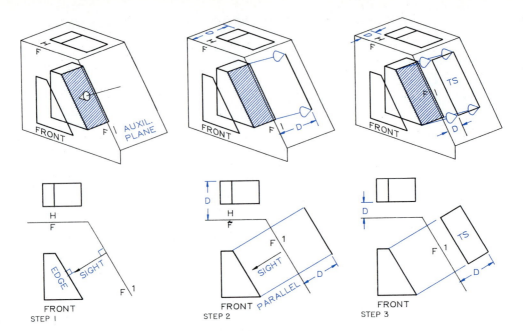

FIGURE 9.10   Auxiliary from the front—folding-line method.

Required   Find the inclined surface true size by an auxiliary view.

Step 1   Draw the line of sight perpendicular to the edge of the plane and draw the F1 reference parallel to the edge. Draw the HF fold line between the top and front views.

Step 2   Project from the edge view of the inclined surface parallel to the line of sight. Use the D dimensions from the top view to locate a line in the auxiliary view.

Step 3   Locate the other corners of the inclined surface by projecting to the auxiliary view. Locate the points by transferring the depth (D) dimensions from the top to the auxiliary view.

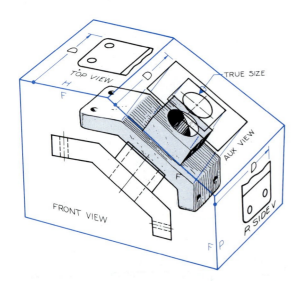

FIGURE 9.11   A pictorial showing the relationship of the projection planes used to find the true-size view of the inclined surface of the object.

9.6

Auxiliaries from the front view—folding-line method

A plane that appears as an edge in the front view (Fig. 9.10) can be found true size in a primary auxiliary view projected from the front view. Fold line F-1 is drawn parallel to the edge view of the inclined plane in the front view at a convenient location.

The line of sight is drawn perpendicular to the edge view of the inclined plane in the front view. Observed from this direction, the frontal plane appears as an edge; thus, the measurements perpendicular to the frontal plane—the depth *(D)* dimensions—are seen true length. Depth dimensions are transferred from the top view to the auxiliary view by using dividers.

The object in Fig. 9.11 is enclosed in a box, and an auxiliary plane is constructed parallel to the inclined plane. When drawn on a sheet of paper, the object appears as shown in Fig. 9.12. The top and side views are drawn as partial views since the auxiliary view eliminates the need for seeing their complete views. The

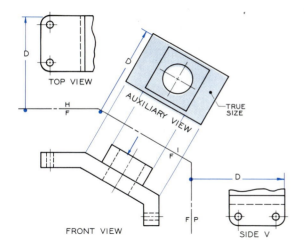

TOP VIEW

AUXILIARY VIEW

TRUE SIZE

FRONT VIEW

SIDE V

FIGURE 9.12 **The layout and construction of an auxiliary view of the object shown in Fig. 9.11.**

COMPUTER METHOD The inclined surface that appears as an edge in the front view (Fig. 9.13) is found true size using the LISP commands PAR-ALLEL and TRANSFER (see Section 20.3). While in AutoCAD's drafting mode, type PARAL-LEL to draw the reference line parallel to edge *AB*. In Step 1, you are prompted for the beginning point of the parallel line (1), its approximate endpoint (2), and the endpoints of the line it is parallel to (3) and (4)

In Step 2, type TRANSFER to obtain prompts for transferring measurements from the top view to the auxiliary view the same manner as you would use a pair of dividers. In Step 3, connect the circular points to complete the auxiliary view.

auxiliary view, which is located by using the depth dimension measured from the edge view of the frontal projection plane, shows the surface's true size.

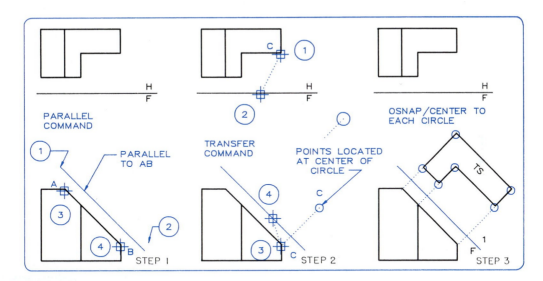

FIGURE 9.13 **Auxiliary view by computer.**

Step 1 While in AutoCAD, type PARALLEL and you are prompted for the first end of the reference line (1) and its approximate second end (2). You are then prompted for the beginning points and endpoints of the line (*AB*) that the reference line is parallel to.

Step 2 Type TRANSFER, and you are prompted to select a point in the top view (1) and its distance from the reference line (2) to be transferred. You are then asked for the front view of the point to be projected from (3) and the reference plane (4). The point is then projected to the auxiliary view and located with a small circle.

Step 3 Continue using TRANS-FER to locate the other corner points of the inclined plane. Connect the points using the CEN-TER option of the OSNAP command to snap to the centers of the circles. Circle points can be ERASEd when their centers have been connected to complete the auxiliary view.

157

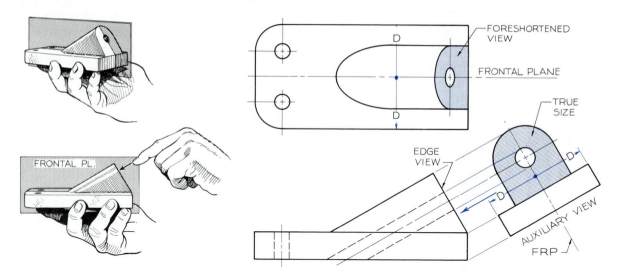

FIGURE 9.14 Since the inclined surface of this part is symmetrical, it is advantageous to use a frontal reference plane (FRP) that passes through the object. The auxiliary view is projected perpendicularly from the edge view of the plane in the front view. The FRP appears as an edge in the auxiliary view, and depth (*D*) dimensions are made on each side of it to locate points on the true-size view of the inclined surface.

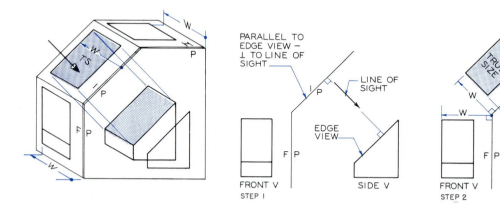

FIGURE 9.15 Auxiliary from the side—folding-line method.

Given An object with an inclined surface.

Required Find the inclined surface true size by an auxiliary view.

Step 1 Draw a line of sight perpendicular to the edge view of the inclined surface. Draw the P-1 fold line parallel to the edge view, and draw the F-P fold line between the given views.

Step 2 Project the corners of the edge parallel to the line of sight. Locate the corners by measuring perpendicularly from the P-1 fold line with width (*W*) dimensions.

9.7

Auxiliaries from the front view—reference-plane method

The object in Fig. 9.14 has an inclined surface that appears as an edge in the front view; therefore, this plane can be found true size in a primary auxiliary view.

It is advantageous to use a reference plane that passes through the center of the symmetrical top view. Because the reference plane is a frontal plane, it is called a **frontal reference plane** (FRP) in the auxiliary view. The FRP is located parallel to the edge view of the inclined plane and through the center of the auxiliary view.

9.8

Auxiliaries from the profile view—folding-line method

Since the inclined surface in Fig. 9.15 appears as an edge in the profile plane, it can be found true size in a primary auxiliary view projected from the profile view. The auxiliary fold line, P-1, is drawn parallel to the edge view of the inclined surface.

A line of sight perpendicular to the auxiliary plane will see the profile plane as an edge. Therefore, width (W) dimensions, which are transferred from the front view to the auxiliary view, will appear true length in the auxiliary view.

9.9

Auxiliaries from the profile— reference-plane method

The object in Fig. 9.16 has two inclined surfaces that appear as edges in the right-side view, the profile view. These inclined surfaces are found true size by using a **profile reference plane** (PRP) that is a vertical edge in the front view.

To find the inclined surface's true size, an auxiliary view is drawn. The profile reference plane is po-

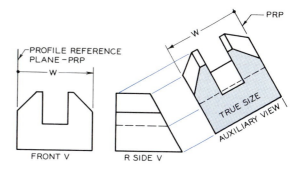

FIGURE 9.16 An auxiliary view is projected from the right-side view by using a profile reference plane (PRP). The auxiliary view shows the true-size view of the inclined surface.

sitioned at the far outside of the auxiliary view instead of between the profile and auxiliary views, as in the folding-line method. By transferring the width (W) dimensions from the edge view of the PRP in the front view to the auxiliary view, the view is found.

9.10

Auxiliaries of curved shapes

When an auxiliary view is drawn to show a curve that is not a true arc, a series of points must be plotted. The cylinder in Fig. 9.17 has a beveled surface that appears as an edge in the front view. This true-size surface is elliptical.

Since the cylinder is symmetrical, it is beneficial to use an FRP through the object so that dimensions can be measured on both sides of it. Points are located about the circular right-side view, and these are projected to the edge view of the surface in the front view. The FRP is located parallel to the edge view of the plane, and the points are projected perpendicularly from the edge view of the plane. Dimensions A and B are shown as examples for plotting points in the auxiliary view. To construct a smooth curve, more points than shown are needed.

An irregular curve is found as an outline of a true-size surface in Fig. 9.18. Points are located on the curve in the top view and are projected to the front view. These points are found in the auxiliary view by plotting each of them using the depth (D) dimensions transferred from the top view.

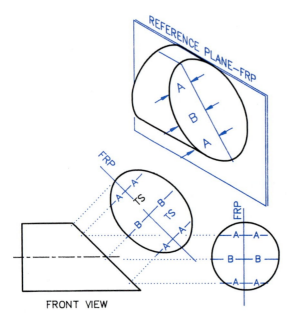

FIGURE 9.17 This auxiliary view requires that a series of points be plotted. Since the object is symmetrical, the reference plane (FRP) is positioned through its center.

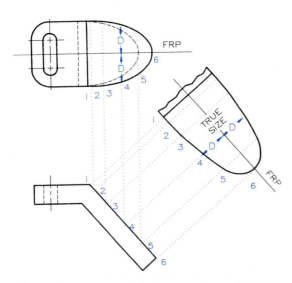

FIGURE 9.18 The auxiliary view of this surface required that a series of points be located in the given view and then projected to the auxiliary view.

9.11
Partial views

An auxiliary view is a supplementary view, so some views of an orthographic arrangement can be drawn as partial views. The object in Fig. 9.19 shows a complete front view and partial auxiliary and side views. The partial views are easier to draw and are more functional without sacrificing clarity.

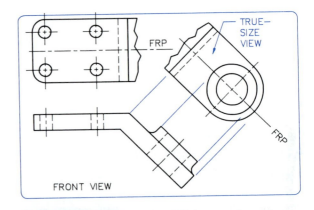

FIGURE 9.19 This computer-drawn combination of views is used to represent an object although the top and side views are partial views. The foreshortened portions of the object have been omitted. The FRP reference line passes through the center of the object in the top view since the object is symmetrical about this line.

A similar example is given in Fig. 9.20, where the front view of the guide bracket is a complete view, and the other two views are partial views. An FRP is passed through the center of the side view. Figure 9.21 shows a photograph of the guide bracket.

9.12
Auxiliary sections

Figure 9.22 shows a section through a part that is projected as an auxiliary view. The section is labeled A-A, and a cutting plane passing through the object is labeled A-A. The cutting plane shows where the sectional view was projected from and where the object was cut to show the section. The auxiliary section pro-

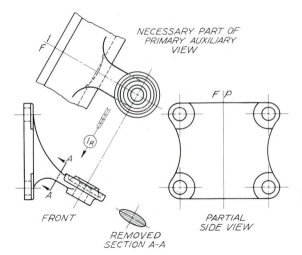

NECESSARY PART OF
PRIMARY AUXILIARY
VIEW

FRONT

PARTIAL
SIDE VIEW

REMOVED
SECTION A-A

FIGURE 9.20 This guide
bracket is represented by partial
views. The hub, which would
appear elliptical, is not shown in
the side view at all.

FIGURE 9.21 A photograph of the
guide bracket drawn in Fig. 9.20.

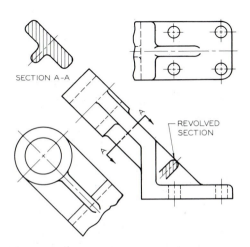

SECTION A-A

REVOLVED
SECTION

FIGURE 9.22 Auxiliary section A-A is projected
from the cutting plane labeled A-A to show the cross
section of the object.

vides a good description of the part that cannot be
readily understood from the given principal views.

9.13
Secondary auxiliary views

> A **secondary auxiliary view** is projected from a
> primary auxiliary view

In Figure 9.23, the inclined plane is found as an edge
view in the primary auxiliary view, and then a line of
sight perpendicular to the edge is drawn. The second
auxiliary view shows the inclined surface true size.
Remember that it is necessary to project from an edge
view of a plane before it can be found true size.

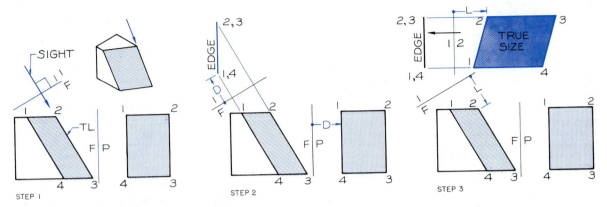

FIGURE 9.23 Secondary auxiliary views.

Step 1 A line of sight is drawn parallel to the true-length view of a line on the oblique surface. The folding line F-1 is drawn perpendicular to the line of sight.

Step 2 The primary auxiliary view of the oblique surface is an edge view. Depth (D) is used to locate a point in the primary auxiliary view.

Step 3 A line of sight is drawn perpendicular to the edge view, and a secondary auxiliary view is projected in this direction. Dimension L is used to locate one of the points in the true-size view.

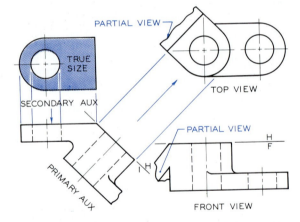

FIGURE 9.25 An example of a secondary auxiliary view projected from a partial auxiliary view.

The problem in Fig. 9.24 is a secondary auxiliary projection in which the point view of a diagonal of a cube is found. When this is found, the three surfaces of the cube are equally foreshortened. The secondary auxiliary view is an **isometric projection,** which is the basis for isometric pictorial drawing.

Since auxiliary views are supplementary views, they can be partial views if all features are sufficiently shown. The object in Fig. 9.25 is shown as a series of orthographic views, all of which are partial views.

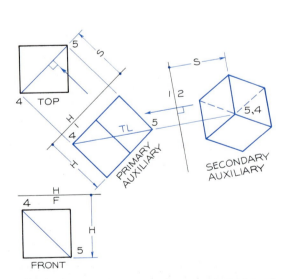

FIGURE 9.24 A point view of a diagonal of a cube is found in the secondary auxiliary view. Line 4–5 is found true length in the primary auxiliary view and then as a point in the secondary auxiliary view, which is an isometric projection of the cube.

9.14
Elliptical features

Occasionally, circular shapes will project as ellipses, which can be drawn using any of the techniques introduced in Chapter 6 once the necessary points have been plotted. The most convenient method of drawing ellipses is with an ellipse guide (template). The angle of the ellipse guide is the angle the line of sight makes with the edge view of the circular feature. In Fig. 9.26, the angle is found to be 45°, so the right-side view is drawn as a 45° ellipse.

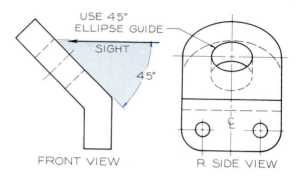

FIGURE 9.26 The ellipse guide angle is the angle that the line of sight makes with the edge view of the circular feature. The ellipse guide for the right-side view is 45°.

Problems

The following problems are to be solved on Size A or Size B sheets, as assigned by your instructor.

1–10. (Fig. 9.27) Using the example layout, change the top and front views by substituting the top views given at the right in place of the one given in the example. The angle of inclination in the front view is 45° for all problems, and the height is 1½ inches in the front view. Construct auxiliary views that show the inclined surface true size. Draw two problems per Size A sheet.

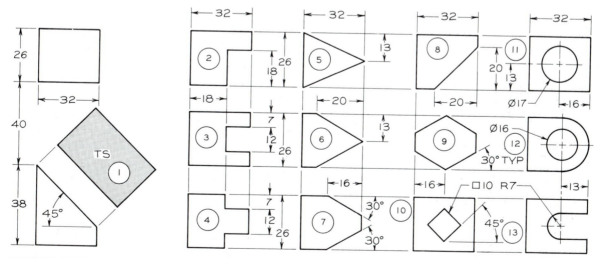

FIGURE 9.27 Problems 1–10. Primary auxiliary views.

11–32. (Figs. 9.28–9.49) Draw the necessary primary and auxiliary views to describe the parts shown. Draw one per Size A or Size B sheet, as assigned. Adjust the scale of each to fit the space on the sheet.

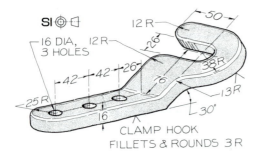

FIGURE 9.28 Problem 11. Clamp hook.

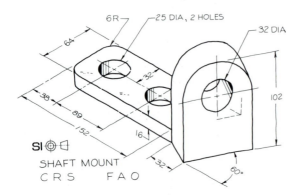

FIGURE 9.29 Problem 12. Shaft mount.

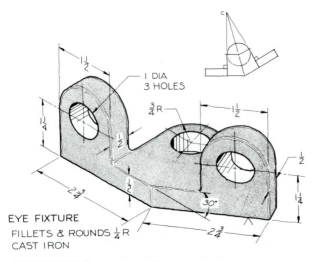

FIGURE 9.30 Problem 13. Eye fixture.

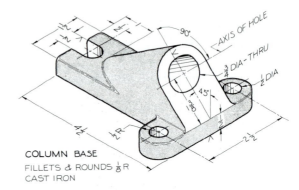

FIGURE 9.31 Problem 14. Column base.

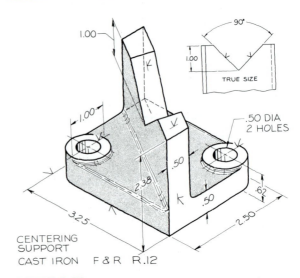

FIGURE 9.32 Problem 15. Centering support.

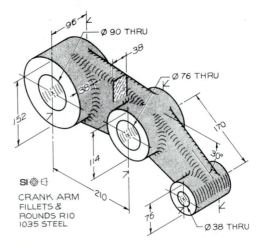

FIGURE 9.33 Problem 16. Crank arm.

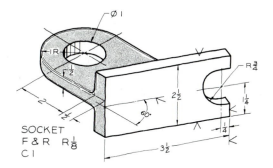

FIGURE 9.34 Problem 17. Socket.

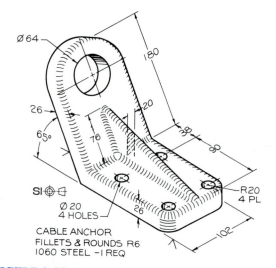

FIGURE 9.35 Problem 18. Cable anchor.

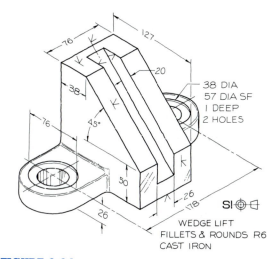

FIGURE 9.36 Problem 19. Wedge lift.

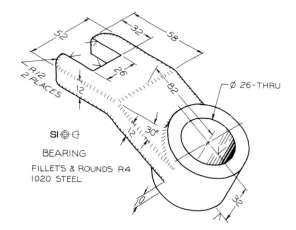

FIGURE 9.37 Problem 20. Bearing.

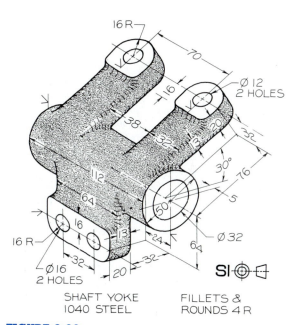

FIGURE 9.38 Problem 21. Shaft yoke.

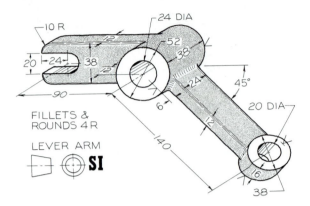

FILLETS &
ROUNDS 4 R

LEVER ARM

FIGURE 9.39 Problem 22. Lever arm.

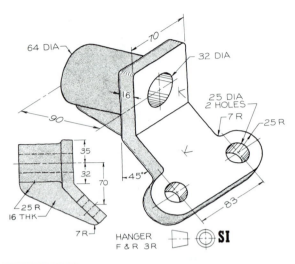

HANGER
F & R 3R

FIGURE 9.41 Problem 24. Hanger.

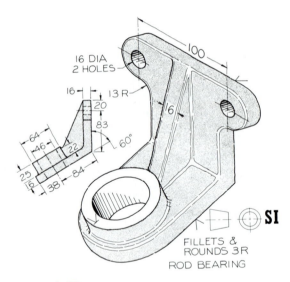

FILLETS &
ROUNDS 3R
ROD BEARING

FIGURE 9.40 Problem 23. Rod bearing.

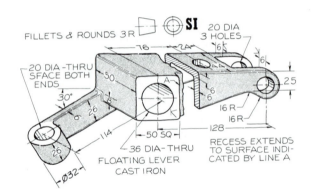

FLOATING LEVER
CAST IRON

RECESS EXTENDS
TO SURFACE INDI-
CATED BY LINE A

FIGURE 9.42 Problem 25. Floating lever.

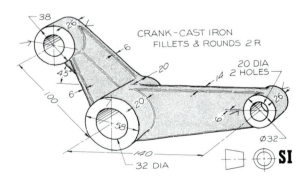

CRANK-CAST IRON
FILLETS & ROUNDS 2 R

FIGURE 9.43 Problem 26. Crank.

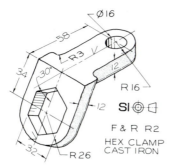

FIGURE 9.44 Problem 27. Hexagon angle.

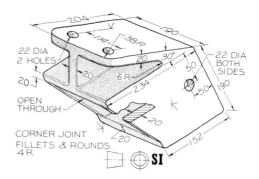

FIGURE 9.47 Problem 30. Corner joint.

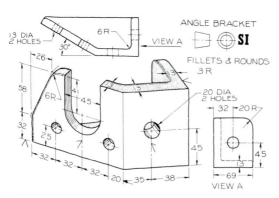

FIGURE 9.45 Problem 28. Angle bracket.

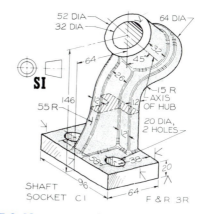

FIGURE 9.48 Problem 31. Shaft socket.

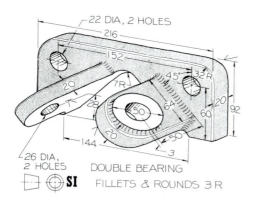

FIGURE 9.46 Problem 29. Double bearing.

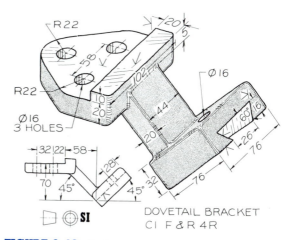

FIGURE 9.49 Problem 32. Dovetail bracket.

33–34. (Figs. 9.50–9.51) Lay out these orthographic views on Size B sheets, and complete the auxiliary and primary views.

35–37. (Figs. 9.52–9.54) Construct orthographic views of the given objects, and using secondary auxiliary views, draw auxiliary views that give the true-size views of the inclined surfaces. Draw one per Size B sheet.

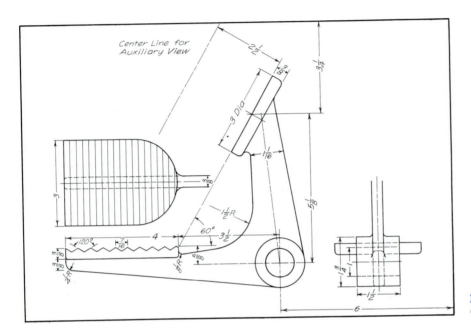

FIGURE 9.50 Problem 33. Clutch pedal.

FIGURE 9.51 Problem 34. Adjustment.

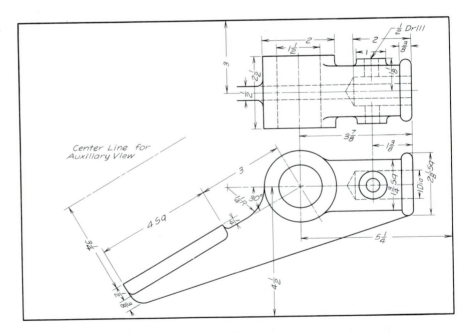

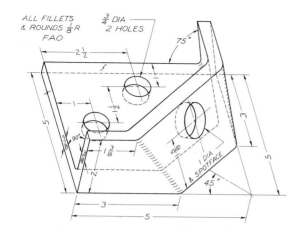

FIGURE 9.52 Problem 35. Corner connector.

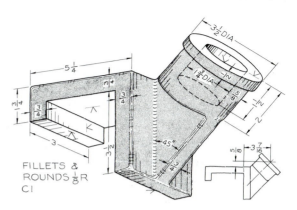

FIGURE 9.53 Problem 36. Shaft bearing.

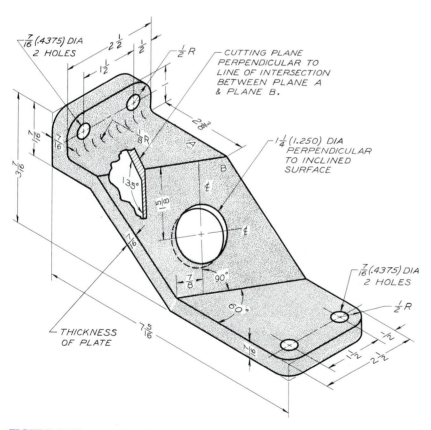

FIGURE 9.54 Problem 37. Oblique support.

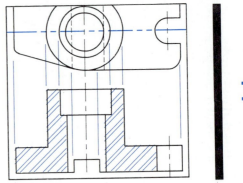

Sections

10.1
Introduction

Standard orthographic views that show all hidden lines may confuse the true details of an object. This shortcoming can often be improved by cutting away part of the object and looking at the cross-sectional view. Such a cutaway view is called a **section.**

A section is shown pictorially in Fig. 10.1A, where an imaginary cutting plane is passed through the object to show its internal features. Figure 10.1A shows the standard top and front views, and Fig. 10.1B shows the method of drawing a section. The front view has been converted to a **full section,** and the cut portion is cross-hatched. Hidden lines have been omitted since they are not needed. The cutting plane is drawn as a heavy line with short dashes at intervals; this can be thought of as a knife-edge cutting through the object.

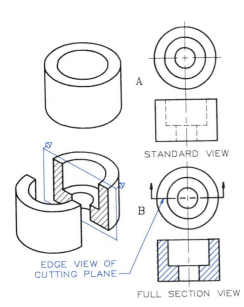

FIGURE 10.1   A comparison of a regular orthographic view with a full-section view showing the internal and external features of the same object.

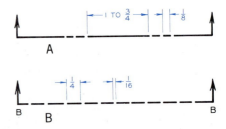

FIGURE 10.2   Typical cutting-plane lines used to represent sections. The cutting plane marked B-B will produce a section labeled B-B.

Figure 10.2 shows two types of cutting planes. Either is acceptable although the example in Fig. 10.2A is more often used. The spacing of the dashes depends on the size of the drawing. The weight of the cutting plane is the same as that of a visible object line. Letters

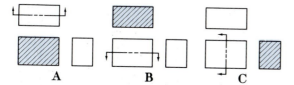

FIGURE 10.3 **The three standard positions of cutting planes that pass through views to result in sectional views in the front, top, and side views. The arrows point in the direction of the line of sight for each section.**

can be placed at each end of the cutting plane to label the sectional view, such as section B-B in Fig. 10.2B.

Figure 10.3 shows the three basic views that appear as sections, with their respective cutting planes. Each cutting plane has perpendicular arrows pointing in the direction of the line of sight for the section. For example, the cutting plane in Fig. 10.3A passes

through the top view, the front of the top view is removed, and the line of sight is toward the remaining portion of the top view. The top view will appear as a section when the cutting plane passes through the front view and the line of sight is downward (Fig. 10.3B). When the cutting plane passes through the front view (Fig. 10.3C), the right-side view will be a section.

10.2
Sectioning symbols

Figure 10.4 shows the symbols used to distinguish between different materials in sections. Although the symbols can be used to indicate the materials within a section, it is advisable to provide supplementary notes specifying the materials to avoid misinterpretation.

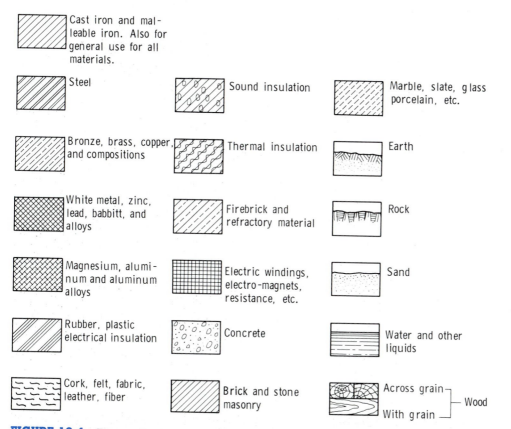

FIGURE 10.4 **The symbols used for lining parts in section. The cast-iron symbol can be used for any material.**

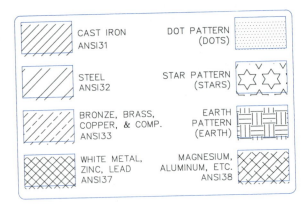

FIGURE 10.5 A few of the sectioning symbols provided by AutoCAD. The pattern scale can be used to vary the spacing of the lines and dashes.

Cast-iron symbols are usually drawn with a 2H pencil with lines slanted at 30°, 45°, or 60° angles and spaced about $\frac{1}{16}$ inch apart.

The **cast-iron symbol** of evenly spaced section lines can be used to represent **any material** and is the most often used sectioning symbol.

COMPUTER METHOD Figure 10.5 shows a few of the many cross-sectional symbols available from AutoCAD. The spacing between the lines and dash lengths may be varied by changing the pattern scale factor.

The proper spacing of section lines is shown in Fig. 10.6A, where the lines are evenly spaced. The other parts of the figure show common errors of section lining.

Extremely thin parts such as sheet metal, washers, and gaskets (Fig. 10.7) are sectioned by completely blacking in the areas rather than by using section lines. Large parts are sectioned with an **outline section** to save time and effort. The section lines are drawn closer together in small parts rather than in larger parts.

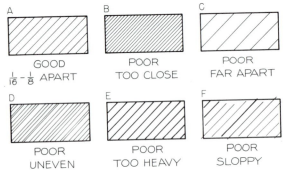

FIGURE 10.6 Good section lines are thin and $\frac{1}{16}$ to $\frac{1}{8}$ inch apart. Some typical lining errors are shown.

Sectioned areas should be lined with symbols that are neither parallel nor perpendicular to the outlines of the parts lest they be confused for serrations or other machining treatments of the surface. (Fig. 10.8).

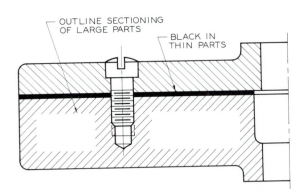

FIGURE 10.7 Thin parts are blacked in, and large areas are section-lined around their outlines to save time and effort.

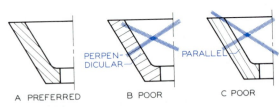

FIGURE 10.8 Section lines should be drawn so that they are neither parallel nor perpendicular to the outline of a part.

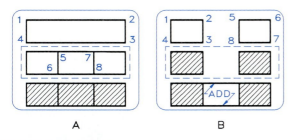

FIGURE 10.9 Sections by computer.

A. If lines were drawn from 1 to 2 to 3 to 4 and division lines were drawn from 5 to 6 and from 7 to 8, the resulting section would ignore lines 5–6 and 7–8 when the area is windowed.

B. If lines are drawn to outline the separate areas, and then windowed, the section lines will fill the areas as intended. The areas can now be connected to complete the sectional view.

> **COMPUTER METHOD** Figure 10.9 shows the method of drawing areas to be section-lined using AutoCAD's HATCH command. The areas to be sectioned must be drawn with lines terminating at each corner point, as Fig. 10.9B shows.

ferent material symbols in an assembly also helps distinguish the materials of the parts. The same part is cross-hatched at the same angle and with the same symbol even though the part may be separated into different areas (Fig. 10.10B). Section lines are effectively used in Fig. 10.11 to identify the parts of the assembly.

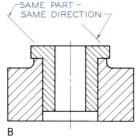

FIGURE 10.11 A typical assembly in section with well-defined parts and correctly drawn section lines.

10.3
Sectioning assemblies

When an assembly of several parts is sectioned, it is important that the section lines be drawn at varying angles to distinguish the parts (Fig. 10.10). Using dif-

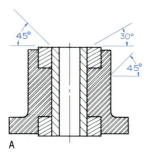

FIGURE 10.10 **A.** Section lines of the same part should be drawn in the same direction. **B.** Section lines of different parts should be drawn at varying angles to distinguish the parts.

10.4
Full sections

> A **full section** is a view formed by passing a cutting plane fully through an object and removing half of it.

In Fig. 10.12, an object is drawn as two orthographic views in which hidden lines are shown. The front view can be drawn as a full section by passing a cutting plane fully through the top view and removing the front portion. The arrows on the cutting plane indicate the direction of sight, and the front view is then section-lined to give the full section.

A full section through a cylindrical part is shown in Fig. 10.13A, where half the object is removed. A common mistake in constructing sectional views is omitting the visible lines behind the cutting plane, as in Fig. 10.13B. Figure 10.13C shows the correctly drawn sectional view. Hidden lines are omitted in all

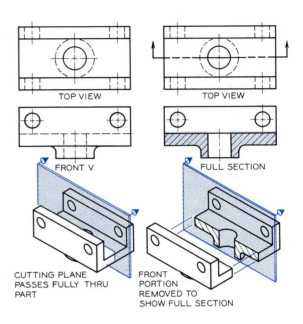

FIGURE 10.12 A full section is formed by a cutting plane that passes completely through the part. The cutting plane is shown passing through the top view, and the direction of sight is indicated by the arrows at each end. The front view is converted to a sectional view to give a clear understanding of the internal features.

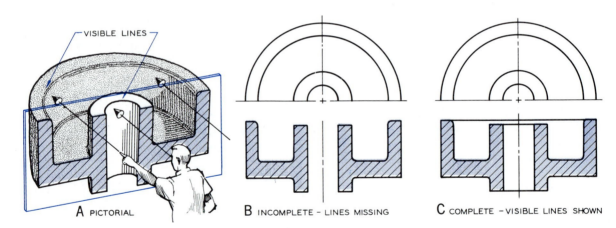

FIGURE 10.13 Full section—cylindrical part.

A. When a full section is passed through an object, you will see lines behind the sectioned area.

B. If only the sectioned area were shown, the view would be incomplete.

C. Visible lines behind the sectioned area must be shown also.

sectional views unless they are considered necessary to provide a clear understanding of the view.

Figure 10.14 is an example of a part whose front view is shown as a full section. Likewise, the part in Fig. 10.15 illustrates a front view that appears as a full section. Lines behind the cutting plane are shown as visible lines.

10.5
Parts not section-lined

Many standard parts like nuts and bolts, rivets, shafts, and set screws, are not section-lined even though the cutting plane passes through them (Fig. 10.16). Since

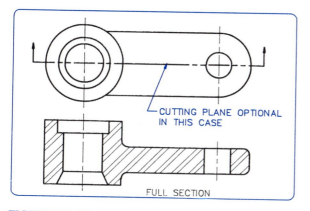

FIGURE 10.14 A full section with the cutting plane shown.

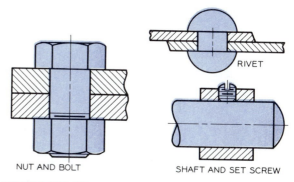

FIGURE 10.16 These parts are not section-lined even though the cutting plane passes through them.

these parts have no internal features, sections through them would be of no value. Other parts not section-lined are roller bearings, ball bearings, gear teeth, dowels, pins, and washers (Fig. 10.17).

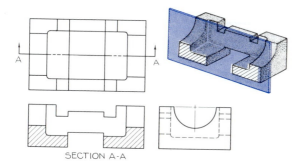

FIGURE 10.15 A full section, section A-A, is used to supplement the given views of the object.

10.6
Ribs in section

Ribs are not section-lined when the cutting plane passes flatwise through them, as in Fig. 10.18A, since this would give a misleading impression of the rib. But a rib is section-lined when the cutting plane passes through it and shows its true thickness (Fig. 10.18B).

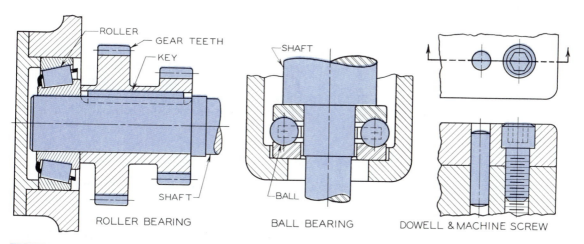

FIGURE 10.17 These parts are not section-lined even though cutting planes pass through them.

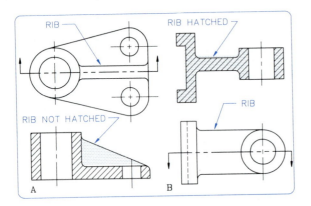

FIGURE 10.18 A rib cut in a flatwise direction by a cutting plane is not section-lined. Ribs are section-lined when cutting planes pass perpendicularly through them, as shown in part B.

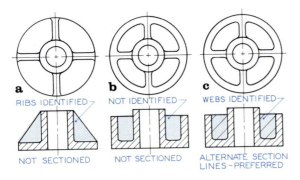

FIGURE 10.19 Outside ribs in section are not section-lined. Poorly identified webs, as in part b, should be identified by alternating section lines with cross-hatching, as in part c.

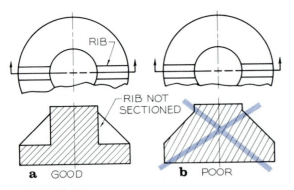

FIGURE 10.20 When ribs are not section-lined, the view is more descriptive of the part. Partial views are used in the top views to save space. The front part of the top view is removed when the front view is a section.

Figure 10.19 shows an alternative method of section-lining webs and ribs. The ribs are not section-lined since the cutting plane passes flatwise through them (Fig. 10.19A). The webs are symmetrically spaced about the hub (Fig. 10.19B). As a rule, webs are not cross-hatched, but this would leave them unidentified; therefore, it is better to use the **alternate sectioning** technique, which extends every other section line through the webs (Fig. 10.19C).

The ribs in Fig. 10.20A are not section-lined and thus afford a more descriptive view of the part. If the ribs had been section-lined, the section would have given the impression that the part was solid and conical, as Fig. 10.20B shows. The top views are partial views, and the portion nearest the sectional view has been omitted.

10.7
Half sections

A **half section** is a view that results from passing a cutting plane halfway through an object and removing a quarter of it to show external and internal features.

A half section is most often used with symmetrical parts, and with cylinders in particular. A cylindrical part in Fig. 10.21A is shown as a pictorial half section in Fig. 10.21B. The method of drawing the orthographic half section is shown in Fig. 10.21C, where both the internal and external features can be seen. Hidden lines are omitted in the sectional view.

The half section in Fig 10.22 has been drawn without showing the cutting plane, which is permissible if it is obvious where the cutting plane was passed through the object. Instead of using an object line, centerlines separate the sectional half from the half that appears as an external view.

10.8
Partial views

Figure 10.23 shows a conventional method of representing symmetrical views. A half view is sufficient when it is drawn adjacent to the sectional view (Fig.

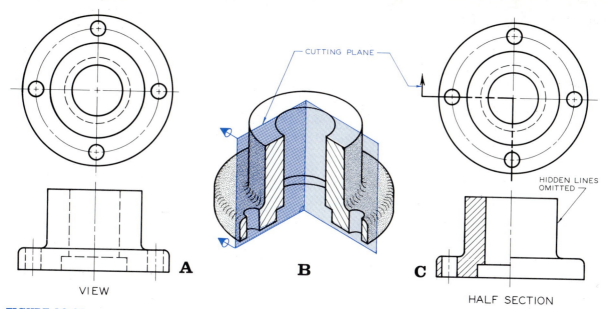

VIEW   A   B   C

CUTTING PLANE

HIDDEN LINES OMITTED

HALF SECTION

FIGURE 10.21 The cutting plane of a half section passes halfway through the object, which results in a sectional view that shows half the outside and half the inside of the object. Hidden lines are omitted unless they are needed to clarify the view.

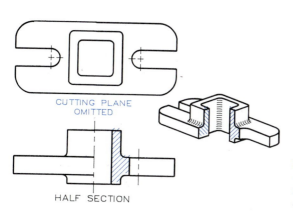

CUTTING PLANE OMITTED

HALF SECTION

FIGURE 10.22 When it is obvious where the cutting plane is located, it is unnecessary to show it, as in this example.

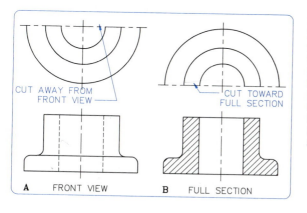

CUT AWAY FROM FRONT VIEW

CUT TOWARD FULL SECTION

A   FRONT VIEW

B   FULL SECTION

FIGURE 10.23 Half views can be used for symmetrical objects to conserve space and drawing time. In part A, the omitted portion of the view is away from the view. In part B, the omitted portion of the view is toward the full section. The omitted half can be toward or away from the section in the case of a half section.

10.23A). In full sections, the removed half is the portion nearest the section (Fig. 10.23B). When drawing half views associated with views (nonsectional views), the removed half of the partial view is the half away from the adjacent view (Fig. 10.23A).

When partial views are drawn with half sections, either the near or the far halves of the partial views can be omitted.

10.9
Offset sections

> An **offset section** is a full section in which the cutting plane is offset to pass through important features.

Figure 10.24 shows an offset section where the plane is offset to pass through the large hole and one of the small holes. The second part of the figure shows the method of drawing the offset section orthographically.

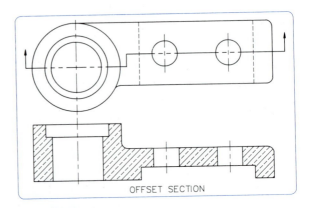

FIGURE 10.25 **An offset section drawn by a computer.**

The cut formed by the offset is not shown in the section since this is an imaginary cut. The computer-drawn object in Fig. 10.25 also lends itself to representation by an offset section.

10.10
Revolved sections

> A **revolved section** is used to describe a cross section of a part by revolving it about an axis of revolution and placing it on the view centered on the axis of revolution.

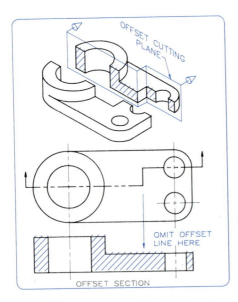

FIGURE 10.24   Offset section.

The cutting plane may need to be offset to pass through features of a part. The offset cutting plane is shown in the top view, and the front view is shown as if it were a full section.

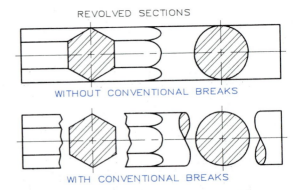

REVOLVED SECTIONS

WITHOUT CONVENTIONAL BREAKS

WITH CONVENTIONAL BREAKS

FIGURE 10.26   Revolved sections can be drawn with or without conventional breaks; either method is acceptable.

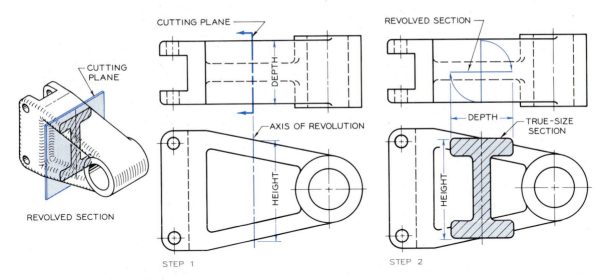

FIGURE 10.27 Revolved section.

Step 1 An axis of revolution is shown in the front view. The cutting plane would appear as an edge in the top view if it were shown.

Step 2 The vertical section in the top view is revolved so that the section can be seen true size in the front view. Object lines are not drawn through the revolved section.

For example, revolved sections are used to indicate cross sections of the parts in Fig. 10.26. Revolved sections are shown with and without conventional breaks. Either method is acceptable.

A more advanced type of revolved section is illustrated in Fig. 10.27, where a cutting plane is passed through the object (Step 1). The plane is imagined to be revolved in the top view to give a true-size revolved section in the front view (Step 2). The object lines do not pass through the revolved section in the front view. It would also have been permissible to use conventional breaks on each side of the revolved section.

Figure 10.28 shows typical revolved sections. These sections provide a method of giving a part's cross section without drawing another complete orthographic view.

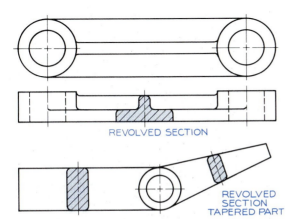

FIGURE 10.28 The revolved sections given here are helpful in describing the cross-sections of the two parts without using additional orthographic views.

10.11
Removed sections

A **removed section** is a revolved section that has been removed from the view where it was revolved (Fig. 10.29).

Centerlines are used as axes of rotation to show where the sections were taken from. Removed sections may be necessary where room does not permit revolution on the given view (Fig. 10.30A); instead, the cross section must be removed from the view (Fig. 10.30B).

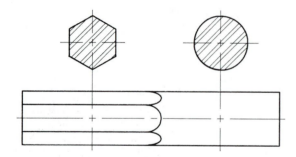

FIGURE 10.29 Removed sections are similar to revolved sections, but they have been removed outside the object along an axis of revolution.

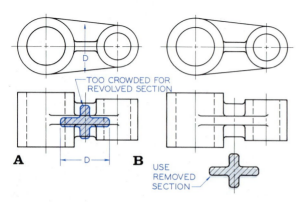

FIGURE 10.30 Removed sections can be used where space does not permit the use of revolved sections.

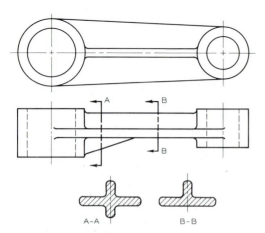

FIGURE 10.31 Sections can be lettered at each end of a cutting plane, such as A-A. This removed section can then be shown elsewhere on the drawing and is designated section A-A.

Removed sections do not have to be positioned directly along an axis of revolution adjacent to the view from where the sections were taken. Instead, cutting planes can be labeled at each end, as Fig. 10.31 shows, to specify the sections. For example, the plane labeled with an A at each end is used to label section A-A; section B-B is similarly found.

FIGURE 10.32 If it is necessary to remove a section to another page in a set of drawings, each end of the cutting plane can be labeled with a letter and a number. The letters refer to section A-A, and the numbers mean this section is on page 7.

When a set of drawings consists of many pages, removed sections may be put on different sheets. In this case, a cutting plane may be labeled as shown in Fig. 10.32. The A at each end indicates section A-A, and the numerals indicate which page the section is on.

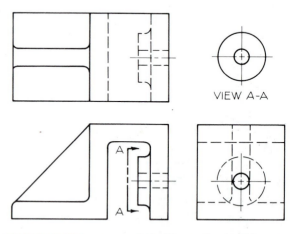

FIGURE 10.33 A removed view (not a section) can also be used to view a part from an unconventional direction.

As Fig. 10.33 shows, removed views can also be used to provide inaccessible orthographic views (non-sectional views).

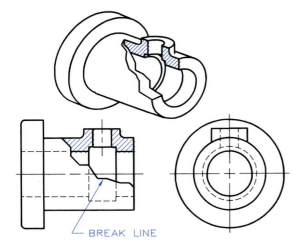

FIGURE 10.34 A broken-out section drawn by computer shows internal features by using a conventional break.

10.12
Broken-out sections

> A **broken-out section** is used to show interior features by breaking away a portion of a view.

A portion of the object in Fig. 10.34 is broken out to reveal details of the wall thickness that better explain the drawing. The irregular lines representing breaks are **conventional breaks.**

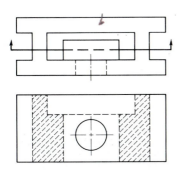

FIGURE 10.35 Phantom sections give an "x-ray" view of an object. The section lines are shown as dashed lines, which makes it possible to show the section without removing the hole in the front of the part.

10.13
Phantom (ghost) sections

> A **phantom** or **ghost section** is used to depict parts as if they were viewed by an x ray.

In Fig. 10.35, the cutting plane is drawn in the usual manner, but the section lines are drawn as dashed lines. If the object had been shown as a regular full section, the circular hole through the front surface could not have been shown in the same view.

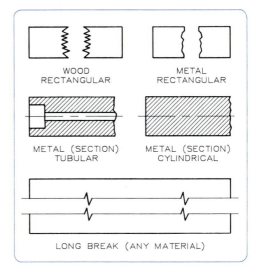

FIGURE 10.36 These conventional breaks indicate that a portion of an object has been broken away.

10.14
Conventional breaks

Figure 10.36 shows examples of **conventional breaks.** The "figure 8" breaks, which are used for cylindrical and tubular parts, can be drawn freehand (Fig. 10.37). They can be drawn with a compass when drawn to a large scale (Fig. 10.38).

One use of conventional breaks is to shorten a long piece that has a uniform cross section. The long

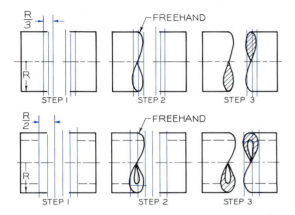

FIGURE 10.37 Conventional breaks in cylindrical and tubular sections can be drawn freehand with the aid of the guidelines shown. The radius, *R*, is used to establish the width of both "figure 8's."

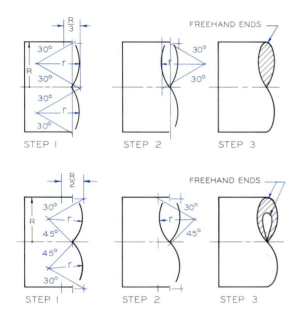

FIGURE 10.38 The steps of constructing conventional breaks of solid and tubular shapes with instruments.

part in Fig. 10.39 has been shortened and drawn at a larger scale for more clarity by using conventional breaks (Fig. 10.39). The dimension specifies the true length of the part, and the breaks indicate that a portion of the length has been removed.

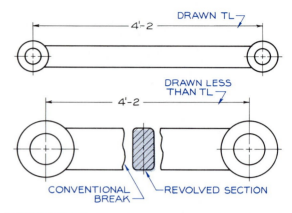

FIGURE 10.39 By using conventional breaks and a revolved section, this part can be drawn at a larger scale that is easier to read.

10.15
Conventional revolutions

Figure 10.40 shows three conventional sections. The center hole is omitted in Fig. 10.40a since it does not pass through the center of the circular plate. However, the hole in Fig. 10.40b does pass through the plate's center and is sectioned accordingly. In Fig. 10.40c, although the cutting plane does not pass through one of

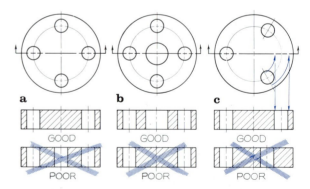

FIGURE 10.40 Symmetrically spaced holes in a circular plate should be revolved in their sectional views to show them at their true radial distance from the center. In part a, no hole is shown at the center, but a hole is shown at the center in part b since the hole is through the center of the plate. In part c, one of the holes is rotated to the cutting plane so that the sectional view will be symmetrical and more descriptive.

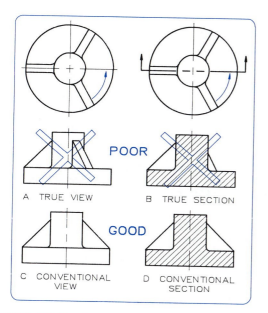

FIGURE 10.41 Symmetrically located ribs are shown revolved in both orthographic and sectional views as a conventional practice.

the symmetrically spaced holes in the top view, the hole is revolved to the cutting plane to show the recommended full section.

When ribs are symmetrically spaced about a hub (Fig. 10.41), it is conventional practice to revolve them to where they will appear true size in either a view or a section. Figure 10.42 shows a full section that shows both ribs and holes revolved to their true-size locations.

The cutting plane can be positioned as Fig. 10.43 shows. Even though the cutting plane does not pass through the ribs and holes in Fig. 10.43a, the sectional view should be drawn as shown in Fig. 10.43b, where the cutting plane is revolved. The cutting plane can be drawn in either position.

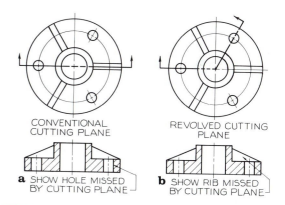

FIGURE 10.43 Symmetrically located ribs are shown in section in revolved positions to show the ribs true size. Foreshortened ribs are omitted.

In the same manner as ribs in section, symmetrically spaced spokes are rotated and not section-lined (Fig. 10.44). Only the revolved, true-size spokes are drawn; the intermediate spokes are omitted.

> If the spokes in Fig. 10.45B had been section-lined, the cross section of the part would be confused with the part in Fig. 10.45A, where there are no spokes but a continuous web.

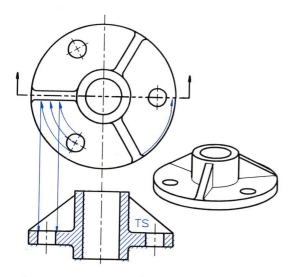

FIGURE 10.42 A part with symmetrically located ribs and holes is shown in section with both ribs and holes rotated to the cutting plane.

The lugs symmetrically positioned about the central hub of the object in Fig. 10.46 are revolved to show they are true size in both views and sections. A more complex object involving the same principle of

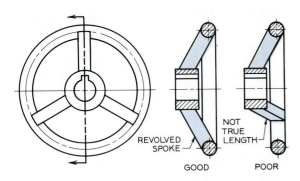

FIGURE 10.44 Symmetrically positioned spokes are revolved to show the spokes true size in section. Spokes are not section-lined in section.

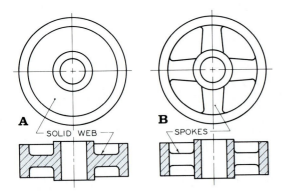

FIGURE 10.45 Solid webs in sections of the type in part A are section-lined. Spokes are not section-lined when the cutting plane passes through them, as in part B.

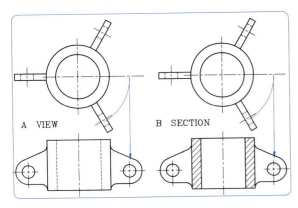

FIGURE 10.46 Symmetrically spaced lugs (flanges) are revolved to show the front view and the sectional view as symmetrical.

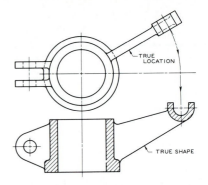

FIGURE 10.47 A part with an oblique feature attached to the circular hub is revolved so that it will appear true shape in the front view, which is a sectional view.

rotation can be seen in Fig. 10.47, where the oblique arm is drawn in the section as if it had been revolved to the centerline in the top view and then projected to the sectional view.

10.16
Auxiliary sections

Auxiliary sections can be used to supplement the principal views used in orthographic projections, as Fig. 10.48 shows. Auxiliary cutting plane A-A is passed through the front view, and the auxiliary view is projected from the cutting plane as indicated by the sight arrows. Section A-A gives the cross-sectional description of the part.

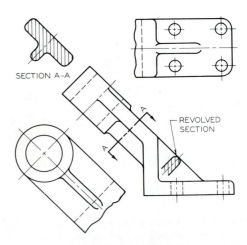

FIGURE 10.48 Sectional views can be shown as auxiliary views for added clarity.

Problems

These problems can be solved on Size A or Size B sheets.

1–24. (Fig. 10.49) Full sections: Draw two of these problems per Size A sheet. Each grid equals 0.20 inch or 5 mm. Complete the front views as full sections.

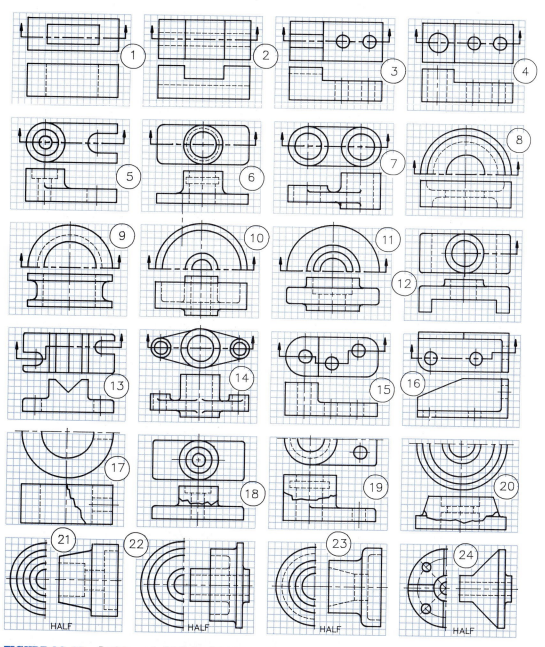

FIGURE 10.49   Problems 1–24. Introductory sections.

25–29. (Figs. 10.50–10.54) Full sections: Complete the drawings as full sections. Draw one problem per Size A sheet. Each grid equals 0.20 inch or 5 mm. Show the cutting planes when they are not given.

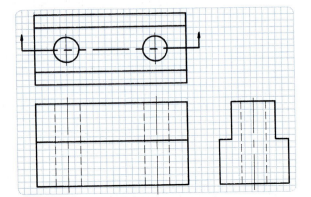

FIGURE 10.50 Problem 25. Full section.

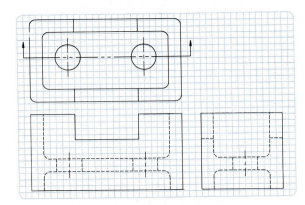

FIGURE 10.51 Problem 26. Full section.

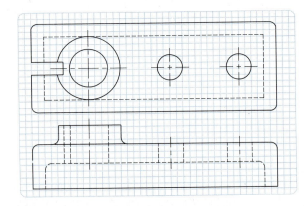

FIGURE 10.52 Problem 27. Full section.

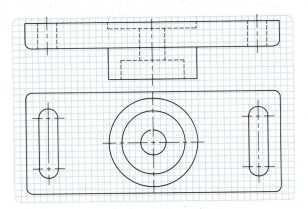

FIGURE 10.53 Problem 28. Full section.

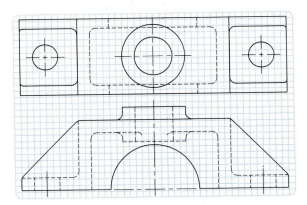

FIGURE 10.54 Problem 29. Full section.

30–32. (Figs. 10.55–10.57) Half sections: Complete the drawings as half sections. Draw one problem per Size A sheet. Each grid equals 0.20 inch or 5 mm. Show the cutting planes when they are not given.

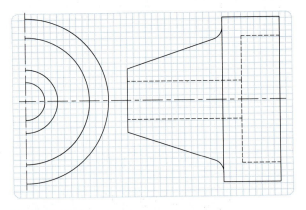

FIGURE 10.55 Problem 30. Half section.

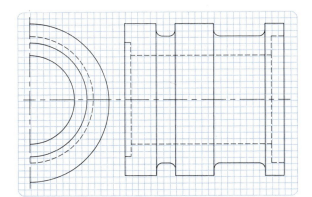

FIGURE 10.56 Problem 31. Half section.

33. (Fig. 10.58) Offset section: Complete the drawing as an offset section. Draw the problem on an Size A sheet. Each grid equals 0.20 inch or 5 mm. Show the cutting planes.

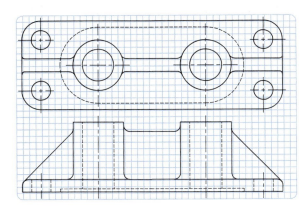

FIGURE 10.57 Problem 32. Half section.

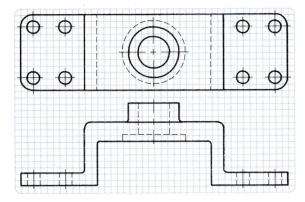

FIGURE 10.58 Problem 33. Offset section.

34. (Fig. 10.59) Full section: Complete the partial view as a full section. Draw the views on a Size A sheet. Each grid equals 0.20 inch or 5 mm. Show the cutting plane.

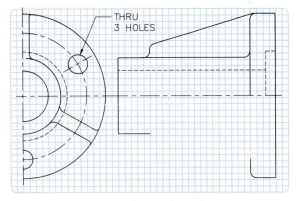

FIGURE 10.59 Problem 34. Full section.

35. (Fig. 10.60) Assembly: Complete the front view as a full section of the assembled parts. Draw the views on a Size A sheet. Each grid equals 0.20 inch or 5 mm. Show the cutting plane.

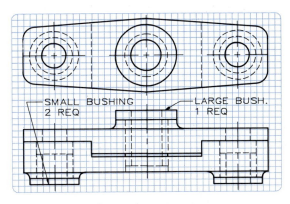

FIGURE 10.60 Problem 35. Full section.

36–38. (Figs. 10.61–10.63) Sections: Draw the necessary views to describe the parts using sections and conventional practices. Draw one problem per Size B sheet. Omit the dimensions.

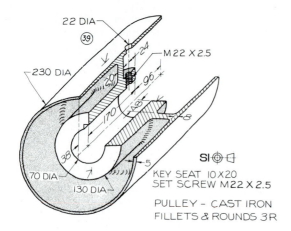

FIGURE 10.61   Problem 36. Section.

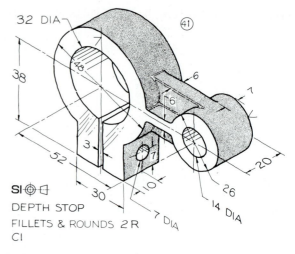

FIGURE 10.63   Problem 38. Section.

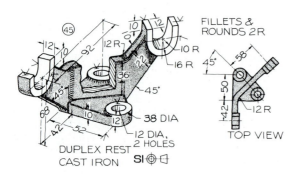

FIGURE 10.62   Problem 37. Section.

Screws, Fasteners, and Springs

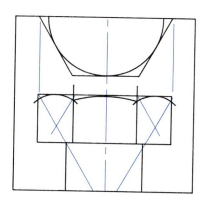

11.1
Threaded fasteners

Screw threads provide a fast and easy method of fastening two parts together and of exerting a force that can be used for adjustment of movable parts. For a screw thread to function, there must be an **internal thread** and an **external thread.** Internal threads may be tapped inside a part such as a motor block or, more commonly, a nut. Whenever possible, the nuts and bolts used in industrial projects should be stock parts that can be obtained from many sources. This reduces manufacturing expenses and improves the interchangeability of parts.

Progress has been made toward establishing standards that will unify threads in this country and abroad by the introduction of metric standards. Other efforts have led to the adoption of the Unified Screw thread by the United States, Britain, and Canada (ABC Standards), which is a modification of both the American Standard thread and the Whitworth thread.

11.2
Definitions of thread terminology

EXTERNAL THREAD is a thread on the outside of a cylinder, such as a bolt (Fig. 11.1).

INTERNAL THREAD is a thread cut on the inside of a part, such as a nut (Fig. 11.1).

MAJOR DIAMETER is the largest diameter on an internal or external thread (Fig. 11.2).

MINOR DIAMETER is the smallest diameter that can be measured on a screw thread (Fig. 11.2).

PITCH DIAMETER is the diameter of an imaginary cylinder passing through the threads at the points where the thread width is equal to the space between the threads (Fig. 11.2).

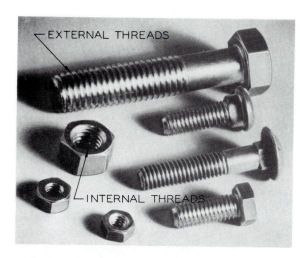

FIGURE 11.1   **Thread terminology.**

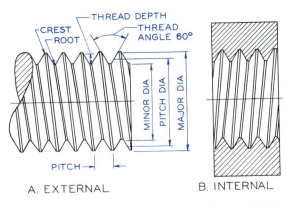

A. EXTERNAL   B. INTERNAL

FIGURE 11.2 **Examples of external threads (bolts) and internal threads (nuts).** (Courtesy of Russell, Burdsall & Ward Bolt and Nut Co.)

LEAD is the distance a screw will advance when turned 360°.

PITCH is the distance between crests of threads. Pitch is found mathematically by dividing 1 inch by the number of threads per inch of a particular thread (Fig. 11.2).

CREST is the peak edge of a screw thread (Fig. 11.2).

THREAD ANGLE is the angle between threads cut by the cutting tool (Fig. 11.2).

ROOT is the bottom of the thread cut into a cylinder (Fig. 11.2).

THREAD FORM is the shape of the thread cut into a threaded part.

THREAD SERIES is the number of threads per inch for a particular diameter, grouped into three series: coarse, fine, extra fine, and there are eight constant-pitch thread series. Coarse series provides rapid assembly, and extra-fine series provides fine adjustment.

THREAD CLASS is a closeness of fit between two mating threaded parts. Class 1 represents a loose fit and Class 3 a tight fit.

RIGHT-HAND THREAD is a thread that will assemble when turned clockwise. A right-hand thread slopes downward to the right on an external thread when the axis is horizontal, and in the opposite direction on an internal thread.

LEFT-HAND THREAD is a thread that will assemble when turned counterclockwise. A left-hand thread slopes downward to the left on an external thread when the axis is horizontal, and in the opposite direction on an internal thread.

11.3
Thread specifications (English system)

Form

Thread form is the shape of the thread cut into a part, as illustrated in Fig. 11.3. The Unified form, a combination of the American National and British Whitworth, is the most widely used because it is a standard in several countries. The Unified form is signified by UN in abbreviations and thread notes, and the American National form is signified by N.

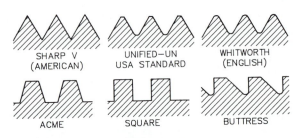

FIGURE 11.3 **Standard thread forms.**

Another thread form, the Unified National Rolled, abbreviated UNR, was introduced into the 1974 ANSI standards. This designation is specified only for external threads—there is no UNR designation for internal threads. The UN form has a flat root (rounded root is optional) in Fig. 11.4A, whereas the UNR thread *must* have a rounded root formed by rolling, as shown in Fig. 11.4B. The rounded root of the UNR thread is designed to reduce the wear of the threading tool and to improve the fatigue strength of the thread.

The transmission of power is achieved by using the **Acme, square,** and **buttress** threads, which are

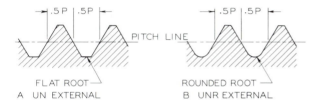

FIGURE 11.4   A. The UN external thread has a flat root (rounded root is optional). B. The UNR has a rounded root formed by rolling. The UNR form does not apply to internal threads.

commonly used in gearing and other pieces of machinery (Fig. 11.4). The **sharp V** thread is used for set screws and in applications where friction in assembly is desired.

Series

Thread series, which is closely related to thread form, designates the type of thread specified for a given application and is abbreviated C, F, and EF.

Eleven standard series of threads are listed under the American National form and the Unified National (UN/UNR) form. There are three series, with abbreviations coarse (C), fine (F), and extra fine (EF), and eight series with constant pitches (4, 6, 8, 12, 16, 20, 28, and 32 threads per inch).

A Unified National form for a coarse-series thread is specified as UNC or UNRC, which is a combination of form and series in a single note. Similarly, an American National form for a coarse thread is written NC. The **coarse-thread** series (UNC/UNRC or NC) is suitable for bolts, screws, nuts, and general use with cast iron, soft metals, or plastics when rapid assembly is desired. The **fine-thread** series (NF or UNF/UNRF) is suitable for bolts, nuts, or screws when a high degree of tightening is required. The **extra-fine** series (UNEF/UNREF or NEF) is suitable for sheet metal, thin nuts, ferrules, or couplings when length of engagement is limited and is used for applications that will have to withstand high stresses.

The 8-thread series (8 UN), 12-thread series, (12 N or 12 UN/UNR), and 16-thread series (16 N or 16 UN/UNR) are threads with a uniform pitch for large diameters. The 8 UN is used as a substitute for the coarse-thread series on diameters larger than 1 inch when a medium-pitch thread is required. The 12 UN is used on diameters larger than $1\frac{1}{2}$ inches, with a thread of a medium-fine pitch as a continuation of the fine-thread series. The 20 UN is used on diameters larger than $1\frac{11}{16}$ inches, with threads of an extra-fine pitch as a continuation of the extra-fine series.

Class of fit

Thread classes are used to indicate the tightness of fit between a nut and bolt or any two mating threaded parts. This fit is determined by the tolerances and allowances applied to threads. Classes of fit are indicated by the numbers 1, 2, or 3 followed by the letters A or B. For UN forms, the letter A represents an external thread, whereas the letter B represents an internal thread. These letters are omitted when the American National form (N) is used.

CLASS 1A AND 1B   threads are used on parts that require assembly with a minimum of binding.

CLASS 2A AND 2B   threads are general-purpose threads for bolts, nuts, screws, and nominal applications in the mechanical field and are widely used in the mass-production industries.

CLASS 3A AND 3B   threads are used in precision assemblies where a close fit is desired to withstand stresses and vibration.

Single and multiple threads

A **single thread** (Fig. 11.5A) is a thread that will advance the distance of its pitch in one full revolution of 360°; in other words, its pitch is equal to its lead. In the drawing of a single thread, the crest line of the thread will slope $\frac{1}{2}P$ since only 180° of the revolution is visible in a single view.

A double thread is composed of two threads, resulting in a lead equal to $2P$, meaning that the threaded part will advance a distance of $2P$ in a single revolution of 360° (Fig. 11.5B). The crest line of a double thread will slope a distance equal to P in the view in which 180° can be seen.

Similarly, a triple thread will advance $3P$ in 360° with a crest line slope of $1\frac{1}{2}P$ in the view in which 180° of the cylinder is visible (Fig. 11.5C). The lead of a double thread is $2P$, and the lead of a triple thread is $3P$. Multiple threads are used wherever quick assembly is required.

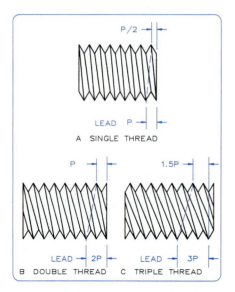

FIGURE 11.5 Single and multiple threads.

Thread notes

Drawings of threads are only symbolic representations and are inadequate unless accompanying notes give the thread specifications (Fig. 11.6). The major diameter is given first, followed by the number of threads per inch, the form and series, the class of fit, and a letter denoting whether the thread is external or internal. For a double or triple thread, the word *DOUBLE* or *TRIPLE* is included in the note; for a left-hand thread, the letters *LH* are included.

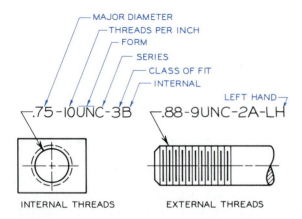

FIGURE 11.6 Parts of a thread note for an external thread.

Figure 11.7A shows the UNR thread note for the external thread. (UNR does not apply to internal threads.) When inches are used as the unit of measurement, thread notes can be written as common fractions, but decimal fractions are preferred.

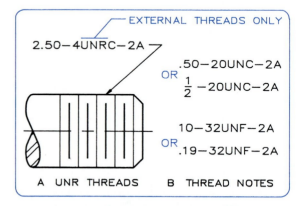

FIGURE 11.7 **A.** The UNR thread notes apply to external threads only. **B.** Notes can be given as decimal fractions or common fractions.

11.4
Using thread tables

Appendix 14 gives the UN/UNR thread table, and Table 11.1 shows part of this table. If an external thread (bolt) with a $1\frac{1}{2}$-inch diameter is to have a "fine" thread, it will have 12 threads per inch. Therefore, the thread note can be written

$$1\frac{1}{2}\text{–12 UNF–2A} \quad \text{or} \quad 1.500\text{–12 UNF–2A.}$$

If the thread were an internal one (nut), the thread note would be the same, but the letter *B* would be used instead of the letter *A*.

A constant-pitch thread series can be selected for the larger diameters. The constant-pitch thread notes are written with the abbreviations C, F, and EF omitted. For example, a $1\frac{3}{4}$-inch diameter bolt with a fine thread could be noted in constant-pitch series as

$$1\frac{3}{4}\text{–12 UN–2A} \quad \text{or} \quad 1.750\text{–12 UN–2A.}$$

TABLE 11.1

AMERICAN NATIONAL STANDARD UNIFIED INCH SCREW THREADS (UN AND UNR THREAD FORM)*

Sizes		Basic Major Diameter	Series with Graded Pitches			Series with Constant Pitches								Sizes
Primary	Second-ary		Coarser UNC	Fine UNF	Extra fine UNEF	UN	6 UN	8 UN	12 UN	16 UN	20 UN	28 UN	32 UN	
1		1.0000	8	12	20	—	—	UNC	UNF	16	UNEF	28	32	1
	$1\frac{1}{16}$	1.0625	—	—	18	—	—	8	12	16	20	28	—	$1\frac{1}{16}$
$1\frac{1}{8}$		1.1250	7	12	18	—	—	8	UNF	16	20	28	—	$1\frac{1}{8}$
	$1\frac{3}{16}$	1.1875	—	—	18	—	—	8	12	16	20	28	—	$1\frac{3}{16}$
$1\frac{1}{4}$		1.2500	7	12	18	—	—	8	UNF	16	20	28	—	$1\frac{1}{4}$
	$1\frac{5}{16}$	1.3125	—	—	18	—	—	8	12	16	20	28	—	$1\frac{5}{16}$
$1\frac{3}{8}$		1.3750	6	12	18	—	UNC	8	UNF	16	20	28	—	$1\frac{3}{8}$
	$1\frac{7}{16}$	1.4375	—	—	18	—	6	8	12	16	20	28	—	$1\frac{7}{16}$
$1\frac{1}{2}$		1.5000	6	12	18	—	UNC	8	UNF	16	20	28	—	$1\frac{1}{2}$
	$1\frac{9}{16}$	1.5625	—	—	18	—	6	8	12	16	20	—	—	$1\frac{9}{16}$

*By using this table, a diameter of $1\frac{1}{2}$ inches that is to be threaded with a fine thread would have the following thread note: $1\frac{1}{2}$-12 UNF-2A.

Source: Courtesy of ANSI; B1.1.

Table 11.1 can also be used for the UNR thread form (for external threads only) by substituting UNR for UN; for example, UNREF for extra fine.

11.5
Metric thread specifications (ISO)

Metric thread specifications are recommended by the ISO (International Organization for Standardization). Thread specifications can be given with a **basic designation,** which is suitable for general applications, or with the **complete designation,** which is used where detailed specifications are needed.

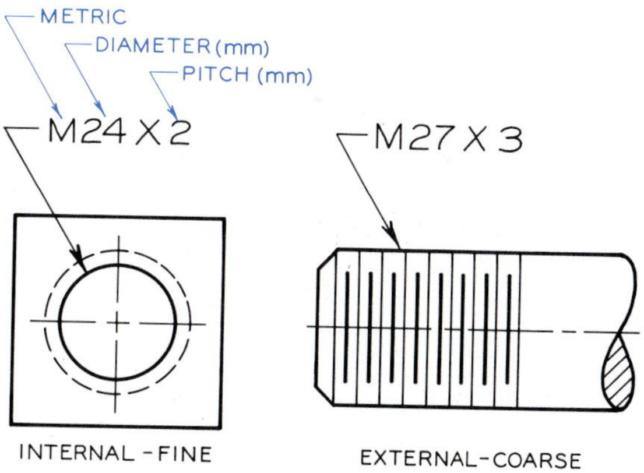

Basic designation

Figure 11.8 shows examples of metric screw thread notes. Each note begins with the letter *M*, which designates the note as a metric note, followed by the diameter in millimeters, and the pitch in millimeters separated by ×, the multiplication sign. The pitch can

FIGURE 11.8 **Basic designations for metric threads.**

be omitted in notes for coarse threads, but U.S. standards prefer that it be shown. Table 11.2 shows the commercially available ISO threads recommended for general use. Appendix 17 gives additional ISO specifications.

TABLE 11.2
BASIC THREAD DESIGNATIONS FOR COMMERCIAL SERIES OF ISO METRIC THREADS

Nominal Size (mm)	Pitch P (mm)	Basic Thread Designation*	Nominal Size (mm)	Pitch P (mm)	Basic Thread Designation*	Nominal Size (mm)	Pitch P (mm)	Basic Thread Designation*
1.6	0.35	M1.6	8	1.25	M8	22	2.5	M22
1.8	0.35	M1.8	8	1	M8 × 1	22	1.5	M22 × 1.5
2	0.4	M2	10	1.5	M10	24	3	M24
2.2	0.45	M2.2	10	1.25	M10 × 1.25	24	2	M24 × 2
2.5	0.45	M2.5	12	1.75	M12	27	3	M27
3	0.5	M3	12	1.25	M12 × 1.25	27	2	M27 × 2
3.5	0.6	M3.5	14	2	M14	30	3.5	M30
4	0.7	M4	14	1.5	M14 × 1.5	30	2	M30 × 2
4.5	0.75	M4.5	16	2	M16	33	3.5	M33
5	0.8	M5	16	1.5	M16 × 1.5	33	2	M33 × 2
6	1	M6	18	2.5	M18	36	4	M36
7	1	M7	18	1.5	M18 × 1.5	36	3	M36 × 3
			20	2.5	M20	39	4	M39
			20	1.5	M20 × 1.5	39	3	M39 × 3

*U.S. practice is to include the pitch symbol even for the coarse pitch series. Basic descriptions shown are as specified in ISO Recommendations.
Source: Courtesy of Greenfield Tap and Die Corp.

Complete designation

For some applications it is necessary to show a complete thread designation (Fig. 11.9). The first part of this note is the same as the basic designation; however, the note also has a tolerance class designation separated by a dash. The 5g represents the pitch diameter tolerance, and 6g represents the crest diameter tolerance.

The numbers 5 and 6 are **tolerance grades** (variations from the basic diameter). Grade 6 is com-

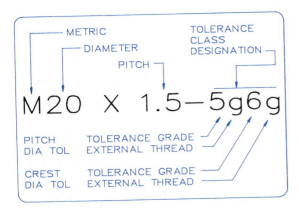

FIGURE 11.9 A complete designation note for metric threads.

TABLE 11.3
TOLERANCE GRADES, ISO THREADS

External Thread		Internal Thread	
Major Diameter (d_1)	Pitch Diameter (d_2)	Minor Diameter (D_1)	Pitch Diameter (D_2)
—	3	—	—
4	4	4	4
—	5	5	5
6	6	6	6
—	7	7	7
8	8	8	8
—	9	—	—

Grade 6 is medium; smaller numbers are finer, and larger numbers are coarser.
Source: Courtesy of ANSI; B1.

monly used for a medium general-purpose thread that is nearly equal to class 2A and 2B of the Unified system. Grades less than 6 are used for fine-quality fits and short lengths of engagement. Grades greater than 6 are recommended for coarse-quality fits and long lengths of engagement. Table 11.3 gives the tolerance grades for internal and external threads for the pitch diameter and the major and minor diameters.

TOLERANCE POSITIONS

EXTERNAL THREADS	INTERNAL THREADS
(Lowercase Letters)	(Uppercase Letters)
e = LARGE ALLOWANCE	G = SMALL ALLOWANCE
g = SMALL ALLOWANCE	H = NO ALLOWANCE
h = NO ALLOWANCE	

LENGTH OF ENGAGEMENT

S = SHORT   N = NORMAL   L = LONG

FIGURE 11.10 **Symbols used to represent tolerance grade, position, and class.**

The letters following the grade numbers designate **tolerance positions** (external or internal). Lowercase letters represent external threads (bolts), as Fig. 11.10 shows. The lowercase letters *e*, *g*, and *h* represent large allowance, small allowance, and no allowance, respectively. (**Allowance** is the variation from the basic diameter.) Uppercase letters designate internal threads (nuts), and lowercase letters designate external threads. On internal threads, *G* designates small allowance, and *H* designates no allowance. The letters are placed after the tolerance grade number. For example, 5g designates a medium tolerance with small allowance for the pitch diameter of an external thread, and 6H designates a medium tolerance with no allowance for the minor diameter of an internal thread.

Tolerance classes are fine, medium, and coarse, as listed in Table 11.4. These classes of fit are combinations of tolerance grades, tolerance positions, and

lengths of engagement—short (S), normal (N), and long (L). The length of engagement can be determined by referring to Appendix 16. Once it has been decided to use a fine, medium, or coarse class of fit for a particular application, the specific designation should be selected, first from the classes shown in large print in Table 11.4, second from the classes shown in medium-size print, and third from the classes shown in small print. Classes shown in boxes are for commercial threads.

Figure 11.11 shows variations in the complete designation thread notes. The tolerance class symbol is written 6H if the crest and pitch diameters have identical grades (Fig. 11.11A). Since an uppercase *H* is

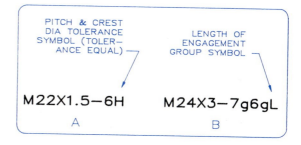

FIGURE 11.11 **A. When both pitch and crest diameter tolerance grades are the same, the tolerance class symbol is shown only once. B. Letters S, N, and L are used to indicate the length of the thread engagement.**

TABLE 11.4
PREFERRED TOLERANCE CLASSES, ISO THREADS*

Quality	External Threads (bolts)									Internal Threads (nuts)					
	Tolerance position e (large allowance)			Tolerance position g (small allowance)			Tolerance position h (no allowance)			Tolerance position G (small allowance)			Tolerance position H (no allowance)		
	Length of engagement			Length of engagement			Length of engagement			Length of engagement			Length of engagement		
	Group S	Group N	Group L	Group S	Group N	Group L	Group S	Group N	Group L	Group S	Group N	Group L	Group S	Group N	Group L
Fine Medium Coarse	6e	7e6e		5g6g	6g / 8g	7g6g / 9g8g	3h4h / 5h6h	4h / 6h	5h4h / 7h6h	5G	6G / 7G	7G / 8G	4H / 5H	5H / 6H / 7H	6H / 7H / 8H

*In selecting tolerance class, select first from the large bold print, second from the medium-size print, and third from the small-size print. Classes shown in boxes are for commercial threads.

used, this is an internal thread. Where considered necessary, the length-of-engagement symbol may be added to the tolerance class designation (Fig. 11.11B).

Designations for the desired fit between mating threads can be specified as shown in Fig. 11.12. A slash is used to separate the tolerance class designations of the internal and external threads.

Additional information about ISO threads may be obtained from *ISO Metric Screw Threads*, a booklet of standards published by ANSI. These standards were used as the basis for most of this section.

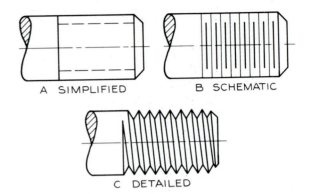

FIGURE 11.13 Three major types of thread representations.

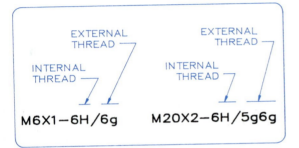

FIGURE 11.12 A slash mark is used to separate the tolerance class designations of mating internal and external threads.

11.6
Thread representation

Three major types of thread representations are the **simplified, schematic,** and **detailed** (Fig. 11.13). The detailed representation is the most realistic approximation of a thread's appearance, and the simplified representation is the least realistic.

11.7
Detailed UN/UNR threads

Figure 11.14 shows examples of detailed representations of internal and external threads. Instead of helical curves, straight lines indicate crest and root lines.

Figure 11.15 shows the construction of a detailed thread representation. The pitch is found by dividing

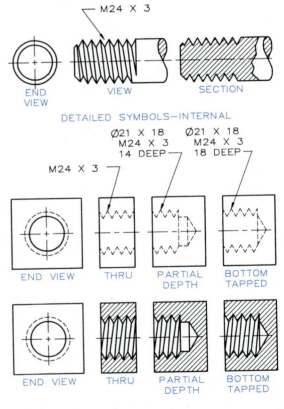

FIGURE 11.14 Detailed thread representations of external and internal threads. Use Appendix 15 to determine tap drill sizes and convert to metric units.

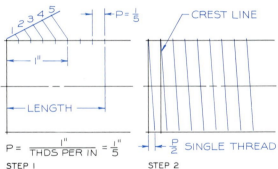
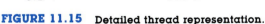

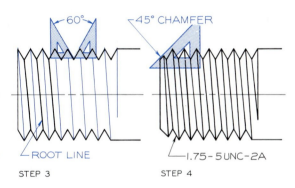

FIGURE 11.15 Detailed thread representation.

Step 1 To draw a detailed representation of a 1.75–5 UNC–2A thread, the pitch is determined by dividing 1″ by the number of threads per inch, 5 in this case. The pitch is laid off the length of the thread. Pitch is usually approximated.

Step 2 Since this is a right-hand thread, the crest lines slope downward to the right equal to ½P. The crest lines will be final lines drawn with an H or F pencil.

Step 3 The root lines are found by constructing 60° vees between the crest lines. The root lines are drawn from the bottom of the vees. Root lines are parallel to each other, but not to crest lines.

Step 4 A 45° chamfer is constructed at the end of the thread from the minor diameter. Strengthen all lines, and add a thread note.

1 inch by the number of threads per inch. This can be done graphically, as shown in Step 1. Usually, the pitch can be approximated and laid off with a scale or dividers. Where threads are close, they should be drawn at a larger spacing to be easier to draw. In Step 4, a 45° chamfer is used to draw a bevel on the threaded end to improve the assembly of the threaded parts.

Metric threads should be drawn in the same manner, using the pitch given in millimeters in the metric thread table, or an expanded pitch, if needed.

> **COMPUTER METHOD** Detailed thread symbols can be drawn by computer (Fig. 11.16) and then duplicated with the `COPY` command set at the `MULTIPLE` option. A typical set of threads is drawn in Step 1 and then copied repetitively in Step 2.

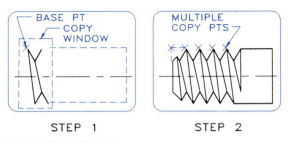

FIGURE 11.16 Detailed threads by computer.

Step 1 A typical detailed thread symbol is drawn at the end of the screw.

Step 2 The typical set of threads are duplicated with the `COPY` command and the `MULTIPLE` option along the predetermined snap points of the screw.

11.8
Detailed square threads

Figure 11.17 shows the method of drawing a detailed representation of a square thread to give an approximation of a square thread.

In Step 1, the major diameter is laid off. The number of threads per inch is taken from Appendix 18. The pitch *(P)* is found by dividing 1 inch by the number of threads per inch, but this pitch can be enlarged if needed. Distances of *P*/2 are marked off with dividers. Steps 2, 3, and 4 are then completed, and a thread note is added.

Square internal threads are drawn in a similar manner (Fig. 11.18). The threads in the section view are drawn in a slightly different way. The thread note

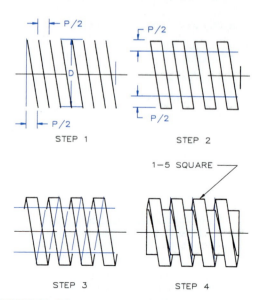

FIGURE 11.17 Drawing the square thread.

Step 1 Lay out the major diameter. Space the crest lines ½P apart. Slope them downward to the right for right-hand threads.

Step 2 Connect every other pair of crest lines. Find the minor diameter by measuring ½P inward from the major diameter.

Step 3 Connect the opposite crest lines with light construction lines. This will establish the profile of the thread form.

Step 4 Connect the inside crest lines with light construction lines to locate the points on the minor diameter where the thread wraps around the minor diameter. Darken the final lines.

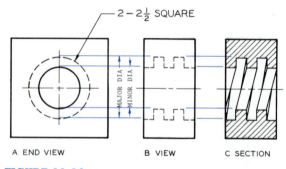

FIGURE 11.18 Internal square threads.

for an internal thread is placed in the circular view whenever possible, with the leader pointing toward the center. When a square thread is long, it need not be drawn continuously but can be represented using the symbol shown in Fig. 11.19.

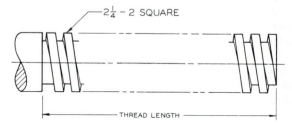

FIGURE 11.19 Conventional method of showing square threads without drawing each thread.

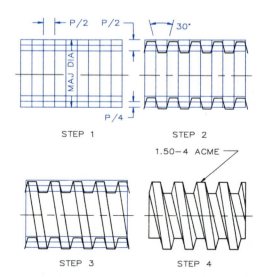

FIGURE 11.20 Drawing the Acme thread.

Step 1 Lay out the major diameter and thread length, and divide the shaft into equal divisions ½P apart. Locate the minor and pitch diameters using distances ½P and ¼P.

Step 2 Draw construction lines at 15° angles with the vertical along the pitch diameter as shown to make a total angle of 30°.

Step 3 Draw the crest lines across the screw.

Step 4 Darken the lines, draw the root lines, and add the thread note to complete the drawing.

11.9
Detailed Acme threads

Figure 11.20 shows the method of preparing detailed drawings of Acme threads. In Step 1, the length and the major diameter are laid off with light construction lines, and the pitch is found by dividing 1 inch by the number of threads per inch. Steps 2, 3, and 4 complete the thread representation, and the thread note is added.

Figure 11.21 shows internal Acme threads. In the section view, left-hand internal threads are sloped so that they look the same as right-hand external threads.

Figure 11.22 shows a shaft that is being threaded with Acme threads on a lathe as the tool travels the length of the shaft.

11.10
Schematic threads

Figure 11.23 shows schematic representations of internal and external threads with parallel nonsloping lines. Since the schematic representation is easy to draw and gives a good symbolic representation of threads, it is the most often used thread symbol. Left-hand threads must be specified in the thread note.

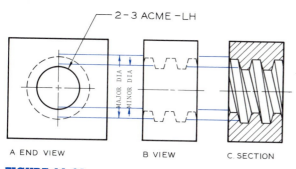

FIGURE 11.21   Internal Acme threads.

FIGURE 11.22   Cutting an Acme thread on a lathe. (Courtesy of Clausing Corp.)

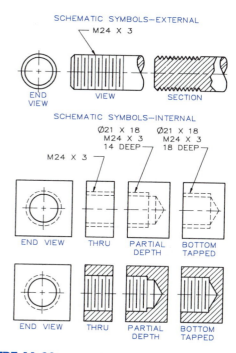

FIGURE 11.23   Schematic representations of external and internal threads as views and sections. Use Appendix 15 to determine tap drill sizes and convert to metric units.

Figure 11.24 shows the method of constructing schematic threads, and Fig. 11.25 shows the method of drawing metric threads using schematic representations. The pitch (in millimeters) given in the metric thread table is used as the approximate distance to separate the crest lines.

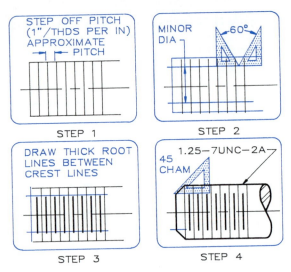

FIGURE 11.24 *Drawing schematic threads.*

Step 1 Lay out the major diameter, and divide the shaft into divisions of a distance of approximately *P* apart. Draw these crest lines as thin lines.

Step 2 Find the minor diameter by drawing a 60° angle between two crest lines on each side.

Step 3 Draw heavy root lines between the crest lines.

Step 4 Chamfer the end of the thread from the minor diameter, and give a thread note.

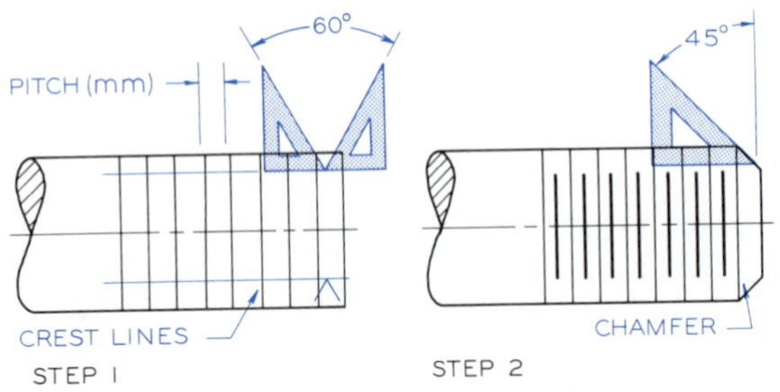

COMPUTER METHOD Schematic threads can be drawn by computer (Fig. 11.26) and then duplicated with the COPY command set at the MULTIPLE option. A typical set of threads is drawn in Step 1 and then copied repetitively in Step 2.

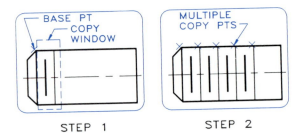

FIGURE 11.26 Schematic threads by computer.

Step 1 Draw the outline of the threaded shaft with a chamfered end. Draw typical minor and major diameters, and window them for a MULTIPLE COPY.

Step 2 Repetitively COPY the threads along the screw at predetermined snap points.

11.11
Simplified threads

Figure 11.27 illustrates the use of simplified representations with notes to specify thread details. Of the three types of thread representations, this is the easiest to draw. Hidden lines can be positioned by eye to approximate the minor diameter. Figure 11.28 shows the steps in constructing a simplified thread drawing.

FIGURE 11.25 *Schematic metric threads.*

Step 1 The pitch of metric threads can be taken directly from the metric tables, which can be used to find the minor diameter.

Step 2 The root lines are drawn in heavy between the crest lines. The end of the thread is chamfered.

11.12
Drawing small threads

Instead of drawing small threads to exact measurements, minor diameters can be drawn smaller to sep-

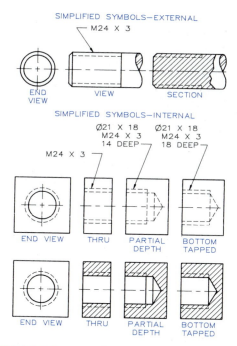

FIGURE 11.27 Simplified thread representations of external and internal threads. Use Appendix 15 to determine tap drill sizes and convert to metric units.

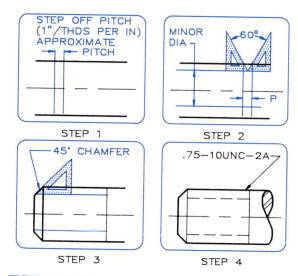

FIGURE 11.28 Drawing simplified threads.

Step 1 Lay out the major diameter. Find the approximate pitch *(P)*, and lay out two lines a distance P apart.

Step 2 Find the minor diameter by constructing a 60° angle between the two lines on both sides.

Step 3 Draw a 45° chamfer from the minor diameter to the major diameter.

Step 4 Show the minor diameter as dashed lines. Add a thread note.

arate the root and crest lines (Fig. 11.29). This procedure makes the thread easier to draw and read. Since the drawing is only a symbolic representation of a thread, exactness is unnecessary. Schematic threads are drawn with crest lines farther apart to prevent crowding of lines. For both internal and external threads, a thread note completes the symbolic drawing by giving the necessary specifications.

11.13
Nuts and bolts

Nuts and bolts come in many forms and sizes for different applications (Fig. 11.30). Figure 11.31 shows the more common types of threaded fasteners. A **bolt** is a threaded cylinder with a head that is used with a nut for holding two parts together (Fig. 11.31A). A **stud** does not have a head but is screwed into one part with a nut attached to the other threaded end (Fig. 11.31B). A **cap screw** is similar to a bolt, but it

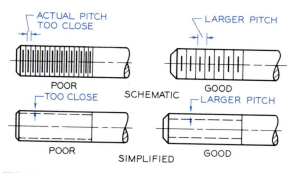

FIGURE 11.29 Simplified and schematic threads should be drawn using approximate dimensions if the actual dimensions would result in lines drawn too close together.

FIGURE 11.30 Examples of nuts and bolts. (Courtesy of Russell, Burdsall & Ward Bolt and Nut Co.)

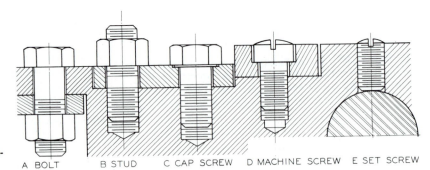

FIGURE 11.31 Types of threaded bolts and screws.

A BOLT   B STUD   C CAP SCREW   D MACHINE SCREW   E SET SCREW

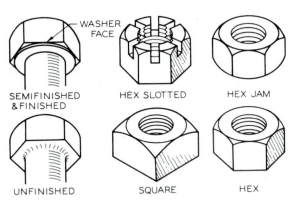

WASHER FACE

SEMIFINISHED & FINISHED   HEX SLOTTED   HEX JAM

UNFINISHED   SQUARE   HEX

FIGURE 11.32 Types of finishes for bolt heads and types of nuts.

does not have a nut; instead, it is screwed into a member with internal threads (Fig. 11.31C). A **machine screw** is similar to, but smaller than, a cap screw (Fig. 11.31D). A **set screw** is used to secure one member with respect to another, usually to prevent a rotational movement (Fig. 11.31E).

Figure 11.32 shows the types of heads used on standard bolts and nuts. These heads are used on both **regular** and **heavy** bolts; the thickness of the head is the primary difference between the two types. Heavy-series bolts have the thicker heads and are used at points where bearing loads are heaviest. Bolts and nuts are either **finished** or **unfinished.** Figure 11.32 shows an unfinished head; that is, none of the surfaces of the head are machined. The finished head has a washer face that is $\frac{1}{64}$ inch thick to provide a circular boss on the bearing surface of the bolt head or the nut.

Other standard forms of bolt and screw heads (Fig. 11.33) are used primarily on cap screws and machine screws. Finished nuts have washer faces for more accurate assembly. A hexagon **jam nut** does not have a washer face, but it is chamfered on both sides.

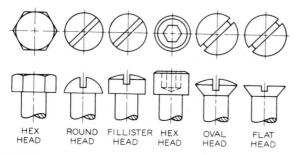

HEX   ROUND   FILLISTER   HEX   OVAL   FLAT
HEAD   HEAD   HEAD   HEAD   HEAD   HEAD

FIGURE 11.33 **Common types of bolt and screw heads.**

Although ANSI tables in Appendixes 19–24 indicate the standard bolt lengths and their corresponding thread lengths, the following can be used as a general guide for square- and hexagon-head bolts:

- Hexagon bolt lengths are available in $\frac{1}{4}$-inch increments up to 8 inches long, in $\frac{1}{2}$-inch increments from 8 to 20-inches long, and in 1-inch increments from 20 to 30 inches long.

- Square-head bolt lengths are available in $\frac{1}{8}$-inch increments from $\frac{1}{2}$ to $\frac{3}{4}$ inch long, in $\frac{1}{4}$-inch increments from $\frac{3}{4}$ to 5 inches long, in $\frac{1}{2}$-inch increments from 5 to 12 inches long, and in 1-inch increments from 12 to 30 inches long.

- The lengths of the threads on both hexagon-head and square-head bolts up to 6 inches long can be found by the formula: Thread length = $2D + \frac{1}{4}$ in., where D is the diameter of the bolt. The threaded length for bolts more than 6 inches long can be found by the formula: Thread length = $2D + \frac{1}{2}$ in.

- The threads for bolts can be coarse, fine, or 8-pitch threads. It is understood that the class of fit for bolts and nuts will be 2A and 2B if no class is specified.

- Standard square-head and hexagon-head bolts are designated by notes in one of the following

forms:

$\frac{3}{8}$–16 × 1$\frac{1}{2}$ SQUARE BOLT—STEEL;

$\frac{1}{2}$–13 × 3 HEX CAP SCREW—SAE GRADE 8—STEEL;

0.75 × 5.00 HEX LAG SCREW—STEEL.

The numbers represent bolt diameter, threads per inch (omit for lag screws), length, name of screw, and material. It is understood that these will have a class 2 fit.

- Nuts are designated by notes in one of the following forms:

$\frac{1}{2}$–13 SQUARE NUT—STEEL;

$\frac{3}{4}$–16 HEAVY HEX NUT, SAE GRADE 5—STEEL;

1.00–8 HEX THICK SLOTTED NUT—CORROSION RESISTANT STEEL.

When nuts are *not* noted as HEAVY, they are assumed to be REGULAR. The class of fit is assumed to be 2B for nuts when not noted.

11.14
Drawing square bolt heads

Detailed tables are available in Appendix 19 for square bolt heads and nuts. Usually, it is sufficient to draw nuts and bolts with only general proportions.

The first step in drawing a bolt head or nut is to determine whether it is to be **across corners** or **across flats**—that is, are the outlines at either side of the view going to represent corners or edge views of flat surfaces of the part? Nuts and bolts should be drawn across corners for the best representation (Fig. 11.34).

11.15
Drawing hexagon bolt heads

Figure 11.35 shows an example of constructing the head of a hexagon-head bolt across corners.

The major diameter of the bolt is D. The thickness of the head is drawn equal to $\frac{2}{3}D$. The top view of the head is drawn as a circle with a radius of $\frac{3}{4}D$. This

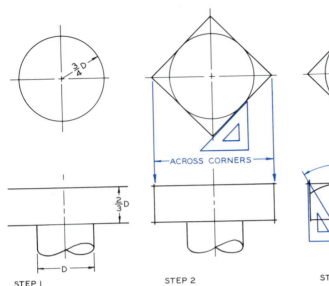

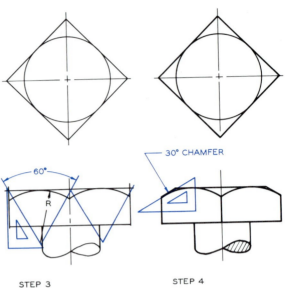

STEP 1

STEP 2

STEP 3

STEP 4

FIGURE 11.34 Drawing the square head.

Step 1 Draw the diameter of the bolt. Use major diameter (D) to establish the head diameter and thickness.

Step 2 Draw the top view of the square head with a 45° triangle to give an across-corners view.

Step 3 Show the chamfer in the front view by using a 30°–60° triangle to find the centers for the radii.

Step 4 Show a 30° chamfer tangent to the arcs in the front view. Strengthen the lines.

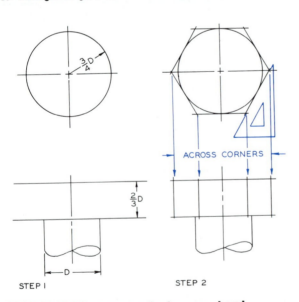

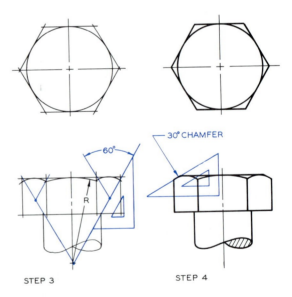

STEP 1

STEP 2

STEP 3

STEP 4

FIGURE 11.35 Drawing the hexagon head.

Step 1 Draw the major (D) diameter of the bolt and use it to establish the head diameter and thickness.

Step 2 Construct a hexagon with a 30°–60° triangle to give an across-corners view.

Step 3 Find arcs in the front view to show the chamfer of the head.

Step 4 Draw a 30° chamfer tangent to the arcs in the front view. Strengthen the lines.

proportionality based on D (the major diameter) is sufficient for drawing bolt heads in most applications.

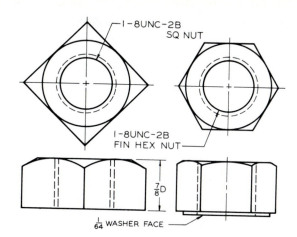

FIGURE 11.37 Drawings of hexagon and square nuts are constructed in the same manner as drawings of bolt heads. Notes are added to give nut specifications.

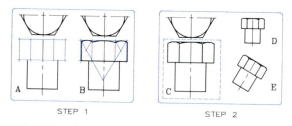

STEP 1   STEP 2

FIGURE 11.36 Bolt head by computer.

Step 1 A hexagon head is drawn using the steps of geometric construction covered in Fig. 11.35. The size of the head is based on a bolt diameter of 1 inch to form a UNIT BLOCK.

Step 2 The drawing is made into a BLOCK by using a window. The block can be INSERTed at any size and positioned as shown at D and E.

11.16
Drawing nuts

The drawing of a square and a hexagon nut across corners is the same as the drawing of a bolt head across corners. The only variation is the thickness of the nut. The REGULAR nut thickness is $\frac{7}{8}D$, and the HEAVY nut thickness is D (D = major diameter).

Figure 11.37 shows square and hexagon nuts drawn across corners. In the front view, hidden lines indicate threads. Since it is understood that nuts are threaded, these hidden lines may be omitted in general applications.

A $\frac{1}{64}$-inch washer face is shown on the hexagon nut. It is usually drawn thicker than $\frac{1}{64}$ inch to make the face more noticeable in the drawing. Thread notes

are placed in the circular views where possible. Since the note for square nut is not labeled HEAVY, it is a REGULAR square nut. The hexagon nut is similar except that it is a finished hexagon nut. The leader from the note is directed toward the center of the circular view, but the arrow stops at the first visible circle it makes contact with.

Nuts can be drawn across flats if doing so improves the drawing (Fig. 11.38). For regular nuts, the

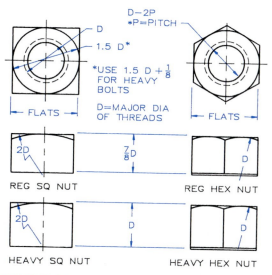

FIGURE 11.38 Examples of hexagon and square nuts drawn across flats. Notes are added to give nut specifications. Square nuts are always unfinished.

205

distance across flats is $1\frac{1}{2} \times D$ (D = major diameter of the thread); for heavy nuts, this distance is $1\frac{5}{8} \times D$. The top views are drawn in the same manner as in across-corners drawings except that they are positioned to give different front views.

A hexagon nut drawn across corners is a better representation than a nut drawn across flats (Fig. 11.37). To complete the representation of the nuts, notes are added with leaders. A washer face should be added to a nut if it is finished—except for the square nut, which is always unfinished.

11.17
Drawing nuts and bolts in combination

The rules followed when drawing nuts and bolts separately also apply when drawing nuts and bolts in assembly (Fig. 11.39). The major diameter, D, of the bolt is used as the basis for other dimensions. The note is added to give the specifications of the nut and bolt. The bolt heads are drawn across corners, and the nuts are drawn across flats. The half-end views show how the front views were found by projection.

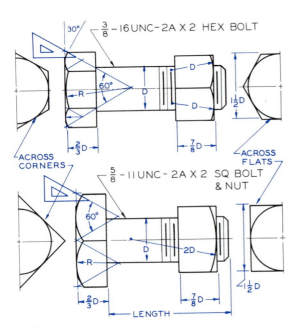

FIGURE 11.39 Construction of nuts and bolts in assembly.

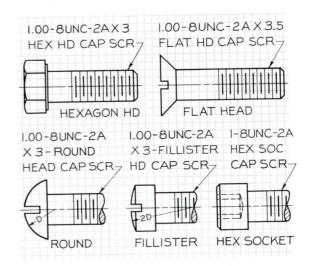

FIGURE 11.40 These proportions of the standard types of cap screws can be used for drawing cap screws of all sizes. Notes provide typical specifications.

11.18
Cap screws

Cap screws are used to hold two parts together without a nut since one of these parts has a threaded cylindrical hole. The other part is drilled with an oversized hole so that the cap screw will pass through it freely. When the cap screw is tightened, the two parts are held securely together.

Figure 11.40 shows the standard types of cap screws. Appendixes 20–22 give the dimensions of several types of cap screws. In Fig. 11.40, the cap screws are drawn on a grid to show their proportions, which can be used for drawing cap screws of all sizes, ranging in diameter from No. 0 (0.060) to No. 12.

11.19
Machine screws

Machine screws are smaller than most cap screws, usually less than 1 inch in diameter. The machine screw is screwed into either another part or a nut. Machine screws are fully threaded when their length is 2 inches or shorter.

Figure 11.41 shows a few of the common machine screws and their notes. Appendix 23 gives the

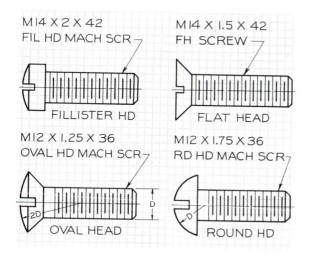

M14 X 2 X 42
FIL HD MACH SCR

FILLISTER HD

M14 X 1.5 X 42
FH SCREW

FLAT HEAD

M12 X 1.25 X 36
OVAL HD MACH SCR

OVAL HEAD

M12 X 1.75 X 36
RD HD MACH SCR

ROUND HD

FIGURE 11.41 **Standard types of machine screws. These proportions can be used for drawing machine screws of all sizes.**

dimensions of round-head machine screws. The four types of machine screws in Fig. 11.41 are drawn on a grid to give proportions of their heads in relation to the major diameter of the screw. Machine screws range in diameter from No. 0 (0.060 inch) to $\frac{3}{4}$ inch.

When slotted-head screws are drawn, conventional practice is to show the slots positioned at a 45° angle in the circular view (Fig. 11.33), and the front view shows the width and depth of the slot.

11.20
Set screws

Set screws or keys are used to secure parts like pulleys and handles on a shaft. Figure 11.42 shows various types of set screws, and Table 11.5 shows their dimensions. Set screws, like other fasteners, can be drawn as approximations.

Set screws are available in combinations of points and heads. The shaft against which the set screw is tightened may have a flat surface to give a good bearing surface for the set screw point; if so, a dog or flat point would be most effective to press against the flat surface. The cup point gives good friction when applied to a round shaft. Appendixes 26–28 give specifications for set screws.

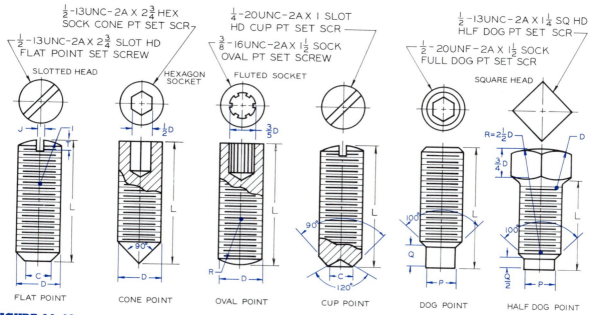

$\frac{1}{2}$-13UNC-2A X 2$\frac{3}{4}$ SLOT HD
FLAT POINT SET SCREW

$\frac{1}{2}$-13UNC-2A X 2$\frac{3}{4}$ HEX
SOCK CONE PT SET SCR

$\frac{3}{8}$-16UNC-2A X 1$\frac{1}{2}$ SOCK
OVAL PT SET SCREW

$\frac{1}{4}$-20UNC-2A X 1 SLOT
HD CUP PT SET SCR

$\frac{1}{2}$-20UNF-2A X 1$\frac{1}{2}$ SOCK
FULL DOG PT SET SCR

$\frac{1}{2}$-13UNC-2A X 1$\frac{1}{4}$ SQ HD
HLF DOG PT SET SCR

SLOTTED HEAD   HEXAGON SOCKET   FLUTED SOCKET   SQUARE HEAD

FLAT POINT   CONE POINT   OVAL POINT   CUP POINT   DOG POINT   HALF DOG POINT

FIGURE 11.42 **Types of set screws. Set screws are available with various combinations of heads and points. Notes give their specifications, and Table 11.5 gives dimensions.**

TABLE 11.5
DIMENSIONS FOR THE SET SCREWS SHOWN IN FIG. 11.42 (DIMENSIONS IN INCHES)

D	I	J	T	R	C		P		Q	q
					Diameter of Cup and Flat Points		Diameter of Dog Point		Length of Dog Point	
Nominal Size	Radius of Headless Crown	Width of Slot	Depth of Slot	Oval Point Radius	Max	Min	Max	Min	Full	Half
5 0.125	0.125	0.023	0.031	0.094	0.067	0.057	0.083	0.078	0.060	0.030
6 0.138	0.138	0.025	0.035	0.109	0.047	0.064	0.092	0.087	0.070	0.035
8 0.164	0.164	0.029	0.041	0.125	0.087	0.076	0.109	0.103	0.080	0.040
10 0.190	0.190	0.032	0.048	0.141	0.102	0.088	0.127	0.120	0.090	0.045
12 0.216	0.216	0.036	0.054	0.156	0.115	0.101	0.144	0.137	0.110	0.055
$\frac{1}{4}$ 0.250	0.250	0.045	0.063	0.188	0.132	0.118	0.156	0.149	0.125	0.063
$\frac{5}{16}$ 0.3125	0.313	0.051	0.076	0.234	0.172	0.156	0.203	0.195	0.156	0.078
$\frac{3}{8}$ 0.375	0.375	0.064	0.094	0.281	0.212	0.194	0.250	0.241	0.188	0.094
$\frac{7}{16}$ 0.4375	0.438	0.072	0.109	0.328	0.252	0.232	0.297	0.287	0.219	0.109
$\frac{1}{2}$ 0.500	0.500	0.081	0.125	0.375	0.291	0.270	0.344	0.344	0.250	0.125
$\frac{9}{16}$ 0.5625	0.563	0.091	0.141	0.422	0.332	0.309	0.391	0.379	0.281	0.140
$\frac{5}{8}$ 0.625	0.625	0.102	0.156	0.469	0.371	0.347	0.469	0.456	0.313	0.156
$\frac{3}{4}$ 0.750	0.750	0.129	0.188	0.563	0.450	0.425	0.563	0.549	0.375	0.188

Source: Courtesy of ANSI; B18.6.2.

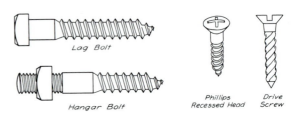

FIGURE 11.44 **Specialty bolts and screws.**

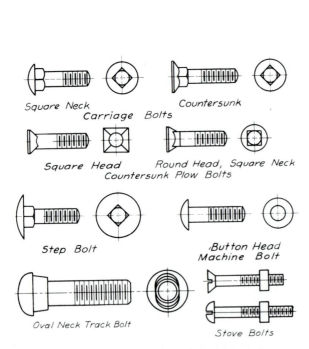

FIGURE 11.43 **Miscellaneous types of bolts.**

11.21
Miscellaneous screws

Figure 11.43 shows a few of the many types of specialty bolts, each of which has its own special application. Figure 11.44 shows further examples of specialty bolts and screws.

Figure 11.45 shows three types of wing screws. They are available in incremental lengths of $\frac{1}{8}$ inch and are used to join parts assembled and disassembled by hand. Figure 11.46 shows two types of thumb screws, which serve the same purpose as wing screws, and Fig. 11.47 shows three types of wing nuts that can be used with various types of screws.

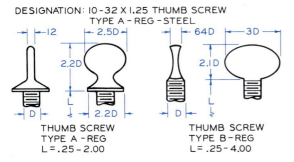

DESIGNATION: .38-16 X 2.00 WING SCREW–TYPE B–STYLE I
STEEL–CADMIUM PLATED

.9D

.5D   .5D

4.5D   4.5D   4.5D

2.6D   2.1D   2.1D

D   L   D   L   D   L

WING SCREW TYPE A
L = .25-4.00

TYPE B
STYLE I
L= .50-4.00

TYPE C
STYLE I
L=.25-1.50

FIGURE 11.45 Wing screw proportions for screw diameters of about $\frac{5}{16}''$. These proportions can be used to draw wing screws of any size diameter. Type A is available in screw diameters of 4, 6, 8, 10, 12, 0.25", 0.313", 0.375", 0.438", 0.50", and 0.625"; Type B in diameters of 10 to 0.625"; and Type C in diameters of 6 to 0.375".

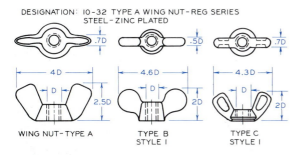

DESIGNATION: 10-32 X 1.25 THUMB SCREW
TYPE A – REG – STEEL

12   2.5D   64D   3D

2.2D   2.1D

D   L   2.2D   D   L

THUMB SCREW
TYPE A – REG
L = .25 - 2.00

THUMB SCREW
TYPE B–REG
L = .25 - 4.00

FIGURE 11.46 Thumb screw proportions for screw diameters of about $\frac{1}{4}''$. These proportions can be used to draw thumb screws of any screw diameter. Type A is available in diameters of 6, 8, 10, 12, 0.25", 0.313", and 0.375". Type B thumb screws are available in diameters of 6 to 0.50".

DESIGNATION: 10-32 TYPE A WING NUT–REG SERIES
STEEL – ZINC PLATED

.7D   .5D   .7D

4D   4.6D   4.3D

D   D   D

2.5D   2D   2D

WING NUT–TYPE A

TYPE B
STYLE I

TYPE C
STYLE I

FIGURE 11.47 Wing nut proportions for screw diameters of $\frac{3}{8}''$. These proportions can be used to draw thumb screws of any size. Type A wing nuts are available in screw diameters (in inches) of 3, 4, 5, 6, 8, 10, 12, 0.25", 0.313", 0.375", 0.438", 0.50", 0.583", 0.625", and 0.75"; Type B nuts are available in sizes from 5 to 0.75"; and Type C nuts in sizes from 4 to 0.50".

11.22
Wood screws

A wood screw is pointed and has a sharp thread of coarse pitch for insertion into wood. Figure 11.48 shows the three most common types of wood screws and their proportions in relation to their major diameters.

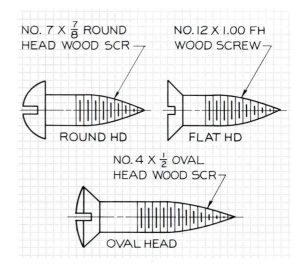

NO. 7 X $\frac{7}{8}$ ROUND
HEAD WOOD SCR

NO.12 X 1.00 FH
WOOD SCREW

ROUND HD   FLAT HD

NO. 4 X $\frac{1}{2}$ OVAL
HEAD WOOD SCR

OVAL HEAD

FIGURE 11.48 Standard types of wood screws. The proportions shown here can be used for drawing wood screws of all sizes.

Sizes of wood screws are specified by single numbers, such as 0, 6, or 16. From 0 to 10, each digit represents a different size (1, 2, 3, and so on). Beginning at 10, only even-numbered sizes are standard, that is, 10, 12, 14, 16, 18, 20, 22, and 24. The following formula relates these numbered sizes to the actual diameter of the screws:

Actual DIA = 0.060 + screw number × 0.013.

For example, the diameter of a No. 5 wood screw is

$$0.060 + 5(0.013) = 0.125.$$

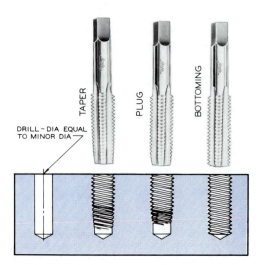

FIGURE 11.49   **Three types of taps for threading internal holes.**

11.23
Tapping a hole

A threaded hole is called a **tapped hole,** and the tool used to cut the threads is called a **tap.** Figure 11.49 shows the types of taps available for threading small holes by hand.

The **taper, plug,** and **bottoming** hand taps are identical in size, length, and measurement; only the chamfered portion of their ends is different. The taper tap has a long chamfer (8 to 10 threads), the plug tap has a shorter chamfer (3 to 5 threads), and the bottoming tap has the shortest chamfer (1 to 1½ threads).

When tapping by hand in open or "through" holes, the taper should be used for coarse threads since it ensures straighter starting. The taper tap is also recommended for the harder metals. The plug tap can be used in soft metals for fine-pitch threads. When a hole is tapped to the very bottom, all three taps—taper, plug, and bottoming—should be used in this order.

Notes can be added to specify the depth of the drilled hole and the depth of the threads. For example, a note reading 7/8 DIA–3 DEEP × 1–8 UNC–2A × 2 DEEP means the hole will be drilled deeper than it is threaded, and the last usable thread will be 2 inches deep in the hole. Note that the drill point has a 120° angle.

11.24
Washers, lock washers, and pins

Washers, called **plain washers,** are used with nuts and bolts to improve the assembly and strength of the fastening. Plain washers are noted on a working drawing in the following manner:

0.938 × 1.750 × 0.134 TYPE A PLAIN WASHER

These numbers represent the washer's inside diameter, outside diameter, and thickness, respectively (see Appendix 36).

A **lock washer** prevents a nut or cap screw from loosening as a result of vibration or movement. Common types of lock washers are the **external-tooth lock washer** and the **helical-spring lock washer** Figure 11.50 shows other types of locking washers and devices.

Appendix 36 gives tables for regular and extra–heavy-duty helical-spring lock washers, which are designated with a note in the following form:

HELICAL-SPRING LOCK WASHER-¼ REGULAR—PHOSPHOR BRONZE.

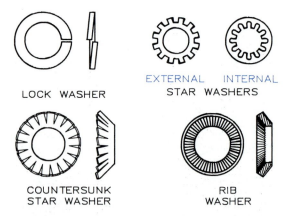

FIGURE 11.50   **Types of lock washers and locking devices.**

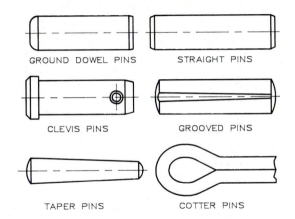

GROUND DOWEL PINS   STRAIGHT PINS

CLEVIS PINS   GROOVED PINS

TAPER PINS   COTTER PINS

FIGURE 11.51 Types of pins used to fix parts together.

(The $\frac{1}{4}$ is the washer's inside diameter). Tooth lock washers are designated with notes in one of the following forms:

INTERNAL-TOOTH LOCK WASHER-$\frac{1}{4}$-TYPE A— STEEL;

EXTERNAL-TOOTH LOCK WASHER-562–TYPE B— STEEL.

Straight pins and taper pins (Fig. 11.51) are used to fix parts together; Appendix 34 gives dimensions for them. Another locking device is the cotter pin; Appendix 37 gives tables of specifications for them.

11.25
Pipe threads

Pipe threads are used in connecting pipes, tubing, and lubrication fittings. The most commonly used pipe thread is tapered at a ratio of 1 to 16, but straight pipe threads are available. Tapered pipe threads will only engage for an effective length determined by the formula:

$$L = (0.80D + 6.8)P,$$

where D is the outside diameter of the pipe, and P is the pitch of the thread. Figure 11.52 shows methods of representing tapered threads.

The following abbreviations are associated with pipe threads:

N = National	G = Grease
P = Pipe	I = Internal

T = Taper		M = Mechanical
C = Coupling		L = Locknut
S = Straight		H = Hose coupling
F = Fuel and oil		R = Railing fittings

Combining these abbreviations gives the following ANSI symbols:

NPT = National pipe taper
NPTF = National pipe thread (dryseal—for pressure-tight joints)
NPS = Straight pipe thread
NPSC = Straight pipe thread in couplings
NPSI = National pipe straight internal thread
NPSF = Straight pipe thread (dryseal)
NPSM = Straight pipe thread for mechanical joints
NPSL = Straight pipe thread for locknuts and locknut pipe threads
NPSH = Straight pipe thread for hose couplings and nipples
NPTR = Taper pipe thread for railing fittings

To specify a pipe thread in note form, the nominal pipe diameter (the internal diameter), the number of threads per inch, and the symbol that denotes the type of thread are given; for example,

$$1\tfrac{1}{4}-11\tfrac{1}{2} \text{ NPT} \quad \text{or} \quad 3-8 \text{ NPTR}$$

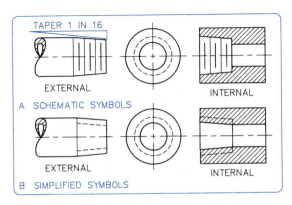

TAPER 1 IN 16

EXTERNAL   INTERNAL

A   SCHEMATIC SYMBOLS

EXTERNAL   INTERNAL

B   SIMPLIFIED SYMBOLS

FIGURE 11.52 **A.** Schematic techniques of representing pipe threads. **B.** Simplified techniques of representing pipe threads.

EXTERNAL NOTES

$\frac{3}{8}$ -18 DRYSEAL NPTF

$\frac{3}{4}$ -14 NPT

INTERNAL NOTES

$\frac{59}{64}$ DIA- $\frac{3}{4}$ -14 NPT

NOMINAL SIZE
THREADS PER INCH
FORM
SERIES

$\frac{1}{8}$ -27 DRYSEAL NPTF

FIGURE 11.53 Typical pipe-thread notes.

Appendix 9 gives these specifications. Figure 11.53 shows examples of external and internal thread notes. Dryseal threads, which may be straight or tapered, are used in applications where a pressure-tight joint is required without the use of a lubricant or sealer. No clearance between the mating parts of the joint is permitted, so the fit is of the highest quality. The tap drill is sometimes given in the internal pipe thread note, but this is optional.

11.26
Keys

Keys are used to attach parts to shafts in order to transmit power to pulleys, gears, or cranks. The four types of keys shown pictorially and orthographically

in Fig. 11.54 are the most commonly used. Notes must be given for the keyway, key, and keyseat, as shown in Fig. 11.54 (parts A, C, E, and G). These notes are typical of those used to give key specifications. Appendixes 31–33 give dimensions for various types of keys.

11.27
Rivets

Rivets are fasteners used to join thin overlapping materials in a permanent joint. The rivet is inserted in a hole slightly larger than the diameter of the rivet, and the headless end is formed into the specified shape by applying pressure to the projecting end. This forming

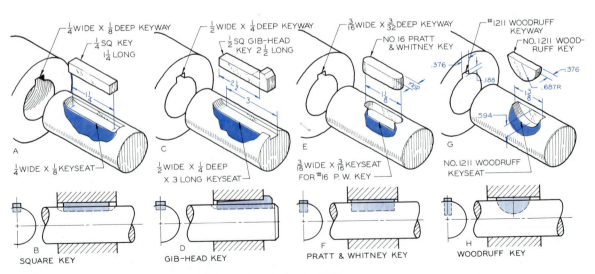

FIGURE 11.54 Standard keys used to hold parts on a shaft.

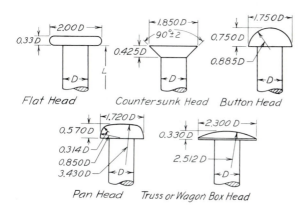

FIGURE 11.55 Types and proportions of small rivets. Small rivets have shank diameters up to ½".

operation is done when the rivets are either hot or cold, depending on the application.

Figure 11.55 shows typical shapes and proportions of small rivets. These rivets vary in diameter from $\frac{1}{16}$ to $1\frac{3}{4}$ inches. Rivets are used extensively in pressure-vessel fabrication, heavy structures such as bridges and buildings, and sheet-metal construction.

Figure 11.56 shows three types of lap joints. The joints are held secure by one, two, or three rivets, as shown in the sectional view. Note that the bodies of the rivets are drawn as hidden circles in the top views.

Figure 11.57 shows the standard symbols recommended by ANSI for representing rivets. Rivets that are driven in the shop are **shop rivets,** and those assembled on the job at the site are **field rivets.**

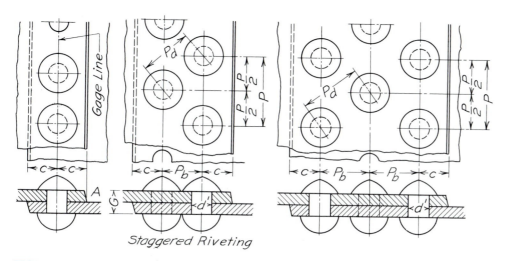

FIGURE 11.56 Examples of lap joints using single rivets, double rivets, and triple rivets.

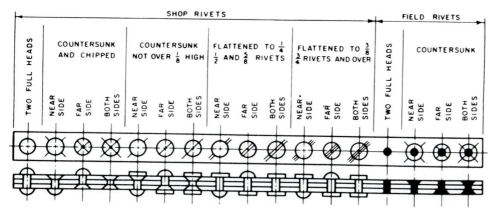

FIGURE 11.57 The symbols used to represent rivets in a drawing. (Courtesy of ANSI 14.14.)

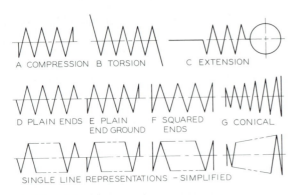

FIGURE 11.58 are: COMPRESSION B TORSION C EXTENSION / D PLAIN ENDS E PLAIN END GROUND F SQUARED ENDS G CONICAL / SINGLE LINE REPRESENTATIONS – SIMPLIFIED

FIGURE 11.58 Single-line representations of various types of springs.

11.28

Springs

Some of the more commonly used types of springs are **compression, torsion, extension, flat,** and **constant force.** Figure 11.58 shows the **single-line** conventional representation of the first three types. Also shown are the types of ends that can be used on compression springs and the simplified single-line representation of coil springs.

Figure 11.59 shows a typical working drawing of a compression spring. The ends of the spring are drawn by using the **double-line** representation, and

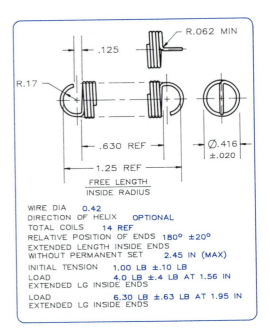

WIRE DIA   0.42
DIRECTION OF HELIX   OPTIONAL
TOTAL COILS   14 REF
RELATIVE POSITION OF ENDS 180° ±20°
EXTENDED LENGTH INSIDE ENDS
WITHOUT PERMANENT SET   2.45 IN (MAX)
INITIAL TENSION   1.00 LB ±.10 LB
LOAD   4.0 LB ±.4 LB AT 1.56 IN
EXTENDED LG INSIDE ENDS
LOAD   6.30 LB ±.63 LB AT 1.95 IN
EXTENDED LG INSIDE ENDS

FIGURE 11.60 A conventional double-line drawing of an extension spring and its specifications.

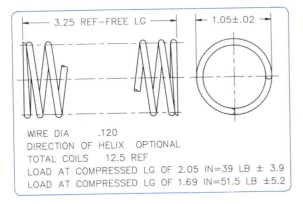

WIRE DIA   .120
DIRECTION OF HELIX   OPTIONAL
TOTAL COILS   12.5 REF
LOAD AT COMPRESSED LG OF 2.05 IN=39 LB ± 3.9
LOAD AT COMPRESSED LG OF 1.69 IN=51.5 LB ±5.2

FIGURE 11.59 A conventional double-line drawing of a compressions spring and its specifications.

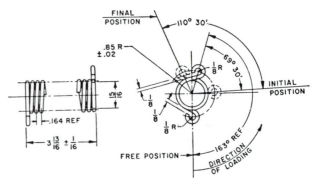

WIRE DIA   .148
DIRECTION OF HELIX   LEFT HAND
TOTAL COILS   20.55 REF
TORQUE   15 LB IN. ± 1.5 LB IN. AT INITIAL POSITION
TORQUE   33 LB IN. ± 3.3 LB IN. AT FINAL POSITION
MAXIMUM DEFLECTION WITHOUT SET BEYOND FINAL POSITION   56°
SPRING RATE   .16 LB IN. / DEG REF

FIGURE 11.61 A conventional double-line drawing of a helical torsion spring and its specifications. (Courtesy of the U.S. Dept. of Defense.)

conventional lines are used to indicate the undrawn portion of the spring. The diameter and free length of the spring are given on the drawing, and the remaining specifications are given a table near the drawing.

A working drawing of an extension spring (Fig. 11.60) is similar to that of a compression spring. In a drawing of a helical torsion spring (Fig. 11.61), angular dimensions must be shown to specify the initial and final positions of the spring as torsion is applied to it. All springs require a table of specifications to describe their details.

11.29
Drawing springs

Springs may be drawn as schematic representations using single lines (Fig. 11.62). Each example is drawn by laying out the diameter of the coils and lengths of the springs, and then the number of active coils are drawn by using the diagonal-line method. In Fig. 11.62B, the two end coils are "dead" coils, and only five are active. Figure 11.62C shows a drawing of an extension spring. Where more realism is desired, a double-line drawing of a thread can be made, as Fig. 11.63 shows.

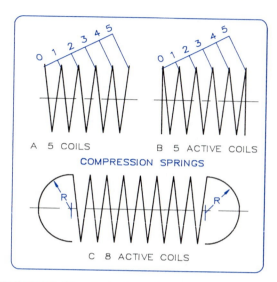

FIGURE 11.62 **A.** A schematic drawing of a spring with four active coils. The diagonal-line method is used to divide it into four equally spaced coils. **B.** A spring with 6 coils, but only four of them are active. **C.** An extension spring with five active coils.

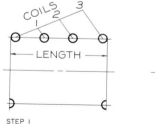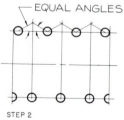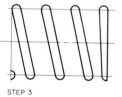

FIGURE 11.63   Detailed drawing of a spring.

Step 1 Lay out the diameter and length of the spring, and locate the coils by the diagonal-line technique.

Step 2 Locate the coils on the lower side along the bisectors of the spaces between the coils on the upper side.

Step 3 Connect the coils on each side. This is a right-hand coil; a left-hand spring would slope in the opposite direction.

Step 4 Construct the back side of the spring and the end coils to complete the detailed drawing of a compression spring.

Problems

These problems are to be drawn and solved on Size A sheets. Each grid equals 0.20 inch or 5 mm.

1. (Fig. 11.64) Draw detailed representations of Acme threads with major diameters of 2 inches. Show both external and internal threads as views and sections. Give a thread note using Appendix 18.

2. Repeat Problem 1, but draw internal and external detailed representations of square threads.

3. Repeat Problem 1, but draw internal and external detailed thread representations of Unified National threads. Give a thread note for a coarse thread with a class 2 fit.

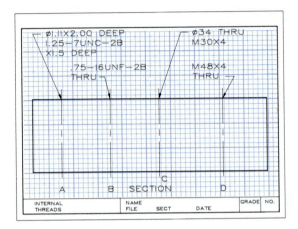

FIGURE 11.65 Problems 4–6. Internal threads.

7. Using the partial views in Fig. 11.66, draw external, internal, and end views of the full-size threaded parts using detailed thread symbols. Apply thread notes for UNC threads with a class 2 fit.

8. Repeat Problem 7, but use schematic thread symbols.

9. Repeat Problem 7, but use simplified symbols.

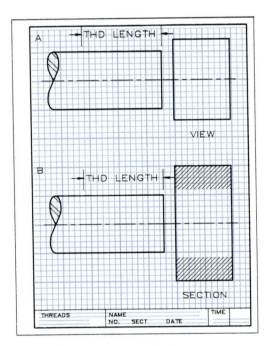

FIGURE 11.64 Problems 1–3. Construction of thread symbols.

4. Using the notes in Fig. 11.65, draw detailed representations of the internal threads and holes in section. Provide thread notes on each as specified.

5. Repeat Problem 4, but use schematic thread symbols.

6. Repeat Problem 4, but use simplified thread symbols.

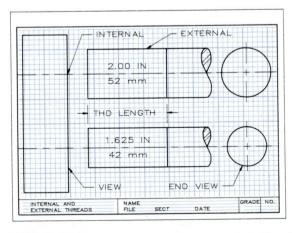

FIGURE 11.66 Problems 7–9. Internal and external threads.

10. (Fig. 11.67) Complete the drawing of the finished hexagon-head bolt and a heavy hexagon nut. The bolt head and nut are to be drawn across corners. Use detailed thread symbols. Provide thread notes in either English or metric forms as assigned.

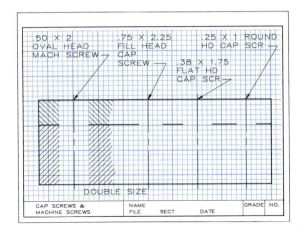

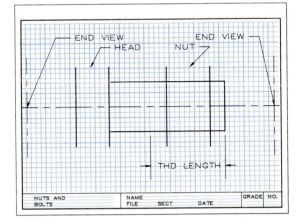

FIGURE 11.67 Problems 10–12. Nuts and bolts in assembly.

FIGURE 11.68 Problems 13–15. Cap screws and machine screws.

11. Repeat Problem 10, but draw the nut and bolt as having unfinished square heads. Use schematic thread symbols.

12. Repeat Problem 10, but draw the bolt with a regular finished hexagon head across flats, using simplified thread symbols. Draw the nut across flats also, and provide thread notes for both.

13. Use the notes in Fig. 11.68 to draw the screws in section and complete the sectional view showing all cross-hatching. Use detailed thread symbols, and apply thread notes to the parts.

14. Repeat Problem 13, but use schematic thread symbols.

15. Repeat Problem 13, but use simplified thread symbols.

16. (Fig. 11.69) The pencil pointer has a $\frac{1}{4}$ in. shaft that fits into a bracket designed to clamp onto a desk top. A set screw holds the shaft in position. Make a drawing of the bracket, estimating its dimensions. Show the details and the method of using the set screw to hold the shaft, and provide a thread note.

FIGURE 11.69 Problem 16. Design involving threaded parts.

17. (Fig. 11.70) On axes *A* and *B,* construct hexagon-head cap screws (across flats) with UNC threads, and a class 2 fit. The cap screws should not reach the bottoms of the threaded holes. Convert the view to a half section.

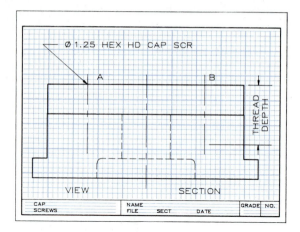

FIGURE 11.70 Problem 17.

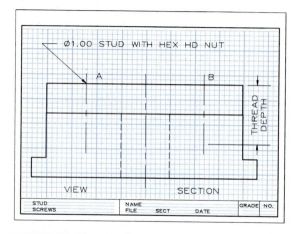

FIGURE 11.71 Problem 18.

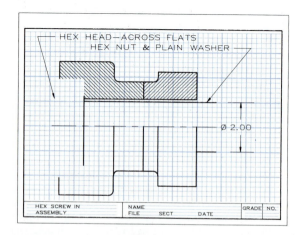

FIGURE 11.72 Problem 19.

18. (Fig. 11.71) On axes *A* and *B*, draw studs with a hexagon-head nut (across flats) that hold the two parts together. The studs are to be fine series with a class 2 fit, and they should not reach the bottom of the threaded hole. Provide a thread note. Show the view as a half section.

19. (Fig. 11.72) Draw a 2.00-inch (50-mm) diameter hexagon-head bolt with its head across flats using schematic symbols. Draw a plain washer and regular nut (across corners) at the right end. Design the size of the opening in the part at the left end to hold the bolt head so that it will not turn. Use a UNC thread with a series 2 fit. Give a thread note.

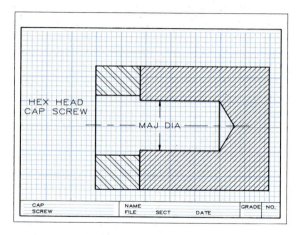

FIGURE 11.73 Problem 20.

20. (Fig. 11.73) Draw a 2.00-inch (50-mm) diameter hexagon-head cap screw that holds the two parts together. Determine the length of the bolt, and show the threads using schematic thread symbols. Give a thread note.

21. (Fig. 11.74) The part at A is held on the shaft by a square key, and the part at B is held on the shaft by gib-head key. Using Appendix 31, complete the drawings and give the necessary notes.

22. Repeat Problem 21, but use Woodruff keys, one with a flat bottom and the other with a round bottom. Using Appendix 32, complete the drawings and give the necessary notes.

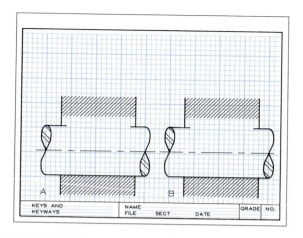

FIGURE 11.74 Problems 21–22.

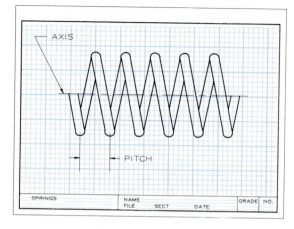

FIGURE 11.75 Problems 23–30.

23–26. (Fig. 11.75) Using Table 11.6, make a double-line drawing of the spring assigned.

27–30. Repeat Problem 23, but draw the springs using single-line representations.

TABLE 11.6
PROBLEMS 23–30

Problem	No. of Turns	Pitch	Size Wire	Inside Diameter	Outside Diameter	
23	4	1	No. 4 = 0.2253	3		RH
24	5	$\frac{3}{4}$	No. 6 = 0.1920		2	LH
25	6	$\frac{5}{8}$	No. 10 = 0.1350	2		RH
26	7	$\frac{3}{4}$	No. 7 = 0.1770		$1\frac{3}{4}$	LH

Gears and Cams

12

12.1
Introduction to gears

Gears are toothed wheels whose circumferences mesh together to transmit force and motion from one gear to the next. In this chapter, we discuss the three most common types—**spur gears, bevel gears,** and **worm gears** (Fig. 12.1).

A   B   C

FIGURE 12.1   **The three basic types of gears are (A) spur gears, (B) bevel gears, and (C) worm gears. (Courtesy of the Process Gear Co.)**

12.2
Spur gear terminology

The spur gear is a circular gear with teeth cut around its circumference. Two mating spur gears can transmit power from a shaft to another, parallel shaft.

When the two meshing gears are unequal in diameter, the smaller gear is called the **pinion,** and the larger one the **gear.**

The following terms describe the parts of a spur gear. Figures 12.2 and 12.3 show many of these features. The corresponding formulas for each feature are also given.

PITCH CIRCLE (PC) is the imaginary circle of a gear if it were a friction wheel without teeth that contacted the pitch circle of another friction wheel.

PITCH DIAMETER (PD) is the diameter of the pitch circle: $PD = N/DP$, where N is the number of teeth, and DP is the diametral pitch.

DIAMETRAL PITCH (DP) is the ratio between the number of teeth on a gear and its pitch diameter. For example, a gear with 20 teeth and a 4-inch pitch diameter will have a diametral pitch of 5, which means there are 5 teeth per inch of a diameter: $DP = N/PD$, where N is the number of teeth.

CIRCULAR PITCH (CP) is the circular measurement from one point on a tooth to the corresponding point on the next tooth measured along the pitch circle: $CP = 3.14/DP$.

CENTER DISTANCE (CD) is the distance from the center of a gear to its mating gear's center: $CD =$

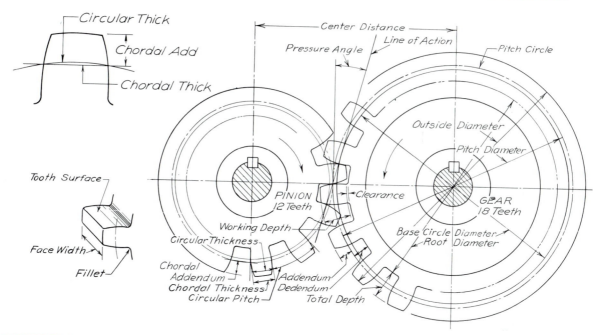

FIGURE 12.2 Gear terminology for spur gears.

$(N_P + N_S)/(2DP)$, where N_P and N_S are the number of teeth in the pinion and spur, respectively.

ADDENDUM (A) is the height of a gear above its pitch circle: $A = 1/DP$.

DEDENDUM (D) is the depth of a gear below the pitch circle: $D = 1.157/DP$.

WHOLE DEPTH (WD) is the total depth of a gear tooth: $WD = A + D$.

WORKING DEPTH (WKD) is the depth to which a tooth fits into a meshing gear: $WKD = 2/DP$, or $WKD = 2A$.

CIRCULAR THICKNESS (CRT) is the circular distance across a tooth measured along the pitch circle: $CRT = 1.57/DP$.

CHORDAL THICKNESS (CT) is the straight-line distance across a tooth at the pitch circle: $CT = PD (\sin 90°/N)$, where N is the number of teeth.

FACE WIDTH (FW) is the width across a gear tooth parallel to its axis. This is a variable dimension, but it is usually three to four times the circular pitch: $FW = 3$ to $4(CP)$.

OUTSIDE DIAMETER (OD) is the maximum diameter of a gear across its teeth: $OD = PD + 2A$.

ROOT DIAMETER (RD) is the diameter of a gear measured from the bottom of its gear teeth: $RD = PD - (2D)$.

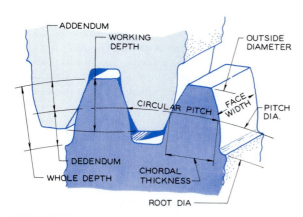

FIGURE 12.3 Gear terminology for spur gears.

PRESSURE ANGLE (PA) is the angle between the line of action and a line perpendicular to the centerline of two meshing gears. Angles of 14.5° and 20° are standard for involute gears.

BASE CIRCLE (BC) is the circle from which an involute tooth curve is generated or developed: $BC = PD (\cos PA)$.

12.3
Tooth forms

The most common gear tooth is an **involute tooth** with a 14.5° pressure angle. The 14.5° angle is the angle of contact between two gears when the tangents of both gears pass through the point of contact. Gears with pressure angles of 20° and 25° are also used. Gear teeth with larger pressure angles are wider at the base and thus stronger than the standard 14.5° teeth.

The standard gear face is an involute that keeps the meshing gears in contact as the gear teeth are revolved. Figure 12.4 shows the principle of constructing an involute.

An involute curve can be thought of as the path of a string that is kept taut as it is unwound from the base arc. It is unnecessary to draw gear teeth as involutes since most detail drawings show only approximations of teeth.

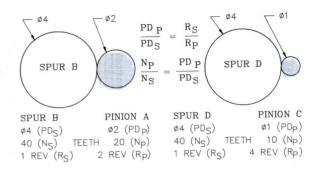

FIGURE 12.5 Ratios between meshing spur gears.

$$\frac{PD_P}{PD_S} = \frac{R_S}{R_P}$$

$$\frac{N_P}{N_S} = \frac{PD_P}{PD_S}$$

SPUR B — ø4 (PD_S), 40 (N_S), 1 REV (R_S)
PINION A — ø2 (PD_P), TEETH 20 (N_P), 2 REV (R_P)
SPUR D — ø4 (PD_S), 40 (N_S), 1 REV (R_S)
PINION C — ø1 (PD_P), TEETH 10 (N_P), 4 REV (R_P)

12.4
Gear ratios

The diameters of two meshing spur gears establish ratios that are important to the function of the gears. (Fig. 12.5).

If the radius of a gear is twice that of its pinion (the small gear), the diameter is twice that of the pinion, and the gear has twice as many teeth as the pinion. The pinion must make twice as many turns as the larger gear; that is, the revolutions per minute (RPM) of the pinion is twice that of the larger gear.

When the diameter of the gear is four times the diameter of the pinion, four times as many teeth must be on the gear as on the pinion, and the number of revolutions of the pinion will be four times that of the larger gear.

The relationship between two meshing spur gears can be developed in a formula by finding the velocity of a point on the small gear that is equal to $\pi PD \times$ RPM of the pinion. The velocity of a point on the large gear equals $\pi PD \times$ RPM of the spur. Since the velocity of points on each gear must be equal, the equation may be written as

$$\pi PD_P(RPM) = \pi PD_S(RPM);$$

therefore,

$$\frac{PD_P}{PD_S} = \frac{RPM_S}{RPM_P}.$$

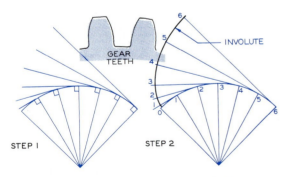

FIGURE 12.4 Construction of an involute.

Step 1 The base arc is divided into equal divisions with radial lines from the center. Tangents are drawn perpendicular to the radial lines on the arc.

Step 2 The chordal distance from 1 to 0 is used as the radius and 1 as the center to find point 1 on the involute curve. The distance from 2 to newly found 1 is revolved to the tangent line through 2 to locate a second point. The process is continued.

If the radius of the pinion is 1 inch, the radius of the spur 4 inches, and the RPM of the pinion 20, then the RPM of the spur can be found as follows:

$$\frac{2(1)}{2(4)} = \frac{RPM_S}{20\ RPM_P}$$

$$RPM_S = \frac{2(20)}{2(4)} = 5\ RPM.$$

The RPM of the spur is 5, or one fourth the RPM of the pinion.

The number of teeth on each gear is proportional to the diameters of a pair of meshing gears. This relationship can be written

$$\frac{N_P}{N_S} = \frac{PD_P}{PD_S},$$

where N_P and N_S are the number of teeth on the pinion and spur, respectively, and PD_P and PD_S are their pitch diameters.

12.5
Spur gear calculations

Before a working drawing of a gear can be started, the drafter must perform calculations to determine the gear's dimensions.

PROBLEM 1 Calculate the dimensions for a spur gear that has a pitch diameter of 5 inches, a diametral pitch of 4, and a pressure angle of 14.5°. (The diametral pitch is the same for meshing gears.)

SOLUTION

Number of teeth: $PD \times DP = 5 \times 4 = 20$.
Addendum: $\frac{1}{4} = 0.25''$.
Dedendum: $= 1.157/4 = 0.2893''$.
Circular thickness: $1.5708/4 = 0.3927''$.
Outside diameter: $(20 + 2)/4 = 5.50''$.
Root diameter: $5 - 2(0.2893) = 4.421''$.
Chordal thickness:
  $5(\sin 90°/20) = 5(0.079) = 0.392''$.
Chordal addendum:
  $0.25 + [0.3927^2/(4 \times 5)] = 0.2577''$.

Face width: $3.5(0.79) = 2.75$.
Circular pitch: $3.14/4 = 0.785''$.
Working depth: $0.6366 \times (3.14/4) = 0.4997''$.
Whole depth: $0.250 + 0.289 = 0.539''$.

These dimensions can be used to draw the spur gear and to provide specifications necessary for its manufacture.

Problem 2 shows the method of determining the design information for two meshing gears when their working ratios are known.

PROBLEM 2 Find the number of teeth and other specifications for a pair of meshing gears with a driving gear that turns at 100 RPM and a driven gear that turns at 60 RPM. The diametral pitch for each is 10, and the center-to-center distance between the gears is 6 inches.

SOLUTION

STEP 1 Find the sum of the teeth on both gears:

$$Total\ teeth = 2 \times (c\text{-to-c distance}) \times DP$$
$$= 2 \times 6 \times 10 = 120\ teeth.$$

STEP 2 Find the number of teeth for the driving gear:

$$\frac{Driver\ RPM}{Driven\ RPM} + 1 = \frac{100}{60} + 1 = 2.667;$$

$$\frac{Total\ teeth}{\frac{100}{60} + 1} = \frac{120}{2.667} = 45\ teeth.^*$$

STEP 3 Find the number of teeth for the driven gear:

$$Total\ teeth - teeth\ on\ driver = teeth\ on\ driven\ gear$$
$$120 - 45 = 75\ teeth.$$

STEP 4 Other specifications for the gears can be calculated as shown in Problem 1 by using the formulas in Section 12.2.

---

*The number of teeth must be a whole number since there cannot be fractional teeth on a gear. It may be necessary to adjust the center distance to yield a whole number of teeth.

12.6
Drawing spur gears

Figure 12.6 shows a conventional drawing of a spur gear. Since it is so time consuming, the teeth need not be drawn. It is possible to omit the circular view and to show only a sectional view of the gear with a table

of dimensions called **cutting data.** Circular centerlines are drawn to represent the root circle, pitch circle, and outside circle of the gear in the circular view.

A table of dimensions is a necessary part of a gear drawing, as Fig. 12.7 shows. These data can be calculated by formula or taken from tables of standards in gear handbooks such as *Machinery's Handbook.*

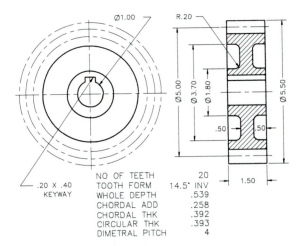

NO OF TEETH	20
TOOTH FORM	14.5° INV
WHOLE DEPTH	.539
CHORDAL ADD	.258
CHORDAL THK	.392
CIRCULAR THK	.393
DIMETRAL PITCH	4

FIGURE 12.6 **A detail drawing of a spur gear with a table of values to supplement the dimensions shown on the drawing.**

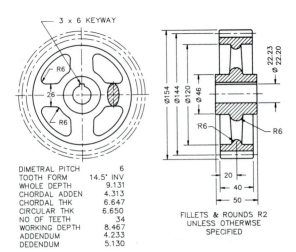

DIMETRAL PITCH	6
TOOTH FORM	14.5° INV
WHOLE DEPTH	9.131
CHORDAL ADDEN	4.313
CHORDAL THK	6.647
CIRCULAR THK	6.650
NO OF TEETH	34
WORKING DEPTH	8.467
ADDENDUM	4.233
DEDENDUM	5.130

FILLETS & ROUNDS R2
UNLESS OTHERWISE
SPECIFIED

FIGURE 12.7 **A computer-drawn detail drawing of a spur gear.**

12.7
Bevel gear terminology

Bevel gears are gears whose axes intersect at angles. Although the angle of intersection is usually 90°, other angles are also used. The smaller of the two bevel gears is the **pinion,** as in spur gearing.

Figure 12.8 illustrates the terminology of bevel gearing. The corresponding formulas for each feature are given below. Gear handbooks can also be used for finding these dimensions.

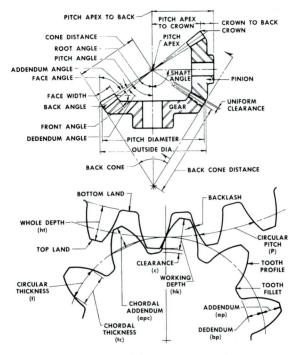

FIGURE 12.8 **The terminology and definitions of bevel gears. (Courtesy of Philadelphia Gear Corp.)**

PITCH ANGLE OF PINION (SMALL GEAR) (PA$_p$):

$$\tan PA_p = \frac{N_p}{N_g},$$

where N_g and N_p are the number of teeth on the gear and pinion, respectively.

PITCH ANGLE OF GEAR (PA$_g$):

$$\tan PA_g = \frac{N_g}{N_p}.$$

PITCH DIAMETER (PD) is the number of teeth (*N*) divided by the diametral pitch *(DP)*: $PD = N/P$.

ADDENDUM (A) is measured at the large end of the tooth: $A = 1/DP$.

DEDENDUM (D) is measured at the large end of the tooth: $D = 1.157/DP$.

WHOLE TOOTH DEPTH (WD): $WD = 2.157/DP$.

THICKNESS OF TOOTH (TT) is measured at the pitch circle: $TT = 1.571/DP$.

DIAMETRAL PITCH: $DP = N/PD$, where N is the number of teeth.

ADDENDUM ANGLE (AA) is the angle formed by the addendum and pitch cone distance:

$$\tan AA = \frac{A}{PCD}.$$

ANGULAR ADDENDUM: $AK = \cos PA \times A$.

PITCH CONE DISTANCE (PCD): $PD/(2 \times \sin PA)$.

DEDENDUM ANGLE (DA) is the angle formed by the dedendum and the pitch cone distance:

$$\tan DA = \frac{D}{PCD}.$$

FACE ANGLE (FA) is the angle between the gear's centerline and the top of its teeth: $FA = 90° - (PCD + AA)$.

CUTTING ANGLE (OR ROOT ANGLE) (CA) is the angle between the gear's axis and the roots of the teeth: $CA = PCD - D$.

OUTSIDE DIAMETER (OD) is the greatest diameter of a gear across its teeth: $OD = PD + 2A$.

APEX TO CROWN DISTANCE (AC) is the distance from the crown of the gear to the apex of the cone measured parallel to the axis of the gear: $AC = OD/(2 \tan FA)$.

CHORDAL ADDENDUM (CA):

$$A + \frac{TT^2 \cos PA}{4(PD)}.$$

CHORDAL THICKNESS (CT) is measured at the large end of the tooth:

$$CT = PD \times \sin\frac{90°}{N}.$$

FACE WIDTH (FW) can vary, but it is recommended that it be approximately equal to the pitch cone distance divided by 3: $FW = PCD/3$.

12.8
Bevel gear calculations

Problem 3 demonstrates how the formulas in Section 12.7 are used. Some of the formulas result in the same specifications that apply to both the gear and pinion.

PROBLEM 3 Two bevel gears intersect at right angles. They have a diametral pitch of 3, 60 teeth on the gear, 45 teeth on the pinion, and a face width of 4 inches. Find the dimensions of the gear.

SOLUTION

Pitch cone angle of gear:

$$\tan PCA = 60/45 = 1.33;$$
$$PCA = 53°7'.$$

Pitch cone angle of pinion: $\tan PCA = 45/60$;
$$PCA = 36°52'.$$

Pitch diameter of gear: 60/3 = 20.00".

Pitch diameter of pinion: 45/3 = 15.00".

The following formulas are the same for both the gear and the pinion:

Addendum: $\frac{1}{3}$ = 0.333".

Dedendum: 1.157/3 = 0.3857".

Whole depth: 2.157/3 = 0.719".

Tooth thickness on pitch circle: 1.571/3 = 0.5237".

Pitch cone distance: 20/(2 sin 53°7') = 12.5015".

Addendum angle: tan AA = 0.333/12.5015 = 1°32'.

Dedendum angle:

$$DA = 0.3857/12.5015 = 0.0308 = 1°46'.$$

Face width: PCD/3 = 4.00".

The following formulas must be applied separately to the gear and pinion:

Chordal addendum of gear:

$$0.333" + \frac{0.5237^2 \times \cos 53°7'}{4 \times 20} = 0.336".$$

Chordal addendum of pinion:

$$0.333" + \frac{0.5237^2 \times \cos 36°52'}{4 \times 15} = 0.338".$$

Chordal thickness of gear:

$$\sin \frac{90°}{60} \times 20" = 0.524".$$

Chordal thickness of pinion:

$$\sin \frac{90°}{45} \times 15" = 0.523".$$

Face angle of gear:

$$90° - (53°7' + 1°32') = 35°21'.$$

Face angle of pinion:

$$90° - (36°52' + 1°32') = 51°36'.$$

Cutting angle of gear:

$$53°7' - 1°46' = 51°21'.$$

Cutting angle of pinion:

$$36°52' - 1°46' = 35°6'.$$

Angular addendum of gear:

$$0.333" \times \cos 53°7' = 0.1999".$$

Angular addendum of pinion:

$$0.333" \times \cos 36°52' = 0.2667".$$

Outside diameter of gear:

$$20" + 2(0.1999") = 20.4000".$$

Outside diameter of pinion:

$$15" + 2(0.2667") = 15.533".$$

Apex-to-crown distance of gear:

$$\frac{20.400"}{2} \times \tan 35°7' = 7.173".$$

Apex-to-crown distance of pinion:

$$\frac{15.533"}{2} \times \tan 51°36' = 9.800".$$

12.9
Drawing bevel gears

The dimensions calculated in Section 12.8 are used to lay out the bevel gears in a detail drawing. Many of the calculated dimensions would be difficult to measure on a drawing within a high degree of accuracy;

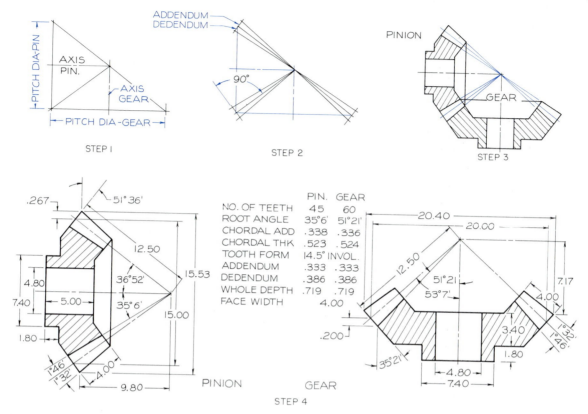

FIGURE 12.9 Construction of bevel gears.

Step 1 Lay out the pitch diameters and axes of the two bevel gears.

Step 2 Draw construction lines to establish the limits of the teeth by using the addendum and dedendum dimensions.

Step 3 Draw the pinion and gear using the specified dimensions or those calculated by formula.

Step 4 Complete the detail drawings of both gears, and provide a table of cutting data.

therefore, it is important to provide a table of cutting data for each gear.

Figure 12.9 shows the steps of drawing the bevel gears. The finished drawings are shown with a combination of dimensions and a table of dimensions.

12.10

Worm gears

A worm gear is composed of a threaded shaft called a **worm** and a circular gear called a **spider** (Fig. 12.10). The worm is revolved, which causes the spider

to revolve about its axis. Figures 12.10 and 12.11 illustrate the following terminology.

Worm specifications and formulas

LINEAR PITCH (P) is the distance from one thread to the next measured parallel to the worm's axis: $P = L/N$, where N is the number of threads: 1 if a single thread, 2 if a double thread, and so on.

LEAD (L) is the distance a thread advances in a turn of 360°.

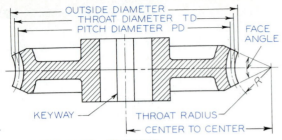

KEYWAY

WORM WHEEL OR SPIDER

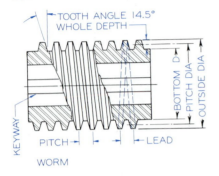

WORM

FIGURE 12.10 The terminology and definitions of worm gears.

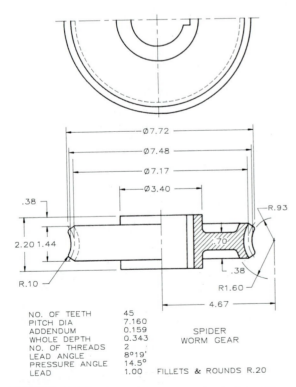

NO. OF TEETH	45
PITCH DIA	7.160
ADDENDUM	0.159
WHOLE DEPTH	0.343
NO. OF THREADS	2
LEAD ANGLE	8°19'
PRESSURE ANGLE	14.5°
LEAD	1.00

SPIDER
WORM GEAR

FILLETS & ROUNDS R.20

FIGURE 12.11 A detail drawing of a worm gear (spider) and the table of cutting data.

228

ADDENDUM OF TOOTH: $AW = 0.3183P.$

PITCH DIAMETER:

$$PDW = OD - 2AW,$$

where OD is the outside diameter.

WHOLE DEPTH OF TOOTH:

$$WDT = 0.6866 \times P.$$

BOTTOM DIAMETER OF WORM:

$$BD = OD - 2WDT.$$

WIDTH OF THREAD AT ROOT: $WT = 0.31P.$

MINIMUM LENGTH OF WORM:

$$MLW = \sqrt{8PDS \times AW},$$

where PDS is the pitch diameter of the spider.

HELIX ANGLE OF WORM:

$$\cot \beta = \frac{3.14(PDW)}{L}.$$

OUTSIDE DIAMETER: $OD = PD + 2A.$

Spider specifications and formulas

PITCH DIAMETER OF SPIDER:

$$PDS = \frac{N(P)}{3.14},$$

where N is the number of teeth on the spider.

THROAT DIAMETER OF SPIDER:

$$TD = PDS + 2A.$$

RADIUS OF SPIDER THROAT:

$$RST = \frac{OD \text{ of worm}}{2} - 2A.$$

FACE ANGLE (FA) may be selected to be between 60° and 80° for the average application.

CENTER-TO-CENTER DISTANCE (CD) is measured between the worm and spider:

$$CD = \frac{PDW + PDS}{2}.$$

OUTSIDE DIAMETER OF SPIDER:

$$ODS = TD + 0.4775P.$$

FACE WIDTH OF GEAR: $FW = 2.38(P) + 0.25.$

12.11
Worm gear calculations

Problem 4 has been solved for a worm gear by using the formulas in Section 12.10.

PROBLEM 4 Calculate the specifications for a worm and worm gear (spider). The gear has 45 teeth, and the worm has an outside diameter of 2.50 inches. The worm has a double thread and a pitch of 0.5 inch.

SOLUTION

Lead: $L = 0.5'' \times 2 = 1''$.
Worm addendum: $AW = 0.3183P = 0.1592''$.
Pitch diameter of worm:

$$PDW = 2.50'' - 2(0.1592'') = 2.1818''.$$

Pitch diameter of gear:

$$PDS = (45'' \times 0.5)/3.14 = 7.166''.$$

Center distance between worm and gear:

$$CD = \frac{(2.182'' + 7.166'')}{2} = 4.674''.$$

Whole depth of worm tooth:

$$WDT = 0.687 \times 0.5'' = 0.3433''.$$

Bottom diameter of worm:

$$BD = 2.50'' - 2(0.3433'') = 1.813''.$$

Helix angle of worm:

$$\cot \beta = \frac{3.14(2.1816)}{1} = 8°19'.$$

Width of thread at root: $WT = 0.31(1) = 0.155''$.
Minimum length of worm:

$$MLW = \sqrt{8(0.1592)\,(7.1656)} = 3.02''.$$

Throat diameter of gear:

$$TD = 7.1656'' + 2(0.1592'') = 7.484''.$$

Radius of gear throat:

$$RST = (2.5/2) - (2 \times 0.1592) = 0.9318''.$$

Face width:

$$FW = 2.38\,(0.5) + 0.25 = 1.44''.$$

Outside diameter of gear:

$$ODS = 7.484 + 0.4775\,(0.5) = 7.723''.$$

12.12
Drawing worm gears

The worm and worm wheel (spider) are drawn and dimensioned as shown in Figs. 12.11 and 12.12. The specifications derived by the formulas in Section 12.11

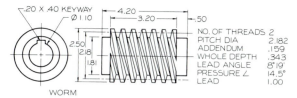

FIGURE 12.12 **A detail drawing of a worm using the dimensions calculated.**

FIGURE 12.13 Examples of machined cams. (Courtesy of Ferguson Machine Co.)

must be used for scaling, laying out the drawings, and providing cutting data.

12.13
Introduction to cams

Plate cams are irregularly shaped machine elements that produce motion in a single plane, usually up and down (Fig. 12.13). As the cam revolves about its center, the variation in the cam's shape produces a rise or fall in the follower that is in contact with it. The shape of the cam is determined graphically before the preparation of manufacturing specifications.

Cams utilize the principle of the inclined wedge, with the surface of the cam causing a change in the slope of the plane, thereby producing the desired motion.

12.14
Cam motion

Cams are designed primarily to produce (1) **uniform** or linear motion (2) **harmonic motion,** (3) **gravity motion,** (uniform acceleration), or (4) **combinations** of these.

Uniform motion

Uniform motion is shown in the displacement diagram in Fig. 12.14A to represent the motion of the cam follower as the cam rotates through 360°. The uniform-

motion curve has sharp corners, which indicates abrupt changes of velocity and causes the follower to bounce. Therefore, uniform motion is usually modified with arcs that smooth this change of velocity. The radius of the modifying arc is varied up to a radius of one half the total displacement, depending on the speed of operation.

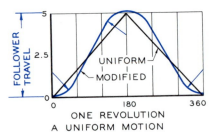

A UNIFORM MOTION

Uniform-motion diagrams are modified with arcs of one fourth to one third total displacement to smooth out the velocity of the follower at these points of abrupt change.

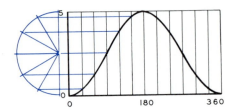

B HARMONIC MOTION

Harmonic motion is plotted by projecting from a semicircle whose diameter is equal to the rise of the follower. The semicircle must be divided into the same number of sectors as the divisions on the *x* axis of the graph to the point of maximum rise.

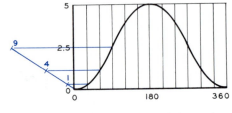

C GRAVITY MOTION

Gravity-motion diagrams are constructed so that the rise of the follower is relative to the square of the units on the *x* axis: 1^2, 2^2, 3^2, and so on to give a uniform acceleration.

FIGURE 12.14 Displacement diagrams.

Harmonic motion

Harmonic motion, plotted in Fig. 12.14B, is a smooth, continuous motion based on the change of position of points on the circumference of a circle. At moderate speeds, this displacement gives a smooth operation.

Gravity motion

Gravity motion (uniform acceleration) is used for high-speed operation (Fig. 12.14C). The variation of displacement is analogous to the force of gravity, with the difference in displacement being 1, 3, 5, 5, 3, 1 based on the square of the number; for instance, $1^2 = 1$; $2^2 = 4$; $3^2 = 9$ to give a uniform acceleration. This motion is repeated in reverse order for the remaining half of the motion of the follower. Intermediate points can be found by squaring fractional increments such as $(2.5)^2$.

Cam followers

Three basic types of cam followers are the **flat surface, roller,** and **knife edge,** (Fig. 12.15). The flat-surface and knife-edge followers are limited to use with slow-moving cams where minor force will be exerted during rotation. The roller follower is used to withstand higher speeds.

12.15
Construction of a plate cam

Plate cam—harmonic motion

Figure 12.16 shows the steps of constructing a plate cam with harmonic motion. Before designing a cam, the drafter must know the motion of the follower, rise of the follower, diameter of the base circle, and direction of rotation. The specifications for the cam pictured in Fig. 12.16 are given graphically in the displacement diagram.

Plate cam—gravity motion

The steps of constructing a cam with gravity motion are the same as in the previous example except for a different displacement diagram and a knife-edge follower. Figure 12.17 shows the graphic layout of the problem.

Plate cam—combination

In Fig. 12.18, a knife-edge follower is used with a plate cam to produce a 4-inch rise with harmonic motion from 0° to 180°, a 4-inch fall with gravity motion (uniform acceleration) from 180° to 300°, and dwell (no follower motion) from 300° to 360°. The drafter must draw the cam that will give this motion from the base circle.

The displacement diagram is drawn with a harmonic curve with a full rise of 4 inches. The curve is then drawn with a uniform acceleration drop of 4 inches. The dwell is a horizontal line to complete the diagram of the 360° rotation of the cam.

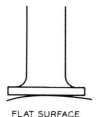

FLAT SURFACE

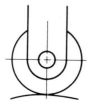

ROLLER

KNIFE-EDGE

FIGURE 12.15 **Three basic types of cam followers are the flat surface, roller, and knife edge.**

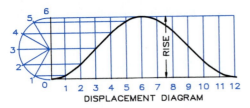

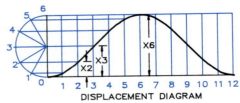

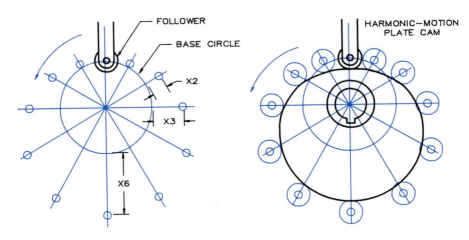

FIGURE 12.16 Construction of a plate cam with harmonic motion.

Step 1 Construct a semicircle whose diameter equals the rise of the follower. Divide the semicircle into the same number of divisions as there are between 0° and 180° on the horizontal axis of the displacement diagram. Plot the displacement curve.

Step 2 Distances of rise and fall (X1, X2, X3, X6) at each interval will be measured from the base circle.

Step 3 Construct the base circle, and draw the follower. Divide the circle into the same number of sectors as there are divisions on the displacement diagram. Transfer distances from the displacement diagram to the respective radial lines of the circle, measuring outward from it.

Step 4 Draw circles to represent the positions of the roller as the cam revolves in a counterclockwise direction. Draw the cam profile tangent to all the rollers to complete the drawing.

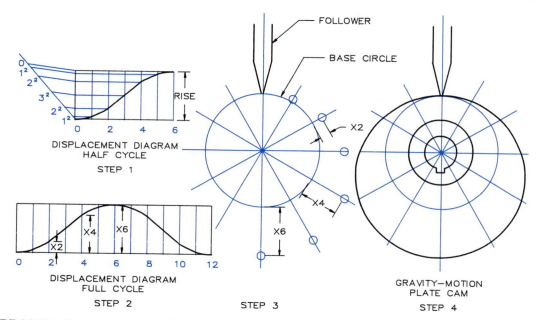

FIGURE 12.17 Construction of a plate cam with uniform acceleration.

Step 1 Construct a displacement diagram to represent the rise of the follower. Divide the horizontal axis into angular increments of 30°. Draw a construction line through point 0; locate the 1^2, 2^2, and 3^2 divisions, and project them to the vertical axis to represent half the rise.

Step 2 Use the same construction to find the right half of the symmetrical curve.

Step 3 Construct the base circle, and draw the knife-edge follower. Divide the circle into the same number of sectors as there are divisions in the displacement diagram. Transfer distances from the displacement diagram to their respective radial lines of the base circle, measuring outward from the base circle.

Step 4 Connect the points found in Step 3 with a smooth curve to complete the cam profile. Show also the cam hub and keyway.

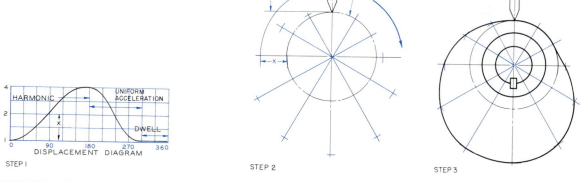

FIGURE 12.18 Construction of a plate cam with combination motions.

Step 1 The cam is to rise 4″ in 180° with harmonic motion, fall 4″ in 120° with uniform acceleration, and dwell for 60°. These motions are plotted on the displacement diagram.

Step 2 Construct the base circle and knife-edge follower. Transfer distances from the displacement diagram to the respective radial lines of the base circle, measuring outward from it.

Step 3 Draw a smooth curve through the points found in Step 2 to complete the profile of the cam.

233

12.16
Construction of a cam with an offset follower

The cam in Fig. 12.19 is designed to produce harmonic motion through 360°. This motion is plotted directly from the follower rather than from a displacement diagram.

A semicircle is drawn with its diameter equal to the total motion of the follower. The base circle is drawn to pass through the center or roller of the follower. The centerline of the follower is extended downward, and a circle is drawn tangent to the extension with its center at the center of the base circle. The small circle is divided into 30° intervals to establish points through which construction lines will be drawn tangent to the circle.

The distances from tangent points to the position points along the path of the follower are laid out along the tangent lines drawn at 30° intervals. These points can be located by measuring from the base circle, as shown in the figure, where point 3 was located distance X from the base circle. The circular roller is drawn in all views, and the profile of the cam is drawn to be tangent to the rollers at all positions.

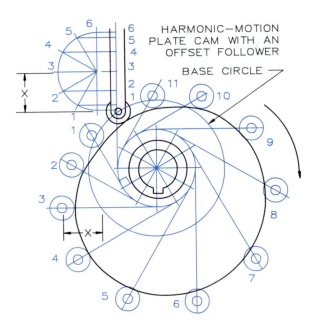

FIGURE 12.19 **Construction of a plate cam with an offset roller follower.**

Problems

Gears

Use Size A sheets ($8\frac{1}{2} \times 11$ inches) for the following gear problems. Select the most appropriate scale so that the drawings will use the available space.

1–5. Calculate the dimensions for the following spur gears, and make a detail drawing of each. Give the dimensions and cutting data for each gear. Provide any other dimensions needed.

Problem	Gear Teeth	Diametral Pitch	14.5° Involute
1	20	5	"
2	30	3	"
3	40	4	"
4	60	6	"
5	80	4	"

6–10. Calculate the gear sizes and number of teeth using the ratios and data below.

Problem	RPM Pinion	RPM Gear	Center to Center	Diametral Pitch
6	100 (driver)	60	6.0"	10
7	100 (driver)	50	8.0"	9
8	100 (driver)	40	10.0"	8
9	100 (driver)	35	12.0"	7
10	100 (driver)	25	14.0"	6

11–20. Make detail drawings of each of the gears for which calculations were made in Problems 6–10. Provide a table of cutting data and other dimensions needed to complete the specifications.

21–25. Calculate the specifications for the bevel gears that intersect at 90°, and make detail drawings of each with the necessary dimensions and cutting data.

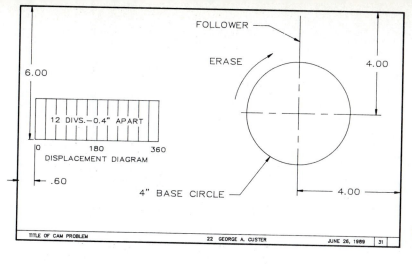

FIGURE 12.20 Layout for Problems 31–41 on Size B sheets.

Problem	Diametral Pitch	No. of Teeth on Pinion	No. of Teeth on Gear
21	3	60	15
22	4	100	40
23	5	100	60
24	6	100	50
25	7	100	30

26–30. Calculate the specifications for worm gears, and make detail drawings of each, with the necessary dimensions and cutting data.

Problem	No. of Teeth in Spider Gear	Outside DIA of Worm	Pitch of Worm	Thread of Worm
26	45	2.50	0.50	double
27	30	2.00	0.80	single
28	60	3.00	0.80	double
29	30	2.00	0.25	double
30	80	4.00	1.00	single

Cams

Use Size B sheets (11 × 17 inches) for the following cam problems. The standard dimensions are base circle, 3.50 in.; roller follower, 0.60-in. diameter; shaft, 0.75-in. diameter; hub, 1.25-in. diameter; direction of rotation, clockwise. The follower is positioned vertically over the center of the base circle except in Problems 40 and 41. Lay out the problems and displacement diagrams as shown in Fig. 12.20.

31. Draw a plate cam with a knife-edge follower for uniform motion and a rise of 1.00 in.

32. Draw a displacement diagram and a cam that will give a modified uniform motion to a knife-edge follower with a rise of 1.7 in. Modify the uniform motion with an arc of one quarter the rise in the displacement diagram.

33. Draw a displacement diagram and a cam that will give a harmonic motion to a roller follower with a rise of 1.60 in.

34. Draw a displacement diagram and a cam that will give a harmonic motion to a knife-edge follower with a rise of 1.00 in.

35. Draw a displacement diagram and a cam that will give uniform acceleration to a knife-edge follower with a rise of 1.70 in.

36. Draw a displacement diagram and a cam that will give a uniform acceleration to a roller follower with a rise of 1.40 in.

37. Draw a displacement diagram and a cam that will give the following motion to a knife-edge follower: rise of 1.25 in. with harmonic motion in 120°; and fall of 1.25 in. with uniform acceleration.

38. Draw a displacement diagram and a cam that will give the following motion to a knife-edge follower: dwell for 70°; rise of 1 in. with a modified uniform motion in 100°; fall of 1 in. with a harmonic motion in 100°; and dwell for 90°.

39. Draw a displacement diagram and a cam that will give the following motion to a roller follower: rise of 1.25 in. with a harmonic motion in 120°; dwell for 120°; and fall of 1.25 in. with a uniform acceleration in 120°.

40. Repeat Problem 32, but offset the follower 0.60 in. to the right of the vertical centerline.

41. Repeat Problem 33, but offset the follower 0.60 in. to the left of the vertical line.

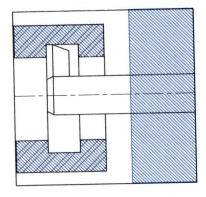

Materials and Processes

13

13.1
Introduction

Metallurgy, the study of metals, is a complex area that is constantly changing as new processes and alloys are developed (Fig. 13.1). The guidelines for designating various types of metals have been standardized by three associations: the American Iron and Steel Institute (AISI), the Society of Automotive Engineers (SAE), and the American Society for Testing Materials (ASTM).

FIGURE 13.1 This furnace operator is pouring an aluminum alloy of manganese into ingots (shown at the right) that will be remelted and cast. (Courtesy of the Aluminum Co. of America.)

13.2
Iron*

Metals that contain iron, even in small quantities, are called **ferrous metals.** The three types of iron are **gray iron, white iron,** and **ductile iron.**

CAST IRON   iron that is melted and poured into a mold to form it, is used in the production of machine parts. Though cheaper and easier to machine than steel, iron does not have steel's ability to withstand shock and force.

GRAY IRON   contains flakes of graphite, which results in low strength and ductility but makes the material easy to machine. Gray iron resists vibrations better than other types of iron. Table 13.1 gives types of gray iron with two designations and their typical applications.

WHITE IRON   contains carbide particles that are very hard and brittle, which enables it to withstand wear and abrasion. There are no designated grades of white iron, but there are differences in composition from one supplier to another. White iron is used for parts on grinding and crushing machines, digging teeth on earthmovers and mining equipment, and wear plates on reciprocating machinery used in textile mills.

---

*This section on iron was developed by Dr. Tom Pollock, a metallurgist at Texas A&M University.

TABLE 13.1
NUMBERING AND APPLICATIONS OF GRAY
IRON

ATSM Grade (1000 psi)	SAE Grade	Typical Uses
ASTM 25 CI	G 2500 CI	Small engine blocks, pump bodies, transmission cases, clutch plates
ASTM 30 CI	G 3000 CI	Auto engine blocks, flywheels, heavy casting
ASTM 35 CI	G 3500 CI	Diesel engine blocks, tractor transmission cases, heavy and high-strength parts
ASTM 40 CI	G 4000 CI	Diesel cylinders, pistons, camshafts

DUCTILE IRON (also called **nodular** or **spheroidized** iron) contains tiny spheres of graphite, making it stronger and tougher than most types of gray iron but also more expensive to produce. The numbering system for ductile iron is given by three sets of numbers, as shown below.

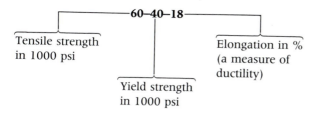

Table 13.2 gives commonly used alloys of ductile iron and their applications.

TABLE 13.2
NUMBERING AND APPLICATIONS OF
DUCTILE IRON

Grade	Typical Uses
60–40–18 CI	Valves, steam fittings, chemical plant equipment, pump bodies
65–45–12 CI	Machine components that are shock loaded, disc brake calipers
80–55–6 CI	Auto crankshafts, gears, rollers
100–70–3 CI	High-strength gears and machine parts
120–90–2 CI	Very high-strength gears, rollers, and slides

MALLEABLE IRON is made from white iron by a heat-treatment process that converts carbides into carbon nodules (similar to ductile iron). The numbering system for designating the grades of malleable iron is shown below.

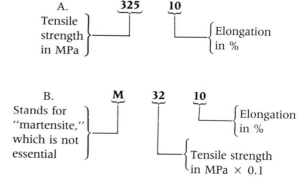

Table 13.3 gives some of the commonly used grades of malleable iron and their applications.

TABLE 13.3
NUMBERING AND APPLICATIONS OF
MALLEABLE IRON

ASTM Grade	Typical Uses
35018 CI	Marine and railroad valves and fittings, "black-iron" pipe fittings (similar to 60–40–18 ductile CI)
45006 CI	Machine parts (similar to 80–55–6 ductile CI)
M3210 CI	Low-stress components, brackets
M4504 CI	Crankshafts, hubs
M7002 CI	High-strength parts, connecting rods, universal joints
M8501 CI	Wear-resistant gears and sliding parts

13.3
Steel

Steel is an alloy of iron and carbon and often contains other constituents such as manganese, chromium, or nickel. Carbon (usually between 0.20% and 1.50%) is the ingredient that has the greatest effect on the grade of the steel. Broadly, the three types of steel are **plain carbon steels, free-cutting carbon steels,** and **al-**

TABLE 13.4
NUMBERING AND APPLICATIONS OF STEEL

Type of Steel	Number	Application
Carbon steels		
Plain carbon	10XX	Tubing, wire, nails
Resulphurized	11XX	Nuts, bolts, screws
Manganese steel	13XX	Gears, shafts
Nickel steel	23XX	Keys, levers, bolts
	25XX	Carburized parts
Nickel-chromium	31XX	Axles, gears, pins
	32XX	Forgings
	33XX	Axles, gears
Molybdenum steel	40XX	Gears, springs
Chromium-molybdenum	41XX	Shafts, tubing
Nickel-chromium	43XX	Gears, pinions
Nickel-molybdenum	46XX	Cams, shafts
	48XX	Roller bearings, pins
Chromium steel	51XX	Springs, gears
	52XX	Ball bearings
Chromium vanadium	61XX	Springs, forgings
Silicon manganese	92XX	Leaf springs

Source: Courtesy of the Society of Automotive Engineers.

loy steels. Table 13.4 gives the types of steels and their designations by four-digit numbers. The first digit indicates the type of steel: 1 is carbon steel, 2 is nickel steel, and so on. The second digit gives the percentage content of the material represented by the first digit. The last two or three digits give the percentage of carbon in the alloy, where 100 equals 1%, and 50 equals 0.50%.

Some frequently used SAE steels are 1010, 1015, 1020, 1030, 1040, 1070, 1080, 1111, 1118, 1145, 1320, 2330, 2345, 2515, 3130, 3135, 3240, 3310, 4023, 4042, 4063, 4140, and 4320.

13.4
Copper

Copper, one of the first metals discovered, can be easily formed and bent without breakage. Because it is highly resistant to corrosion and highly conductive, it is used in the manufacture of pipes, tubing, and electrical wiring. It is also an excellent roofing and screening material since it withstands the weather well.

Copper has several alloys, including brasses, tin bronzes, nickel silvers, and copper nickels. **Brass** is an alloy of copper and zinc, and **bronze** is an alloy of copper and tin. Copper and copper alloys can be easily finished by buffing or plating. These alloys can be

joined by soldering, brazing, or welding and can be easily machined and used for casting.

Wrought copper has properties that permit it to be formed by hammering. A few of the numbered designations of wrought copper are C11000, C11100, C11300, C11400, C11500, C11600, C10200, C12000, and C12200.

13.5
Aluminum

Aluminum is a corrosion-resistant, lightweight metal that has applications for many industrial products. Most materials called aluminum are actually aluminum alloys, which are stronger than pure aluminum.

The types of wrought aluminum alloys are designated by four digits, as Table 13.5 shows. The first digit, from 2 through 9, indicates the alloying element that is combined with aluminum. The second digit indicates modifications of the original alloy or impurity limits. The last two digits identify the other alloying materials or indicate the aluminum purity.

Table 13.6 shows a four-digit numbering system used to designate types of cast aluminum and aluminum alloys. The first digit indicates the alloy group. The next two digits identify the aluminum alloy or

TABLE 13.5
NUMBERING DESIGNATIONS FOR WROUGHT ALUMINUM AND ALUMINUM ALLOYS

Composition	Alloy Number	Applications
Aluminum (99% pure)	1XXX	Tubing, tank cars
Aluminum alloys		
Copper	2XXX	Aircraft parts, screws, rivets
Manganese	3XXX	Tanks, siding, gutters
Silicon	4XXX	Forging, wire
Magnesium	5XXX	Tubes, welded vessels
Magnesium and silicon	6XXX	Auto body, pipe
Zinc	7XXX	Aircraft structures
Other elements	8XXX	

TABLE 13.6
ALUMINUM CASTING AND
INGOT DESIGNATIONS

Composition	Alloy Number
Aluminum (99 % pure)	1XX.X
Aluminum alloys	
Copper	2XX.X
Silicon with copper and/or magnesium	3XX.X
Silicon	4XX.X
Magnesium	5XX.X
Magnesium and silicon	6XX.X
Zinc	7XX.X
Tin	8XX.X
Other elements	9XX.X

Source: Courtesy of the Society of Automotive Engineers.

aluminum purity. The number 1 to the right of the decimal point represents the aluminum form: XX.0 indicates castings, XX.1 indicates ingots with a specified chemical composition, and XX.2 indicates ingots with a specified chemical composition other than the XX.1 ingot, whereas 0 represents aluminum for casting. **Ingots** are blocks of cast metal to be remelted, and **billets** are castings of aluminum to be formed by forging.

13.6
Magnesium

Magnesium is a light metal available in an inexhaustible supply since it is extracted from seawater and natural brines. Approximately half the weight of aluminum, magnesium is an excellent material for aircraft parts, clutch housing, crankcases for air-cooled engines, and applications where lightness is desirable.

Magnesium is used for die and sand castings, extruded tubing, sheet metal, and forging. Magnesium and its alloys can be joined by bolting, riveting, or welding. Some numbered designations of magnesium alloys are M10100, M11630, M11810, M11910, M11912, M12390, M13320, M16410, and M16620.

13.7
Properties of materials

All materials have properties that designers must use to their best advantage. The following terms describe these properties:

DUCTILITY is a softness present in some materials, such as copper and aluminum, that permits them to be formed by stretching (drawing) or hammering without breaking. Wire is made of ductile materials that can be drawn through a die.

BRITTLENESS is a characteristic of metals that will not stretch without breaking, such as cast irons and hardened steels.

MALLEABILITY is the ability of a metal to be rolled or hammered without breaking.

HARDNESS is the ability of a metal to resist being dented when it receives a blow.

TOUGHNESS is the property of being resistant to cracking and breaking while remaining malleable.

ELASTICITY is the ability of a metal to return to its original shape after being bent or stretched.

13.8
Heat treatment of metals

The properties of metals can be changed by various forms of heat treating. Steels are affected to a greater extent by heat treating than are other materials.

HARDENING is performed by heating steel to a prescribed temperature and then quenching it in oil or water.

QUENCHING is the process of rapidly cooling heated metal by immersing it in liquids, gases, or solids (such as sand, limestone, or asbestos).

TEMPERING is the process of reheating previously hardened steel and then cooling it, usually by air. This increases the steel's toughness.

ANNEALING is the process of heating and cooling metals to soften them, release their internal stresses, and make them easier to machine.

NORMALIZING is achieved by heating metals and letting them cool in air to relieve their internal stresses.

CASE HARDENING is the process of hardening a thin outside layer of a metal. The outer layer is placed in contact with carbon or nitrogen compounds that are absorbed by the metal as it is heated; afterward, the metal is quenched.

FLAME HARDENING is the method of hardening by heating a metal to within a prescribed temperature range with a flame and then quenching the metal.

13.9
Castings

Two major methods of forming shapes are **casting** and **pressure forming.** Casting involves the preparation of a mold into which is poured molten metal that cools and forms the part. The types of casting, which differ in the way the molds are made, are **sand casting, permanent-mold casting, die casting,** and **investment casting.**

Sand casting

In the first step of sand casting, a wood or metal form or pattern is made that is representative of the final part to be cast. The pattern is placed in a metal box called a **flask,** and molding sand is packed around the pattern. When the pattern is withdrawn from the

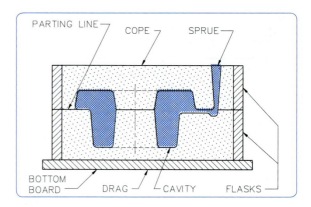

FIGURE 13.3 **A two-section sand mold.**

sand, it leaves a void forming the mold. Molten metal is poured into the mold through sprues, or gates. After cooling, the casting is removed and cleaned (Fig. 13.2).

Cores, parts formed in sand, are placed within a mold to leave holes or hollow portions within the finished casting. Once the casting has been formed, the cores can be broken apart and removed, leaving behind the desired void within the casting. Cores add to the cost of a casting and should not be used unless their expense is offset by savings in materials.

Since the patterns are placed in sand and then withdrawn before the metal is poured, the sides of the patterns must be tapered for ease of withdrawal from the sand (Fig. 13.3). This taper is called **draft.** The amount of draft depends on the depth of the pattern in the sand; for most applications, it varies from 2° to 8°. To compensate for shrinkage that will occur when the metal cools, patterns are made oversize.

Because the sand casting has a rough surface that ought not to be a contact surface with other moving parts, it is common practice to machine portions of a casting by drilling, grinding, or shaping (Fig. 13.4). The pattern should be made larger in these areas to compensate for removal of metal.

Fillets and rounds are used at intersections to increase the strength of a casting and because it is difficult to form square corners by the sand-casting process.

Permanent-mold casting

Permanent molds are made for the mass production of parts. They are generally made of cast iron and coated to prevent fusing with the molten metal poured into them (Fig. 13.5).

FIGURE 13.2 **A large casting of an aircraft's landing-gear mechanism is being removed from its mold. (Courtesy of Cameron Iron Works.)**

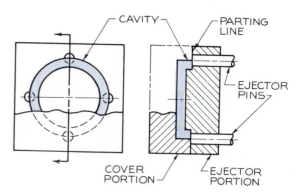

FIGURE 13.6 A die for casting a simple part. Unlike the sand casting, the metal is forced into the die to form the casting.

FIGURE 13.4 This casting of the outer cylinder of an aircraft's landing gear is being bored on a horizontal boring mill. (Courtesy of Cameron Iron Works.)

Die casting

Die castings are used for the mass production of parts made of aluminum, magnesium, zinc alloys, copper, or other materials. Made by forcing molten metal into

dies (or molds) under pressure, die castings can be produced at a low cost, at close tolerances, and with good surface qualities. The same general principles recommended for sand castings—using fillets and rounds, allowing for shrinkage, and specifying draft angles—apply to die castings (Fig. 13.6).

Investment casting

Investment casting is a process used to produce complicated parts that would be difficult to form by any other method. This technique is used to form the intricate shapes of artistic sculptures.

Since a new pattern must be used for each investment casting, a mold or die is made to cast a wax master pattern. The wax pattern, which will be identical to the finished casting, is placed inside a container, and plaster or sand is poured (or invested) around the pattern. Once the investment has cured, the wax pattern is melted, leaving a hollow cavity that will serve as the mold for the molten metal. When filled and set, the plaster or sand is broken away from the finished investment casting.

13.10
Forgings

Forging is the process of shaping or forming heated metal by hammering or squeezing it into a die. Drop forges and press forges are used to hammer the metal (called billets) into the forging dies by multiple blows. The resulting forging possesses high strength and a resistance to loads and impacts.

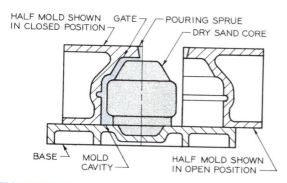

FIGURE 13.5 Permanent molds are made of metal for repetitive usage. Here, a sand core made from another mold is placed in the permanent mold to give a void within the casting.

FIGURE 13.7 Three stages of manufacturing a turbine fan. **A.** The blank is formed by forging. **B.** It is machined. **C.** The fan blades are attached to their machined slots. (Courtesy of Avco Lycoming.)

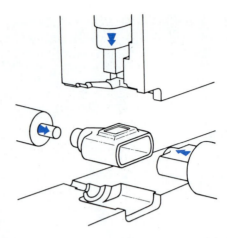

FIGURE 13.9 Auxiliary rams can be used to form internal features on a part. (Courtesy of General Motors Corp.)

When preparing forging drawings, the following must be considered: (1) draft angles and parting lines, (2) fillets and rounds, (3) forging tolerances, (4) extra material for machining, and (5) heat treatment of the finished forging (Fig. 13.7). Some of the standard steels used for forging are designated by the SAE numbers 1015, 1020, 1025, 1045, 1137, 1151, 1335, 1340, 4620, 5120, and 5140. Iron, copper, and aluminum can also be forged.

Figure 13.8 shows examples of dies. A single-impression die gives an impression on one side of the parting line between the mating dies; a double-impression die gives an impression on both sides of the parting line; and the interlocking dies give an

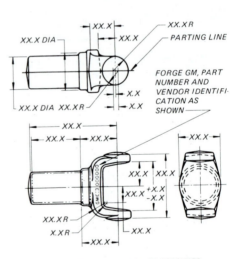

UNLESS OTHERWISE SPECIFIED:
DRAFT ANGLES X°.
ALL FILLETS X.XR, CORNERS X.XR.
+X.X– X.X TOLERANCES ON
FORGING DIM.

SNAG AND REMOVE SCALE.

SAMPLE FORGINGS ARE TO BE
APPROVED BY METALLURGICAL
AND ENGRG DEPTS FOR GRAIN
FLOW STRUCTURE.

FORGING DRAWING

FIGURE 13.10 A drawing of a forging. The blank is forged oversize to allow for machining operations that will remove metal from it. (Courtesy of General Motors Corp.)

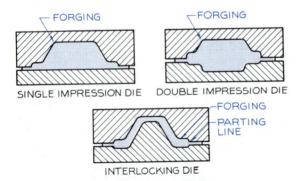

FIGURE 13.8 Three types of forging dies. **A.** A single-impression die. **B.** A double-impression die. **C.** An interlocking die.

impression that may cross the parting line on either side. Figure 13.9 shows an object that is forged with auxiliary rams to hollow the forging, and Fig. 13.10 shows a drawing of a forged part.

Rolling

Rolling is a type of forging in which the stock is rolled between two rollers to give it a desired shape. Rolling can be done at right angles to the axis of the part or parallel to its axis (Fig. 13.11). If a high degree of shaping is required, the stock is usually heated before rolling. If the forming requires only a slight change in configuration, the rolling can be performed when the metal is cold; this is called **cold rolling.**

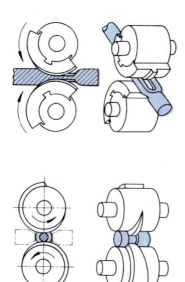

FIGURE 13.11 Features on parts may be formed by rolling. In these examples, parts are being rolled parallel and perpendicular to their axes. (Courtesy of General Motors Corp.)

13.11

Stamping

Stamping is a method of forming flat metal stock into three-dimensional shapes. The first step of stamping is to cut out the shapes, called **blanks,** to be bent. Blanks are formed into shape by bending and pressing

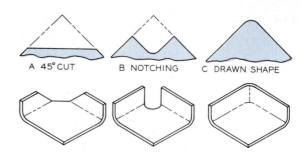

FIGURE 13.12 Box-shaped parts formed by stamping. **A.** A corner cut of 45° permits folding flanges and may require no further trim. **B.** Notching has the same effect as the 45° cut, but it is often more attractive. **C.** A continuous corner flange requires that the blank be developed so that it can be drawn into shape.

them against forms. Figure 13.12 shows examples of box-shaped parts, and Fig. 13.13 shows a flange stamping. Holes in stampings are made by punching, extruding, or piercing (Fig. 13.14).

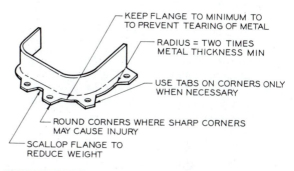

FIGURE 13.13 A sheet metal flange design with notes calling attention to design details.

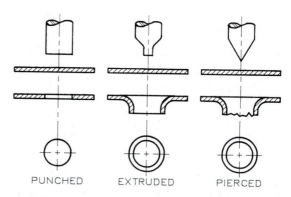

FIGURE 13.14 Three methods of forming holes in sheet metal. (Courtesy of General Motors Corp.)

E=EXCELLENT
G=GOOD
F=FAIR
P=POOR
A=ADHESIVES

	MACHINABILITY	FORMABILITY	CASTABILITY	WELDABILITY	CORROSION RES.	ABRASION RES.	LB/CU FT	YIELD: 1000 PSI	TYPICAL APPLICATIONS
NYLON	E	G	G		G	E	73	15	HELMETS, GEARS, DRAWER SLIDES, HINGES, BEARINGS
ABS PLASTIC	G	G	G		E	G	66	66	LUGGAGE, BOAT HULLS, TOOL HANDLES, PIPE FITTINGS
POLYETHLENE	G	F	G	A	F	F	58	2	CHEMICAL TUBING, CONTAINERS, ICE TRAYS, BOTTLES
POLYPROPYLENE	G	G	G	A	E	G	56	5.3	CARD FILES, COSMETIC CASES, AUTO PEDALS, LUGGAGE
POLYSTYRENE	G	E	G	A	P	G	67	7	JUGS, CONTAINERS, FURNITURE, LIGHTED SIGNS
POLYURETHANE	G	G	G	A	G	E	74	6	SOLID TIRES, GASKETS, BUMPERS, SYNTHETIC LEATHERS
POLYVINYLE CHLORIDE	E	E	G	A	G	G	78	4.8	RIGID PIPE AND TUBING, HOUSE SIDING, PACKAGING
ACRYLIC	G	G	E	A	E	F	74	9	AIRCRAFT WINDOWS, TV PARTS, LENSES, SKYLIGHTS
EPOXY	F	G	G		E	G	69	17	CIRCUIT BOARDS, BOAT BODIES, COATINGS FOR TANKS
SILICONE	F	G	G		G	G	109	28	FLEXIBLE FUEL HOSES, HEART VALVES, GASKETS
GLASS	F	G			F	F	160	10+	BOTTLES, WINDOWS, TUMBLERS, CONTAINERS
FIBERGLASS	G		E	A	G		109	20+	BOATS, SHOWER STALLS, AUTO BODIES, CHAIRS, SIGNS

FIGURE 13.15 Specifications for commonly used plastics and materials.

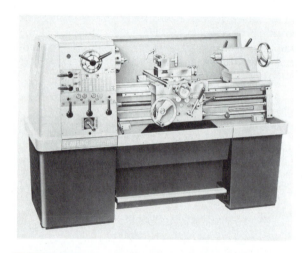

FIGURE 13.16 A typical metal lathe that holds and rotates the work piece between the centers of the lathe. (Courtesy of the Clausing Corp.)

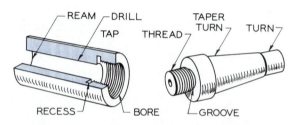

FIGURE 13.17 The fundamental operations performed on a lathe are illustrated on the two parts above.

13.12
Plastics and miscellaneous materials

Figure 13.15 shows commonly used plastics and materials. The weights and yields of the materials are given along with examples of their applications.

13.13
Machining operations

After the metal has been formed, machining operations must be performed to complete the part. The following machines are often used: **lathe, drill press, milling machine, shaper,** and **planer.**

The lathe

The lathe shapes cylindrical parts while rotating the work piece between the centers of the lathe (Fig. 13.16). The more fundamental operations performed on the lathe are **turning, facing, drilling, boring, reaming, threading,** and **undercutting** (Fig. 13.17).

TURNING forms a cylinder by a tool that advances against and moves parallel to the cylinder being turned (Fig. 13.18).

FACING forms flat surfaces perpendicular to the axis of rotation of the part being rotated by the lathe.

DRILLING is performed by mounting a drill in the tail stock of the lathe and rotating the work while the bit is advanced into the part (Fig. 13.19).

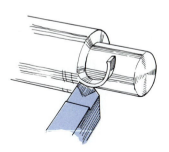

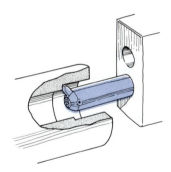

FIGURE 13.20 Boring is the method of enlarging holes that are larger than available drill bits. The cutting tool is attached to the boring bar on the lathe.

FIGURE 13.18 Turning is the most basic operation performed on the lathe. A continuous chip is removed by a cutting tool as the part is rotated.

BORING makes large holes that are too big to be drilled. Large holes are bored by enlarging smaller drilled holes (Fig. 13.20).

REAMING removes only a few thousandths of an inch of material inside a drilled hole to bring it to its required level of tolerance. Conical and cylindrical reaming can be performed on the lathe (Fig. 13.21).

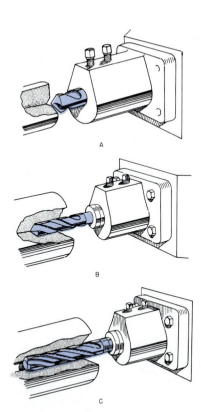

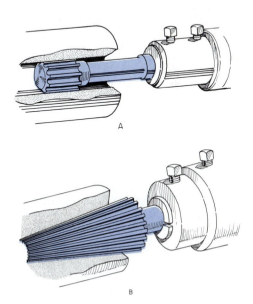

FIGURE 13.19 Three steps of drilling a hole in the end of a cylinder: **A.** start drilling, **B.** twist drilling, and **C.** core drilling, which enlarges the previously drilled hole to the required size.

FIGURE 13.21 Fluted reamers can be used to finish inside cylindrical and conical holes within a few thousandths of an inch.

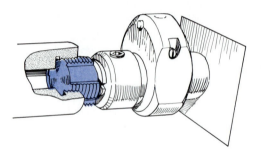

FIGURE 13.22 External and internal threads (shown here) can be cut on a lathe. The die used to cut the threads is called a tap. A recess has been formed at the end of the threaded hole prior to threading.

FIGURE 13.24 A turret lathe that performs a sequence of operations.

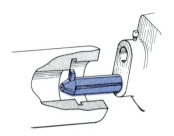

FIGURE 13.23 A recess (undercut) can be formed by using the boring bar with the cutting tools attached as shown. As the boring bar is moved off center of the axis, the tool will form the recess.

THREADING of external and internal holes can be done on the lathe. The die used for cutting internal holes is called a **tap** (Fig. 13.22).

UNDERCUTTING cuts a recess inside a cylindrical hole with a tool mounted on a boring bar. The groove is cut as the tool advances from the center of the axis of revolution into the part (Fig. 13.23).

The turret lathe is a programmable lathe that can perform sequential operations, such as drilling a series of holes, boring them, and then reaming them. The turret is mounted to rotate each tool into position for its particular operation (Fig. 13.24).

The drill press

The drill press is used to drill small- and medium-sized holes into stock that is held on the bed of the press by a fixture or clamp (Fig. 13.25). The drill press can be

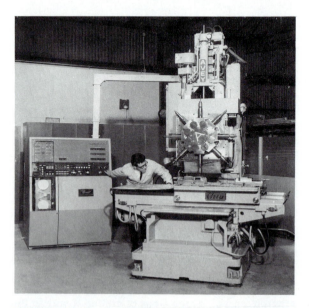

FIGURE 13.25 A multiple-head drill press that can be programmed to perform a series of operations in a desired sequence.

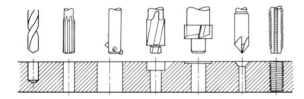

FIGURE 13.26 The basic operations that can be performed on the drill press are (left to right): drilling, reaming, boring, counterboring, spotfacing, countersinking, and tapping (threading).

used for **counterdrilling, countersinking, counterboring, spotfacing,** and **threading** (Fig. 13.26).

BROACHING Cylindrical holes can be converted into square or hexagonal holes by using a **broach.** The broach has a series of teeth along its axis, beginning with teeth that are nearly the size of the hole to be broached and tapering to the size of the finished hole to be broached. The broach is forced through the

hole, with each tooth cutting more from the hole as it passes through.

The milling machine

The milling machine uses a variety of cutting tools, rotated about a shaft (Fig. 13.27), to form different grooved slots, threads, and gear teeth. The milling machine can cut irregular grooves in cams and finish surfaces on a part within a high degree of tolerance.

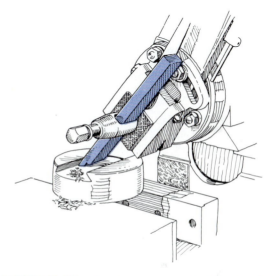

FIGURE 13.28 The shaper moves back and forth across the part, removing metal as it advances. It can be used to finish surfaces, cut slots, and for many other operations.

FIGURE 13.27 This milling machine is being used to cut a groove in the work piece. (Courtesy of the Brown & Sharpe Mfg. Co.)

The shaper

The shaper is a machine that holds the work piece stationary while the cutter passes back and forth across the work to finish the surface or to cut a groove one stroke at a time (Fig. 13.28). With each stroke of the cutting tool, the material is shifted slightly so as to align the part for the next overlapping stroke.

The planer

Unlike the shaper, which holds the work piece stationary, the planer passes the work under the cutters

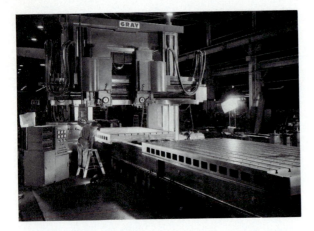

FIGURE 13.29 The planer has stationary cutters, and the work is fed past them to finish larger surfaces. This planer has a 30-foot bed. (Courtesy of Gray Corp.)

(Fig. 13.29). Like the shaper, the planer can cut grooves or slots and finish surfaces that must meet tolerance specifications.

13.14
Surface finishing

Surface finishing is the process of finishing a surface to the desired uniformity. It may be accomplished by several methods, including **grinding, polishing, lapping, buffing,** and **honing.**

GRINDING finishes of a flat surface by holding it against a rotating abrasive wheel (Fig. 13.30). Grinding is used to smooth surfaces and to sharpen edges used for cutting, such as drill bits.

POLISHING is performed in the same manner as grinding except the polishing wheel is flexible since it is made of felt, leather, canvas, or fabric.

LAPPING produces very smooth surfaces. The surface to be finished is held against a **lap,** a large, flat surface coated with a fine abrasive powder. As the lap rotates, the surface is finished. Lapping is done only after the surface has been previously finished by a less accurate technique like grinding or polishing. Cylindrical parts can be lapped by using a lathe with the lap.

BUFFING removes scratches from a surface with a rotating buffer wheel made of wool, cotton, or other fabric. Sometimes the buffer is a cloth or felt belt that is applied to the surface being buffed. To enhance the buffing, an abrasive mixture is applied to the buffed surface from time to time.

HONING finishes the outside or inside of holes within a high degree of tolerance. As it is passed through the holes, the honing tool is rotated to produce the sort of finishes found in gun barrels, engine cylinders, and other products requiring a high degree of smoothness.

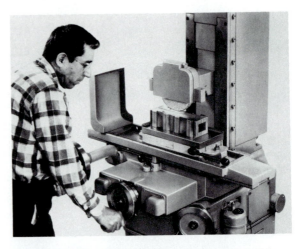

FIGURE 13.30 The upper surface of this part is being ground to a smooth finish by a grinding wheel. (Courtesy of the Clausing Corp.)

Dimensioning

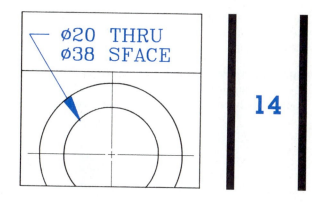

14

14.1
Introduction

Working drawings are dimensioned drawings used to describe the details of a part or project so that construction can be performed according to specifications. When properly applied, dimensions and notes will supplement the drawings so that they can be used as legal contracts for construction.

The techniques of dimensioning presented in this chapter are based primarily on the standards of the American National Standards Institute (ANSI), especially Y14.5M, *Dimensioning and Tolerancing for Engineering Drawings*. Various industrial standards from companies such as General Motors Corporation have also been used.

14.2
Dimensioning terminology

The guide slide in Fig. 14.1 illustrates some of the terms of dimensioning.

DIMENSION LINES are thin lines (2H–4H pencil) with arrows at each end. Numbers placed near their midpoints specify a part's size.

EXTENSION LINES are thin lines (2H–4H pencil) that extend from a view of an object for dimensioning

the part. The arrowheads of dimension lines end at these lines.

CENTERLINES are thin lines (2H–4H pencil) used to locate the centers of cylindrical parts such as cylindrical holes.

LEADERS are thin lines (2H–4H pencil) drawn from a note to a feature the note applies to.

ARROWHEADS are placed at the ends of dimension lines and leaders to indicate their endpoints. Ar-

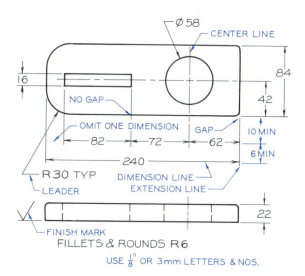

FIGURE 14.1 This typical working drawing is dimensioned in millimeters.

rowheads are drawn the same length as the height of the letters or numerals, usually ⅛ inch. Figure 14.2 shows the form of the arrowhead.

DIMENSION NUMBERS are placed near the middle of the dimension line and are usually ⅛ inch high; units of measurement (″, IN, or mm) are omitted.

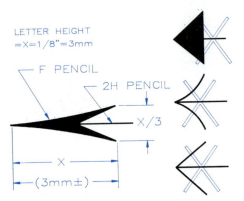

FIGURE 14.2 Arrowheads are drawn as long as the height of the letters used on a drawing. They are one third as wide as they are long.

14.3
Units of measurement

The two most commonly used units of measurement are the decimal inch, in the **English** (imperial) system, and the millimeter, in the **metric** (SI) system.

The inch in its common fraction form can be used, but it is preferable to give fractions in decimal form. Common fractions make arithmetic hard to perform. Figure 14.3 compares dimensions in millimeters with those in inches.

Examples are shown in Fig. 14.4, where the units are given in millimeters, decimal inches, and fractional inches.

> Dimensions in millimeters are usually rounded off to **whole numbers** without decimal fractions.

When a metric dimension is less than a millimeter, a zero precedes the decimal point, but no zero precedes a decimal point when inches are used. When using

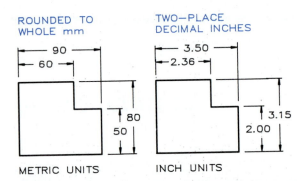

FIGURE 14.3 In the metric system, millimeters are rounded to the nearest whole number. In the English system, inches are carried to two decimal places even for whole numbers like 1.00.

decimal inches, show all dimensions with two-place decimal fractions even if the last numbers are zeros.

> Units of measurement are omitted from the dimension numbers since they are normally understood to be in millimeters or inches. For example, 112, not 112 mm, and 67, not 67″ or 5′–7″.

Architects use a combination of feet and inches, but the inch units are omitted (for example, 7′–2). Engineers use feet and decimal fractions of feet to dimension large-scale projects such as road designs (for example, 252.7′ where feet units *are* shown).

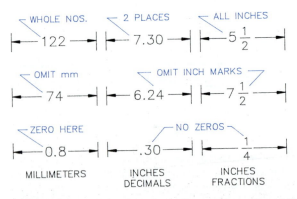

FIGURE 14.4 A comparison of the application of SI units with English units with decimal and common fractions.

14.4
English/metric conversions

Dimensions in inches can be converted to millimeters by multiplying by 25.4. Similarly, dimensions in millimeters can be converted to inches by dividing by 25.4.

For most applications, the millimeter does not need more than a one-place decimal. When millimeters are found by conversion from inches

- The last digit retained in a conversion of either mm or inches is unchanged if it is followed by a number less than 5; for example, 34.43 is rounded off to 34.4.

- The last digit to be retained is increased by 1 if it is followed by a number greater than 5; 34.46 is rounded off to 34.5.

- The last digit to be retained is unchanged if it is even and followed by exactly 5; 34.45 is rounded off to 34.4.

- The last digit to be retained is increased by 1 if it is odd and followed by exactly 5; 34.75 is rounded off to 34.8.

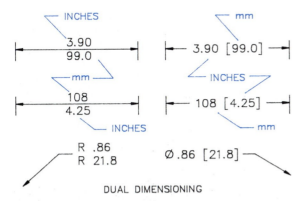

DUAL DIMENSIONING

FIGURE 14.5 If the drawing was originally made in inches, the equivalent measurement in millimeters is placed under or to the right of the inches in brackets in dual-dimensioning. If the drawing was originally made in millimeters, the equivalent measurement in inches is placed under or to the right. When inches are converted to millimeters, the millimeters may need to be written as decimal fractions.

14.5
Dual dimensioning

Some drawings require that both metric and English units be given on each dimension. This **dual dimensioning** can be shown by placing the inch equivalent of millimeters either under or over the other units (Fig. 14.5). If the drawing was originally dimensioned in inches, the inch dimensions are placed on top, and the equivalent millimeters are given underneath. If the drawing was originally dimensioned in millimeters and then converted to inches, the millimeters would be placed over the equivalent in inches.

A second method of dual dimensioning uses brackets placed on either side of the converted dimensions (Fig. 14.5). Do not mix these two methods on the same drawing.

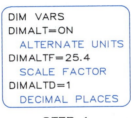 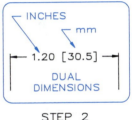

STEP 1 STEP 2

FIGURE 14.6 Alternate (dual) dimensions.

Step 1 Set the Dim Vars to DIMALT to set dual dimensions to ON, assign the scale factor (DIMALTF), and specify the number of decimal places (DIMALTD).

Step 2 Linear dimensions are found using the same steps as shown in Fig. 14.29. The dimension in brackets is the metric equivalent to the inch dimensions.

COMPUTER METHOD As Fig. 14.6 shows, the dimensioning variable DIMALT must be set to ON to obtain alternate (dual) dimensions in brackets following the primary units used. The variable DIMALTF (scale factor) is set to a value representing the multiplier by which the first dimension is changed. DIMALTD is used to assign the desired number of decimal places for the second dimension.

Dimensions are then selected using the DIM command in the same way single-value dimensions are found (see Fig. 14.29).

14.6
Metric designation

The metric system is the Système Internationale d'Unités and is denoted by the abbreviation SI (Fig. 14.7). This system uses the first-angle of projection, which locates the front view over the top view and the right-side view to the left of the front view.

When drawings are made for international circulation, it is customary to use one of the symbols shown in Figs. 14.7C and 14.7D to designate the angle of projection. Either the letters *SI* or the word *METRIC*, written prominently on the drawing or in the title block, indicates the measurements are metric.

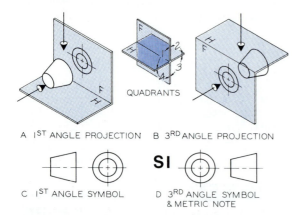

FIGURE 14.7 **A.** The European system of orthographic projection places the top view under the front view. **B.** The American system places the front view under the top view. **C.** These symbols indicate first-angle projection. **D.** These symbols indicate third-angle projection. SI indicates that metric units are used.

14.7
Aligned and unidirectional numbers

The two methods of positioning dimension numbers on a dimension line are the **aligned** and **unidirectional** methods. The unidirectional system is more widely accepted since it is easier to apply numerals positioned horizontally (Fig. 14.8A). The aligned system aligns the numerals with the dimension lines (Fig. 14.8B).

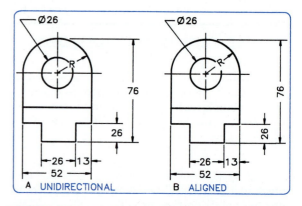

FIGURE 14.8 **A.** When dimensions are positioned so that all of them read from the bottom of the page, they are unidirectional. **B.** Aligned dimensions are positioned to read from the bottom and right side of the sheet.

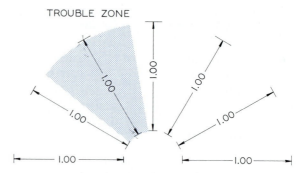

FIGURE 14.9 Numbers on angular dimension lines should not be placed in the trouble zone. To do so would cause the numbers to be read from the left side of the sheet rather than from the right side and bottom.

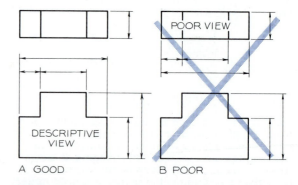

FIGURE 14.10 Place the dimensions on the most descriptive views where the true contour of the object can be seen.

The numbers must be readable from the bottom or right side of the page.

Figure 14.9 shows examples of aligned dimensions on angular dimension lines. Avoid placing aligned dimensions in the "trouble zone" since these numerals would read from the left instead of the right side of the sheet.

COMPUTER METHOD A mode of DIMTIH (dimensioning text inside dimension lines is horizontal) under DIM Vars of the DIM command of AutoCAD must be set to OFF for aligned dimensions and to ON for unidirectional dimensions (Fig. 14.8). The DIMTOH mode controls the positioning of text that lies outside the dimension line in cases where the numerals do not fit within a short dimension line. When DIMTOH is ON, the dimensions will be horizontal; when OFF, the dimensions will be aligned with the direction of the dimension line.

14.8
Placement of dimensions

It is good practice to dimension the most descriptive views.

The front view in Fig. 14.10 is more descriptive than the top view, so the front view should be dimensioned. Dimensions should be applied to views in an organized manner (Fig. 14.11). Locate the dimension

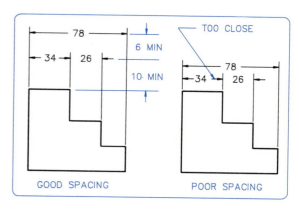

FIGURE 14.12 The first row of dimensions should be placed at least 0.40 in. (10 mm) from the view, and successive rows should be at least 0.25 in. (6 mm) from the first row. If greater spaces are used, these proportions should still be maintained.

lines by beginning with the smaller ones to avoid crossing dimension and extension lines.

Always leave at least 0.40 in. (10 mm) between the object and first row of dimensions (Fig. 14.12). Successive rows of dimensions should be equal and at least 0.25 in. (6 mm) apart.

If greater spaces are used, apply these same general proportions. The Braddock-Rowe lettering guide triangle can be used to space the dimension lines (Fig. 14.13).

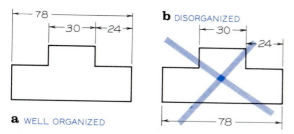

FIGURE 14.11 Dimensions should be placed on the views in a well-organized manner to make them as readable as possible.

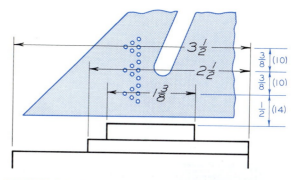

FIGURE 14.13 When common fractions are used, the center holes of the triangular arrangements on the Braddock-Rowe triangle are aligned with the dimension lines to automatically space the lines.

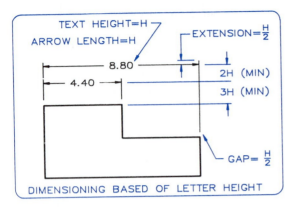

FIGURE 14.14 The proportions for placing dimensions on a drawing are based on the letter height, usually 0.125" or 3 mm.

The positioning of dimensions is based on the letter height, as Fig. 14.14 shows. These proportions are the **minimum** values.

COMPUTER METHOD Figure 14.15 shows dimensioning variables and their minimum settings. Computer variables can be set at these proportions, and all of them can be changed at once using DIMSCALE. For example, DIMSCALE = 25.4 would convert inch proportions to millimeter proportions.

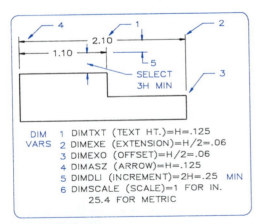

FIGURE 14.15 The assignment of dimensioning variables (Dim Vars) to be applied by AutoCAD is shown here. Once set, these variables remain set with the file in use. DIMSCALE can be used to enlarge or reduce these proportions.

Figure 14.16 shows other dimensioning variables. Chapter 23 covers variables not covered in this chapter.

When a row of dimensions is placed on a drawing, one of the dimensions is omitted since the overall

DIM VARS	DEFAULT	
DIMSCALE	1.0000	Overall scale factor
DIMASZ	0.1800	Arrow size
DIMCEN	0.0900	Center mark size
DIMEXO	0.0625	Extension line origin offset
DIMDLI	0.3800	Dimension line increment
DIMEXE	0.1800	Extension above dimen. line
DIMTP	0.0000	Plus tolerance
DIMTM	0.0000	Minus tolerance
DIMTXT	0.1800	Text height
DIMTSZ	0.0000	Tick size
DIMRND	0.0000	Rounding value
DIMDLE	0.0000	Dimension line extension
DIMTOL	OFF	Generate dimension tolerances
DIMLIM	OFF	Generate dimension limits
DIMTIH	ON	Text inside extensions horiz.
DIMTOH	ON	Text outside extensions horiz.
DIMSE1	OFF	Suppress first extension line
DIMSE2	OFF	Suppress 2nd extension line
DIMTAD	OFF	Place text above dimen. line
DIMZIN	0	Zero inches/feet control
DIMALT	OFF	Alternate units selected
DIMALTF	25.4000	Alternate unit scale factor
DIMALTD	2	Alternate unit decimal places
DIMLFAC	1.0000	Linear unit scale factor
DIMBLK		Arrow block name
DIMASO	ON	Create associative dimensions
DIMSHO	OFF	Update dimen. while dragging
DIMPOST		Default suffix for dimen. text
DIMAPOST		Default suffix for alternate text

FIGURE 14.16 The dimensioning variables (Dim Vars) shown here are covered partly in this chapter and partly in Chapter 23.

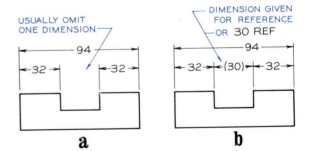

FIGURE 14.17 **A.** One intermediate dimension is customarily omitted since the overall dimension provides this measurement. **B.** If all the intermediate dimensions are given, one should be placed in parentheses to indicate that it has been given as a reference dimension.

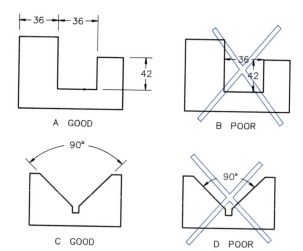

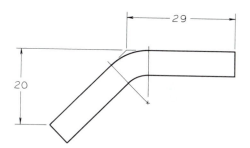

FIGURE 14.20 A curved surface is dimensioned by locating the theoretical point of intersection with extension lines.

FIGURE 14.18 Dimensioning rules.

A. Placing the dimensions outside a part is the preferred practice.

B. Dimension lines should not be used as extension lines.

C. Angles are dimensioned with arcs and extension lines.

D. The angle should not be placed inside the angular cut.

dimension supplements the omitted dimension (Fig. 14.17A). A dimension that needs to be given as a reference dimension is either placed in parentheses or followed by the abbreviation REF to indicate it is a reference dimension.

Figure 14.18 shows the recommended techniques of dimensioning features, and Fig. 14.19 shows examples of the placement of extension lines. Extension lines may cross other extension lines or object lines; they are also used to locate theoretical points outside curved surfaces (Fig. 14.20).

14.9
Dimensioning in limited spaces

Figure 14.21 shows examples of dimensioning in limited spaces. Regardless of space limitations, the numerals should not be drawn smaller than they appear elsewhere on the drawing. Where dimension lines are

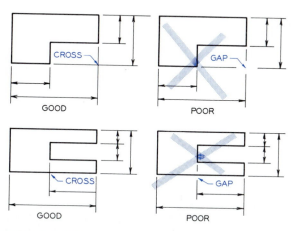

FIGURE 14.19 Extension lines extend from the edge of an object, leaving a small gap. They do not have gaps where they cross object lines or other extension lines.

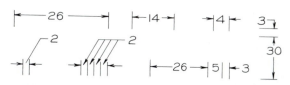

FIGURE 14.21 Where room permits, numerals and arrows should be placed inside the extension lines. Other placements are shown as the spacing becomes smaller.

closely grouped, the numerals should be staggered to make them more readable (Fig. 14.22).

14.10
Dimensioning symbology

Figure 14.23 shows several symbols used in dimensioning. The sizes of the symbols are based on the height of the lettering used in dimensioning an object,

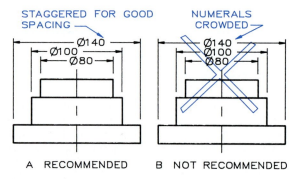

FIGURE 14.22 Dimensioning numerals should be staggered when close spacing tends to crowd them.

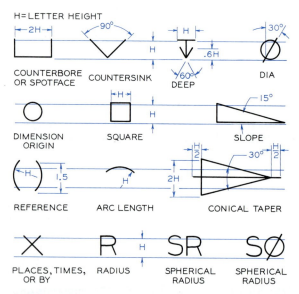

FIGURE 14.23 These symbols can be used to dimension parts. The proportions of the symbols are based on the letter height, *H*, which is usually ⅛ inch.

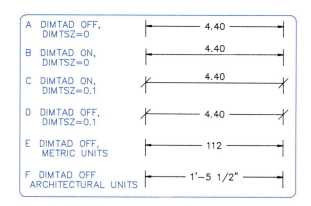

FIGURE 14.24 The dimension lines illustrate the effects of using the different DIM VARS (dimensioning varibles) in AutoCAD.

usually ⅛ inch. Symbols reduce the time in preparing notes and add to the clarity of a dimension.

14.11
Computer dimensioning

The rules of dimensioning must be followed whether you are using computer graphics or manual techniques. The AutoCAD program offers many features to use in applying the rules. For example, Fig. 14.24 shows several combinations of DIM VARS, which are modes of the DIM command.

Text can be placed inside the dimension line (DIMTAD: OFF) or above the dimension line (DIMTAD: ON). Arrowheads can be placed at the ends of dimension lines DIMASZ>0, or tick marks (slashes) can be used when DIMTSZ is set to a value greater than 0, usually about half the letter height of text. UNITS can be selected to be architectural (feet and inches), metric (no decimal fractions), decimal inches (two or more decimal fractions), or engineering units (feet and decimal inches).

An important DIM VARS in dimensioning is DIMSCALE, which can be used to change all variables of dimensioning to allow for changes in scales (Fig. 14.25). DIMSCALE is set at a default of 1.00, which contains the lengths of arrows, text size, and extension line offsets. When invoked, DIMSCALE will not change previously drawn dimensions, only those applied afterward.

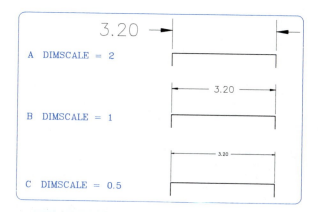

DIMSCALE, a subcommand under DIM:, can be used to change dimensioning variables by inputting a single factor. The text sizes, arrows, offsets, and extensions are all changed at one time.

The text STYLE that was last used will be used as the dimensioning text. If you had set the text to a specified height, this height would be used in dimensioning whether or not DIMSCALE was used. For DIMSCALE to change text height, you should set the text height for the STYLE being used at 0 (zero). You will be prompted for the letter height each time, which allows you either to accept the default last used or to assign another height. In this case, DIMSCALE will enlarge or reduce text height along with the other variables.

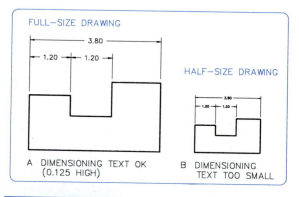

FIGURE 14.26 When dimensioning a drawing, you should be aware of its final plotted size for the dimensioning variables to be sized to compensate for reduction or enlargement. DIMSCALE is the most efficient command for assigning a scale to dimensioning variables.

You must apply DIMSCALE to a dimensioned drawing when it will be plotted at a different scale (Fig. 14.26). For example, if a drawing is to be plotted half size, you would want to use a DIMSCALE of 2.00 so that the variables would appear full size when reduced.

AutoCAD automatically performs many dimensioning operations, but occasionally you will want to modify the placement of arrows and text. Figure 14.27 shows an edited dimension where MOVE and ERASE commands are used.

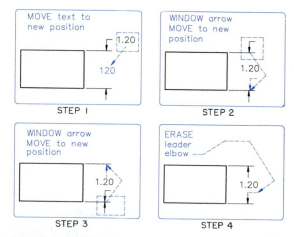

FIGURE 14.27 Editing dimensions.

Step 1 You may wish to MOVE a dimension text within the extension lines.

Step 2 Arrows can be MOVEd inside extension lines also.

Step 3 A second arrow is windowed and MOVEd inside the extension lines.

Step 4 The unneeded leader extension is ERASEd to complete the modified dimension.

14.12
Dimensioning prisms

Figure 14.28 illustrates the following rules for dimensioning prisms:

1. Dimensions should extend from the most **descriptive views** (Fig. 14.28A).

2. Dimensions that apply to two views should be placed **between** them (Fig. 14.28A).

3. The first row of dimension lines should be placed at **least 0.40 in.** (10 mm) from the view. Successive rows should be placed at **least 0.25** in. (6 mm) apart.

4. Extension lines may **cross** each other and other lines, but dimension lines should **not cross** unless absolutely necessary.

5. To dimension a part in its most descriptive view, dimensions may have to be placed in **more** than one view (Fig. 14.28B).

6. Dimension lines may be placed **inside** a notch to improve clarity.

7. Whenever possible, dimensions should be applied to **visible** lines rather than to hidden lines (Fig. 14.28D).

8. Dimensions should **not be repeated,** and unnecessary information should not be given.

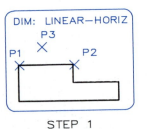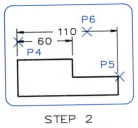

STEP 1   STEP 2

FIGURE 14.29 **Linear dimensioning.**

Step 1 Command: <u>DIM</u> **(CR)**
Dim: HORIZONTAL **(CR)**
First extension line origin or RETURN to select: **(Select P1.)**
Second extension line origin: **(Select P2.)**
Dimension line location: **(Select P3.)**
Dimension text <current>: **(CR) to select default value or input correct value from the keyboard.**

Step 2 Dim: **(CR)**
HORIZONTAL
First extension line origin or RETURN to select: **(Select P4.)**
Second extension line origin: **(Select P5.)**
Dimension line location: **(Select P6.)**
Dimension text <current>: **(CR) to select default or input correct value from the keyboard.**

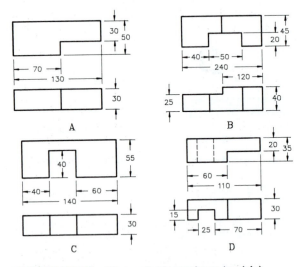

FIGURE 14.28 **Dimensioning prisms (metric).**

A. **Dimensions should extend from the most descriptive view and be placed between the views they apply to.**

B. **One intermediate dimension is not given. Extension lines may cross object lines.**

C. **It is permissible to dimension a notch inside the object if this improves clarity.**

D. **Whenever possible, dimensions should be placed on visible lines, not hidden lines.**

COMPUTER METHOD The part in Fig. 14.29 is dimensioned by entering the DIM command and selecting the LINEAR and HORIZONTAL options. Select the start of the first and second extension lines, and locate the dimension line to obtain each horizontal dimension.

The extension lines can be obtained automatically by responding to the First extension line origin prompt with a (CR). You will get the prompt Select line, arc, or circle. Point to the line to be measured, and locate the dimension line when prompted.

If the dimensioning variable DIMASO is set to ON before dimensions are applied to a part, the dimensions will be associative, that is, they will change in value as the size of the part is STRETCHed (Fig. 14.30). Associative dimensions can also be erased as a single unit (extension lines, dimension lines, arrows, and numerals).

When DIMSHO is ON, the dimensioning numerals will be recomputed dynamically as the part is dragged to a new size.

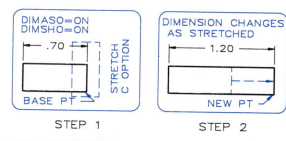

FIGURE 14.30 Associative dimensions.

Step 1 Set Dim Vars, DIMASO, and DIMSHO, to
ON, and the dimensions will be associated with the
size of the part they are applied to. Use the C option
of the STRETCH command to window the part and
its dimension.

Step 2 A base point is selected, and a new point is
selected. As the size of the part is dragged, the dimensions will be recalculated dynamically during
the dragging.

14.13
Dimensioning angles

Angles can be dimensioned either by using coordinates to locate the ends of angular lines or planes or
by using angular measurements in degrees (Fig.
14.31). The two methods should not be mixed when
dimensioning the same angle since they may not
agree.

Units for angular measurements are degrees, minutes, and seconds (Fig. 14.31C). There are 60 minutes
in a degree and 60 seconds in a minute. Seldom will
angular measurements need to be measured to the
nearest second.

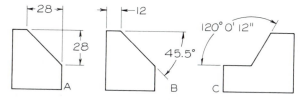

FIGURE 14.31 **A.** Angular planes can be
dimensioned by using coordinates. **B.** Angular
planes can also be dimensioned by using an angle
measured in degrees from the located vertex. Angles
can be measured in decimal fractions. **C.** Angles
can also be measured in degrees, minutes, and
seconds.

COMPUTER METHOD By selecting the AN-
GULAR option under the DIM command, an angle
can be dimensioned as shown in Fig. 14.32. If
room is not available for the arrows between the
extension lines, they will be drawn outside the extension lines. You will be prompted to locate the
position of numerals.

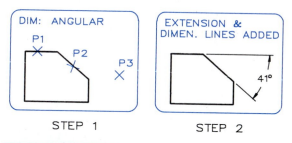

FIGURE 14.32 Angular dimensions.

Step 1 Command: DIM (CR)
Dim: ANGULAR (CR)
Select first line: (Select pt. P1)
Second line: (Select pt. P2)
Enter dimension line arc location: (Se-
lect pt. P3)
Dimension text <41>: (CR) to accept default
value 41.

Step 2 Enter text location: (Select pt. P4)
Dim: (CR) to continue angular dimensioning or Ctrl-
E to return to a Command prompt.

14.14
Dimensioning cylinders

Cylinders that are dimensioned may be either solid
cylinders or cylindrical holes.

Solid cylinders are dimensioned in their **rectangular views** by using **diameters,** not radii.

All diametral dimensions should be preceded by
the symbol ∅ (Fig. 14.33), which indicates the dimension is a diameter. The English system often uses the
abbreviation DIA following the diametral dimension.

Parts having several cylinders, which are concentric, are dimensioned with diameters, beginning with
the smallest cylinder (Fig. 14.33C). A cylindrical part

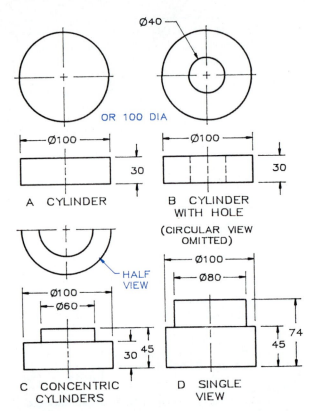

FIGURE 14.33 caption labels: OR 100 DIA; A CYLINDER; B CYLINDER WITH HOLE (CIRCULAR VIEW OMITTED); C CONCENTRIC CYLINDERS; D SINGLE VIEW; HALF VIEW

FIGURE 14.33 **A.** It is preferred that cylinders be dimensioned in their rectangular views using a diameter rather than a radius. **B.** Dimensions should be placed between the views when possible, and holes should be dimensioned with leaders. **C.** Dimension concentric cylinders beginning with the smallest one. **D.** The circular view can be omitted if DIA or ∅ is used with the diametral dimensions.

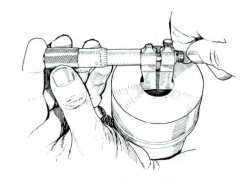

FIGURE 14.34 An internal micrometer caliper for measuring internal cylindrical diameters.

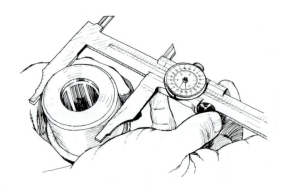

FIGURE 14.35 An external micrometer caliper for measuring the diameter of a cylinder.

may be sufficiently dimensioned with only one view if ∅ or DIA is used with the diametral dimension (Fig. 14.33D).

14.15
Measuring cylindrical parts

Cylindrical parts are dimensioned with diameters rather than radii because diameters are easier to measure. An internal cylindrical hole is measured with an internal micrometer caliper (Fig. 14.34). Likewise, an external micrometer caliper can be used for measuring the outside diameters of a part (Fig. 14.35). Measuring

a **diameter** rather than a **radius** makes it possible to measure during machining when the part is held between centers on a lathe.

14.16
Cylindrical holes

Cylindrical holes may be dimensioned by one of the methods shown in Fig. 14.36. The preferred method of dimensioning cylindrical holes is to draw a **leader** from the **circular view,** and then add the dimension preceded by ∅ (Fig. 14.37) or followed by DIA. Sometimes the note DRILL or BORE is added to specify the shop operation, but current standards prefer the use of DIA instead.

To illustrate various methods of dimensioning, a part containing cylindrical features is dimensioned in Fig. 14.38 in both the circular and rectangular views.

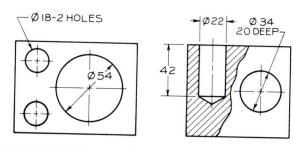

FIGURE 14.36 Several acceptable methods of dimensioning cylindrical holes and shapes. The symbol ∅ is placed in front of the diametral dimension.

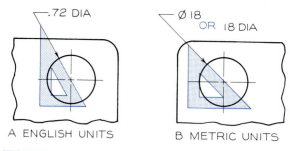

FIGURE 14.37 The preferable method of dimensioning cylindrical holes is with a leader, the symbol ∅, and a dimension to indicate that the dimension is a diameter. Previous standards recommended the abbreviation DIA after the dimension.

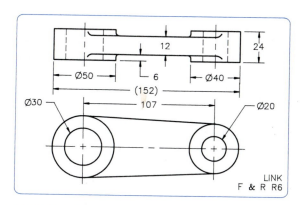

FIGURE 14.38 This part, composed of cylindrical features, has been dimensioned using several approved methods. (F&R R6 means that fillets and rounds have a 6 radius.)

Figure 14.39 shows three methods of dimensioning holes. The depth of a hole is its **usable** depth, not the depth to the point left by the drill bit.

> **COMPUTER METHOD** Circles can be dimensioned in their circular views as shown in Fig. 14.40. The dimension lines will begin with the point selected on the circle and pass through the center. The diameter symbol will be placed in front of the dimension numerals. Small circles will be dimensioned with the arrows inside the circle and the dimension numerals outside connected by a leader.
>
> Even smaller circles will have the arrows and dimension outside the circle.

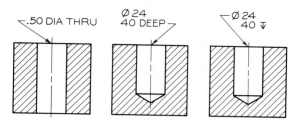

FIGURE 14.39 Three methods of dimensioning holes with combinations of notes and symbols.

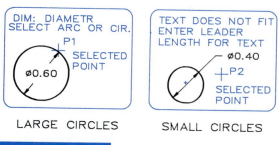

LARGE CIRCLES   SMALL CIRCLES

FIGURE 14.40

Step 1 Select the Diametr option under the DIM command to dimension a circle. Select P1 on the arc (an endpoint of the dimension), and the dimension is drawn. You have the option to replace the dimension measured by the computer with a different value.

Step 2 When text does not fit, you receive the prompt "Text does not fit, Enter leader length for text." Select a point, P2, and the leader and dimension are drawn.

14.17
Pyramids, cones, and spheres

Figures 14.41A–C show three methods of dimensioning pyramids. The **pyramids** in Figs. 14.41B and 14.41C are truncated (that is, the apex has been cut off and replaced by a plane). In all three examples, the apex of the pyramid is in the rectangular view.

Figures 14.41D and 14.41E show two acceptable methods of dimensioning **cones.**

A **sphere,** if it is complete, is dimensioned by using its diameter (Fig. 14.41F); a radius is used if it is not a complete sphere (Fig. 14.41G). Only one view is necessary to describe a sphere.

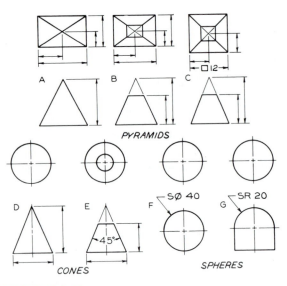

FIGURE 14.41   Methods of dimensioning pyramids, cones, and spheres.

14.18
Leaders

Leaders are used to apply notes and dimensions to a feature they describe. As illustrated in Fig. 14.37, leaders are drawn at a standard angle of a triangle. Figure 14.42 shows examples of notes using leaders. The leader should be drawn from either the first word of the note or the last word of the note and begin with a short horizontal line from the note.

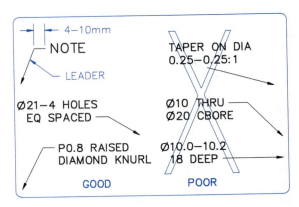

FIGURE 14.42   Leaders from notes should begin with a horizontal bar from the first or last word of the note, not from the middle of the note.

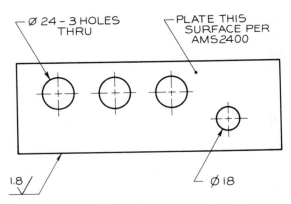

FIGURE 14.43   Examples of notes with leaders applied to a part.

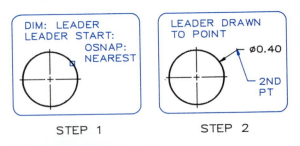

STEP 1   STEP 2

FIGURE 14.44   Leaders.

Step 1 Command: DIM (CR)
Dim: LEADER (CR)
Leader start: OSNAP to P1 using NEAREST option.

Step 2 To point: (Select next point.) (CR)
Dimension text <0.40>: (CR) to accept default value. (The leader and dimension are drawn.)
Dim: (CR) to continue leader dimensioning or Ctrl-C to return to a Command prompt.

Figure 14.43 shows applications of leaders on a part. A dot is used instead of an arrowhead when the note applies to a surface that does not appear as an edge.

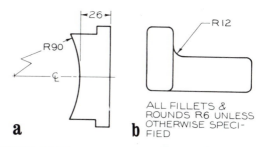

> **COMPUTER METHOD** By selecting the LEADER option under the DIM command, a leader can be drawn that begins with the arrow end (Fig. 14.44). To ensure that the arrow touches the circle, use OSNAP and NEAREST to snap to the circumference. You will be prompted "To point" until (CR) is pressed. Then you will be shown the diameter of the circle, which you can accept by pressing (CR) or override by typing in a different value.
>
> If the previous option was DIAMETR, the dimension will be preceded with the diameter symbol. If the previous option was RADIUS, the dimension will be preceded with an R, the radius symbol.

FIGURE 14.46 a. When a radius is very long, it may be shown with a false radius and a zigzag to indicate that it is not true length. It should end on the centerline of the true center. b. Fillets and rounds may be noted to reduce repetitive dimensions of small arcs.

standards recommended that the R follow the dimension, such as 10R. Thus both methods are seen in practice.

When the arc being dimensioned is long, it may be dimensioned with a false radius, as Fig. 14.46a shows. A zigzag is placed in the radius to indicate that it is not a true radius. The false center of the arc should lie on the extended centerline of the true center.

14.19
Dimensioning arcs

Cylindrical parts **less than a full circle** are dimensioned with **radii,** as Fig. 14.45 shows. Current standards recommend that radii be dimensioned with an **R preceding** the dimension, such as R10. Previous

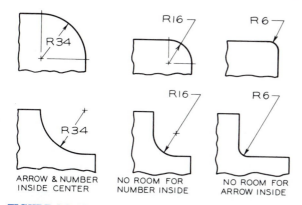

FIGURE 14.45 When space permits, the dimension and arrow should be placed between the center and the arc. If room is not available for the number, the arrow is placed between the center and the arc with the number on the outside. If there is no room for the arrow, both the dimension and arrow are placed outside the arc with a leader.

> **COMPUTER METHOD** By using the RADIUS option of the DIM command, arcs can be dimensioned by selecting a point on the arc, which will be the starting point of the arrow (Fig. 14.47). If room permits, the dimension will be placed between the center and the arrow.
>
> For smaller arcs, the arrows will be placed inside, and dimensions will be placed outside. When room is not available for arrows inside, both the arrow and dimension will be placed outside. Decimal values will be preceded with a zero unless you override them by typing in the value without a preceding zero. If this is done, the R must be typed in also.

14.20
Fillets and rounds and TYP

Fillets and **rounds** are rounded corners conventionally used on castings. A fillet is an internal rounding, and a round is an external rounding.

A note may be placed on the drawing to eliminate the need for repetitive dimensioning of fillets and

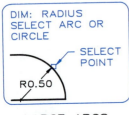
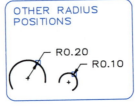

LARGE ARCS   SMALLER ARCS

FIGURE 14.47 By using the RADIUS option of the DIM command, select a point on the arc, and the dimension is drawn with its arrow located at the point selected. The dimension is preceded by an R. Smaller arcs are dimensioned with the dimensions outside the arc, or because of limited space, with the arcs and dimensions outside the arc.

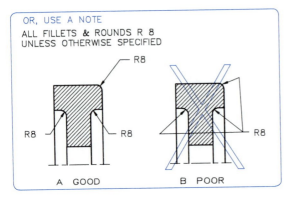

FIGURE 14.48 When several arcs are dimensioned, using separate leaders is preferable to extending the leaders.

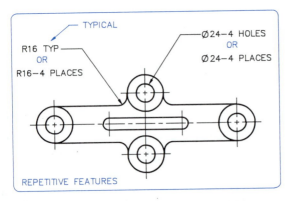

FIGURE 14.49 Notes can be used to indicate that similar features and dimensions are repeated on drawings without having to dimension them individually.

rounds. The note may read ALL FILLETS AND ROUNDS R6. If most, but not all, of the fillets and rounds have equal radii, the note may read ALL FILLETS AND ROUNDS R6 UNLESS OTHERWISE SPECIFIED. In this case, only the fillets and rounds of different radii are dimensioned (Fig. 14.46b). The notes may be abbreviated, for example, **F&R10.**

Fillets and rounds should be dimensioned with short, simple leaders (Fig. 14.48A) rather than with long, confusing leaders (Fig. 14.48B).

Repetitive features on a drawing may be noted as shown in Fig. 14.49. The note TYPICAL or TYP means although only one of these features is dimensioned, the dimensions are typical of those undimensioned. The note PLACES is sometimes used to specify the number of places that a similar feature appears. The number of holes sharing the same dimension may be similarly indicated.

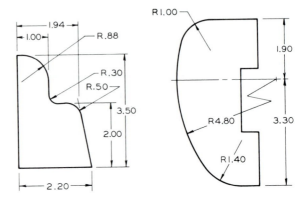

FIGURE 14.50 Examples of dimensioned parts composed of a series of tangent parts.

14.21
Curved surfaces

An irregular shape composed of several tangent arcs of varying sizes (Fig. 14.50) can be dimensioned by using a series of radii.

When the curve is irregular rather than composed of arcs (Fig. 14.51), the coordinate method can be used to locate a series of points along the curve from two datum lines. The drafter must use judgment to determine the proper spacing for the points. Extension lines may be placed at an angle to provide additional space for showing dimensions.

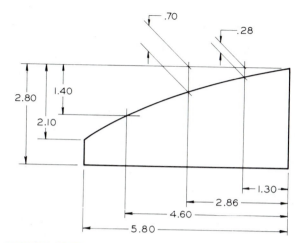

FIGURE 14.51 This object with an irregular curve is dimensioned by using coordinates.

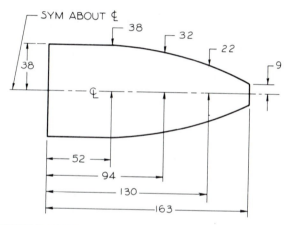

FIGURE 14.52 This symmetrical part is dimensioned about its centerline.

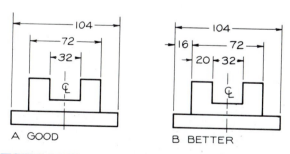

FIGURE 14.53 **A.** Symmetrical parts may be dimensioned about their centerlines as shown here. **B.** The better way of dimensioning symmetrical parts is shown here.

An irregular curve that is symmetrical about an axis is shown dimensioned in Fig. 14.52. Dimension lines are used as extension lines, which is an acceptable violation of rules.

14.22
Symmetrical objects

Symmetrical objects may be dimensioned as shown in Fig. 14.53A, where it is assumed that the dimensions are each centered about the centerline, abbreviated CL. The better method is shown in Fig. 14.53B, where the assumption is eliminated. All dimensions are located with respect to each other.

14.23
Finished surfaces

Many parts are formed as castings in a mold that gives the parts' exterior surfaces a rough finish. If the part is designed to come in contact with another surface, the rough finish must be machined by grinding, shaping, lapping, or similar process.

To indicate that a surface is to be finished, **finish marks** are drawn on the surface where it appears as an **edge** (Fig. 14.54).

> Finish marks should be repeated in **every view** where the finished surface appears as an edge even if it is a **hidden line.**

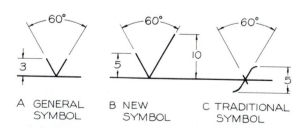

FIGURE 14.54 Finish marks indicate that a surface has been machined to a smooth surface. **A.** The traditional V can be used for general applications. **B.** The new finish mark is related to surface texture. **C.** The *f* is also used to indicate finished surfaces.

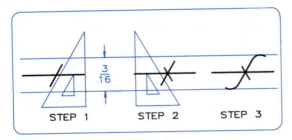

FIGURE 14.55 The steps of drawing the f finish mark.

Figure 14.54 shows three methods of drawing finish marks. The simple V mark is used in general cases. The uneven V (Fig. 14.54), a newly recommended symbol, is related to surface texture; we discuss it in Chapter 15. Figure 14.55 shows the steps of constructing the traditional F finish mark. When an object is finished on all surfaces, the note FINISHED ALL OVER (abbreviated **FAO**) is placed on the drawing.

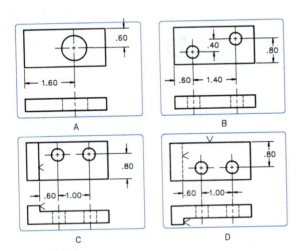

FIGURE 14.56 Location dimensions.

A. Cylindrical holes should be located in their circular views from two surfaces of the object.

B. When more than one hole is to be located, they should be located from center to center.

C. Holes should be located from finished surfaces.

D. Holes should be located in the circular view and from finished surfaces even if the finished surfaces are hidden.

14.24
Location dimensions

Location dimensions are used to locate the **positions, not the sizes,** of geometric elements, such as cylindrical holes (Fig. 14.56). The sizes of the holes are omitted for clarity. The centers of the holes are located with coordinates in the circular view when possible.

> Holes should be **located** from **finished surfaces** since holes can be located more accurately from a smooth, machined surface than from a rough, unfinished one.

This rule is followed even if the finished surface is a hidden line, as in Fig. 14.57.

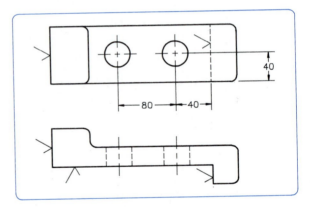

FIGURE 14.57 Cylindrical holes are located from center to center in their circular view and from a finished surface if one is available.

Location dimensions should be placed on views where both dimensions can be shown (Fig. 14.58). Cylinders are located in their circular views (Figs. 14.58B and 14.59).

14.25
Location of holes

When holes must be located accurately, the dimensions should originate from a common datum plane on the part to reduce errors in measurement (Figs.

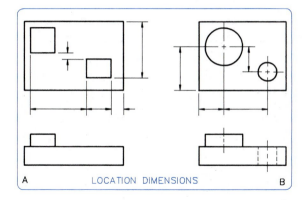

FIGURE 14.58 Prisms and cylinders are located with coordinates in the view where both coordinates can be seen.

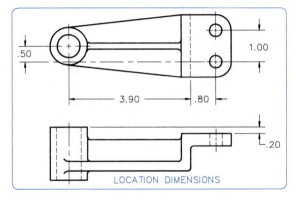

FIGURE 14.59 Location dimensions applied to a part to locate its geometric features.

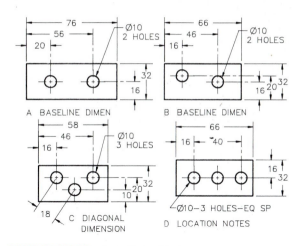

FIGURE 14.60 Location of holes.

A and B. Holes can be more accurately located if measured from common datum.

C. A diagonal dimension can be used to locate a hole of this type from another hole's center.

D. A note can be used to specify the spacing between the centers of equally spaced holes.

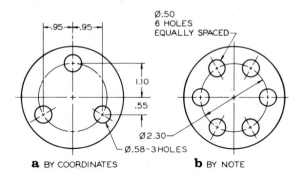

FIGURE 14.61 Holes may be located in circular plates by coordinates or notes.

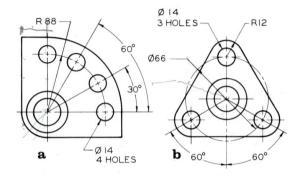

FIGURE 14.62 Methods of locating cylindrical holes on concentric arcs.

14.60A and 14.60B). When several holes in a series are to be equally spaced, as in Fig. 14.60C, a note specifying as much can be used to locate the holes. The first and last holes of the series are determined by the usual location dimensions.

Holes through circular plates may be located by coordinates or a note (Fig. 14.61). When a note is used, the diameter of the imaginary circle passing through the centers of the holes must be given. This circle is called the **bolt circle** or **circle of centers.**

A similar method of locating holes is the **polar system** illustrated in Fig. 14.62. The radial distances from the point of concurrency and their angular measurements (in degrees) between the holes are used to locate the centers.

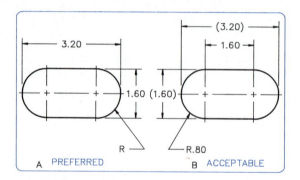

FIGURE 14.63 **A. The preferred method of dimensioning objects with rounded ends. B. The less desirable method.**

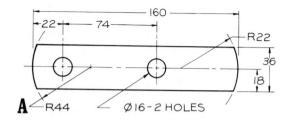

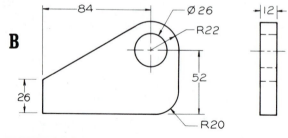

FIGURE 14.64 Examples of dimensioned parts with rounded ends and cylindrical features.

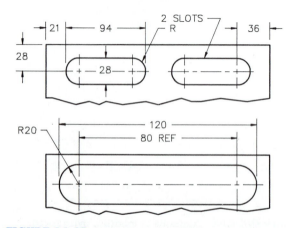

FIGURE 14.65 Methods of dimensioning slots.

14.26
Objects with rounded ends

Objects with rounded ends should be dimensioned from end to end (Fig. 14.63A). The radius, shown as R without a dimension, specifies the end is an arc. Since the height is given, the radius is understood to be half the height.

If the object is dimensioned from center to center (Fig. 14.63B), the overall dimension should be given as a reference dimension (3.40) to eliminate calculating the overall dimension. In this case, the radii must be given.

A part with partially rounded ends is dimensioned in Fig. 14.64A. The overall dimension and radii are given so that their centers may be located. When an object has a rounded end that is less than a semicircle (Fig. 14.64B), location dimensions must be used to locate the center of the arc.

Slots with rounded ends are dimensioned in Fig. 14.65. Only one slot is dimensioned in Fig. 14.65A, with a note indicating there are two slots. The slot in Fig. 14.65B is dimensioned giving the overall dimension and the two arcs, which are understood to apply to both ends. The distance between the centers is given as a reference dimension.

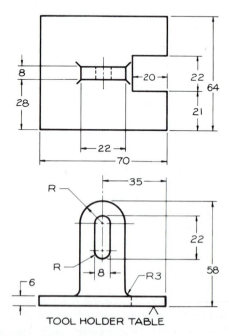

FIGURE 14.66 Examples of arcs and slots.

The dimensioned views of the tool holder table in Fig. 14.66 show examples of arcs and slots. To prevent dimension lines from crossing, it is often necessary to place dimensions in a view less descriptive than might be desired.

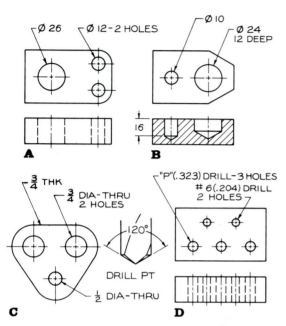

FIGURE 14.67 A. and B. Cylindrical holes may be dimensioned by either of these methods. C. and D. When only one view is given, it is necessary to note the THRU holes.

14.27
Machined holes

Machined holes are made or refined by a machine operation, such as drilling or boring (Fig. 14.67). It is preferable to give the diameter of the hole with the symbol $\emptyset$ in front of the dimension ($\emptyset$ 32); however, the note 32 DRILL may be used in some cases. You will also see diameters dimensioned as XX DIA since this was the standard previously recommended.

DRILLING is the most common method of machining holes. The depth of a drilled hole can be specified in the note, or it may be dimensioned in the rectangular view. The depth of a drilled hole is **dimensioned** as the **usable part** of the hole; the conical point is disregarded (Fig. 14.67B).

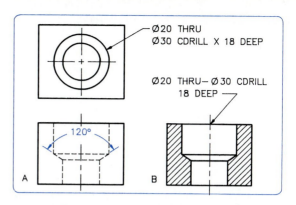

FIGURE 14.68 Counterdrilling notes gives the specifications for a larger hole drilled inside a smaller hole. The 120° angle is not required as a dimension since this is the standard angle of a drill point.

COUNTERDRILLING is drilling a large hole inside a smaller drilled hole to enlarge it (Fig. 14.68). The 120° dimension indicates the angle of the drill point; it need not be dimensioned on the drawing.

COUNTERSINKING is the process of forming a conical hole and is often used with flat-head screws (Fig. 14.69). The diameter of the countersunk hole

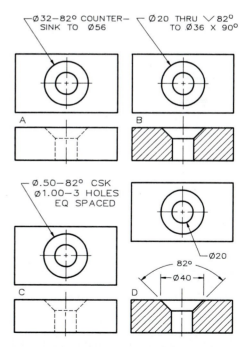

FIGURE 14.69 Examples of notes and symbols specifying countersunk holes.

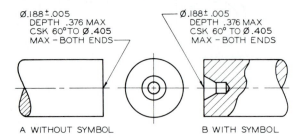

A WITHOUT SYMBOL   B WITH SYMBOL

FIGURE 14.70 Methods of specifying countersinks in the ends of cylinders for mounting them on a lathe between centers.

(the maximum diameter on the surface) and the angle of the countersink are given in the notes. Countersinking is also used to provide center holes in shafts, spindles, and other cylindrical parts held between the centers of a lathe (Fig. 14.70).

SPOTFACING is a machining process used to finish the surface around the top of a hole to provide a level seat for a washer or fastener head (Figs. 14.71A and

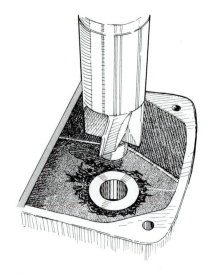

FIGURE 14.72 This tool is spotfacing the cylindrical boss to provide a smooth seat for a bolt head.

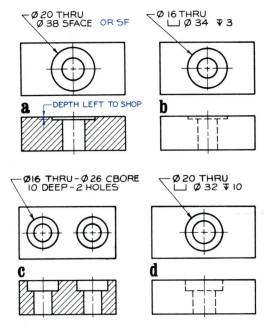

FIGURE 14.71 a. and b. Spotfaces can be specified as shown here. The depth of the spotface can be specified if needed. c. and d. Counterbores are dimensioned by giving the diameters of both holes and the depth of the larger hole.

FIGURE 14.73 Boring a large hole on a lathe with a boring bar. (Courtesy of Clausing Corp.)

14.71B). The spotfacing tool in Fig. 14.72 has spot-faced a boss (a raised cylindrical element).

BORING   is a machine operation for making large holes. It is usually performed on a lathe with a boring bar (Fig. 14.73).

COUNTERBORING   is the process of enlarging the diameter of a drilled hole (Fig. 14.74). The bottoms of counterbored holes are flat with no taper as in counterdrilled and countersunk holes.

REAMING   is the operation of finishing or slightly enlarging holes that have been drilled or bored. This operation uses a ream similar to a drill bit.

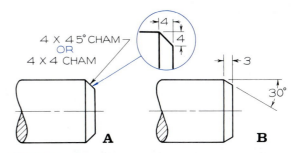

FIGURE 14.74   **A. Chamfers can be dimensioned by using either type of note when the angle is 45°. B. If the angle is other than 45°, it should be dimensioned as shown here.**

14.28
Chamfers

Chamfers are beveled edges that are made on cylindrical parts, such as shafts and threaded fasteners. They eliminate sharp edges and facilitate the assembly of parts.

When a chamfer angle is 45°, a note can be used in either of the forms shown in Fig. 14.74A. When the chamfer angle is other than 45°, the angle and length are given, as Fig. 14.74B shows. Chamfers can also be specified at the openings of holes (Fig. 14.75).

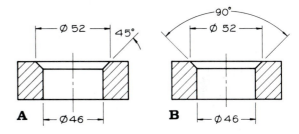

FIGURE 14.75   **Inside chamfers are dimensioned by using one of these methods.**

14.29
Keyseats

A **keyseat** is a slot cut into a shaft for aligning the shaft with a pulley or collar mounted on it. Figure 14.76 shows the method of dimensioning a keyway and keyseat. The double dimensions are tolerances, which we discuss in Chapter 15.

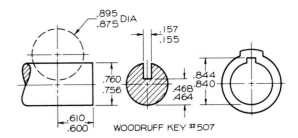

FIGURE 14.76   **Methods of dimensioning slots in a shaft and a slot for a Woodruff key that will hold the part on the shaft. Appendix 33 gives the dimensions for these features.**

14.30
Knurling

Knurling is the operation of cutting diamond-shaped or parallel patterns on cylindrical surfaces for gripping, decoration, or a press fit between two parts that will be permanently assembled.

A **diamond knurl** and **straight knurl** are drawn and dimensioned in Fig. 14.77. Knurls should be dimensioned with specifications that give type, pitch, and diameter.

The abbreviation DP in Fig. 14.77 means diametral pitch, the ratio of the number of grooves on the circumference *(N)* to the diameter *(D)*, which is found by the equation $DP = N/D$. The preferred diametral

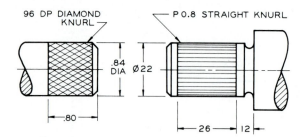

FIGURE 14.77 **A.** A diamond knurl with a diametral pitch of 96. **B.** A straight knurl where the linear pitch (P) is 0.8 mm. Pitch is the distance between the grooves on the circumference.

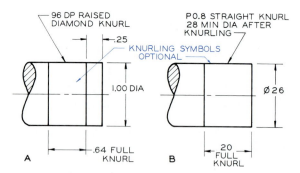

FIGURE 14.78 Knurls need not be drawn if they are dimensioned as shown here.

pitches for knurling are 64 DP, 96 DP, 128 DP, and 160 DP. For diameters of 1 inch, knurling of 64 DP, 96 DP, 128 DP, and 160 DP will have 64, 96, 128, and 160 teeth, respectively, on the circumference. The note P 0.8 means the knurling grooves are 0.8 mm apart. Calculations for knurling must be made using inches, with conversion to millimeters made afterward.

Knurls for press fits are specified with diameters before knurling and with the minimum diameter after knurling. A simplified method of representing knurls is shown in Fig. 14.78, where notes are used and the knurls are not drawn.

14.31
Necks and undercuts

A **neck** is a recess cut into a cylindrical part. Where cylinders of different diameters join (Fig. 14.79), a neck ensures that the part assembled on the smaller shaft will fit flush against the shoulder of the larger cylinder.

Undercuts, which are similar to necks (Fig. 14.80A), ensure that a part fitting in the corner of the part will fit flush against both surfaces; they also permit space for trash to drop out of the way when entrapped in the corner. An undercut could also be a recessed neck inside a cylindrical hole. A thread relief

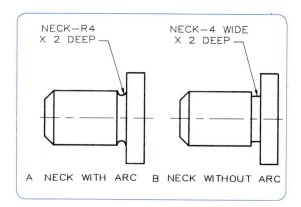

FIGURE 14.79 Necks are recesses in cylinders that are used where cylinders of different sizes join together.

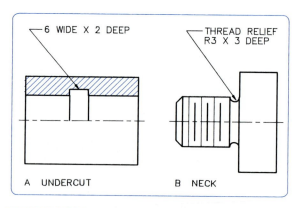

FIGURE 14.80 **A.** An undercut can be dimensioned as shown here. **B.** A thread relief, which is a type of neck, is dimensioned here.

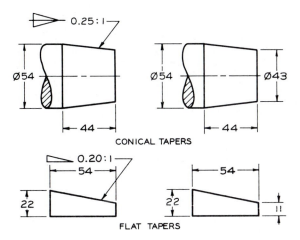

CONICAL TAPERS

FLAT TAPERS

FIGURE 14.81 **Examples of conical and flat tapers. Taper is the ratio of the diameters (or heights) at each end of a sloping surface to the length of the taper.**

(Fig. 14.80B) is used to square the threads where they intersect a larger cylinder.

14.32
Tapers

Tapers can be either conical surfaces or flat planes (Fig. 14.81). A taper can be dimensioned by the diameter or width at each end of the taper, the length of the tapered feature, or the rate of taper.

Taper is the ratio of the difference in the diameters of a cone to the distance between two diameters (Fig. 14.81). Taper can be expressed as inches per inch (.25 per inch), inches per foot (3.00 per foot), or millimeters per millimeter (0.25:1).

Flat taper is the ratio of the difference in the heights at each end of a feature to the distance between the heights. Flat taper can be expressed as inches per inch (.20 per inch), inches per foot (2.40 per foot), or millimeters per millimeter (0.20:1), as Fig. 14.81 shows.

14.33
Dimensioning sections

Sections are dimensioned in the same manner as regular views (Fig. 14.82). Most of the principles of dimensioning covered in this chapter have been applied to the part shown in this figure.

FIGURE 14.82 **An example of a computer-drawn part dimensioned with a variety of dimensioning principles.**

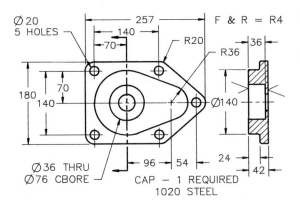

CAP — 1 REQUIRED
1020 STEEL

14.34
Miscellaneous notes

A variety of notes are used on detail drawings to provide information and specifications that would otherwise be difficult to represent (Figs. 14.83 and 14.84).

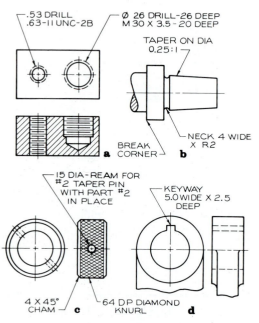

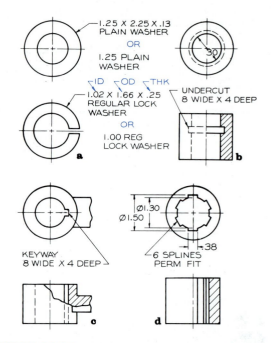

FIGURE 14.84 **a.** Methods of dimensioning washers and lock washers. These dimensions can be found in Appendixes 35 and 36. **b.** An undercut is dimensioned. **c.** A keyway is dimensioned. **d.** A spline inside a hole is dimensioned.

FIGURE 14.83 **a.** and **b.** Threaded holes are sometimes dimensioned by giving the tap drill size in addition to the thread specifications. The tap drill size is not required, but it is permissible. **b.** This part is dimensioned to indicate a neck, taper, and break corner, which is a slight round to remove the sharpness from a corner. **c.** This collar has a knurl note, chamfer note, and note indicating the insertion of a #2 taper pin. **d.** This part has a note for dimensioning a keyway.

Notes are placed horizontally on the sheet; if the notes lie on the same line, short dashes should be used between them. The abbreviations shown in these notes can be used to save space. Appendix 1 lists the standard abbreviations.

Problems

1–24. (Figs. 14.85–14.86) These problems are to be solved on Size A paper, one per sheet, if they are drawn full size. If drawn double size, use Size B paper. The views are drawn on a 0.20″ (5 mm) grid.

You will need to vary the spacing between the views to provide adequate room for the dimensions. It would be a good idea to sketch the views and dimensions to determine the required spacing before laying out the problems with instruments.

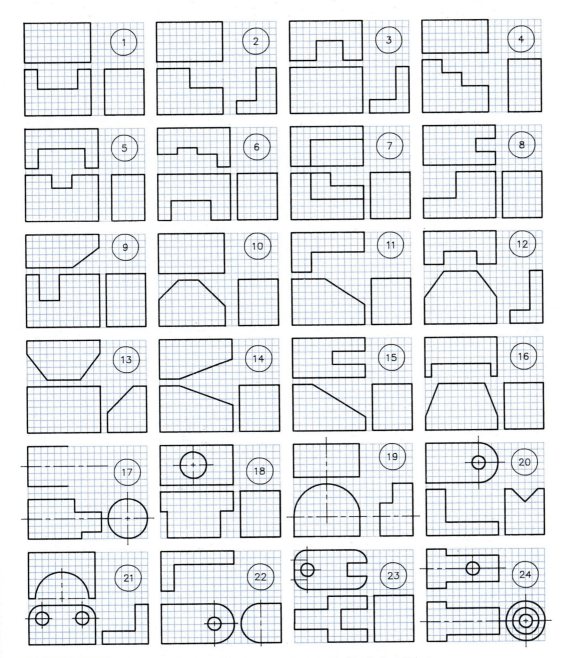

FIGURE 14.85   Problems 1–24(a). Lay out the views and dimension them.

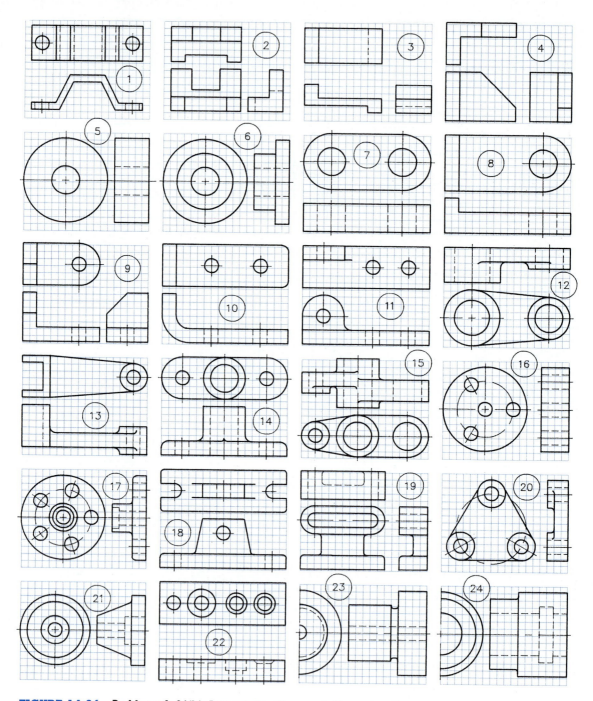

FIGURE 14.86 Problems 1–24(b). Lay out the views and dimension them.

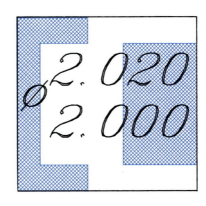

Tolerances

15

15.1
Introduction

Today's technologies require increasingly exact dimensions. What is more, many of today's parts are made by different companies in different locations; therefore, these parts must be specified so that they will be interchangeable.

The techniques of dimensioning parts to ensure interchangeability is called **tolerancing.** Each dimension is allowed a certain degree of variation within a specified zone, or a **tolerance.** For example, a part's dimension might be expressed as 100 ± 0.50, which yields a tolerance of 1.00 mm.

> Dimensions should be given as **large a tolerance** as possible without interfering with the function of the part to reduce production costs. Manufacturing to close tolerances is expensive.

15.2
Tolerance dimensions

Figure 15.1 shows several acceptable methods of specifying tolerances. When plus-and-minus tolerancing is used, tolerances are applied to a **basic dimension.** When dimensions allow variation in only one direction, the tolerancing is **unilateral.** Tolerancing that

permits variation in either direction from the basic dimension is **bilateral.**

Tolerances may also be given in the form of **limits;** that is, two dimensions are given that represent

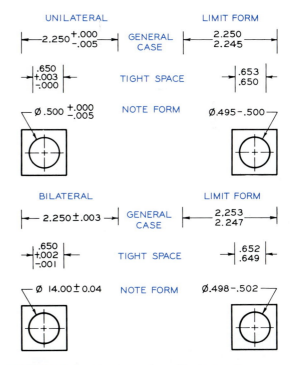

FIGURE 15.1 **Methods of positioning and indicating tolerances in unilateral, bilateral, and limit forms.**

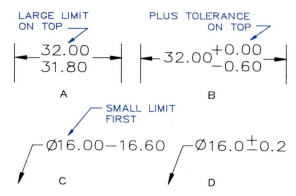

LARGE LIMIT
ON TOP

PLUS TOLERANCE
ON TOP

32.00
31.80

A

$32.00^{+0.00}_{-0.60}$

B

SMALL LIMIT
FIRST

⌀16.00—16.60   ⌀16.0±0.2

C   D

FIGURE 15.2 When limit dimensions are given, the large limits are placed either above or to the right of the small limits. In plus-and-minus tolerancing, the plus limits are placed above the minus limits.

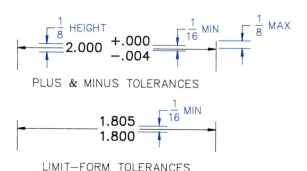

$\frac{1}{8}$ HEIGHT

$2.000 \begin{array}{c} +.000 \\ -.004 \end{array}$

$\frac{1}{16}$ MIN $\frac{1}{8}$ MAX

PLUS & MINUS TOLERANCES

$\frac{1.805}{1.800}$ $\frac{1}{16}$ MIN

LIMIT—FORM TOLERANCES

FIGURE 15.3 Positioning and spacing of numerals used to specify tolerances.

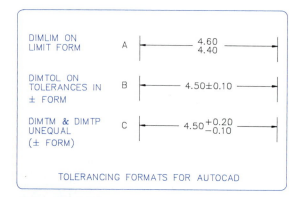

DIMLIM ON
LIMIT FORM   A $\frac{4.60}{4.40}$

DIMTOL ON
TOLERANCES IN   B   4.50±0.10
± FORM

DIMTM & DIMTP
UNEQUAL   C $4.50^{+0.20}_{-0.10}$
(± FORM)

TOLERANCING FORMATS FOR AUTOCAD

FIGURE 15.4 AutoCAD will give toleranced dimensions in the forms shown here. In all cases, the DIMTOL mode must be ON. If DIMLIM is ON, tolerances will be applied in limit form; when OFF, tolerances will be in plus-and-minus form.

the largest and smallest sizes permitted for a feature of the part.

Figure 15.2 shows the customary methods of indicating toleranced dimensions, and Fig. 15.3 shows the positioning and spacing of numerals of toleranced dimensions.

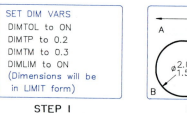

SET DIM VARS
DIMTOL to ON
DIMTP to 0.2
DIMTM to 0.3
DIMLIM to ON
(Dimensions will be
in LIMIT form)

STEP 1

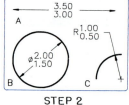

$\frac{3.50}{3.00}$

A

$R^{1.00}_{0.50}$

$⌀^{2.00}_{1.50}$

B   C

STEP 2

FIGURE 15.5 Limit tolerances by computer.

Step 1 Set DIM: variables (DIM VARS) as shown for tolerances to be given in LIMIT form.

Step 2 Linear dimensions will be given in limit form, diametral dimensions will be given as limits preceded by ⌀, and radial dimensions will be preceded by R.

TOLERANCES BY COMPUTER Using AutoCAD, toleranced dimensions can be shown in limit form or plus-and-minus form (Fig. 15.4). Both DIMLIM and DIMTOL must be turned ON to obtain dimensions in limit form. The tolerances are assigned to DIMTM and DIMTP modes under the DIM: command.

In addition to linear dimensions, diametral and radial dimensions are automatically given with either a circle symbol or an R preceding the dimensions as Figs. 15.5 and 15.6 show.

Angular measurements can be toleranced using the plus-and-minus form or the limit form as shown in Fig. 15.7. The computer measures the angle and automatically computes the upper and lower limits.

You will want to edit a toleranced dimension when you change the limits given by the automatic process. By a series of erasures and moves (Fig. 15.8), you will be able to make the desired changes.

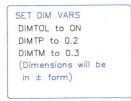

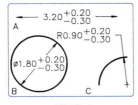

FIGURE 15.6 Plus-and-minus tolerances by computer.

Step 1 Set DIM: variables (DIM VARS) as shown with DIMLIM set to OFF.

Step 2 Linear dimensions will be given as a basic diameter followed by plus-and-minus tolerances, diametral dimensions will be given as linear dimensions preceded by ⌀, and radial dimensions will be preceded by R.

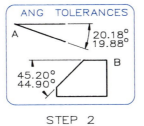

FIGURE 15.7 Angular tolerances.

Step 1 Set DIM VARS as shown above. Be sure the UNITS command is used to assign the desired number of decimal places for angular units beforehand.

Step 2 Select the lines forming the angles, and the toleranced measurements will be given in limit form. Angular measurements can be specified by the UNITS command as decimal degrees, minutes and seconds, grads, radians, or surveyor's units.

15.3
Mating parts

Mating parts are parts that fit together within a prescribed degree of accuracy (Fig. 15.9). The upper piece is dimensioned with two measurements that indicate the upper and lower limits of the size. The notch is slightly larger, allowing the parts to be assembled with a clearance fit.

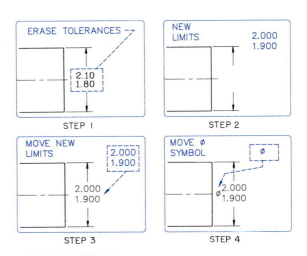

FIGURE 15.8 Editing tolerances by computer.

Step 1 When you wish to change a toleranced dimension, begin by erasing the given tolerances.

Step 2 Using TEXT, draw the new limits of tolerance in a convenient location.

Step 3 Using a window, MOVE the new limits to the dimension line.

Step 4 Locate a diameter symbol, ⌀, by typing %%C, and move it in front of the two new limits.

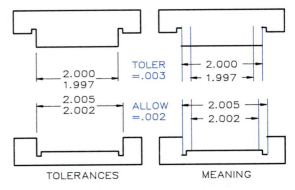

FIGURE 15.9 Each of these mating parts has a tolerance of 0.003″ (variation in size). The allowance between the assembled parts (tightest fit) is 0.002″.

Figure 15.10A shows an example of mating cylindrical parts, and Fig. 15.10B illustrates the meaning of the tolerance dimensions. The size of the shaft can vary in diameter from 1.500 in. (maximum size) to 1.498 in. (minimum size). The difference between

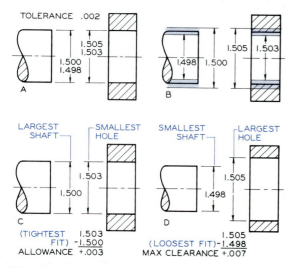

FIGURE 15.10 The allowance (tightest fit) between these assembled parts is +0.003″. The maximum clearance is 0.007″.

these limits on a single part is a tolerance of 0.002 in. The dimensions of the hole in Fig. 15.10A are given with limits of 1.503 and 1.505, for a tolerance of 0.002 (the difference between the limits as shown in Fig. 15.10B).

15.4
Terminology of tolerancing

The meaning of most of the terms used in tolerancing can be seen by referring to Fig. 15.10.

TOLERANCE is the difference between the limits prescribed for a single part. The tolerance of the shaft in Fig. 15.10 is 0.002 in.

LIMITS OF TOLERANCE are the extreme measurements permitted by the maximum and minimum sizes of a part. The limits of tolerance of the shaft in Fig. 15.10 are 1.500 and 1.498.

ALLOWANCE is the tightest fit between two mating parts. The allowance between the largest shaft and the smallest hole in Fig. 15.10 is 0.003 (negative for an interference fit).

NOMINAL SIZE is an approximate size that is usually expressed with common fractions. The nomi-

nal sizes of the shaft and hole in Fig. 15.10 are 1.50 in. or $1\frac{1}{2}$ in.

BASIC SIZE is the exact theoretical size from which limits are derived by the application of plus-and-minus tolerances. There is no basic diameter if this is expressed with limits.

ACTUAL SIZE is the measured size of the finished part.

FIT signifies the type of fit between two mating parts when assembled. There are four types of fit: clearance, interference, transition, and line.

CLEARANCE FIT is a fit that gives a clearance between two assembled mating parts. The fit between the shaft and the hole in Fig. 15.10 is a clearance fit that permits a minimum clearance of 0.003 in. and a maximum clearance of 0.007 in.

INTERFERENCE FIT is a fit that results in an interference between the two assembled parts. The shaft in Fig. 15.11A is larger than the hole, so it requires a force or press fit, which has an effect similar to welding the two parts.

TRANSITION FIT can result in either an interference or a clearance. The shaft in Fig. 15.11B can be either smaller or larger than the hole and still be within the prescribed tolerances.

LINE FIT can result in a contact of surfaces or a clearance between them. The shaft in Fig. 15.11C can have contact or clearance when the limits are approached.

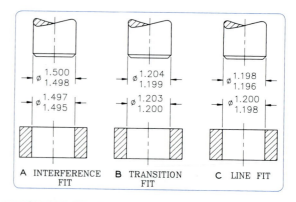

FIGURE 15.11 Types of fits between mating parts. The clearance fit is not shown.

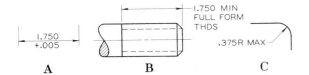

FIGURE 15.12 **Single tolerances can be given in some applications in MAX or MIN form.**

SELECTIVE ASSEMBLY is a method of selecting and assembling parts by trial and error. Using this method, parts can be assembled that have greater tolerances and produced at a reduced cost. This hand-assembly process is a compromise between a high degree of manufacturing accuracy and an ease of assembly of interchangeable parts.

SINGLE LIMITS are dimensions designated by either MIN (minimum) or MAX (maximum), not by both (Fig. 15.12). Depths of holes, lengths, threads, corner radii, chamfers, and so on are sometimes dimensioned in this manner. Caution should be taken to prevent substantial deviations from the single limit.

15.5
Basic hole system

Widely used by industry, the basic hole system of dimensioning holes and shafts gives the required allowance between two assembled parts. The smallest hole is taken as the basic diameter from which the limits of tolerance and allowance are applied. It is advantageous to use the hole diameter as the basic dimension because many of the standard drills, reamers, and machine tools are designed to give standard hole sizes.

If the smallest diameter of a hole is 1.500 in., the allowance (0.003 in this example) can be subtracted from this diameter to find the diameter of the largest shaft (1.497 in.). The smallest limit for the shaft can then be found by subtracting the tolerance from 1.497 in.

15.6
Basic shaft system

Some industries use the basic shaft system of applying tolerances to dimensions of shafts since many shafts

come in standard sizes. In this system, the largest diameter of the shaft is used as the basic diameter from which the tolerances and allowances are applied.

If the largest permissible shaft is 1.500 in., the allowance can be added to this dimension to yield the smallest possible diameter of the hole into which the shaft must fit. Therefore, if the parts are to have an allowance of 0.004 in., the smallest hole would have a diameter of 1.504 in.

15.7
Metric limits and fits

In this section, we cover the metric system as recommended by the International Standards Organization (ISO), which has been presented in ANSI B4.2. These fits usually apply to cylinders—holes and shafts. These tables can also be used to determine the fits between any parallel surfaces, such as a key in a slot.

Metric definitions of limits and fits

Figure 15.13 illustrates some of the definitions given below.

BASIC SIZE is the size from which the limits or deviations are assigned. Basic sizes, usually diameters,

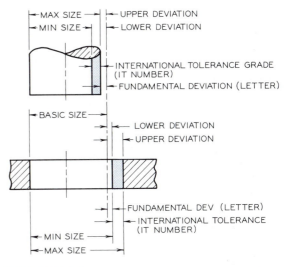

FIGURE 15.13 **Terms related to metric fits and limits.**

TABLE 15.1
PREFERRED SIZES

Basic Size (mm)		Basic Size (mm)		Basic Size (mm)	
First Choice	Second Choice	First Choice	Second Choice	First Choice	Second Choice
1		10		100	
	1.1		11		110
1.2		12		120	
	1.4		14		140
1.6		16		160	
	1.8		18		180
2		20		200	
	2.2		22		220
2.5		25		250	
	2.8		28		280
3		30		300	
	3.5		35		350
4		40		400	
	4.5		45		450
5		50		500	
	5.5		55		550
6		60		600	
	7		70		700
8		80		800	
	9		90		900
				1000	

should be selected from Table 15.1 under the First Choice column.

DEVIATION   is the difference between the hole or shaft size and the basic size.

UPPER DEVIATION   is the difference between the maximum permissible size of a part and its basic size.

LOWER DEVIATION   is the difference between the minimum permissible size of a part and its basic size.

FUNDAMENTAL DEVIATION   is the deviation closest to the basic size. In the note 40H7, the H (an uppercase letter) represents the fundamental deviation for a hole. In the note 40g6, the g (a lowercase letter) represents the fundamental deviation for a shaft.

TOLERANCE   is the difference between the maximum and minimum allowable sizes of a single part.

INTERNATIONAL TOLERANCE (IT) GRADE is a group of tolerances that vary in accordance with the basic size and provide a uniform level of accuracy

within a given grade. In the note 40H7, the 7 represents the IT grade. There are 18 IT grades: IT01, IT0, IT1, . . . , IT16.

TOLERANCE ZONE   is the zone that represents the tolerance grade and its position in relation to the basic size. This is a combination of the fundamental deviation (represented by a letter) and the international tolerance grade (IT number). In note 40H8, the H8 indicates the tolerance zone.

HOLE BASIS   is a system of fits based on the minimum hole size as the basic diameter. The fundamental deviation for a hole basis system is an uppercase letter, *H,* for example (Fig. 15.14).

SHAFT BASIS   is a system of fits based on the maximum shaft size as the basic diameter. The fundamental deviation for a shaft basis system is a lowercase letter, *f,* for example (Fig. 15.14).

CLEARANCE FIT   is a fit that results in a clearance between two assembled parts under all tolerance conditions.

INTERFERENCE FIT   is a fit between two parts that requires they be forced together when assembled.

TRANSITION FIT   is a fit that results in either a clearance or an interference fit between two assembled parts.

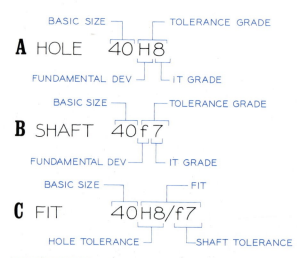

FIGURE 15.14 Symbols and their definitions as applied to holes and shafts.

$$40H8 \quad 40H8 \begin{pmatrix} 40.039 \\ 40.000 \end{pmatrix} \quad \begin{matrix} 40.039 \\ 40.000 \end{matrix} (40H8)$$

A   B   C

FIGURE 15.15 Three methods of giving tolerance symbols. The numbers in parentheses are for reference.

TOLERANCE SYMBOLS are notes used to communicate the specifications of tolerance and fit (Fig. 15.14). The **basic size** is the primary dimension the tolerances are determined from; therefore, it is the first part of the symbol. It is followed by the fundamental deviation letter and the IT number to give the tolerance zone. **Uppercase letters** are used to indicate the fundamental deviation for **holes,** and **lowercase letters** are used for **shafts.**

Figure 15.15 shows three methods of specifying tolerance information. Parenthetical information is for reference only. Appendixes 44–47 give the upper and lower limits.

15.8
Preferred sizes and fits

Table 15.1 shows the preferred basic sizes for computing tolerances. Under the First Choice heading, each number increases by about 25% of the preceding number. Each number in the Second Choice column increases by about 12%. To reduce expenses, you should, where possible, select basic diameters from the first column since these correspond to standard stock sizes for round, square, and hexagonal metal products.

Preferred fits for clearance, transition, and interference fits are shown in Table 15.2 for hole basis and shaft basis fits. The tables in Appendixes 44–47 correspond to these fits.

TABLE 15.2
DESCRIPTION OF PREFERRED FITS

	ISO Symbol		Description	
	Hole Basis	Shaft Basis		
Clearance fits	H11/c11	C11/h11	**Loose running fit** for wide commercial tolerances or allowances on external members	More clearance
	H9/d9	D9/h9	**Free running fit** not for use where accuracy is essential, but good for large temperature variations, high running speeds, or heavy journal pressures	
	H8/f7	F8/h7	**Close running fit** for running on accurate machines and for accurate location at moderate speeds and journal pressures	
	H7/g6	G7/h6	**Sliding fit** not intended to run freely but to move and turn freely and locate accurately	
Transition fits	H7/h6	H7/h6	**Locational clearance fit** provides snug fit for locating stationary parts but can be freely assembled and disassembled	
	H7/k6	K7/h6	**Locational transisition fit** for accurate location; a compromise between clearance and interference	
	H7/n6	N7/h6	**Locational transition fit** for more accurate location where greater interference is permissible	More interference
Interference fits	H7/p6*	P7/h6	**Locational interference fit** for parts requiring rigidity and alignment with prime accuracy of location but without special bore pressure requirements	
	H7/s6	S7/h6	**Medium drive fit** for ordinary steel parts or shrink fits on light sections; the tightest fit usable with cast iron.	
	H7/u6	U7/h6	**Force fit** suitable for parts that can be highly stressed or for shrink fits where the heavy pressing forces required are impractical	

*Transition fit for basic sizes in range from 0 through 3 mm.

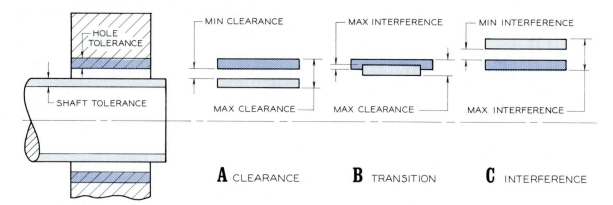

A CLEARANCE   **B** TRANSITION   **C** INTERFERENCE

FIGURE 15.16 Types of fits.

A. A clearance fit.

B. A transition fit where there can be an interference or a clearance.

C. An interference fit where the parts must be forced together.

PREFERRED FITS—HOLE BASIS SYSTEM Figure 15.16 illustrates the symbols used to show the possible combinations of fits when using the hole basis system. There is a **clearance fit** between the two parts, a **transition fit,** and an **interference fit.** This technique of representing fits is used in Fig. 15.17 to show a series of fits for a hole basis system. Note that the lower deviation of the hole is zero; in other words, the smallest size of the hole is the basic size. The different sizes of the shafts give a variety of fits from c11

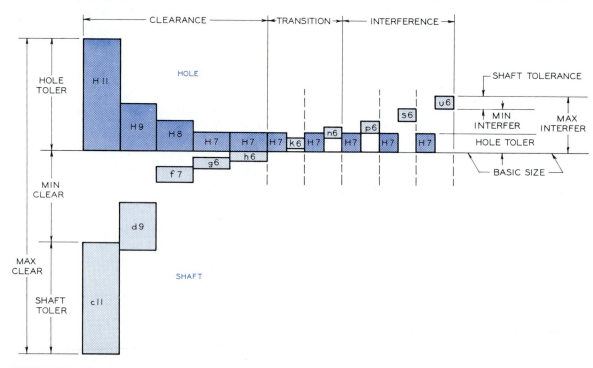

FIGURE 15.17 The preferred fits for a shaft basis system. These fits correspond to those given in Table 15.2.

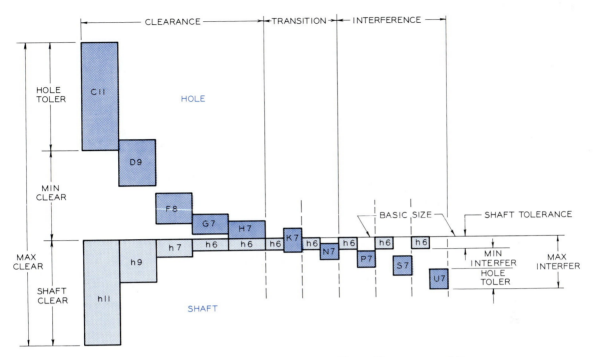

FIGURE 15.18 The preferred fits for a shaft basis system. These fits correspond to those given in Table 15.2.

to u6, where there is a maximum of interference. These fits correspond to those given in Table 15.2.

PREFERRED FITS—SHAFT BASIS SYSTEM Figure 15.18 shows the preferred fits based on the shaft basis system, where the largest shaft size is the basic diameter. The variation in the fit between the parts is caused by varying the size of the holes, which results in a range from a clearance fit of C11/h11 to an interference fit of U7/h6.

15.9
Example problems—metric system

The following problems are given and solved as examples of determining the sizes and limits and the applications of the proper symbols to mating parts. The solution of these problems requires using the tables in Appendix 44, the table of preferred sizes (Table 15.1), and the table of preferred fits (Table 15.2).

EXAMPLE 1 (Fig. 15.19)

Given: Hole basis system, close running fit, basic diameter = 39 mm.

Solution: Use a basic diameter of 40 mm (Table 15.1) and fit of H8/f7 (Table 15.2).

Hole: Find the upper and lower limits of the hole in Appendix 44 under H8 and across from 40 mm. These limits are 40.000 and 40.039 mm.

Shaft: The upper and lower limits of the shaft are found under f7 and across from 40 mm in Appendix 44. These limits are 39.950 and 39.975 mm.

Symbols: Figure 15.19 shows the methods of noting the drawings. Any of these methods is appropriate.

EXAMPLE 2 (Fig. 15.20)

Given: Hole basis system, locational transition fit, basic diameter = 57 mm.

Solution: Use a basic diameter of 60 mm (Table 15.1) and a fit of H7/k6 (Table 15.2).

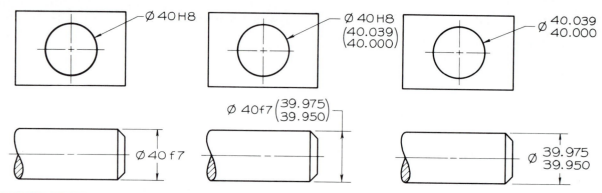

FIGURE 15.19 Any of these three methods can be used to apply symbols to a detail drawing of two mating parts.

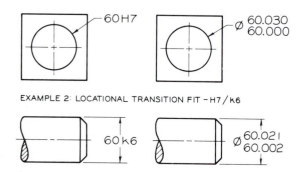

EXAMPLE 2: LOCATIONAL TRANSITION FIT – H7/k6

FIGURE 15.20 Two methods of applying tolerance symbols to a transition fit.

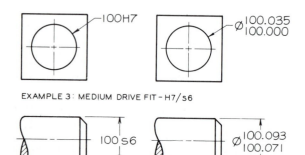

EXAMPLE 3: MEDIUM DRIVE FIT – H7/s6

FIGURE 15.21 Two methods of applying tolerance symbols to an interference fit.

Hole: Find the upper and lower limits of the hole in Appendix 45 under H7 and across from 60 mm. These limits are 60.000 and 60.030 mm.

Shaft: The upper and lower limits of the shaft are found under k6 and across from 60 mm in Appendix 45. These limits are 60.021 and 60.002 mm.

Symbols: Figure 15.20 shows two methods of applying the tolerance symbols to a drawing.

EXAMPLE 3 (Fig. 15.21)

Given: Hole basis system, medium drive fit, basic diameter = 96 mm.

Solution: Use a basic diameter of 100 mm (Table 15.1) and a fit of H7/s6 (Table 15.2).

Hole: Find the upper and lower limits of the hole in Appendix 45 under H7 and across from 100 mm. These limits are 100.035 and 100.000 mm.

Shaft: The upper and lower limits of the shaft are found under s6 and across from 100 mm in Appendix 45. These limits are found to be 100.093 and 100.071 mm. From the appendix, the tightest fit is an interference of 0.093 mm, and the loosest fit is an interference of 0.036 mm. An interference is indicated by a minus sign in front of the numbers.

EXAMPLE 4 (Fig. 15.22)

Given: Shaft basis system, loose running fit, basic diameter = 116 mm.

Solution: Use a basic diameter 120 mm (Table 15.1) and a fit of C11/h11 (Table 15.2).

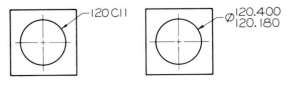

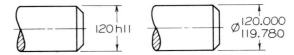

EXAMPLE 4: LOOSE RUNNING FIT – C11/h11

FIGURE 15.22 **Two methods of applying tolerance symbols to a clearance fit.**

Hole: Find the upper and lower limits of the hole in Appendix 46 under C11 and across from 120 mm. These limits are 120.400 and 120.180 mm.

Shaft: The upper and lower limits of the shaft are found under h11 and across from 120 mm in Appendix 46. These limits are 119.780 and 120.000 mm.

15.10
Preferred metric fits— nonpreferred sizes

Limits of tolerances for preferred fits, shown in Table 15.2, can be calculated for nonstandard sizes. Limits of tolerances appear in Appendix 48 for nonstandard hole sizes and in Appendix 49 for nonstandard shaft sizes.

The hole and shaft limits for an H8/f7 fit and a 45-mm DIA are calculated in Fig. 15.23. The tolerance limits of 0.000 and 0.039 mm for an H8 hole are taken from Appendix 48 across from the size range of 40–50 mm. The tolerance limits of −0.025 and −0.050 mm

FIT: H8/f7   Ø 45 BASIC

FROM APPENDIX		HOLE LIMITS	45.039
			45.000
HOLE	SHAFT		
H8	f7		
0.039	−0.025	SHAFT LIMITS	44.975
0.000	−0.050		44.950

FIGURE 15.23 **The limits of a nonstandard diameter, 45 mm, and an H8/f7 fit are calculated by using values from Appendixes 48 and 49.**

are taken from Appendix 49. The limits of sizes for the hole and shaft are calculated by applying these limits of tolerance to the 45-mm basic diameter.

15.11
Standard fits—English units

The ANSI B4.1 standard specifies a series of fits between cylindrical parts that are based on the basic hole system in inches. The types of fit covered in this standard are

RC—Running and sliding fits

LC—Clearance locational fits

LT—Transition locational fits

LN—Interference locational fits

FN—Force and shrink fits

Appendixes 38–42 list these five types of fit, each of which has several classes.

RUNNING AND SLIDING FITS (RC) are fits which provide a similar running performance, with suitable lubrication allowance throughout the range of sizes. The clearance for the first two classes (RC 1 and RC 2), used chiefly as slide fits, increases more slowly with diameter size than that of other classes so that accurate location is maintained even at the expense of free relative motion.

LOCATIONAL FITS (LC, LT, LN) are intended to determine only the location of the mating parts; they may provide rigid or accurate location (interference fits) or some freedom of location (clearance fits). Locational fits are divided into three groups: **clearance fits** (LC), **transition fits** (LT), and **interference fits** (LN).

FORCE FITS (FN) are special types of interference fits, typically characterized by maintenance of constant bore pressures throughout the range of sizes. The interference therefore varies almost directly with diameter, and the difference between its minimum and maximum values is small enough to maintain the resulting pressures within reasonable limits.

Figure 15.24 illustrates how to use the values from the tables in Appendix 38 for an RC 9 fit. The basic diameter for the hole and shaft is 2.5000 in., which is between 1.97 and 3.15 in. given in the last column of the table. Since all limits are in thou-

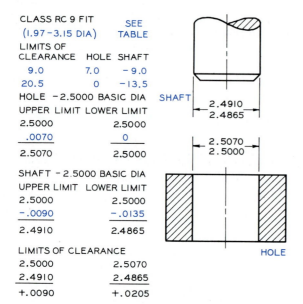

CLASS RC 9 FIT	SEE	
(1.97–3.15 DIA)	TABLE	
LIMITS OF		
CLEARANCE	HOLE	SHAFT
9.0	7.0	– 9.0
20.5	0	–13.5

HOLE – 2.5000 BASIC DIA

UPPER LIMIT	LOWER LIMIT
2.5000	2.5000
.0070	0
2.5070	2.5000

SHAFT – 2.5000 BASIC DIA

UPPER LIMIT	LOWER LIMIT
2.5000	2.5000
–.0090	–.0135
2.4910	2.4865

LIMITS OF CLEARANCE

2.5000	2.5070
2.4910	2.4865
+.0090	+.0205

FIGURE 15.24 The method of calculating limits and allowances for an RC 9 fit between a shaft and a hole. The basic diameter is 2.5000 inches.

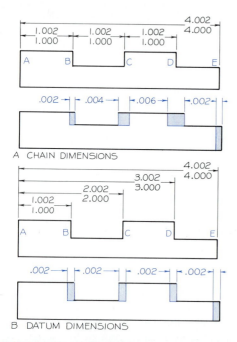

A CHAIN DIMENSIONS

B DATUM DIMENSIONS

FIGURE 15.25 When dimensions are given as chain dimensions, the tolerances can accumulate to give a variation of 0.006″ at *D* instead of 0.002″. When dimensioned from a single datum, the variations of *X* and *Y* cannot deviate more than the specified 0.002″ from the datum.

sandths, the values can be converted by moving the decimal point three places to the left; for example, +0.7 is +0.0007 in.

The upper and lower limits of the shaft (2.4910 and 2.4865 in.) are found by subtracting the two limits (−0.0090 and −0.0135 in.) from the basic diameter. The upper and lower limits of the hole (2.5007 and 2.5000 in.) are found by adding the two limits (+0.007 and 0.000 in.) to the basic diameter.

When the two parts are assembled, the tightest fit (+0.0090 in.) and the loosest fit (+0.0205 in.) are found by subtracting the maximum and minimum sizes of the holes and shafts. These values are provided in the first column of the table as a check on the limits

The same method (but different tables) is used for calculating the limits for all types of fit. Plus values indicate clearance, and minus values indicate interference between the assembled parts.

To convert these values to millimeters multiply inches by 25.4, or use metric tables, instead.

15.12
Chain dimensions

When parts are dimensioned to locate surfaces or geometric features by a chain of dimensions (Fig. 15.25A), variations may occur that exceed the tolerances specified. As successive measurements are made, with each based on the preceding one, the tolerances may accumulate, as Fig. 15.25A shows. For example, the tolerance between surfaces *A* and *B* is 0.002; between *A* and *C*, 0.004; and between *A* and *D*, 0.006.

This accumulation of tolerances can be eliminated by measuring from a single plane called a **datum plane.** A datum plane is usually on the object, but it could be on the machine used to make the part. Since each of the planes in Fig. 15.25B was located with respect to a single datum, the tolerances between the intermediate planes are a uniform 0.002, which represents the maximum tolerance.

15.13
Origin selection

Sometimes there is a need to specify a surface as the origin for locating another surface. An example is

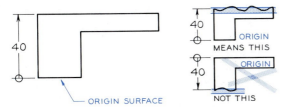

FIGURE 15.26 This method is used to indicate the origin surface for locating one feature of a part with respect to another.

shown in Fig. 15.26, where the origin surface is the shorter one at the base of an object. The result is that the angular variation permitted is less for the longer surface than it would be if the origin plane had been the longer surface.

15.14
Conical tapers

Taper is a ratio of the difference in the diameters of two circular sections of a cone to the distance between the sections. Figure 15.27 shows a method of specifying a conical taper by giving a basic diameter and basic taper. The basic diameter of 20 mm is located midway in the length of the cone with a toleranced dimension. Figure 15.27 also shows how to find the radial tolerance zone.

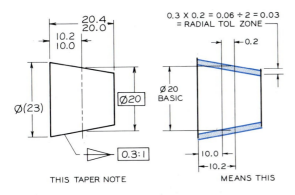

FIGURE 15.27 Taper is indicated with a combination of tolerances and taper symbols. The variation in diameter at any point is 0.06 mm, or 0.03 mm in radius.

15.15
Tolerance notes

All dimensions on a drawing are toleranced either by the rules previously discussed or by a note placed in or near the title block. For example, the note TOLERANCE $\pm\frac{1}{64}$ (or its decimal equivalent, 0.40 mm) might be given on a drawing for less critical dimensions.

Some industries may give dimensions in inches where decimals are carried out to two, three, and four places. A note for dimensions with two and three decimal places might be given on the drawing as TOLERANCES XX.XX ±0.10; XX.XXX ±0.005. Tolerances of four places would be given directly on the dimension lines.

The most common method of noting tolerances is to give as large a tolerance as feasible in a note, such as TOLERANCES ±0.05 (±1 mm when using metrics), and to give the tolerances for the mating dimensions that require smaller tolerances on the dimension lines.

Angular tolerances should be given in a general note in or near the title block, such as ANGULAR TOLERANCES $\pm0.5°$ or $\pm30'$. Angular tolerances less than this should be given on the drawing where these angles are dimensioned. Figure 15.28 shows techniques of tolerancing angles.

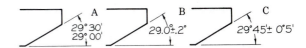

FIGURE 15.28 Angles can be toleranced by any of these techniques using limits or the plus-and-minus method.

15.16
General tolerances—metric

All dimensions on a drawing are understood to have tolerances, and the amount of tolerance must be noted. In this section, which is based on the metric system as outlined in ANSI B4.3, we use the **millimeter** as the unit of measurement.

Tolerances may be specified by applying them directly to the dimensions, giving them in specification documents, or giving them in a general note on the drawing.

FIGURE 15.29 International tolerance (IT) grades and their applications.

LINEAR DIMENSIONS may be toleranced by indicating ± one half of an international tolerance (IT) grade as given in Appendix 43. The appropriate IT grade can be selected from the graph in Fig. 15.29.

TABLE 15.3

IT GRADES AND THEIR RELATIONSHIP TO MACHINING PROCESSES

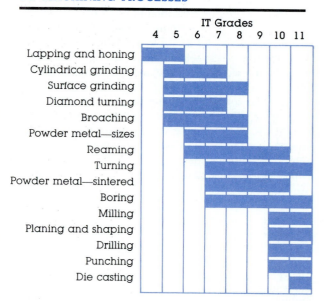

IT grades for mass-produced items range from IT12 through IT16. When the machining process is known, the IT grades can be selected from Table 15.3.

General tolerances using IT grades may be expressed in a note as follows:

UNLESS OTHERWISE SPECIFIED ALL

UNTOLERANCED DIMENSIONS ARE $\pm\dfrac{\text{IT}14}{2}$.

This means a tolerance of ±0.700 mm is allowed for a dimension between 315 and 400 mm. The value of the tolerance is listed in Appendix 43 as 1.400.

Table 15.4 shows recommended tolerances for **fine, medium,** and **coarse series** for dimensions of graduated sizes. A medium tolerance, for example, can be specified by the following note:

GENERAL TOLERANCES SPECIFIED IN
ANSI B4.3 MEDIUM SERIES APPLY.

This same information can be given on the drawing in a table by selecting the grade—medium in this example—from Table 15.4 and giving it as shown in Fig. 15.30.

General tolerances can be expressed in a table that gives the tolerances for dimensions with one or no decimal places (Fig. 15.31).

Another method of giving general tolerances is a note in the following form:

UNLESS OTHERWISE SPECIFIED ALL
UNTOLERANCED
DIMENSIONS ARE ±0.8 mm.

TABLE 15.4

FINE, MEDIUM, AND COARSE SERIES: GENERAL TOLERANCE—LINEAR DIMENSIONS

Basic Dimensions (mm)		Variations (mm)						
		0.5 to 3	Over 3 to 6	Over 6 to 30	Over 30 to 120	Over 120 to 315	Over 315 to 1000	Over 1000 to 2000
Permissible variations	Fine series	±0.05	±0.05	±0.1	±0.15	±0.2	±0.3	±0.5
	Medium series	±0.1	±0.1	±0.2	±0.3	±0.5	±0.8	±1.2
	Coarse series		±0.2	±0.5	±0.8	±1.2	±2	±3

TABLE 15.5
GENERAL TOLERANCE—ANGLES AND TAPERS

Length of the Shorter Leg (mm)		Up to 10	Over 10 to 50	Over 50 to 120	Over 120 to 400
Permissible variations	In degrees and minutes	±1°	±0°30′	±0°20′	±0°10′
	In millimeters per 100 mm	±1.8	±0.9	±0.6	±0.3
	In Milliradians	±18	±9	±6	±3

DIMENSIONS IN mm

GENERAL TOLERANCE
UNLESS OTHERWISE SPECIFIED THE FOLLOWING TOLERANCES ARE APPLICABLE

LINEAR	OVER TO	0.5 6	6 30	30 120	120 315	315 1000	1000 2000
TOL	±	0.1	0.2	0.3	0.5	0.8	1.2

FIGURE 15.30 This table is a *medium* series of values taken from Table 15.4. It is placed on the working drawing to provide the tolerances for various ranges of sizes.

DIMENSIONS IN mm

GENERAL TOLERANCE
UNLESS OTHERWISE SPECIFIED THE FOLLOWING TOLERANCES ARE APPLICABLE

LINEAR	OVER TO	− 120	120 315	315 1000	1000 −
TOL	ONE DECIMAL ±	0.3	0.5	0.8	1.2
	NO DECIMALS ±	0.8	1.2	2	3

FIGURE 15.31 This table of tolerances can be placed on a drawing to indicate the tolerances for dimensions with one or no decimal places.

This method should be used only where the dimensions on a drawing have slight differences in size.

ANGULAR TOLERANCES are expressed (1) as an **angle** in decimal degrees or in degrees and minutes, (2) as a taper expressed in **percentage** (number of millimeters per 100 mm), or (3) as **milliradians.** A milliradian is found by multiplying the degrees of an

angle by 17.45. The suggested tolerances for each of these units are shown in Table 15.5 and are based on the length of the shorter leg of the angle.

General angular tolerances may be given on the drawing with a note in the following form:

UNLESS OTHERWISE SPECIFIED THE GENERAL TOLERANCES IN ANSI B4.3 APPLY.

A second method shows a portion of Table 15.5 on the drawing using the units desired (Fig. 15.32). A third method is a note with a single tolerance such as:

UNLESS OTHERWISE SPECIFIED THE GENERAL ANGULAR TOLERANCES ARE ±0°30′ (or ±0.5°).

15.17
Geometric tolerances

Geometric tolerancing is a term used to describe tolerances that specify and control form, profile, orientation, location, and runout on a dimensioned part. The basic principles of this area of tolerancing are stan-

ANGULAR TOLERANCE				
LENGTH OF SHORTER LEG - mm	UP TO 10	OVER 10 TO 50	OVER 50 TO 120	OVER 120 TO 400
TOL	±1°	±0° 30′	±0° 20′	±0° 10′

FIGURE 15.32 This table can be placed on a drawing to indicate the general tolerances for angles that were extracted from Table 15.5.

	TOLERANCE	CHARACTERISTIC	SYMBOL
INDIVIDUAL FEATURES	FORM	STRAIGHTNESS	—
		FLATNESS	▱
		CIRCULARITY	○
		CYLINDRICITY	⌭
INDIVIDUAL OR RELATED FEATURES	PROFILE	PROFILE OF A LINE	⌒
		PROFILE OF A SURFACE	⌓
RELATED FEATURES	ORIENTATION	ANGULARITY	∠
		PERPENDICULARITY	⊥
		PARALLELISM	//
	LOCATION	POSITION	⌖
		CONCENTRICITY	◎
	RUNOUT	CIRCULAR RUNOUT	↗
		TOTAL RUNOUT	↗↗

FIGURE 15.33 These symbols are used to specify the geometric characteristics of a dimensioned part.

dardized by the ANSI Y14.5M–1982 Standards and the Military Standards (Mil-Std) of the U.S. Department of Defense.

These standards are based on the metric system with the millimeter as the unit of measurement. Inch units with decimal fractions can be used instead of millimeters if needed.

15.18
Symbology of geometric tolerances

Figure 15.33 shows the various symbols used to specify geometric characteristics of dimensioned drawings. Additional features and their proportions are shown in Fig. 15.34, where the letter height *(H)* is used as the basis of the proportions. On most drawings, a ⅛-in. or 3-mm letter height is recommended. Figure 15.35 shows other examples of feature control symbols.

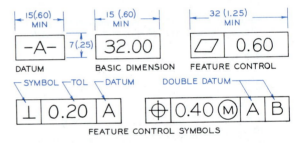

FIGURE 15.35 Examples of symbols used to indicate datum planes, basic dimensions, and feature control symbols.

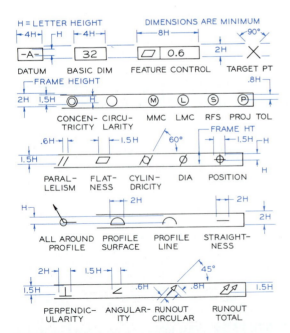

FIGURE 15.34 The general proportions of notes and symbols used in feature control symbols.

15.19
Limits of size

Three terms used to specify the limits of size of a part when applying geometric tolerances are **maximum material condition (MMC), least material condition (LMC),** and **regardless of feature size (RFS).**

MMC indicates a part is made with the maximum amount of material. For example, the shaft in Fig. 15.36 is at MMC when it has the largest permitted diameter of 24.6 mm. The hole is at MMC when it has the most material or the smallest diameter of 25.0 mm.

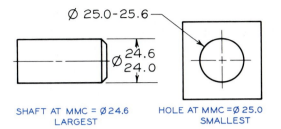

FIGURE 15.36 A shaft is at MMC when it is at the largest size permitted by its tolerance. A hole is at MMC when it is at the smallest size.

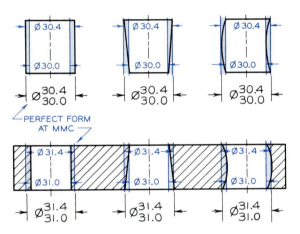

FIGURE 15.37 When only a tolerance of size is specified on a part, the limits prescribe the form of the part, as shown in these shafts and holes with the same limits of tolerance.

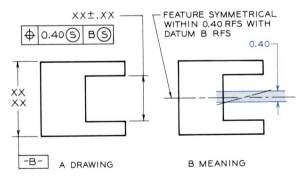

FIGURE 15.38 Tolerances of position should include a note of M, S, or L to indicate maximum material condition, regardless of feature size or least material condition.

LMC indicates a part has the least amount of material. The shaft in Fig. 15.36 is at LMC when it has the smallest diameter of 24.0 mm. The hole is at LMC when it has the largest diameter of 25.6 mm.

RFS indicates tolerances apply to a geometric feature regardless of the size it may be, from MMC to LMC.

15.20
Three rules of tolerances

Three general rules of tolerancing geometric features should be followed in this type of dimensioning.

RULE 1 (INDIVIDUAL FEATURE SIZE) When only a tolerance of size is specified on a part, the limits of size prescribe the amount of variation permitted in its geometric form. In Fig. 15.37, the forms of the shaft and hole are permitted to vary within the tolerance of size indicated by the dimensions.

RULE 2 (TOLERANCES OF POSITION) When a tolerance of position is specified on a drawing, RFS, MMC, or LMC must be specified with respect to the tolerance, datum, or both. The specification of symmetry of the part in Fig. 15.38 is based on a tolerance at RFS from a datum at RFS.

RULE 3 (ALL OTHER GEOMETRIC TOLERANCES) RFS applies for all other geometric tolerances for individual tolerances and datum references if no modifying symbol is given in the feature control symbol. If a feature is to be at MMC, it must be specified.

15.21
Three-datum plane concept

A datum plane is used as the origin of a part's dimensioned features that have been toleranced. Datum planes are usually associated with manufacturing equipment, such as machine tables, or with locating pins.

Three mutually perpendicular datum planes are used to dimension a part accurately. For example, the part in Fig. 15.39 is placed in contact with the primary datum plane at its base where three points must make contact with the datum. The part is further related to the secondary plane with two contacting points. The third (tertiary) datum is contacted by a single point on the object.

The priority of these datum planes is noted on the drawing of the part by feature control symbols, as Fig. 15.40 shows. The primary datum is surface *P*, the secondary is surface *S*, and the tertiary is surface *T*. Examples of feature control symbols are given in Fig. 15.41, where the primary, secondary, and tertiary datum planes are listed in order of priority.

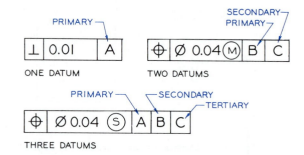

FIGURE 15.41 Feature control symbols may indicate from one to three datum planes listed in order of priority.

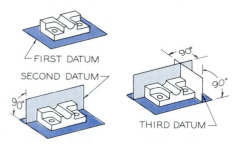

FIGURE 15.39 When an object is referenced to a primary datum plane, it contacts the plane with its three highest points. The vertical surface contacts the secondary vertical datum plane with two points. The third datum plane is contacted by one point on the object. The datum planes are listed in this order in the feature control symbol.

15.22
Cylindrical datum features

Figure 15.42 illustrates a part with a cylindrical datum feature that is the axis of a true cylinder. Primary datum *K* establishes the first datum. Datum *m* is associated with two theoretical planes—the second and third in a three-plane relationship.

The two theoretical planes are represented on the drawing by perpendicular centerlines. The intersection of the centerlines coincides with the datum axis. All dimensions originate from the datum axis, which is perpendicular to datum *K;* the two intersecting datum planes indicate the direction of measurements in an *x* and *y* direction.

The sequence of the datum reference in the feature control symbol is significant to the manufacturing

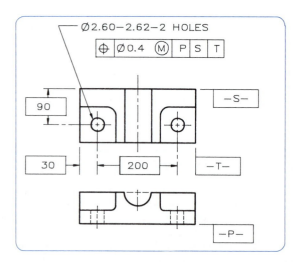

FIGURE 15.40 The sequence of the three-plane reference system (shown in Fig. 15.39) is labeled where the planes appear as edges. Note that the primary datum plane, *P*, is listed first in the feature control symbol; the secondary plane, *S*, is next; and the tertiary plane, *T*, is last.

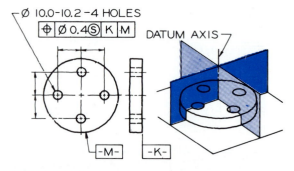

FIGURE 15.42 These true-position holes are located with respect to primary datum *K* and datum *M*. Since datum *M* is a circle, this implies that the holes are located about two intersecting datum planes formed by the crossing centerlines in the circular view satisfying the three-plane concept.

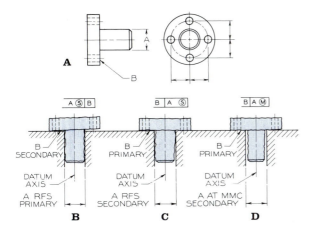

FIGURE 15.43 Three examples that illustrate the effects of selection of the datum planes in order of priority and the effect of RFS and MMC.

and inspection processes. The part in Fig. 15.43 is dimensioned with an incomplete feature control symbol; it does not specify the primary and secondary datum planes. The schematic drawing in Fig. 15.43B illustrates the effect of specifying the diameter A as the primary datum plane and surface B as the secondary datum plane. This means the part is centered about cylinder A by mounting the part in a chuck, mandrel, or centering device on the processing equipment, which centers the part at RFS. Surface B is assembled to contact at least one point of the third datum plane.

If surface B were specified as the primary datum feature, it would be assembled to contact datum plane

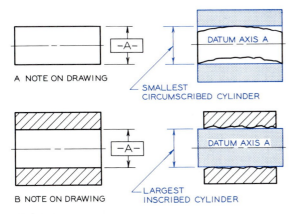

FIGURE 15.44 The datum axis of a shaft is the smallest circumscribed cylinder that contacts the shaft. The datum axis of a hole is the centerline of the largest inscribed cylinder that contacts the hole.

B in at least three points. The axis of datum feature A will be gauged by the smallest true cylinder that is perpendicular to the first datum that will contact surface A at RFS.

In Fig. 15.43D, plane B is specified as the primary datum feature, and cylinder A is specified as the secondary datum feature at MMC. The part is mounted on the processing equipment where at least three points on feature B are in contact with datum B. The second and third planes intersect at the datum axis to complete the three-plane relationship. Using the modifier to specify MMC gives a more liberal tolerance zone than otherwise would be acceptable when RFS was specified.

15.23
Datum features at RFS

When dimensions of size are applied to a part at RFS, the datum is established by contacting surfaces on the processing equipment with surfaces of the part. Variable machine elements, such as chucks or center devices, are adjusted to fit the external or internal features of a part and thereby establish datums.

PRIMARY DIAMETER DATUMS For an external cylinder (shaft), the datum axis is the axis of the smallest circumscribed cylinder that contacts the cylindrical feature of the part. That is, the largest diameter of the part will make contact with the smallest contacting cylinder of the machine element that holds the part (Fig. 15.44).

For an internal cylinder (hole), the datum axis is the axis of the largest inscribed cylinder that contacts the inside of the hole. That is, the smallest diameter of the hole will make contact with the largest cylinder of the machine element inserted in the hole (Fig. 15.44).

PRIMARY EXTERNAL PARALLEL DATUMS The datum for external features is the center plane between two parallel planes, at their minimum separation, that contact the planes of the object (Fig. 15.45A). These are planes of a viselike device that holds the part; therefore, the planes of the part are at maximum separation, whereas the planes of the device are at minimum separation.

PRIMARY INTERNAL PARALLEL DATUMS The datum for internal features is the center plane between two parallel planes, at their maximum separa-

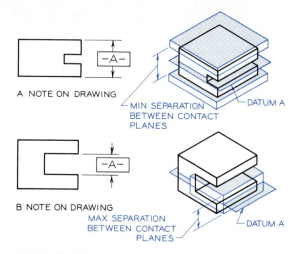

A NOTE ON DRAWING

MIN SEPARATION
BETWEEN CONTACT
PLANES

DATUM A

B NOTE ON DRAWING

MAX SEPARATION
BETWEEN CONTACT
PLANES

DATUM A

FIGURE 15.45 The datum plane for external parallel surfaces is the center plane between two contacting parallel planes at their minimum separation. The datum plane for internal parallel surfaces is the center plane between two contacting parallel surfaces at their maximum separation.

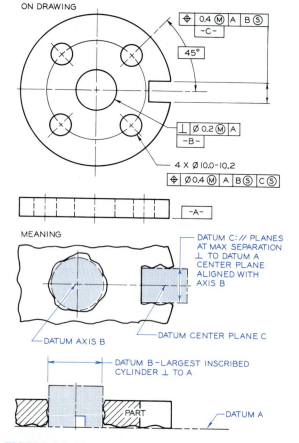

ON DRAWING

MEANING

DATUM C: // PLANES
AT MAX SEPARATION
⊥ TO DATUM A
CENTER PLANE
ALIGNED WITH
AXIS B

DATUM AXIS B

DATUM CENTER PLANE C

DATUM B—LARGEST INSCRIBED
CYLINDER ⊥ TO A

PART

DATUM A

FIGURE 15.46 A part located with respect to primary, secondary, and tertiary datum planes.

tion, that contact the planes of the object (Fig. 15.45B). This is the condition in which the slot is at its smallest opening size.

SECONDARY DATUMS The secondary datum (axis or center plane) for both external and internal diameters or distances between parallel planes is found as covered in the previous two paragraphs but with an additional requirement: The contacting cylinder of the contacting parallel planes must be perpendicularly oriented to the primary datum. Fig. 15.46 illustrates how datum B is the axis of a cylinder. This principle also can be applied to parallel planes.

TERTIARY DATUMS The third datum (axis or center plane) for both external and internal features is found as covered in the previous three paragraphs but with an additional requirement: The contacting cylinder or parallel planes must be angularly oriented to the secondary datum. Datum C in Fig. 15.46 is the tertiary datum plane.

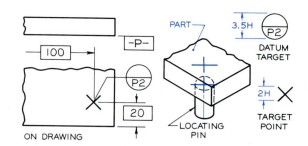

FIGURE 15.47 Target points from which a datum point is established are located with an X and a target symbol.

15.24
Datum targets

Instead of using a plane surface as a datum, specified datum targets are indicated on the surface of a part where the part is supported by spherical or pointed locating pins. The symbol X indicates target points that are supported by locating pins at specified points (Fig. 15.47). Datum target symbols are placed outside the outline of the part with a leader directed toward the target. When the target is on the near (visible) surface, the leaders are solid lines; when the target is on the far (invisible) surface (see Fig. 15.47), the leaders are hidden.

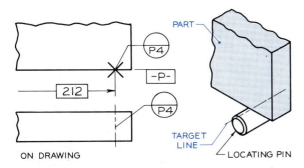

FIGURE 15.48 An X and a phantom line are used to locate target lines on a drawing.

Three target points are required to establish the primary datum plane, two for the secondary, and one for the tertiary. The target symbol in Fig. 15.47 is labeled *P2* to match the designation of the primary datum, *P*. Were the other two points shown, they would be labeled *P1* and *P2* to establish the primary datum.

A datum target line is specified in Fig. 15.48 for a part supported on a datum line instead of on a datum point. An X and a phantom line are used to locate the line of support.

Target areas are specified for cases where spherical or pointed locating pins are inadequate to support a part. The diameters of the targets are specified with cross-hatched circles surrounded by phantom lines, as Fig. 15.49 shows. The target symbols give both the diameter of the targets and their number designations. The X could also be used to indicate target areas.

The part is located on its datum plane by placing

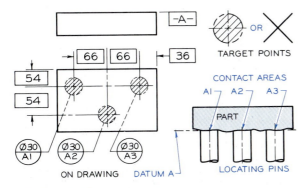

FIGURE 15.49 Target points with areas are located with basic dimensions and target symbols that give the diameters of the targets. Hidden leaders indicate that the targets are on the hidden side of the plane.

it on the three locating pins with 30-mm diameters (Fig. 15.49). Leaders from the target areas to the target symbols are hidden, indicating the targets are on the hidden side of the object.

15.25
Tolerances of location

Tolerances of location deal with **position, concentricity,** and **symmetry.**

Position tolerancing

Whereas toleranced location dimensions give a square tolerance zone for locating the center of a hole, **true-position dimensions** locate the exact position of a hole's center, about which a circular tolerance zone is given (Fig. 15.50). **Basic dimensions** are exact untoleranced dimensions used to locate true positions indicated by boxes drawn around them (Fig. 15.50B).

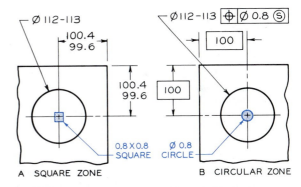

FIGURE 15.50 **A.** These dimensions give a square tolerance zone for the axis of the hole. **B.** Basic dimensions locate the true center of the circle about which a circular tolerance zone of 0.8 mm is specified.

In both methods, the diameters of the holes are toleranced by notes. The true-position method (Fig. 15.50B) uses a feature control symbol to specify the diameter of the circular tolerance zone inside which the center of the hole must lie. A circular zone gives a more uniform tolerance of the hole's true position than a square.

In Fig. 15.51, you can see an enlargement of the square tolerance zone that results from using toler-

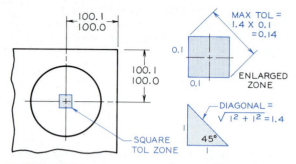

FIGURE 15.51 The toleranced-coordinate method of dimensioning gives a square tolerance zone. The diagonal of the square exceeds the specified tolerance by a factor of 1.4.

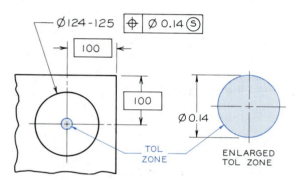

FIGURE 15.52 The true-position method of tolerancing gives a circular tolerance zone with equal variations in all directions from the true axis of the hole.

anced coordinates to locate a hole's center. The diagonal across the square zone is greater than the specified tolerance by a factor of 1.4. The true-position method shown in Fig. 15.52 can have a larger circular tolerance zone by a factor of 1.4 and still have the same degree of accuracy of position as the specified 0.1 square zone.

If the toleranced coordinate method could accept a variation of 0.014 across the diagonal of the square tolerance zone, the true-position tolerance should be acceptable with a circular zone of 0.014, which is a greater tolerance than the square zone permitted (Fig. 15.51). True-position tolerances can be applied by symbol, as Fig. 15.52 shows.

The circular tolerance zone specified in the circular view of a hole is assumed to extend the full depth of the hole. Therefore, the tolerance zone for the centerline of the hole is a cylindrical zone inside which

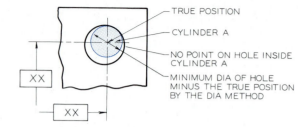

FIGURE 15.53 When a hole is located at true position at MMC, no element of the hole will be inside the imaginary cylinder *A*.

the axis must lie. Since the size of the hole and its position are both toleranced, these two tolerances are used to establish the diameter of a gauge cylinder used to check for the conformance of the holes to the specifications (Fig. 15.53).

By subtracting the true-position tolerance from the hole at MMC (the smallest permissible hole), the circle is found. This zone represents the least favorable condition when the part is gauged or assembled with a mating part. When the hole is not at MMC, it is larger and permits greater tolerance and easier assembly.

Gauging a two-hole pattern

Gauging is a technique of checking dimensions to determine whether they have met the specifications of tolerance (Fig. 15.54). The two holes, which are positioned 26.00 mm apart with a basic dimension, have

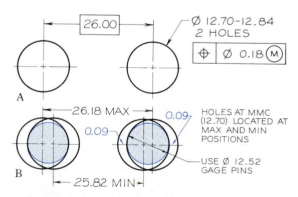

FIGURE 15.54 When two holes are located at true position at MMC, they may be gauged with pins 12.52 mm in diameter that are located 26.00 mm apart.

limits of 12.70 and 12.84 for a tolerance of 0.14 and are located at true position within a diameter of 0.18. The gauge pin diameter is calculated to be 12.52 mm (the smallest hole's size minus the true-position tolerance), as Fig. 15.54B shows. This means two pins with diameters of 12.52 mm spaced exactly 26.00 mm apart could be used to check the diameters and positions of the holes at MMC, the most critical size. If the pins can be inserted in the holes, the holes are properly sized and located.

When the holes are not at MMC—that is, when they are larger than their minimum size—these gauge pins will permit a greater range of variation (Fig. 15.55). When the holes are at their maximum size of 12.84 mm, they can be located as close as 25.68 mm from center to center or as far apart as 26.32 mm from center to center.

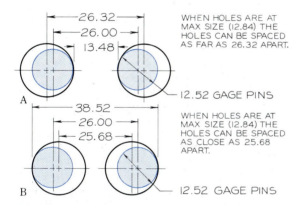

FIGURE 15.55 **A.** When two holes are at their maximum size, the centers of the holes can be spaced as far as 26.32 mm apart and still be acceptable. **B.** The holes can be placed as close as 25.68 mm apart when they are at maximum size.

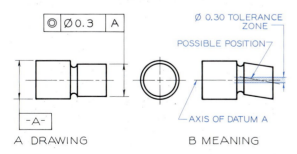

FIGURE 15.56 Concentricity is a tolerance of location. The feature control symbol specifies that the smaller cylinder should be concentric to cylinder A, within 0.3 mm about the axis of A.

Concentricity

Concentricity is a feature of location because it specifies the relationship of one cylinder with another since both share the same axis. In Fig. 15.56, the large cylinder is flagged as datum A, which means the large diameter is used as the datum for measuring the variation of the smaller cylinder's axis.

Feature control symbols will be used to specify concentricity and other geometric characteristics throughout the remainder of this chapter (Fig. 15.57).

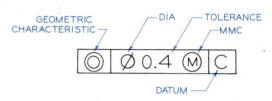

FIGURE 15.57 A typical feature control symbol. This one indicates that a surface is concentric to datum C within a diameter of 0.4 mm at MMC.

Symmetry

Symmetry is also a feature of location. A part or feature is symmetrical when it has the same contour and size on opposite sides of a central plane. A symmetry tolerance locates features with respect to a datum plane (Fig. 15.58). The feature control symbol notes that the notch is symmetrical about datum B within a zone of 0.6 mm.

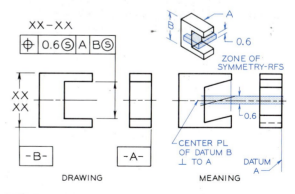

FIGURE 15.58 Symmetry is a tolerance of location that specifies a part's feature be symmetrical about the center plane between parallel surfaces of the part.

15.26
Tolerances of form

Flatness

A surface is flat when all its elements are in one plane. A feature control symbol is used to specify flatness within a 0.4 mm zone in Fig. 15.59. No point on the surface may vary more than 0.40 from the highest to the lowest point on the surface.

Straightness

A surface is straight if all its elements are straight lines. A feature control symbol is used to specify straightness of a cylinder in Fig. 15.60. A total of 0.12 mm is permitted as the elements are gauged in a vertical plane parallel to the axis of the cylinder.

Roundness

A surface of revolution (a cylinder, cone, or sphere) is round when all points on the surface intersected by a plane are equidistant from the axis. A feature control

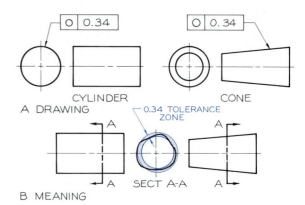

FIGURE 15.61   Roundness is a tolerance of form that indicates a cross section through a surface of revolution is round and lies within two concentric circles.

symbol is used to specify roundness of a cone and cylinder in Fig. 15.61. This symbol permits a tolerance of 0.34 mm on the radius of each part. Figure 15.62 specifies the roundness of a sphere.

Cylindricity

A surface of revolution is cylindrical when all its elements form a cylinder. A cylindricity tolerance zone is specified in Fig. 15.63, where a tolerance of 0.54 mm is permitted on the radius of the cylinder. Cylindricity is a combination of tolerances of roundness and straightness.

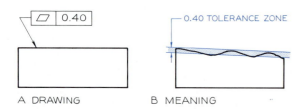

FIGURE 15.59   Flatness is a tolerance of form that specifies two parallel planes inside of which the object's surface must lie.

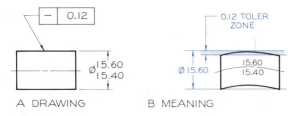

FIGURE 15.60   Straightness is a tolerance of form that indicates the elements of a surface are straight lines. The symbol must be applied where the elements appear as straight lines.

15.27
Tolerances of profile

Profile tolerancing is used to specify tolerances about a contoured shape formed by arcs or irregular curves. Profile can apply to a surface or a single line.

The surface in Fig. 15.64 is given a unilateral profile tolerance because it can only be smaller than the points located. Figure 15.64B shows examples of specifying bilateral and unilateral tolerance zones.

A profile tolerance for a single line can be specified as shown in Fig. 15.65, where the curve is formed by tangent arcs whose radii are given as basic dimensions. The radii are permitted to vary by plus or minus 0.10 mm about the basic radii.

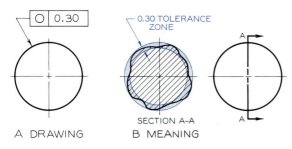

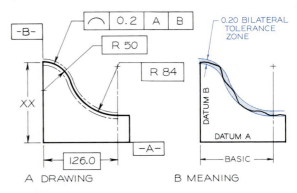

FIGURE 15.62 Roundness of a sphere is indicated in this manner, which means any cross section through it is round within the specified tolerance in the symbol.

15.28
Tolerances of orientation

Tolerances of orientation include **parallelism, perpendicularity,** and **angularity.**

Parallelism

A surface or line is parallel when all its points are equidistant from a datum plane or axis. Two types of

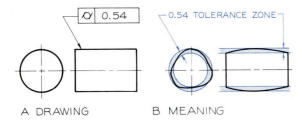

FIGURE 15.63 Cylindricity is a tolerance of form that indicates the surface of a cylinder lies within an envelope formed by two concentric cylinders.

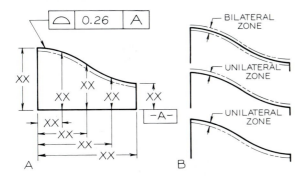

FIGURE 15.64 Profile is a tolerance of form used to tolerance irregular curves of planes. **A.** The curving plane is located by coordinates. **B.** The tolerance is located by any of these methods.

FIGURE 15.65 Profile of a line is a tolerance of form that specifies the variation allowed from the path of a line. Here, the line is formed by tangent arcs. The tolerance zone may be either bilateral or unilateral, as Fig. 15.64B shows.

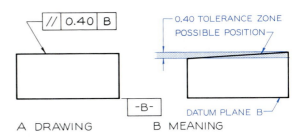

FIGURE 15.66 Parallelism is a tolerance of form that specifies a plane is parallel to another within specified limits. Plane *B* is the datum plane in this figure.

parallelism are

1. A tolerance zone between planes parallel to a datum plane within which the axis or surface of the feature must lie (Fig. 15.66). This tolerance also controls flatness.

2. A cylindrical tolerance zone parallel to a datum feature within which the axis of a feature must lie (Fig. 15.67).

The effect of specifying parallelism at MMC can be seen in Fig. 15.68, where the modifier *M* is given in the feature control symbol. Tolerances of form apply RFS when not specified. Specifying parallelism at MMC means the axis of the cylindrical hole must vary no more than 0.20 mm when the holes are the smallest permissible size.

As the hole approaches its upper limit of 30.30, the tolerance zone increases until it reaches 0.50 DIA. Therefore, a greater variation is given at MMC than at RFS.

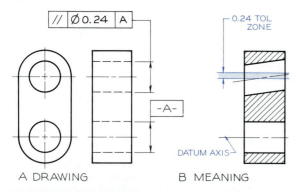

A DRAWING   B MEANING

FIGURE 15.67 Parallelism of one centerline to another can be specified by using the diameter of one of the holes as the datum.

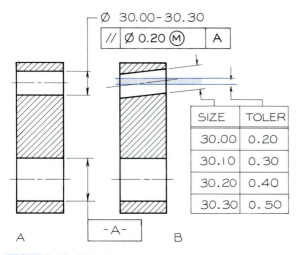

SIZE	TOLER
30.00	0.20
30.10	0.30
30.20	0.40
30.30	0.50

FIGURE 15.68 The most critical tolerance will exist when features are at MMC. Here, the upper hole must be parallel to the hole used as datum A within 0.20 DIA. As the hole approaches its maximum size of 30.30 mm, the tolerance zone approaches 0.50 mm.

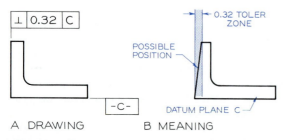

A DRAWING   B MEANING

FIGURE 15.69 Perpendicularity is a tolerance form that gives a tolerance zone for a plane perpendicular to a specified datum plane.

Perpendicularity

Figure 15.69 specifies the perpendicularity of two planes. Note that datum plane *C* is flagged, and the feature control symbol is applied to the perpendicular surface. A hole is specified as perpendicular to a surface in Fig. 15.70, where surface *A* is indicated as the datum plane.

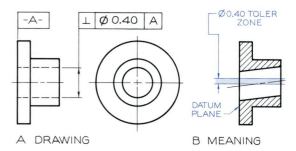

A DRAWING   B MEANING

FIGURE 15.70 Perpendicularity can apply to the axis of a feature, such as the centerline of a cylinder.

Angularity

A surface or line is angular when it is at a specified angle (other than 90°) from a datum or an axis. The angularity of a surface is specified in Fig. 15.71, where the angle is given a basic dimension of 30°. The angle is permitted to vary within a tolerance zone of 0.25 mm about the angle.

15.29
Tolerances of runout

Runout tolerance is a means of controlling the functional relationship between one or more parts to a common datum axis. The features controlled by runout are surfaces of revolution about an axis and surfaces perpendicular to the axis.

The datum axis is established by using a functional cylindrical feature that rotates about the axis, such as diameter *B* in Fig. 15.72. When the part is rotated about this axis, the features of rotation must fall within the prescribed tolerance at **full indicator movement (FIM).**

The two types of runout are circular runout and total runout. One arrow in the feature control symbol

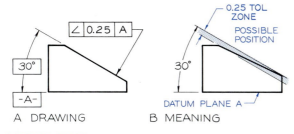

FIGURE 15.71 Angularity is a tolerance of form that specifies the tolerance for an angular surface with respect to a datum plane. The 30° angle is a true angle, a basic angle. The tolerance of 0.25 mm is applied to this basic angle.

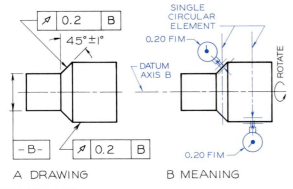

FIGURE 15.72 Runout tolerance of a surface, a composite of several tolerance-of-form characteristics, is used to specify concentric cylindrical parts. The part is mounted on the datum axis and is gauged as it is rotated about it.

indicates circular runout, and two arrows indicate total runout.

CIRCULAR RUNOUT (one arrow) is measured by rotating an object about its axis for 360° to determine whether a circular cross section at any point exceeds the permissible runout tolerance. This same technique is used to measure the amount of wobble existing in surfaces perpendicular to the axis of rotation.

TOTAL RUNOUT (two arrows) is used to specify cumulative variations of circularity, straightness, coaxiality, angularity, taper, and profile of a surface (Fig. 15.73). Total runout is applied to all circular and profile positions as the part is rotated 360°. When applied to surfaces perpendicular to the axis, total runout controls variations in perpendicularity and flatness.

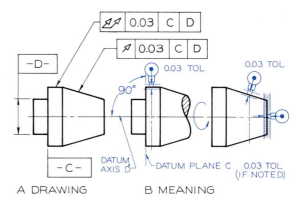

FIGURE 15.73 The runout tolerance in this example is measured by mounting the object on the primary datum plane, surface C, and the secondary datum plane, cylinder D. The cylinder and conical surface is gauged to determine if it conforms to a tolerance zone of 0.03 mm. The end of the cone could have been noted to specify its runout (perpendicularity to the axis).

The part shown in Fig. 15.74 is dimensioned by using a composite of several techniques of geometric tolerancing.

15.30
Surface texture

Because the surface texture of a part will affect its function, it must be precisely specified. The finish mark V does not elaborate on the finish desired. Figure 15.75 illustrates most of the terms of surface texture defined below.

SURFACE TEXTURE is the variation in the surface, including roughness, waviness, lay, and flaws.

ROUGHNESS describes the finest of the irregularities in the surface. These are usually caused by the manufacturing process used to smooth the surface.

ROUGHNESS HEIGHT is the average deviation from the mean plane of the surface. It is measured in microinches (μin) or micrometers (μm), respectively millionths of an inch and of a meter.

ROUGHNESS WIDTH is the width between successive peaks and valleys that forms the roughness measured in microinches or micrometers.

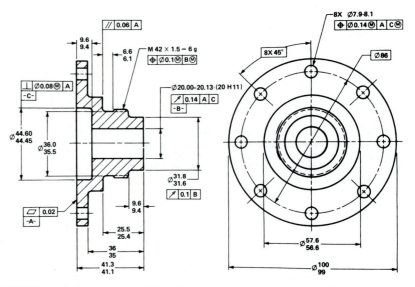

FIGURE 15.74 A part dimensioned with a combination of notes and symbols to describe its geometric features. (Courtesy of ANSI Y14.5M–1982.)

ROUGHNESS WIDTH CUTOFF is the largest spacing of repetitive irregularities that includes average roughness height (measured in inches or millimeters). When not specified, a value of 0.8 mm (0.030 in.) is assumed.

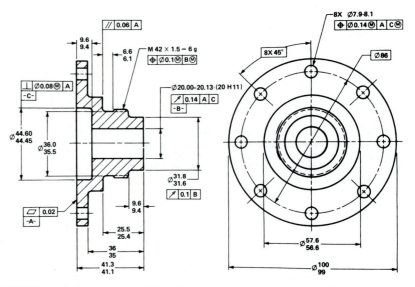

FIGURE 15.75 **Characteristics of surface texture.**

WAVINESS is a widely spaced variation that exceeds the roughness width cutoff. Roughness may be regarded as superimposed on a wavy surface. Waviness is measured in inches or millimeters.

WAVINESS HEIGHT is the peak-to-valley distance between waves. It is measured in inches or millimeters.

WAVINESS WIDTH is the spacing between peaks or wave valleys measured in inches or millimeters.

LAY is the direction of the surface pattern and is determined by the production method used.

FLAWS are irregularities or defects that occur infrequently or at widely varying intervals on a surface. These include cracks, blow holes, checks, ridges, scratches, and the like. Unless otherwise specified, the effect of flaws is not included in roughness height measurements.

CONTACT AREA is the surface that will make contact with its mating surface.

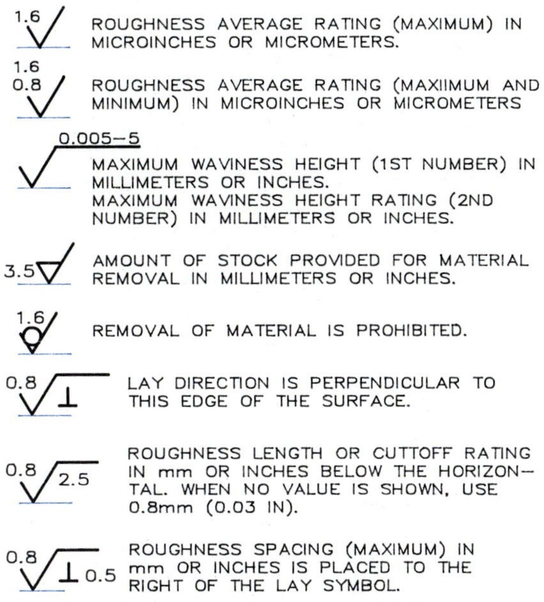

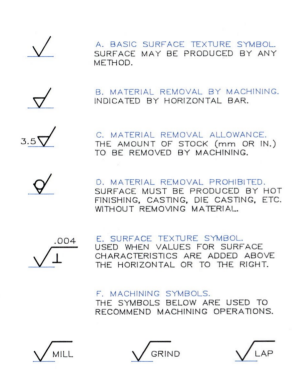

ROUGHNESS AVERAGE RATING (MAXIMUM) IN MICROINCHES OR MICROMETERS.

ROUGHNESS AVERAGE RATING (MAXIIMUM AND MINIMUM) IN MICROINCHES OR MICROMETERS

MAXIMUM WAVINESS HEIGHT (1ST NUMBER) IN MILLIMETERS OR INCHES.
MAXIMUM WAVINESS HEIGHT RATING (2ND NUMBER) IN MILLIMETERS OR INCHES.

AMOUNT OF STOCK PROVIDED FOR MATERIAL REMOVAL IN MILLIMETERS OR INCHES.

REMOVAL OF MATERIAL IS PROHIBITED.

LAY DIRECTION IS PERPENDICULAR TO THIS EDGE OF THE SURFACE.

ROUGHNESS LENGTH OR CUTTOFF RATING IN mm OR INCHES BELOW THE HORIZON—TAL. WHEN NO VALUE IS SHOWN, USE 0.8mm (0.03 IN).

ROUGHNESS SPACING (MAXIMUM) IN mm OR INCHES IS PLACED TO THE RIGHT OF THE LAY SYMBOL.

FIGURE 15.77 Values can be added to surface control symbols for more precise specifications. These may be in combinations other than those shown here.

A. BASIC SURFACE TEXTURE SYMBOL. SURFACE MAY BE PRODUCED BY ANY METHOD.

B. MATERIAL REMOVAL BY MACHINING. INDICATED BY HORIZONTAL BAR.

C. MATERIAL REMOVAL ALLOWANCE. THE AMOUNT OF STOCK (mm OR IN.) TO BE REMOVED BY MACHINING.

D. MATERIAL REMOVAL PROHIBITED. SURFACE MUST BE PRODUCED BY HOT FINISHING, CASTING, DIE CASTING, ETC. WITHOUT REMOVING MATERIAL.

E. SURFACE TEXTURE SYMBOL. USED WHEN VALUES FOR SURFACE CHARACTERISTICS ARE ADDED ABOVE THE HORIZONTAL OR TO THE RIGHT.

F. MACHINING SYMBOLS. THE SYMBOLS BELOW ARE USED TO RECOMMEND MACHINING OPERATIONS.

MILL   GRIND   LAP

FIGURE 15.76 Surface control symbols for specifying surface finish.

Figure 15.76 shows the symbols used to specify surface texture. The point of the V must touch the edge view of the surface, an extension line from the surface, or a leader pointing to the surface being specified.

In Fig. 15.77, values of surface texture that can be applied to surface texture symbols, individually or in combination, are given. The roughness height values are related to manufacturing processes used to finish the surface (Fig. 15.78).

Lay symbols that indicate the direction of texture of a surface (Fig. 15.79) can be incorporated into surface texture symbols (Fig. 15.80). Figure 15.81 shows a part with a variety of surface texture symbols.

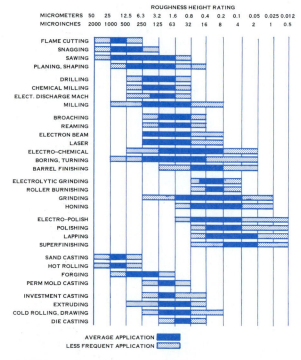

ROUGHNESS HEIGHT RATING

| | MICROMETERS | 50 | 25 | 12.5 | 6.3 | 3.2 | 1.6 | 0.8 | 0.4 | 0.2 | 0.1 | 0.05 | 0.025 | 0.012 |
| | MICROINCHES | 2000 | 1000 | 500 | 250 | 125 | 63 | 32 | 16 | 8 | 4 | 2 | 1 | 0.5 |

FLAME CUTTING
SNAGGING
SAWING
PLANING, SHAPING

DRILLING
CHEMICAL MILLING
ELECT. DISCHARGE MACH
MILLING

BROACHING
REAMING
ELECTRON BEAM
LASER
ELECTRO–CHEMICAL
BORING, TURNING
BARREL FINISHING

ELECTROLYTIC GRINDING
ROLLER BURNISHING
GRINDING
HONING

ELECTRO–POLISH
POLISHING
LAPPING
SUPERFINISHING

SAND CASTING
HOT ROLLING
FORGING
PERM MOLD CASTING

INVESTMENT CASTING
EXTRUDING
COLD ROLLING, DRAWING
DIE CASTING

AVERAGE APPLICATION
LESS FREQUENT APPLICATION

FIGURE 15.78 The surface roughness heights produced by various types of production methods are shown here in micrometers (microinches). (Courtesy of the General Motors Corp.)

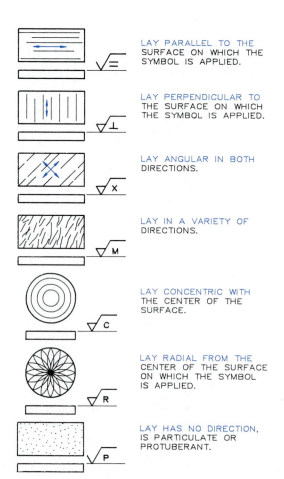

LAY PARALLEL TO THE SURFACE ON WHICH THE SYMBOL IS APPLIED.

LAY PERPENDICULAR TO THE SURFACE ON WHICH THE SYMBOL IS APPLIED.

LAY ANGULAR IN BOTH DIRECTIONS.

LAY IN A VARIETY OF DIRECTIONS.

LAY CONCENTRIC WITH THE CENTER OF THE SURFACE.

LAY RADIAL FROM THE CENTER OF THE SURFACE ON WHICH THE SYMBOL IS APPLIED.

LAY HAS NO DIRECTION, IS PARTICULATE OR PROTUBERANT.

FIGURE 15.79 These symbols are used to indicate the direction of lay with respect to the surface where the control symbol is placed.

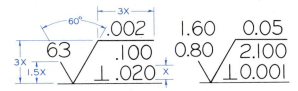

FIGURE 15.80 Examples of fully specified surface control symbols.

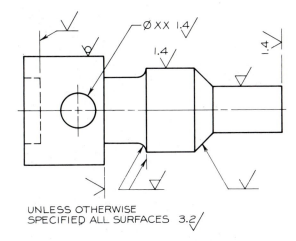

UNLESS OTHERWISE SPECIFIED ALL SURFACES 3.2

FIGURE 15.81 This drawing illustrates the techniques of applying surface texture symbols to a part.

Problems

These problems can be solved on Size A sheets. The problems are laid out on a grid of 0.20 inches (5 mm).

Cylindrical fits

1. (Fig. 15.82) Construct the drawing of a shaft and hole as shown (it need not be drawn to scale), give the limits for each diameter, and complete the table of values. Use a basic diameter of 1.00 in. (25 mm) and a class RC 1 fit or a metric fit of H8/f7.

2. Same as Problem 1, but use a basic diameter of 1.75 in. (45 mm) and a class RC 9 fit or a metric fit of H11/c11.

3. Same as Problem 1, but use a basic diameter of 2.00 in. (51 mm) and a class RC 5 fit or a metric fit of H9/d9.

4. Same as Problem 1, but use a basic diameter of 12.00 in. (305 mm) and a class LC 11 fit or a metric fit of H7/h6.

5. Same as Problem 1, but use a basic diameter of 3.00 in. (76 mm) and a class LC 1 fit or a metric fit of H7/h6.

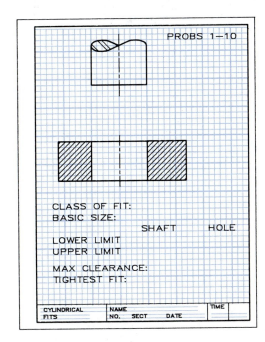

FIGURE 15.82 Problems 1–10.

6. Same as Problem 1, but use a basic diameter of 8.00 in. (203 mm) and a class LC 1 fit or a metric fit of H7/k6.

7. Same as Problem 1, but use a basic diameter of 102 in. (2591 mm) and a class LN 3 fit or a metric fit of H7/n6.

8. Same as Problem 1, but use a basic diameter of 11.00 in. (279 mm) and a class LN 2 fit or a metric fit of H7/p6.

9. Same as Problem 1, but use a basic diameter of 6.00 in. (152 mm) and a class FN 5 fit or a metric fit of H7/s6.

10. Same as Problem 1, but use a basic diameter of 2.60 in. (66 mm) and a class FN 1 fit or a metric fit of H7/u6.

Tolerances of position

11. (Fig. 15.83) On Size A paper, make an instrument drawing of the part shown. Locate the two holes with a size tolerance of 1.00 mm and a position tolerance of 0.50 DIA. Show the proper symbols and dimensions for this arrangement.

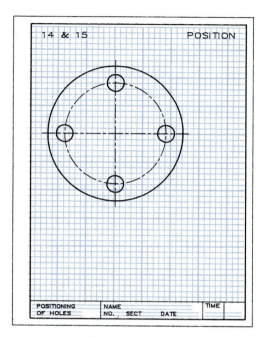

FIGURE 15.84   **Problems 14 and 15.**

12. Same as Problem 11, but locate three holes using the same tolerances for size and position.

13. Give the specifications for a two-pin gauge that can be used to gauge the correctness of the two holes specified in Problem 11. Make a sketch of the gauge and show the proper dimensions on it.

14. (Fig. 15.84) Using positioning tolerances, locate the holes and properly note them to provide a size tolerance of 1.50 mm and a locational tolerance of 0.60 DIA.

15. Same as Problem 14, but locate six equally spaced, equally sized holes using the same tolerances of position.

16. (Fig. 15.85) Using a feature control symbol and the necessary dimensions, indicate that the notch is symmetrical to the left-hand end of the part within 0.60 mm.

17. (Fig. 15.85) Using a feature control symbol and the necessary dimensions, indicate that the small cylinder is concentric with the large one (the datum cylinder) within a tolerance of 0.80.

18. (Fig. 15.86) Using a feature control symbol and the necessary dimensions, indicate that the ele-

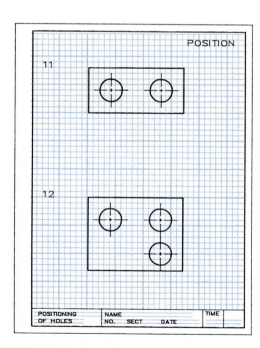

FIGURE 15.83   **Problems 11–13.**

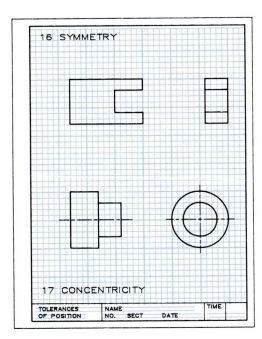

FIGURE 15.85 Problems 16 and 17.

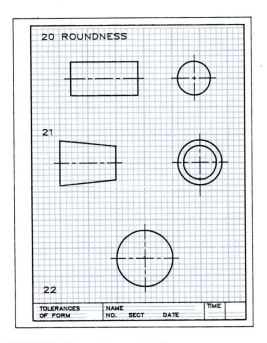

FIGURE 15.87 Problems 20–22.

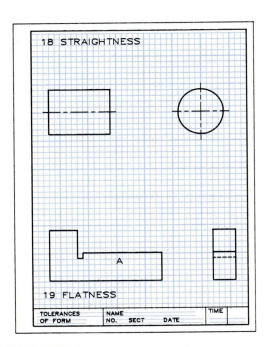

FIGURE 15.86 Problems 18 and 19.

ments of the cylinder are straight within a tolerance of 0.20 mm.

19. (Fig. 15.86) Using a feature control symbol and the necessary dimensions, indicate that surface *A* of the object is flat within a tolerance of 0.08 mm.

20–22. (Fig. 15.87) Using feature control symbols and the necessary dimensions, indicate that the cross sections of the cylinder, cone, and sphere are round within a tolerance of 0.40 mm.

23. (Fig. 15.88) Using a feature control symbol and the necessary dimensions, indicate that the profile of the irregular surface of the object lies within a bilateral or unilateral tolerance zone of 0.40 mm.

24. (Fig. 15.88) Using a feature control symbol and the necessary dimensions, indicate that the profile of the line formed by tangent arcs lies within a bilateral or unilateral tolerance zone of 0.40 mm.

25. (Fig. 15.89) Using a feature control symbol and the necessary dimensions, indicate that the cylindricity of the cylinder is 0.90 mm.

26. (Fig. 15.89) Using a feature control symbol and the necessary dimensions, indicate that the angularity tolerance of the inclined plane is 0.7 mm from the bottom of the object, the datum plane.

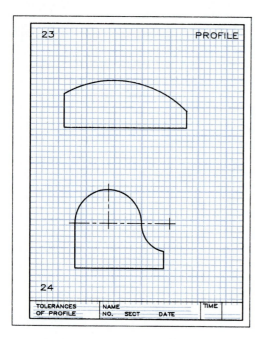

FIGURE 15.88 Problems 23 and 24.

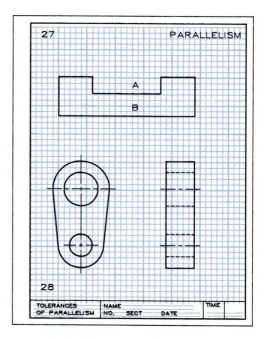

FIGURE 15.90 Problems 27 and 28.

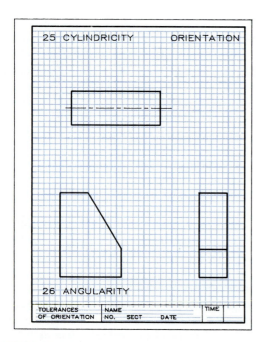

FIGURE 15.89 Problems 25 and 26.

27. (Fig. 15.90) Using a feature control symbol and the necessary dimensions, indicate that surface A of the object is parallel to datum B within 0.30 mm.

28. (Fig. 15.90) Using a feature control symbol and the necessary dimensions, indicate that the small hole is parallel to the large hole, the datum, within a tolerance of 0.80 mm.

29. (Fig. 15.91) Using a feature control symbol and the necessary dimensions, indicate that the vertical surface B is perpendicular to the bottom of the object, the datum C, within a tolerance of 0.20 mm.

30. (Fig. 15.91) Using a feature control symbol and the necessary dimensions, indicate that the hole is perpendicular to datum A within a tolerance of 0.08 mm.

31. (Fig. 15.92) Using a feature control symbol and cylinder A as the datum, indicate that the conical feature has a runout of 0.80 mm.

32. (Fig. 15.92) Using a feature control symbol with cylinder B as the primary datum and surface C as the secondary datum, indicate that surfaces D, E, and F have a runout of 0.60 mm.

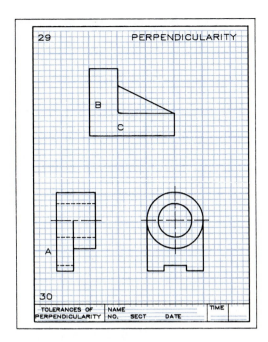

FIGURE 15.91 Problems 29 and 30.

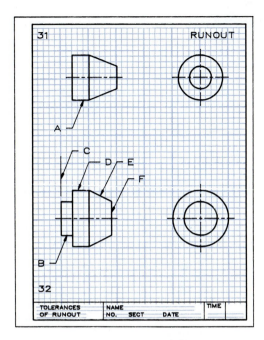

FIGURE 15.92 Problems 31 and 32.

Welding

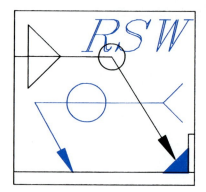

16

16.1
Introduction

Welding is the process of permanently joining metal by heating a joint to a suitable temperature with or without applying pressure and with or without using filler material.

The welding practices in this chapter comply with the standards developed by the American Welding Society and the American National Standards Institute (ANSI). We also refer to the drafting standards used by General Motors Corporation.

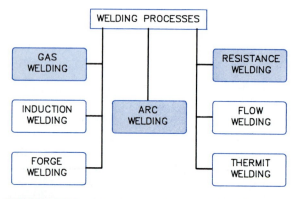

FIGURE 16.1 **The three main types of welding processes are gas welding, arc welding, and resistance welding.**

Advantages of welding over other methods of fastening include (1) simplified fabrication, (2) economy, (3) increased strength and rigidity, (4) ease of repair, (5) creation of gas- and liquid-tight joints, and (6) reduction in weight and size.

Figure 16.1 shows various types of welding processes, but the three main types are **gas welding, arc welding,** and **resistance welding.**

GAS WELDING is a process in which gas flames are used to melt and fuse metal joints. Gases like acetylene or hydrogen are mixed in a welding torch and burned with air or oxygen (Fig. 16.2). The oxyacetylene method is widely used for repair work and field construction.

Most oxyacetylene welding is done manually with a minimum of equipment. Filler material in the form of welding rods is used to deposit metal at the joint as it is heated. Most metals, except for low- and medium-carbon steels, require fluxes to aid the process of melting and fusing the metals.

ARC WELDING is a process that uses an electric arc to heat and fuse the joints (Fig. 16.3). Pressure is sometimes required in addition to heat. The filler material is supplied by a consumable or nonconsumable electrode through which the electric arc is transmitted. Metals well-suited to arc welding are wrought iron, low- and medium-carbon steels, stainless steel, copper, brass, bronze, aluminum, and some nickel alloys.

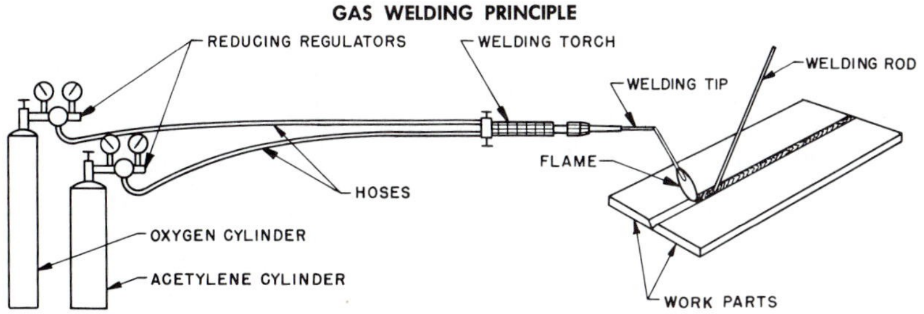

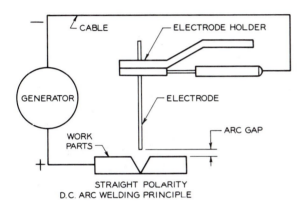

FIGURE 16.2 The gas welding process burns gases like oxygen and acetylene in a torch to apply heat to a joint. The welding rod supplies the filler material. (Courtesy of General Motors Corp.)

Flash welding is a form of arc welding, but it is similar to resistance welding in that both pressure and electric current are used to join two pieces (Fig. 16.4). The two pieces are brought together, and an electric current causes heat to build up between them. As the metal burns, the current is turned off, and the pressure between the pieces is increased to fuse them together.

RESISTANCE WELDING is a group of processes where metals are fused both by the heat produced from the resistance of the parts to the flow of an electric current and by the pressure applied. Fluxes and filler materials are normally not used. All resistance welds are either lap- or butt-type welds.

FIGURE 16.3 In arc welding, either AC or DC current is passed through an electrode to heat the joint.

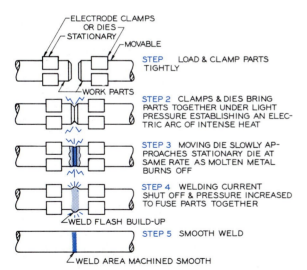

FIGURE 16.4 Flash welding, a type of arc welding, uses a combination of electric current and pressure to fuse two parts.

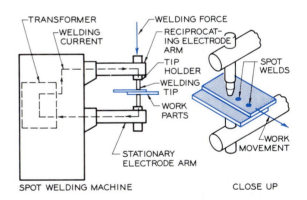

FIGURE 16.5 Resistance spot welding can be used to join lap and butt joints.

Fig. 16.5 illustrates how resistance spot welding is performed on a lap joint. The two parts are lapped and pressed together, and an electric current fuses the parts where they join. A series of spots spaced at intervals, called **spot welds,** are used to secure the parts. Table 16.1 suggests the welding processes that can be used for different materials.

TABLE 16.1
RECOMMENDED RESISTANCE WELDING
PROCESSES

Material	Spot Welding	Flash Welding
Low-carbon mild steel		
SAE 1010	R*	R
SAE 1020	R	R
Medium-carbon steel		
SAE 1030	R	R
SAE 1050	R	R
Wrought alloy steel		
SAE 4130	R	R
SAE 4340	R	R
High-alloy austenitic stainless steel		
SAE 30301–30302	R	R
SAE 30309–30316	R	R
Ferritic and martensitic stainless steel		
SAE 51410–51430	S	S
Wrought heat-resisting alloys†		
19–9–DL	S	S
16–25–6	S	S
Cast iron	NA	NR
Gray iron	NA	NR
Aluminum and aluminum alloys	R	S
Nickel and nickel alloys	R	S

*R—Recommended   NR—Not recommended
S—Satisfactory   NA—Not applicable
†For composition, see *American Society of Metals Handbook.*
Source: Courtesy of General Motors Corp.

16.2
Weld joints

Figure 16.6 shows five standard weld joints. The **butt joint** can be joined with the square groove, V-groove, bevel groove, U-groove, and J-groove welds. The **corner joint** can be joined with these welds and with the fillet weld. The **lap joint** can be joined with the bevel groove, J-groove, fillet, slot, plug, spot, projection, and seam welds. The **edge joint** uses the same welds as the lap joint along with the square groove, V-groove, U-groove, and seam welds. The **tee joint** can be joined by the bevel groove, J-groove, and fillet welds.

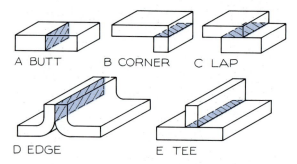

A BUTT   B CORNER   C LAP
D EDGE   E TEE

FIGURE 16.6 The standard weld joints.

16.3
Welding symbols

The specification of welds on a working drawing is done by symbols. If a drawing has a general note such as ALL JOINTS WELDED or WELDED THROUGHOUT, the designer has transferred the design responsibility to the welder. Welding is too important to be left to chance; it must be thoroughly specified.

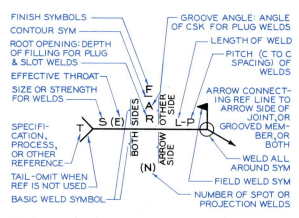

FIGURE 16.7 The welding symbol. Usually it is modified to a simpler form.

A welding symbol is used to provide specifications on a drawing (Fig. 16.7). This example gives the symbol in its complete form, which is seldom needed. The symbol is usually modified to a simpler form when all the specifications are not needed.

The scale of the welding symbol is shown in Fig. 16.8, where it is drawn on a $\frac{1}{8}$ in. (3 mm) grid. Its size can be scaled using these same proportions. The lettering used is the standard height of $\frac{1}{8}$ in. or 3 mm.

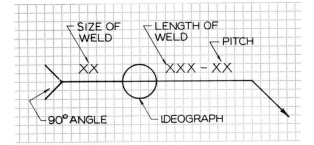

FIGURE 16.8 When the grid is drawn full size (⅛" or 3 mm), the size of the welding symbol can be determined.

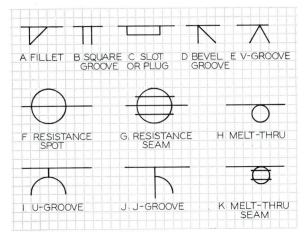

FIGURE 16.9 The sizes of the ideographs are shown on the ⅛" (3 mm) grid. These sizes are proportionately equal to the size of the welding symbol shown in Fig. 16.8.

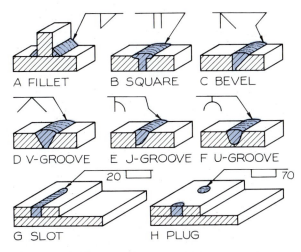

FIGURE 16.10 The standard welds and their corresponding ideographs.

> The **ideograph** is the symbol that denotes the type of weld desired. Generally, the ideograph depicts the cross section of the weld.

Figure 16.9 shows the most often used ideographs. They are drawn to scale on the ⅛ in. (3 mm) grid to represent their full size when added to the welding symbol.

16.4
Types of welds

Figure 16.10 shows commonly used welds, along with their corresponding ideographs. The fillet weld is a built-up weld at the angular intersection between two surfaces. The square, bevel, V-groove, J-groove, and U-groove welds all have grooves, and the weld is placed inside these grooves. Slot and plug welds have intermittent holes or openings where the parts are welded. Holes are unnecessary when resistance welding is used.

16.5
Application of symbols

In Fig. 16.11A, the fillet ideograph is placed below the horizontal line of the symbol, indicating the weld is to be at the joint on the **arrow side,** the side of the arrow.

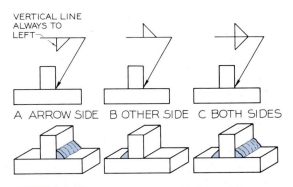

FIGURE 16.11 Fillet welds are indicated by abbreviated symbols. When the ideograph is below the horizontal line, it refers to the arrow side; when it is above the line, it refers to the other side.

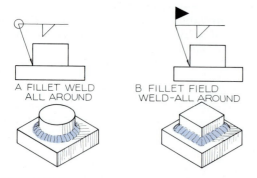

FIGURE 16.12 **Symbols for indicating fillet welds all around two types of parts.**

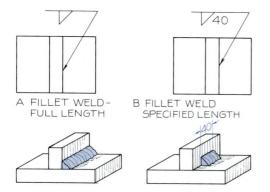

FIGURE 16.13 **A. Symbol for indicating full-length fillet welds. B. Symbol for indicating fillet welds less than full length.**

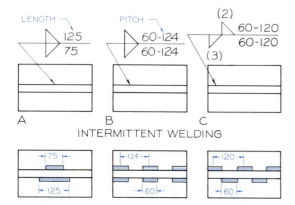

FIGURE 16.14 **Symbols for specifying varying and intermittent welds.**

The vertical leg of the ideograph is **always** on the left side.

Placing the ideograph above the horizontal line indicates that the weld is to be on the **other side**—that is, the joint on the other side of the part away from the arrow. When the part is to be welded on both sides, the ideograph shown in Fig. 16.11C is used. It is permissible to omit the tail and other specifications from the symbol when detailed specifications are given elsewhere.

A single arrow is often used to specify a weld that is to be all around two joining parts (Fig. 16.12); a circle, 6 mm in diameter, placed at the bend in the leader of the symbol denotes this. If the welding is to be done in the field (on the site rather than in the shop), a black circle, 3 mm in diameter, can be used to denote this joint.

A fillet weld that is to be the full length of the two parts may be specified as Fig. 16.13 shows. Since the ideograph is on the lower side of the horizontal line, the weld will be on the arrow side. A fillet weld that is to be less than full length may be specified as shown in Fig. 16.13B, where 40 represents the weld's length in millimeters.

Fillet welds of different lengths and positioned on both sides may be specified as Fig. 16.14A shows. The dimension on the lower side of the horizontal gives the length of the weld on the arrow side, and the dimension on the upper side gives the length on the other side.

Intermittent welds are welds of a given length that are spaced uniformly apart from center to center by a distance called the **pitch.** In Fig. 16.14B, the welds are on both sides, are 60 mm long, and have pitches of 124 mm; this can be indicated by the symbol shown. Intermittent welds that are to be staggered to alternate positions on both sides can be specified by the symbol shown in Fig. 16.14C.

16.6
Groove welds

Figure 16.15 shows the more standard groove welds. When the depth of the grooves, angle of the chamfer, and root openings are not given on a symbol, they must be specified elsewhere on the drawings or in supporting documents. In Figs. 16.15B and 16.15E,

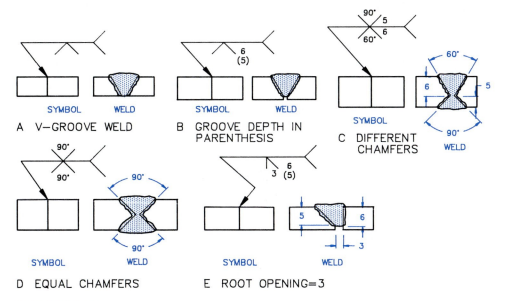

FIGURE 16.15 Types of groove welds and their general specifications.

the depth of the chamfer of the prepared joint is given in parentheses under the size dimension of the weld, which takes into account the penetration of the weld beyond this chamfer. If the size of the joint equals the depth of the prepared joint, only one number is given.

When the chamfer is different on each side of the joint, it can be noted with the symbol shown in Fig. 16.15C. If the spacing between the two parts, the **root opening,** is to be specified, this is done by placing its dimensions between the groove angle number and the weld ideograph, as Fig. 16.15E shows.

A bevel weld is a groove beveled from one of the parts being joined; therefore, the symbol must indicate which part is to be beveled (Fig. 16.15E). To call attention to this operation, the leader from the symbol is bent and aimed toward the part to be beveled. This practice also applies to J-welds where one side is grooved and the other is not (Fig. 16.16).

16.7
Surface contoured welds

Contour symbols are used to indicate which of the three types of contours, **flush, concave,** or **convex,** is desired on the surface of the weld. Flush contours are smooth with the surface or flat across the hypotenuse of a fillet weld. Concave contours bulge inward with a curve, and convex contours bulge outward with a curve (Fig. 16.17).

It is often necessary to finish the weld by a supplementary process to bring it to the desired contour. These processes, which may be indicated by their ab-

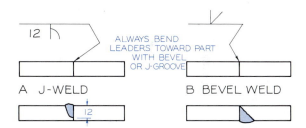

FIGURE 16.16 J-welds and bevel welds are specified by bent arrows pointing to the side of the joint to be grooved.

TYPE	SYMBOL	EXAMPLE
FLUSH		
CONCAVE		
CONVEX		

FIGURE 16.17 The contour symbols used to specify the surface finish of a weld.

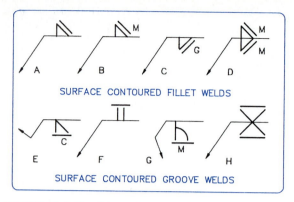

FIGURE 16.18 Examples of contoured symbols and letters of finishing applied to them.

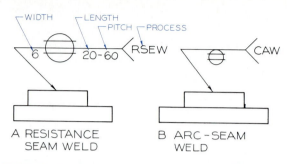

FIGURE 16.19 The process used for resistance seam welds and arc-seam welds is indicated in the tail of the symbol. The arc weld must specify arrow side or other side in the symbol.

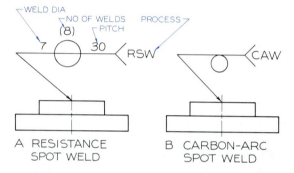

FIGURE 16.20 The process used for resistance spot welds and arc spot welds is indicated in the tail of the symbol. The arc weld must specify arrow side or other side in the symbol.

breviations, are chipping (C), grinding (G), hammering (H), machining (M), rolling (R), and peening (P). Figure 16.18 shows examples of these.

16.8
Seam welds

A seam weld joins two lapping parts with either a continuous weld or a series of closely spaced spot welds. The process used for seam welds must be given by abbreviations placed in the tail of the weld symbol. The ideograph for a resistance weld is about 12 mm in diameter and is placed with the horizontal line of the symbol through its center. Figure 16.19 shows the weld's width, length, and pitch.

When the seam weld is made by arc welding, the diameter of the ideograph is about 6 mm and is placed on the upper or lower side of the symbol's horizontal line to indicate whether the seam will be applied to the arrow side or other side (Fig. 16.19B). When the length of the weld is omitted from the symbol, it is understood that the seam weld extends between abrupt changes in the seam or as it is dimensioned.

Spot welds are similarly specified with ideographs, and specifications are given by diameter, number of welds, and pitch between the welds. The process, resistance spot welding (RSW), is noted in the tail of the symbol (Fig. 16.20A). For arc welding, the arrow side or other side must be indicated by a symbol (Fig. 16.20B). Also note the abbreviation of the welding process. Table 16.2 gives the abbreviations of various welding processes.

TABLE 16.2
WELDING PROCESS SYMBOLS

CAW	Carbon arc welding	FRW	Friction welding	PGW	Pressure gas welding
CW	Cold welding	FW	Flash welding	RB	Resistance brazing
DB	Dip brazing	GMAW	Gas metal arc welding	RPW	Projection welding
DFW	Diffusion welding	GTAW	Gas tungsten welding	RSEW	Resistance seam welding
EBW	Electron beam welding	IB	Induction brazing	RSW	Resistance spot welding
ESW	Electroslag welding	IRB	Infrared brazing	RW	Resistance welding
EXW	Explosion welding	OAW	Oxyacetylene welding	TB	Torch brazing
FB	Furnace brazing	OHW	Oxyhydrogen welding	UW	Upset welding
FOW	Forge welding				

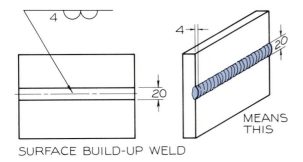

SURFACE BUILD-UP WELD

FIGURE 16.21 The method of applying a symbol to a built-up weld on a surface.

16.9
Built-up welds

When the surface of a part is to be enlarged, or **built-up,** by welding, this can be indicated by a symbol (Fig. 16.21). The width of the built-up weld is dimensioned in the view, and the height of the weld above the surface is specified in the symbol to the left of the ideograph. The radius of the circular segment is 6 mm.

16.10
Welding standards

Figure 16.22 (pages 320 and 321) gives an overview of the welding symbols and specifications discussed in the previous paragraphs. The chart can be used as a reference for most general types of welding and their symbols.

16.11
Brazing

Like welding, brazing is a method of joining pieces of metal. The process entails heating the joints above 800°F, and distributing by capillary action a nonferrous filler material, with a melting point below the base materials, between the closely fit parts.

Before brazing, the parts must be cleansed and the joints fluxed. The brazing filler is added before or just as the joints are heated beyond the filler's melting point. After the filler material has melted, it is allowed to flow between the parts to form the joint. As Fig. 16.23 shows, there are two basic brazing joints: lap joints and butt joints.

Brazing is used to hold parts together, to provide gas- and liquid-tight joints, to ensure electrical conductivity, and to aid in repair and salvage. Brazed

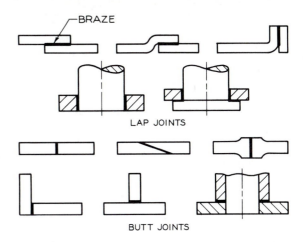

BUTT JOINTS

FIGURE 16.23 Two basic types of brazing joints: lap joints and butt joints.

joints will withstand more stress, higher temperature, and more vibration than soft-soldered joints.

16.12
Soft soldering

Soldering is the process of joining two metal parts with a third metal that melts below the temperature of the metals being joined. Solders are alloys of nonferrous metals that melt below 800°F. Widely used in the automotive and electrical industries, soldering is one of the basic techniques of welding and is often done by hand with a soldering iron like the one shown in Fig. 16.24. The iron is placed on the joint to heat it and to melt the solder. Figure 16.24 shows the method of indicating a soldered joint.

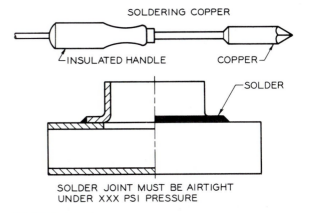

SOLDER JOINT MUST BE AIRTIGHT UNDER XXX PSI PRESSURE

FIGURE 16.24 A typical hand-held soldering iron used to soft-solder two parts together and the method of indicating a soldered joint on a drawing.

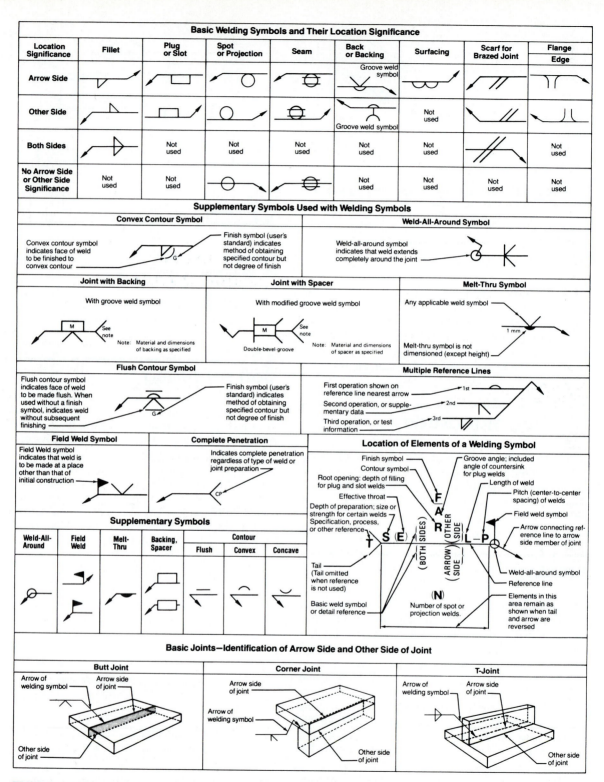

FIGURE 16.22 The American Welding Society Standard Welding Symbols. (Courtesy of the American Welding Society of Miami, FL.)

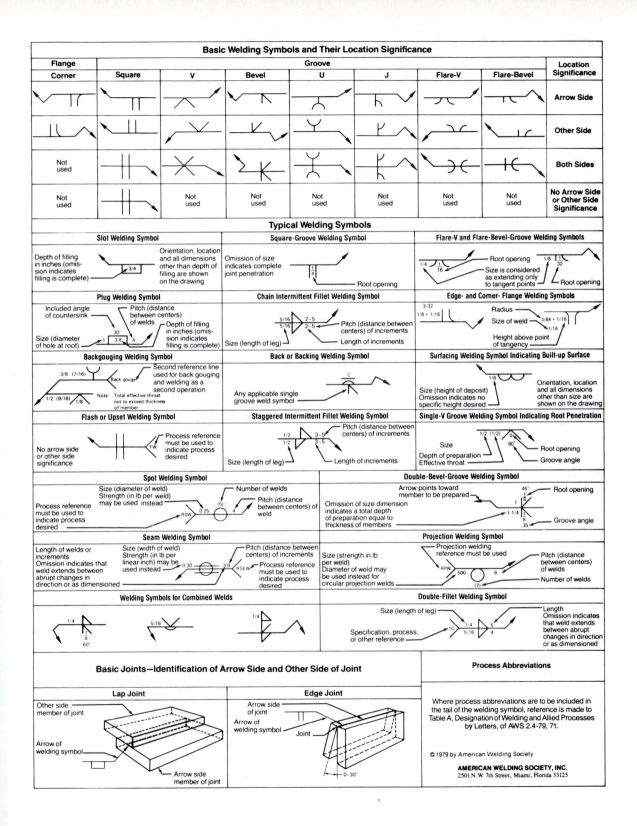

Basic Welding Symbols and Their Location Significance

Flange	Groove								Location Significance
Corner	Square	V	Bevel	U	J	Flare-V	Flare-Bevel		
									Arrow Side
									Other Side
Not used									Both Sides
Not used		Not used	Not used	Not used	Not used	Not used	Not used		No Arrow Side or Other Side Significance

Typical Welding Symbols

Slot Welding Symbol

Depth of filling in inches (omission indicates filling is complete)

3/4

Orientation, location and all dimensions other than depth of filling are shown on the drawing

Square-Groove Welding Symbol

Omission of size indicates complete joint penetration

1/4

Root opening

Flare-V and Flare-Bevel-Groove Welding Symbols

1/4 1/16   1/8 1/32

Size is considered as extending only to tangent points

Root opening

Plug Welding Symbol

Included angle of countersink

Pitch (distance between centers) of welds

Depth of filling in inches (omission indicates filling is complete)

30° 1 3/8 4

Size (diameter of hole at root)

Chain Intermittent Fillet Welding Symbol

5/16 5/16   2-5 2-5

Pitch (distance between centers) of increments

Length of increments

Size (length of leg)

Edge- and Corner- Flange Welding Symbols

3/32 1/8 + 1/16

Radius

Size of weld   3/64 + 1/16 1/16

Height above point of tangency

Backgouging Welding Symbol

3/8 (7/16) Back gouge

Second reference line used for back gouging and welding as a second operation

1/2 (9/16) 1/8

Note: Total effective throat not to exceed thickness of member

Back or Backing Welding Symbol

C

Any applicable single groove weld symbol

Surfacing Welding Symbol Indicating Built-up Surface

1/8

Size (height of deposit) Omission indicates no specific height desired

Orientation, location and all dimensions other than size are shown on the drawing

Flash or Upset Welding Symbol

No arrow side or other side significance

FW

Process reference must be used to indicate process desired

Staggered Intermittent Fillet Welding Symbol

1/2 1/2   3-5 3-5

Pitch (distance between centers) of increments

Size (length of leg)

Length of increments

Single-V Groove Welding Symbol Indicating Root Penetration

1/2 (1/2) 0

90°

Size

Depth of preparation Effective throat

Root opening

Groove angle

Spot Welding Symbol

Size (diameter of weld) Strength (in lb per weld) may be used instead

(5)

Number of welds

Pitch (distance between centers) of weld

RSW 0.25   4

Process reference must be used to indicate process desired

Double-Bevel-Groove Welding Symbol

Arrow points toward member to be prepared

45° 1/8

1 1-1/4

1/8 35

Root opening

Omission of size dimension indicates a total depth of preparation equal to thickness of members

Groove angle

Seam Welding Symbol

Length of welds or increments Omission indicates that weld extends between abrupt changes in direction or as dimensioned

Size (width of weld) Strength (in lb per linear inch) may be used instead

0.30   3.9

Pitch (distance between centers) of increments

RSEW

Process reference must be used to indicate process desired

Projection Welding Symbol

Projection welding reference must be used

RPW 500   6

(7)

Size (strength in lb per weld) Diameter of weld may be used instead for circular projection welds

Pitch (distance between centers) of welds

Number of welds

Welding Symbols for Combined Welds

1/4   5/16   1/4

8 60°

Double-Fillet Welding Symbol

Size (length of leg)

1G 1/4 6 4
   5/16

Specification, process, or other reference

Length Omission indicates that weld extends between abrupt changes in direction or as dimensioned

Basic Joints—Identification of Arrow Side and Other Side of Joint

Lap Joint

Other side member of joint

Arrow of welding symbol

Arrow side member of joint

Edge Joint

Arrow side of joint

Arrow of welding symbol

Joint

0–30°

Process Abbreviations

Where process abbreviations are to be included in the tail of the welding symbol, reference is made to Table A, Designation of Welding and Allied Processes by Letters, of AWS 2.4-79, 71.

© 1979 by American Welding Society

AMERICAN WELDING SOCIETY, INC.
2501 N. W. 7th Street, Miami, Florida 33125

Working Drawings

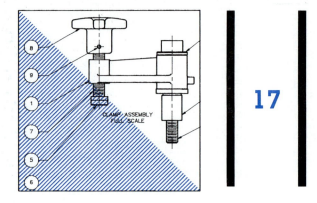

17

17.1
Introduction

Working drawings are the drawings a design is constructed from. Depending on the complexity of the project, a set of working drawings may contain any number of sheets, from only one to more than one hundred. It is important to give the number of sheets in the set on each sheet, for example, sheet 2 of 6, sheet 3 of 6, and so on.

The written instructions that accompany working drawings are called **specifications.** When the project can be represented on several sheets, the specifications are often written on the drawings to consolidate the information into a single format. Much of the work in preparing working drawings is done by the drafter, but the designer, who is usually an engineer, is responsible for their correctness.

> A working drawing is often called a **detail drawing** because it describes and gives the dimensions of the details of the parts being presented.

All the principles of orthographic projection and all the techniques of graphical presentation are used to communicate the details in a working drawing.

FIGURE 17.1 This pulley and setscrew are detailed in the working drawing in Fig. 17.2.

17.2
Working drawings—
inch system

The inch is the basic unit of the English system, which is being superseded by the metric system.

The pulley (Fig. 17.1) is represented by the working drawings shown in Fig. 17.2. Dimensions and notes give the information needed to construct the pieces. This drawing is dimensioned in millimeters.

The clamp in Fig. 17.3 is represented by computer-drawn working drawings in Figs. 17.4–17.6.

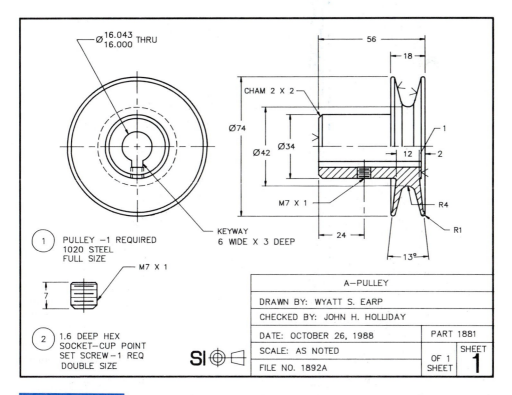

$\varnothing \frac{16.043}{16.000}$ THRU

CHAM 2 X 2

$\varnothing 74$

$\varnothing 42$

$\varnothing 34$

M7 X 1

56

18

1

2

12

R4

R1

24

13°

KEYWAY
6 WIDE X 3 DEEP

1 PULLEY −1 REQUIRED
1020 STEEL
FULL SIZE

M7 X 1

7

2 1.6 DEEP HEX
SOCKET−CUP POINT
SET SCREW −1 REQ
DOUBLE SIZE

SI

A−PULLEY	
DRAWN BY: WYATT S. EARP	
CHECKED BY: JOHN H. HOLLIDAY	
DATE: OCTOBER 26, 1988	PART 1881
SCALE: AS NOTED	SHEET
FILE NO. 1892A	OF 1 SHEET / 1

FIGURE 17.2 A computer-drawn working drawing of a pulley and setscrew dimensioned using the metric system.

FIGURE 17.3 A revolving clamp assembly manufactured to hold parts while they are being machined.

Figure 17.3 is dimensioned with decimal inches, which are preferable to common fractions although both systems are still widely used. Using decimal inches makes it possible to handle arithmetic with greater ease than is possible with fractions.

Inch marks are omitted from dimensions on a working drawing since it is understood that the units are in inches. Several dimensioned parts are shown on each sheet as orthographic views. The arrangement of these parts on the sheet has no relationship to how they fit together; they are simply positioned to take advantage of the available space. Each part is given a number and name for identification. The material that each part is made of is indicated along with other notes to explain any necessary manufacturing procedures.

The orthographic assembly given on sheet 3 (Fig. 17.6) explains how the parts fit together. The parts are numbered to correspond to the part numbers in the parts list, which serves as a bill of materials.

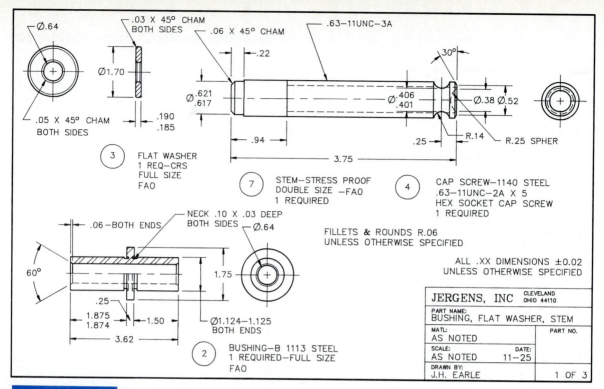

FLAT WASHER
1 REQ-CRS
FULL SIZE
FAO

STEM-STRESS PROOF
DOUBLE SIZE -FAO
1 REQUIRED

CAP SCREW-1140 STEEL
.63-11UNC-2A X 5
HEX SOCKET CAP SCREW
1 REQUIRED

FILLETS & ROUNDS R.06
UNLESS OTHERWISE SPECIFIED

ALL .XX DIMENSIONS ±0.02
UNLESS OTHERWISE SPECIFIED

BUSHING-B 1113 STEEL
1 REQUIRED-FULL SIZE
FAO

JERGENS, INC   CLEVELAND
OHIO 44110

PART NAME:
BUSHING, FLAT WASHER, STEM

MATL:
AS NOTED   PART NO.

SCALE: DATE:
AS NOTED   11-25

DRAWN BY:
J.H. EARLE   1 OF 3

FIGURE 17.4 A set of three computer-drawn detail drawings showing parts of the clamp assembly. (Courtesy of Jergens, Inc.)

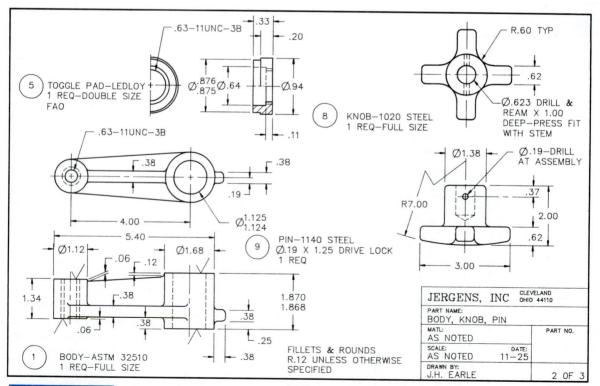

TOGGLE PAD-LEDLOY
1 REQ-DOUBLE SIZE
FAO

KNOB-1020 STEEL
1 REQ-FULL SIZE

PIN-1140 STEEL
Ø.19 X 1.25 DRIVE LOCK
1 REQ

BODY-ASTM 32510
1 REQ-FULL SIZE

FILLETS & ROUNDS
R.12 UNLESS OTHERWISE
SPECIFIED

JERGENS, INC   CLEVELAND
OHIO 44110

PART NAME:
BODY, KNOB, PIN

MATL:
AS NOTED   PART NO.

SCALE: DATE:
AS NOTED   11-25

DRAWN BY:
J.H. EARLE   2 OF 3

FIGURE 17.5 A continuation of Fig. 17.4. (Courtesy of Jergens, Inc.)

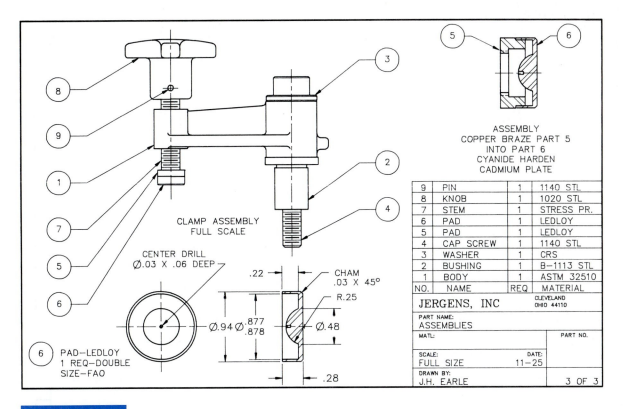

ASSEMBLY
COPPER BRAZE PART 5
INTO PART 6
CYANIDE HARDEN
CADMIUM PLATE

9	PIN	1	1140 STL
8	KNOB	1	1020 STL
7	STEM	1	STRESS PR.
6	PAD	1	LEDLOY
5	PAD	1	LEDLOY
4	CAP SCREW	1	1140 STL
3	WASHER	1	CRS
2	BUSHING	1	B−1113 STL
1	BODY	1	ASTM 32510
NO.	NAME	REQ	MATERIAL

JERGENS, INC   CLEVELAND OHIO 44110

PART NAME:
ASSEMBLIES

MATL: PART NO.

SCALE: DATE:
FULL SIZE   11−25

DRAWN BY:
J.H. EARLE   3 OF 3

CLAMP ASSEMBLY
FULL SCALE

CENTER DRILL
Ø.03 X .06 DEEP

CHAM
.03 X 45°
R.25

Ø.94 Ø .877/.878   Ø.48

.22   .28

6 PAD−LEDLOY
1 REQ−DOUBLE
SIZE−FAO

FIGURE 17.6 A computer-drawn detail drawing and an orthographic assembly of the clamp assembly.

17.3
Working drawings— metric system

FIGURE 17.7 The left-end handcrank detailed in the working drawings in Figs. 17.8 and 17.9.

The left-end handcrank (Fig. 17.7) is detailed on two Size B sheets in Figs. 17.8 and 17.9. Since the millimeter is the basic unit of the metric system, all dimensions are measured to the nearest whole millimeter, with no decimal fractions except where tolerances are shown. Metric abbreviations after the numerals are omitted from dimensions on a working drawing since it is understood from the SI symbol near the title block that the units are metric.

If you have trouble relating to the length of a millimeter, remember that the fingernail of your index finger is about 10 mm wide.

Figure 17.9, an orthographic assembly of the left-end handcrank, shows how the parts are put together.

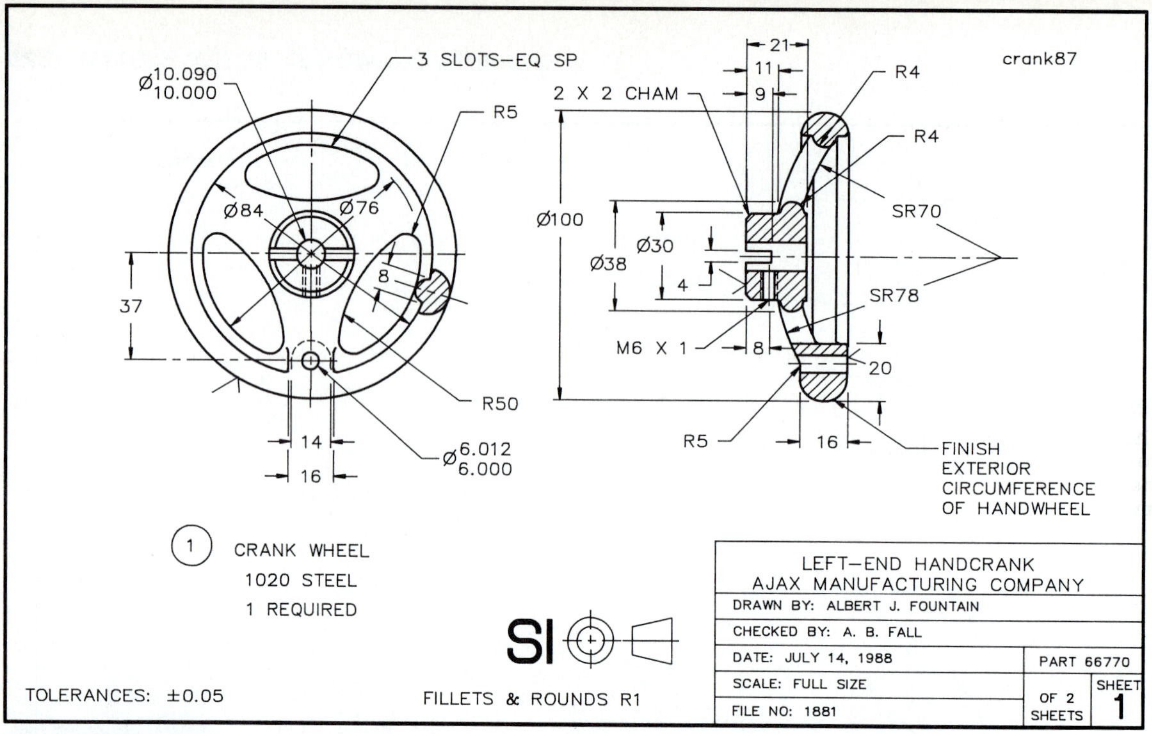

FIGURE 17.8 A computer-drawn working drawing of the crank wheel of the left-end handcrank. Dimensions are in millimeters.

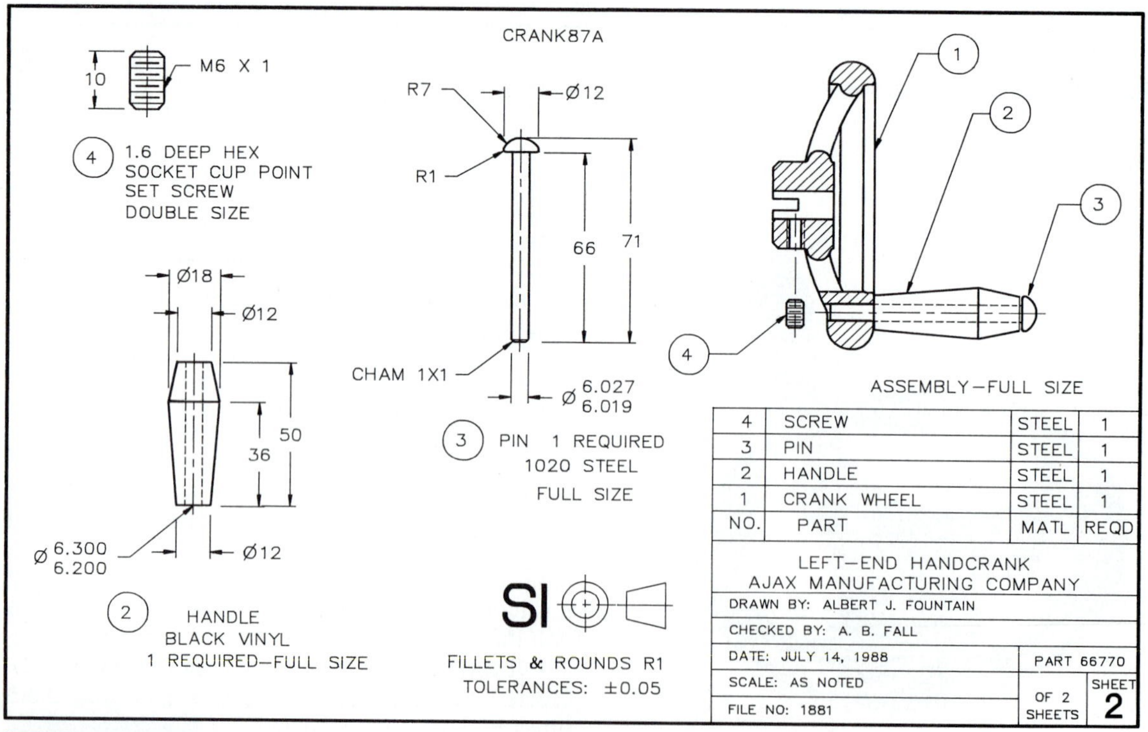

NO.	PART	MATL	REQD
4	SCREW	STEEL	1
3	PIN	STEEL	1
2	HANDLE	STEEL	1
1	CRANK WHEEL	STEEL	1

FIGURE 17.9 The second sheet of the working drawings of the left-end handcrank, including an assembly drawing.

FIGURE 17.10 **A Lev-L-ine lifting device used to level heavy machinery. This product is the basis of the working drawings in Figs. 17.11 and 17.12. (Courtesy of Unisorb Machinery Installation Systems.)**

The numbers in the balloons refer to the numbers in the parts list, which is above the title block.

Figure 17.10 shows a lifting device used to level heavy equipment such as lathes and milling machines, and Figs. 17.11 and 17.12 show working drawings of the parts of this product. The SI symbol indicates that the dimensions are in millimeters, and the truncated cone indicates that the views are drawn using third-angle projection. Figure 17.12 shows the parts in assembly.

17.4
Working drawings— dual dimensions

Some working drawings are dimensioned in both inches and millimeters. Figure 17.13 shows an example, where the dimensions in parentheses or brackets are millimeters. The units may also be given in inches

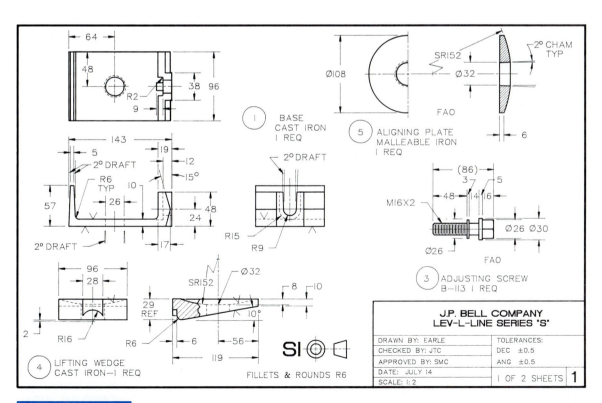

FIGURE 17.11 **A working drawing of the lifting device dimensioned in SI units. (Courtesy of Unisorb Machinery Installation Systems.)**

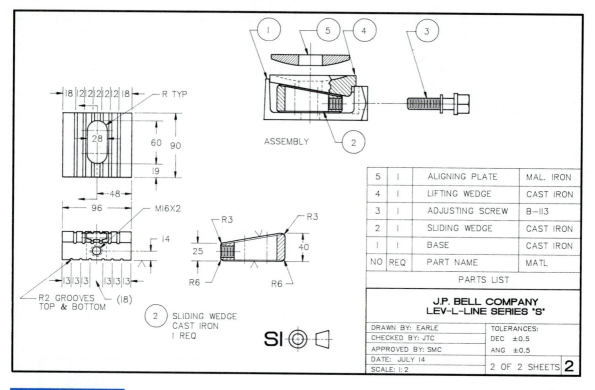

FIGURE 17.12 **A working drawing and assembly of the lifting device dimensioned in SI units. (Courtesy of Unisorb Machinery Installation Systems.)**

first and then converted to millimeters. Converting from one unit to the other will result in fractional units that must be rounded off. An explanation of the system used should be noted in the title block.

17.5
Laying out a working drawing

A typical working drawing is laid out by computer by beginning with the border (if a printed border is not available), as shown in Fig. 17.14. A margin of at least 0.25 in. (7 mm) should be allowed between the edge of the sheet and the borderlines. Space for the title block in the lower right-hand corner of the sheet is outlined to ensure that the drawing does not occupy this area.

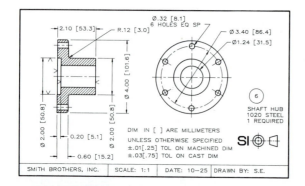

FIGURE 17.13 **A dual-dimensioned drawing. The dimensions are given in millimeters, and their equivalents in inches are given in brackets.**

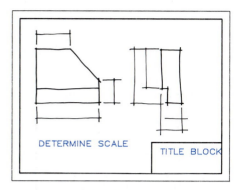

1 SKETCH THE VIEWS AND DETERMINE THE
   DIMENSIONS THAT WILL BE NEEDED.

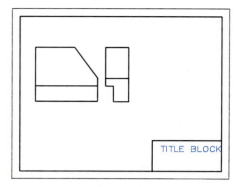

2 PLOT THE VIEWS CLOSE TOGETHER TO
   MAKE PROJECTION EASIER.

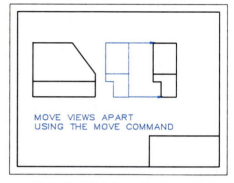

3 SEPARATE THE VIEWS TO MAKE ROOM
   FOR THE DIMENSIONS.

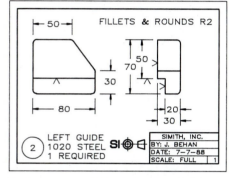

4 APPLY THE DIMENSIONS AND NOTES,
   COMPLETE THE TITLE BLOCK.

FIGURE 17.14 **The steps of laying out a working drawing begin with a freehand sketch to determine the views and dimensions needed.**

A freehand sketch is made in Step 1 to determine the necessary views, the number of dimensions, and their placement so that the proper scale can be selected. In Step 2, the views are drawn close together since this makes projection from view to view easier.

In Step 3, the views are MOVEd to make room for notes and dimensions. In Step 4, the dimensions, notes, SI symbol, and title block are added to complete the drawing.

When making a drawing by pencil on paper or film, it is more efficient to first lay out the views and dimensions on a different sheet of paper, and then overlay the drawing with vellum or film for tracing the final drawing. Guidelines for lettering must be drawn for each dimension and note.

Figure 17.15 shows the standard sheet sizes for working drawings. Papers, films, cloths, and reproduction materials are available in these modular sizes.

DRAWING SHEET SIZES				
ENGLISH SIZES		METRIC SIZES		
A	11 X 8.5	A4	297 X 210
B	17 X 11	A3	420 X 297
C	22 X 17	A2	594 X 420
D	34 X 22	A1	841 X 594
E	44 X 34	A0	1189 X 841

FIGURE 17.15 **The standard sheet sizes for working drawings dimensioned in inches.**

17.6
Title blocks and parts lists

Figure 17.16 shows a parts list and title block for student assignments. Title blocks are usually in the lower right-hand corner of the drawing sheet against the borders and usually contain the **title** or **part name, drafter, date, scale, company,** and **sheet** number. Other information such as **tolerances, checkers,** and **materials** may also be shown. The parts list should be placed directly over the title block.

Figure 17.17 shows a title block used by General Motors. A note to the left of the title block lists John F. Brown as the inventor. Two associates were asked to date and sign the drawings as witnesses of his work.

REVISIONS	COMPANY NAME COMPANY ADDRESS		
CHG HEIGHT	TITLE: LEFT—END BEARING		
FAO	DRAWN BY: JOHNNY RINGO		
	CHECKED BY: FRED DODGE		
	DATE: JULY 14, 1989		
	SCALE HALF SIZE	SHEET OF 3 SHEETS 1	

FIGURE 17.18   An example of a title block with a revision block.

This procedure establishes the ownership of the ideas and dates their development in case this becomes an issue in obtaining a patent.

Another example of a title block (Fig. 17.18) is typical of those used by various industries. Revision blocks list any modifications that will improve the design.

Figure 17.19 shows a computer shortcut for filling in a title block that will be used on several sheets. The title block can be drawn only once and filled in with dummy values to establish the positions of the text. The CHANGE command can be used to replace the dummy entries with updated values when the block is inserted or copied into a new drawing.

A similar shortcut is the design of a title block in which ATTRIBUTES are used (see Section 23.82). Here, you will be prompted for the entries into the title block at the time of its INSERTion.

2	SHAFT	2	1020 STL
1	BASE	1	CI
NO	PART NAME	REQ	MATERIAL

5" APPROXIMATELY
.125" LETTERS

TITLE	
BY:	SECT:
DATE:	SHEET:
SCALE:	OF SHEETS

FIGURE 17.16   A typical parts list and title block suitable for most student assignments.

INVENTOR: JACK OMOHUNDRO MAY 2, 1990 WITNESS: J. B. HICKOK MAY 2, 1990	DRAFT
	DATE:
	MATER
	FEATU
	SHEET

INVENTOR'S SIGNATURE AND DATE
WITNESS

FIGURE 17.17   A General Motors title block witnessed by an associate. (Courtesy of General Motors Corp.)

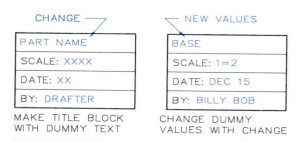

CHANGE
PART NAME
SCALE: XXXX
DATE: XX
BY: DRAFTER

MAKE TITLE BLOCK WITH DUMMY TEXT

NEW VALUES
BASE
SCALE: 1=2
DATE: DEC 15
BY: BILLY BOB

CHANGE DUMMY VALUES WITH CHANGE

FIGURE 17.19   Title block by computer

Step 1   Draw the title block and add dummy text values using the desired text style and size. Make a BLOCK of the title block.

Step 2   Insert the title block against the lower bottom and right borders. EXPLODE the BLOCK and, using the CHANGE command, change the dummy values to the desired ones.

17.7
Scale specification

If all drawings are the same scale, the scale of a working drawing should be indicated in or near the title block. If several drawings are made at different scales, the scales should be indicated on the drawings.

Figure 17.20 shows several methods of indicating scales. When the colon is used (for example, 1:2), the metric system is implied; when the equal sign is used (for example, 1 = 2), the English system is implied. The SI symbol or METRIC designation on a drawing specifies that the units of measurement are millimeters.

In some cases, a graphical scale with calibrations is given that permits the interpretation of linear units by transferring dimensions with your dividers from the drawing to the scale.

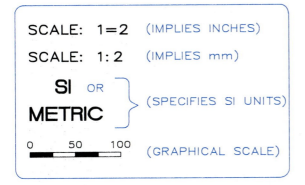

FIGURE 17.20 Methods of specifying scales and metric units on working drawings.

17.8
Tolerances

General notes can be given on working drawings to specify the tolerances of dimensions. Figure 17.21 shows a table of values where a check can be made to indicate whether the units will be in inches or millimeters. As the figure shows, plus-or-minus tolerances are given in the blanks, under the number of digits, and under each decimal fraction. For example, this ta-

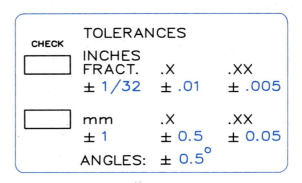

FIGURE 17.21 General tolerance notes given on working drawings to specify the tolerances permitted on dimensions.

ble specifies that each dimension with two-place decimals will have a tolerance of ±0.01 in.

Angular tolerances can also be given in general notes. (See Chapter 15 for more detailed examples of using tolerance notes.)

17.9
Part names and numbers

Each part should be given a name and number (Fig. 17.22). The letters and numbers should be $\frac{1}{8}$ in. (3 mm) high. The part numbers are placed inside circles, called **balloons,** which are drawn approximately four times the height of the numbers.

On the working drawings, the part numbers should be placed near the parts to clarify which part they are associated with. On assembly drawings, balloons are especially important since the numbers of the parts refer to the same numbers in the parts list.

FIGURE 17.22 Each part of a working drawing should be named and numbered for listing in the parts list.

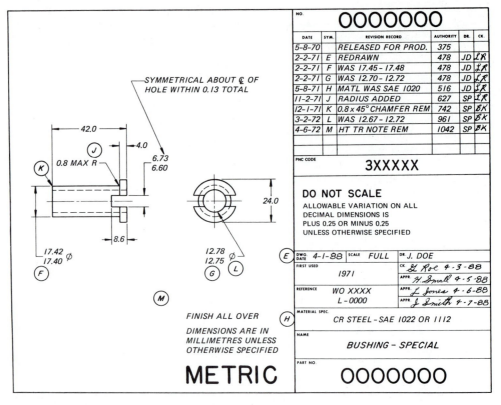

FIGURE 17.23 The modifications to this working drawing are noted near the details to be revised. The letters in balloons are cross-referenced in the revision table. (Courtesy of General Motors Corp.)

17.10
Checking a drawing

All drawings must be checked before they are released for production since a mistake could prove expensive. The people who check drawings have special qualifications enabling them to suggest revisions and modifications that will result in a better product at less cost. The checker may be a chief drafter experienced in this type of work or the engineer or designer who originated the project. In larger companies, the drawings are reviewed by the various shops involved to determine whether the most efficient production methods are specified for each part.

Checkers never check the original drawing; instead, they note corrections with a colored pencil on a

diazo print (a blue-line print). The print is returned to the drafter for revision of the original drawing, and another print is made for approval.

In Fig. 17.23, the various modifications made by checkers are labeled with letters that are circled and placed near the revisions. Changes are listed and dated in the revision record by the drafter.

Checkers check for the soundness of the design and its functional characteristics. They are also responsible for the drawing's completeness and quality, its readability and clarity, and its lettering and drafting techniques. Quality of lettering is especially important since the shop person must rely on lettered notes and dimensions.

The best way for students to check their drawings is to make a rapid scale drawing of the part from the working drawings. Figure 17.24 shows a grading scale

	Max Value	Points Earned
TITLE BLOCK		
Student's name	1	_____
Checker	1	_____
Date	1	_____
Scale	1	_____
Sheet number	1	_____
REPRESENTATION OF DETAILS		
Selection of views	5	_____
Assembly drawings	10	_____
Positioning of views	5	_____
DRAFTING PRACTICES		
Line quality	8	_____
Lettering	8	_____
Proper dimensioning	10	_____
Proper use of sections	5	_____
Proper use of auxiliary views	5	_____
DESIGN INFORMATION		
Indication of tolerances	5	_____
General tolerance notes	5	_____
Fillets and rounds notes	5	_____
Finish marks	5	_____
Parts list	8	_____
Thread notes and symbols	5	_____
PRESENTATION		
Properly trimmed	2	_____
Properly folded	2	_____
Properly stapled	2	_____
	100	_____

grade ⬜

FIGURE 17.24 A checklist for evaluating a working-drawing assignment.

for checking the working drawings prepared by students. This list can be used as an outline for reviewing working drawings to ensure that the major requirements have been met.

17.11
Drafter's log

Drafters should keep a **log,** a record that shows all changes made during the project. As the project progresses, the changes, dates, and people involved should be recorded for reference and later review. Calculations are often made during a drawing's preparation. If they are lost or poorly done, it may be necessary to do them again; therefore, they should be made a permanent part of the log.

17.12
Assembly drawings

After the parts have been made according to the specifications of the working drawings, they will be assembled (Fig. 17.25) by following the directions of an **assembly drawing.**

FIGURE 17.25 An assembly drawing is used to explain how the parts of a product such as this Ford tractor are assembled. (Courtesy of Ford Motor Co.)

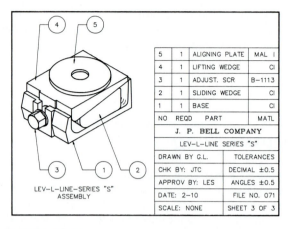

5	1	ALIGNING PLATE	MAL I
4	1	LIFTING WEDGE	CI
3	1	ADJUST. SCR	B-1113
2	1	SLIDING WEDGE	CI
1	1	BASE	CI
NO	REQD	PART	MATL

J. P. BELL COMPANY

LEV-L-LINE SERIES "S"

DRAWN BY G.L.	TOLERANCES
CHK BY: JTC	DECIMAL ±0.5
APPROV BY: LES	ANGLES ±0.5
DATE: 2-10	FILE NO. 071
SCALE: NONE	SHEET 3 OF 3

LEV-L-LINE-SERIES "S"
ASSEMBLY

FIGURE 17.26 An isometric assembly drawing that shows the parts of the lifting device fully assembled. Dimensions are usually omitted from an assembly drawing, and a parts list is given.

Two general types of assembly drawings are **orthographic assemblies** and **pictorial assemblies.** Dimensions are usually omitted from assembly drawings.

The lifting device in Fig. 17.10 is shown as an isometric assembly in Fig. 17.26. Each part is numbered with a balloon and leader to cross-reference them to the parts list, where more information about each part is given.

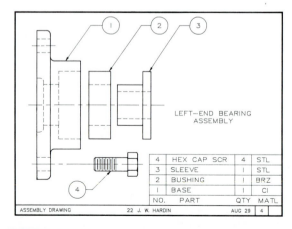

LEFT-END BEARING
ASSEMBLY

4	HEX CAP SCR	4	STL
3	SLEEVE	1	STL
2	BUSHING	1	BRZ
1	BASE	1	CI
NO.	PART	QTY	MATL

ASSEMBLY DRAWING · 22 J. W. HARDIN · AUG 29 · 4

FIGURE 17.27 An exploded orthographic assembly illustrating how parts are put together.

Figure 17.27 shows an **orthographic exploded assembly.** In many applications, the assembly of parts can be more easily understood when the parts are shown "exploded" along their centerlines. In this example, the views are shown as regular views with some lines shown as hidden lines.

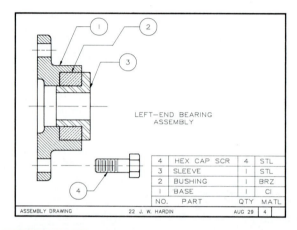

LEFT-END BEARING
ASSEMBLY

4	HEX CAP SCR	4	STL
3	SLEEVE	1	STL
2	BUSHING	1	BRZ
1	BASE	1	CI
NO.	PART	QTY	MATL

ASSEMBLY DRAWING · 22 J. W. HARDIN · AUG 29 · 4

FIGURE 17.28 A sectioned orthographic assembly that shows the parts in their assembled positions except for the exploded bolt.

The same assembly of the part is shown in Fig. 17.28 as an orthographic assembly where the parts are in their assembled positions. The views have been sectioned as full sections to make them more easily understandable.

The **outline assembly** in Fig. 17.29 shows how various components are connected. Each part is composed of subassemblies not shown in detail. Only general dimensions are given that might be of value in connecting the pump into its overall system.

The brake-pedal assembly in Fig. 17.30 is an **exploded pictorial assembly** that shows how the parts fit together. Special balloons are used to give a variety of information in a standard manner.

17.13
Freehand working drawings

A freehand sketch can serve the same purpose as an instrument drawing, provided the part is simple and the essential dimensions are given (Fig. 17.31). The

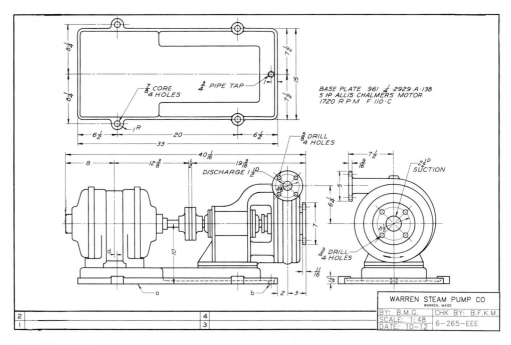

FIGURE 17.29 An outline assembly showing the general relationship of the parts of the assembly and their overall dimensions.

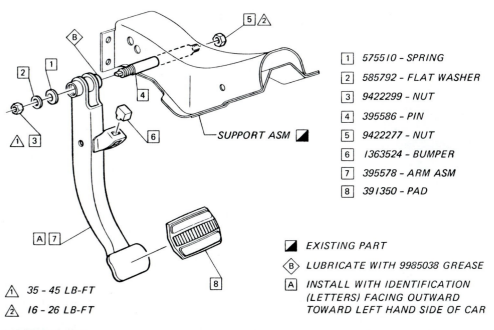

1	575510 – SPRING
2	585792 – FLAT WASHER
3	9422299 – NUT
4	395586 – PIN
5	9422277 – NUT
6	1363524 – BUMPER
7	395578 – ARM ASM
8	391350 – PAD

◤ EXISTING PART

Ⓑ LUBRICATE WITH 9985038 GREASE

△1 35 - 45 LB-FT

△2 16 - 26 LB-FT

Ⓐ INSTALL WITH IDENTIFICATION (LETTERS) FACING OUTWARD TOWARD LEFT HAND SIDE OF CAR

FIGURE 17.30 An exploded pictorial assembly of a pedal assembly. The balloons are cross-referenced to a legend that gives a variety of information. (Courtesy of General Motors Corp.)

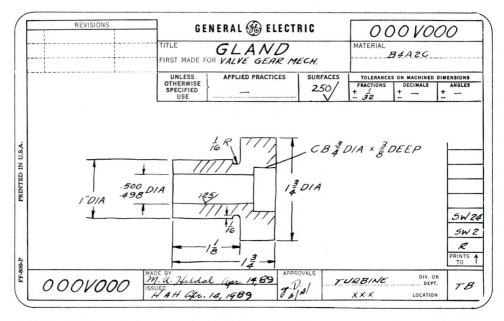

FIGURE 17.31 A freehand working drawing with the essential dimensions can be as adequate as an instrument-drawn detail drawing. (Courtesy of General Electric Co.)

same principles of working-drawing construction should be followed when instruments are used.

17.14
Castings and forged parts

The two parts shown in Fig. 17.32 illustrate the difference between a forged part and a machined part. A **forging** is a rough, oversize form made by hammering the metal into shape or pressing it between two forms (called **dies**). The forging is then machined to its finished dimensions and tolerances.

A **casting,** like a forging, is a general shape that must be machined so that it will fit with other parts in an assembly. The casting is formed by pouring molten metal into a mold formed by a pattern made slightly larger than the finished part (Fig. 17.33). For the pattern to be removable from the sand that forms the mold, its sides must be tapered. This taper of from 5° to 10° is called the **draft** (Fig. 17.33).

In some industries, casting and forging drawings are made separate from machine drawings (Fig. 17.34). But more often, these drawings are combined

FIGURE 17.32 The top part is a "blank" that has been forged. When the forging has been machined, it will look like the bottom part.

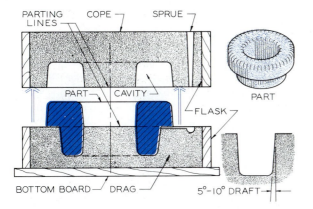

FIGURE 17.33 A two-part sand mold is used to produce a casting. A draft of from 5° to 10° is needed to permit withdrawal of the pattern from the sand. The casting must be machined to size it within specified tolerances.

into one drawing with the understanding that the forgings and castings must be made oversize to allow for the removal of material by the machining operations.

17.15
Sheet metal drawings

Parts made of sheet metal are formed by bending or stamping. The flat metal patterns for these parts must be developed graphically. Figure 17.35 shows an example of a sheet metal part, where the angles and radii of the bends are given. The note B.D. means to bend downward, and B.U. means to bend upward.

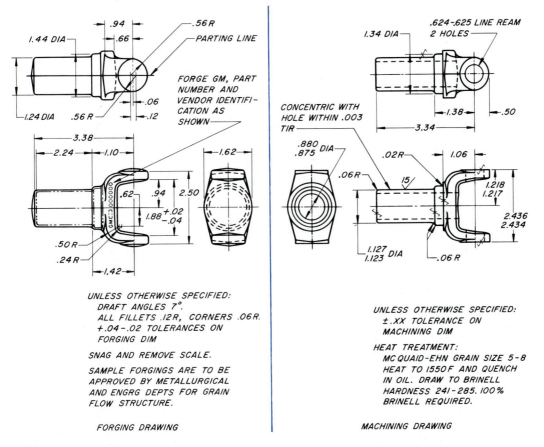

FORGING DRAWING

MACHINING DRAWING

FIGURE 17.34 Two separate working drawings, a forging drawing and a machining drawing, are used to give the details of the same part. Often, this information is combined into a single drawing. (Courtesy of General Motors Corp.)

Problems

The following problems (Figs. 17.36–17.89) are to be drawn on the sheet sizes assigned or those suggested. Each problem should be drawn with the appropriate dimensions and notes to fully describe the parts and assemblies being drawn.

Working drawings may be made on film or tracing vellum in ink or pencil. Select a suitable title block, and complete it using good lettering practices. Some problems will require more than one sheet to show all the parts properly.

Assemblies should be prepared with a parts list where there are several parts. These may be orthographic or pictorial assemblies, either exploded, partially exploded, or assembled.

The dimensions given in the problems do not always represent good dimensioning practices because of space limitations; however, the dimensions given are usually adequate for you to complete the detail drawings. In some cases, there may be omitted dimensions that you must approximate using your own judgment. When making the detail drawings, strive to provide all the necessary information, notes, and dimensions to describe the views completely. Use any of the previously covered principles, conventions, and techniques to present the views with the maximum clarity and simplicity.

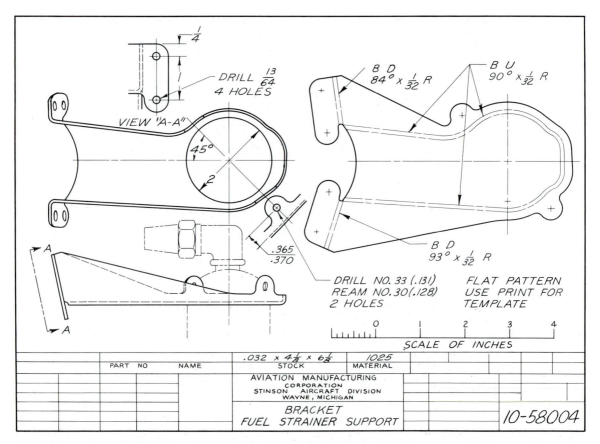

FIGURE 17.35 A detail drawing of a sheet metal part shown as a flat pattern and as orthographic views when bent into shape.

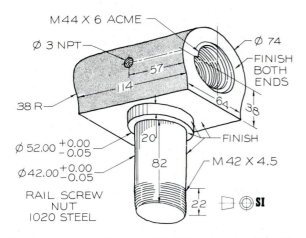

FIGURE 17.36 Make a detail drawing on a Size B sheet.

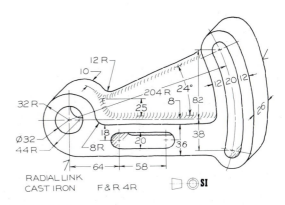

FIGURE 17.37 Make a detail drawing on a Size B sheet.

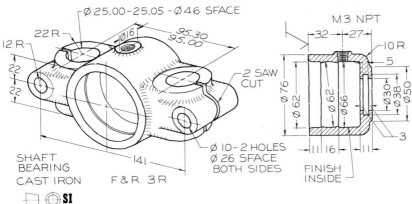

FIGURE 17.38 Make a detail drawing on a Size C sheet.

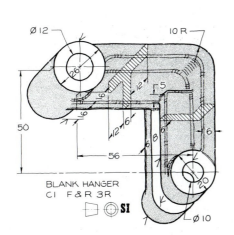

FIGURE 17.39 Make a detail drawing on a Size B sheet.

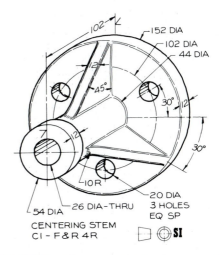

FIGURE 17.40 Make a detail drawing on a Size B sheet.

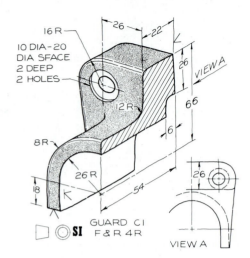

FIGURE 17.41   **Make a detail drawing on a Size B sheet.**

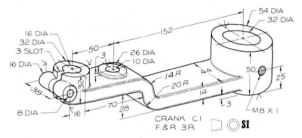

FIGURE 17.42   **Make a detail drawing on a Size B sheet.**

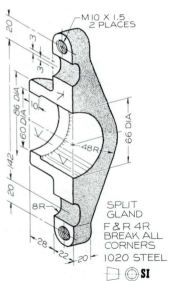

FIGURE 17.43   **Make a detail drawing on a Size B sheet.**

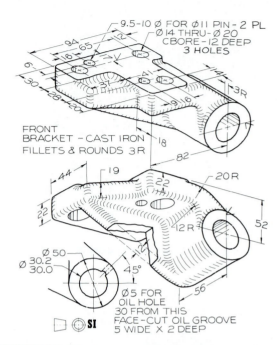

FIGURE 17.44   **Make a detail drawing on a Size B sheet.**

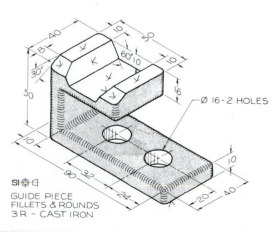

FIGURE 17.45   **Make a detail drawing on a Size B sheet.**

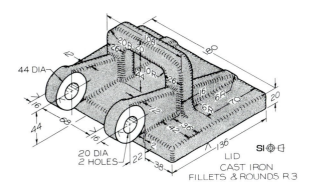

FIGURE 17.46 Make a detail drawing on a Size B sheet.

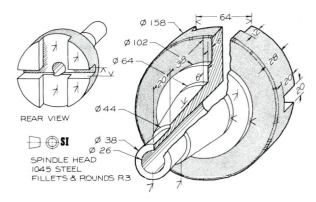

FIGURE 17.49 Make a detail drawing on a Size B sheet.

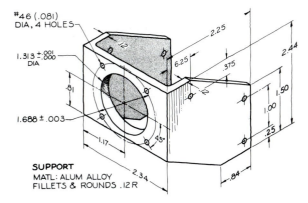

FIGURE 17.47 Make a detail drawing on a Size B sheet using (A) decimal inches or (B) inches converted to millimeters.

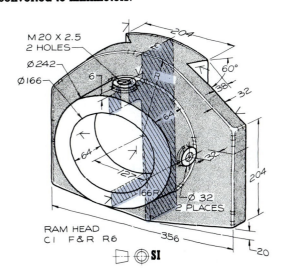

FIGURE 17.48 Make a detail drawing on a Size B sheet.

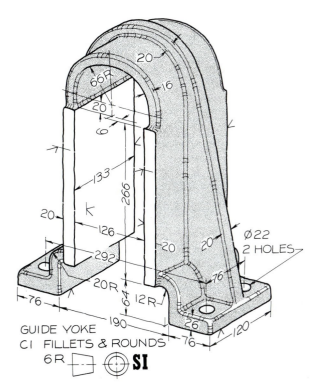

FIGURE 17.50 Make a detail drawing on a Size B sheet.

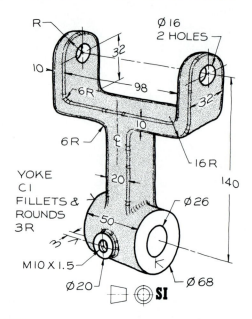

Ø16
2 HOLES

R

32

10

6R

98

32

10

6R

C

16R

YOKE
CI
FILLETS &
ROUNDS
3R

20

Ø26

140

50

3

M10 X 1.5

Ø20

Ø68

SI

FIGURE 17.51 Make a detail drawing on a Size B sheet.

20

3

Ø3-3 HOLES

Ø2
4 HOLES

6

6

18R

6R

30

Ø44

TUBE
BRACKET CI

FILLETS &
ROUNDS IR

SI

38

32

Ø48

3 RIB

Ø38

6R

6

Ø6

6

36

11R

6

3

9

16

4

FIGURE 17.52 Make a detail drawing on a Size B sheet.

FIGURE 17.53 Make a detail drawing on a Size B sheet.

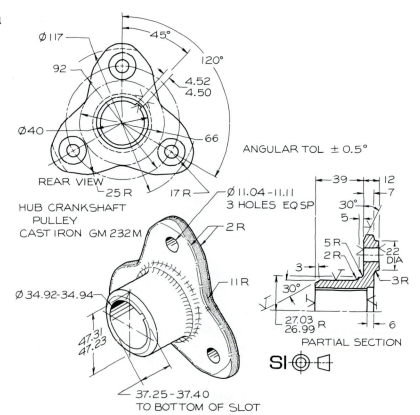

Ø117

45°

120°

92

4.52
4.50

Ø40

66

ANGULAR TOL ± 0.5°

REAR VIEW

25 R

17 R

Ø11.04-11.11
3 HOLES EQ SP

39

12

7

30°

HUB CRANKSHAFT
PULLEY
CAST IRON GM 232M

2 R

5

5R

2R

22
DIA

3R

Ø34.92-34.94

11R

3

30°

6

47.31
47.23

30°

27.03
26.99

R

PARTIAL SECTION

37.25-37.40
TO BOTTOM OF SLOT

SI

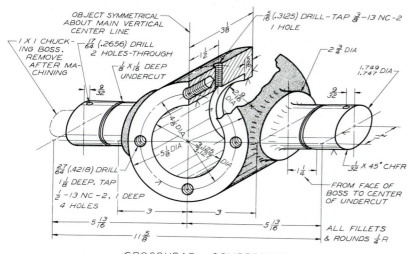

OBJECT SYMMETRICAL ABOUT MAIN VERTICAL CENTER LINE

$\frac{5}{16}$ (.3125) DRILL-TAP $\frac{3}{8}$-13 NC-2 I HOLE

I X I CHUCK-ING BOSS. REMOVE AFTER MACHINING

$\frac{17}{64}$ (.2656) DRILL 2 HOLES-THROUGH

$\frac{1}{8}$ X $\frac{1}{16}$ DEEP UNDERCUT

2 $\frac{3}{4}$ DIA

$\frac{1.749}{1.747}$ DIA

$\frac{9}{32}$

$\frac{9}{32}$

$3\frac{3}{8}$

$1\frac{1}{2}$

$\frac{27}{64}$ (.4218) DRILL $1\frac{1}{8}$ DEEP. TAP $\frac{1}{2}$-13 NC-2, I DEEP 4 HOLES

$\frac{1}{32}$ X 45° CHFR

$1\frac{1}{4}$

FROM FACE OF BOSS TO CENTER OF UNDERCUT

3

3

$5\frac{13}{16}$

$5\frac{13}{16}$

$11\frac{5}{8}$

ALL FILLETS & ROUNDS $\frac{1}{4}$ R

CROSSHEAD — COMPRESSOR

FIGURE 17.54 Make a detail drawing on a Size B sheet using (A) decimal inches or (B) inches converted to millimeters.

FIGURE 17.55 Make a detail drawing on a Size B sheet using (A) decimal inches or (B) inches converted to millimeters.

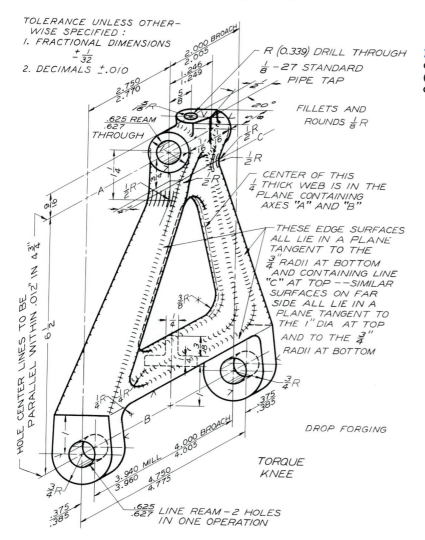

TOLERANCE UNLESS OTHER-WISE SPECIFIED :
1. FRACTIONAL DIMENSIONS $\pm\frac{1}{32}$
2. DECIMALS $\pm$.010

2.000 BROACH
2.005

1.246
1.249

2.750
2.770

$\frac{5}{16}$ R

$\frac{5}{8}$

R (0.339) DRILL THROUGH $\frac{1}{8}$ - 27 STANDARD PIPE TAP

20°

.625 REAM .627 THROUGH

FILLETS AND ROUNDS $\frac{1}{8}$ R

$\frac{1}{2}$ R

$\frac{1}{2}$ R

$\frac{1}{4}$

$\frac{1}{2}$ R

CENTER OF THIS $\frac{1}{4}$ THICK WEB IS IN THE PLANE CONTAINING AXES "A" AND "B"

THESE EDGE SURFACES ALL LIE IN A PLANE TANGENT TO THE $\frac{3}{4}$" RADII AT BOTTOM AND CONTAINING LINE "C" AT TOP --SIMILAR SURFACES ON FAR SIDE ALL LIE IN A PLANE TANGENT TO THE I" DIA AT TOP AND TO THE $\frac{3}{4}$" RADII AT BOTTOM

$\frac{3}{8}$ R

$\frac{1}{4}$

$\frac{3}{4}$ R

.375
.385

DROP FORGING

$\frac{3}{4}$ R

HOLE CENTER LINES TO BE PARALLEL WITHIN .012" IN 4$\frac{3}{4}$"

$6\frac{1}{2}$

$9\frac{1}{16}$

$\frac{1}{4}$

$\frac{1}{4}$ R $\frac{1}{4}$ R

B

TORQUE KNEE

3.940 MILL
3.960

4.000 BROACH
4.005

4.750
4.775

$\frac{3}{4}$ R

.375
.385

.625 LINE REAM - 2 HOLES .627 IN ONE OPERATION

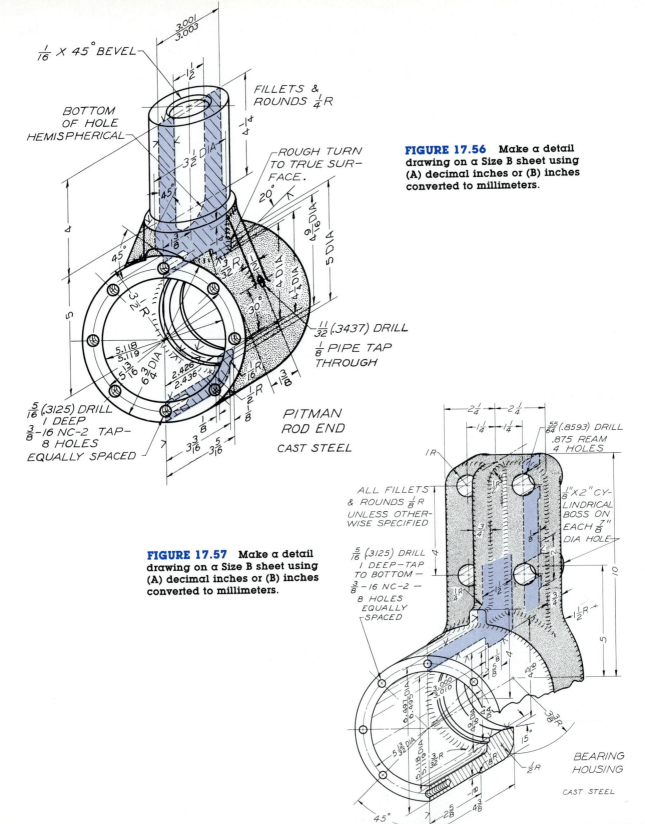

$\frac{1}{16} \times 45°$ BEVEL

$\frac{3.001}{3.003}$

$1\frac{1}{2}$

BOTTOM
OF HOLE
HEMISPHERICAL

FILLETS &
ROUNDS $\frac{1}{4}$R

$1\frac{1}{4}$

$3\frac{1}{2}$ DIA

ROUGH TURN
TO TRUE SUR-
FACE.

45°

20°

$4\frac{9}{16}$ DIA

45°

$3\frac{1}{8}$

$4\frac{1}{4}$ DIA

5 DIA

4 DIA

45°

$\frac{3}{32}$ R

$\frac{11}{32}$ (.3437) DRILL
$\frac{1}{8}$ PIPE TAP
THROUGH

30°

5

$3\frac{1}{2}$ R

$\frac{5.118}{5.119}$

$5\frac{13}{16}$ DIA

$6\frac{3}{4}$ DIA

$\frac{2.426}{2.436}$

$\frac{5}{16}$ R

$\frac{3}{8}$

$\frac{1}{2}$ R

$2\frac{1}{8}$

$\frac{5}{16}$ (3125) DRILL
1 DEEP
$\frac{3}{8}$ - 16 NC-2 TAP-
8 HOLES
EQUALLY SPACED

$\frac{1}{8}$

$3\frac{3}{16}$

$3\frac{5}{16}$

PITMAN
ROD END

CAST STEEL

FIGURE 17.56 Make a detail drawing on a Size B sheet using (A) decimal inches or (B) inches converted to millimeters.

FIGURE 17.57 Make a detail drawing on a Size B sheet using (A) decimal inches or (B) inches converted to millimeters.

$2\frac{1}{4}$

$2\frac{1}{4}$

$1\frac{1}{4}$

$1\frac{1}{4}$

$\frac{55}{64}$ (.8593) DRILL
.875 REAM
4 HOLES

1R

1R

ALL FILLETS
& ROUNDS $\frac{1}{8}$R
UNLESS OTHER-
WISE SPECIFIED

$\frac{3}{4}$

$1''$ X $2''$ CY-
LINDRICAL
BOSS ON
EACH $\frac{7}{8}''$
DIA HOLE

$\frac{5}{16}$ (3125) DRILL
1 DEEP-TAP
TO BOTTOM —
$\frac{3}{8}$ - 16 NC-2 —
8 HOLES
EQUALLY
SPACED

1R

4

$\frac{1}{2}$

$\frac{3}{4}$

2

10

$1\frac{1}{2}$ R

5

$\frac{3.000}{3.010}$

$\frac{6.497}{6.495}$ DIA

45°

$\frac{3}{8}$ R

15

$\frac{1}{2}$ R

$5\frac{13}{32}$ DIA

$\frac{5.118}{5.119}$ DIA

$\frac{3}{32}$ R

$\frac{5}{8}$ R

45°

$2\frac{5}{8}$

$4\frac{3}{8}$

BEARING
HOUSING

CAST STEEL

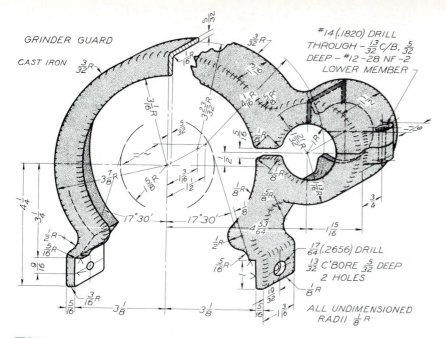

FIGURE 17.58 Make a detail drawing on a Size B sheet using (A) decimal inches or (B) inches converted to millimeters.

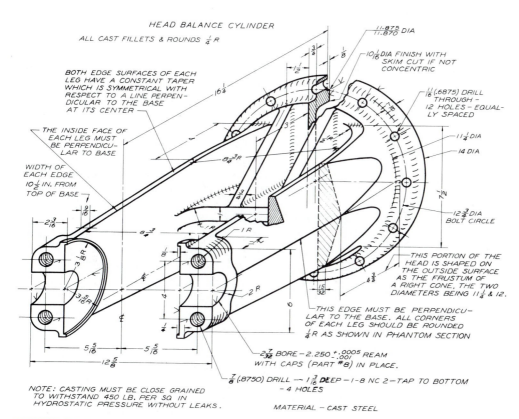

FIGURE 17.59 Make a detail drawing on a Size B sheet using (A) decimal inches or (B) inches converted to millimeters.

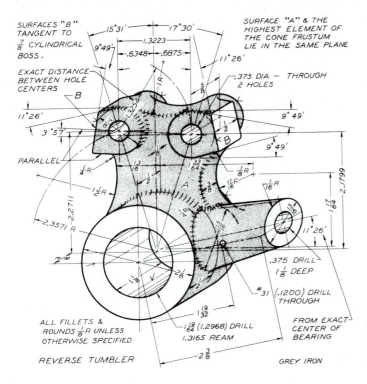

FIGURE 17.60 Make a detail drawing on a Size C sheet using (A) decimal inches or (B) inches converted to millimeters.

FIGURE 17.61 Make a detail drawing on a Size C sheet using (A) decimal inches or (B) inches converted to millimeters.

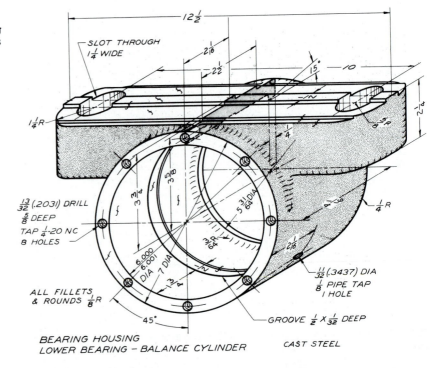

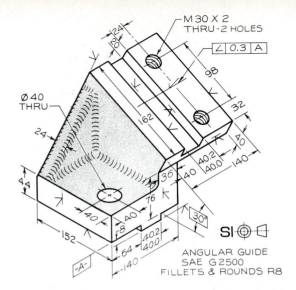

FIGURE 17.62 Make a detail drawing on a Size B sheet using (A) decimal inches or (B) millimeters.

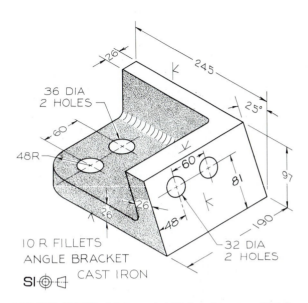

FIGURE 17.65 Make a detail drawing on a Size A sheet.

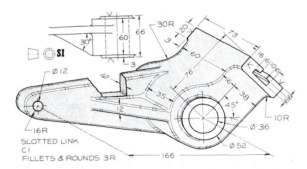

FIGURE 17.63 Make a detail drawing on a Size B sheet using (A) decimal inches or (B) millimeters.

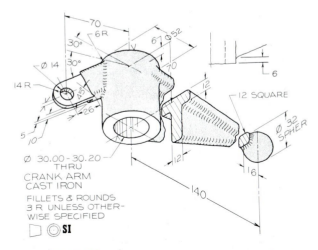

FIGURE 17.64 Make a detail drawing on a Size B sheet.

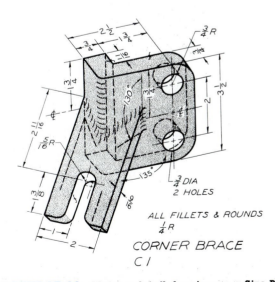

FIGURE 17.66 Make a detail drawing on a Size B sheet. Convert the fractional inches to (A) decimal inches or (B) millimeters.

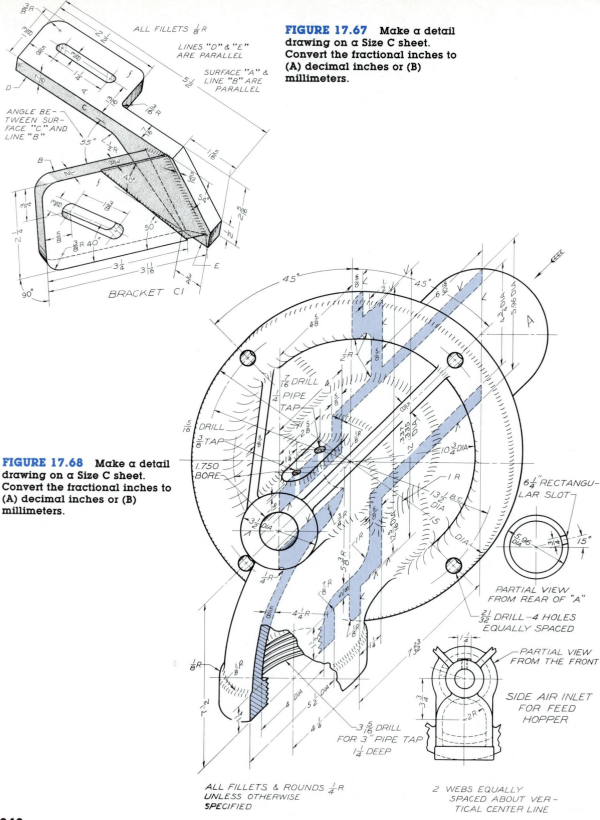

FIGURE 17.67 Make a detail drawing on a Size C sheet. Convert the fractional inches to (A) decimal inches or (B) millimeters.

ALL FILLETS $\frac{1}{8}$R

LINES "D" & "E" ARE PARALLEL

SURFACE "A" & LINE "B" ARE PARALLEL

ANGLE BE-TWEEN SUR-FACE "C" AND LINE "B"

BRACKET C1

FIGURE 17.68 Make a detail drawing on a Size C sheet. Convert the fractional inches to (A) decimal inches or (B) millimeters.

$6\frac{1}{4}$" RECTANGU-LAR SLOT

PARTIAL VIEW FROM REAR OF "A"

$\frac{21}{32}$ DRILL–4 HOLES EQUALLY SPACED

PARTIAL VIEW FROM THE FRONT

SIDE AIR INLET FOR FEED HOPPER

ALL FILLETS & ROUNDS $\frac{1}{4}$R UNLESS OTHERWISE SPECIFIED

2 WEBS EQUALLY SPACED ABOUT VER-TICAL CENTER LINE

348

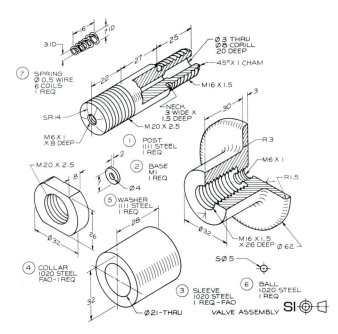

FIGURE 17.69 Make a detail drawing of the parts of the valve assembly on Size B sheets. Draw an assembly and provide a parts lists.

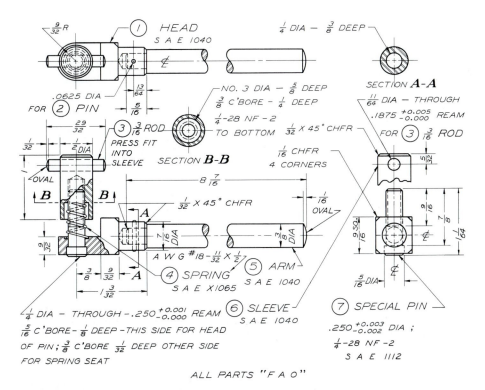

FIGURE 17.70 Make a detail drawing of the parts of the cut-off crank on Size B sheets. Convert the fractional inches to (A) decimal inches or (B) millimeters. Draw an assembly and provide a parts list.

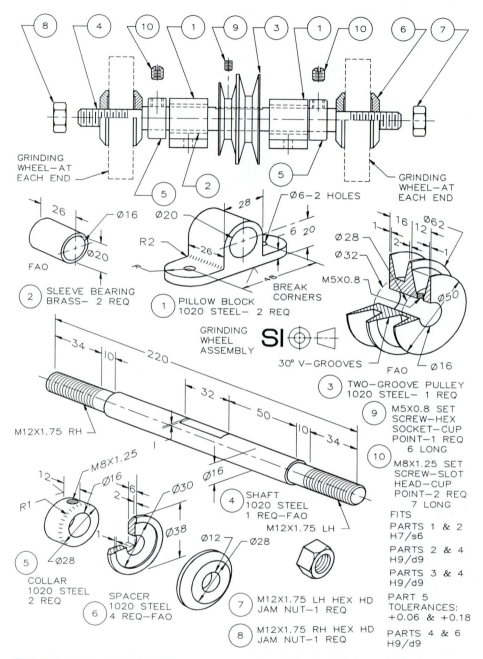

FIGURE 17.71 Make working drawings of the pulley assembly on two Size B sheets. Draw an assembly with a parts list on a third Size B sheet. Scale: full size.

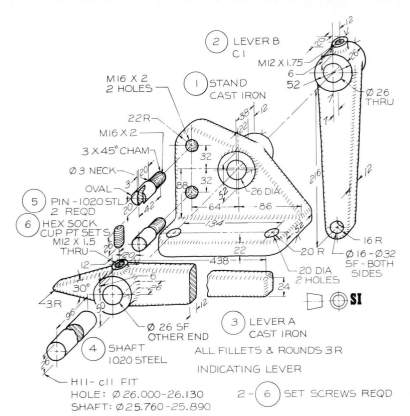

(2) LEVER B
C I

(1) STAND
CAST IRON

M16 X 2
2 HOLES

22R

M16 X 2

3 X 45° CHAM

Ø 3 NECK

OVAL

(5) PIN -1020 STL.
2 REQD

(6) HEX SOCK
CUP PT SET S
M12 X 1.5
THRU

(3) LEVER A
CAST IRON

(4) SHAFT
1020 STEEL

H11- c11 FIT
HOLE: Ø 26.000-26.130
SHAFT: Ø25.760-25.890

M12 X 1.75
6
52

Ø 26
THRU

16 R

Ø 16 - Ø 32
SF - BOTH
SIDES

20 R

20 DIA
2 HOLES

Ø 26 SF
OTHER END

ALL FILLETS & ROUNDS 3 R

INDICATING LEVER

2 - (6) SET SCREWS REQD

SI

FIGURE 17.72 Make a detail drawing of the parts of the indicating lever on Size B sheets. Draw an assembly and provide a parts list.

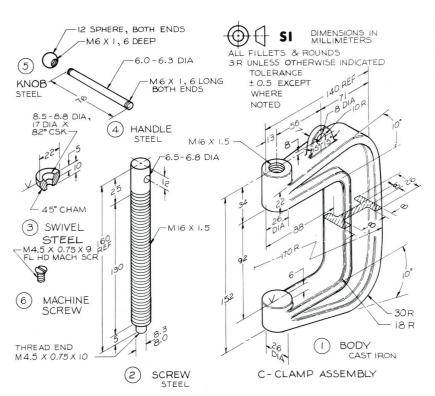

12 SPHERE, BOTH ENDS
M6 X 1, 6 DEEP

(5) KNOB
STEEL

6.0 - 6.3 DIA

M6 X 1, 6 LONG
BOTH ENDS

(4) HANDLE
STEEL

8.5 - 8.8 DIA,
17 DIA X,
82° CSK

6.5 - 6.8 DIA

45° CHAM

(3) SWIVEL
STEEL
M4.5 X 0.75 X 9
FL HD MACH SCR

(6) MACHINE
SCREW

THREAD END
M4.5 X 0.75 X 10

(2) SCREW
STEEL

M16 X 1.5

SI DIMENSIONS IN MILLIMETERS

ALL FILLETS & ROUNDS
3 R UNLESS OTHERWISE INDICATED
TOLERANCE
± 0.5 EXCEPT
WHERE
NOTED

M16 X 1.5

140 REF

8 DIA

10°

10 R

56

M16 X 1.5

170 R

10°

30R
18 R

(1) BODY
CAST IRON

26
DIA

C- CLAMP ASSEMBLY

FIGURE 17.73 Make a detail drawing of the parts of the C-clamp assembly on Size B sheets. Draw an assembly and provide a parts list.

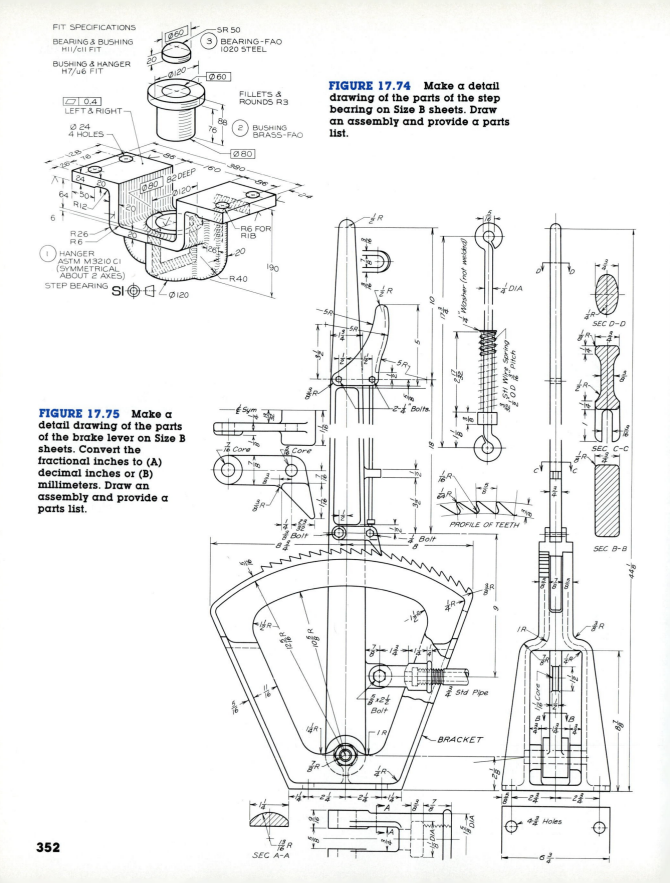

FIGURE 17.74 Make a detail drawing of the parts of the step bearing on Size B sheets. Draw an assembly and provide a parts list.

FIGURE 17.75 Make a detail drawing of the parts of the brake lever on Size B sheets. Convert the fractional inches to (A) decimal inches or (B) millimeters. Draw an assembly and provide a parts list.

352

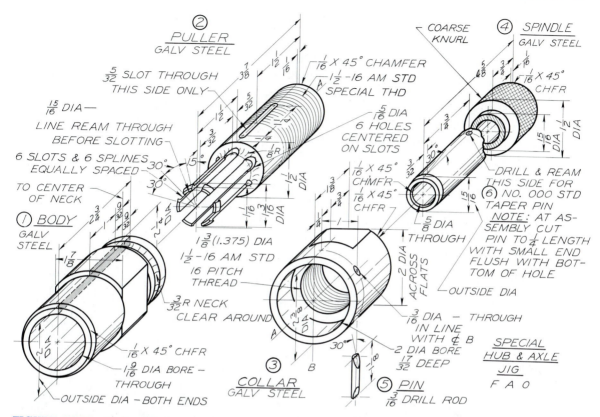

② PULLER
GALV STEEL

$\frac{5}{32}$ SLOT THROUGH
THIS SIDE ONLY

$\frac{15}{16}$ DIA —
LINE REAM THROUGH
BEFORE SLOTTING

6 SLOTS & 6 SPLINES
EQUALLY SPACED

TO CENTER
OF NECK

① BODY
GALV
STEEL

$\frac{1}{16} \times 45°$ CHFR

$1\frac{9}{16}$ DIA BORE —
THROUGH

OUTSIDE DIA – BOTH ENDS

$\frac{1}{16} \times 45°$ CHAMFER
$1\frac{1}{2}$-16 AM STD
SPECIAL THD

$\frac{5}{16}$ DIA
6 HOLES
CENTERED
ON SLOTS

$1\frac{3}{8}$(1.375) DIA
$1\frac{1}{2}$-16 AM STD
16 PITCH
THREAD

$\frac{3}{32}$R NECK
CLEAR AROUND

③ COLLAR
GALV STEEL

$\frac{1}{16} \times 45°$ CHMFR
$\frac{1}{16} \times 45°$ CHFR

2 DIA
ACROSS
FLATS

$\frac{3}{16}$ DIA – THROUGH
IN LINE
WITH ₵ B
2 DIA BORE
$\frac{17}{32}$ DEEP

⑤ PIN
$\frac{3}{16}$ DRILL ROD

④ SPINDLE
GALV STEEL

COARSE KNURL

$\frac{1}{16} \times 45°$ CHFR

DRILL & REAM
THIS SIDE FOR
⑥ NO. 000 STD
TAPER PIN
NOTE: AT AS-
SEMBLY CUT
PIN TO $\frac{1}{4}$ LENGTH
WITH SMALL END
FLUSH WITH BOT-
TOM OF HOLE

OUTSIDE DIA

SPECIAL
HUB & AXLE
JIG
F A O

FIGURE 17.76 Make detail drawings of the parts of the hub and axle jig on Size B sheets. Convert the fractional inches to (A) decimal inches or (B) millimeters. Draw an assembly of the parts and provide a parts list.

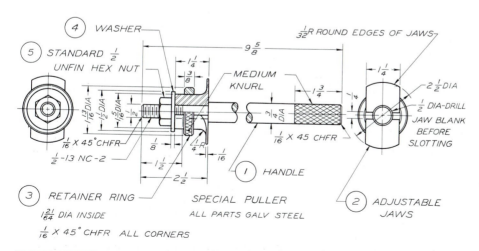

④ WASHER
⑤ STANDARD $\frac{1}{2}$ UNFIN HEX NUT

MEDIUM KNURL

$\frac{1}{32}$R ROUND EDGES OF JAWS

$9\frac{5}{8}$

$2\frac{1}{2}$DIA
$\frac{1}{2}$ DIA-DRILL
JAW BLANK
BEFORE
SLOTTING

$\frac{1}{16} \times 45°$ CHFR
$\frac{1}{2}$-13 NC-2

③ RETAINER RING
$1\frac{21}{64}$ DIA INSIDE
$\frac{1}{16} \times 45°$ CHFR ALL CORNERS

① HANDLE

SPECIAL PULLER
ALL PARTS GALV STEEL

② ADJUSTABLE JAWS

FIGURE 17.77 Make a detail drawing of the parts of the special puller on Size B sheets. Convert the fractional inches to (A) decimal inches or (B) millimeters. Draw a pictorial assembly of the parts and provide a parts list.

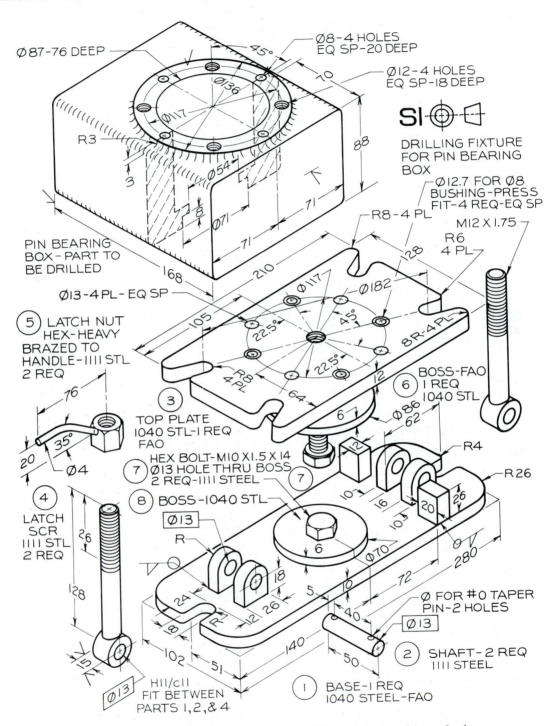

FIGURE 17.78 Make detail drawings of the parts of the drilling jig and crank pin bearing box. Draw an assembly of the parts and provide a parts list.

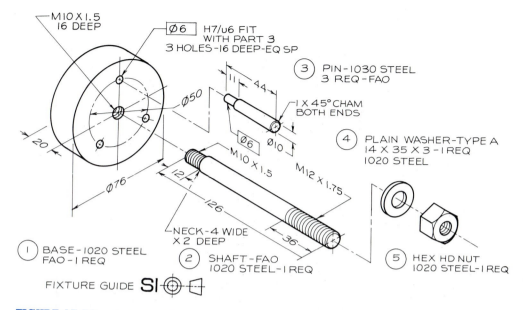

M10 X 1.5
16 DEEP

∅6 H7/u6 FIT
WITH PART 3
3 HOLES-16 DEEP-EQ SP

③ PIN-1030 STEEL
3 REQ-FAO

1 X 45° CHAM
BOTH ENDS

∅50

20

∅76

11

44

∅6

∅10

M10 X 1.5

M12 X 1.75

④ PLAIN WASHER-TYPE A
14 X 35 X 3 -1 REQ
1020 STEEL

12

126

36

NECK-4 WIDE
X 2 DEEP

① BASE-1020 STEEL
FAO-1 REQ

② SHAFT-FAO
1020 STEEL-1 REQ

⑤ HEX HD NUT
1020 STEEL-1 REQ

FIXTURE GUIDE SI ⊕ ⊏⊐

FIGURE 17.79 Make detail drawings of the parts of the fixture guide. Draw an
assembly and provide a parts list.

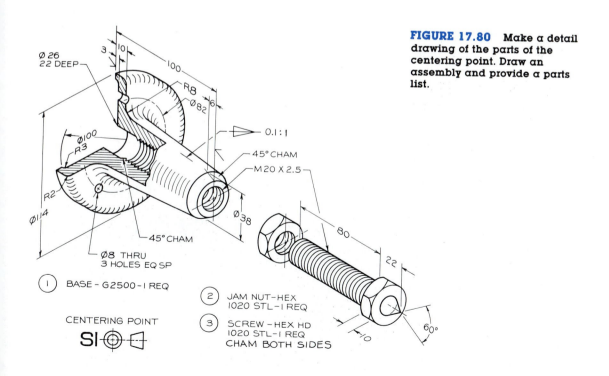

FIGURE 17.80 Make a detail
drawing of the parts of the
centering point. Draw an
assembly and provide a parts
list.

∅ 26
22 DEEP

3 10

100

R8

∅82

6

0.1:1

45° CHAM

M20 X 2.5

∅100
R3

R2

∅114

∅38

45° CHAM

∅8 THRU
3 HOLES EQ SP

① BASE - G2500-1 REQ

② JAM NUT-HEX
1020 STL-1 REQ

③ SCREW - HEX HD
1020 STL-1 REQ
CHAM BOTH SIDES

80

22

10

60°

CENTERING POINT
SI ⊕ ⊏⊐

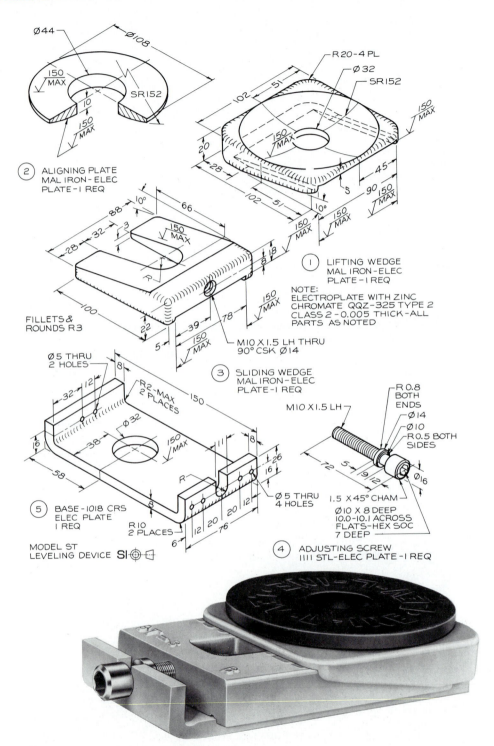

Ø44
Ø108
150 MAX
SR152
10
150 MAX
150 MAX

(2) ALIGNING PLATE
MAL IRON - ELEC
PLATE - I REQ

R 20-4 PL
Ø 32
SR152
102
51
20
28
102
51
10°
150 MAX
150 MAX
45
5
90
150 MAX
150 MAX

(1) LIFTING WEDGE
MAL IRON - ELEC
PLATE - I REQ

10°
88
66
150 MAX
28 32 3
18
8
150 MAX
R
100
22
5
150 MAX
39 78
150 MAX

FILLETS & ROUNDS R3

M10 X 1.5 LH THRU
90° CSK Ø14

(3) SLIDING WEDGE
MAL IRON - ELEC
PLATE - I REQ

NOTE:
ELECTROPLATE WITH ZINC
CHROMATE QQZ-325 TYPE 2
CLASS 2 - 0.005 THICK - ALL
PARTS AS NOTED

Ø 5 THRU
2 HOLES
8
32 12
R2 - MAX
2 PLACES
150
Ø 32
38
150 MAX
11
8
6
26
16
58
8
12 20 12
R
R 10
2 PLACES
12 20 76
6
Ø 5 THRU
4 HOLES

(5) BASE - 1018 CRS
ELEC PLATE
I REQ

MODEL ST
LEVELING DEVICE SI

M10 X 1.5 LH
R 0.8 BOTH ENDS
Ø14
Ø10
R 0.5 BOTH SIDES
72
5 9 12
Ø16
1.5 X 45° CHAM
Ø10 X 8 DEEP
10.0-10.1 ACROSS
FLATS - HEX SOC
7 DEEP

(4) ADJUSTING SCREW
1111 STL - ELEC PLATE - I REQ

FIGURE 17.81 Make a detail drawing of the parts of the leveling device. Draw an assembly and provide a parts list.

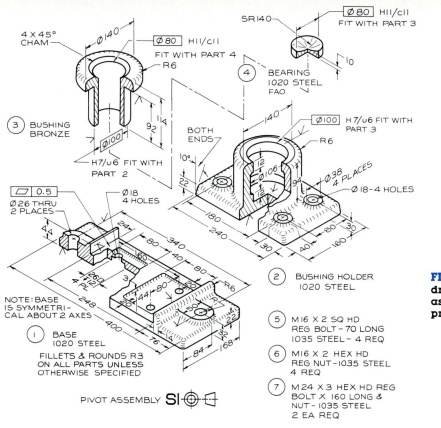

4 X 45° CHAM

Ø140

Ø80 H11/c11
FIT WITH PART 4

R6

③ BUSHING
BRONZE

Ø100

H7/u6 FIT WITH
PART 2

92   114

BOTH
ENDS

SR140

Ø80 H11/c11
FIT WITH PART 3

10

④ BEARING
1020 STEEL
FAO

140

Ø100 H7/u6 FIT WITH
PART 3

R6

10°

22

12

Ø106

Ø38
4 PLACES

94

12

Ø18-4 HOLES

180

240

30

30   40   160

② BUSHING HOLDER
1020 STEEL

⑤ M16 X 2 SQ HD
REG BOLT - 70 LONG
1035 STEEL - 4 REQ

⑥ M16 X 2 HEX HD
REG NUT -1035 STEEL
4 REQ

⑦ M24 X 3 HEX HD REG
BOLT X 160 LONG &
NUT - 1035 STEEL
2 EA REQ

0.5

Ø26 THRU
2 PLACES

Ø18
4 HOLES

24   80   340   160   40

4   4

26

12
4 PL

3   44   80   R6

248

400

76   84   168

32

22

NOTE: BASE
IS SYMMETRI-
CAL ABOUT 2 AXES

① BASE
1020 STEEL

FILLETS & ROUNDS R3
ON ALL PARTS UNLESS
OTHERWISE SPECIFIED

PIVOT ASSEMBLY SI ⊕ ◁

FIGURE 17.82 Make detail
drawings of the parts of the pivot
assembly. Draw an assembly and
provide a parts list.

FIGURE 17.83 Make detail
drawings of the parts of the drill
press vise. Draw an assembly
and provide a parts list.

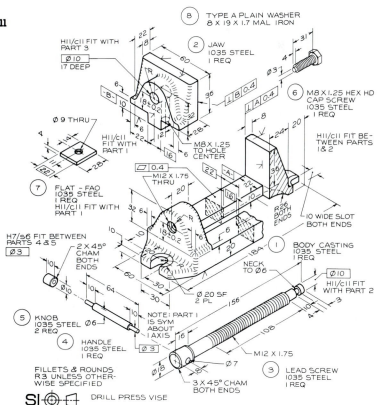

⑧ TYPE A PLAIN WASHER
8 X 19 X 1.7 MAL IRON

② JAW
1035 STEEL
1 REQ

H11/c11 FIT WITH
PART 3

Ø10
17 DEEP

22
8   60

R

36

32

18±0.2

6
10

6
22

12

6

M8 X 1.25
TO HOLE
CENTER

⊥ B 0.4

⊥ A 0.4

31

Ø37   4

⑥ M8 X 1.25 HEX HD
CAP SCREW
1035 STEEL
1 REQ

8

24   20

H11/c11 FIT BE-
TWEEN PARTS
1 & 2

Ø 9 THRU

14

22   28

⑦ FLAT - FAO
1035 STEEL
1 REQ
H11/c11 FIT WITH
PART 1

0.4

M12 X 1.75
THRU

22   -A-

16   10

36

R26
BOTH
ENDS

10 WIDE SLOT
BOTH ENDS

H7/s6 FIT BETWEEN
PARTS 4 & 5

Ø3

2 X 45°
CHAM
BOTH
ENDS

10

Ø10

10

64

18±0.2

32   6

0

20

60

20

18A

① BODY CASTING
1035 STEEL
1 REQ

NECK
TO Ø6

Ø10
H11/c11 FIT
WITH PART 2

⑤ KNOB   Ø6
1035 STEEL
2 REQ

④ HANDLE
1035 STEEL
1 REQ

Ø3

FILLETS & ROUNDS
R3 UNLESS OTHER-
WISE SPECIFIED

10

Ø 20 SF
2 PL

NOTE: PART 1
IS SYM
ABOUT
1 AXIS

30

16

156

108

3

10

M12 X 1.75

Ø18   Ø7

8

3 X 45° CHAM
BOTH ENDS

③ LEAD SCREW
1035 STEEL
1 REQ

SI ⊕ ◁ DRILL PRESS VISE

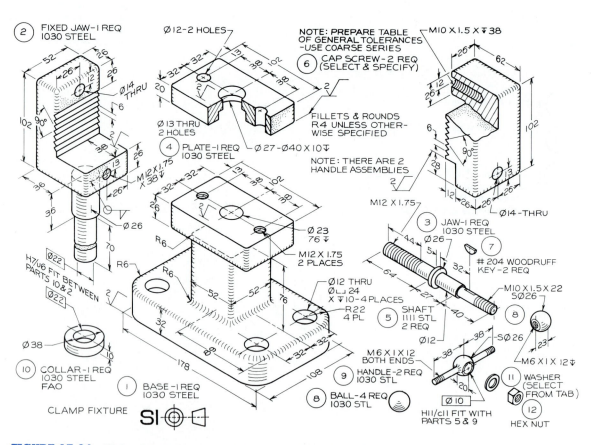

FIGURE 17.84 Make detail drawings of the parts of the clamp fixture. Draw an assembly and provide a parts list.

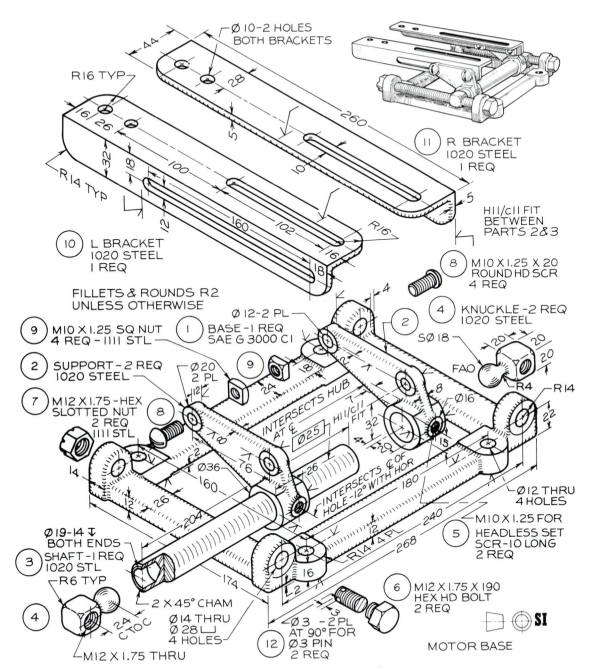

Ø 10–2 HOLES
BOTH BRACKETS

R16 TYP

44

28

260

R14 TYP

16 26

32 18

100

5

10

160 102

R16

16

18

11 R BRACKET
1020 STEEL
1 REQ

5

H11/c11 FIT
BETWEEN
PARTS 2&3

10 L BRACKET
1020 STEEL
1 REQ

8 M10 X 1.25 X 20
ROUND HD SCR
4 REQ

FILLETS & ROUNDS R2
UNLESS OTHERWISE

4 KNUCKLE –2 REQ
1020 STEEL

9 M10 X 1.25 SQ NUT
4 REQ – 1111 STL

Ø 12–2 PL

1 BASE –1 REQ
SAE G 3000 C1

2

SØ 18

FAO

20 20 20

R4 R14

2 SUPPORT – 2 REQ
1020 STEEL

Ø20
2 PL

9

24

12

INTERSECTS HUB
AT ℄

Ø25

H11/c11
FIT

32

Ø16

7 M12 X1.75 –HEX
SLOTTED NUT
2 REQ
1111 STL

8

8

4

20

15

22

14

Ø36

76

26

160

26

INTERSECTS ℄ OF
HOLE –12° WITH HOR

180

Ø12 THRU
4 HOLES

12

204

12 240

M10 X 1.25 FOR
HEADLESS SET
SCR –10 LONG
2 REQ

Ø19–14 ↧
BOTH ENDS

3 SHAFT –1 REQ
1020 STL

R6 TYP

4

174

R14 4 PL

268

16

2

5

2 X 45° CHAM

24
C TO C

Ø14 THRU
Ø 28 ⌴
4 HOLES

M12 X 1.75 THRU

Ø3 –2 PL
AT 90° FOR

12 Ø3 PIN
2 REQ

6 M12 X1.75 X 190
HEX HD BOLT
2 REQ

SI

MOTOR BASE

FIGURE 17.85 Make a detail drawing of the parts of the motor base. Draw an
assembly and provide a parts list.

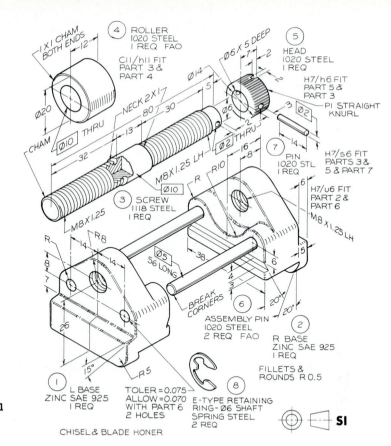

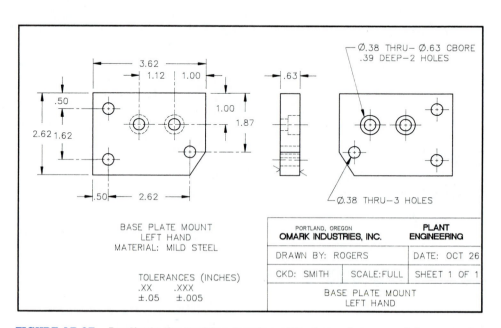

FIGURE 17.86 Make a detail drawing of the parts of the chisel and blade honer. Draw an assembly and give a parts list.

FIGURE 17.87 Duplicate the working drawing of the base plate mount, by computer or drafting instruments, on a Size B sheet.

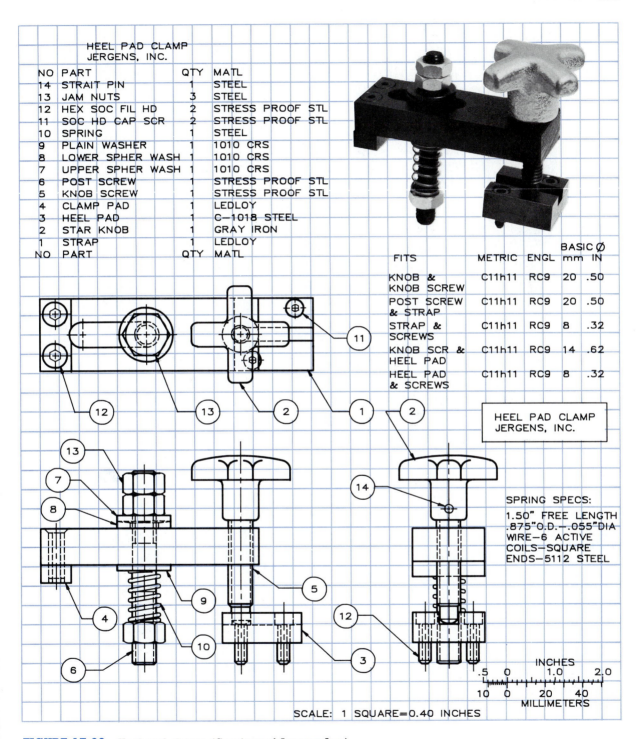

HEEL PAD CLAMP
JERGENS, INC.

NO	PART	QTY	MATL
14	STRAIT PIN	1	STEEL
13	JAM NUTS	3	STEEL
12	HEX SOC FIL HD	2	STRESS PROOF STL
11	SOC HD CAP SCR	2	STRESS PROOF STL
10	SPRING	1	STEEL
9	PLAIN WASHER	1	1010 CRS
8	LOWER SPHER WASH	1	1010 CRS
7	UPPER SPHER WASH	1	1010 CRS
6	POST SCREW	1	STRESS PROOF STL
5	KNOB SCREW	1	STRESS PROOF STL
4	CLAMP PAD	1	LEDLOY
3	HEEL PAD	1	C—1018 STEEL
2	STAR KNOB	1	GRAY IRON
1	STRAP	1	LEDLOY
NO	PART	QTY	MATL

FITS	METRIC	ENGL	mm	IN
KNOB & KNOB SCREW	C11h11	RC9	20	.50
POST SCREW & STRAP	C11h11	RC9	20	.50
STRAP & SCREWS	C11h11	RC9	8	.32
KNOB SCR & HEEL PAD	C11h11	RC9	14	.62
HEEL PAD & SCREWS	C11h11	RC9	8	.32

HEEL PAD CLAMP
JERGENS, INC.

SPRING SPECS:
1.50" FREE LENGTH
.875"O.D.—.055"DIA
WIRE—6 ACTIVE
COILS—SQUARE
ENDS—5112 STEEL

INCHES
MILLIMETERS

SCALE: 1 SQUARE=0.40 INCHES

FIGURE 17.88 Heel pad clamp. (Courtesy of Jergens Inc.)

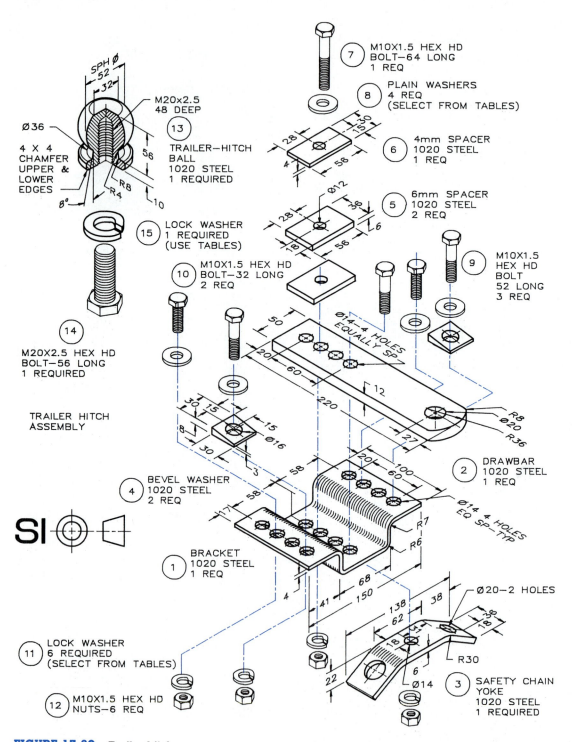

FIGURE 17.89 Trailer hitch.

Reproduction Methods and Drawing Shortcuts

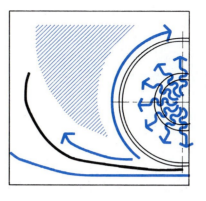

18

18.1
Introduction

So far, we have discussed the processes of preparing drawings and specifications through the working-drawing stage where a detailed drawing is completed on tracing film or paper. Now the drawing must be reproduced, folded, and prepared for filing or transmittal to the drawing's users.

18.2
Reproduction of working drawings

A drawing made by a drafter is of little use in its original form. It would be impractical for the original to be handled by checkers and, even more so, by workers in the field or shop. The drawing would quickly be damaged or soiled, and no copy would be available as a permanent record of the job. Therefore, reproduction of drawings is necessary so that copies can be available for use by the people concerned.

The most often used processes of reproducing engineering drawings are (1) **diazo printing,** (2) **blueprinting,** (3) **microfilming,** (4) **xerography,** and (5) **photostatting.**

Diazo printing

The diazo print is more correctly called a whiteprint or blue-line print than a blueprint, since it has a white background and blue lines. Other colors of lines are available depending on the type of paper used. The white background makes notes and corrections more clearly visible than does the blue background of the blueprint.

> Both blueprinting and diazo printing require that the original drawing be made on semitransparent tracing paper, cloth, or film that will allow light to pass through the drawing.

The paper the copy is made on, the diazo paper, is chemically treated so that it has a yellow tint on one side. To prevent spoilage, this paper must be stored away from heat and light.

The tracing paper or film drawing is placed face up on the yellow side of the diazo paper and is run through the diazo-process machine, which exposes the drawing to a built-in light. The light passes through the tracing paper and burns out the yellow chemical on the diazo paper except where the drawing lines have shielded the paper from the light. After exposure, the diazo paper is a duplicate of the original drawing except that the lines are light yellow and not permanent.

FIGURE 18.1   **A typical wheelprinter that operates on the diazo process. (Courtesy of Blu-Ray, Inc., Essex, CT)**

The diazo paper is then passed through the developing unit of the diazo machine where the yellow lines are developed into permanent blue lines by exposure to ammonia fumes. Figure 18.1 shows a typical diazo printer-developer, sometimes called a white printer.

The speed at which the drawing passes under the light determines the darkness of the copy. A slow speed burns out more of the yellow and produces a clear white background; however, some of the lighter lines of the drawing may be lost. Most diazo copies are made at a somewhat faster speed to give a light tint of blue in the background and stronger lines in the copy. Ink drawings give the best reproductions.

Blueprinting

Blueprints are made with paper that is chemically treated on one side. As in the diazo process, the tracing-paper drawing is placed in contact with the chemically treated side of the paper and exposed to light. The exposed blueprint paper is washed in clear water and coated with a solution of potassium dichromate. Then the print is washed again and dried. The wet sheets can be hung on a line to dry, or they can be dried by equipment made for this purpose.

Microfilming

Microfilming is a photographic process that converts large drawings into film copies—either aperture cards or roll film. Drawings must be photographed on either 16-mm or 35-mm film. Figure 18.2 shows a camera and copy table.

The roll film or aperture cards can be placed in a microfilm enlarger-printer (Fig. 18.3), where the individual drawings can be viewed on a built-in screen. The selected drawings can then be printed from the film to give standard-size drawings. Microfilm copies

are usually smaller than the original drawings to save paper and make the drawings easier to use.

Microfilming makes it possible to eliminate large, bulky files of drawings since hundreds of drawings can

FIGURE 18.2   **The Micro-Master 35-mm camera and copy table are used for microfilming engineering drawings. (Courtesy of Keuffel & Esser Co., Morristown, NJ)**

FIGURE 18.3   **The Bruning 1200 microfilm enlarger-printer makes drawings up to 18″ × 24″ from aperture cards and roll film. (Courtesy of Bruning Co.)**

FIGURE 18.4 This Xerox 7080 Engineering Print system accepts original drawings sized A to E; makes prints sized A to C as fast as 58 per minute; and stamps, folds, and sorts prints automatically.

be stored in miniature size on a small amount of film. The aperture cards shown in Fig. 18.3 are data processing cards that can be cataloged and recalled by a computer to make them accessible with a minimum of effort.

Xerography

Xerography is an electrostatic process of duplicating drawings on ordinary, unsensitized paper. Originally developed for business and clerical uses, xerography has more recently been used for the reproduction of engineering drawings.

One advantage of the xerographic process is its ability to make reduced copies of drawings (Fig. 18.4). The Xerox 2080 can reduce a 24-×-36-inch drawing to 8 × 10 inches.

Photostatting

Photostatting is a method of enlarging or reducing drawings using a camera. Figure 18.5 shows a combination camera and processor used for photographing drawings and producing high-contrast copies.

The drawing or artwork is placed under the glass of the exposure table, which is lit by built-in lamps. The image can be seen on the glass inside the darkroom where it is exposed on photographically sensi-

tive paper. The negative paper that has been exposed to the image is placed in contact with receiver paper, and the two are fed through the developing solution to obtain a photostatic copy.

These high-contrast reproductions are often used to prepare artwork that is to be printed by offset printing presses. Photostatting can also be used to make reproductions on transparent films and for reproducing halftones (photographs with tones of gray).

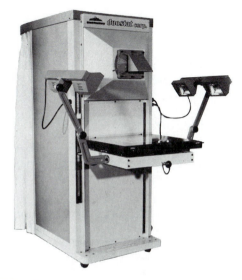

FIGURE 18.5 A camera-processor for enlarging and reducing drawings to be reproduced as photostats. (Courtesy of the Duostat Corp.)

18.3
Folding the drawing

Once the prints have been finished, the original drawings should be stored in a flat file for future use and updating. The original drawings should not be folded, and their handling should be kept to a minimum.

The printed drawings, on the other hand, are usually folded for transmittal from office to office. Figure 18.6 shows the methods of folding Sizes B, C, D, and E sheets; in each case, the final size after folding is $8\frac{1}{2} \times 11$ inches (or 9 × 12 inches). The drawings are folded so that the title blocks are positioned at the top and lower right of the drawing to allow them to be easily retrieved from a file.

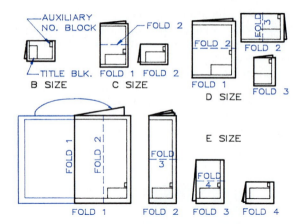

FIGURE 18.6 **Standard folds for engineering drawing sheets. The final size in each case is $8\frac{1}{2}'' \times 11''$.**

18.4
Overlay drafting techniques

Valuable drafting time can be saved by taking advantage of current processes and materials that use a series of overlays to separate parts of a single drawing. Engineers and architects often work from a single site plan or floor plan. For example, the floor plan of a building will be used for the electrical plan, furniture arrangement plan, air-conditioning plan, floor materi-

FIGURE 18.7 **In the pin system, separate overlays are aligned by seven pins mounted on metal strips. (Courtesy of Keuffel & Esser Co., Morristown, NJ)**

als plan, and so on. It would be expensive to retrace the plan for each application.

A series of overlays can be used in a system referred to as **pin drafting,** where accurately spaced holes are punched in the polyester drafting film at the top edge of the sheets. These holes are aligned on pins attached to a metal strip that match the holes punched in the film (Fig. 18.7). The pins ensure accurate alignment of registration of a series of sheets, and the polyester ensures stability of the material since it does not stretch or sag with changes in humidity.

The set of overlays could be attached by the alignment pins or taped together and run through a diazo machine for full-size prints. Another reproduction operation is the use of a flat-bed process camera (Fig. 18.8), which photographs and reduces the drawings to $8\frac{1}{2} \times 11$ inches.

When many prints or multicolor prints are needed, offset lithography is the reproduction process often used.

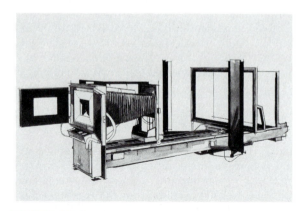

FIGURE 18.8 **The process camera used to reduce and enlarge engineering drawings is the heart of the pin system. (Courtesy of Keuffel & Esser Co., Morristown, NJ)**

18.5
Paste-on photos

When several repetitive drawings are needed, it is often more economical to use photographic reproductions of the drawings on transparent film. These features can be pasted into position on the master drawing. (Although the term pasted is commonly used to describe this attachment, tape is actually used when

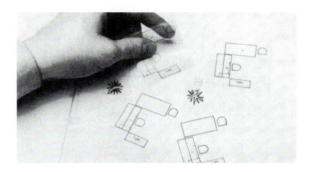

FIGURE 18.9 When drawings are to be used repetitively, it may be more economical to reproduce them than to redraw them. (Courtesy of Eastman Kodak Co.)

FIGURE 18.10 Photographically reproduced drawings are taped in position to complete the overall drawing. (Courtesy of Eastman Kodak Co.)

the drawings are reproduced on transparent film. If the reproductions are made as opaque photostats, rubber cement may be used.)

The office arrangements shown in Fig. 18.9 are transparencies that have been photographically duplicated from a single drawing. The architect shown in Fig. 18.10 is composing an entire drawing sheet with paste-on images of repetitive features that were previously drawn.

18.6
Stick-on materials

Several companies market stick-on symbols, screens, and lettering that can be applied to drawings to save time and improve the appearance of drawings. The three standard types of materials are stick-ons, burnish-ons, and tape-ons.

Stick-on symbols or letters are printed on thin plastic sheets. The symbols are cut out with a razor-sharp blade and affixed to the drawing (Fig. 18.11). This material is available in glossy and matte finishes.

Burnish-on symbols are applied by placing the entire sheet over the drawing and burnishing the desired symbol into place with a rounded-end object such as the end of a pen cap.

Tape-ons are colored adhesive tapes in varying widths used in the preparation of graphs and charts. Tapes are also used to represent wide lines on large drawings.

Sheets of symbols can be custom-printed for users who have repetitive needs for trademarks and other

FIGURE 18.11 Stick-on lettering can be applied to a drawing by cutting the letters from the plastic sheet, applying them to the drawing in alignment with a guideline, and burnishing them to the drawing. (Courtesy of Graphic Products Corp.)

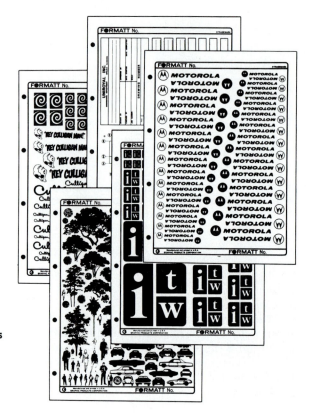

FIGURE 18.12 Stick-on symbols are available in a wide range of styles, and they can be custom-designed to suit the client's needs. (Courtesy of Graphic Products Corp.)

often-used symbols (Fig. 18.12). Title blocks are sometimes printed in this manner to reduce drawing time. Figure 18.13 shows several symbols used on architectural plans.

Also available is a matte finished sheet with an adhesive back that can be typed on with a standard typewriter and then transferred to the drawing.

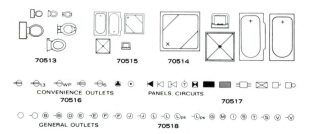

FIGURE 18.13 Examples of stick-on architectural symbols. (Courtesy of Zip-a-Tone Inc.)

18.7
Photo drafting

To clarify assembly details, it is sometimes worthwhile to build a model for photographing. This is especially true in the piping industry, where complex refineries are built as models, then photographed, noted, and reproduced as photodrawings.

Figure 18.14 shows the steps of preparing a photodrawing, where it is desired to specify certain parts of a sprocket-and-chain assembly. In Step 1, a halftone print of the photograph is made. (This is the process of screening the photograph or representing it by a series of dots that give varying tones of gray.) In Step 2, the halftone is taped to a white drawing sheet and then photographed to give a negative. The negative is used to produce a positive on polyester drafting film. The notes can be lettered on this drawing film to complete the master drawing (Step 4). The master drawing can then be used to make diazo prints, or it can be microfilmed or reproduced photographically to the desired size.

Step 1: Make a halftone print of the photograph.

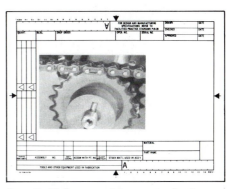

Step 3: Make a positive reproduction on matte film.

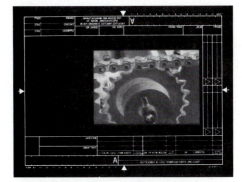

Step 2: Tape the halftone print to a drawing form, and photograph it to produce a negative.

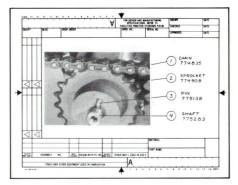

Step 4: Now draw in your callouts—and the job is done.

FIGURE 18.14 The steps in making a photodrawing. (Courtesy of Eastman Kodak Co.)

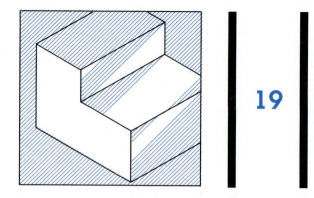

Pictorials

19

19.1
Introduction

A **pictorial** is an effective means of communicating an idea. Pictorials are especially helpful if a design is unique or the person to whom it is being explained has difficulty interpreting multiview drawings.

Pictorials, sometimes called **technical illustrations,** are widely used to describe various products in catalogs, parts manuals, and maintenance publications.

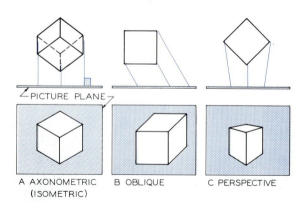

FIGURE 19.1 Types of projection systems for pictorials. **A.** Axonometric pictorials are formed by parallel projectors perpendicular to the picture plane. **B.** Obliques are formed by parallel projectors oblique to the picture plane. **C.** Perspectives are formed by converging projectors that make varying angles with the picture plane.

19.2
Types of pictorials

The four commonly used types of pictorials are (1) **obliques,** (2) **isometrics,** (3) **axonometrics,** and (4) **perspectives** (Fig. 19.1).

OBLIQUE PICTORIALS are three-dimensional pictorials on a plane of paper drawn by projecting from the object with parallel projectors that are oblique to the picture plane (Fig. 19.1B).

ISOMETRIC AND AXONOMETRIC PICTORIALS are three-dimensional pictorials on a plane of paper drawn by projecting from the object to the picture plane (Fig. 19.1A). The parallel projectors are perpendicular to the picture plane.

PERSPECTIVE PICTORIALS are drawn with projectors that converge at the viewer's eye and make varying angles with the picture plane (Fig. 19.1C).

19.3
Oblique pictorials

Although seldom used, **oblique pictorials** are the basis of oblique drawings. In Fig. 19.2a, a number of lines of sight are drawn through point 2 of line 1–2. Each line of sight makes a 45° angle with the picture plane, which creates a cone with its apex at 2, and each element on the cone makes a 45° angle with the plane. A variety of projections of line 1–2' on the pic-

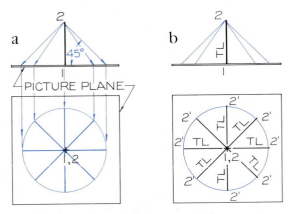

FIGURE 19.2 The underlying principle of the cavalier oblique can be seen here, where a series of projectors form a cone. Each element makes a 45° angle with the picture plane. Thus the projected lengths of 1–2′ are equal in length to line 1–2, which is perpendicular to the picture plane.

ture plane can be seen in the front view (Fig. 19.2b). Each of these projections of 1–2′ is equal in length to the true length of 1–2. This is a **cavalier oblique projection** because all projectors make a 45° angle with the picture plane, and measurements along the receding axis can be made true length and in any direction.

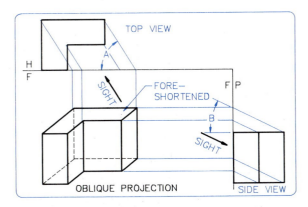

FIGURE 19.3 An oblique projection can be drawn at any angle of sight to obtain an oblique pictorial. However, the line of sight should not make an angle less than 45° with the picture plane. This would result in a receding axis longer than true length, thereby distorting the pictorial.

The top and side views are given as orthographic views, and the front view is drawn as an oblique projection by using the two views of a selected line of sight (Fig. 19.3). Projectors are drawn from the object parallel to the lines of sight to locate their respective points in the oblique view.

19.4
Oblique drawings

Oblique projections are seldom used in the manner just illustrated.

> Instead, based on these principles are three basic types of **oblique drawings:** (1) **cavalier,** (2) **cabinet,** and (3) **general** (Fig. 19.4).

In each case, the angle of the receding axis can be at any angle between 0° and 90° (Fig. 19.4). Measurements along the receding axes of the cavalier oblique are true length (full scale). The cabinet oblique has measurements along the receding axes reduced to half length. The general oblique has measurements along the receding axes reduced to between half and full length.

Figure 19.5 shows three examples of cavalier obliques of a cube. Each has a different angle for the receding axes, but the measurements along the receding axes are true length. Figure 19.6 compares cavalier with cabinet obliques.

19.5
Constructing obliques

An oblique should be drawn by constructing a box using the overall dimensions of height, width, and depth with light construction lines. In Fig. 19.7, the front view is drawn true size in Step 1. This will be a cavalier oblique. True measurements can be made parallel to the three axes. To complete the oblique, the notches are removed from the blocked-in construction box. These measurements are transferred from the given orthographic views with your dividers.

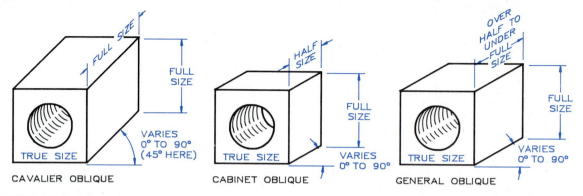

FIGURE 19.4 Types of obliques.

A. The cavalier oblique can be drawn with a receding axis at any angle, but the measurements along this axis are true length.

B. The cabinet oblique can be drawn with a receding axis at any angle, but the measurements along this axis are half size.

C. The general oblique can be drawn with a receding axis at any angle, but the measurements along this axis can vary from half to full size.

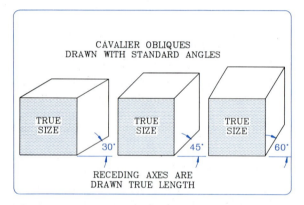

FIGURE 19.5 A cavalier oblique is usually drawn with the receding axis at the standard angles of the drafting triangles. Each gives a different view of a cube.

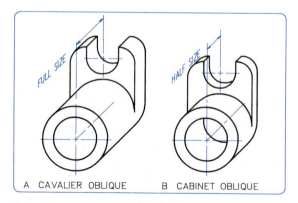

FIGURE 19.6 Measurements along the receding axis of a cavlier oblique are full size; those in a cabinet oblique are half size.

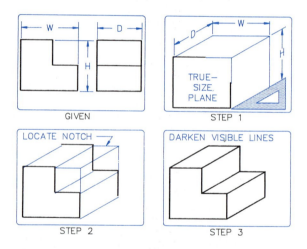

FIGURE 19.7 Oblique drawing construction.

Step 1 The front surface of the oblique is drawn as a true-size plane. The receding axis is drawn at a convenient angle, and depth is found by using the true distance of *D* taken from the given views.

Step 2 The notch in the front plane is drawn and projected to the rear plane.

Step 3 The lines are darkened to complete the cavalier oblique.

19.6

Angles in oblique

Angular measurements can be made on the true-size plane of an oblique. However, angular measurements

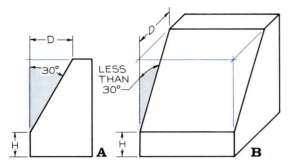

FIGURE 19.8 Angles in oblique must be located by using coordinates; they cannot be measured true size except on a true-size plane.

will not be true size on the other two planes of the oblique.

To construct an angle in an oblique, coordinates must be used (Fig. 19.8). The sloping surface of 30° must be found by locating the vertex of the angle *H* distance from the bottom. The inclination is found by measuring the distance of *D* along the receding axis to establish the slope. This angle is not equal to the 30° angle given in the orthographic view.

You can see in Fig. 19.9 that a true angle can be measured on a true-size surface. In Fig. 19.9B, angles along the receding planes are either smaller or larger than their true angles.

It requires less effort and gives a better appearance when obliques are drawn where angles will appear true size, as shown in Fig. 19.9A rather than in Fig. 19.9B.

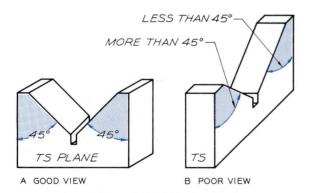

FIGURE 19.9 The best view takes advantage of the ease of construction offered by oblique drawings. The view in part B is less descriptive and more difficult to construct than the view in part A.

19.7
Cylinders in oblique

The major advantage of obliques is that circular features can be drawn as true circles when parallel to the picture plane (Fig. 19.10).

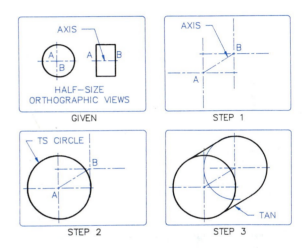

FIGURE 19.10 An oblique drawing of a cylinder.

Step 1 Draw axis *AB*, and locate the centers of the circular ends of the cylinder at *A* and *B*. Since the axis is drawn true length, this will be a cavalier oblique.

Step 2 Draw a true-size circle with its center at *A* using a compass or computer-graphics techniques.

Step 3 Draw the other circular end with its center at *B*, and connect the circles with tangent lines parall to the axis, *AB*.

The centerlines of the circular end at and the receding axis is drawn at any The end at *B* is located by measuring Circles drawn at each end using ce connected with tangent lines paral

These same principles are object with semicircular feat oblique is positioned to take of drawing circular featur and *C* are located for dra

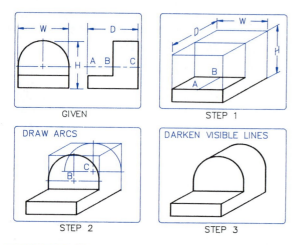

FIGURE 19.11 Circular features in oblique.

Step 1 Block in the overall dimensions of the cavalier oblique with light construction lines, and ignore the semicircular features.

Step 2 Locate centers B and C, and draw arcs with a compass or computer tangent to the sides of the construction boxes.

Step 3 Connect the arcs with lines tangent to each arc and parallel to axis BC; then darken the lines.

19.8
Circles in oblique

Although circular features will be true size on a true-size plane of an oblique pictorial, circular features on ~ other two planes will appear as ellipses (Fig.

A more frequently used technique of drawing elliptical views in oblique is the four-center ellipse method (Fig. 19.13). A rhombus is drawn that would be tangent to the circle at four points. Perpendicular construction lines are drawn from where the centerlines cross the sides of the rhombus. As Steps 2 and 3 show, this construction is made to locate four centers that are used to draw four arcs that form the ellipse in oblique.

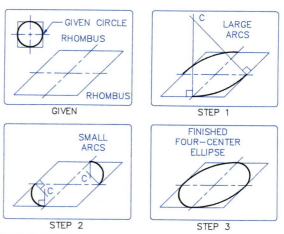

FIGURE 19.13 Four-center ellipse in oblique.

Given The circle to be drawn in oblique is blocked in with a square that is tangent to the circle at four points. This square will appear as a rhombus on the oblique plane.

Step 1 Construction lines are drawn perpendicularly from the points of tangency to locate the centers for drawing two segments of the ellipse.

Step 2 The centers for the two remaining arcs are located with perpendiculars drawn from adjacent tangent points.

Step 3 When the four arcs have been drawn, the final result is an approximate ellipse.

The four-center ellipse method will not work for the cabinet or general oblique. In these cases, coordinates must be used to locate a series of points on the curve. Figure 19.14 shows coordinates in orthographic view and cabinet oblique. The coordinates parallel to the receding axis are reduced to half size; horizontal coordinates are drawn full size. The ellipse can be drawn with an irregular curve or ellipse template that approximates the plotted points.

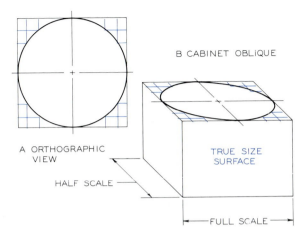

FIGURE 19.14   The four-center ellipse technique cannot be used to locate circular shapes on the foreshortened surface of a cabinet oblique. These ellipses must be plotted with coordinates.

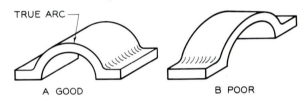

FIGURE 19.15   An oblique should be positioned to enable circular features to be drawn most easily.

Whenever possible, oblique drawings of objects with circular features should be positioned with circles on true-size planes so that they can be drawn as true circles. The view in Fig. 19.15A is better than the one in Fig. 19.15B because it gives a more descriptive view of the part and is easier to draw.

19.9
Curves in oblique

Irregular curves in oblique must be plotted point by point using coordinates (Fig. 19.16). The coordinates are transferred from the orthographic view to the oblique view, and the curve is drawn through these points with an irregular curve.

If the object has a uniform thickness, the lower curve can be found by projecting vertically downward from the upper points a distance equal to the height of the object.

In Fig. 19.17, the elliptical feature on the inclined surface was found by using a series of coordinates to locate the points along its curve. These points are then connected by using an irregular curve or ellipse template of approximately the same size.

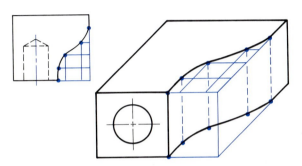

FIGURE 19.16   Coordinates are used to establish irregular curves in oblique. The lower curve is found by projecting the points downward a distance equal to the height of the oblique.

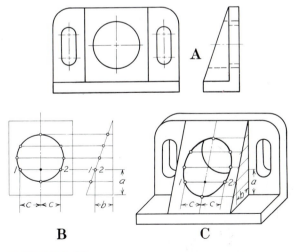

FIGURE 19.17   The construction of a circular feature on an inclined surface must be found by plotting points using three coordinates, a, b, and c, to locate points 1 and 2 in this example. The plotted points are connected to complete the elliptical feature.

19.10
Oblique sketching

Understanding the mechanical principles of oblique construction is essential for sketching obliques. As Fig. 19.18 shows, guidelines are helpful in developing a sketch. These guidelines should be drawn lightly so that they will not need to be erased when the finished lines of the sketch are darkened. When sketching on tracing vellum, it is helpful if a printed grid is placed under the vellum to provide guidelines.

19.11
Dimensioned obliques

A dimensioned full-section oblique is given in Fig. 19.19, where the interior features and dimensions are shown.

In oblique pictorials, numerals and lettering should be applied using either the aligned method, in which the numerals are aligned with the dimensioned lines, or the unidirectional method, in which the numerals are all positioned in a single direction regard-

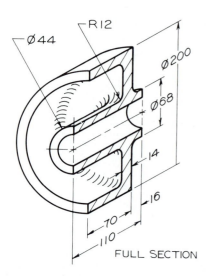

FIGURE 19.19 Oblique pictorials can be drawn as sections and dimensioned to serve as working drawings.

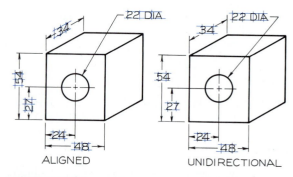

FIGURE 19.20 Oblique pictorials can be dimensioned by using either of these methods of applying numerals to the dimension lines.

less of the direction of the dimension lines (Fig. 19.20). Notes connected with leaders are positioned horizontally in both methods.

19.12
Isometric pictorials

An **isometric pictorial** is a type of axonometric projection in which parallel projectors are perpendicular to the picture plane, and the diagonal of a cube is seen

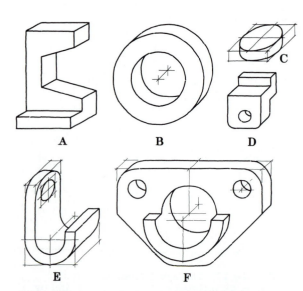

FIGURE 19.18 Obliques may be drawn as freehand sketches by using the same principles used for instrument pictorials. Use light construction lines to locate the more complex features.

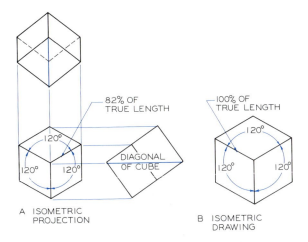

A ISOMETRIC PROJECTION

B ISOMETRIC DRAWING

FIGURE 19.21 A true isometric projection is found by constructing a view that shows the diagonal of a cube as a point. An isometric drawing is not a true projection since the dimensions are drawn true size rather than reduced as in a projection.

as a point (Fig. 19.21). The three axes are spaced 120° apart, and the sides are foreshortened to 82% of their true length.

> **Isometric,** which means equal measurement, is used to describe this type of pictorial since the planes are equally foreshortened.

An **isometric drawing** is similar to an **isometric projection** except that it is not a true axonometric projection but an approximate method of drawing a

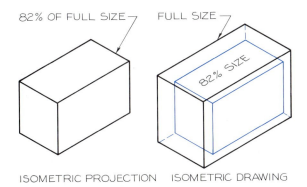

ISOMETRIC PROJECTION   ISOMETRIC DRAWING

FIGURE 19.22 The isometric projection is foreshortened to 82% of full size. The isometric drawing is drawn full size for convenience.

pictorial. Instead of reducing the measurements along the axes 82%, they are drawn true length (Fig. 19.21B).

Figure 19.22 compares an isometric projection with an isometric drawing. By using the isometric drawing instead of the isometric projection, pictorials can be measured using standard scales, the only difference being the 18% increase in size. Isometric drawings are used more often than isometric projections.

The axes of isometric drawings are separated by 120° (Fig. 19.23). Although one of the axes is usually drawn vertically, this is not necessary.

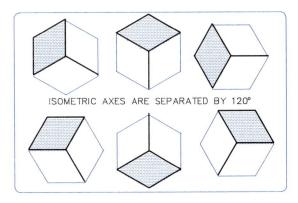

ISOMETRIC AXES ARE SEPARATED BY 120°

FIGURE 19.23 Isometric axes are spaced 120° apart, but they can be revolved into any position.

Constructing isometric drawings

An isometric drawing is begun by drawing three axes 120° apart. Lines parallel to these axes are called **isometric lines** (Fig. 19.24A). True measurements can be made along isometric lines but not along nonisometric lines (Fig. 19.24B).

The three surfaces of a cube in isometric are called **isometric planes** (Fig. 19.24C). Planes parallel to these planes are isometric planes, and planes not parallel to them are nonisometric planes.

To draw an isometric, you will need a scale and 30°–60° triangle (Fig. 19.25). Begin by constructing a plane of the isometric using the dimensions of height (H) and depth (D). In Step 3, the third dimension, width (W), is used to complete the isometric drawing.

All isometric drawings should be blocked in using light guidelines (Fig. 19.26) and overall dimensions of W, D, and H. Other dimensions can be taken from the given views and measured along the isometric axes to

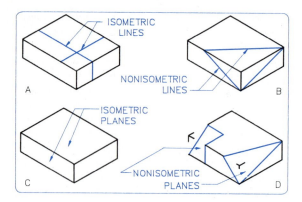

FIGURE 19.24 **A.** True measurements can be made along isometric lines (parallel to the three axes). **B.** Nonisometric lines cannot be measured true length. **C.** Three isometric planes are indicated; no nonisometric planes are in this drawing. **D.** Nonisometric planes are planes inclined to any of the three isometric planes of a cube.

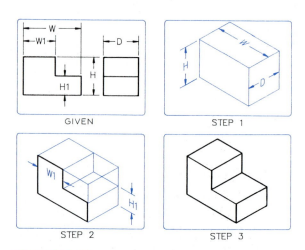

FIGURE 19.26 Layout of a simple isometric drawing.

Step 1 Construct an isometric drawing of a box by using the overall dimensions W, D, and H taken from the given view.

Step 2 Locate the notch in the box by using dimensions W1 and H1 taken from the given orthographic views.

Step 3 Darken the lines to complete the isometric drawing.

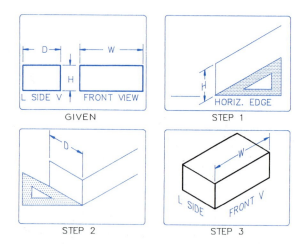

FIGURE 19.25 An isometric drawing of a box.

Step 1 Use a 30°–60° triangle and a horizontal straight edge to construct a vertical line equal to the height (H), and draw two isometric lines through each end.

Step 2 Draw two 30° lines, and locate the depth (D) by transferring this dimension from the given views.

Step 3 Locate the width (W) of the object, and complete the surfaces of the isometric box.

locate notches and portions removed from the blocked-in drawing.

Figure 19.27 shows a more complex isometric. Again, the object is blocked in using H, W, and D, and portions of the block are removed to complete the isometric drawing.

19.13
Angles in isometric

Angles cannot be measured true size in an isometric drawing since the surfaces of an isometric are not true size. Angles must be located by using coordinates measured along isometric lines (Fig. 19.28). Lines AD and BC are equal in length in the orthographic view, but they are shorter and longer than true length in the isometric drawing.

A similar example can be seen in Fig. 19.29, where two angles are drawn in isometric. The equal-size angles in orthographic are less than and greater than true size in the isometric drawing.

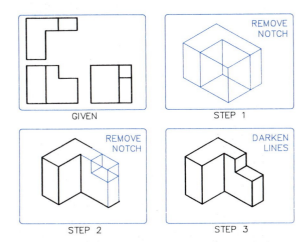

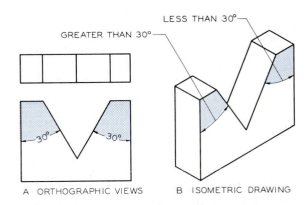

FIGURE 19.27 Layout of an isometric drawing.

Step 1 The overall dimensions of the object are used to block in the object with light lines. The notch is removed.

Step 2 The second notch is removed.

Step 3 The lines are darkened to complete the isometric drawing.

An isometric drawing of an object with an inclined surface is drawn in three steps in Fig. 19.30. The object is blocked in pictorially, and portions are removed.

19.14
Circles in isometric

Three methods of constructing circles in isometric drawings are (1) **point plotting,** (2) **four-center ellipse construction,** and (3) **ellipse templates.**

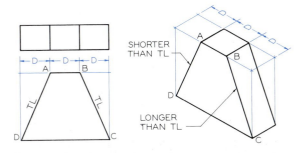

FIGURE 19.28 Inclined surfaces must be found by using coordinates measured along the isometric axes. The lengths of angular lines will not be true length in isometric.

FIGURE 19.29 Angles in isometric must be found by using coordinates. Angles will not appear true size in isometric.

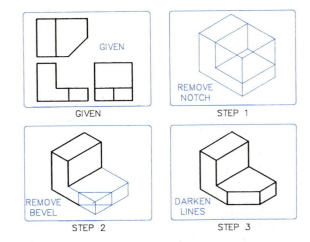

FIGURE 19.30 Inclined planes in isometric.

Step 1 The object is lightly blocked in using the overall dimensions, and the notch is removed.

Step 2 The ends of the inclined plane are located using measurements parallel to the isometric axes.

Step 3 The lines are darkened to complete the isometric drawing.

Point plotting

Point plotting is a method where a series of points on a circle are located by coordinates in the X and Y directions. The coordinates are transferred to the isometric drawing to locate the points on the ellipse one at a time. A series of points located on a circle can be

379

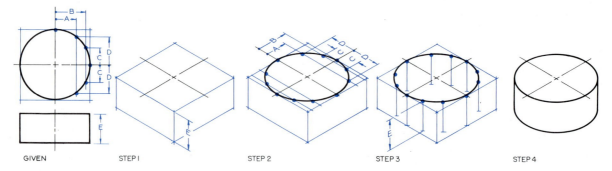

GIVEN   STEP 1   STEP 2   STEP 3   STEP 4

FIGURE 19.31 Plotting circles.

Step 1 The cylinder is blocked in using the overall dimensions. The centerlines locate the points of tangency of the ellipse.

Step 2 Coordinates are used to locate points on the circumference of the circle.

Step 3 The lower ellipse is found by dropping each point a distance equal to the height of the cylinder, *E*.

Step 4 The two ellipses can be drawn with an irregular curve and connected with tangent lines to complete the cylinder.

located in an isometric drawing by using **coordinates** parallel to the isometric axes (Fig. 19.31).

The cylinder is blocked in and drawn pictorially, with the centerlines added in Step 1. Coordinates *A, B, C,* and *D* are used in Step 2 to locate points on the ellipse and then are connected using an irregular curve.

The lower ellipse located on the bottom plane of the cylinder can be found by using a second set of coordinates. The most efficient method is by measuring the distance, *E*, vertically beneath each point that was located on the upper ellipse (Step 3). A plotted

ellipse is a true ellipse; it is equivalent to a 35° ellipse on an isometric plane. An example of a design composed of circular features drawn in isometric is the handwheel shown in Fig. 19.32.

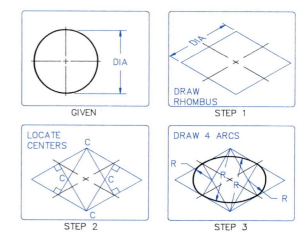

GIVEN   STEP 1

LOCATE CENTERS   STEP 2

DRAW 4 ARCS   STEP 3

FIGURE 19.33 The four-center ellipse.

Step 1 The diameter of the given circle is used to draw a rhombus and centerlines.

Step 2 Light construction lines are drawn perpendicularly from the midpoints of each side to locate four centers.

Step 3 Four arcs are drawn from each center to represent an ellipse tangent to the rhombus.

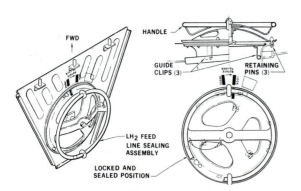

FIGURE 19.32 An example of parts drawn by using ellipses in isometric to represent circles. This is a handwheel proposed for use in an orbital workshop to be launched into space.

Four-center ellipse construction

The **four-center ellipse** method can be used to construct an approximate ellipse in isometric by using four arcs drawn with a compass (Fig. 19.33). The four-center ellipse is drawn by blocking in the orthographic view of the circle with a square tangent to the circle at four points. The four centers are found by constructing perpendiculars to the sides of the rhombus at the midpoints of the sides (Step 2). The four arcs are drawn to give the completed four-center ellipse (Step 3). This method can be used to draw ellipses on any of the three isometric planes, since each is equally foreshortened (Fig. 19.34).

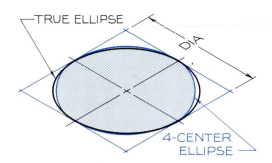

FIGURE 19.35 The four-center ellipse is not a true ellipse but an approximate ellipse.

Figure 19.35 shows that the four-center ellipse is only an approximate ellipse when compared with a true ellipse.

Ellipse templates

A specially designed **ellipse template** can be used for drawing ellipses in isometric (Fig. 19.36).

The diameters of the ellipses on the template are measured along the direction of the isometric lines since this is how diameters are measured in an isometric drawing (Fig. 19.37). The maximum diameter across the ellipse is the **major diameter,** which is a true diameter. Thus the size of the diameter marked on the template is less than the ellipse's major diameter, the true diameter, since isometric drawings are drawn 18% larger than true projections.

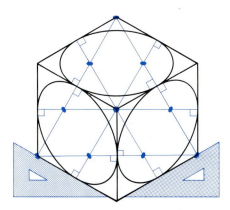

FIGURE 19.34 Four-center ellipses can be drawn on all three surfaces of an isometric drawing.

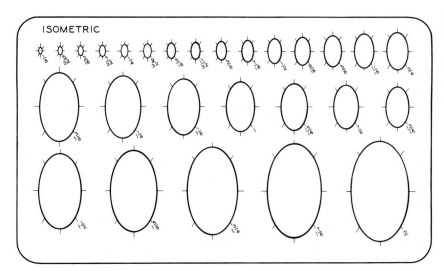

FIGURE 19.36 The isometric template is designed to reduce drafting time. The isometric diameters of the ellipses are not the major diameters of the ellipses but are diameters parallel to the isometric axes.

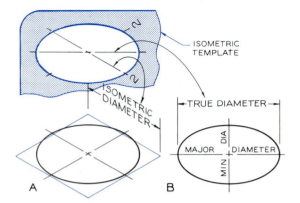

FIGURE 19.37 The diameter of a circle in isometric is measured along the direction of the isometric axes. Therefore, the major diameter of an isometric ellipse is greater than the measured diameter. The minor diameter is perpendicular to the major diameter.

The isometric ellipse template can be used to draw an ellipse by constructing the centerlines of the ellipse in isometric and aligning the ellipse template with these isometric lines (Fig. 19.37A).

19.15

Cylinders in isometric

A cylinder can be drawn in isometric by using the four-center ellipse method (Fig. 19.38). A rhombus is drawn at each end of the cylinder's axis with the cen-

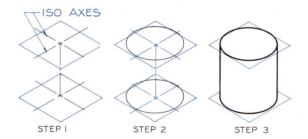

FIGURE 19.38 Cylinder: four-center method.

Step 1 A rhombus is drawn in isometric at each end of the cylinder's axis.

Step 2 A four-center ellipse is drawn within each rhombus.

Step 3 Lines are drawn tangent to each rhombus to complete the isometric drawing.

terlines drawn as isometric lines (Step 1). The ellipses are drawn using the four-center ellipse method at each end (Step 2). The ellipses are then connected with tangent lines, and the lines are darkened (Step 3).

A cylinder can also be drawn using the ellipse template (Fig. 19.39). The axis of the cylinder is drawn, and perpendiculars are constructed at each end (Step 1). Since the axis of a right cylinder is perpendicular to the major diameter of its elliptical end, the ellipse template is positioned as shown (Step 2). The ellipses are drawn at each end and are connected with tangent lines (Step 3).

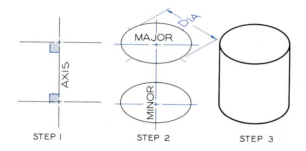

FIGURE 19.39 Cylinder: ellipse template.

Step 1 The axis of the cylinder is drawn to its proper length, and perpendiculars are drawn at each end.

Step 2 The elliptical ends are drawn by aligning the major diameter with the perpendiculars at the ends of the axis. The isometric diameter of the isometric ellipse template is given along the isometric axis.

Step 3 The ellipses are connected with tangent lines to complete the isometric drawing. Hidden lines are omitted.

To construct a cylindrical hole in the block (Fig. 19.40), begin by locating the center of the hole on the isometric plane. The axis of the cylinder is drawn parallel to the isometric axis that is perpendicular to this plane through its center (Step 2). The ellipse template is aligned with the major and minor diameters to complete the elliptical view of the cylindrical hole (Step 3).

19.16

Partial circular features

When an object has a semicircular end, as in Fig. 19.41, the four-center ellipse method can be used with only two centers to draw half the circle (Step 2). To

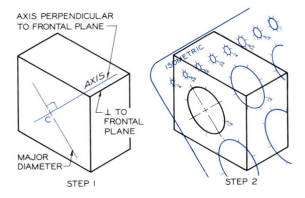

STEP I

STEP 2

FIGURE 19.40 Cylinders in isometric.

Step 1 The center of the hole with a given diameter is located on a face of the isometric drawing. The axis of the cylinder is drawn from the center parallel to the isometric axis that is perpendicular to the plane of the circle. The major diameter is drawn perpendicular to the axis.

Step 2 The $1\frac{3}{8}''$ ellipse template is used to draw the ellipse by aligning the major and minor diameters with the guidelines on the template.

draw the lower ellipse at the bottom of the object, the centers are projected downward a distance of H, the height of the object. These centers are used with the same radii that were used on the upper surface to draw the arcs.

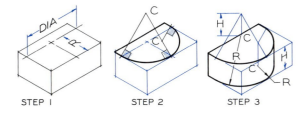

STEP I　　　STEP 2　　　STEP 3

FIGURE 19.41 Semicircular features.

Step 1 Objects with semicircular features can be drawn by blocking in the objects as if they had square ends. The centerlines are drawn to locate the centers and tangent points.

Step 2 Perpendiculars are drawn from each point of tangency to locate two centers. These are used to draw half a four-center ellipse.

Step 3 The lower surface can be drawn by lowering the centers by the distance of H, the thickness of the part. The same radii are used with these new centers to complete the isometric.

In Fig. 19.42, an object with rounded corners is blocked in, and the centerlines are located at each rounded corner (Step 1). An ellipse template is used to construct the rounded corners (Step 2). The rounded corners could have been constructed by the four-center ellipse method or by plotting points on the arcs.

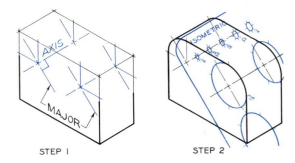

STEP I　　　STEP 2

FIGURE 19.42 Rounded corners.

Step 1 To construct rounded corners of an object, the centerlines of the ellipse are drawn.

Step 2 The elliptical corners are drawn with an ellipse template.

A similar drawing involving the construction of ellipses is the conical shape in Fig. 19.43. The ellipses are blocked in at the top and bottom surfaces (Step 1), and by using a template or the four-center method, the half ellipses are drawn (Step 2).

19.17
Measuring angles

Angles in isometric may be located by coordinates (Fig. 19.44) since angles will not appear true size in an isometric.

A second method of measuring and locating angles is the ellipse template method shown in Fig. 19.45. Since the hinge line is perpendicular to the path of revolution, an ellipse is drawn in Step 1 with the major diameter perpendicular to the hinge line. A true circle is drawn with a diameter that is equal to the major diameter of the ellipse.

In Step 2, point A is located on the ellipse and is projected to the circle to locate the direction of a horizontal line. From this line, the angle of revolution of the hinged part can be measured true size, 120° in this

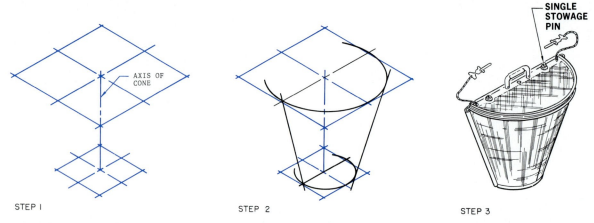

STEP 1   STEP 2   STEP 3

FIGURE 19.43 Construction of a cone in isometric.

Step 1 The axis of the cone is constructed. Each circular end of the cone is blocked in.

Step 2 An ellipse guide is used for constructing the circles in isometric at each end. These ends are connected to give the outline of the object.

Step 3 The remaining details of the screen storage provisions are added to complete the isometric. (Courtesy of the National Aeronautics and Space Administration.)

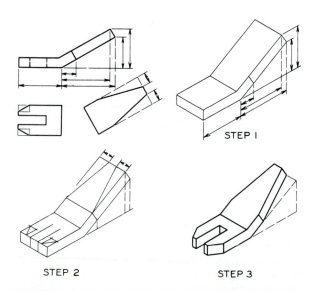

STEP 1

STEP 2   STEP 3

FIGURE 19.44 Inclined surfaces in isometric must be located by using coordinates laid off parallel to the isometric axes. True angles cannot be measured in isometric drawings.

example (Step 3). Point B is projected to the ellipse to locate a line at 120° in isometric.

The thickness of the revolved part is found by drawing line C perpendicular to line A and projecting back to the ellipse. A smaller ellipse through point D is drawn to locate the thickness of the revolved part in the isometric. The remaining lines are drawn parallel to these key lines.

19.18
Curves in isometric

Irregular curves must be plotted point by point, using coordinates to locate each point. Points A through F are located in the orthographic view with coordinates of width and depth (Fig. 19.46). These coordinates are transferred to the isometric view of the blocked-in part (Step 1) and are connected with an irregular curve.

Each point on the upper curve is projected downward for a distance of H, the height of the part, to locate points on the lower curve. The points are connected with an irregular curve to complete the isometric.

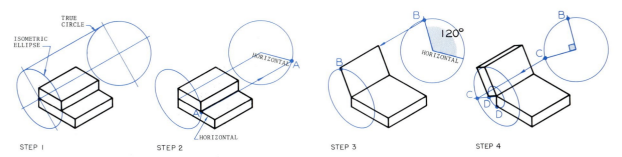

FIGURE 19.45 Measuring angles with an ellipse template.

Step 1 An ellipse is drawn with the major diameter perpendicular to the hinge line of the two parts. Any size of ellipse could be used. A true circle is drawn with its center on the projection of the hinge line and with a diameter equal to the major diameter of the ellipse.

Step 2 Point A is projected to the circle to locate the direction of the horizontal in the circular view.

Step 3 The position of rotation is measured 120° from the horizontal to locate point B, which is then projected to the ellipse to locate the position of the revolved surface.

Step 4 To locate the perpendicular to the surface, a 90° angle is drawn in the circular view, and point C is projected to the ellipse; a line is drawn from point C on the ellipse to the center of the ellipse. A smaller ellipse is drawn to pass through point D on the lower part of the object. The point where this ellipse intersects the line from C to the center of the ellipse establishes the thickness of the revolved part.

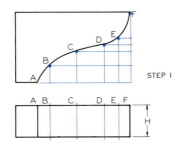

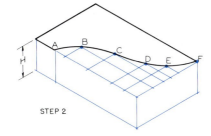

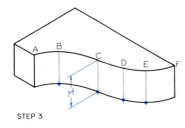

FIGURE 19.46 Plotting irregular curves.

Step 1 Draw two coordinates to locate a series of points on the irregular curve. These coordinates must be parallel to the standard W, D, and H dimensions.

Step 2 Block in the shape using overall dimensions. Locate points on the irregular curve using the coordinates from the orthographic views.

Step 3 Since the object has a uniform thickness, the lower curve can be found by projecting downward the distance H from the upper points.

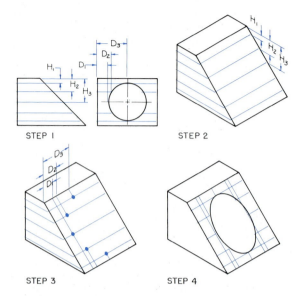

FIGURE 19.47 Construction of ellipses on an inclined plane.

Step 1 Coordinates are established in the orthographic views.

Step 2 One set of coordinates is transferred to the isometric view.

Step 3 The second set of coordinates is transferred to the isometric to establish points on the ellipse.

Step 4 The plotted points are connected with an elliptical curve. An ellipse template can usually serve as a guide for connecting the points.

19.19
Ellipses on nonisometric planes

When an ellipse lies on a nonisometric plane, like the one shown in Fig. 19.47, points on the ellipse can be plotted to locate it. Coordinates are located in the orthographic views and then transferred to the isometric as shown in Steps 1 and 2. The plotted points can be connected with an irregular curve, or an ellipse template can be selected that will approximate the plotted points.

19.20
Machine parts in isometric

Figure 19.48 shows orthographic and isometric views of a spotface, countersink, and boss. The isometric drawings of these features can be drawn by point-plotting the circular features, using the four-center method, or using an ellipse template, which is the easiest method.

A threaded shaft can be drawn in isometric, as Fig. 19.49 shows, by first drawing the cylinder in isometric (Step 1). Next, the major diameters of the crest lines of the thread are drawn separated at a distance

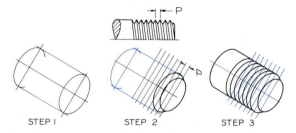

FIGURE 19.49 Threads in isometric.

Step 1 Draw the cylinder to be threaded by using an ellipse template.

Step 2 Lay off perpendiculars that are spaced by a distance equal to the pitch of the thread, *P.*

Step 3 Draw a series of ellipses to represent the threads. The chamfered end is drawn by using an ellipse whose major diameter is equal to the root diameter of the threads.

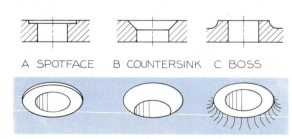

A SPOTFACE   B COUNTERSINK   C BOSS

FIGURE 19.48 Examples of circular features drawn in isometric. These can be drawn by using ellipse templates.

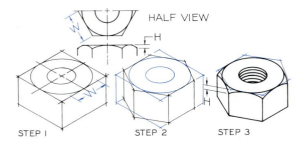

FIGURE 19.50 Construction of a nut.

Step 1 The overall dimensions of the nut are used to block in the nut.

Step 2 The hexagonal sides are constructed at the top and bottom.

Step 3 The chamfer is drawn with an irregular curve. Threads are drawn to complete the isometric.

of P, the pitch of the thread (Step 2). Ellipses are then drawn by aligning the major diameter of the ellipse template with the perpendiculars to the cylinder's axis (Step 3). Note that the 45° chamfered end is drawn using a smaller ellipse at the end.

A hexagon-head nut (Fig. 19.50) is drawn in three steps using an ellipse template. The nut is blocked in, and an ellipse drawn tangent to the rhombus. The hexagon is constructed by locating the distance across a flat, W, parallel to the isometric axes. The other sides of the hexagon are found by drawing lines tangent to the ellipse (Step 2). The distance H is laid off at each corner to establish the amount of chamfer at each corner (Step 3).

A hexagon-head bolt is drawn in two positions in Fig. 19.51. The washer face can be seen on the lower side of the head, and the chamfer on the upper side of the head.

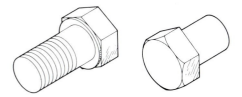

FIGURE 19.51 Isometric drawings of the upper and lower sides of a hexagon-head bolt.

To draw a sphere in isometric, three ellipses are drawn as isometric planes with a common center (Fig. 19.52). The center is used to construct a circle that will be tangent to each ellipse, as shown in Step 3.

A portion of a sphere is used to draw a round-head screw in Fig. 19.53. A hemisphere is constructed in Step 1. The centerline of the slot is located along one of the isometric planes. The thickness of the head is measured at a distance of E from the highest point on the sphere.

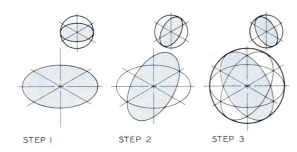

FIGURE 19.52 An isometric sphere.

Step 1 The three intersecting isometric axes are drawn. The ellipse template is used to draw the horizontal elliptical section.

Step 2 The isometric ellipse template is used to draw one of the vertical elliptical sections.

Step 3 The third vertical elliptical section is drawn, and the center is used to draw a circle tangent to the three ellipses.

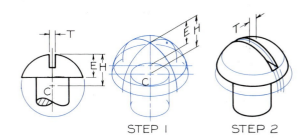

FIGURE 19.53 Spherical features.

Step 1 An isometric ellipse template is used to draw the elliptical features of a round-head screw.

Step 2 The slot in the head is drawn, and the lines are darkened to complete the isometric of the head.

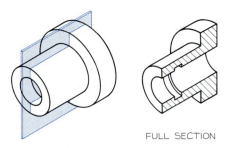

FIGURE 19.54 **Parts can be shown in isometric sections to clarify internal features.**

19.21
Isometric sections

A full section can be drawn in isometric to clarify internal details that might otherwise be overlooked (Fig. 19.54). Half sections can also be used, as Fig. 19.55

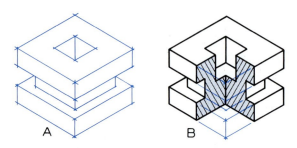

FIGURE 19.55 **An isometric drawing of a half section can be constructed.**

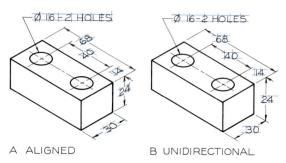

FIGURE 19.56 **Dimensions can be placed on isometric drawings by using either technique shown here. Guidelines should always be used for lettering.**

shows. Figure 19.56 shows the same as a full section and a half section.

19.22
Dimensioned isometrics

When it is advantageous to dimension and note a part shown in isometric, either the aligned or the unidirectional method can be used to apply the notes (Fig. 19.57). In both cases, notes connected with leaders are positioned horizontally. Always use guidelines for your lettering and numerals.

19.23
Fillets and rounds

Fillets and rounds in isometric can be represented by either of the methods shown in Fig. 19.57 to give added realism to a pictorial drawing. Figures 19.57A and 19.57B show how intersecting guidelines are drawn equal in length to the radii of the fillets and rounds, and how arcs are drawn tangent to these lines. These arcs can be drawn freehand or with an ellipse template. The method in Fig. 19.57C uses freehand lines drawn parallel to or concentric with the directions of the fillets and rounds. Figure 19.58 shows an example of these two methods of showing fillets and rounds . The stipple shading was applied by using an adhesive overlay film.

When fillets and rounds are illustrated as shown in Fig. 19.59 and dimensions are applied, it is much easier to understand the features of the part than when it is represented by orthographic views.

19.24
Isometric assemblies

Assemblies are used to explain how parts are assembled. Figure 19.60A shows common mistakes in applying leaders to an assembly; whereas, Fig. 19.60B shows the more acceptable techniques. The numbers in the circles ("balloons") refer to the number given to each part listed in the parts list.

Figure 19.61 shows an assembly from a parts manual along with its parts list. This assembly is exploded, which makes it clear how the parts are to be assembled.

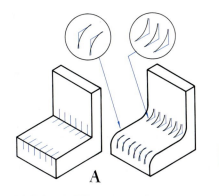
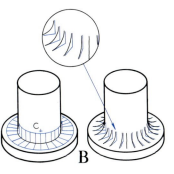
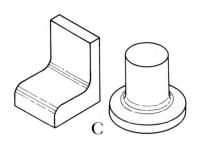

FIGURE 19.57 Representation of fillets and rounds.

A. Fillets and rounds can be represented by segments of an isometric ellipse if guidelines are constructed at intervals.

B. Fillets and rounds can be represented by elliptical arcs by constructing radial guidelines.

C. Fillets and rounds can be represented by lines that run parallel to the fillets and rounds.

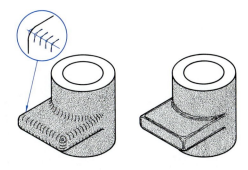

FIGURE 19.58 Two methods of representing fillets and rounds on a part.

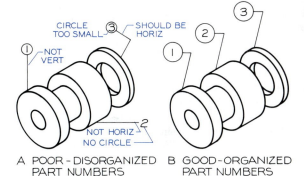

A POOR—DISORGANIZED PART NUMBERS

B GOOD—ORGANIZED PART NUMBERS

FIGURE 19.60 **A.** Common mistakes in applying part numbers to an assembly. **B.** Acceptable techniques of applying part numbers.

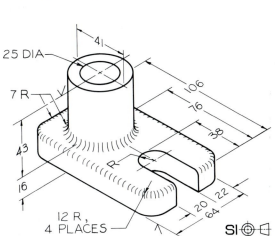

FIGURE 19.59 An isometric drawing with complete dimensions and fillets and rounds represented.

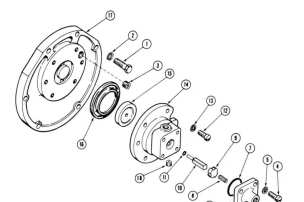

FIGURE 19.61 An exploded isometric assembly.

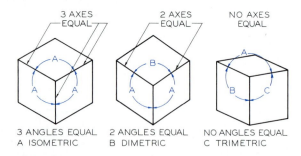

FIGURE 19.62 **Three types of axonometric projection.**

19.25
Axonometric pictorials

An **axonometric pictorial** is a form of orthographic projection in which the pictorial view is projected perpendicularly onto the picture plane with parallel projectors. The object is positioned in an angular position with the picture plane so that its pictorial projection will be a three-dimensional view. Three types of axonometric pictorials are possible: (1) **isometric,** (2) **dimetric,** or (3) **trimetric.**

The **isometric projection** is the view where the diagonal of a cube is viewed as a point. The planes will be equally foreshortened and the axes equally spaced 120° apart (Fig. 19.62A). The measurements along the three axes will be equal but less than true length since this is true projection.

A **dimetric projection** is the view where two planes are equally foreshortened, and two of the axes are separated by equal angles (Fig. 19.62B). The measurements along two of the axes are equal.

A **trimetric projection** is the view where all three planes are unequally foreshortened, and the angles between the three axes are different (Fig. 19.62C).

19.26
Perspective pictorials

A **perspective pictorial** is a view normally seen by the eye or a camera; it is the most realistic form of pictorial. In a perspective, all parallel lines converge at infinite vanishing points as they recede from the observer.

The three basic types of perspectives are (1) **one point,** (2) **two point,** and (3) **three point,** depending on the number of vanishing points in their construction (Fig. 19.63).

The **one-point perspective** has one surface of the objective parallel to the picture plane; therefore, it is true shape. The other sides vanish to a single point on the horizon, called a *vanishing point* (Fig. 19.63A).

A **two-point perspective** is a pictorial positioned with two sides at an angle to the picture plane; this requires two vanishing points (Fig. 19.63B). All horizontal lines converge at the vanishing points, but vertical lines remain vertical and have no vanishing point.

The **three-point perspective** has three vanishing points since the object is positioned so that all sides of it are at an angle with the picture plane (Fig. 19.63C). The three-point perspective is used in drawing larger objects such as buildings.

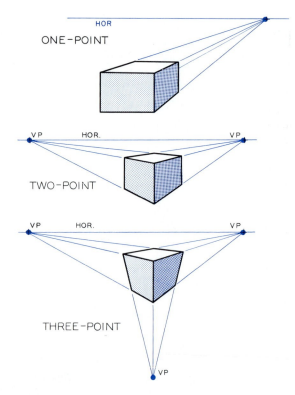

FIGURE 19.63 **A comparison of one-point, two-point, and three-point perspectives.**

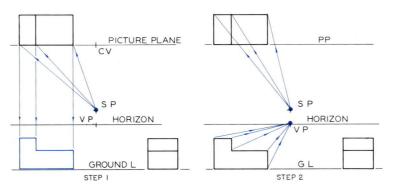

FIGURE 19.64 Construction of a one-point perspective.

Step 1 Since the object is parallel to the picture plane, there will be only one vanishing point, located on the horizon below the station point. Projections from the top and side views establish the front plane. This surface is true size, since it lies in the picture plane.

Step 2 Draw projectors from the station point to the rear points of the object in the top view and from the front view to the vanishing point on the horizon. In a one-point perspective, the vanishing point is the front view of the station point.

Step 3 Construct vertical projectors from the top view to the front view from the points where the projectors cross the picture plane. These projectors intersect the lines leading to the vanishing point. This is a one-point perspective, since the lines converge at a single VP.

19.27
One-point perspectives

The steps of drawing a one-point perspective are shown in Fig. 19.64, which shows the top and side views of the object, picture plane, station point, horizon, and ground line.

PICTURE PLANE is the plane the perspective is projected on. It appears as an edge in the top view.

STATION POINT (SP) is the location of the observer's eye in the plan view. The front view of the station point will always lie on the horizon.

HORIZON is a horizontal line in the front view that represents an infinite horizontal, such as the surface of the ocean.

GROUND LINE is an infinite horizontal line in the front view that passes through the base of the object being drawn.

CENTER OF VISION (CV) is a point that lies on the picture plane in the top view and on the horizon in the front view. In both cases, it is on the line from the station point that is perpendicular to the picture plane.

When drawing any perspective, the station point should be far enough away from the object so that the perspective can be contained in a cone of vision not more than 30° (Fig. 19.65). If a larger cone of vision is required, the perspective will be distorted.

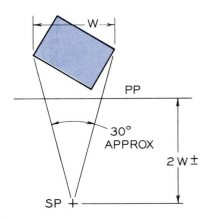

FIGURE 19.65 The station point (SP) should be far enough away from the object to permit the cone of vision to be less than 30° to reduce distortion.

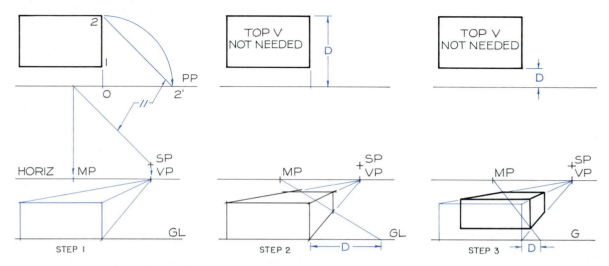

FIGURE 19.66 One-point perspective—measuring points.

Step 1 Line 0–2 is revolved into the picture plane to locate point 2'. A line is drawn parallel to 2'–2 through the SP to the PP. The measuring point is located on the horizon.

Step 2 Distance *D* is laid off along the ground line from the front corner of the perspective. This distance is projected to the measuring point to locate the rear corner of the perspective.

Step 3 The front of the object is located by laying off distance *D* from the corner of the perspective and projecting to the measuring point. This locates the front surface of the object.

Measuring points

An additional vanishing point used to locate measurements along the receding lines that vanish to the horizon is the **measuring point.** In Fig. 19.66, the measuring point of a one-point perspective is found by revolving line 0–2 into the picture plane to 0–2' and drawing a construction line from the station point to the picture plane parallel to 2–2'. The measuring point is located on the horizon by projection from the picture plane (Step 1).

Since the distance 0–2 equals 0–2', depth dimensions can be laid off along the ground line and then projected to the measuring point. This locates the real corner of the one-point perspective (Step 2). The depth from the picture plane to the front of the object is similarly found.

The use of the measuring-point method eliminates the need for placing the top view in the customary top-view position; instead, the dimensions can be more conveniently transferred to the ground line.

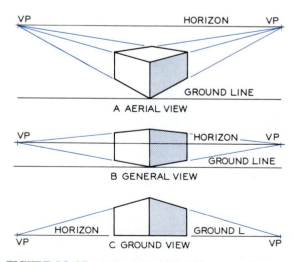

FIGURE 19.67 Different perspectives can be obtained by locating the horizontal over, under, and through the object.

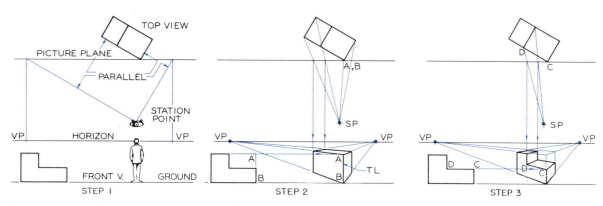

FIGURE 19.68 Construction of a two-point perspective.

Step 1 Construct projectors that extend from the top view of the station point to the picture plane parallel to the forward edges of the object. Project these points vertically to the horizon in the front view to locate vanishing points. Draw the ground line below the horizon, and construct the side view on the ground line.

Step 2 Since all lines in the picture plane are true length, line *AB* is true length. Thus, line *AB* is projected from the side view to determine its height. Then project each end of *AB* to the vanishing points. Draw projectors from the SP to the exterior edges of the top view. Project the intersections of these projectors with the picture plane to the front view.

Step 3 Determine point *C* in the front view by projecting from the side view to line *AB*. Draw a projector from point *C* to the left vanishing point. Point *D* will lie on this projector beneath the point where a projector from the station point to the top view of point *D* crosses the picture plane. Complete the notch by projecting to the respective vanishing points.

19.28
Two-point perspectives

If two surfaces of an object are positioned at an angle to the picture plane, two vanishing points will be required to draw it as a perspective. Different views can be obtained by changing the relationship between the horizontal and the ground line (Fig. 19.67).

An **aerial view** is obtained when the horizon is placed above the ground line and the top of the object in the front view. When the ground line and horizon coincide in the front view, a **ground-level view** is obtained. A **general view** is obtained when the horizon is placed above the ground line and through the object, usually equal to the height of a person.

Figure 19.68 shows the steps of constructing a two-point perspective.

Since line *AB* lies in the picture plane, it will be true length in the perspective. All height dimensions must originate at this vertical line because it is the only true-length line in the perspective. Points *C* and *D* are found by projecting to *AB* and then projecting toward the vanishing points.

Figure 19.69 shows a typical two-point perspective. By referring to Fig. 19.68, you will be able to understand the development of the construction used.

The object in Fig. 19.70 does not contact the picture plane in the top view as in the previous examples. To draw a perspective of this object, the lines of the object must be measured where the extended plane intersects the picture plane. The height is measured, and the infinite plane is drawn to the vanishing point. The corner of the object can be located on this infinite plane by projecting the corner to the picture plane in the top view with a projector from the station point.

Measuring points

Measuring points are found in Fig. 19.71 to aid in constructing two-point perspectives. The use of measuring points eliminates the need to have the top view in the top-view position after the vanishing points have been found.

For two-point perspectives, two vanishing points are used to locate dimensions along the receding planes that vanish to the horizon (Step 2). Depth dimensions are laid off along the ground line from point

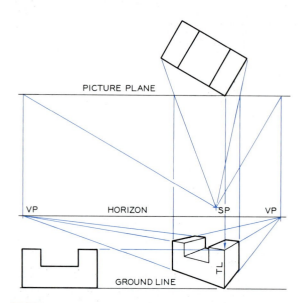

FIGURE 19.69 A two-point perspective of an object.

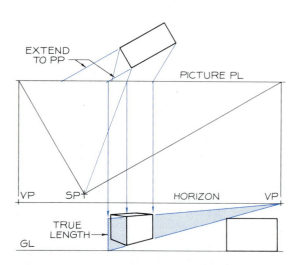

FIGURE 19.70 A two-point perspective of an object that is not in contact with the picture plane.

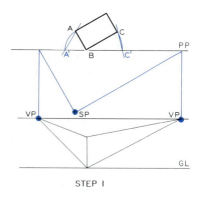

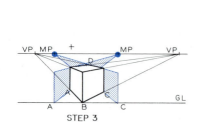

FIGURE 19.71 Construction of measuring points.

Step 1 In the top view, revolve lines *AB* and *BC* into the picture plane using point *B* as the center of revolution. Draw construction lines through points *A–A'* and points *C–C'*. These lines represent edge views of vertical planes passing through the corners *A* and *C*.

Step 2 Draw lines from the station point parallel to lines *A–A'* and *C–C'* to the picture plane. Project these points of intersection to the horizon to locate two measuring points. Distances *AB* and *BC* can be laid off true length on the ground line (GL), since they have been revolved into the picture plane in the top view.

Step 3 Extend planes from points *A* and *C* on the ground line to their respective measuring points. These planes intersect the infinite planes, which are extended to their vanishing points, to locate two corners of the block. True measurements can be laid off on the ground line and projected to measuring points to find corner points.

B, and secondary planes are passed from these points to their measuring points.

Measuring points are used in Fig. 19.72 to construct a two-point perspective. No top view is needed since this system is used to locate measurements along the receding lines.

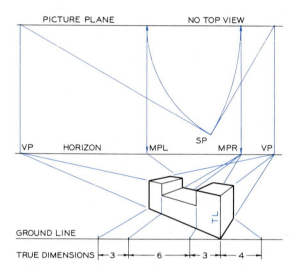

FIGURE 19.72 A two-point perspective drawn with the use of measuring points.

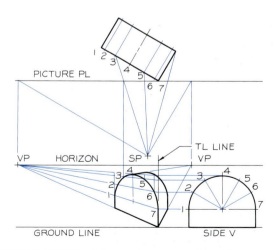

FIGURE 19.73 A two-point perspective of an object with semicircular features.

Arcs in perspective

Arcs in two-point perspectives must be found by using coordinates to locate points along the curves in perspective (Fig. 19.73). Points 1 through 7 are located along the semicircular arc in the orthographic view. These same points are found in the perspective by projecting coordinates from the top and side views. The points do not form a true ellipse but an egg-shaped oval. An irregular curve is used to connect the points.

19.29
Axonometric pictorials by computer

An ISOMETRIC style grid and snap can be used as an aid in making isometric drawings. The STYLE subcommand of the SNAP command can be used to change the rectangular GRID (called STANDARD) (S) to ISOMETRIC (I) where the dots are plotted vertically and at 30° to the horizontal (Fig. 19.74). While in this mode, the cursor's cross hairs, which can be made to snap to the grid points, will align with two axes of the isometric.

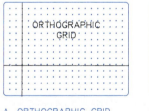

A ORTHOGRAPHIC GRID

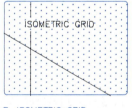

B ISOMETRIC GRID

FIGURE 19.74 The SNAP command permits you to rotate the orthographic grid (Standard) or to select the isometric grid option that can be used as an aid in drawing isometric pictorials.

Although an isometric drawing can be made by using the grid and snap features, this method is not truly a three-dimensional system; instead, points must be located manually, and elliptical features must be inserted individually, as Fig. 19.75 shows.

Isometric ellipses can be automatically drawn by using the ISOCIRCLE option under the ELLIPSE

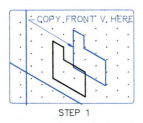

STEP 1

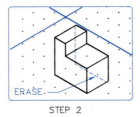

STEP 2

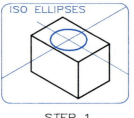

STEP 1

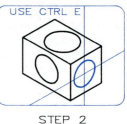

STEP 2

FIGURE 19.75 Isometric drawings—computer.

Step 1 When the isometric grid is on the screen, the cursor can be made to SNAP to it. The front view is drawn as an isometric plane and duplicated at the back of the object by using the COPY command.

Step 2 The hidden lines are ERASEd to complete the isometric drawing.

FIGURE 19.77 Isometric ellipses.

Step 1 While in the Isometric-grid mode, draw isometric ellipses as follows:
```
Command: ELLIPSE (CR)
<Axis endpoint 1>:/Center/Isocircle:
I (CR)
Center of Circle: (Select with cursor.)
<Circle radius>/Diameter: (Select radius
with cursor.)
```

Step 2 Change the orientation of the cursor for drawing isometric ellipses on the other two planes by pressing CTRL-E. Repeat the process in Step 1.

command. When in this command, you can change the position of the cursor to be aligned with each of the three isometric planes by pressing CTRL-E on the keyboard (Fig. 19.76). When the cursor is aligned with the proper axes of an isometric plane, you can select the center of the isometric ellipse or its diameter (Fig. 19.77).

The ISOPLANE command can be used to change the position of the cursor in the same manner as CTRL-E. When using ISOPLANE, you will be prompted to select from Left/Top/Right/ <Toggle>: options. The Toggle option can be used by pressing (CR) to successively move the cursor position from plane to plane. The other options are self-explanatory.

AutoCAD 3-D

AutoCAD contains features that provide limited three-dimensional capabilities for drawing axonometric pictorials with three commands: ELEV, VPOINT, and HIDE. These commands enable you to draw the plan view of an object, specify its height, select a desired viewpoint, obtain its axonometric projection, and remove hidden lines from it. Each point of the object is rotated as a three-dimensional point in space when each viewpoint is selected, and circular features are converted to ellipses.

A box with a cylindrical hole through it is drawn as follows:

```
Command: ELEV (CR)
New current elevation <.20
current elevation>: 0 (CR)
New current thickness <0.000>:
4 (CR)
```

Thickness is the distance the object extends above its current elevation. A minus thickness results in downward extrusion of the plane of the object drawn on elevation 0. We can now draw the box:

```
Command: LINE (CR)
From point: (Draw plan view of box.)
```

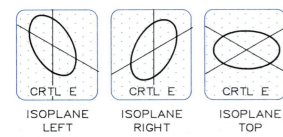

ISOPLANE LEFT

ISOPLANE RIGHT

ISOPLANE TOP

FIGURE 19.76 Three isometric ellipses can be used under the ELLIPSE command and the ISOCIRCLE option when drawing isometrics. By pressing CTRL-E, the isometric ellipses can be alternatively positioned to fit the three isometric planes. The ISOPLANE can also be used for the same purpose.

Since the elevation and thickness of the cylindrical hole will be the same as that of the box, these values need not be changed, and the cylinder is located in the plan view as a circle:

```
Command: CIRCLE (CR)
3P/2P/<Centerpoint>: (Specify center.)
Diameter/<Radius>: 1.0 (CR)
```

The VPOINT command can now be used to select a viewpoint of the object that you want to see in three dimensions. Respond by entering the desired location of your viewpoint, such as 1, −1, 1, which locates the point of vision at $X=1$, $Y=-1$, and $Z=1$ from the origin of 0, 0, 0. By not giving coordinates but selecting AXES from the Root Menu, the X, Y, and Z axes will be shown on the screen (Fig. 19.78). The circles represent the compass on a globe. The centerpoint is the north pole (0, 0, 1), the inner circle is the equator (n, n, 1), and the outer circle is the south pole (0, 0, −1). When the cursor is moved about the compass, the tripod axes will change accordingly to show the axes of the axonometric that will be drawn.

The viewpoint can be selected in the following manner:

```
Command: VPOINT(CR)
Enter view point <current X, Y, Z
view points>: 1, −1, 1(CR)
```

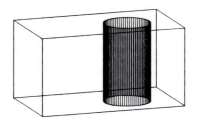

FIGURE 19.79 **This axonometric drawing is a wire diagram of a cylindrical hole in a box.**

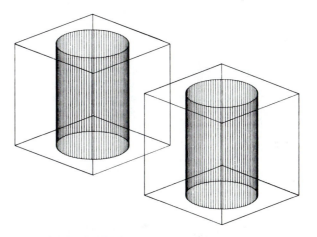

FIGURE 19.80 **Two cubes with cylindrical holes through them.**

Press CR, and a three-dimensional wire diagram will appear on the screen (Fig. 19.79).

The Root Menu will show subcommands PLAN, AXES, and HIDE under the VPOINT command. By selecting the PLAN command, the drawing will return to its plan view for modification if needed.

The HIDE command is illustrated in Fig. 19.80, where two boxes with cylindrical holes through them have been drawn. The one at the left was drawn using the DRAW and CIRCLE commands, and the one at the right was drawn the same way, but the SOLID command was used to fill in the plan view (over the circle). The viewpoint was selected, (CR) pressed, and a wire diagram of each object was drawn on the screen. The HIDE command is selected from the Root Menu, and AutoCAD calculates the visibility for each line and redraws the objects on the screen (Fig. 19.81). The object at the left shows the box as an empty box, whereas the one at the right shows the box as a solid.

```
Layer  VISIBLE  Ortho  Snap        4.00,5.00    AutoCAD
                                                ****
COMPASS/GLOBE ────                              3DLINE:
              Z                                 3DFACE
                                                ELEV:
                                                CHANGE:
                                                VPOINT:
AXES ────                                       rotate
                 Y                              axes
                        CURSOR ─                plan
                      X
                                                HIDE:

COMMAND:
```

FIGURE 19.78 **When using the AXES option under the VPOINT command, the screen will show the three axes and compass/globe that can be used to select a viewpoint for an axonometric projection of the object being drawn. As the cursor is moved about the globe, the axes will change on the screen to show the current relationship of the axes.**

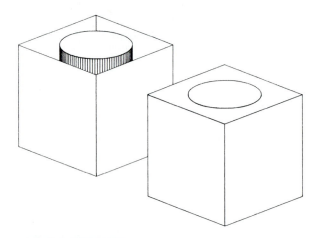

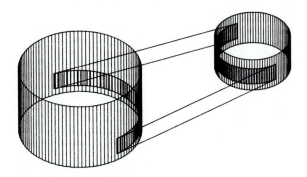

HIDE command is used, the objects will appear as solids rather than as empty perimeters. A wire-diagram drawing is often better than one where visibility has been shown since all features can be seen better.

Surfaces can be drawn on multiple elevations and thicknesses to represent a part (Fig. 19.83). The cylinders have the same base elevation, 0, and thicknesses of 2 and 1, respectively. The rib has an elevation of 0.2 and a thickness of 0.5.

All pictorials using this option of AutoCAD will be axonometrics with projection lines parallel to the picture plane. Perspective drawings cannot be drawn with this system. However, you may obtain an infinite number of views of each part and the view you wish to have by selecting VPOINTS.

FIGURE 19.81 When the HIDE command is applied to the cubes drawn in Fig. 19.80, visibility for each is shown differently. The cube at the right was filled in with the SOLID command when it was drawn in the PLAN mode.

To plot or print the axonometric projections with the hidden lines removed, you must respond during plotter configuration when prompted as follows:

Remove hidden lines? <N>: Y (CR)

This response will remove the hidden lines at the time of plotting.

Irregular shapes can also be drawn as axonometrics by using the PLINE command (Fig. 19.82). The Root Menu will show subcommands PLAN, AXES, and HIDE under the VPOINT command.

Objects drawn with TRACE lines, wide POLYLINES, or SOLIDS will have solid tops and bottoms when they are extruded. Therefore, when the

FIGURE 19.83 A three-dimensional drawing with varying elevations can be constructed with AutoCAD's 3-D option.

TEXT and BLOCKS can be inserted in three-dimensional drawings when the drawing is in its PLAN mode. When converted to axonometric, the text will lie on the ELEVation it was inserted at. BLOCKS will be inserted on the current layer and extruded to the thickness of the part.

19.30
Perspectives by computer

Design Board Professional®, a software package produced by MegaCADD, Inc., illustrates how perspectives can be constructed in three-dimensional space. The MegaCADD program offers three basic modes for developing perspective drawings of objects: CRE-

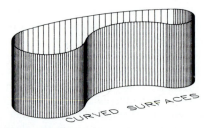

FIGURE 19.82 This object with curved surfaces was drawn using a combination of ARC and PLINE commands.

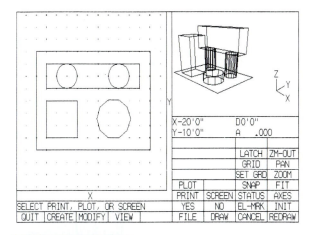

FIGURE 19.84 The CREATE screen of MegaCADD's Design Board Professional® is used to create the orthographic views of the objects that will be converted into perspective views.

ATE, VIEW, and MODIFY. The CREATE mode (Fig. 19.84) gives a screen on the monitor on which a perspective can be drawn by establishing three-dimensional shapes in their plan views. The X and Y axes are given at the perimeters of the plan view with a grid for locating measurements.

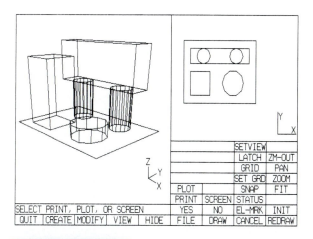

FIGURE 19.85 The VIEW screen of DBPro® shows the plan view of the model and its corresponding perspective view. The viewpoint can be changed by selecting the observer's position and the target position in the plan view.

The VIEW mode gives a screen on which you may select a viewpoint from which to look at the orthographic views (Fig. 19.85). The orthographic views of the drawing will appear at the right of the screen, and the corresponding perspective view will be shown in the large area at the left. By selecting CREATE, the screen will return to the mode shown in Fig. 19.84, where the perspective axes are shown.

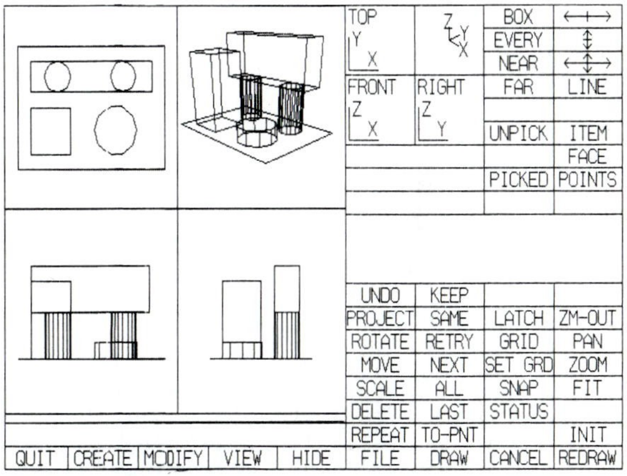

FIGURE 19.86 The MODIFY screen of DBPro® shows three orthographic views and the perspective of the model created. Modifications of the model can be made in this mode.

The MODIFY mode (Fig. 19.86) gives three orthographic views (top, front, and right-side views) of the model as well as its perspective. Parts of the drawing can be deleted, changed, moved, enlarged, and edited in several ways. Once the model is completed and the viewpoint selected, the DRAW command can be used to obtain a drawing on the screen or a hard copy by a printer or plotter (Fig. 19.87). The plotted drawing in Fig. 19.87 is a wire diagram. To show visibility, hidden lines can be omitted before plotting (Fig. 19.88). Figure 19.89 shows another wire-diagram view of the model from a different position. Views can be obtained from any position, from any distance around the model, and from any height.

MegaCADD's Prohouse is viewed from different locations in Fig. 19.90, where the hidden lines have been removed. A series of views can be generated to represent "walk-through" views of the house by locating viewpoints and target points within the house

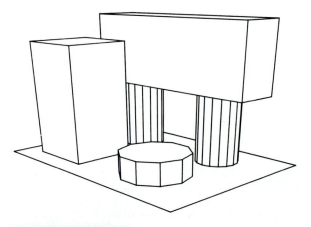

FIGURE 19.88 **The HIDE command is used to remove hidden lines from the wire diagram plotted in the previous figure.**

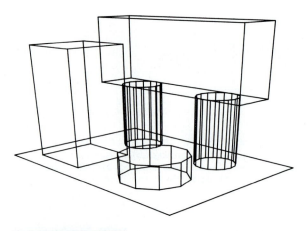

FIGURE 19.87 **A wire diagram of several three-dimensional solids drawn by DBPro®.**

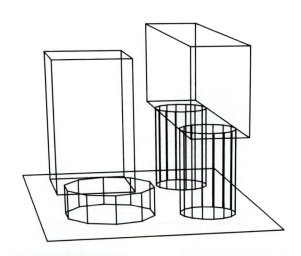

FIGURE 19.89 **An infinite number of views may be obtained from each model.**

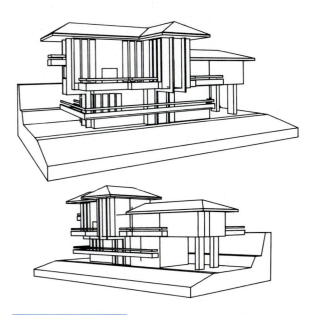

FIGURE 19.90   **An example of different views obtained from the same house developed by DBPro.**

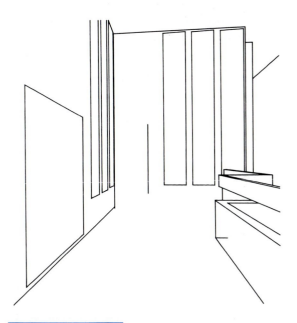

FIGURE 19.92   **The hidden lines are omitted to give a "walk-through" view of the house.**

(Fig. 19.91). The hidden lines can be removed and the view can be enlarged to give a perspective from this vantage point (Fig. 19.92). Since the house was drawn in six hours, some details were omitted; however, the true value of the views lies in their giving a quick means of analyzing the major shapes within the structure.

A different application of this powerful software program is the representation of an entire downtown area for comparing and studying a group of buildings (Fig. 19.93). Like the house, viewpoints can be selected at sky or street level to simulate the view from any position. Views also can be generated that represent a walk through the town.

MegaCADD offers methods of drawing curved surfaces, domes, inclined surfaces, free forms, orthographic views, and stereographic views that can be used for representing perspectives for viewing models in three dimensions.

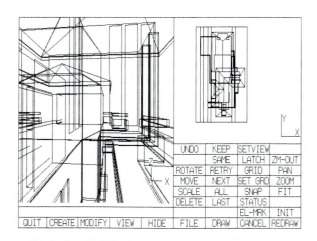

FIGURE 19.91   **A viewpoint and target position can be located in the plan view at the right to obtain the perspective shown at the left when the screen is in the** VIEW **mode.**

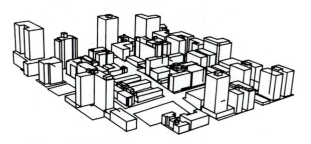

FIGURE 19.93   **A perspective view of MegaCADD's Urban City.**

Problems

The following problems (Fig. 19.94) are to be shown on Size A or B sheets as assigned. Select the appropriate scale that will take advantage of the space available on each sheet. By setting each square to 0.20 inch (about 5 mm), two problems can be drawn on each Size A sheet. When each square is set to 0.40 inch (about 10 mm), one problem can be drawn on each Size B sheet.

Obliques

1–24. Construct either cavalier, cabinet, or general obliques of the parts assigned in Fig. 19.94.

Isometrics

1–24. Construct isometric drawings of the objects assigned in Fig. 19.94.

Axonometrics

1–24. Construct axonometric scales, and find the ellipse template angles for each surface by using a cube rotated into positions of your choice. Calibrate the scales to represent $\frac{1}{2}$-in. intervals. Overlay the scales, and draw axonometrics of the objects assigned from Fig. 19.94.

Perspectives

1–24. Lay out perspective views of the assigned parts in Fig. 19.94 on Size B sheets.

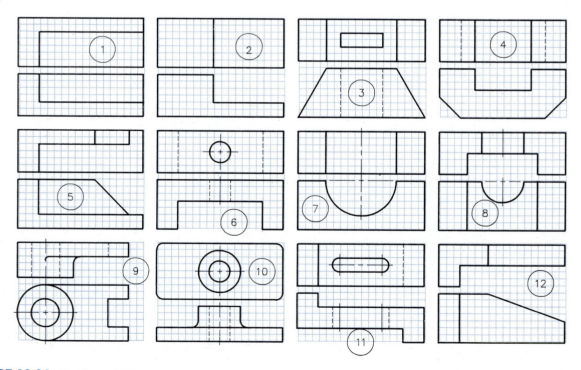

FIGURE 19.94 Problems 1–24.

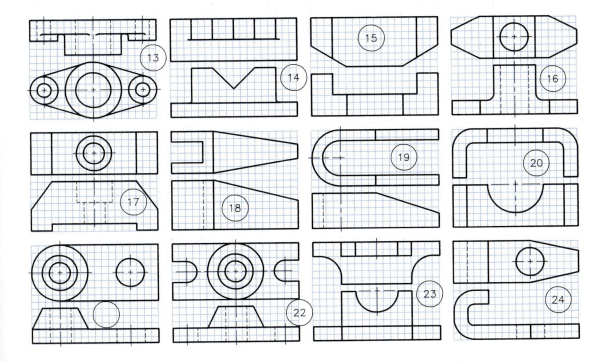

Descriptive Geometry

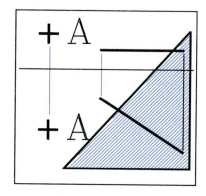

20

20.1
Introduction

Descriptive geometry is the projection of three-dimensional figures onto a two-dimensional plane of paper in a manner that allows geometric manipulations to determine lengths, angles, shapes, and other information by means of graphics. Orthographic projection is used for laying out descriptive geometry problems.

In the first part of this chapter, we cover the representation of points, lines, and planes by orthographic projection. In the next part, we deal with primary and secondary auxiliary views, and in the last part, we introduce intersections and developments.

Whereas the first several chapters of this book discuss the preparation of working drawings and pictorials as a means of detailing three-dimensional objects, descriptive geometry is a problem-solving tool for developing a design and determining its geometry. Descriptive geometry's most useful application in the design process is in the refinement step.

20.2
Techniques of labeling points, lines, and planes

Points, lines, and planes are the basic geometric elements used in graphics and descriptive geometry. Figure 20.1 illustrates the following rules of solving and labeling descriptive geometry problems:

LETTERING should be labeled using $\frac{1}{8}$-in. letters with guidelines. Lines should be labeled at each end, and planes should be labeled at each corner with either letters or numerals.

POINTS should be designated by two perpendicular dashes approximately $\frac{1}{8}$ in. long that form a cross, not

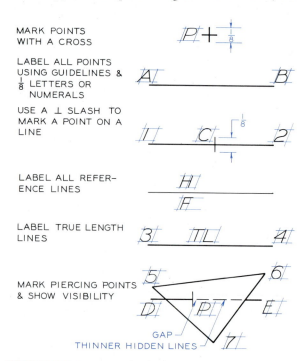

FIGURE 20.1 Standard practices for labeling points, lines, and planes.

a dot. Label the points or corners with letters or numerals.

POINTS ON A LINE should be marked with a perpendicular dash about $\frac{1}{8}$ in. long, not a dot. Label the point with a letter or numeral.

REFERENCE LINES are thin (about as thin as they can be drawn) lines that should be labeled according to the instructions in Section 20.3.

OBJECT LINES are used to represent points, lines, and planes. They should be drawn about twice as thick as reference lines with an H or F pencil. Hidden lines are drawn thinner than visible object lines.

TRUE-LENGTH LINES should be labeled by the note, TRUE LENGTH, or by the abbreviation, TL.

TRUE-SIZE PLANES should be labeled by the note, TRUE SIZE, or by the abbreviation, TS.

PROJECTION LINES should be precisely drawn with a 4H pencil as thin, gray lines. They should be drawn just dark enough to be visible and need not be erased after the problem is completed.

20.3
Descriptive geometry by computer

Four useful commands for solving descriptive geometry problems are covered in this section and documented in Appendix 51. The commands are PERPLINE, PARALLEL, TRANSFER, COPYDIST, and BISECT.*

While making a drawing in AutoCAD's Drawing Editor, any of these commands can be executed by typing its name. For example, Command: PERPLINE will enter the command for constructing a line perpendicular to the line of your choice by giving you a series of prompts to follow.

PERPLINE In Fig. 20.2, line 3–4 is drawn perpendicular to line 1–2 by locating the starting point, and next, the line to which the line is to be perpendicular.

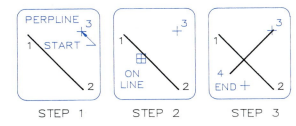

FIGURE 20.2 The PERPLINE **command.**

Step 1 Command: PERPLINE (CR) Select START point of perpendicular line: **(Select pt. 3.)**

Step 2 Select ANY point on line to which perp'lr: **(Select point on line.)**

Step 3 Select END point of desired perpendicular (for length only): **(Select pt. 4.) (3–4 is drawn.)**

The third point is the endpoint in the general area where the line is to end.

PARALLEL In Fig. 20.3, line 3–4 is drawn parallel to line 1–2 by locating the starting point of the parallel and the general location of its endpoint. Next, select the endpoints of line 1–2 in the same order that you want the parallel to be drawn; in this case, in the direction from 1 to 2.

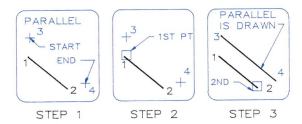

FIGURE 20.3 The PARALLEL **command.**

Step 1 Command: PARALLEL (CR) Select START point of parallel line: **(Select pt. 3.)** Select END point of parallel line: **(Select pt. 4.)**

Step 2 Select 1st point on line for parallelism: **(Select pt. 1.)**

Step 3 Select 2nd point on line for parallelism: **(Select pt. 2.) (3–4 is drawn.)**

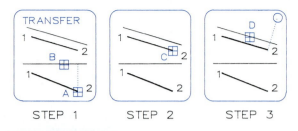

FIGURE 20.4 The TRANSFER **command.**

Step 1 Command: <u>TRANSFER</u> **(CR)**
Select start of transfer distance:
(Select pt. A.)
Select the reference plane: **(Select pt. B.)**

Step 2 Select point to be projected:
(Select pt. C.)

Step 3 Select other reference plane:
(Select pt. D.) (Point is located with a circle.)

TRANSFER The TRANSFER command (Fig. 20.4) is used to transfer distances from reference lines to solve descriptive geometry problems in the same manner that you would transfer the distances with your dividers. First select endpoint 2 in the front view, and then select the reference line. Select endpoint 2 in the top view and the auxiliary reference line, and the endpoint is located with a small circle in the auxiliary view.

COPYDIST A distance can be copied from a given position to another position as Fig. 20.5 shows. Select

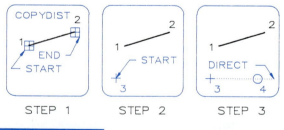

STEP 1   STEP 2   STEP 3

FIGURE 20.5 The COPYDIST **command.**

Step 1 Command: <u>COPYDIST</u> **(CR)**
Select start point of line distance
to be copied: **(Select End 1.)**
End point? : **(Select End 2.)**

Step 2 Start point of new distance location: **(Select pt. 3.)**

Step 3 Which direction?: **(Select with cursor, and endpoint 4 is located with a circle.)**

the endpoints of the line to be copied, and locate its beginning point in the new position. Next, locate the direction of the copied line, and the endpoint of the line will be scaled and located with a small circle.

BISECT An angle can be easily bisected (Fig. 20.6) by selecting the vertex of the angle, the first line, and the second line. The second line must be counter-clockwise from the first line selected. Locate the end-point when prompted, and the bisector will be drawn.

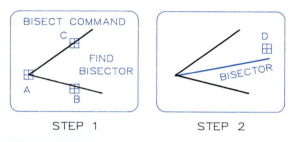

STEP 1   STEP 2

FIGURE 20.6 The BISECT **command.**

Step 1 Command: <u>BISECT</u> **(CR)**
Select corner of angle: **(Select pt. A.)**
Select first side (remember CCW):
(Select pt. B.)
Select other side: **(Select pt. C.)**

Step 2 Select endpoint of bisecting line (for length only): **(Select pt. D.)**
(The bisector is drawn.)

20.4
Orthographic projection of a point

A point is a theoretical location in space; it has no dimension. But a series of points can establish areas, volumes, and lengths.

A point must be projected perpendicularly onto at least two principal planes to establish its position (Fig. 20.7). When the planes of the projection box (Fig. 20.7A) are opened into the plane of the drawing surface (Fig. 20.7B), the projectors from each view of point 2 are perpendicular to the reference lines between the planes. The letters *H*, *F*, and *P* represent the horizontal, frontal, and profile planes, the three principal projection planes.

A point can be located from verbal descriptions with respect to the principal planes. For example,

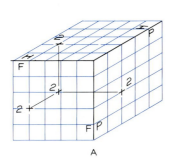
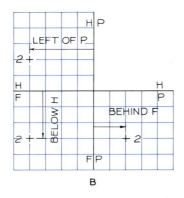
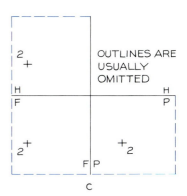

A   B   C

FIGURE 20.7   **A.** The three projections of point 2 are shown pictorially. **B.** The three projections are shown orthographically. The projection planes are opened into the plane of the drawing paper. Point 2 is four units to the left of the profile, three below the horizontal, and two behind the frontal. **C.** The outlines of the projection are usually omitted in orthographic projection.

point 2 in Fig. 20.7 can be described as (1) four units left of the profile plane, (2) three units below the horizontal plane, and (3) two units behind the frontal plane.

When looking at the front view, the horizontal and profile planes appear as edges. The frontal and profile planes appear as edges in the top view, and the frontal and horizontal planes appear as edges in the side view.

20.5
Lines

A line is a straight path between two points in space. A line can appear as (1) a foreshortened line, (2) a true-length line, or (3) a point (Fig. 20.8).

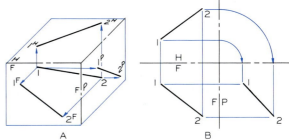

FIGURE 20.9   **A.** A pictorial of the orthographic projection of a line. **B.** A standard orthographic projection of a line.

OBLIQUE LINES   are neither parallel nor perpendicular to a principal projection plane (Fig. 20.9). When line 1–2 is projected onto the horizontal, frontal, and profile planes, it appears foreshortened in each view.

PRINCIPAL LINES   are parallel to at least one of the principal projection planes. A principal line is true length in the view where the principal plane it is parallel to appears true size. The three types of principal lines are horizontal, frontal, and profile lines.

A **horizontal line** is shown in Fig. 20.10A, where it appears true length in the horizontal view, the top view. It may be shown in an infinite number of positions in the top view and still appear true length provided it is parallel to the horizontal plane.

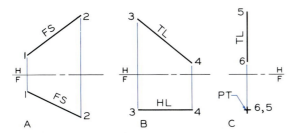

A   B   C

FIGURE 20.8   A line in orthographic projection can appear as a point (PT), foreshortened (FS), or true length (TL).

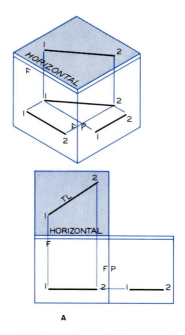

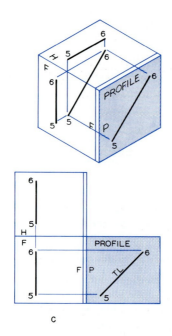

FIGURE 20.10 Principal lines.

A. The horizontal line is true length in the horizontal view (the top view). It appears parallel to the edge view of the horizontal plane in the front and side views.

B. The frontal line is true length in the front view. It appears parallel to the edge view of the frontal plane in the top and side views.

C. The profile line is true length in the profile view (the side view). It appears parallel to the edge view of the profile plane in the top and front views.

An observer cannot tell whether the line is horizontal when looking at the top view. This must be determined from looking at the front or side views where a horizontal line will be parallel to the edge view of the horizontal, which is the *H-F* fold line. A line that projects as a point in the front view is a combination horizontal and profile line.

A **frontal line** is parallel to the frontal projection plane, and it appears true length in the front view since the observer's line of sight is perpendicular to it in this view. Line 3–4 in Fig. 20.10B is determined to be a frontal line by observing its top and side views where the line is parallel to the edge view of the frontal plane.

A **profile line** is parallel to the profile projection planes, and it appears true length in the side views, the profile views. To tell whether a line is a profile line, it is necessary to look at a view adjacent to the profile view. In Fig. 20.10C, line 5–6 is parallel to the edge view of the profile plane in the top and side views.

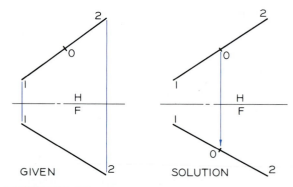

FIGURE 20.11 A point on a line can be found in the front view by projection. The direction of the projection is perpendicular to the reference line between the two views.

20.6
Location of a point on a line

Figure 20.11 shows the top and front views of line 1–2. Point 0 is located on the line in the top view, and it is required that the front view of the point be found.

Since the projectors between the views are perpendicular to the *H-F* fold line in orthographic projection, point 0 is found by projecting in this same direction from the top view to the front view of the line.

If a point is to be located at the midpoint of a line, it will be at the line's midpoint whether the line appears true length or foreshortened.

20.7
Intersecting and nonintersecting lines

Lines that intersect have a point of intersection that lies on both lines and is common to both. Point 0 in Fig. 20.12a is a point of intersection since it projects to a common crossing point in the three views.

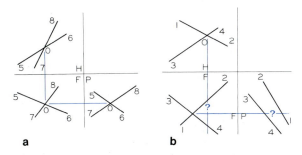

FIGURE 20.12 **a.** The lines intersect because the point of intersection, 0, projects as a point of intersection in all views. **b.** The lines cross in the top and front views, but they do not intersect. A common point of intersection does not project from view to view.

On the other hand, the crossing point of the lines in Fig. 20.12b in the front view is not a point of intersection. Point 0 does not project to a common crossing point in the top view. Therefore, the lines do not intersect although they do cross.

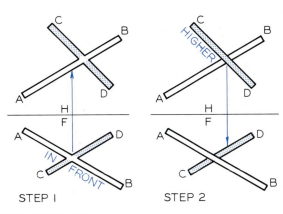

FIGURE 20.13 Visibility of lines.

Required Find the visibility of the lines in both views.

Step 1 Project the point of crossing from the front to the top view. This projector strikes *AB* before *CD*; therefore, line *AB* is in front and is visible in the front view.

Step 2 Project the point of crossing from the top view to the front view. This projector strikes *CD* before *AB*; therefore, *CD* is above *AB* and is visible in the top view.

20.8
Visibility of crossing lines

The visibility of nonintersecting lines *AB* and *CD* is found in Fig. 20.13. Select a crossing point in one of the views, the front view in Step 1, and project it to the top view to determine which line is in front of the other. This process of determining visibility is done by analysis rather than visualization. If only one view were available, it would be impossible to determine visibility.

20.9
Visibility of a line and a plane

The principle of visibility of intersecting lines is used in determining the visibility for a line and a plane. In Step 1 of Fig. 20.14, the intersections of *AB* and lines 1–3 and 2–3 are projected to the top view to determine that the lines of the plane are in front of *AB* in

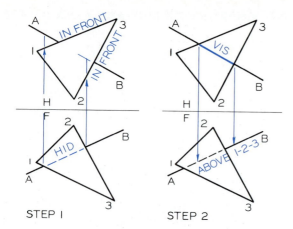

STEP 1   STEP 2

FIGURE 20.14   **Visibility of a line and plane.**

Required   Find the visibility of the plane and line in both views.

Step 1   Project the points where *AB* crosses the plane in the front view to the top view. These projectors encounter lines 1–3 and 2–3 of the plane first; therefore, the plane is in front of the line, making the line invisible in the front view.

Step 2   Project the points where *AB* crosses the plane in the top view to the front view. These projectors encounter *AB* first; therefore, the line is higher than the plane, and the line is visible in the top view.

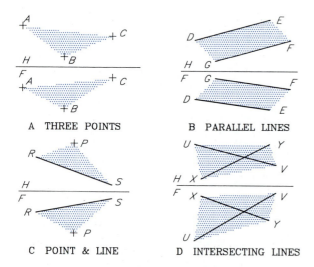

A   THREE POINTS   B   PARALLEL LINES

C   POINT & LINE   D   INTERSECTING LINES

FIGURE 20.15   **Representations of a plane.**

A.   Three points not on a straight line.

B.   Two parallel lines.

C.   A line and a point not on the line or its extension.

D.   Two intersecting lines.

the front view. Therefore, the line is shown as a dashed line in the front.

Similarly, the two intersections on *AB* in the top view are projected to the front view, where line *AB* is found to be above the two lines of the plane, 2–3 and 1–3. Since it is above the plane, *AB* is drawn as visible in the top view.

20.10
Planes

A plane can be represented by any of the four methods shown in Fig. 20.15. Planes in orthographic projection can appear as (1) an edge, (2) a true-size plane, and (3) a foreshortened plane (Fig. 20.16).

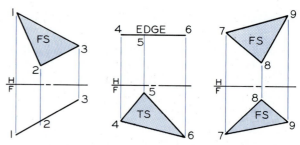

FIGURE 20.16   **A plane in orthographic projection can appear as an edge, true size (TS), or foreshortened (FS). If a plane is foreshortened in all principal views, it is an oblique plane.**

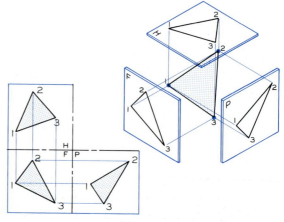

FIGURE 20.17   **An oblique plane is neither parallel nor perpendicular to a principal plane; it can be called a general-case plane.**

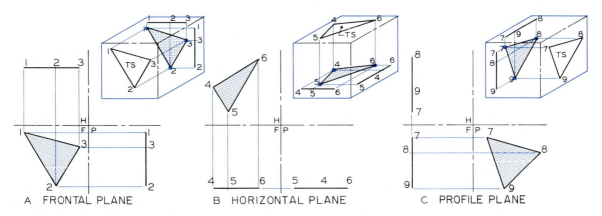

FIGURE 20.18   Principal planes.

A. The frontal line is true length in the front view. It will appear parallel to the edge view of the frontal plane in the top and side views.

B. The horizontal line is true length in the horizontal view (the top view). It will appear parallel to the edge view of the horizontal plane in the front and side views.

C. The profile line is true length in the profile view (the side view). It will appear parallel to the edge view of the profile plane in the top and front views.

OBLIQUE PLANES are neither parallel nor perpendicular to principal projection planes in any view (Fig. 20.17).

PRINCIPAL PLANES are parallel to projection planes (Fig. 20.18). The three types of principal planes are frontal, horizontal, and profile planes.

A **frontal plane** is parallel to the frontal projection plane, and it appears true size in the front view. To tell that the plane is frontal, you must observe the top or side views where its parallelism to the edge view of the frontal plane can be seen.

A **horizontal plane** is parallel to the horizontal projection plane, and it is true size in the top view. To tell that the plane is horizontal, you must observe the front or side views where its parallelism to the edge view of the horizontal plane can be seen.

A **profile plane** is parallel to the profile projection plane, and it is true size in the side view. To tell that it is a profile plane, you must observe the top or front views where its parallelism to the edge view of the profile plane can be seen.

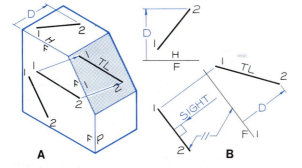

FIGURE 20.19   **A.** A pictorial of line 1–2 is shown inside a projection box where a primary auxiliary plane is established perpendicular to the frontal plane and parallel to the line. **B.** The orthographic arrangement shows the auxiliary view projected from the front view to find 1–2 true length.

20.11

Primary auxiliary view of a line

The top and front views of line 1–2 are shown pictorially and orthographically in Fig. 20.19. Since the line is not a principal line, it is not true length in the principal views. A primary auxiliary view, a view projected from a principal view, must be used to find its true-length view.

In Fig. 20.19B, the line of sight is drawn perpendicular to the front view of the line, and reference line F-1 is drawn parallel to its frontal view. You can see in the pictorial that the auxiliary plane is parallel to

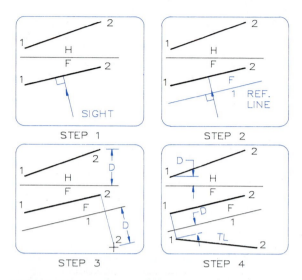

FIGURE 20.20 True length of a line.

Step 1 To find 1–2 true length, it must be viewed with a line of sight perpendicular to one of its views.

Step 2 The F-1 reference line is drawn parallel to the line and perpendicular to the line of sight.

Step 3 Project point 2 perpendicularly from the front view. Distance D from the top view locates point 2.

Step 4 Point 1 is similarly located. Line 1–2 is true length in the auxiliary view.

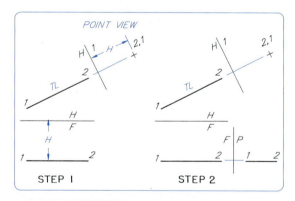

FIGURE 20.21 Point view of a line.

Step 1 The point view of a line can be found in a primary auxiliary view that is projected from the true-length view of the line.

Step 2 An auxiliary view projected from a foreshortened view of a line will result in a foreshortened view of the line, not a point view.

the line and perpendicular to the frontal plane, which accounts for its being labeled F-1.

The auxiliary view is found by projecting parallel to the line of sight and perpendicular to the F-1 reference line. Point 2 is found by transferring distance D with your dividers to the auxiliary view since the frontal plane appears as an edge in both the top and auxiliary views. Point 1 is located in the same manner, and the points are connected to find the true-length view of the line.

Figure 20.20 shows the steps required to find the true length of an oblique line. It is best to letter all reference planes using the notation suggested in Fig. 20.1.

A point view of a line can be found in a primary auxiliary view projected from a true-length view of a line (Fig. 20.21). The auxiliary view projected from the front view of the line does not give a point view since the line is foreshortened in the front view.

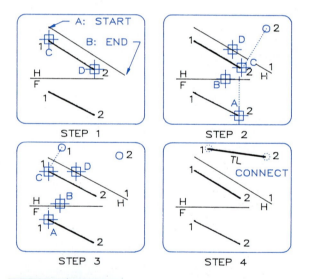

FIGURE 20.22 True length by computer.

Step 1 Command: <u>PARALLEL</u> **(CR)** (By following the prompts as covered in Section 19.2, draw the reference line parallel to CD.)

Step 2 Command: <u>TRANSFER</u> **(CR)** (Follow the prompts as covered in Section 20.2 to locate point 2 in the auxiliary view.)

Step 3 Locate point 1 in the auxiliary view using TRANSFER.

Step 4 Draw a line from the centers of the circles found in Steps 2 and 3 by using the CENTER option of the OSNAP command. The circles can be erased after this step.

COMPUTER METHOD Figure 20.22 illustrates how the PARALLEL and TRANSFER commands can be used to find a true-length view of a line by an auxiliary view. A line is drawn parallel to the top view of line 1–2 in Step 1. The endpoints of the line are TRANSFERred in Steps 2 and 3. To draw the ends of line 1–2 from the centers of the circles (Step 4), use the CENTER option of the OSNAP command. The circles can be erased afterward if you wish to remove them.

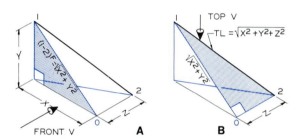

FIGURE 20.24 **A three-dimensional line that is foreshortened in a principal view can be found true length by the Pythagorean theorem in two steps. A. The frontal projection, 1–0, is found using the X and Y coordinates. B. The hypotenuse of the right triangle 1–0–2 is found using X, Y, and Z coordinates.**

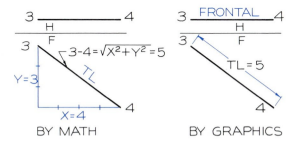

BY MATH   BY GRAPHICS

FIGURE 20.23 **A line that appears true length in a view (the front view in this case) can have its length calculated by applying the Pythagorean theorem. Since the line is a frontal line and is true length in the front view, its length can be measured graphically.**

20.13
The true-length diagram

A true-length diagram is constructed with two perpendicular lines to find a line of true length (Fig. 20.25). This method does not give a direction for the line but merely its true length.

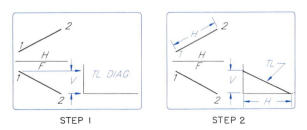

STEP 1   STEP 2

FIGURE 20.25 **True-length diagram.**

Step 1 **Transfer the vertical distance between the ends of 1–2 to the vertical leg of the TL diagram.**

Step 2 **Transfer the horizontal length of the line in the top view to the horizontal leg of the TL diagram.**

20.12
True length by analytical geometry

In Fig. 20.23, you can see that the length of a frontal line can be found in the front view by applying analytical geometry (mathematics) and the Pythagorean theorem, which states that the hypotenuse of a right triangle equals the square root of the sum of the squares of the other two sides.

The line shown pictorially in Fig. 20.24A can be found true length with analytical geometry by determining the length of the front view where the X and Y coordinates form a right triangle. In Fig. 20.24B, a second right triangle, 1–0–2, is solved to find its hypotenuse, which is the true length of 1–2. The true length of an oblique line is the square root of the sum of the squares of the X, Y, and Z coordinates that correspond to width, height, and depth.

The two measurements laid out on the true-length diagram can be transferred from any two adjacent orthographic views. One measurement is the distance between the endpoints in one of the views. The other measurement, taken from the adjacent view, is measured between the endpoints in a direction perpendicular to the reference line between the two views.

20.14
Slope of a line

> **Slope** is the angle a line makes with the horizontal plane.

Slope may be specified by any of the three methods shown in Fig. 20.26: **slope angle, percent grade,** or **slope ratio**.

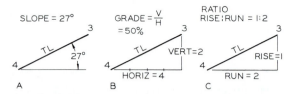

FIGURE 20.26 **The inclination of a line with the horizontal can be measured and expressed by any of three methods. A. Slope angle. B. Percent grade. C. Slope ratio.**

Slope angle

> The **slope** of a line can be measured in a view where the line is true length and the horizontal plane appears as an edge.

Thus the slope of line 3–4 in Fig. 20.26A can be measured in the front view, where Θ is found to be 27°.

Percent grade

The percent grade of a line is found in the view where the line is true length and the horizontal plane appears as an edge.

> **Grade** is the ratio of the vertical (rise) divided by the horizontal (run) between the ends of a line expressed as a percentage.

The percent grade of *AB* is found in Fig. 20.27 by using a combination of mathematics and graphics. Ten units are laid off parallel to the horizontal from *A* using a convenient scale with decimal units (Step 1).

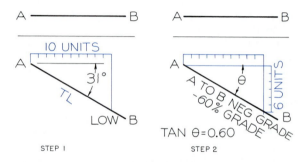

FIGURE 20.27 **Percent grade of a line.**

Step 1 **The percent grade of a line can be measured in the view where the horizontal appears as an edge and the line is true length (the front view here). Ten units are laid off parallel to the horizontal from the end of the line.**

Step 2 **A vertical distance from *A* to the line is measured to be six units. The percent grade is 6 divided by 10, or 60%. This is negative when the direction is from *A* to *B*. The tangent of this slope angle is $\frac{6}{10}$ or 0.60.**

The vertical drop of the line at ten units along the horizontal is measured as six units (Step 2). The tangent of the angle is 0.60, which is easily converted into a −60% grade. The grade is negative from *A* to *B* since this is downhill; it would be positive from *B* to *A* since it would be uphill. Line *CD* has a +50% grade from *C* to *D* in Fig. 20.28A.

Slope ratio

The first number of the slope ratio is the rise, and the second number is the run (see Fig. 20.26). In the slope ratio, the rise is always 1 (for example, 1:10, 1:200).

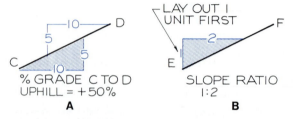

FIGURE 20.28 **A. The percent grade of a line is positive if uphill, and negative if downhill. B. The slope ratio is expressed as 1:XX, where 1 is the rise and XX is the horizontal distance. The 1 unit must be drawn before the horizontal distance.**

The graphical method of finding the slope ratio is shown in Fig. 20.28B, where the rise of 1 is laid off on the true-length view of *EF.* The corresponding horizontal is found to be 2, which results in a slope ratio of 1:2.

OBLIQUE LINES An oblique line does not appear true length in the front view; therefore, it must be found true length by projecting from the top view to find its slope. This auxiliary view shows the horizontal as an edge and the line true length, making it possible to measure the slope angle (Fig. 20.29A).

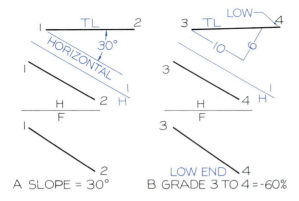

FIGURE 20.29 Slope of an oblique line.

A. The slope angle can be measured in a view where the horizontal appears as an edge and the line is true length. The slope of 30° is found in an auxiliary view projected from the top view.

B. The percent grade can be measured in a true-length view of the line projected from the top view. Line 3–4 has a −60% grade from 3 to the low end at 4.

Similarly, an auxiliary view projected from the top view must be used to find the percent grade of an oblique line (Fig. 20.29B). Ten units are laid off horizontally, parallel to the H-1 reference line, and the vertical distance is found to be 6, or a −60% grade downhill from 3 to 4.

20.15
Compass bearing of a line

Two types of bearings of a line's direction are **compass bearings** and **azimuth bearings** (Fig. 20.30).

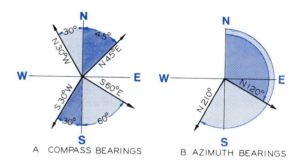

A COMPASS BEARINGS   B AZIMUTH BEARINGS

FIGURE 20.30 A. Compass bearings are measured with respect to north and south directions on the compass. B. Azimuth bearings are measured with respect to north in a clockwise direction up to 360°.

> Compass bearings always begin with the north or south directions, and the angles with north and south are measured toward east or west.

The line in Fig. 20.30A that makes 30° with north has a bearing of N 30° W. A line making 60° with south toward the east has a compass bearing of south 60° east, or S 60° E. Since a compass can be read only when held horizontally, the compass bearings of a line can be determined only in the top, or horizontal view.

> An azimuth bearing is measured from north in clockwise direction to 360° (Fig. 20.30B).

Azimuth bearings of a line are written N 120°, N 210°, and so forth, with this notation indicating that the measurements are made from the north.

> The compass bearing (direction) of a line is assumed to be toward the low end of the line unless otherwise specified.

For example, line 2–3 in Fig. 20.31 has a bearing of N 45° E since the line's low end is point 3. It can be seen in the front view that point 3 is the lower end.

The compass bearing and slope of a line are found in Fig. 20.32. This information can be used to verbally describe the line as having a compass bearing of S 60° E and a slope of 30° from 5 to 6.

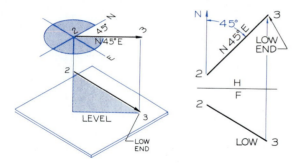

FIGURE 20.31 The compass bearing of a line is measured in the top view toward its low end (unless specified toward the high end). Line 2–3 has a bearing of N 45° E toward the low end at 3.

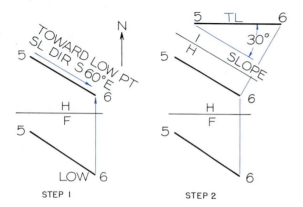

FIGURE 20.32 Slope and bearing of a line.

Step 1 Slope bearing can be found in the top view toward its low end. Direction of slope is S 60° E.

Step 2 The slope angle of 30° is found in an auxiliary view projected from the top view where the line is found true length.

20.16

Edge view of a plane

> The edge view of a plane can be found in a view where any line on the plane appears as a point.

A line can be found as a point by projecting from a true-length view of the line (Fig. 20.33).

A true-length line can be found on any plane by drawing the line parallel to one of the principal planes

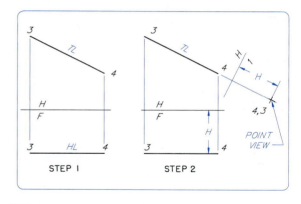

FIGURE 20.33 Point view of a line.

Step 1 Line 3–4 is horizontal in the front view and is therefore true length in the top view.

Step 2 The point view of 3–4 is found by projecting from the top view to the auxiliary view.

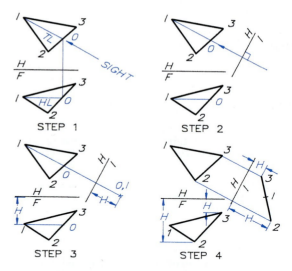

FIGURE 20.34 Edge view of a plane.

Step 1 By finding the point view of a line on plane 1–2–3, you will find its edge view. Draw a horizontal line 1–0 on the plane in the front view, and project it to the top view where it is true length. The line of sight is parallel to the TL line.

Step 2 Draw the H-1 reference line perpendicular to the TL line.

Step 3 The point view of line 1–0 is found by projecting from the top view. Locate 1 by transferring dimension H from the front view.

Step 4 Locate points 2 and 3 by transferring their height dimensions from the front view.

and projecting it to the adjacent view, as shown in Fig. 20.34 (Step 1). Since line 3–4 is true length in the top view, its point view may be found, and the plane will appear as an edge in this auxiliary view.

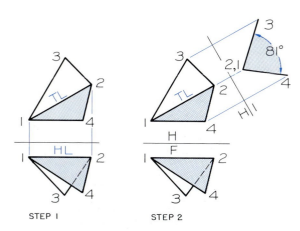

FIGURE 20.35 Dihedral angle.

Step 1 The line of intersection between the planes, 1–2, is true length in the top view.

Step 2 The angle between the planes (the dihedral angle) can be found in the auxiliary view where the line of intersection appears as a point.

20.17
Dihedral angles

> A **dihedral angle,** the angle between two planes, can be found in the view where the line of intersection between two planes appears as a point.

The line of intersection, line 1–2, between the two planes in Fig. 20.35 is true length in the top view. This makes it possible to find the point view of line 1–2 and the edge view of both planes in a primary auxiliary view.

20.18
Piercing points by auxiliary views

The piercing point of a line and plane can be found by an auxiliary view (Fig. 20.36). Piercing point P can be seen in Step 2, where the plane is found as an edge. Point P is projected back to the line in the top and front views from the auxiliary view in Step 3. The location

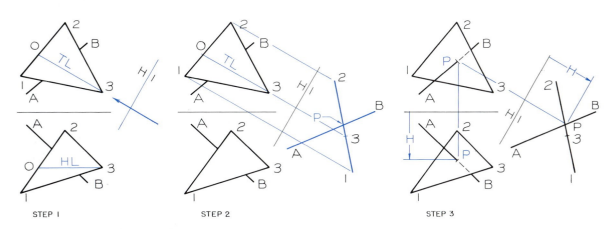

FIGURE 20.36 Piercing points by auxiliary view.

Step 1 Draw a horizontal line on the plane in the front view, project it to the top view to find TL line 0–3 on the plane. The line of sight is drawn parallel to the TL line.

Step 2 Find the edge view of the plane, and project line AB to this view. Point P is the piercing point in the auxiliary view.

Step 3 Point P is projected to the top and front views. Point A of the auxiliary view is the highest end, and AP will be visible in the top view. Since end B in the top view is the farthest back, it is hidden in the front view.

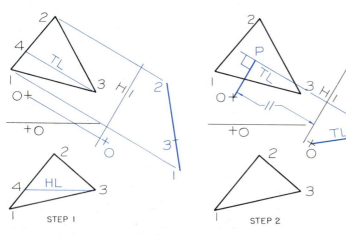

FIGURE 20.37 Line perpendicular to a plane.

Step 1 Find the edge view of the plane by finding the point view of a line on it, 3–4. Project from either view. Project point 0 also.

Step 2 Line 0P is drawn perpendicular to the edge view of the plane. Since 0P is TL in the auxiliary view, it will be parallel to the H-1 line in the top view and perpendicular to a TL line on the plane.

Step 3 Piercing point P is found in the front view by projecting from the top view. Point P is accurately located by transferring dimension H from the auxiliary view to the front view.

of *P* in the front view is checked by transferring dimension *H* from the auxiliary view with your dividers.

Visibility is easily determined for the top view since it can be seen that *AP* is higher than the plane in the auxiliary view and is therefore visible in the top view. Analysis of the top view shows that endpoint *A* is the most forward point; therefore, *AP* is visible in the front view.

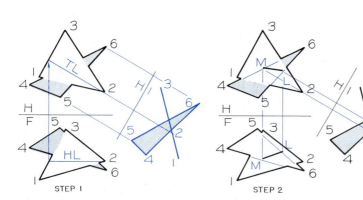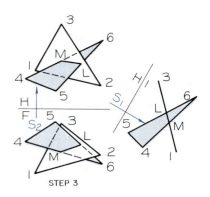

FIGURE 20.38 Intersection by auxiliary view.

Step 1 Find the edge view of one of the planes, and project the other plane to this view also.

Step 2 Piercing points *L* and *M* can be seen on the edge view of the plane. *LM* is projected to the top and front views.

Step 3 The line of sight from the top view strikes 1–5 first in the auxiliary view, which is 1–5 means visible in the top view. Line 4–5 is farthest forward in the top view and is visible in the front view.

20.19
Perpendicular to a plane

In Fig. 20.37, it is required that a line be drawn from point 0 perpendicular to the plane.

> A perpendicular line will appear true length and perpendicular to a plane where the plane appears as an edge.

The true-length perpendicular is drawn in Step 2 to locate piercing point *P*. Point *P* is found in the top view by drawing line *0P* parallel to the H-1 reference line. It must lie in this direction since *0P* is true length in the auxiliary view. Line *0P* will also be perpendicular to a true-length line in the top view of the plane.

20.20
Intersections by auxiliary view

The intersection between planes can be found by finding the edge view of one of the planes, as shown in Fig. 20.38 (Step 1). Piercing points *L* and *M* are projected from the auxiliary view to their respective lines, 5–6 and 4–6, in the top view (Step 2).

The visibility of plane 4–5–6 in the top view is apparent by inspecting the auxiliary view, where sight line S_1 has an unobstructed view of the 4–*L*–*M* portion of the plane. Plane 4–5–*L*–*M* is visible in the front view since sight line S_2 has an unobstructed view of the top view of this portion of the plane.

20.21
Slope of a plane

Planes can be established by using verbal specifications of slope and direction of slope of a plane.

SLOPE is the angle the plane's edge view makes with the edge of the horizontal plane.

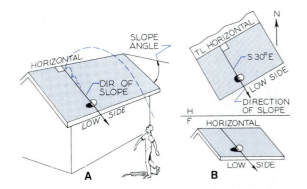

FIGURE 20.39 The direction of slope of a plane is the compass bearing of a line on the plane. This is measured in the top view toward the low side of the plane.

DIRECTION OF SLOPE is the compass bearing of a line perpendicular to a true-length line in the top view of a plane toward its low side. This is the direction in which a ball would roll on the plane. It can be seen in Fig. 20.39A that a ball would roll perpendicular to all horizontal lines of the roof toward the low side. The slope is seen when the roof and horizontal are edges in a single view.

Figure 20.40 gives the steps of finding the direction of slope and the slope angle of a plane. Under-

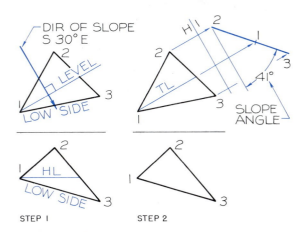

FIGURE 20.40 Slope and direction of slope of a plane.

Step 1 Slope direction is perpendicular to a true-length level line in the top view toward the low side of the plane, S 30° E in this case.

Step 2 Slope is measured in an auxiliary view where the horizontal is an edge and the plane is an edge, 41° in this case.

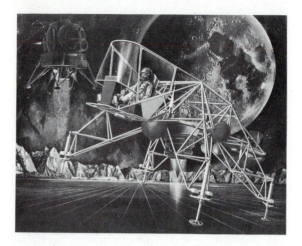

FIGURE 20.41 **The lengths of each structural member and the angles between them must be determined by descriptive geometry during the design process. (Courtesy of Bell Aerosystems Company.)**

standing these terms enables you to verbally describe a sloping plane.

20.22
Successive auxiliary views

A design cannot be detailed with complete specifications unless its complete geometry has been determined, which usually requires applying descriptive geometry. The drawing of a lunar vehicle in Fig. 20.41

reveals many lengths and shapes that must be determined by descriptive geometry before the vehicle can be designed and built.

> A **secondary auxiliary view** is an auxiliary view projected from a primary auxiliary view. A **successive auxiliary view** is an auxiliary view of a secondary auxiliary view.

20.23
Point view of a line

When a line appears true length, its point view can be found by projecting an auxiliary view from it. In Fig. 20.42, line 1–2 is true length in the top view since it is horizontal in the front view. Its point view is found in the primary auxiliary view by constructing reference line H-1 perpendicular to the true-length line. The height dimension, H, is transferred to the auxiliary view to locate the point view of line 1–2.

Since the line in Fig. 20.43 is not true length in either view, the line must be found true length by a primary auxiliary view before its point view can be found. By projecting from the front view, the line is found true length. This view could have been pro-

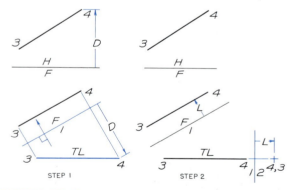

FIGURE 20.43 **Point view of an oblique line.**

Step 1 A line of sight is drawn perpendicular to one of the views, the front view in this example. Line 3–4 is found true length by projecting perpendicularly from the front view.

Step 2 A secondary reference line, 1–2, is drawn perpendicular to the true-length view of 3–4. The point view is found by transferring dimension L from the front view to the secondary auxiliary view.

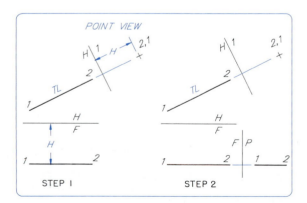

FIGURE 20.42 **The point of view of a line can be found by projecting an auxiliary view from the true-length view of the line.**

jected from the top as well. The point view of the line is found by projecting from the true-length line to a secondary auxiliary view and is labeled 4, 3 since point 4 is seen first.

20.24
Angle between planes

> The angle between two planes is a **dihedral angle.** The dihedral angle can be found in the view where the line of intersection appears as a point.

Since this view results in the point view of a line that lies on both planes, both will appear as edges.

The two planes in Fig. 20.44 represent a special case since the line of intersection, 1–2, is true length in the top view. This permits you to find its point view in a primary auxiliary view where the true angle can be measured.

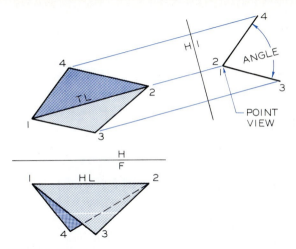

FIGURE 20.44 The angle between two planes can be found in the view where the line of intersection between them projects as a point. Since the line of intersection, 1–2, is true length in the top view, it can be found as a point in a view projected from the top view.

A more typical case is shown in Fig. 20.45, where the line of intersection between the two planes is not true length in either view. The line of intersection, 1–2, is found true length in a primary auxiliary view,

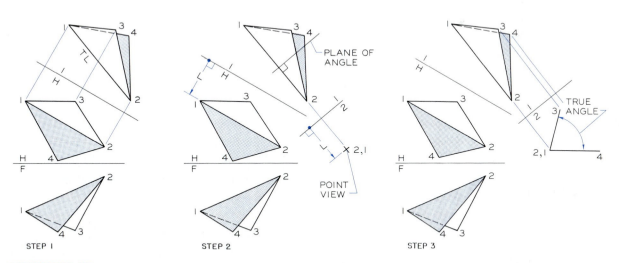

FIGURE 20.45 Angle between two planes.

Step 1 The angle between two planes can be measured in a view where the line of intersection appears as a point. The line of intersection is found true length by projecting a primary auxiliary view perpendicularly from the top view.

Step 2 The point view of the line of intersection is found in the secondary auxiliary view by projecting parallel to the true length view of 1–2. The plane of the dihedral angle is an edge and is perpendicular to the true-length line of intersection.

Step 3 The edge views of the planes are completed in the secondary auxiliary view by locating points 3 and 4. The angle between the planes, the dihedral angle, can be measured in this view.

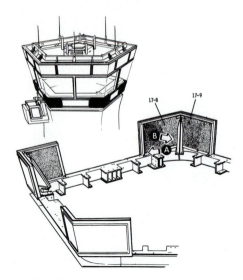

FIGURE 20.46 **Determining the angle between the planes of the corner panels of the control tower uses principles of descriptive geometry. (Courtesy of the Federal Aviation Agency.)**

and the point view of the line is then found in the secondary auxiliary view, where the dihedral angle is measured.

This principle must be used to determine the angles between side panels of a control tower (Fig. 20.46) so that corner braces can be designed that will hold the structure together.

20.25
True size of a plane

A plane can be found true size in a view projected perpendicularly from an edge view of a plane.

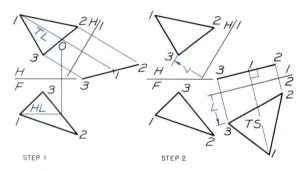

FIGURE 20.48 **True size of a plane.**

Step 1 **The edge view of plane 1–2–3 is found by finding the point view of TL line, 1–0, in a primary auxiliary view.**

Step 2 **A true-size view is found by projecting a secondary auxiliary view perpendicularly from the edge view of the plane. Dimension L is a typical measurement used to complete the TS view.**

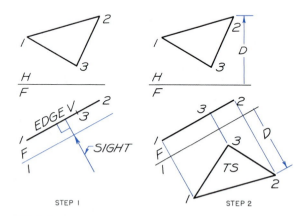

FIGURE 20.47 **True size of a plane.**

Step 1 **Since plane 1–2–3 appears as an edge in the front view, the line of sight is drawn perpendicular to the edge. The F–1 reference line is drawn parallel to the edge.**

Step 2 **The plane will appear true size in the primary auxiliary by locating the vertex points with depth *(D)* dimensions.**

Plane 1–2–3 appears as an edge in the front view (Fig. 20.47). It can be found true size in a primary auxiliary view projected perpendicularly from the edge view.

In Fig. 20.48, the true size of plane 1–2–3 is found by first finding the edge view of the plane (Step 1). The secondary auxiliary view is then projected perpendicularly from the edge view (Step 2) to find the true-size view of the plane where each angle is true size.

This principle can be used to find the angle between lines, such as bends in a fuel line of an aircraft

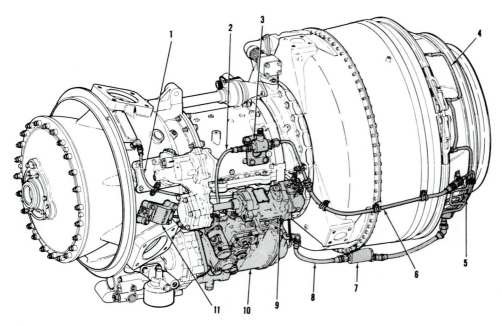

FIGURE 20.49 The angles of bend in the fuel line were found by applying the principle of finding the angle between two lines. (Courtesy of Avco Lycoming.)

engine (Fig. 20.49). This type of problem is shown in Fig. 20.50, where the top and front views of intersecting lines are given. It is required that the angles of bend be determined and a radius of curvature shown.

Angle 1–2–3 is found as an edge in the primary auxiliary view and true-size in the secondary view, where the angle can be measured and the radius of curvature drawn.

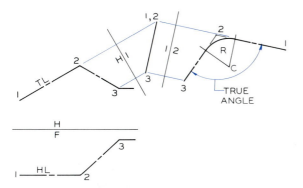

FIGURE 20.50 The angle between two lines can be found by finding the plane of the lines true size.

20.26

Shortest distance from a point to a line

> The shortest distance from a point to a line can be measured in the view where the line appears as a point.

The shortest distance from point 3 to line 1–2 is found in a primary auxiliary view in Fig. 20.51 (Step 1). Since the distance from point 3 to the line is true length where the line is a point, it must be parallel to reference line F-1 in the front view.

This type of problem is solved in Fig. 20.52 by finding the line 1–2 true length in a primary auxiliary. The point view of the line is found in the secondary auxiliary view, where the distance from point 3 is true length. Since line 0–3 is true length in this view, it must be parallel to the 1–2 reference line in the preceding view, the primary auxiliary view. It is also perpendicular to the true-length view of line 1–2 in the primary auxiliary view.

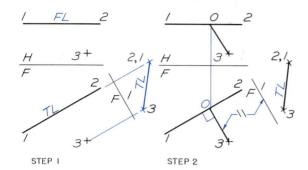

STEP 1   STEP 2

FIGURE 20.51 Shortest distance from a point to a line.

Step 1 The shortest distance from a point to a line will appear TL where the line appears as a point. The TL line is found in the primary auxiliary view.

Step 2 For the connecting line to be TL in the auxiliary view, it must be parallel to the F-1 reference line in the preceding view, the front view. Line 3–0 is found and projected to the top view.

20.27
Shortest distance between skewed lines—line method

> Randomly positioned (nonparallel) lines are **skewed lines.** The shortest distance between two skewed lines can be measured in the view where one of the lines appears as a point.

The shortest distance between two lines is perpendicular to both lines. The location of the shortest distance is both functional and economical, as demonstrated by the connector between two pipes in Fig. 20.53 since a standard connector is a 90° tee.

A problem of this type is solved by the **line method.** In Fig. 20.54, line 3–4 is found as a point in the secondary auxiliary view, where the shortest distance is drawn perpendicular to line 1–2. Since the distance is true length in the secondary auxiliary view, it must be parallel to the 1–2 reference line in the pri-

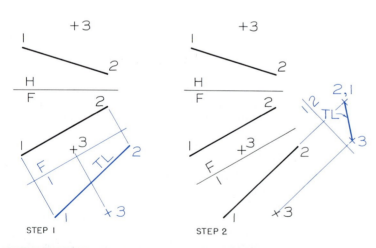
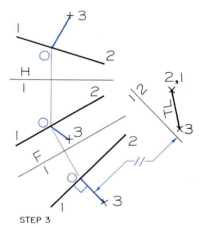

STEP 1   STEP 2   STEP 3

FIGURE 20.52 Shortest distance from a point to a line.

Step 1 The shortest distance from a point to a line can be found in the view where the line appears as a point. Line 1–2 is found true length by projecting from the front view.

Step 2 Line 1–2 is found as a point in a secondary auxiliary view projected from the true-length view of 1–2. The shortest distance appears true length in this view.

Step 3 Since 3–0 is true length in the secondary auxiliary view, it is parallel to the 1–2 line in the primary auxiliary view and perpendicular to the line. The front and top views of 3–0 are found by projection.

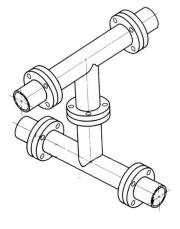

FIGURE 20.53 The shortest distance between two lines is a line perpendicular to both. This is the most economical and functional connection since perpendicular connectors are standard.

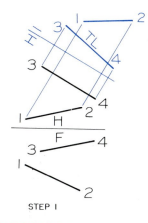

STEP I

STEP 2

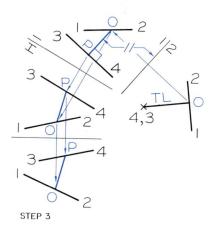

STEP 3

FIGURE 20.54 Shortest distance between skewed lines—line method.

Step 1 The shortest distance between two skewed lines can be found in the view where one of the lines appears as a point. Line 3–4 is found true length by projecting from the top view along with line 1–2.

Step 2 The point view of line 3–4 is found in a secondary auxiliary view projected from the true-length view of 3–4. The shortest distance between the lines is drawn perpendicular to line 1–2.

Step 3 Since the shortest distance is TL in the secondary auxiliary view, it must be parallel to the reference line in the preceding view. Points 0 and P are projected to the given view.

mary auxiliary view. Point 0 is found by projection, and 0P is drawn perpendicular to line 3–4. The line is projected back to the given principal views.

This principle of the shortest distance between two skewed lines was applied to the design of powerlines to ensure that they were properly separated to provide the necessary clearance (Fig. 20.55).

20.28
Angular distance to a line

Standard connectors used to connect pipes and structural members are available in two standard angles— 90° and 45°. It is far more economical to incorporate

auxiliary view. The angle can be measured in this view where the plane of the line and point is true size.

The 45° connector is drawn from 0 to the line toward point 2 if it slopes downhill or toward point 1 if it slopes uphill. Slope can be determined by referring to the front view where height can be easily seen. Line 0P is projected back to the given views.

20.29
Angle between a line and a plane—plane method

The angle between a line and a plane can be measured in the view where the plane appears as an edge and the line appears true length.

In Fig. 20.57, the edge view of the plane is found in a primary auxiliary view projected from any primary view. The plane is then found true size in Step 2, where the line is foreshortened. Line *AB* can be found true length in a third auxiliary view projected perpendicularly from the secondary auxiliary view of *AB*. The line appears true length, and the plane appears as an edge in the third successive auxiliary view.

these angles into a design than to design specially made connectors.

In Fig. 20.56, it is required to locate the point of intersection on line 1–2 of a line drawn from point 0 at a 45° angle to the line. The plane of the line and point, 1–2–0, is found as an edge in the primary auxiliary view and as a true-size plane in the secondary

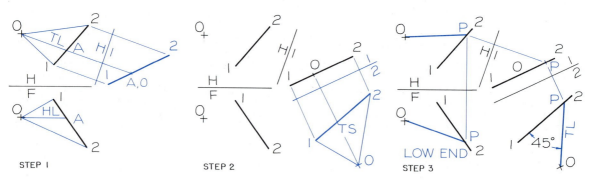

STEP 1   STEP 2   STEP 3

FIGURE 20.56 Line through a point with a given angle to a line.

Step 1 Connect 0 to each end of the line to form a plane 1–2–0 in both views. Draw a horizontal line in the front view of the plane, and project it to the top view, where it is TL. Find the edge view of the plane by obtaining the point view of *A*0.

Step 2 Find the true size of plane 1–2–0 by projecting perpendicularly from the edge view of the plane in the primary auxiliary view. The plane can be omitted in this view and only line 1–2 and point 0 are shown.

Step 3 Line 0P is constructed at the specific angle with the line 1–2, 45°. If the angle is toward point 2, the line slopes downhill; if toward point 1, it slopes uphill. Point P is projected back to the other views in sequence.

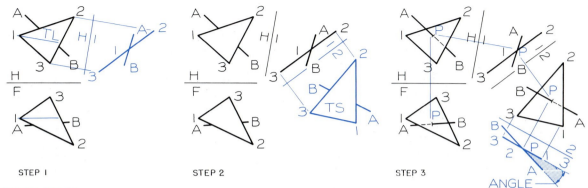

FIGURE 20.57 Angle between a line and a plane—plane method.

Step 1 The angle between a line and a plane can be measured in the view where the plane is an edge and the line is TL. The plane is found as an edge by projecting off the top view. The line is not true length in this view.

Step 2 The plane is found true size by projecting perpendicularly from the edge view of the plane. A view projected in any direction from a TS plane will show the plane as an edge.

Step 3 A third successive auxiliary view is projected perpendicularly from line *AB*. The line appears as an edge in this view where the angle is measured. The piercing points and visibility are shown by projecting back in sequence to all views.

20.30

Intersections and developments

In this section, we discuss the methods of finding lines of **intersections** between parts that join. Usually these parts are made of sheet metal or plywood if used to form concrete.

Once the intersections have been determined, **developments** can be found. These flat patterns can be laid out on the sheet metal and cut to conform to the desired shape. You can see many examples of intersections and developments in Fig. 20.58, where a refinery is under construction.

FIGURE 20.58 This storage tank facility involves many applications and principles of intersections and developments. (Courtesy of Phillips Petroleum Company.)

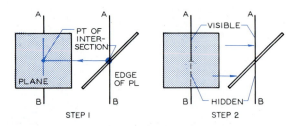

FIGURE 20.59 Intersection of a line and a plane.

Step 1 The point of intersection can be found in the view where the plane appears as an edge, the side view in this example.

Step 2 Visibility in the front view is determined by looking from the front view to the right-side view.

20.31
Intersections of lines and planes

Figure 20.59 shows the steps of finding the intersection between a line and plane. This is a special case where the point of intersection can be easily seen since the plane appears as an edge (Step 1). It is projected to the front view, and the visibility of the line is found (Step 2).

This principle is used in Fig. 20.60 to find the line of intersection between two planes. Since *EFGH* ap-

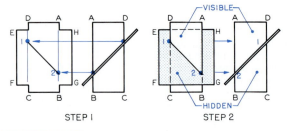

FIGURE 20.60 Intersection between planes.

Step 1 The points where lines *AB* and *DC* intersect plane *EFGH* are found where the plane appears as an edge. These points are projected to the front view.

Step 2 Line 1–2 is the line of intersection. Visibility is determined by looking from the front view to the right-side view.

pears as an edge in the side view, points of intersection 1 and 2 can be found, projected to the front view, and the visibility determined (Step 2). The intersection was found by locating the piercing points of lines *AB* and *DC* and connecting these points.

The intersection of a plane at a corner of a prism results in a line of intersection that bends around the corner (Fig. 20.61). Piercing points 2′ and 1′ and found in Step 1.

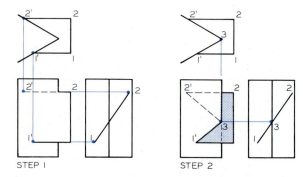

FIGURE 20.61 Intersection of a plane at a corner.

Step 1 The intersecting plane appears as an edge in the side view. Intersection points 1′ and 2′ are projected from the top and side views to the front view.

Step 2 The line of intersection from 1′ to 2′ must bend around the vertical corner at 3′ in the top and side views. Point 3′ is projected to the front view to locate line 1′–2′–3′.

Corner point 3 is seen in the side view of Step 2, where the vertical corner pierces the plane. Point 3 is projected to the corner in the front view. Point 2′ is hidden in the front view since it is on the back side.

The intersection of a plane and prism is found in Fig. 20.62, where the plane appears as an edge. The points of intersection are found for each corner line and are connected; visibility is shown to complete the line of intersection.

An intersection between a plane and prism is shown in Fig. 20.63, where the vertical corners of the prism are true length in the front view, and the plane appears foreshortened in both views. Vertical cutting planes are passed through the planes of the prism in the top view to find the piercing points of the corners

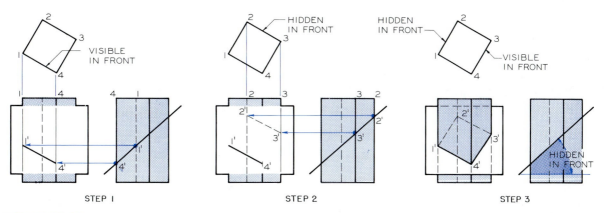

FIGURE 20.62 Intersection of a plane and a prism.

Step 1 Vertical corners 1 and 4 intersect the edge view of the plane in the side view at 1′ and 4′. These points are projected to the front view and are connected with a visible line.

Step 2 Vertical corners 2 and 3 intersect the edge of the inclined plane at 2′ and 3′ in the side view. These points are connected in the front view with a hidden line.

Step 3 Lines 1′–2′ and 3′–4′ are drawn as hidden and visible lines, respectively. Visibility is determined by inspecting the top and side views.

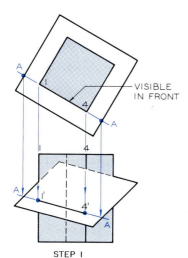

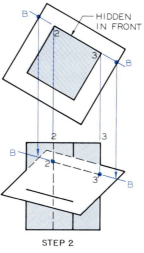

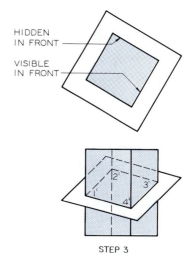

FIGURE 20.63 Intersection of an oblique plane and a prism.

Step 1 Vertical cutting plane A–A is passed through the vertical plane, 1–4, in the top view and projected to the front view. Piercing points 1′ and 4′ are found in this view.

Step 2 Vertical plane B–B is passed through the top view of plane 2–3 and projected to the front view where piercing points 2′ and 3′ are found. Line 2′–3′ is a hidden line.

Step 3 The line of intersection is completed by connecting the four points in the front view. Visibility in the front view is found by inspecting the top view.

in the front view. The points are connected, and the visibility is determined to complete the solution.

The intersection between a foreshortened plane and an oblique prism is found in Fig. 20.64. The plane is found as an edge in a primary auxiliary view. The piercing points of the corners of the prism are located in the auxiliary view and are projected back to the given views.

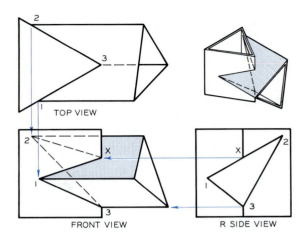

FIGURE 20.65 Three views of intersecting prisms. The points of intersection can be seen where intersecting planes appear as edges.

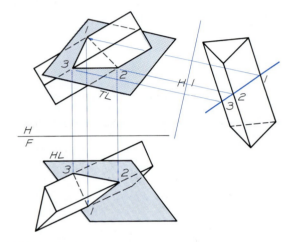

FIGURE 20.64 The intersection between a plane and a prism can be found by constructing a view in which the plane appears as an edge.

Points 1, 2, and 3 are projected from the auxiliary view to the given views as examples. Visibility is determined by analysis of crossing lines.

In Fig. 20.66, an inclined prism intersects a vertical prism. The end view of the inclined prism is found by an auxiliary view (Step 1). In the auxiliary view, you can see where plane 2–3 bends around corner AB at point X (Step 2). Points of intersection 1' and 2' are projected from the top to the front view. The line of intersection 2'–X–3' can be drawn to complete this portion of the line of intersection. The remaining lines, 1'–3' and 2'–3', are connected to complete the solution (Step 3).

> The conduit connector in Fig. 20.67 is an example of intersecting planes and prisms.

20.32
Intersections between prisms

The same principles used to find the intersection between a plane and line are used to find the intersection between two prisms in Fig. 20.65. Piercing points 1, 2, and 3 are found in the front view by projecting from the side and top views. Point X is located in the side view where line of intersection 1–2 bends around the vertical corner of the other prism. Points 1, X, and 2 are connected in sequence, and the visibility is determined.

20.33
Intersection of a plane and cylinder

The intersections of the components of the gas transmission system shown in Fig. 20.68 offer many applications of the principles of intersections.

The intersection between a plane and cylinder is found in Fig. 20.69, where the plane appears as an edge in one of the views. Cutting planes are passed

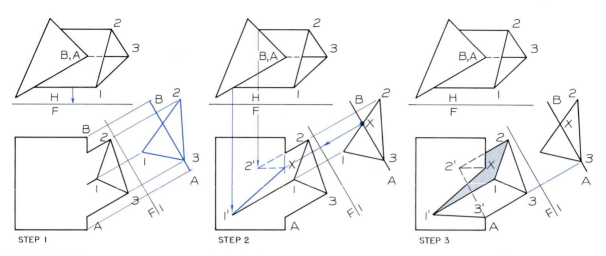

FIGURE 20.66 Intersection between two prisms.

Step 1 Construct the end view of the inclined prism by projecting an auxiliary view from the front view. Show only line *AB* of the vertical prism in the auxiliary view.

Step 2 Locate piercing points 1′ and 2′ in the top and front views. Intersection line 1′–2′ will bend around corner *AB* at point *X*, which is projected from the auxiliary view.

Step 3 Intersection lines from 2′ and 1′ to 3′ do not bend around the corner. Therefore, these are drawn as straight lines. Line 1′–3′ is visible, and line 2′–3′ is invisible.

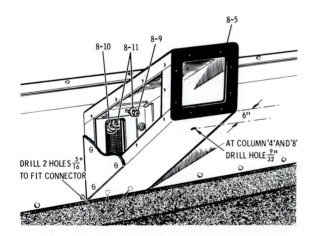

FIGURE 20.67 This conduit connector was designed using the principles of the intersection of a plane and prism. (Courtesy of the Federal Aviation Administration.)

FIGURE 20.68 This complex of pipes and vessels contains many applications of intersections. (Courtesy of Phillips Petroleum Company.)

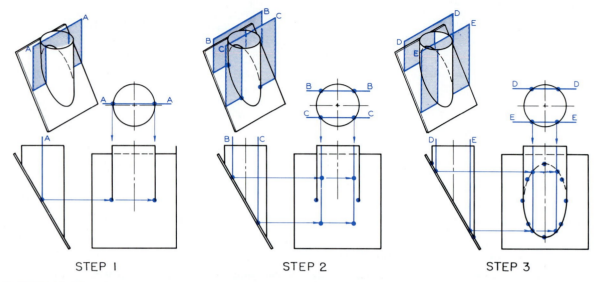

STEP 1   STEP 2   STEP 3

FIGURE 20.69 Intersection between a cylinder and a plane.

Step 1 A vertical cutting plane, A–A, is passed through the cylinder parallel to its axis to find two points of intersection.

Step 2 Cutting planes, B–B and C–C, are used to find four additional points in the top and left-side views; these points are projected to the front view.

Step 3 Additional cutting planes are used to find more points. These points are connected to give an elliptical line of intersection.

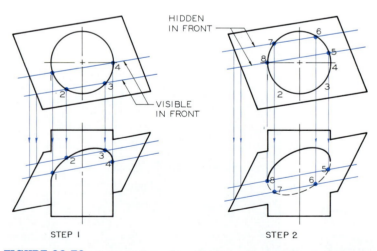

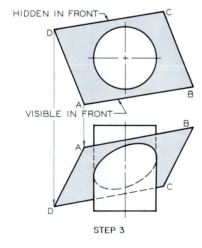

STEP 1   STEP 2   STEP 3

FIGURE 20.70 Intersection of a cylinder and an oblique plane.

Step 1 Vertical cutting planes are passed through the cylinder in the top view to establish elements on its surface and lines on the oblique plane. Piercing points 1, 2, 3, and 4 are projected to the front view of their respective lines and are connected with a visible line.

Step 2 Additional cutting planes are used to find other piercing points—5, 6, 7, and 8—which are projected to the front view of their respective lines on the oblique plane. They are connected with a hidden line.

Step 3 Visibility of the plane and cylinder is completed in the front view. Line AB is found to be visible by inspecting the top view, and line CD is found to be hidden.

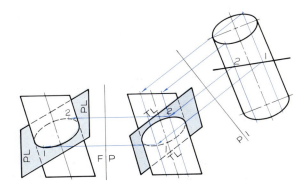

FIGURE 20.71 The intersection between an oblique cylinder and an oblique plane can be found by constructing a view that shows the plane as an edge.

vertically through the top view of the cylinder to establish elements on the cylinder and their piercing points. The piercing points are projected to each view to find the line of intersection, which is an ellipse.

A more general problem is solved in Fig. 20.70, where the cylinder is vertical, but the plane is oblique. Vertical cutting planes are passed through the cylinder and the plane in the top view to find piercing points

of the cylinder's elements on the plane. These points are projected to the front view to complete the line of intersection, an ellipse. The more cutting planes used, the more accurate the line of intersection.

The general case of the intersection between a plane and cylinder is solved in Fig. 20.71, where both are oblique in the given views. The edge view of the plane is found in an auxiliary view. Cutting planes are passed through the cylinder parallel to the cylinder's axis in the auxiliary view to find the piercing points. The piercing points of the elements are connected to give elliptical lines of intersection in the given views. Visibility is determined by analysis.

20.34
Intersections between cylinders and prisms

A series of vertical cutting planes is used in Fig. 20.72 to establish lines that lie on the surfaces of the cylinder and prism. A primary auxiliary view is drawn to show the end view of the inclined prism. Also shown in this view are the vertical cutting planes, which are spaced the same distance apart as in the top view (Step 1).

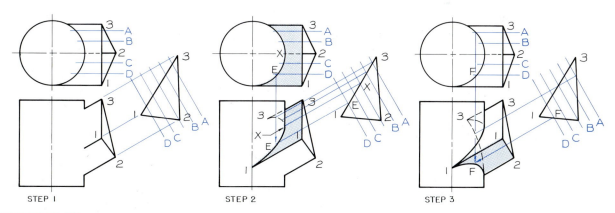

FIGURE 20.72 Intersection between a cylinder and a prism.

Step 1 Project an auxiliary view of the triangular prism from the front view to show three of its surfaces as edges. Pass frontal cutting planes through the top view of the cylinder, and project them to the auxiliary view. The spacing between the planes is equal in both views.

Step 2 Locate points along the line of intersection 1–3 in the top view, and project them to the front view. Example: Point E on cutting plane D is found in the top and auxiliary views and projected to the front view where the projectors intersect. Visibility changes in the front view at point X.

Step 3 Determine the remaining points of intersection by using the same cutting planes. Point F is shown in the top and auxiliary views and is projected to the front view of 1–2. Connect the points, and determine visibility. Space the cutting planes so that they will produce the most accurate line of intersection.

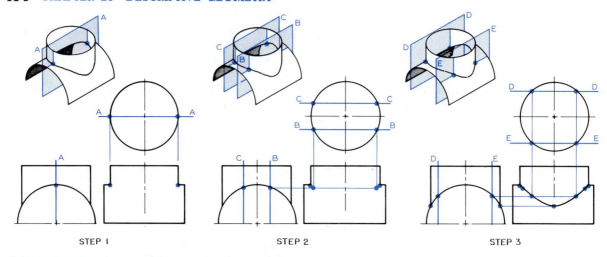

STEP 1   STEP 2   STEP 3

FIGURE 20.73   Intersection between two cylinders.

Step 1   Cutting plane *A–A* is passed through the cylinders parallel to the axes of both. Two points of intersection are found.

Step 2   Cutting planes *C–C* and *B–B* are used to find four additional points of intersection

Step 3   Cutting planes *D–D* and *E–E* locate four more points. Points found in this manner give the line of intersection.

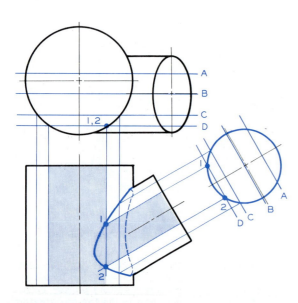

FIGURE 20.74   The intersection between these cylinders is found by locating the end view of the inclined cylinder in an auxiliary view. Vertical cutting planes are used to find the piercing points of the elements of the cylinder and the line of intersection.

The line of intersection from 1 to 3 is projected from the auxiliary view to the front view (Step 2), where the intersection is an elliptical curve. The change of visibility of this line is found at point *X* in the top and auxiliary views, and it is projected to the front view. The process is continued to find the lines of intersection of the other two planes of the prism. (Step 3).

20.35
Intersections between two cylinders

The line of intersection between two perpendicular cylinders can be found by passing cutting planes through the cylinders parallel to the centerlines of each (Fig. 20.73). Each cutting plane locates the piercing points of two elements of one cylinder on an element of the other cylinder. The points are connected, and visibility is determined to complete the solution.

The intersection between nonperpendicular cylinders is found in Fig. 20.74 by a series of vertical cutting planes. Each cutting plane is passed through the cylinders parallel to the centerline of each. As examples of points on the line of intersection, points 1 and

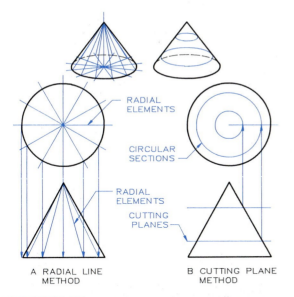

FIGURE 20.75 **A.** Intersections on conical surfaces can be found with radial cutting planes that pass through the cone's centerline and perpendicular to its base. **B.** A second method shows cutting planes that are parallel to the cone's base.

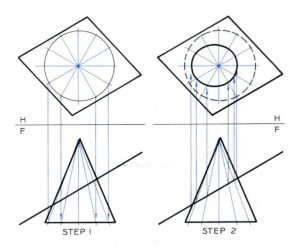

FIGURE 20.76 Intersection of a plane and a cone.

Step 1 Divide the base into even divisions in the top view, and connect these points with the apex to establish elements on the cone. Project these elements to the front view.

Step 2 The piercing point of each element on the edge view is projected to the top view to the same elements, where they are connected to form the line of intersection.

2 are labeled on cutting plane *D*. Other points are found in the same manner. Although the auxiliary view is not required for the solution, it helps you visualize the problem. Points 1 and 2 are shown on cutting plane *D* in the auxiliary view, where they can be projected to the front view as a check on the solution found when projecting from the top view.

20.36
Intersections between planes and cones

To find points of intersection on a cone, cutting planes can be used that are (1) perpendicular to the cone's axis or (2) parallel to the cone's axis. Horizontal cutting planes are shown in Fig. 20.75A, where they are labeled H_1 and H_2. The horizontal planes cut circular sections that appear true size in the top view.

The cutting planes in Fig. 20.75B are passed radially through the top view to establish elements on the surface of the cone that are projected to the front view. Points 1 and 2 are found on these elements in both views by projection.

A series of radial cutting planes is used to find elements on the cone in Fig. 20.76. These elements cross the edge view of the plane in the front view to locate piercing points that are projected to the top view of the same elements to form the line of intersection.

20.37
Intersections between cones and prisms

A primary auxiliary view is used to find the end view of the inclined prism that intersects the cone in Fig. 20.77 (Step 1). Cutting planes that radiate from the apex of the cone in the top view are drawn in the auxiliary view to locate elements on the cone's surface that intersect the prism. These elements are found in the front view by projection.

Wherever the edge view of plane 1–3 intersects an element in the auxiliary view, the piercing points are projected to the same element in the front and top views (Step 2). An extra cutting plane is passed

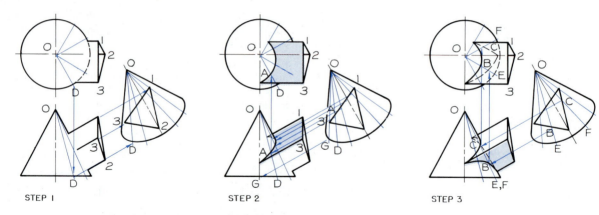

STEP 1   STEP 2   STEP 3

FIGURE 20.77   Intersection between a cone and a prism.

Step 1   Construct an auxiliary view to obtain the edge views of the lateral surfaces of the prism. In the auxiliary view, pass cutting planes through the cone that radiates from the apex to establish elements on the cone. Project the elements to the front and auxiliary views.

Step 2   Locate the piercing points of the cone's elements with the edge view of plane 1–3 in the primary view, and project them to the front and top views. Example: Point A lies on element OD in the primary auxiliary view, so it is projected to the front and top views of element OD.

Step 3   Locate the piercing points where the conical elements intersect the edge views of the planes of the prism in the auxiliary view. Example: Point B is found on OE in the primary auxiliary view and projected to the front and top views of OE.

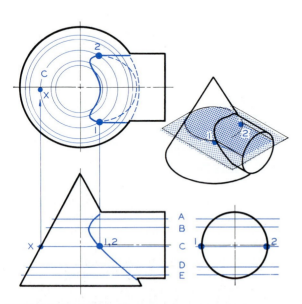

FIGURE 20.78   Horizontal cutting planes are used to find the intersection between the cone and cylinder. The cutting planes form circles in the top view.

FIGURE 20.79   This electrically operated distributor is an application of intersections between a cone and a series of cylinders. (Courtesy of GATX.)

through point 3 in the auxiliary view to locate an element that is projected to the front and top views. Piercing point 3 is projected to this element in sequence from the auxiliary view to the top view.

This same procedure is used to find the piercing points of the other two planes of the prism (Step 3). All projections of points of intersection originate in the auxiliary view, where the planes of the prism appear as edges.

In Fig. 20.78, horizontal cutting planes are passed through the cone and intersecting perpendicular cylinder to locate the line of intersection. A series of circular sections are found in the top view. Points 1 and 2 are found on cutting plane *C* in the top view as examples and are projected to the front view. Other points are found in this same manner.

This method is feasible only when the centerline of the cylinder is perpendicular to the axis of the cone so that circular sections can be found in the top view, rather than elliptical sections that would be difficult to draw.

The distributor housing in Fig. 20.79 is an example of an intersection between cylinders and a cone.

20.38
Intersections between pyramids and prisms

The intersection between an inclined prism and a pyramid is solved in Fig. 20.80. The end view of the inclined prism is found in a primary auxiliary view; the pyramid is shown in this view also (Step 1). Radial lines 0*B* and 0*A* are passed through corners 1 and 3 in the auxiliary view (Step 2). The radial lines are projected from the auxiliary view to the front and top views. Intersecting points 1 and 3 are located on 0*B* and 0*A* in each of these views by projection. Point 2 is the point where line 1–3 bends around corner 0*C*. Lines of intersection 1–4 and 4–3 are then found (Step 3).

Figure 20.81 shows a prism parallel to the base of a pyramid. Its lines of intersection are found by using a series of horizontal cutting planes that pass through the pyramid parallel to its base to form triangular sections in the top view.

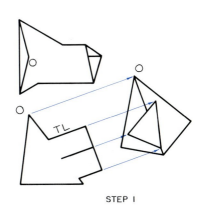

STEP 1

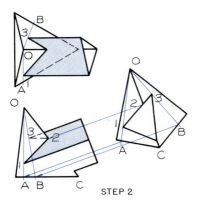

STEP 2

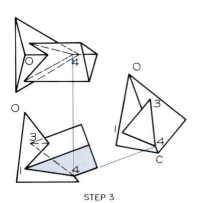

STEP 3

FIGURE 20.80 Intersection between a prism and a pyramid.

Step 1 Find the edge view of the surfaces of the prism by projecting an auxiliary view from the front view. Project the pyramid into this view also. Only the visible surfaces need be shown in this view.

Step 2 Pass planes *A* and *B* through apex *O* and points 1 and 3 in the auxiliary view. Project lines *OA* and *OB* to the front and top views. Project points 1 and 3 to *OA* and *OB* in the principal views. Point 2 lies on line *OC*. Connect points 1, 2, and 3 to give the intersection of the upper plane.

Step 3 Point 4 lies on line *OC* in the auxiliary view. Project this point to the principal views. Connect point 4 to points 3 and 1 to complete the intersections. Visibility is indicated. These geometric shapes are assumed to be hollow as though constructed of sheet metal.

The same cutting planes are passed through the corner lines of the prism in the front and auxiliary views. Each corner edge is extended in the top view to intersect the triangular section formed by the cutting plane passing through it. Point 1 is given as an example.

Corner point X is found by passing cutting plane B through it in the auxiliary view where it crosses the corner line. This is where the line of intersection of this plane bends around the corner.

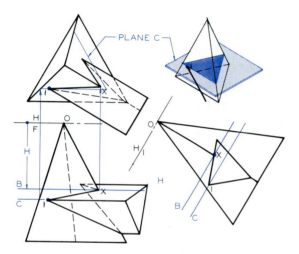

FIGURE 20.81 The intersection between this pyramid and prism is found by locating the end view of the prism in an auxiliary view. Horizontal cutting planes are passed through the fold lines of the prism to find the piercing points and the line of intersection.

FIGURE 20.82 Almost all the surfaces shown in this refinery were made from flat stock fabricated to form these irregular shapes. These flat patterns are called developments. (Courtesy of Phillips Petroleum Company.)

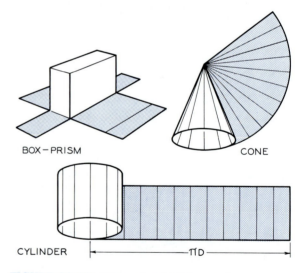

FIGURE 20.83 Three standard types of developments: the box, cylinder, and cone.

20.39

Principles of developments

The processing plant shown in Fig. 20.82 illustrates examples of sheet metal shapes designed using the principles of developments. That is, the patterns were laid out on a flat stock and then formed to the proper shape by bending and seaming the joints.

Figure 20.83 shows the development of the surfaces of three typical shapes into a flat pattern. The sides of a box are imagined to be unfolded into a common plane. The cylinder is rolled out for a distance equal to its circumference. The pattern of a right cone is developed using the length of an element as a radius.

Patterns of shapes with parallel elements, such as the prisms and cylinders shown in Figs. 20.84A and 20.84B, are begun by constructing stretch-out lines parallel to the edge view of the right section of the parts. The distance around the right section is laid off along the stretch-out line. The prism and cylinder in Figs. 20.84C and 20.84D are inclined; thus the right

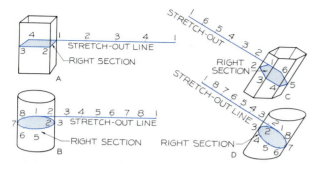

FIGURE 20.84 **A. and B.** The developments of
right prisms and cylinders are found by rolling out
the right sections along a stretch-out line. **C. and D.**
When these figures are oblique, the right sections
are found to be perpendicular to the sides of the
prism and cylinder. The development is laid out
along the stretch-out line parallel to the edge view
of the right section.

sections must be drawn perpendicular to their sides,
not parallel to their bases.

> An inside pattern (development) is more desirable
> than an outside pattern because most bending ma-
> chines are designed to fold metal so that markings
> are folded inward and because markings and scrib-
> ings are hidden when the pattern is assembled.

The method of denoting a pattern is labeled by a series
of lettered or numbered points about its layout. All
lines on a development must be true length.

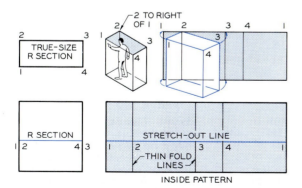

FIGURE 20.85 The development of a rectangular
prism to give an inside pattern. The stretch-out line
is parallel to the edge view of the right section.

Seam lines (lines where the pattern is joined)
should be the shortest lines so that the expense
of riveting or welding the pattern is the least possible.

20.40
Development of prisms

A pattern for a prism is developed in Fig. 20.85. Since
the edges of the prism are vertical in the front view,
its right section is perpendicular to these sides. The top
view shows the right section true size. The stretch-out
line is drawn parallel to the edge view of the right
section, beginning with point 1.

If an inside pattern is drawn and it is to be laid
out to the right, point 2 will be to the right of point 1.
This is determined by looking from inside the top
view, where 2 is seen to the right of 1.

To locate the fold lines on the pattern, lines 1–2,
2–3, 3–4, and 4–1 are transferred with your dividers
from the right section in the top view to the stretch-
out line. The length of each fold line is found by projecting its true length from the front view. The ends of
the fold lines are connected to form the limits of the
developed surface. Fold lines are drawn as thin lines,
and the outside lines are drawn as visible object lines.

The complex installation in Fig. 20.86 is composed of many developments ranging from simple
prisms to complicated shapes.

FIGURE 20.86 Several examples of developments
fabricated from sheet metal are shown in this sheet
metal assembly. (Courtesy of Gar-Bro Mfgr. Co.)

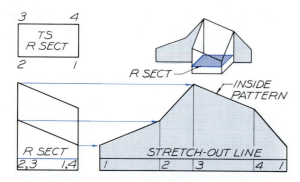

FIGURE 20.87 The development of a rectangular prism with a beveled end to give an inside pattern. The stretch-out line is parallel to the right section.

The development of the prism in Fig. 20.87 is similar to the example in Fig. 20.85 except that one end is beveled rather than square. The stretch-out line is drawn parallel to the edge view of the right section in the front view. The true-length distances around the right section are laid off along the stretch-out line,

and the fold lines are located. The lengths of the fold lines are found by projecting from the front view of these lines.

20.41
Development of oblique prisms

The prism in Fig. 20.88 is inclined to the horizontal plane, but its fold lines are true length in the front view. The right section is drawn as an edge perpendicular to these fold lines, and the stretch-out line is drawn (Step 1). A true-size view of the right section is found in the auxiliary view.

In Step 2, the distances between the fold lines are transferred from the true-size right section to the stretch-out line. The lengths of the fold lines are found by projecting from the front view.

In Step 3, the ends of the prism are found and attached to the pattern so that they can be folded into position.

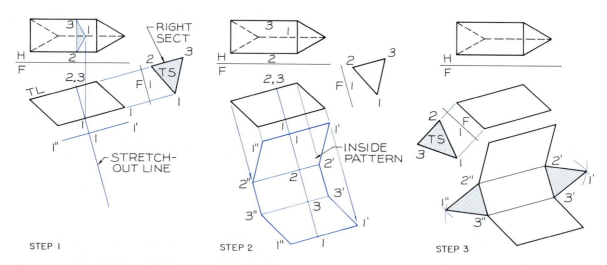

FIGURE 20.88 Development of an oblique prism

Step 1 The edge view of the right section is perpendicular to the true-length axis of the prism in the front view. Determine the true-size view of the right section by an auxiliary view. Draw the stretch-out line parallel to the edge view of the right section. Line 1'–1" is the first line of the development.

Step 2 Since the pattern is developed toward the right, from line 1'–1", the next point is line 2'–2" by referring to the auxiliary view. Transfer true-length lines 1–2, 2–3, and 3–1 from the right section to the stretch-out line to locate the elements. Determine the lengths of the bend lines by projection.

Step 3 Find the true-size views of the end pieces by projecting auxiliary views from the front view. Connect these surfaces to the development of the lateral sides to form the completed pattern. Fold lines are drawn with thin lines, and outside lines are drawn as object lines.

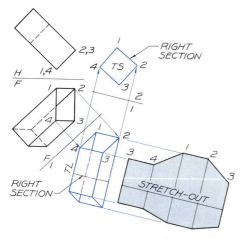

FIGURE 20.89 The development of an oblique prism is found by locating an auxiliary view in which the fold lines are true length and a secondary auxiliary view in which the right section appears true size. The stretch-out line is drawn parallel to the right section.

A prism whose fold lines do not project true length in either view can be developed as shown in Fig. 20.89. The fold lines are found true length in an auxiliary view projected from the front view. The right section will appear as an edge perpendicular to the fold lines in the auxiliary view. The true size of the right section is found in a secondary auxiliary view.

The stretch-out line is drawn parallel to the edge view on the right section. The fold lines are located on the stretch-out line by measuring around the right section in the secondary auxiliary view. The lengths of the fold lines are then projected to the development from the primary auxiliary view.

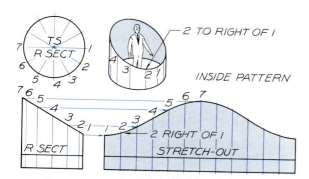

FIGURE 20.90 The development of a right cylinder's inside pattern. The stretch-out line is parallel to the right section. Point 2 is to the right of point 1 for an inside pattern.

20.42
Development of cylinders

The development of a cylinder is found in Fig. 20.90. Since the elements of the cylinder are true length in the front view, the right section will appear as an edge in this view and true size in the top view. The stretch-out line is drawn parallel to the edge view of the right section, and point 1 is chosen as the beginning point since it is the shortest element. To draw an inside pattern, assume you are standing on the inside looking at point 1, and you will see that point 2 is to the right of point 1; therefore, the pattern is laid out with point 2 to the right of point 1.

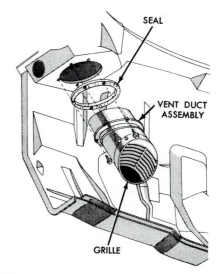

FIGURE 20.91 This ventilator air duct was designed using development principles. (Courtesy of Ford Motor Co.)

The spacing between the elements in the top view can be conveniently done by drawing radial lines at 15° or 30° intervals. Using this technique, the elements will be equally spaced, making it convenient to lay them out along the stretch-out line. The lengths of the elements are found by projecting from the front view to complete the pattern.

An application of a developed cylinder with a beveled end is the air-conditioning duct from an automobile shown in Fig. 20.91.

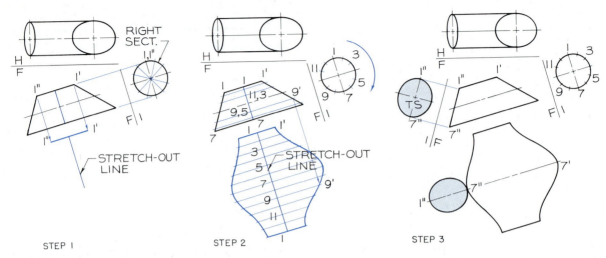

FIGURE 20.92 Development of an oblique cylinder.

Step 1 The right section is an edge in the front view, in which it is perpendicular to the true-length axis. Construct an auxiliary view to determine the true size of the right section. Draw a stretch-out line parallel to the edge view of the right section. Locate element 1′–1″. Divide the right section into equal divisions.

Step 2 Project these elements to the front view from the right section. Transfer measurements between the points in the auxiliary view to the stretch-out line to locate the elements in the development. Determine the lengths of the elements by projection.

Step 3 The development of the end pieces will require auxiliary views that project these surfaces as ellipses, as shown for the left end. Attach this true-size ellipse to the pattern. The beginning line for the pattern was line 1′–1″, the shortest element, for economy.

20.43
Development of oblique cylinders

The development of an oblique cylinder (Fig. 20.92) is found as in the previous examples except that the right section must first be found true-size in an auxiliary view. In Step 1, a series of equally spaced elements is located around the right section in the auxiliary view and is projected back to the true-length view. The stretch-out line is drawn parallel to the edge view of the right section in the front view.

In Step 2, the spacing between the elements is laid out along the stretch-out line, and the elements are drawn through these points perpendicular to the stretch-out line. The lengths of the elements are found by projecting from the front view.

The ends of the cylinder are found in Step 3 to complete the pattern. Only one end pattern is shown as an example.

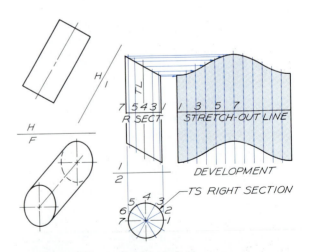

FIGURE 20.93 The development of an oblique cylinder is found by constructing an auxiliary view in which the elements appear true length. The right section is found true size in a secondary auxiliary view.

A more general case is the oblique cylinder in Fig. 20.93, where the elements are not true length in the given views. A primary auxiliary view is used to find a view where the elements are true length, and a secondary auxiliary view is drawn to find the true-size view of the right section. The stretch-out line is drawn parallel to the edge view of the right section in the primary auxiliary view, and the elements are located along this line by transferring their distances apart from the true-size section.

The elements are drawn perpendicular to the stretch-out line. The length of each element is found by projecting from the primary auxiliary view. The endpoints are connected with a smooth curve to complete the pattern.

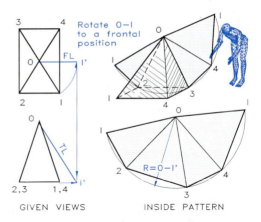

FIGURE 20.95 Development of a right pyramid.

20.44
Development of pyramids

All lines used to draw a pattern must be true length. Pyramids have only a few lines that are true length in the given views; for this reason, the sloping corner lines must be found true length at the outset.

The corner lines of a pyramid can be found true length by revolution, as Fig. 20.94 shows. Line 0–5 is revolved in the frontal position of 0–5′ in the top view

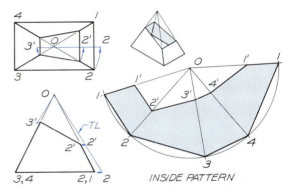

FIGURE 20.96 The development of an inside pattern of a truncated pyramid. The corner lines are found true length by revolution.

(Step 1). Since 0–5′ is a frontal line, it will be true length in the front view (Step 2).

Figure 20.95 shows the development of a pyramid. Line 0–2 is revolved into the frontal plane in the top view to find its true length in the front view. Since this is a right pyramid, all bends are equal in length. Line 0–2′ is used as a radius to construct the base circle for drawing the development. Distance 3–2 is transferred from the base in the top view to the development, where it forms a chord on the base circle. Lines 2–1, 1–4, and 4–3 are found in the same manner and in sequence. The bend lines are drawn as thin lines from the base to the apex, point 0.

A variation of this problem is given in Fig. 20.96, where the pyramid has been truncated, or cut at an angle to its axis. The development of the inside pattern is found as in the previous example; however, to es-

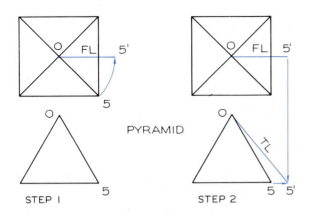

FIGURE 20.94 True length by revolution.

Step 1 Corner 0–5 of a pyramid is found true length by revolving it into the frontal plane in the top view, 0–5′.

Step 2 Point 5′ is projected to the front view where 0–5′ is true length.

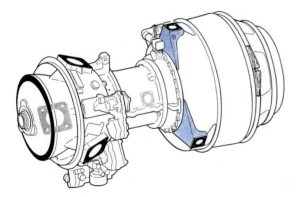

FIGURE 20.97 **Examples of pyramid shapes in the design of mounting pads for an engine. (Courtesy of Avco Lycoming.)**

tablish the upper lines of the development, an additional step is required. The true-length lines from the apex to points 1′, 2′, 3′, and 4′ are found by revolution. These distances are located on their respective lines of the pattern to find the upper limits of the pattern.

The mounting pads in Fig. 20.97 are sections of pyramids that intersect an engine body.

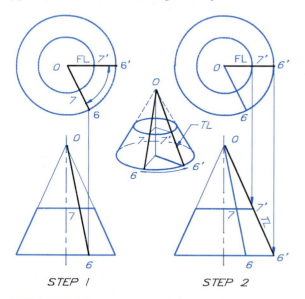

FIGURE 20.98 **True length by revolution.**

Step 1 An element of a cone, 0–6, is revolved into a frontal plane in the top view.

Step 2 Point 6′ is projected to the front view where it is the outside element of the cone and is true length. Line 0–7′ is found TL by projecting to the outside element in the front view.

20.45
Development of cones

All elements of a right cone are equal in length, as illustrated in Fig. 20.98, where 0–6 is found true length by revolution. When revolved to 0–6′ position, it is a frontal line and is therefore true length in the front view where it is an outside element of the cone. Point 7′ is found by projecting horizontally to element 0–6′.

The right cone in Fig. 20.99 is developed by dividing the base into equally spaced elements in the top view and projecting them to the base in the front view. These elements radiate to the apex at 0. The outside elements in the front view, 0–10 and 0–4, are true length.

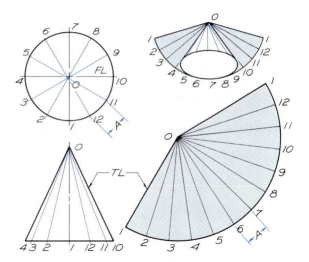

FIGURE 20.99 **The development of an inside pattern of a right cone. In this case, all elements are equal in length.**

Using element 0–10 as a radius, draw the base circle of the development. The elements are located along the base circle equal to the chordal distances between them around the base in the top view. You can see this is an inside pattern by inspecting the top view, where point 2 is to the right of point 1 when viewed from the inside.

The sheet metal conical vessel in Fig. 20.100 is an example of a large vessel that was designed using principles of development.

Figure 20.101 shows the development of a truncated cone. The pattern is found by laying out the total

FIGURE 20.100 An example of a conical shape formed from steel panels by applying principles of developments.

cone, ignoring the portion removed from it. The removed upper portion can be found by using true-length line 0–7' as the radius in the pattern view. The hyperbolic sections through the front view of the cone can be found on their respective elements in the top and front views. Lines 0–2' and 0–3' are projected horizontally to the true-length element 0–1 in the front view, where they will appear true length. These distances and others are measured off along their respective elements in the development to establish a smooth curve on the development.

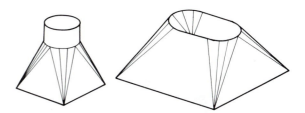

FIGURE 20.102 Examples of transition pieces that join parts having different cross sections.

20.46
Development of transition pieces

A transition piece is a form that transforms the section at one end to a different shape at the other (Fig. 20.102). Huge transition pieces can be seen in the industrial installation in Fig. 20.103.

Figure 20.104 shows the development of a transition piece. In Step 1, radial elements are drawn from each corner to the equally spaced points on the circular end of the piece. Each of these lines is found true length by revolution.

In Step 2, the true-length lines are used with the true-length chordal distance in the top view to lay out a series of adjacent triangles to form the pattern beginning with element A–2.

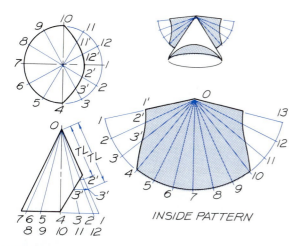

INSIDE PATTERN

FIGURE 20.101 The development of a conical surface with a side opening.

446

FIGURE 20.103 Transition-piece developments are used to join a circular shape with a rectangular section. (Courtesy of Western Precipitation Group, Joy Manufacturing Co.)

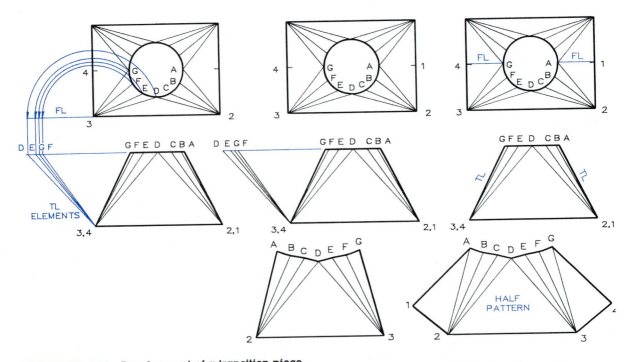

FIGURE 20.104 Development of a transition piece.

Step 1 Divide the circular edge of the surface into equal parts in the top view. Connect these points with bend lines to the corner points, 2 and 3. Find the true length (TL) of these lines by revolving them into a frontal plane and projecting them to the front view. These lines represent elements on the surface of a cone.

Step 2 Using the TL lines found in the TL diagram and the lines on the circular edge in the top view, draw a series of triangles, which are joined together at common sides. Examples: Arcs 2D and 2C are drawn from point 2. Point C is found by drawing arc DC from point D. Line DC is TL in the top view.

Step 3 Construct the remaining planes, A–1–2 and G–3–4, by triangulation to complete the inside half pattern of the transition piece. Draw the fold lines as thin lines at the places where the surface is to be bent slightly. The line of departure for the pattern is chosen along A–1, the shortest possible line.

In Step 3, the triangles *A*–1–2 and *G*–3–4 are added at each end of the pattern to complete the development of half pattern.

20.47
Solution of descriptive geometry problems

Figure 20.105 illustrates the techniques of labeling and solving a descriptive geometry problem. Some of the lettering and numbering is aligned with inclined lines and reference lines to which the labeling applies, whereas other lettering is parallel to the edge of the paper. You may use either technique or a combination of the two. Always use guidelines, and ⅛-inch lettering is best for most drawings. Observe the difference in the line qualities used in the problem solution.

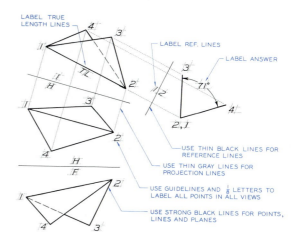

FIGURE 20.105 **Rules that should be followed in solving descriptive geometry problems.**

Problems

Use Size A sheets for the following problems, and lay out the problems using instruments. Each square on the grid is equal to 0.20 in. (about 5 mm). The problems can be laid out on grid or plain paper. Label all reference planes and points in each problem with ⅛-inch (3mm) letters or numbers, and use guidelines.

1. (Fig. 20.106) **A.** through **D.** Find the true-length views of the lines as indicated by the given lines of sight by an auxiliary view.

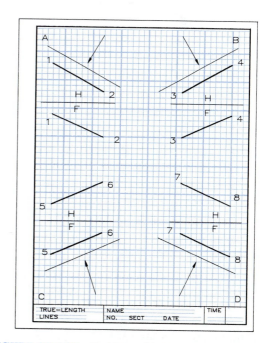

FIGURE 20.106 Problems 1A–1D.

2. (Fig. 20.107) **A.** through **D.** Find the angles that these lines make with the respective principal planes indicated by the given auxiliary reference lines.

3. (Fig. 20.108) **A.** and **B.** Find the true-length views of the lines by the true-length diagram method, and use the same diagram for both lines. **C.** and **D.** Find the point views of the lines.

4. (Fig. 20.109) **A.** through **D.** Find the slope angle, tangent of the slope angle, and the percent grade of the four lines.

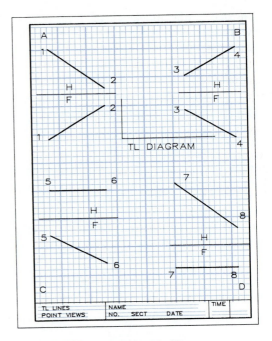

FIGURE 20.108   Problems 3A–3D.

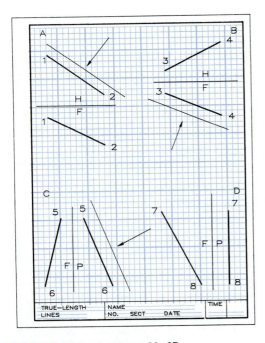

FIGURE 20.107   Problems 2A–2D.

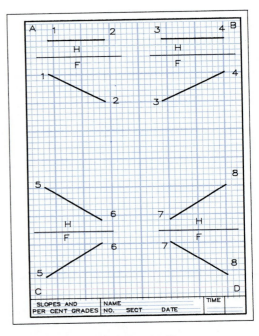

FIGURE 20.109   Problems 4A–4D.

5. (Fig. 20.110) **A.** and **B.** Find the edge views of the two planes.

6. (Fig. 20.111) **A.** Find the angle between the planes. **B.** and **C.** Find the piercing points by auxiliary views.

7. (Fig. 20.112) **A.** Construct a 0.50-in. line perpendicular to the plane on its lower side through point 0 on the plane by the auxiliary view method. **B.** Construct a line perpendicular to the plane from point 0. Show the line and its visibility in all views.

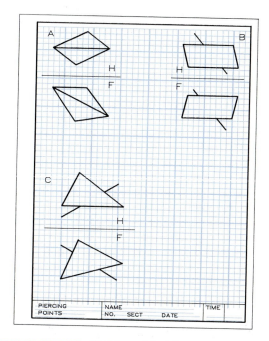

FIGURE 20.111 Problems 6A–6C.

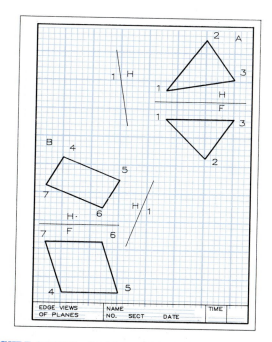

FIGURE 20.110 Problems 5A and 5B.

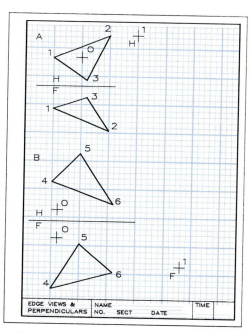

FIGURE 20.112 Problems 7A and 7B.

8. (Fig. 20.113) **A.** Find the line of intersection between the planes by the auxiliary view method, and show visibility in all views. **B.** Find the angle between the planes by an auxiliary view.

9. (Fig. 20.114) **A.** and **B.** Find the slope angle and direction of slope of the planes.

10. (Fig. 20.115) **A.** and **B.** Find the point views of the lines. **C.** and **D.** Find the angles between the planes.

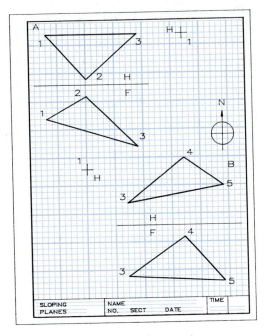

FIGURE 20.114 Problems 9A and 9B.

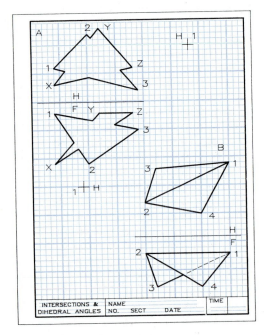

FIGURE 20.113 Problems 8A and 8B.

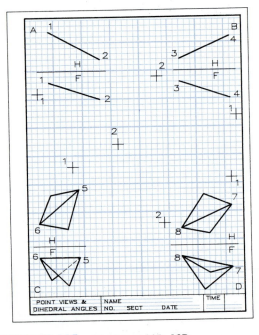

FIGURE 20.115 Problems 10A–10D.

11. (Fig. 20.116) **A.** and **B.** Find the true-size views of the planes.

12. (Fig. 20.117) **A.** and **B.** Find the angles between the intersecting lines.

13. (Fig. 20.118) **A.** and **B.** Find the shortest distances from the points to the lines, and show the distances in all views. Scale: full size.

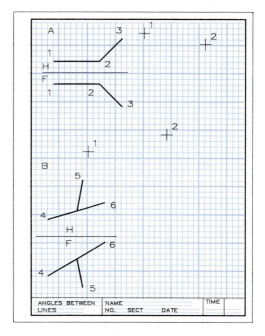

FIGURE 20.117 Problems 12A and 12B.

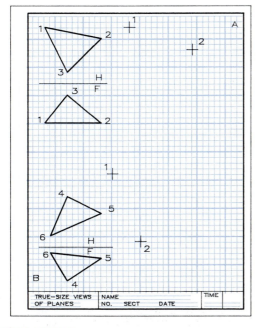

FIGURE 20.116 Problems 11A and 11B.

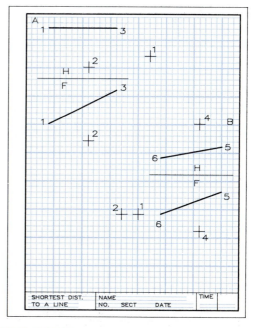

FIGURE 20.118 Problems 13A and 13B.

14. (Fig. 20.119) **A.** and **B.** Find the shortest distances between the skewed lines. The first reference line should pass through point 1, and the secondary reference line should pass through point 2. Show the lines in all views.

15. (Fig. 20.120) Find the connector from point 1 that makes 60° with the given line. Show the connector in all views.

16. (Fig. 20.121) Find the angle between the line and plane, and show the visibility in all views.

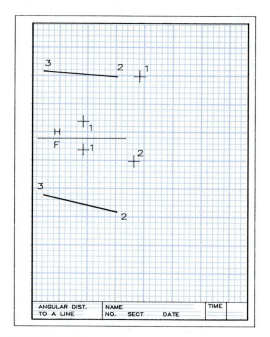

FIGURE 20.120 Problem 15.

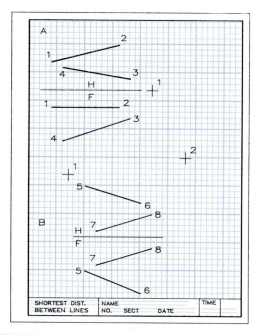

FIGURE 20.119 Problems 14A and 14B.

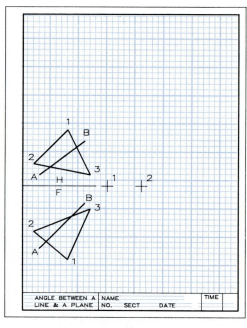

FIGURE 20.121 Problem 16.

The following problems are scaled so that two problems will fit on a Size A sheet when the grid is assigned a value of 0.20″ or 5 mm. When the grid is assigned a value of 0.40″ or 10 mm, only one problem will fit on a Size A sheet.

Intersections

17–40. (Fig. 20.122) Lay out the problems assigned, and find the intersections needed to complete the views.

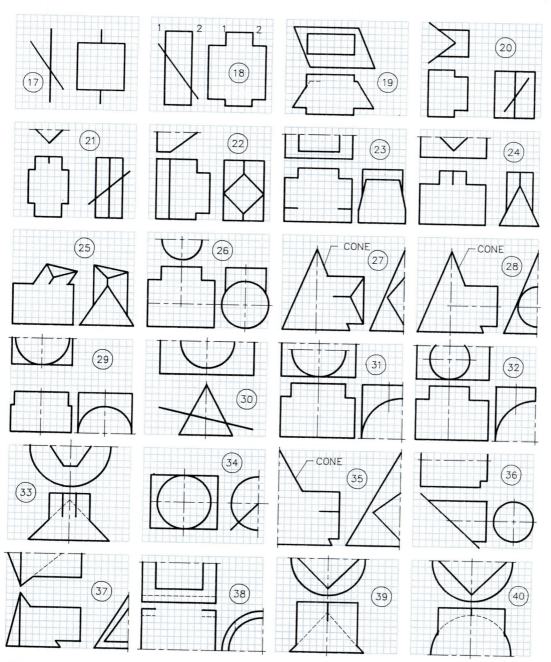

FIGURE 20.122 Problems 17–40.

Developments

41–64. (Fig. 20.123) Lay out the problems assigned on Size A sheets. Place the long side of the sheet horizontally to permit room at the right of the given views for drawing the development.

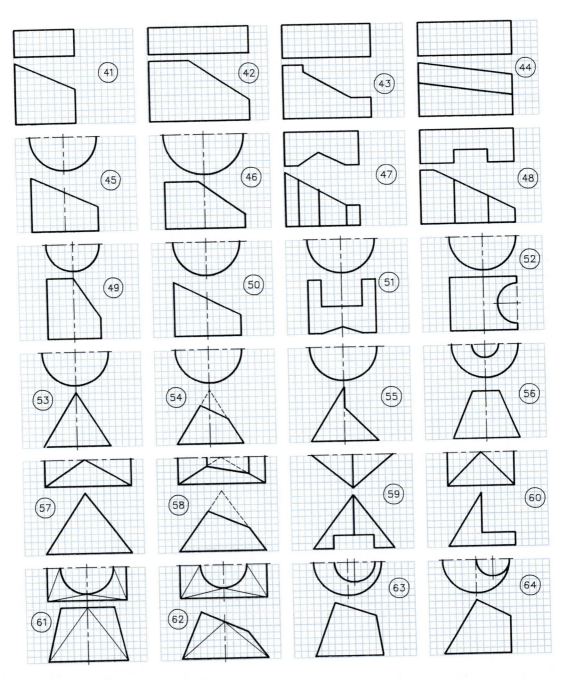

FIGURE 20.123 Problems 41–64.

Civil Engineering Applications

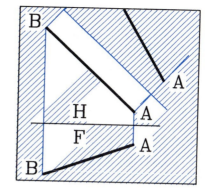

21

21.1
Introduction

The ultimate foundation for all structures and most projects is the earth. Thus the engineer and technologist must understand how the surface of the earth is depicted in a technical drawing. Likewise, they must know how to specify changes and modifications to be made to the surface of the earth.

The principles of **topography** (the graphical representation of the top of the earth) are used and applied more by civil engineers than other types of engineers, but all engineering branches should have a general understanding of topography. Besides civil engineers, these principles are used extensively by the mining and petroleum industry, geologists, architects, environmentalists, and builders.

In this chapter, we introduce the fundamentals of (1) topography, (2) profiles, (3) cut and fill, (4) strike and dip, (5) dam design, and (6) outcrop.

21.2
Plot plans

A **plot plan** is the top view of the portion of the earth shown in a drawing. The locations of key points of the plot plan are taken from surveyors' notes or scaled aerial photographs, which may be converted into drawings to more clearly show significant details and dimensions.

COMPASS BEARINGS The directions of the sides of a plot plan are specified by compass directions, or bearings. Two types of bearings are **compass bearings** and **azimuth bearings** (Fig. 21.1).

Compass bearings always begin with the north or south directions, and the angles with north and south are measured toward east or west. The line in Fig. 21.1A that makes a 30° angle with north has a bearing of N 30° W. A line making 60° with south toward the east has a bearing of south 60° east, or S 60° E. Since a compass can be read only when held horizontally, the compass bearing of the line must be determined in the top view.

An azimuth bearing is measured from north in a clockwise direction to 360° (Fig. 21.1). Azimuth bearings of a line are written N 120°, N 210°, and so forth, with this notation indicating that the measurements are made from the north.

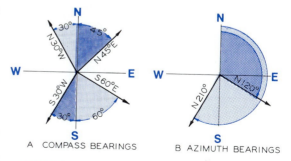

A COMPASS BEARINGS   B AZIMUTH BEARINGS

FIGURE 21.1 **A.** Compass bearings are measured with respect to north and south directions on the compass. **B.** Azimuth bearings are measured with respect to north in a clockwise direction up to 360°.

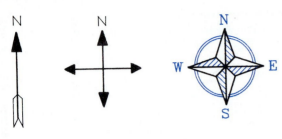

NORTH ARROWS

FIGURE 21.2 North arrows should be placed on plot plans to relate the plot to the compass. Arrows can be drawn to differ from the simple to the complex.

NORTH ARROW Since directions on a topographical drawing or plot plan are made with respect to compass directions, a north arrow must be placed on the drawing. The north arrow can vary from simple to complex (Fig. 21.2).

SCALE INDICATION Plot plans and topographical maps should have a scale provided to allow measurements to be made on them. Figure 21.3 shows examples of numerical and graphical scales.

A typical plot plan is shown in Fig. 21.4, where the north arrow is parallel to the vertical side of the drawing sheet. This is a rare example since the north

arrow points only generally toward the top of the sheet; the shape of the plot often requires that the drawing be rotated to efficiently fit on the sheet.

The sides of the plot plan are drawn with lines that have long dashes separated by two short dashes. Each side is labeled with its length and compass direction. The internal angles at each corner are given in degrees and minutes. The total of the interior angles must add up to the number of degrees found by the formula

$$\text{Interior sum of angles} = (N - 2) \times 180°$$

where N is the number of sides. This four-sided traverse (a series of connected sides of a surveyed plot) must close, therefore, the sum of the interior angles must be 360°. If they do not add up to 360°, there has been an error in surveying the plot.

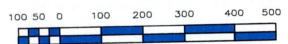

FIGURE 21.3 The scale of a plot plan can be indicated with numbers or a graphical scale.

FIGURE 21.4 This plot plan is drawn with its north arrow pointing vertically on the drawing sheet. Given on the plan are the lengths and bearings of the sides, interior angles, north arrow, and scale.

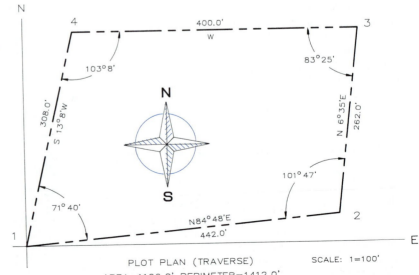

PLOT PLAN (TRAVERSE)   SCALE: 1=100'
AREA=1166.0' PERIMETER=1412.0'

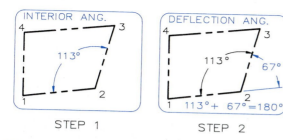

FIGURE 21.5 Compass angles.

Step 1 The interior angles are usually shown on plot plans. The sum of the interior angles = $(N - 2) \times 180°$, where N = number of sides.

Step 2 The deflection angles are the exterior angles at the corner points showing the angles that must be turned by the surveyor's level. The sum of the deflection angles is always 360°.

SURVEYING ANGLES The surveyor is concerned with both interior angles and deflection angles when a plot is surveyed or drawn after surveying. The **interior angle** is the angle between two intersecting sides of the plot. The **deflection angle** is the exterior angle that must be turned from the current corner to

the next corner of the plot by the surveyor (Fig. 21.5). The sum of the deflection angles will always be 360° regardless of the number of sides to a plot.

COMPUTER METHOD A general case plot plan (Fig. 21.6) shows a drawing in which the north arrow is at a slight angle to the right on the drawing. AutoCAD provides an option under the UNITS command for aligning a plot plan with the north arrow. When using this command, you are prompted

```
System of angle              (Example)
measure:
  1. Decimal degrees         45.0000
  2. Degree/
     minutes/seconds         45d0'0"
  3. Grids                   50.0000g
  4. Radians                 0.7854r
  5. Surveyor's
     units                   N45d0'0"E
Enter choice, 1 to 5
<default>: 5
```

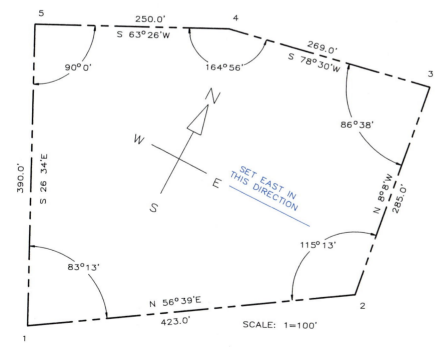

FIGURE 21.6 Most plot plans are drawn where the north arrow is not vertical on the drawing sheet but at an angle since the plot must be positioned to fit the drawing sheet.

PLOT PLAN: GENERAL CASE
AREA=1,627.0 SQ FT PERIMETER=1,617.0'

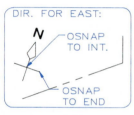

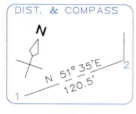

STEP 1   STEP 2

FIGURE 21.7 Setting compass direction.

Step 1 First, draw the compass arrow on the plot plan. When in the UNITS command, select 5. Surveyor's units, and you will be prompted for the angle of east. Press F1, and then with the cursor, select two points on the east line (west to east) to calibrate the compass direction.

Step 2 Using the DIST command, select the end-points of each side in a counterclockwise direction about the traverse. Label the sides with direction inside and lengths outside the plot.

By entering 5, you obtain surveyor's units that will give directional angles with respect to north or south in an east or west direction, such as N 30d45'10", where d = degrees, ' = minutes, and " = seconds. Interior angles measured with the DIM command will give the angles in the same form but without reference to points on the compass such as 152d34'17".

Next, the UNITS command will give a prompt for setting the direction of east to establish the relationship of the drawing to the directional north arrow:

```
Direction for angle 0:
    East      3 o'clock = 0
    North    12 o'clock = 90
    West      9 o'clock = 180
    South     6 o'clock = 270
Enter direction for angle 0
<current>: (F1 and align)
```

When you press F1, your screen returns to the drawing you were working on. With the cursor, locate a point on the east line of the compass arrow, and then locate a second point on the line toward the east (Fig. 21.7). By picking these two points, you have indexed the geometry of your drawing to the compass directions being used. The last prompt

reads

```
Do you want angles measured
clockwise? <N>: N (CR)
```

By selecting a counterclockwise direction, angles are measured in this direction, the standard direction. By using the DIST command, you can find the lengths and compass directions of lines by selecting points at the ends of the lines in a counterclockwise direction about the traverse (Fig. 21.7).

21.3
Contour maps and profiles

A **contour map** is a plot plan that shows the elevations (heights) of points on its surfaces. A contour map is a two-dimensional drawing of a three-dimensional surface (Fig. 21.8).

CONTOUR LINES are horizontal lines that represent constant elevations from a horizontal datum, such as sea level. The contours in Fig. 21.8 are labeled to show that they are spaced 10 ft apart vertically. Figure 21.9 shows a contour map of the Boulder Dam area. This figure also shows a section, or profile, through the cableways.

PROFILES are vertical sections through a contour map that are used to show the earth's surface at any desired location (Fig. 21.8). Contour lines in profiles are the edge views of equally spaced horizontal planes. The true representation of a profile is drawn with the vertical scale equal to the scale of the contour map; however, the vertical scale is often increased to emphasize changes in elevation that would be overlooked if drawn at the same scale.

STATION NUMBERS are used to locate distance on a contour map. Since the civil engineer uses a chain (metal tape) 100 ft long, stations are located 100 ft apart (Fig. 21.10). Station 7 is 700 ft from the beginning point, station 0. A point 32 ft from station 7 toward station 8 is labeled station 7 + 32.

The contour lines on a plot plan can be approximated if the elevations of a series of points on the plot are located (Fig. 21.11). Since contour lines are spaced at even intervals such as 2 ft, 20 ft, or some other

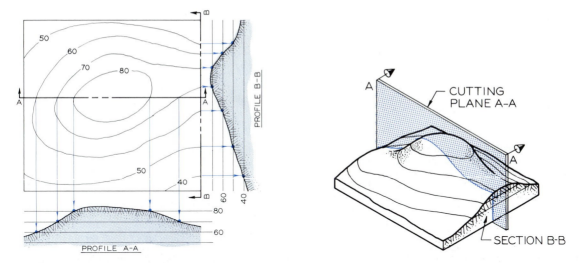

FIGURE 21.8 A contour map uses contour lines to show variations in elevation on an irregular surface. Vertical sections taken through a contour map are called profiles.

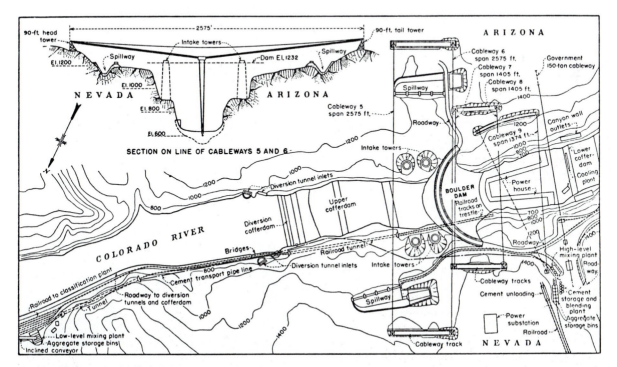

FIGURE 21.9 A contour map of the Boulder Dam area. The inset shows a vertical section through the cableways. (Courtesy of U.S. Department of the Interior.)

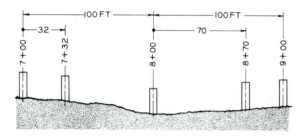

FIGURE 21.10 Station points are located 100 ft apart. For example, station 7 is 700 ft from the beginning point. A point 32 ft beyond station 7 is labeled station 7 + 32. A point 870 ft from the origin is labeled station 8 + 70.

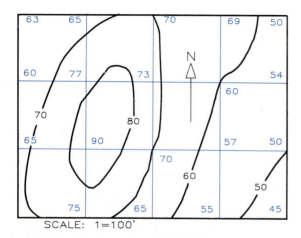

FIGURE 21.11 Once the surveyor has located a grid of elevation points on a plot, contour lines can be found by interpolation. Contour lines are level (horizontal) lines. The steeper the terrain, the closer the contour lines are to each other.

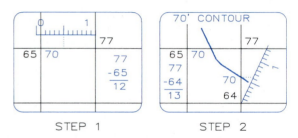

FIGURE 21.12 Locating contour lines.

Step 1 If elevations are located on a 12-ft grid and the elevation difference between 77 ft and 65 ft is 12 ft, a scale with 12 units can be used to locate the 70-ft contour line 5 units from the 65-ft elevation.

Step 2 The 70-ft elevation line is located between the 77-ft elevation and 64-ft elevation by positioning the scale where 13 units extend between the horizontal lines of the grid. The 70-ft contour line is 6 units from the 64-ft elevation.

line between 64 ft and 77 ft. This procedure can be continued to locate other contour lines. Figure 21.13 shows a model of a dam depicting contour lines.

In many applications, the plot plan is also a contour map.

21.4
Profiles

The vertical section through the contour map in Fig. 21.8 is an example of a **profile.** A profile is used in Fig. 21.14 to show an underground pipe known to have an elevation of 90 ft at point 1 and 60 ft at point

functional spacing, using a method of interpolation to locate the contour lines on the plot is necessary. For example, the 70-ft contour line between two elevation points of 64' and 75' will be located between the two points.

A more general example is shown Fig. 21.12, where a scale is used to interpolate between known elevation points on the plan. In Step 1, the vertical distance between 77' and 65' is 12'. A scale is positioned with 12 units between elevations 65 and 77, and 5 units are counted off from the 65' elevation to locate the point of the 70-ft elevation line. In Step 2, the same procedure is used to locate the 70-ft contour

FIGURE 21.13 A model of a dam that shows contour lines as layers of a thickness equal to the difference between the contour lines. (Courtesy of U.S. Department of the Interior.)

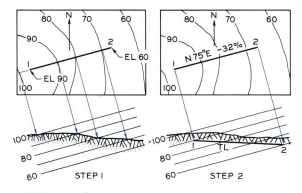

FIGURE 21.14 Vertical sections (profiles).

Step 1 An underground pipe is known to have elevations of 90 ft and 60 ft at each end. An auxiliary view is projected perpendicularly from the top view, and contours are drawn at 10-ft intervals to correspond to the plan view. The top of the ground is found by projecting from the contour lines in the plan view.

Step 2 Points 1 and 2 are located at elevations of 90 ft and 60 ft in the vertical section (profile). Since 1–2 is TL in the section, its slope or percent grade can be measured. The compass direction and percent grade are labeled in the top view of the line.

2. An auxiliary view is projected perpendicularly from the top view, contour lines are located, and the top of the earth over the pipe is found in a profile.

To measure the true lengths of angles of slope in the profile, the scale used to draw the contour map is used to draw the profile. The percent grade and compass bearings of the line are labeled on the contour map.

21.5
Plan profiles

A **plan profile** is a drawing that includes a plan with contours and a vertical section called a profile. A plan profile is used to show an underground drainage system from manhole 1 to manhole 3 in Figs. 21.15 and 21.16.

The profile is drawn with an exaggerated vertical scale to emphasize the variations in the earth's surface and the grade of the pipe. Although the vertical scale is usually increased, it can be drawn at the same scale used in the plan if desired.

Manhole 1 is projected to the profile using orthographic projection, but the other points are not ortho-

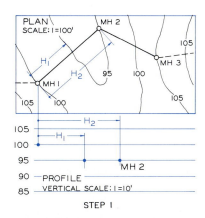

STEP 1

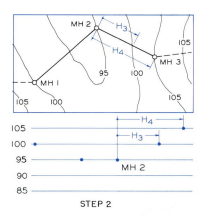

STEP 2

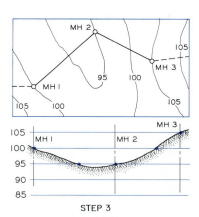

STEP 3

FIGURE 21.15 Plan profile.

Required Find the profile of the earth over the underground drainage system.

Step 1 Distances H_1 and H_2 from manhole 1 are transferred to their respective elevations in the profile. This is not an orthographic projection.

Step 2 Distances H_3 and H_4 are measured from manhole 2 in the plan and transferred to their respective elevations in the profile. These points represent elevations of points on the earth above the pipe.

Step 3 The five points are connected with a freehand line, and the drawing is cross-hatched to represent the earth's surface. Centerlines are drawn to show the locations of the three manholes.

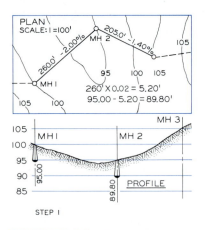

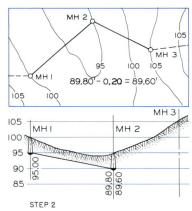

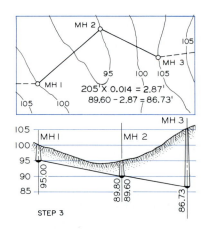

STEP I   STEP 2   STEP 3

FIGURE 21.16 Plan profile—manhole location.

Step 1 The horizontal distance from MH1 to MH2 is multiplied by the percent grade. The elevation of the bottom of manhole 2 is calculated by subtracting from the elevation of manhole 1.

Step 2 The lower side of manhole 2 is 0.20' lower than the inlet side to compensate for loss of head (pressure) due to the turn in the pipeline. The lower side is found to be 89.60' and is labeled.

Step 3 The elevation of manhole 3 is calculated to be 86.73' since the grade is 1.40% from manhole 2 to manhole 3. The flow line of the pipeline is drawn from manhole to manhole, and the elevations are labeled.

graphic projections (Fig. 21.15). The points where the contour lines cross the top view of the pipe are transferred to their respective elevations in the profile with dividers. These points are connected to show the surface of the earth over the pipe and the location of manhole centerlines (Fig. 21.16). The drop from manhole 1 to manhole 2 is found to be 5.20 ft by multiplying the horizontal distance of 260.00 ft by a −2.00% grade (Fig. 21.16).

Since the pipes intersect at manhole 2 at an angle, the flow of the drainage is disrupted at the turn; thus a drop of 0.20 ft is given from the inlet across the floor of the manhole to compensate for the loss of pressure (head) through the manhole.

The true lengths of the pipes cannot be accurately measured in the profile when the vertical scale is different from the horizontal scale. For this computation, trigonometry must be used.

21.6
Cut and fill

A level roadway routed through irregular terrain or the embankment of a fill used to build a dam involves the principles of cut and fill (Fig. 21.17). **Cut and fill** is the process of cutting away equal amounts of the high ground to fill the lower areas.

In Fig. 21.18, it is required that a level roadway of an elevation of 60 ft be constructed about the given centerline in the contour map using the specified angles of cut and fill.

FIGURE 21.17 The road across the top of this dam was built by applying the principles of cut and fill. (Courtesy of the Bureau of Reclamation, U.S. Department of the Interior.)

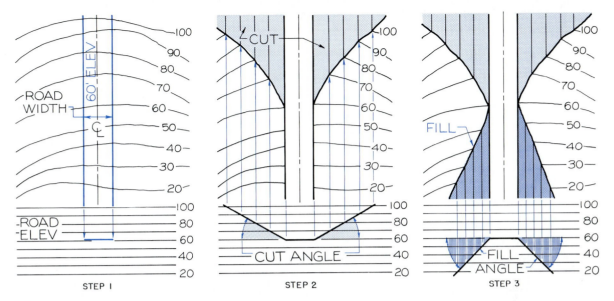

FIGURE 21.18 Cut and fill of a level roadway.

Step 1 Draw a series of elevation planes in the front view at the same scale as the map, and label them to correspond to the contours on the map. Draw the width of the roadway in the top view and in the front view at the given elevation, 60′ in this case.

Step 2 Draw the cut angles on the upper sides of the road in the front view according to the given specifications. The points of intersection between the cut angles and the contour planes in the front view are projected to their respective contour lines in the top view to determine the limits of cut.

Step 3 Draw the fill angles on the lower sides of the road in the front view. The points in the front view where the fill angles cross the contour lines are projected to their respective contour lines in the top view to give the limits of the fill. Contour lines are changed in the cut-and-fill areas to indicate the new contours parallel to the roadway.

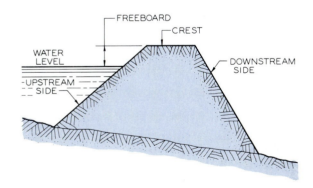

FIGURE 21.19 Terms and symbols used in the construction of a dam.

In Step 1, the roadway is drawn in the top view, and the contour lines in the profile view are drawn 10 ft apart since the contours in the top view are this far apart. In Step 2, the cut angles are measured and

drawn on both sides of the roadway on the upper side. The points on the elevation lines crossed by the cut embankments are projected to the top view to find the limits of cut in this view.

In Step 3, the fill angles are laid off in the profile from given specifications. The crossing points on the profile view of the elevation lines are projected to the top view to find the limits of fill. New contour lines are drawn inside the areas of cut and fill to indicate they have been changed.

21.7
Design of a dam

Some of the terms used in the drawing of a dam are (1) **crest,** the top of the dam; (2) **water level;** and (3) **freeboard,** the height of the crest above the water level (Fig. 21.19).

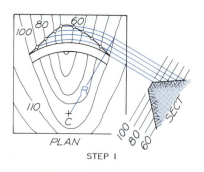
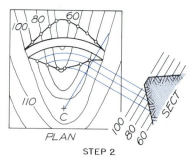
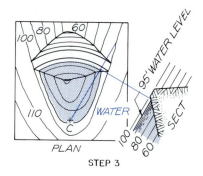

STEP 1   STEP 2   STEP 3

FIGURE 21.20   Graphical design of a dam.

Step 1   A dam in the shape of an arc with its center at *C* has an elevation of 100′. Draw radius *R* from *C* and project perpendicularly from this line. Then draw a section through the dam from specifications. The downstream side of the dam is projected to radial line *R*. Using the radii from *C*, locate points on their respective contour lines.

Step 2   The elevations of the dam on the upstream side of the section are projected to the radial line, *R*. Using center *C* and your compass, locate points on their respective contour lines in the plan view as they are projected from the section.

Step 3   The elevation of the water level is 95′ and is drawn in the section. The point where the water intersects the dam is projected to the radial line in the plan view and drawn as an arc using center *C*. The limit of the water is drawn between the 90′ and 100′ contour lines in the top view.

FIGURE 21.21   The Hoover Dam and Lake Mead, which were built from 1931 to 1935. (Courtesy of the Bureau of Reclamation, U.S. Department of the Interior.)

An earthen dam is located on the contour map in Fig. 21.20. It makes an arc with its center at point *C*. The top of the dam is to be level to provide a roadway. Figure 20.20 shows the method of drawing the top view of the dam and indicating the level of the water held by the dam.

These same principles were used in the design and construction of the 726-ft Hoover Dam, built in the 1930s. Since this dam was made of concrete instead of earth, the dam was built in the shape of an arch bowed toward the water to take advantage of the compressive strength of concrete (Fig. 21.21).

21.8
Strike and dip

Strike and dip are terms used in geological and mining engineering to describe strata of ore under the surface of the earth.

STRIKE   is the compass bearing of a level line in the top view of a plane. It has two possible compass bearings since it is the direction of a level line.

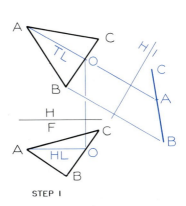

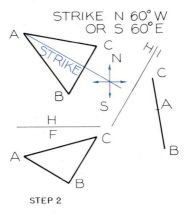

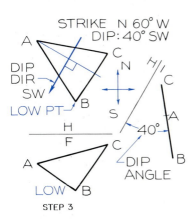

STEP 1   STEP 2   STEP 3

FIGURE 21.22   Strike and dip of a plane.

Step 1   Find the edge view of the plane by projecting from the top view.

Step 2   Strike is the compass direction of a level line on the plane and is measured in the top view. The strike of the plane is either N 60° W or S 60° E.

Step 3   Dip is the angle a plane makes with the horizontal (40° in the auxiliary view) plus the general compass direction toward the low side in the top view (SW). Dip direction is perpendicular to a TL line in the top view. Dip is written 40° SW.

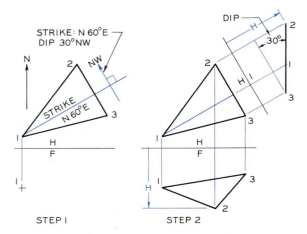

STEP 1   STEP 2

FIGURE 21.23   Working from strike and dip specifications.

Step 1   Strike is drawn on the top view of the plane, which is a TL horizontal line. The direction of dip is drawn perpendicular to the strike toward the NW, as specified.

Step 2   A point view of the strike line is found in the auxiliary view to locate point 1. The edge view of the plane is drawn through 1 at a dip of 30°, according to the specifications. The front view is completed by transferring height dimensions from the auxiliary to the front view.

DIP   is the angle the edge view of a plane makes with the horizontal plus its general compass direction, such as NW or SW. The dip angle is found in the primary auxiliary view projected from the top view, and its dip direction is measured perpendicular to a level line in the top view toward the low side.

Figure 21.22 gives the steps of finding the strike and dip of a plane. Strike can be measured in the top view by finding a true-length line on the plane in this view. The dip angle is found in an auxiliary view projected from the top view that shows the horizontal and plane as edges.

A plane can be constructed from strike and dip specifications (Fig. 21.23). The strike is drawn to represent a true-length horizontal line on the plane. Dip direction is perpendicular to the strike (Step 1). The edge view of the plane is then drawn through point 1 at a dip of 30° (Step 2). Points 2 and 3 are located in the front view.

21.9
Distances from a point to a plane

Descriptive geometry principles can be used to find various distances from a point to a plane. Figure 21.24 shows an example where the distance from point 0 on the ground to an underground ore vein is found.

Three points are located on the top plane of an ore vein. Point 0 is the point on the earth from which the tunnels are to be drilled to the vein. Point 4 is a point on the lower plane of the vein.

The edge view of plane 1–2–3 is found by projecting from the top view. The lower plane is drawn

parallel to the upper plane through point 4. The horizontal distance from point 0 to the plane is drawn parallel to the H-1 reference line. The vertical distance is perpendicular to the H-1 line. The shortest distance to the plane is perpendicular to the plane. Each of these lines is true length in this view where the ore vein appears as an edge.

Finding the distance from a point to a plane or vein is a technique often used in solving mining and geological problems; for example, test wells are drilled into coal zones to learn more about them (Fig. 21.25).

21.10
Outcrop

Strata of ore formations usually approximate planes of a uniform thickness. This assumption is used in analyzing the orientation of underground ore veins. A vein of ore may be inclined to the surface of the earth and may outcrop on its surface. Outcrops permit open-surface mining operations at minimum expense.

Figure 21.26 shows the steps of finding the outcrop of an ore vein. The locations of sample drillings, A, B, and C, are shown on the contour map, and their elevations are located on the contours of the profile. These points are known to lie on the upper surface of the vein; point D is known to lie on the lower plane of the vein.

The edge view of the ore vein can be found in an auxiliary view projected from the top view (Step 1). The points on the upper surface are projected back to their respective contour lines in the top view (Step 2). The points on the lower surface of the vein are then projected to the top view (Step 3). If the ore vein continues uniformly at its angle of inclination to the surface, the space between these two lines will be the outcrop of the vein on the surface of the earth.

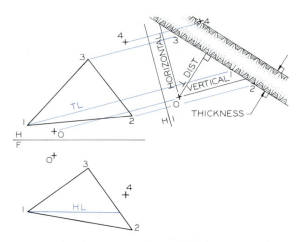

FIGURE 21.24   **The vertical, horizontal, and perpendicular distances from a point to an ore vein can be found in an auxiliary view projected from the top view. The thickness of an ore vein is perpendicular to the upper and lower planes of the vein.**

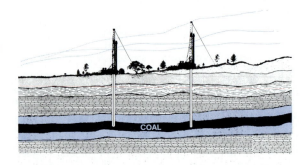

FIGURE 21.25   Test wells are drilled into coal zones to determine which coal seams will contribute significantly to the production of gas. (Courtesy of Texas Eastern News.)

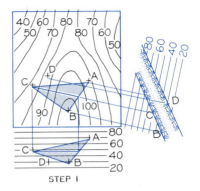

STEP 1

STEP 2

STEP 3

FIGURE 21.26 Ore vein outcrop.

Step 1 Using points *A*, *B*, and *C* on the upper surface of the plane, find its edge view by projecting an auxiliary off the top view. The lower surface of the plane is drawn parallel to the upper surface through point *D*, a point on the lower surface.

Step 2 Points of intersection between the upper surface and the contour lines in the auxiliary view are projected to their respective contour lines in the top view to find one line of the outcrop.

Step 3 Points from the lower surface in the auxiliary view are projected to their respective contour lines in the top view to find the second line of outcrop. Cross-hatch this area to indicate the outcrop of the vein.

Problems

Solve the following problems on Size A sheets with instruments. Each square on the grid equals 0.20 in. (about 5 mm). The problems can be laid out on a grid or plain paper, by hand or by computer. Label all reference planes and points in each problem with $\frac{1}{8}$-in. (3 mm) letters or numbers, and use guidelines.

1. (Fig. 21.27) Draw the plot plan, label the lengths of the sides, and give their compass directions. Give the interior angles of each corner. Add other information usually required on a plot plan.

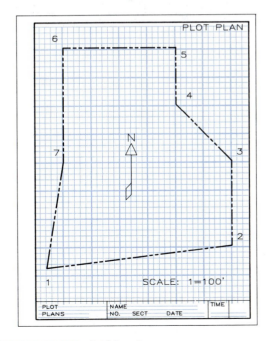

FIGURE 21.27 Problem 1.

2. (Fig. 21.28) Same as problem 1. Notice that the north arrow is at a slight angle to edges of the drawing.

3. (Fig. 21.29) Draw the plot plan, label the lengths of the sides, and give their compass directions. Show both the interior angles and deflection angles for each corner. Give the angular values in the table.

4. (Fig. 21.30) Draw the contour map with its contour lines. Give the lengths of the sides, their compass directions, interior angles, scale, and north arrow.

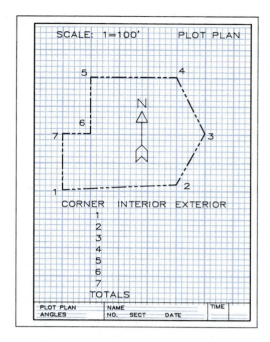

FIGURE 21.29 Problem 3.

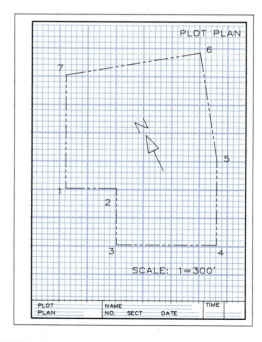

FIGURE 21.28 Problem 2.

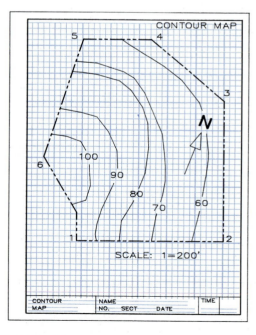

FIGURE 21.30 Problem 4.

5. (Fig. 21.31) Draw the contour map, and construct the profile (vertical section) as indicated by the cutting-plane line in the plan view. Notice that the profile scale is different from the plan scale.

6. (Fig. 21.32) By using the elevations shown on the grid, construct contour lines every 10 ft, for example, 40', 50', 60', and so forth.

7. (Fig. 21.33) Construct a vertical section (profile) through line 1–2 that slopes downward from point 1 at 25%. Point 1 lies on the surface of the earth.

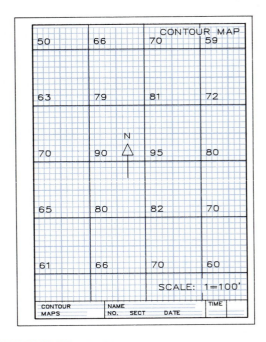

FIGURE 21.32 Problem 6.

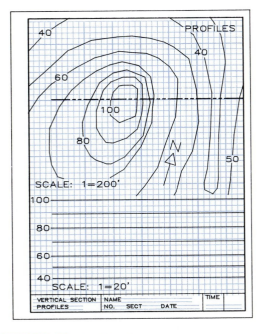

FIGURE 21.31 Problem 5.

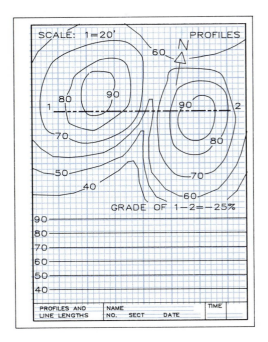

FIGURE 21.33 Problem 7.

8. (Fig. 21.34) Draw the contour map, and construct a plan profile through the three manholes. Use the grade specifications given for each line.

9. (Fig. 21.35) Draw the contour map through the level road, and construct the profile that shows the cut (35°) and fill (40°). Show the limits of the cut and fill in the plan view.

10. (Fig. 21.36) Find the strike and dip of the two planes.

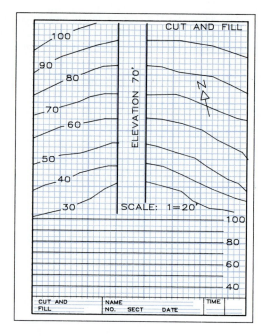

FIGURE 21.35 Problem 9.

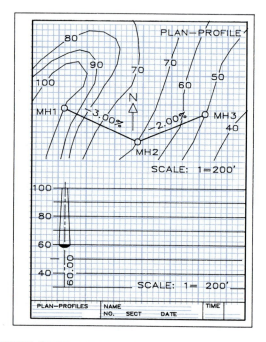

FIGURE 21.34 Problem 8.

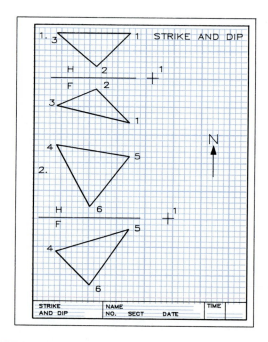

FIGURE 21.36 Problem 10.

11. (Fig. 21.37) Find the shortest distance, horizontal distance, and vertical distance from a point on the surface of the earth to the underground ore vein represented by the triangle. Point *B* is on the lower plane of the vein. Find the thickness of the vein.

12. (Fig. 21.38) Find the outcrop of the ore vein represented by the plane in the contour-map view and profile view. Point *B* is on the lower surface of the vein.

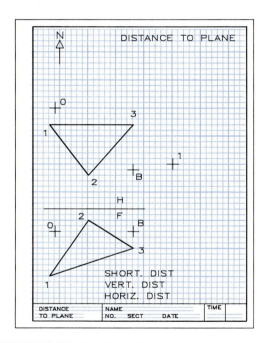

FIGURE 21.37   Problem 11.

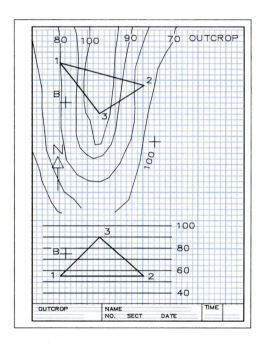

FIGURE 21.38   Problem 12.

Graphs

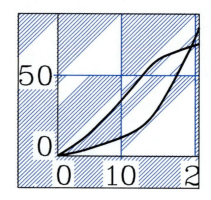

22

22.1
Introduction

Data and information expressed as numbers and words are often difficult to analyze or evaluate unless transcribed into graphical form. **Graphs** (charts is an acceptable term but is more appropriate when referring to maps, a specialized form of graphs) are a popular means of briefing other people on trends that might otherwise be difficult to communicate (Fig. 22.1). The trends of a plotted curve on a graph can be compared to the expressions on a person's face, which is a graph of sorts that reveals a person's feelings. For example, a flat curve shows no change, whereas an

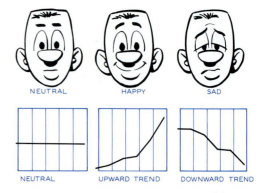

FIGURE 22.2 Curves on a graph are similar to expressions on a face.

upwardly inclined curve indicates a positive increase. A downward curve, on the other hand, represents a negative result (Fig. 22.2).

In this chapter, we deal with the more commonly used graphs. The basic types are

1. Pie graphs
2. Bar graphs
3. Linear coordinate graphs
4. Logarithmic coordinate graphs
5. Semilogarithmic coordinate graphs
6. Polar graphs
7. Schematics and diagrams

FIGURE 22.1 Graphs are helpful in presenting technical data to one's associates.

22.2
Size proportions of graphs

Graphs may be used to illustrate technical reports reproduced in quantity, as well as for projection by slide or overhead projectors. In all cases, the proportion of the graph must be determined so that it will match the proportion of the space or the format of the visual aid.

If a graph is to be photographed by a 35-mm camera, the graph must conform to the standard size of the 35-mm film used. This proportion is approximately 2 × 3 (Fig. 22.3). The proportions of the area in which the graph is to be drawn can be enlarged or reduced by using the diagonal-line method.

22.3
Pie graphs

> Pie graphs compare the relationship of parts to a whole when there are only a few parts.

Figure 22.4 shows the types of computer graphics software used in industry.

Figure 22.5 shows the method of drawing a pie graph. The tabular data does not give as good an impression of the comparisons as does the pie graph even though the data is quite simple.

To facilitate lettering within narrow spaces, the

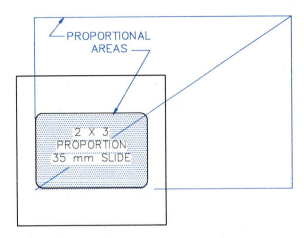

FIGURE 22.3 **This diagonal-like method can be used for construction areas proportional to those of a 35-mm slide.**

thin sectors should be placed as nearly horizontal as possible to provide more room for the label. The actual percentage should be given, and it may also be desirable to give the actual numbers or values in each sector.

22.4
Bar graphs

Since they are well understood by the general public, bar graphs are effective to compare values (Fig. 22.6). In this example, the bars show not only the overall

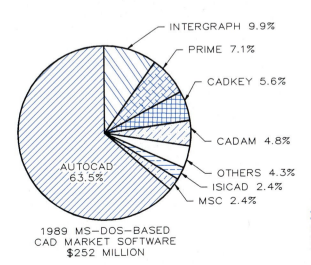

INTERGRAPH 9.9%
PRIME 7.1%
CADKEY 5.6%
CADAM 4.8%
OTHERS 4.3%
ISICAD 2.4%
MSC 2.4%
AUTOCAD 63.5%

1989 MS−DOS−BASED
CAD MARKET SOFTWARE
$252 MILLION

FIGURE 22.4 **A pie graph shows the relationship of the parts to a whole. It is effective when there are only a few parts.**

PRODUCT
DEVELOPMENT
COST PER UNIT

LABOR	$40	40%×360	=144°
RESEARCH	30	30%×360	=108°
MATERIALS	20	20%×360	= 72°
OVERHEAD	10	10%×360	= 36°
	$100		360°

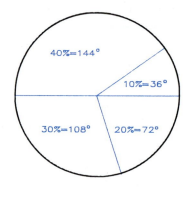

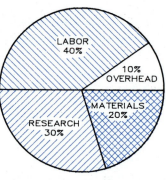

NEW PRODUCT DEVELOPMENT
COST PER UNIT

FIGURE 22.5   Drawing pie graphs.

Step 1   The total sum of the parts is found, and the percentage of each is found. Each percentage is multiplied by 360° to find the angle of each sector of the pie graph.

Step 2   The circle is drawn to represent the pie graph. Each sector is drawn using the degrees found in Step 1. The smaller sectors should be placed as nearly horizontal as possible.

Step 3   The sectors are labeled with their proper names and percentages. In some cases, it might be desirable to include the exact numbers in each sector as well.

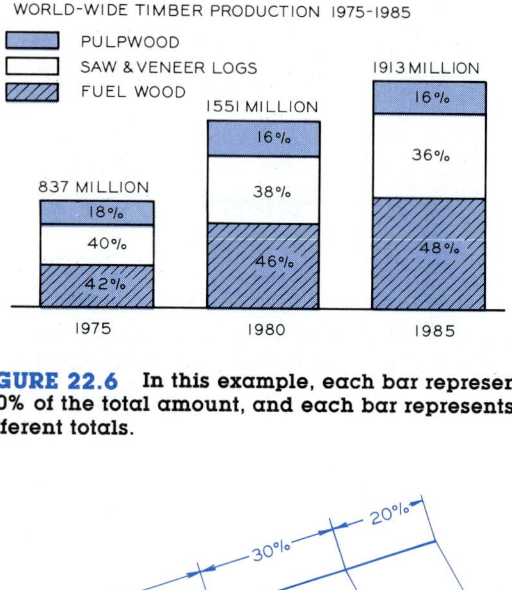

WORLD-WIDE TIMBER PRODUCTION 1975-1985

- PULPWOOD
- SAW & VENEER LOGS
- FUEL WOOD

FIGURE 22.6   In this example, each bar represents 100% of the total amount, and each bar represents different totals.

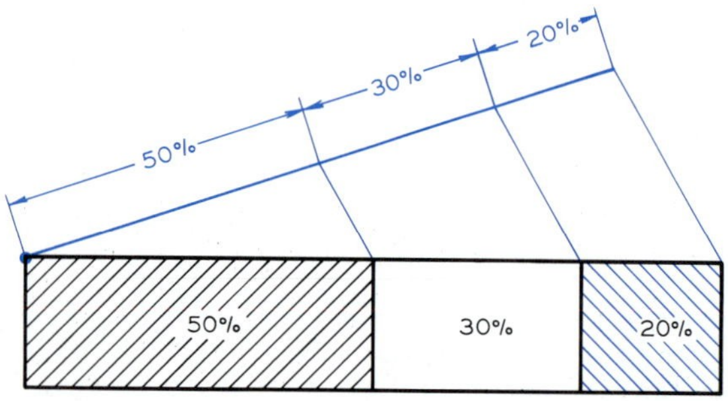

production of timber (the total lengths of the bars) but also the portions of the total devoted to the three uses of the timber.

A bar graph can be composed of a single bar (Fig. 22.7). The total length of the bar is 100%, and the bar is divided into lengths proportional to the percentages represented by each of the three parts of the bar.

Figure 22.8 shows the method of constructing a bar graph. In this case, the title of the graph is placed inside the graph where space is available. Titles are often placed under or over the graph.

> The data should be sorted by arranging the bars in ascending or **descending** order since it is desirable to know how the data represented by the bars rank from category to category (Fig. 22.9B).

An alphabetical or numerical arrangement of bars results in a graph more difficult to evaluate (Fig. 22.9A).

If the data are sequential and involve time, such as sales per month, it would be less effective to rank the data in ascending order because it is important to see variations in the data over time.

Bars in a bar graph may be horizontal (Fig. 22.10) or vertical (Fig. 22.11). To show a true comparison of data, it is desirable that the bars begin at zero.

FIGURE 22.7   The method of constructing a single bar where the sum of all the parts will be 100%.

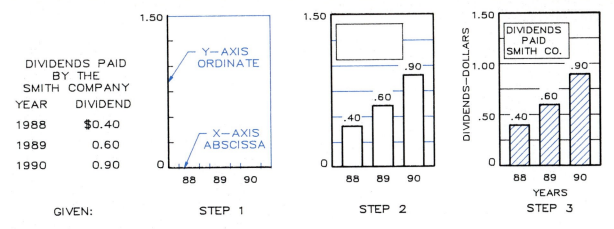

FIGURE 22.8 Construction of a bar graph.

Given These data are to be plotted as a bar graph.

Step 1 Lay off the vertical and horizontal axes so that the data will fit on the grid. Make the bars begin at zero.

Step 2 Construct and label the bars. The width of the bars should be different from the space between them. Grid lines should not pass through them.

Step 3 Strengthen lines, title the graph, label the axes, and cross-hatch the bars.

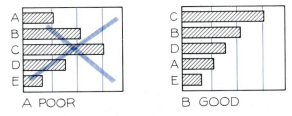

FIGURE 22.9 The bars at A are arranged alphabetically. The resulting graph is not as easy to evaluate as the one at B, where the bars have been arranged in descending order.

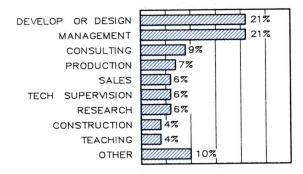

EMPLOYMENT FUNCTIONS OF ENGINEERS

FIGURE 22.10 A horizontal bar graph that is arranged in descending order to show where engineers are employed.

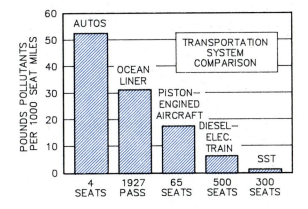

FIGURE 22.11 This bar graph has the bars arranged in descending order to compare several sources of pollution.

22.5
Linear coordinate graphs

Figure 22.12 shows a typical coordinate graph, with notes explaining its important features. The axes are linear if the divisions along the axes are equally spaced.

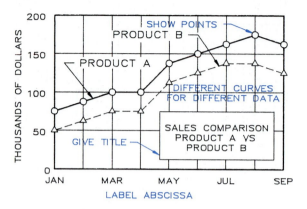

FIGURE 22.12 The basic linear coordinate graph with the important features identified.

Points are plotted on the grid by using two measurements, called **coordinates**, made along each axis. The plotted points are indicated by using symbols, such as circles, that can be drawn with a template.

> The horizontal scale of the graph is called the **abscissa** or *X* axis, and the vertical scale is called the **ordinate** or *Y* axis.

Once the points have been plotted, the curve is drawn from point to point. (The line drawn to represent the plotted points is called a curve whether it is a smooth or broken line.) The curve should not close up the plotted points; rather, they should be left as open circles or symbols.

The curve must be drawn as a heavy prominent line since it is the most important part of the graph. In Fig. 22.12, there are two curves; therefore, drawing them as different types of lines and labeling them with notes and leaders is helpful. The title of the graph is placed in a box inside the graph.

Units are given along the *X* and *Y* axis with labels that designate the units of the graph.

Broken-line graphs

Figure 22.13 shows the steps involved in drawing a linear coordinate graph. Because the data points are one year apart on the *X* axis, it is impossible to assume the change in the data is a smooth, continual progression from point to point. Therefore, the points are connected with a broken-line curve.

For the best appearance, the plotted points should not be crossed by the curve or grid lines of the graph (Fig. 22.14). Each circle or symbol used to plot points should be about $\frac{1}{8}$ in. (5 mm) in diameter. Figure 22.15 shows several approved symbols and lines.

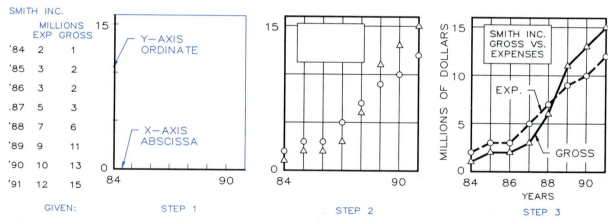

FIGURE 22.13 Construction of a broken-line graph.

Given A record of the Smith Company's gross and expenses.

Step 1 The vertical and horizontal axes are laid off to provide space for the largest values.

Step 2 The points are plotted over the respective years. Different symbols are used for each curve.

Step 3 The data points are connected with straight lines, the axes are labeled, the graph is titled, and the lines are strengthened.

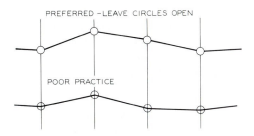

PREFERRED –LEAVE CIRCLES OPEN

POOR PRACTICE

FIGURE 22.14 The curve of a graph should be drawn from point to point, but it should not close up the symbols used to locate the plotted points.

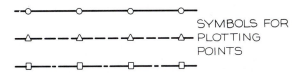

SYMBOLS FOR PLOTTING POINTS

FIGURE 22.15 Any of these symbols or lines can be effectively used to represent different curves on a single graph. The symbols are about $\frac{1}{8}''$ (3 mm) in diameter.

COMPUTER METHOD The points on a graph can be drawn as CIRCLEs, DONUTs (open and closed), or POLYGONs. The grid lines and curves that pass through the open symbols can be easily removed by using the TRIM command as shown in Fig. 22.16.

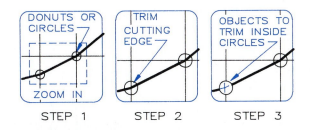

DONUTS OR CIRCLES

ZOOM IN

STEP 1

TRIM CUTTING EDGE

STEP 2

OBJECTS TO TRIM INSIDE CIRCLES

STEP 3

FIGURE 22.16   Editing data points.

Step 1 Open data points can be plotted on a graph as CIRCLEs, DONUTs, or POLYGONs. To remove lines from inside the open points, ZOOM in on several points.

Step 2 Command: TRIM (CR)
Select cutting edges (s):...
Select objects: (Select the circle.) (CR)

Step 3 Select object to trim: (Select the lines inside the circle, and they will be removed.) Continue this process for all points.

The title of a graph can be placed in any of the positions shown in Fig. 22.17. The title should never be as meaningless as ''graph'' or ''coordinate graph.'' Instead, it should explain the graph by giving the important information such as the company, date, source of data, and general comparisons being shown.

The proper calibration and labeling of axes is important to the appearance and use of a graph. Figure 22.18A shows a properly executed axis. The axis in Fig. 22.18B has too many grid lines and too many

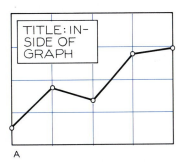

TITLE: IN-SIDE OF GRAPH

A

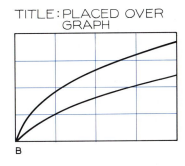

TITLE: PLACED OVER GRAPH

B

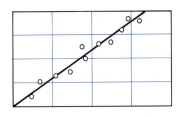

TITLE: PLACED UNDER GRAPH

C

FIGURE 22.17 Title placement on a graph.

A. The title of a graph can be lettered inside a box placed within the area of the graph. The perimeter lines of the box should not coincide with grid lines within the graph.

B. The title can be placed over the graph. The title should be drawn in $\frac{1}{8}''$ letters or slightly larger.

C. The title can be placed under the graph. It is good practice to be consistent when a series of graphs is used in the same report.

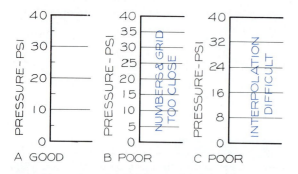

FIGURE 22.18 The scale at A is the best. It has about the right number of grid lines and divisions, and the numbers are given in well-spaced, easy-to-interpolate form. The numbers at B are too close, and there are too many grid lines. The units at C make interpolation difficult by eye.

units labeled along the axis. The units selected in Fig. 22.18C make it difficult to interpolate between the labeled values; for example, it is more difficult to locate a value such as 22 by eye on this scale than on the one in Fig. 22.18A.

Smooth-line graphs

The strength of cement as related to curing time is plotted in Fig. 22.19. Since you know the strength of cement changes gradually in relation to curing time, the data points are connected with a smooth curve rather than a broken-line curve. Even if the data points do not lie on the curve, you can be reasonably

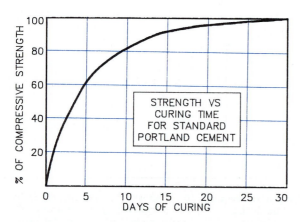

FIGURE 22.19 When the process graphed involves gradual, continuous changes in relationships, the curve should be drawn as a smooth line.

certain the deviation is due to errors of measurement or the methods used in collecting the data.

Similarly, the strength of clay tile, as related to its absorption characteristics, is an example of data that yield a smooth curve (Fig. 22.20). Note that the plotted data do not lie on the curve. Since you know this relationship should be represented by a smooth curve, the **best curve** is drawn to interpret the data to give an average representation of the points.

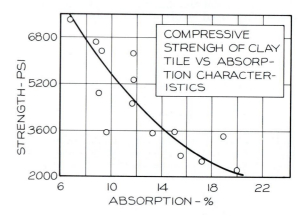

FIGURE 22.20 If it is known that a relationship plotted in a graph should yield a smooth gradual curve, a smooth-line best curve is drawn to represent the average of the plotted points. You must use your judgment and knowledge of the data in cases of this type.

There is a smooth-line curve relationship between miles per gallon and the speed at which a car is driven. In Fig. 22.21, two engines are compared with two smooth-line curves. Figure 22.22 compares the effect of speed on several automotive characteristics.

When a smooth-line curve is used to connect data points, the implication is that you can **interpolate** between the plotted points to estimate other values. Points connected by a broken-line curve imply you cannot interpolate between the plotted points.

Straight-line graphs

Some graphs have neither broken-line curves nor smooth-line curves but straight-line curves as Fig. 22.23 shows. Using this graph, you can determine a

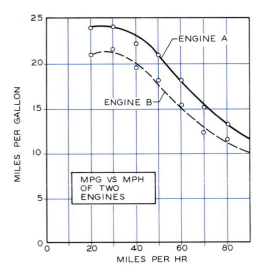

FIGURE 22.21 These are best curves that approximate the data without necessarily passing through each data point. Inspecting the data tells you this curve should be a smooth-line curve rather than a broken-line curve.

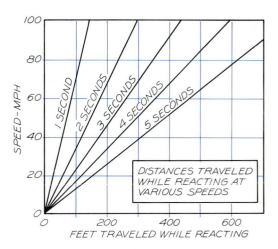

FIGURE 22.23 A graph can be used to determine a third value when two variables are known. Taking this information from a graph is easier than computing each answer separately.

third value from the two given values. For example, if you are driving 70 miles per hour and it takes 5 seconds to react and apply your brakes, you will have traveled 500 feet in this time.

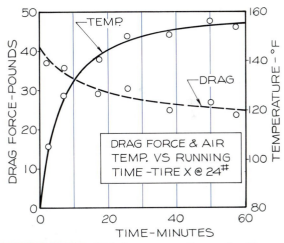

FIGURE 22.24 This is a composite graph with different scales along each Y axis. The curves are labeled so that reference can be made to the applicable scale.

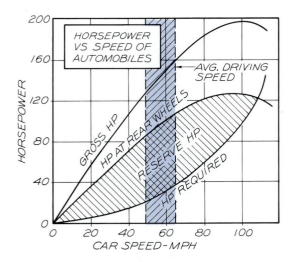

FIGURE 22.22 A linear coordinate graph is used here to analyze data affecting the design of an automobile's power system.

Two-scale coordinate graphs

Graphs can be drawn with different scales in combination, such as the one shown in Fig. 22.24. The vertical scale at the left is in units of pounds, and the one at the right is in degrees of temperature. Both curves are drawn using their respective Y axes, and each curve is labeled.

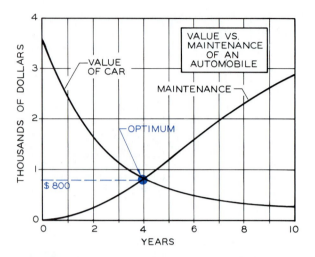

FIGURE 22.25 This graph shows the optimum time to sell a car based on the intersection of two curves that represent the depreciation of the car's value and its increasing maintenance costs.

With graphs of this type, care must be taken to avoid confusing the reader. These graphs are effective when comparing related variables, such as the drag force and air temperature of a tire, as shown in this example.

Optimization graphs

Figure 22.25 shows the optimization of the depreciation of an automobile and its increase in maintenance

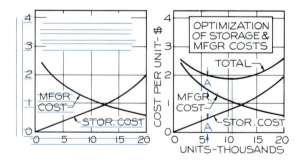

FIGURE 22.26 Optimization graphs.

Step 1 Lay out the graph, and plot the given curves.

Step 2 Add the two curves to find a third curve. Distance A is shown transferred to locate a point on the third curve. The lowest point of the "total" curve is the optimum point of 11,000 units.

costs. These two sets of data are plotted, and the curves cross at an X axis value of four years. At this time, the cost of maintenance is equal to the value of the car, indicating this might be a desirable time to exchange it for a new one.

Another optimization graph is constructed in Fig. 22.26. The manufacturing cost per unit is reduced as more units are made, but the warehousing cost increases. By adding the two curves, a third curve is found in Step 2. The "total" curve tells you the optimum number to manufacture at a time is about 11,000 units. When more or fewer units are manufactured, the expense per unit is greater.

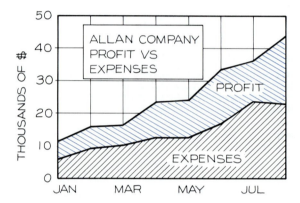

FIGURE 22.27 This graph is a combination of a coordinate graph and an area graph. The upper curve represents the total of two values plotted, one above the other.

Composite graphs

The graph in Fig. 22.27 is a composite (or combination) of an area graph and a coordinate graph. The lower curve is plotted first. The upper curve is found by adding the values to the lower curve so that the two areas represent the data. The upper curve is equal to the sum of the two Y values.

Figure 22.28 shows a combination of a coordinate graph and a bar graph used to illustrate the Dow Jones industrial stock average. The bars represent the daily ranges in the index. The broken-line curve connects the points where the market closed for each day.

Break-even graphs

Break-even graphs help evaluate the marketing and manufacturing costs used to determine the selling cost

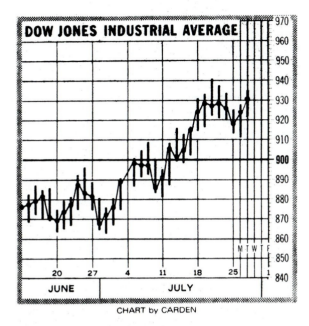

FIGURE 22.28 This graph is a combination of a coordinate graph and a bar graph. The bars represent the ranges of selling during a day, and the broken-line curve connects the points at which the market closed each day.

of a product. The break-even graph in Fig. 22.29 reveals that 10,000 units must be sold at $3.50 each to cover the costs of manufacturing and development. Sales in excess of 10,000 result in profit.

A second type of break-even graph (Fig. 22.30) uses the cost of manufacturing per unit versus the number of units produced. In this example, the development costs must be incorporated into the unit costs. The manufacturer can determine how many units must be sold to break even at a given price or the price per unit if a given number is selected. In this example, a sales price of $0.80 requires that 8400 units be sold to break even.

22.6
Logarithmic coordinate graphs

Both scales of a logarithmic grid are calibrated into divisions equal to the logarithms of the units represented. Commercially printed logarithmic grid paper is available in many variations for graphing data.

The graph in Fig. 22.31 has a logarithmic grid and shows the geometry of standard railroad cars as they relate to the tracks in order to not exceed projection

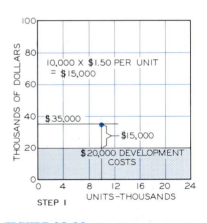
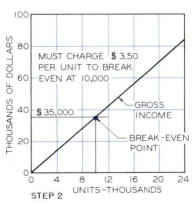
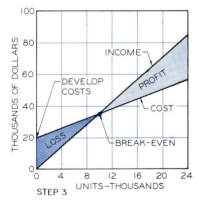

FIGURE 22.29 Break-even graph.

Step 1 The graph is drawn to show the cost ($20,000 in this case) of developing the product. Each unit would cost $1.50 to manufacture. This is a total investment of $35,000 for 10,000 units.

Step 2 To break even at 10,000, the units must be sold for $3.50 each. Draw a line from zero through the break-even point for $35,000.

Step 3 The loss is $20,000 at zero units and becomes progressively less until the break-even point is reached. The profit is the difference between the cost and income to the right of the break-even point.

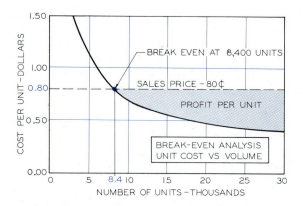

FIGURE 22.30 The break-even point can be found on a graph that shows the relationship between the cost per unit, which includes the development cost, and the number of units produced. The sales price is a fixed price. The break-even point is reached when 8400 units have been sold at $0.80 each.

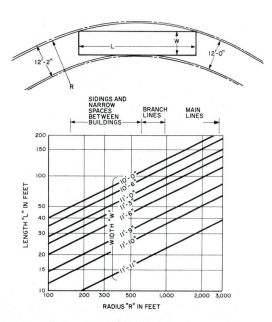

FIGURE 22.31 This logarithmic graph shows the maximum load projection of 12 ft in relation to the length of a railroad car and the radius of the curve. (Courtesy of Plant Engineering.)

width of 12 ft around curves. Extremely large values can be shown on logarithmic grids since the lengths are considerably compressed.

22.7
Semilogarithmic coordinate graphs

Semilogarithmic graphs are called **ratio graphs** because they give graphical representations of ratios.

One scale, usually the vertical scale, is logarithmic, and the other is linear (divided into equal divisions). Parallel curves on a semilogarithmic graph have equal percentage increases.

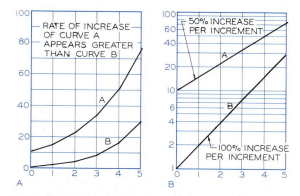

FIGURE 22.32 When plotted on a standard grid, curve A appears to be increasing at a greater rate than curve B. However, the true rate of change can be seen when the same data are plotted on a semilogarithmic graph in part B.

Figure 22.32 shows the same data plotted on a linear grid and on a semilogarithmic grid. The semilogarithmic graph reveals that the percent of change from 0 to 5 is greater for curve B than for curve A since curve B is steeper. This comparison was not apparent in the plot on the linear grid.

Figure 22.33 shows the relationship between the linear scale and the logarithmic scale. Equal divisions along the linear scale have unequal ratios, and equal divisions along the log scale have equal ratios.

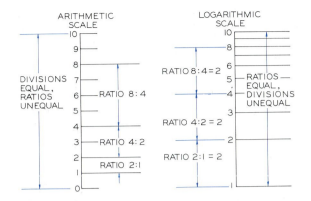

FIGURE 22.33 The spacings on an arithmetic scale are equal, with unequal ratios between points. The spacings on logarithmic scales are unequal, but equal spaces represent equal ratios.

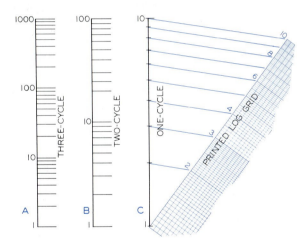

FIGURE 22.34 Logarithmic paper can be either bought or drawn using several cycles. Three-, two-, and one-cycle scales are shown here. Calibrations can be drawn on a scale of any length by projecting from a printed scale, as shown in part C.

Log scales can be drawn to have one or many cycles. Each cycle increases by a factor of 10. For example, the scale in Fig. 22.34A is a three-cycle scale, and the one in Fig. 22.34B is a two-cycle scale. When these must be drawn to a special length, commercially printed log scales can be used to graphically transfer the calibrations to the scale being drawn (Fig. 22.34C).

In Fig. 22.35, the calibrations along the log scale are separated by the difference in their logarithms. The logarithms are laid off using a scale calibrated in decimal divisions. It can be seen in Fig. 22.35B that parallel straight-line curves yield equal ratios of increase. Figure 22.36 is an example of a semilogarithmic graph used to present industrial data.

Semilog graphs are sometimes misunderstood by people who do not realize they are different from linear coordinate graphs; also, zero values cannot be shown on log scales.

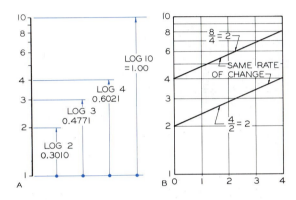

FIGURE 22.35 **A.** A number's logarithm is used to locate its position on a log scale. **B.** This makes it possible to see the true rate of change at any location on a semilogarithmic graph.

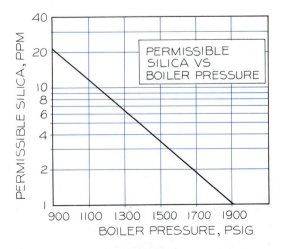

FIGURE 22.36 A semilogarithmic graph is used to compare the permissible silica (parts per million) in relation to the boiler pressure.

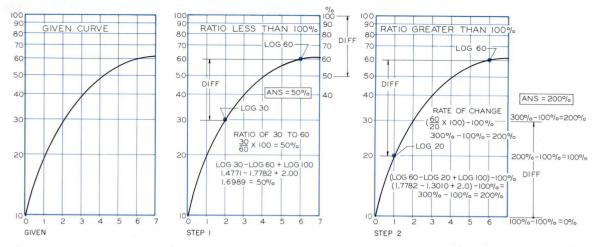

FIGURE 22.37   Percentage graphs.

Given   The data are plotted on a semilogarithmic graph to enable you to determine percentages and ratios in much the same manner that you use a slide rule.

Step 1   In finding the percent that a smaller number is of a larger number, you know that the percent will be less than 100%. The log of 30 is subtracted from the log of 60 with dividers and is transferred to the percent scale at the right, where 30 is found to be 50% of 60.

Step 2   To find the percent of increase, a smaller number is divided into a larger number to give a value greater than 100%. The difference between the logs of 60 and 20 is found with dividers and is measured upward from 100% at the right to find that the percent of increase is 200%.

Percentage graphs

The percent that one number is of another, or the percent increase of one number that is greater than the other, can be determined by using a semilogarithmic graph (Fig. 22.37).

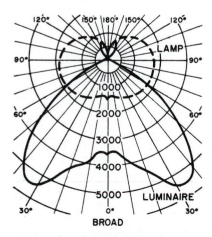

FIGURE 22.38   A polar graph is used to show the illumination characteristics of luminaires.

Data plotted in Step 1 are used to find the percent that 30 is of 60, two points on the curve. The vertical distance between them is equal to the difference of their logarithms. This distance is subtracted from the log of 100 at the right of the graph to give a value of 50% as a direct reading.

In Step 2, the percent of increase between two points is transferred from the grid to the lower end of the log scale and measured upward since the increase is greater than zero. These methods can be used to find percent increases or decreases of any set of points on the grid.

22.8
Polar graphs

Polar graphs are drawn with a series of concentric circles with the origin at the center. Lines are drawn from the center toward the perimeter of the graph, where the data can be plotted through 360° by measuring values from the origin. For example, the illumination of a lamp is shown in Fig. 22.38, where the maximum lighting of the lamp is 550 lumens at 35° from the vertical.

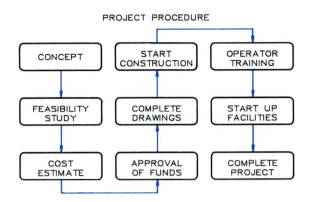

PROJECT PROCEDURE

FIGURE 22.39 This schematic shows a block diagram of the steps required to complete a project.

This type of graph is used to plot the areas of illumination of all types of lighting fixtures and other applications. Polar graph paper is available commercially for drawing graphs of this type.

22.9
Schematics

The **block diagram** in Fig. 22.39 shows the steps required to complete a construction project. Each step is blocked in and connected with arrows to explain the sequence of events.

The organization of a company or group of people can be depicted in an **organizational chart** like the one shown in Fig. 22.40. The offices represented by

the blocks in the lower part of the graph are responsible to the offices represented by the blocks above them. The lines of authority connecting the blocks suggest the routes for communication from one office to another in an upward or downward direction.

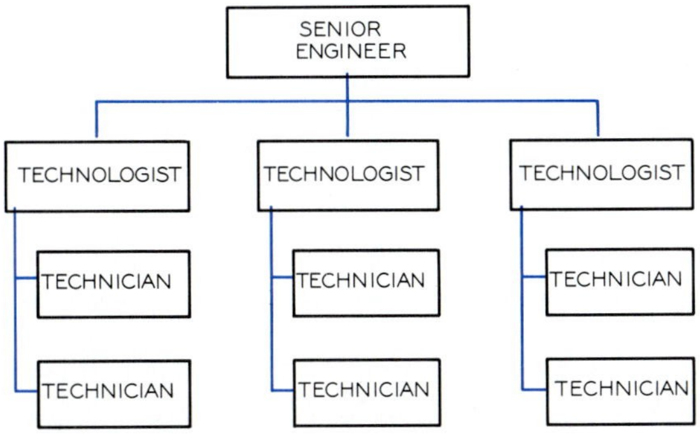

The drawing in Fig. 22.41 is not a graph, nor is it a true view of the apparatus; instead, it is a **schematic** that effectively shows how the parts and their functions relate to one another.

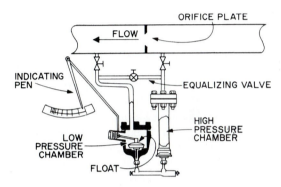

FIGURE 22.41 A schematic showing the components of a gauge that measures the flow in a pipeline. (Courtesy of Plant Engineering.)

Geographical graphs are used to combine maps and other relationships such as weather (Fig. 22.42). Different symbols represent the annual rainfall in various areas of the nation.

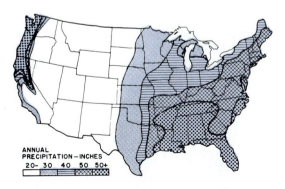

ANNUAL PRECIPITATION—INCHES
20- 30   40   50 50+

FIGURE 22.42 A map chart that shows the weather characteristics of various geographical areas. (Courtesy of the Structural Clay Products Institute.)

FIGURE 22.40 This schematic shows the organization of a design team in a block diagram.

Problems

The problems below are to be drawn on Size A sheets in pencil or ink, as specified. Follow the techniques covered in this chapter and the examples given as you solve the problems.

	Percent of Labor Force	
Ages	Graduates	Dropouts
16–17	18	22
18–19	12.5	17.5
20–21	8	13
22–24	5	9

Pie graphs

1. Draw a pie graph that compares the employment of male youth between the ages of 16 and 21: Operators–25%; Craftsmen–9%; Professionals, technicians, and managers–6%; Clerical and sales–17%; Service–11%; Farm workers–13%; and Laborers–19%.

2. Draw a pie graph that shows the relationship between the following members of the technological team: Engineers–985,000; Technicians–932,000; Scientists–410,000.

3. Construct a pie graph of the following percentages of the employment status of graduates of two-year technician programs one year after graduation: Employed–63%, Continuing full-time study–23%; Considering job offers–6%; Military–6%; Other–2%.

4. Construct a pie graph that shows the relationship between the types of degrees held by engineers in aeronautical engineering: Bachelor–65%; Master–29%; Ph.D.–6%.

5. Draw a pie graph for the following average annual expenditures of a state on public roads: Construction–$13,600,000; Maintenance–$7,100,000; Purchase and upkeep of equipment–$2,900,000; Bonds–$5,600,000; Engineering and administration–$1,600,000.

6. Draw a pie graph that shows the data given in Problem 10.

7. Draw a pie graph that shows the data given in Problem 11.

Bar graphs

8. Draw a bar graph that depicts the unemployment rate of high school graduates and dropouts in various age categories given in the following table:

9. Draw a single bar that represents 100% of a die casting alloy. The proportional parts of the alloy are as follows: Tin–16%; Lead–24%; Zinc–38.8%; Aluminum–16.4%; Copper–4.8%.

10. Draw a bar graph that compares the number of skilled workers employed in various occupations. Arrange the graph for ease of interpretation and comparison of occupations. Use the following data: Carpenters–82,000; All-round machinists–310,000; Plumbers–350,000; Bricklayers–200,000; Appliance servicers–185,000; Automotive mechanics–760,000; Electricians–380,000; Painters–400,000.

11. Draw a bar graph that represents the flow of a river in cubic feet per second (cfs) as shown in the following table. Show bars that represent the data for ten days in the first month. Omit the second month.

Day of Month	Rate of Flow (in 100 cfs)	
	1st Month	2nd Month
1	19	19
2	130	70
3	228	79
4	129	33
5	65	19
6	32	14
7	17	15
8	13	11
9	22	19
10	32	27

12. Draw a bar graph that shows the airline distances in statute miles from New York to the cities listed below. Arrange the bars in ascending or descending order.

TABLE 22.1

	1890	1900	1910	1920	1930	1940	1950	1960	1970	1980	1990
Supply	80	90	110	135	155	240	270	315	380	450	460
Demand	35	35	60	80	110	125	200	320	410	550	570

Berlin 3965

Buenos Aires 5300

Honolulu 4960

London. 3465

Manila 8510

Mexico City 2090

Moscow 4665

Paris. 3634

Tokyo 6740

13. Draw a bar graph that compares the corrosion resistance of the materials listed in the table below.

	Loss in Weight (%)	
	In Atmosphere	In Sea Water
Common steel	100	100
10% nickel steel	70	80
25% nickel steel	20	55

14. Draw a bar graph using the data in Problem 1.

15. Draw a bar graph using the data in Problem 2.

16. Draw a bar graph using the data in Problem 3.

17. Construct a rectangular grid graph to show the accident experience of Company A. Plot the numbers of disabling accidents per million person-hours of work on the *Y*-axis. Years will be plotted on the *X*-axis. Data: 1973–0.63; 1974–0.76; 1975–0.99; 1976–0.95; 1977–0.55; 1978–0.76; 1979–0.68; 1980–0.55; 1981–0.73; 1982–0.52; 1983–0.46; 1984–0.53; 1985–0.49; 1986–0.55.

Linear coordinate graphs

18. Using the data given in Table 22.1, draw a linear coordinate graph that compares the supply and demand of water in the United States from 1890 to 1990 in billions of gallons of water per day.

19. Present the data in Table 22.2 in a linear coordinate graph to decide which lamps should be selected to provide economical lighting for an industrial plant. The table gives the candlepower directly under the lamps (0°) and at various angles from the vertical when the lamps are mounted at a height of 25 ft.

20. Construct a linear coordinate graph that shows the relationship in energy costs (mills per kilowatt-hour) and the percent capacity of two types of power plants. Plot energy costs along the *Y* axis, and the capacity factor along the *X* axis. The plotted curve will compare the costs of a nuclear plant with a gas- or oil-fired plant. Gas-fired plant data: 17 mills, 10%; 12 mills, 20%; 8 mills, 40%; 7 mills, 60%; 6 mills, 80%; 5.8 mills, 100%. Nuclear-plant data: 24 mills, 10%; 14 mills, 20%; 7 mills, 40%; 5 mills, 60%; 4.2 mills, 80%; 3.7 mills, 100%.

21. Plot the data from Problem 17 as a linear coordinate graph.

22. Construct a linear coordinate graph to show the relationship between the transverse resilience in inch-pounds (*Y* axis) and the single-blow impact in foot-pounds (*X* axis) of gray iron. Data: 21 fp, 375 ip; 22 fp, 350 ip; 23 fp, 380 ip; 30 fp, 400 ip; 32 fp, 420 ip; 33 fp, 410 ip; 38 fp, 510 ip; 45 fp, 615 ip; 50 fp, 585 ip; 60 fp, 785 ip; 70 fp, 900 ip; 75 fp, 920 ip.

TABLE 22.2

Angle with vertical	0°	10°	20°	30°	40°	50°	60°	70°	80°	90°
Candlepower (thous.) 2–400W	37	34	25	12	5.5	2.5	2	0.5	0.5	0.5
Candlepower (thous.) 1–1000W	22	21	19	16	12.3	7	3	2	0.5	0.5

23. Draw a linear coordinate graph to compare the two sets of data in the following table: capacity vs. diameter, and capacity vs. weight of a brine cooler. The horizontal scale is to be tons of capacity, and the vertical scales are to be outside diameter on the left and weight (cwt) on the right.

Tons Refrigerating Capacity	Outside Diameter (in)	Weight (cwt)
15	22	25
30	28	46
50	34	73
85	42	116
100	46	136
130	50	164
160	58	215
210	60	263

Use 20 × 20 graph paper 8½" × 11". Horizontal scale of 1" = 40 tons. Vertical scales of 1" = 10" of outside DIA and 1" = 40 cwt (hundred weight).

24. Draw a linear coordinate graph that shows the voltage characteristics for a generator as given in the following table of values—abscissa-armature current in amperes (I_a); ordinate-terminal voltage in volts (E_t):

I_a	E_t	I_a	E_t	I_a	E_t
0	288	31.1	181.8	41.5	68
5.4	275	35.4	156	40.5	42.5
11.8	257	39.7	108	39.5	26.5
15.6	247	40.5	97	37.8	16
22.2	224.5	40.7	90	13.0	0
26.2	217	41.4	77.5		

25. Draw a linear coordinate graph for the centrifugal pump test data in the table below. The units along the X axis are to be gallons per minute. There will be four curves to represent the variables given.

Gallons per Minute	Discharge Pressure	Water HP	Electric HP	Efficiency (%)
0	19.0	0.00	1.36	0.0
75	17.5	0.72	2.25	32.0
115	15.0	1.00	2.54	39.4
154	10.0	1.00	2.74	36.5
185	5.0	0.74	2.80	26.5
200	3.0	0.63	2.83	22.2

26. Draw a linear coordinate graph that compares two of the values shown in Table 22.3—ultimate strength and elastic limit—with degrees of temperature labeled along the X axis.

27. Draw a linear coordinate graph that compares two of the values shown in the table in Problem 26—percent of elongation and percent of reduction of area of the cross section—with the degrees of temperature that will be represented along the X axis.

Break-even graphs

28. Construct a break-even graph that shows the earnings for a new product that has a development cost of $12,000. The first 8000 will cost $0.50 each to manufacture, and you wish to break even at this quantity. What would be the profit at volumes of 20,000 and 25,000?

29. Same as Problem 28 except that the development costs are $80,000, the manufacturing cost of the first 10,000 is $2.30 each, and the desired break-even point is 10,000. What would be the profit at volumes of 20,000 and 30,000?

30. A manufacturer has incorporated the manufacturing and development costs into a cost-per-unit estimate. He wishes to sell the product at $1.50 each. On the Y axis, plot cost per unit in dollars; on the X axis, plot number of units in thousands. Data: 1000, $2.55; 2000, $2.01; 3000, $1.55; 4000, $1.20; 5000, $0.98; 6000, $0.81; 7000, $0.80; 8000, $0.75; 9000, $0.73; 10,000, $0.70. How many must be sold to break even? What will be the total profit when 9000 are sold?

31. The cost per unit to produce a product by a manufacturing plant is given below. Construct a break-even graph with the cost per unit plotted on the Y axis and the number of units on the X axis. Data: 1000, $5.90; 2000, $4.50; 3000, $3.80; 4000, $3.20; 5000, $2.85; 6000, $2.55; 7000, $2.30; 8000, $2.17; 9000, $2.00; 10,000, $0.95.

Logarithmic graphs

32. Using the data given in Table 22.4, construct a logarithmic graph where the vibration amplitude (A) is plotted as the ordinate, and the vibration frequency (F) is plotted as the abscissa. The data for curve 1 represent the maximum limits of ma-

TABLE 22.3

°F	Ultimate Strength	Elastic Limit	Elongation (%)	Reduction of Area (%)	Brinell Hardness No.
400	257,500	208,000	10.8	31.3	500
500	247,000	224,500	12.5	39.5	483
600	232,500	214,000	13.3	42.0	453
700	207,500	193,500	15.0	47.5	410
800	180,500	169,000	17.0	52.5	358
900	159,500	146,500	18.5	56.5	313
1000	142,500	128,500	20.3	59.2	285
1100	126,500	114,000	23.0	60.8	255
1200	114,500	96,500	26.3	67.8	230
1300	108,000	85,500	25.8	58.3	235

chinery in good condition with no danger from vibration. The data for curve 2 are the lower limits of machinery that is being vibrated excessively to the danger point. The vertical scale should be three cycles, and the horizontal scale should be two cycles.

33. Plot the data below on a two-cycle log graph to show the current in amperes (Y axis) versus the voltage in volts (X axis) of precision temperature-sensing resistors. Data: 1 volt, 1.9 amps; 2 volts, 4 amps; 4 volts, 8 amps; 8 volts, 17 amps; 10 volts, 20 amps; 20 volts, 30 amps; 40 volts, 36 amps; 80 volts, 31 amps; 100 volts, 30 amps.

34. Plot the data from Problem 18 as a logarithmic graph.

35. Plot the data from Problem 24 as a logarithmic graph.

Semilogarithmic graphs

36. Construct a semilogarithmic graph with the Y axis a two-cycle log scale from 1 to 100 and the X axis a linear scale from 1 to 7. Plot the data below to show the survivability of a shelter at varying distances from a one-megaton bomb exploding in air. The data consists of overpressure in psi along the Y axis, and distance from ground zero in miles

along the X axis. The data points represent an 80% chance of survival of the shelter. Data: 1 mile, 55 psi; 2 miles, 11 psi; 3 miles, 4.5 psi; 4 miles, 2.5 psi; 5 miles, 2.0 psi; 6 miles, 1.3 psi.

37. The growth of two divisions of a company, Division A and Division B, is given in the data below. Plot the data on a rectilinear graph and on a semilog graph. The semilog graph should have a one-cycle log scale on the Y axis for sales in thousands of dollars, and a linear scale on the X axis showing years for a six-year period. Data in dollars: 1 yr, A = \$11,700 and B = \$44,000; 2 yr, A = \$19,500 and B = \$50,000; 3 yr, A = \$25,000 and B = \$55,000; 4 yr, A = \$32,000 and B = \$64,000; 5 yr, A = \$42,000 and B = \$66,000; 6 yr, A = \$48,000 and B = \$75,000. Which division has the better growth rate?

38. Draw a semilog chart showing probable engineering progress. Use the following indices: 40,000 B.C. = 21; 30,000 B.C. = 21.5; 20,000 B.C. = 22; 16,000 B.C. = 23; 10,000 B.C. = 27; 6000 B.C. = 34; 4000 B.C. = 39; 2000 B.C. = 49; 500 B.C. = 60; A.D. 1900 = 100. Horizontal scale 1" = 10,000 years. Height of cycle = about 5". Two-cycle printed paper may be used if available.

39. Plot the data from Problem 24 as a semilogarithmic graph.

TABLE 22.4

F	100	200	500	1000	2000	5000	10,000
A(1)	0.0028	0.002	0.0015	0.001	0.0006	0.0003	0.00013
A(2)	0.06	0.05	0.04	0.03	0.018	0.005	0.001

40. Plot the data from Problem 26 as a semilogarithmic graph.

Percentage graphs

41. Plot the data given in Problem 18 on a semilog graph to determine the percentages and ratios of the data. What is the percent of increase in the demand for water from 1890 to 1920? What percent of demand is the supply for 1900, 1930, and 1970?

42. Using the graph plotted in Problem 37, determine the percent of increase of Division A and Division B from Year 1 to Year 4. What percent of sales of Division A are the sales of Division B at the end of Year 2? At the end of Year 6?

43. Plot two values from Problem 26—water horsepower and electric horsepower—on semilog paper compared with gallons per minute along the X axis. What percent of electric horsepower is water horsepower when 75 gallons per minute are being pumped? What is the percent increase of the electric horsepower from 0 to 185 gallons per minute?

Organizational charts

44. Draw an organizational chart for a city government organized as follows: The electorate elects school board, city council, and municipal court officers; the city council is responsible for the civil service commission, city manager, and city planning board; the city manager's duties cover finance, public safety, public works, health and welfare, and law.

45. Draw an organizational chart for a manufacturing plant. The sales manager, chief engineer, treasurer, and general manager are responsible to the president. The general manager has three department heads: master mechanic, plant superintendent, and purchasing agent. The plant superintendent has charge of the shop foremen, under whom are the working forces, and direct charge of the shipping, tool and die, inspection, order, and stores and supplies departments.

Polar graphs

46. Construct a polar graph of the data given in Problem 19.

47. Construct a polar graph of the following illumination, in lumens at various angles, emitted from a luminaire. The zero-degrees position is vertically under the overhead lamp. Data: 0°, 12,000; 10°, 15,000; 20°, 10,000; 30°, 8000; 40°, 4200; 50°, 2500; 60°, 1000; 70°, 0. The illumination is symmetrical about the vertical.

AutoCAD Computer Graphics

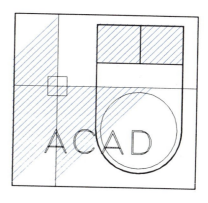

23.1

Introduction

This chapter is devoted to the coverage of computer graphics using AutoCAD software (Release 11) on an IBM AT 286 or compatible computer with a 20 MB hard disk, a mouse (or tablet) and an A-B plotter (Fig. 23.1). AutoCAD is used as a basis for presenting microcomputer graphics because it is the most widely used software for the microcomputer. Although CAD software differs, much of what is learned on one system is transferrable to another system.

The coverage of computer graphics has been introduced in previous chapters to demonstrate how it can be applied to specific types of problems. However, these last two chapters appear at the end of the book so that any last-minute updates from AutoCAD can be included before the book goes to press.

Although commands and data can be input through the keyboard when using AutoCAD, the drafter will be more productive if a mouse or tablet is used. AutoCAD operates on a two-button mouse (Fig. 23.2), where the left button is used to select entities, and the right button acts as a carriage return, which is abbreviated throughout the text as (CR).

Most of AutoCAD's principles and commands are covered in this chapter and previous chapters. However, a great many more sophisticated applications are possible through the application of specialized commands and LISP programming. Several AutoCAD manuals accompany the software that can be used as references for these specialized applications.

FIGURE 23.1 View of an Engineering Design Graphics computer laboratory at Texas A&M University.

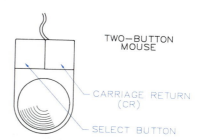

FIGURE 23.2 The left button on the mouse is the select button, and the right button is the carriage return (CR) button.

Release 11

The AutoCAD Release 11 covered in this chapter is an upgrade of Release 10 that can be used in the two-dimensional and three-dimensional manner as Release 10 was used.

A major addition revision is the introduction of paper space (PSPACE) for two-dimensional drawings and notations, and model space (MSPACE) for three-dimensional drawings. Both types of drawings can be merged together in a single drawing and plotted on paper at the same time.

A second software package, and addition to Release 11 (386 version), is Advanced Modeling Extrusion (AME), which is used for solid modeling. An introduction to AME is given in Chapter 24.

23.2
Starting up

Begin AutoCAD by booting up the system and typing ACAD. The Main Menu will appear on the screen (Fig. 23.3). The abbreviation (CR) is used throughout the text to represent the carriage return or the ENTER key. The number 1 was entered to specify that a new drawing, called DRAW1, is to be drawn. Names of drawings can be no longer than eight characters with

no blank spaces, exclusive of drive prefix (A:, for example). AutoCAD automatically adds a file type of .DWG following the assigned name, which will be filed as A:DRAW1.DWG. The Main Menu leaves the screen, and the drawing area with the Root Menu ready for making a drawing is displayed.

Had you wished to edit a previous drawing, you could have responded:

> Enter selection: 2 (Edit an existing drawing.) (CR)
> Enter NAME of drawing: A:DRAW2 (CR)

The drawing, A:DRAW2, would then be displayed on the screen as it appeared when last exited from. You may now edit (change) the drawing as a continuation of the last drawing session.

If you have forgotten the names of the drawing files in memory, enter selection 6 from the Main Menu and enter 1 for a listing of them. Type 0 to return to the Main Menu, and enter 2 to name the drawing to edit.

23.3
Experimenting

The beginner who has not used any version of computer graphics, or AutoCAD in particular, can benefit from turning the machine on, loading the system by typing ACAD, and responding to the screen prompts to see how much could be understood from the prompts alone. For example, to draw a line, you would select DRAW from the Root Menu, and LINE from the submenu. By experimenting, you will become familiar with the system, its prompts, and menus. But you will soon find the need for specific instructions.

23.4
Introduction to plotting

You may wish to make a plot of your drawing before ending your first session on the computer. To do so, see that a pen is inserted properly in your plotter and that a sheet of paper is loaded. (You may need help from your instructor on how your plotter works.)

```
Main Menu
  0.   AutoCAD
  1.   Begin a NEW drawing
  2.   Edit and EXISTING drawing
  3.   Plot a drawing
  4.   Printer Plot a drawing

  5.   Configure AutoCAD
  6.   File Utilities
  7.   Compile shape/font description file
  8.   Convert old drawing file
  9.   Recover damaged file

Enter selection:
```

FIGURE 23.3 **After booting up AutoCAD, the Main Menu will appear on the screen. Enter 1 to begin a new drawing, A:DRAW1.**

Type the command <u>PLOT</u>, and press (CR). You will be prompted at the command line as follows:

```
What to plot-Display, Extents,
Limits, View, or Window <D>: E
```
(E for Extents.)

The graphics on the screen will be replaced with text that gives the current plotting specifications. For now, look at the current PLOT ORIGIN and the SCALE. If they are not 0,0 and F, respectively, type <u>Y</u> in response to the last question:

```
Do you want to change any-
thing? <N>: Y
```

The current settings will be shown, one at a time on the screen, and each setting can be changed by typing new values after each prompt. But for now, change only the ORIGIN to 0,0 and SCALE to F for FIT, which means the plot will be sized to fit the extents of the paper size. Continue by pressing (CR) at the other prompts until the plotter begins to plot your drawing.

There is more to know about plotting, but this will introduce you to how your plotter works. In Section 23.81, we give a more detailed coverage of plotting.

23.5
Shutting down

The first step in turning off the machine is to return the Main Menu, and select FILE from the pull-down menu, which gives SAVE, END, QUIT, PLOT, FILES and PRINT.

END By selecting END, your drawing is saved under the name it was given at the start of the session; your drawing leaves the screen, and the Main Menu reappears. If you were editing a previous drawing and had selected END, the last version of the drawing would become an automatic back-up file (A:DRAW1.BAK), and the version displayed on the screen would become the saved drawing file (A:DRAW1.DWG).

SAVE The SAVE command shows a dialogue box on the screen (Fig. 23.4), which lists the drawing on

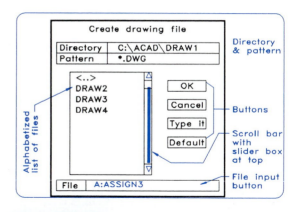

FIGURE 23.4 This dialogue box appears when you SAVE a drawing. It shows the files on the current drive and enables you to save the default file or to type in the name of the file by selecting Type It.

the drive (disk) being used. The directory box shows the directory and the pattern box shows the file pattern (file extension).

By selecting the File box with the cursor, you can type the name of the file to be saved, or, you can use the cursor to select the desired name from the list of files. You can select a file from another directory by cursor-selecting the Directory box and entering the directory name, or by picking the directory from the listing (in angle brackets).

If you would rather type the information, select the Type It box and the dialogue box disappears, returning you to type at the command line at the bottom of the screen. Select the Default box to return the dialogue box to its initial settings. The Cancel box is selected to cancel the current command and return to the command prompt. Select the OK box to accept the information in the dialogue box, execute the command, and close the box.

By setting the system variable FILEDIA to 0 (off) instead of 1 (on), this dialogue box will be disabled, requiring that all responses be typed at the command line.

If a selected file has the same name as an existing file, a warning box will appear, permitting you to replace it or to cancel the command.

The drawing file is saved under its current name shown in angles (<A:DRAW1>, for example) in the File box after OK is selected, without leaving the Drawing Editor. If you wish to save a duplicate copy of the drawing under a second name (A:DRAW1A,

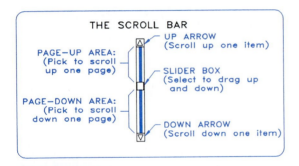

Many dialogue boxes have scroll bars to permit you to scroll through a list slowly or quickly.

for example), type A:DRAW1A into the File box when prompted. Enter QUIT.

Notice that the Create drawing file dialogue box in Fig. 23.4 has a Scroll Bar at its side, as shown in Fig. 23.5. By cursor-selecting the top or bottom arrows, you can scroll through the list of files one at a time. You can scroll up or down one page at a time by selecting bars on each side of the slider box; or, you can pick the slider box and drag it up and down for viewing the list. Slider bars appear on several dialogue boxes within AutoCAD.

QUIT If you are editing a drawing, and have not SAVEd during the session, and wish to return to the Main Menu discarding all changes made during the current session, type QUIT. You will be prompted, Do you really want to discard all changes to the drawing? By responding Yes or Y you will be returned to the Main Menu, and the drawing will be left unchanged.

EXIT Type 0 to exit AutoCad and return to DOS (disk operating system). Turn the monitor and computer off to end the session. Do not turn off the computer until returning the Main Menu to the screen.

WARNING! When working from your own disk, not the hard disk, do not remove the disk until the Main Menu has returned to the screen.

23.6
Drawing layers

AutoCAD provides an infinite number of layers to make a drawing on. Each layer is assigned its own name, color, and line type. For example, you may have a layer name HIDDEN that appears on the screen in yellow, and the line types drawn on it will be dashed lines to represent hidden lines.

Layers can be used to reduce the duplication of effort. For example, architects commonly use the same floor plan for several different applications: one for dimensions, one for furniture arrangement, one for floor finishes, one for electrical details, and so forth. The same basic plan is used for all these applications by turning on the needed layers and turning off others.

SETTING LAYERS The layers shown in Fig. 23.6 are sufficient for most working drawings. Layers are given names that correspond with their line types, and each is assigned a different color to distinguish it on a color monitor. The 0 (zero) layer is the system default layer, which can be turned off but not deleted.

LAYERS By selecting LAYERs from the Root Menu, you will receive a subcommand of LAYER, which is selected from the keyboard or by mouse. The following is AutoCAD's dialogue with you while you are setting layers:

```
Command: LAYER (CR)
?/Set/New/On/Off/Color/Ltype/
```

Layer Name	State	Color		LineType
0	On	7	(white)	CONTINUOUS
VISIBLE	On	1	(red)	CONTINUOUS
HIDDEN	On	2	(yellow)	HIDDEN
CENTER	On	3	(green)	CENTER
HATCH	On	4	(cyan)	CONTINUOUS
DIMEN	On	5	(blue)	CONTINUOUS
CUT	On	6	(magenta)	PHANTOM
WINDOW	On	7	(white)	CONTINUOUS

The following layers and their settings are typical of those needed on a format file.

LINETYPES

```
CONTINUOUS _____
DASHED  _  __  __  __  __  __  __  __
HIDDEN  _ _ _ _ _ _ _ _ _ _ _ _ _ _ _
CENTER  ____  _  ____  ____  __  _
PHANTOM  _____  _  _  _____  _  _
DOT  . . . . . . . . . . . . . . . .
DASHDOT  ____  .  ____  .  ____  .
BORDER  __ __ . __ __ . __ __ . __ .
DIVIDE  ____  _  ____  _  ____  _
```

FIGURE 23.7 **These are the** LINETYPES **available in AutoCAD.**

```
Freeze/Thaw: New or N (CR)
Layer name(s): VISIBLE,
HIDDEN,CENTER,HATCH,DIMEN,
CUT,WINDOW (CR)
```

This series of responses has created seven new layers by name, and each will have a white default color and a continuous line type.

COLOR   To set the COLOR of each layer, we must respond in the following manner:

```
?/Set/NEW/ON/OFF/COLOR/Ltype/
Freeze/Thaw: COLOR or C (CR)
Color: RED (or 1) (CR)
Layer name(s) for color 1
(red) <VISIBLE>: VISIBLE (CR)
```

The VISIBLE layer has been assigned the color red, which could have been assigned with the number 1 instead of the word, RED. The colors of the other layers must be assigned in the same manner, one at a time.

LINE TYPES   are assigned to layers in the following manner:

```
?/Set/On/Off/Color/Ltype/
Freeze/Thaw: Ltype (or L) (CR)
Linetype (or ?) <CONTINUOUS>:
HIDDEN (CR)
```

```
Layer name(s) for linetype
HIDDEN <0>: HIDDEN (CR)
```

The lines on the HIDDEN layer will now be drawn with dashed (hidden) lines. By typing LINETYPE, and pressing (CR) twice, the line types available from AutoCAD will be displayed (Fig. 23.7).

WARNING!   If you wish to set LAYERS by using a dialogue box (shown in Fig. 23.13), do not use the defaults of BYLAYER and BYBLOCK. These formats assume the characteristics of the current layer rather than those of the named layers.

SET   We must SET a layer to make it the current layer in order to draw on it by responding in the following manner:

```
?/Set/New/On/Off/Ltype/Freeze/
Thaw: SET (or S) (CR)
New current layer <0>:
VISIBLE (CR) (CR) (Two carriage returns.)
```

ON/OFF   Any line drawn will appear on the VISIBLE layer in red with continuous lines. Once we have defined a number of layers and have drawn on each, they can be turned on or off by the ON/OFF command in the following manner:

```
?/Set/New/On/Off/Ltype/Freeze/
Thaw: ON (or OFF)
Layer name(s) to turn on:
VISIBLE (or * for all layers)
(Or VISIBLE, HIDDEN, CENTER, to turn
these layers ON or OFF.) (CR)
```

Even though several layers are on, you can draw on only one layer, the **current layer.** By selecting the question-mark option (?), the screen will display the current listing of the layers, their line types, colors, and on/off status. Press function key F1 to change the screen back to the graphics editor.

FREEZE AND THAW   FREEZE and THAW are options under the LAYER command. By FREEZE-ing a specified layer, it will not be redrawn or plotted until it has been THAWed; thus a drawing can be regenerated much faster on the screen than when OFF is used. This option allows unneeded layers to be turned off much like the layer option, OFF. To save your file of layers and their specifications, use the command END, which will return you to the Main Menu.

But before we END, let's set additional parameters as shown in the next section.

23.7
Setting screen parameters

To set screen parameters, you must become familiar with the commands under SETTINGS of the Root Menu which are LIMITS, GRID, UNITS, SNAP, AXIS, ORTHO, DRAGMODE, BLIPMODE, LTSCALE, and the function keys.

LIMITS The LIMITS command is used to establish the size of the screen area that represents the area of the paper a drawing will be plotted on. A full-size drawing that will fit on a size A sheet (11 × 8.5 inches) will have a drawing area of about 10.1 × 7.8 inches for most plotters. If millimeters are used, the limits will be approximately 257, 198. LIMITS are set as follows:

```
Command: LIMITS (CR)
ON/OFF Lower Left corner:
<0.00,0.00>: (CR) (To accept default
value.)
Upper right corner:
<36.00,24.00>: 10,7.8 (CR)
```

GRID A grid on the screen can be set in the following manner:

```
Command: GRID (CR)
On/Off/Value (X)/Aspect
<0.00>: .2 (CR)
```

This command paints the screen with a square pattern of dots that are each 0.2 in. apart. To make the newly assigned limits fill the working area of the screen, use the ZOOM/ALL command from the DISPLAY command on the Root Menu.

UNITS The UNITS of dimensions used in a drawing must be assigned before the LIMITS can be set. By typing the UNITS command, AutoCAD will respond with a list of formats to select from (Fig. 23.8). Select the type of units, and respond to the following

```
Command: UNITS

Report formats:
    1.  Scientific        1.66E+01
    2.  Decimal           16.60
    3.  Engineering       1'—4.60"
    4.  Architectural     1'—4 5/8"
    5.  Fractional        16 5/8

Enter choice, 1 to 5 <default>: 2
```

FIGURE 23.8 **The UNITS command is used to assign the form of the dimensioning units used for measurements.**

prompts to specify the details desired, such as the number of decimal places. Decimal units should be used for the metric system (millimeters) and the English system (inches). To return to the Drawing Editor, press function key F1.

SNAP The SNAP command can be used to make the cursor on the screen snap to a visible or invisible grid. If your grid is set at spacings of 0.2 in., the cursor can be made to snap to the grid or to an invisible 0.1-in. grid between the visible grid points (Fig. 23.9). Press function key F9 to turn SNAP on or off.

```
Command: SNAP (CR)
On/Off/Value/Aspect/
Rotate/Style: 0.1 (CR)
```

AXIS The AXIS command is used to display ruler lines of a specified spacing on the screen across the

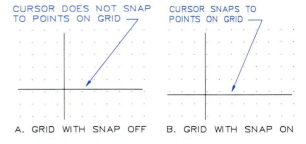

CURSOR DOES NOT SNAP TO POINTS ON GRID

CURSOR SNAPS TO POINTS ON GRID

A. GRID WITH SNAP OFF   B. GRID WITH SNAP ON

FIGURE 23.9 **The SNAP command can be used to make the cursor stop on visible or invisible grids on the screen.**

bottom and right sides in the following manner:

> Command: AXIS (CR)
> On/Off/Tick spacing
> (X)/Aspect: 0.20 (CR)

This response sets the ruler with tick marks 0.20 units apart. By responding with 5X, the tick marks are spaced at a multiple of five times the snap resolution, which places tick marks at every fifth snap point. The ASPECT option allows you to set different values on each scale of the axis.

ORTHO   The ORTHO command forces all lines to be either vertical or horizontal, which aids in making orthographic views where most lines meet at right angles.

> Command: ORTHO (Select ON or OFF from the screen menu.) (CR)

ORTHO can be turned off by using Ctrl O or by pressing function key F8 (Fig. 23.10). Function keys can be used to turn screen parameters on and off once they have been set. **Running coordinates** can be obtained at the top of the screen to give the coordinates of the cursor's position by pressing F6. The **status line** (Fig. 23.11) at the top of the screen gives the current layer by name; it shows when SNAP and ORTHO are turned on by displaying these words.

DRAGMODE   When an object is moved, it can be dynamically moved across the screen by setting the DRAGMODE command to ON. If this command was

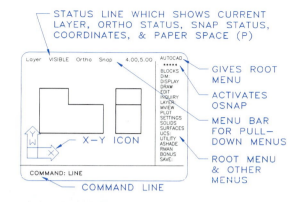

FIGURE 23.11   **The menu area of the screen when AutoCAD is in use. When * * * * is selected, the** OSNAP **mode is activated.**

set to OFF, the object would be moved, but its image would appear only after its destination point had been selected. By setting DRAGMODE to AUTO, all objects that can be dragged are dragged automatically without prompting.

BLIPMODE   By setting BLIPMODE to ON, a temporary blip (+ sign) will appear on the screen when a point is selected with the cursor. Blips are removed by pressing function key F7 twice. If this command is set to OFF, no blips will appear on the screen.

LTSCALE   Hidden lines, centerlines, phantom lines, and other noncontinuous lines can be drawn with varying sizes of dashes and spaces by changing the LTSCALE value to a larger or smaller number.

FORMAT FILE   If the previous parameters were specified, they would be saved when you ENDed the file. This file should be saved with its parameters as an ''empty'' file with no drawing, and named A:FORMAT so that it can be called up as a drawing to begin a new drawing on. Once the drawing is completed in this FORMAT file, SAVE the drawing, and give it a name different from A:FORMAT; then QUIT, and respond YES when asked if you wish to discard all changes in the drawing. In this way, you have returned to the original file, A:FORMAT, without disturbing its parameters, allowing it to be used again, and you have saved a drawing that was made from it.

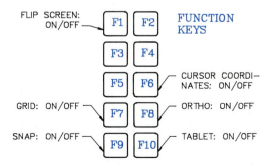

FIGURE 23.10   **The function keys on the IBM (or compatible) keyboard can be used to activate six functions of AutoCAD.**

ROOT MENU By placing the cursor on AUTO-CAD, the Root Menu can be recalled. The four stars, ****, can be selected to activate the OSNAP command (Fig. 23.11).

Several submenus are longer than will fit on the screen at one time, so the NEXT command is used to obtain them.

MENUS Commands can be executed from the **keyboard,** the **root menu,** or the **pull-down menu bar.** The keyboard is used like a typewriter to type in commands that are shown at the **command line** as they are typed (Fig. 23.11). Your mouse or digitizer can be used to select commands from the Root Menu at the right of the screen by moving to the desired command and pressing the left mouse button.

Pull-down menus

Many pull-down menus and dialogue boxes can be selected by selecting the heading on the menu bar at the top of the screen with a mouse, as shown in Fig. 23.12. Only one heading can be pulled down at a time. Commands under the headings are given in Fig.

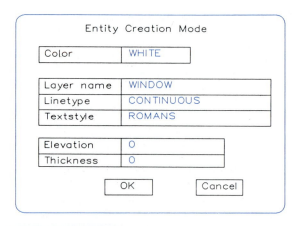

FIGURE 23.13 **This dialogue box is located under the** Entity Creation **heading under the pulldown menu heading "Options."**

23.12. Commands with an angle, >, at the right have additional subheadings that will be displayed when they are selected.

Figure 23.13 shows a dialogue box that is filed under the headings OPTIONS and ENTITY CRE-

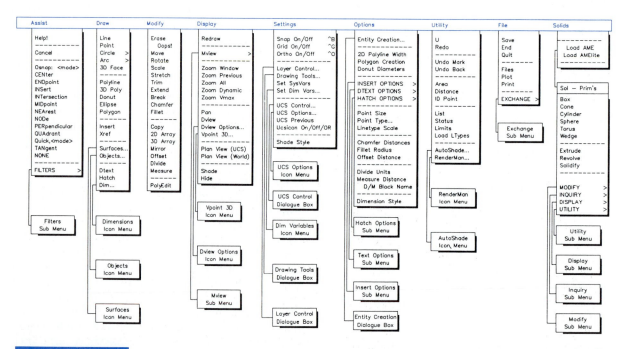

FIGURE 23.12 **The pull-down menu bar at the top of the screen lists the commands that can be selected with your cursor.**

tttthithithithithithith

ATION. By selecting the boxes with the cursor, you can type in the data to create a layer. It is best to assign colors and line types BYLAYER rather than using the BYBLOCK setting offered as a default.

The Layer Control dialogue box (Fig. 23.14) located under the SETTINGS heading of the Menu Bar can be used to change the status of a layer. It can be used to turn layers on or off, freeze or thaw them, assign the current layer, or create a new one.

The Drawing Aids dialogue box (Fig. 23.15) is also found under the SETTINGS heading of the Menu Bar. It can be used to set SNAP specifications,

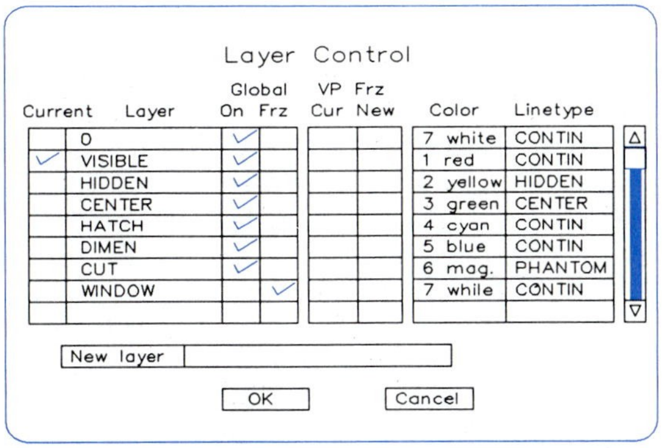

FIGURE 23.14 **This dialogue box is located under the** Layer Control **heading under the pull-down menu** SETTINGS **heading. It is used to control the layers in the current file.**

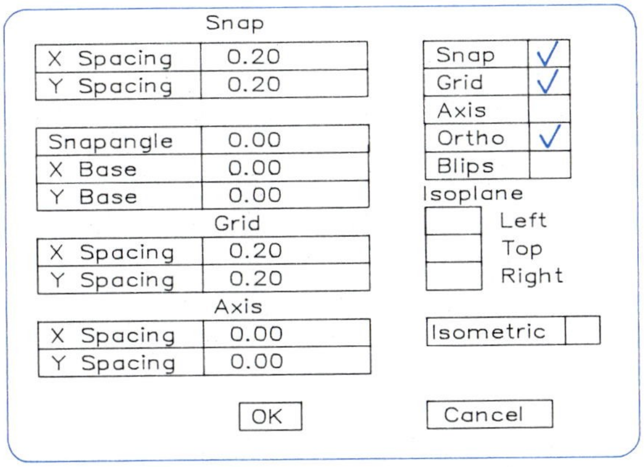

FIGURE 23.15 **The** Drawing Aids **dialogue box can be used to change a variety of settings.**

turn ON or OFF the GRID, AXIS, ORTHO, BLIP-MODE, or ISOMETRIC features.

Dynamic dialogue boxes (DDboxes)

The dialogue boxes mentioned above that are used by the pull-down menu are also available by typing at the command line (Example: Command: DDEDIT. This will give a dialogue box on the screen for editing text.) Each dialogue box begins with DD, an abbreviation for dynamic dialogue, and some are preceded by an apostrophe, such as 'DDEMODES. The apostrophe means that the dialogue box can be used "transparently," which means it can be used while in another command.

Although there are dialogue boxes within dialogue boxes to aid you in making selections, the basic dialogue box commands are

DDATTE (Dynamic Dialog Attribute Edit) is used to edit ATTRIBUTES. (See Section 23.82.)

DDEDIT (Dynamic Dialogue Edit) gives a dialogue box that can be used for editing TEXT that has been previously drawn on the screen.

'DDEMODES (Dynamic Dialogue Entity Modes) is used to control the color, line type, layer, elevation, and thickness of entities.

'DDLMODES (Dynamic Dialogue Layer Modes) is used to control the LAYER command.

'DDRMODES (Dynamic Drawing Modes) allows you to set drawing aids such as AXIS, GRID, ISOPLANE, ORTHO, SNAP, and BLIPMODE.

DDUCS (Dynamic Dialog User Coordinate System) enables you to manage the UCS command by a dialogue box. (See Section 23.63.)

Control-C

To exit from an existing operation or a command, such as removing the Help text from the screen, press and hold down the Ctrl key and then press the C key, as shown in Fig. 23.16. This will give the command: prompt at the bottom of the screen ready for another new command. Control-C is also used to abort a print or a plot operation, but it does not stop the drawing immediately since the buffer must empty its data first.

STEP 1   STEP 2

FIGURE 23.16 **By holding down the** `Ctrl` **button and pressing the** `C` **key, you can end the current operation and go back to the command mode for entering a new command.**

```
Command: FILES

    0.  Exit File Utility Menu
    1.  List Drawing files
    2.  List user specified files
    3.  Delete files
    4.  Rename files
    5.  Copy file
    6.  Unlock file

Enter selection (0 to 6) <0>: 1
```

FIGURE 23.17 **While in the drawing editor, you can use the** `FILES` **option to perform any of these operations on files.**

23.8
Utility commands

Utility commands can be used to control the operation of AutoCAD and make changes in the files that are developed.

HELP The HELP (`?`) command gives a list of commands on the screen. Press `Ctrl C` to abort the listing process. Press `F1` to return to the graphics mode on the screen. To obtain additional information on a specific command, respond to the prompts as follows:

```
Command: HELP (or ?) (CR)
Command name (Return for
list): LINE (CR)
```

The screen will give information about the `LINE` command.

RENAME The RENAME command is used to change the name of a block, layer, linetype, text style, named view, `UCS`, or viewport. The command prompts you for the `Old name` and the `New name`. Continuous lines and layer 0 cannot be RE-NAMEd.

FILES The `FILES` command permits you to list, delete, rename, and copy files from the **Drawing Editor.** By typing `FILES`, the listing will appear on the screen, as shown in Fig. 23.17. Option 1, `List Drawing files`, is used to list the files on a

specified drive. Enter A: if that is the disk drive you wish to have searched.

Option 2, `List user specified files`, can be used to display the files that you specify. For example, by responding to `ENTER FILE SEARCH SPECIFICATIONS:` with `A:*.BAK`, you will get a listing of all drawing files with a suffix `.BAK` that are stored on disk drive A.

Option 3, `Delete files`, allows you to eliminate unneeded files on your disks. You can specify a file such as `A:DRAW1.DWG`, or you can use wild cards such as `A:*.DWG` to delete all files with a `DWG` suffix. When using wild cards, each file that meets the specifications will be listed one at a time, and you are given the option to delete them by entering `Y` or `N` as they are listed.

Option 4, `Rename files`, is used to change the name of a file by responding to the prompts as follows:

```
Enter current file name:
A:DRAW1.DWG (CR)
Enter new file name:
A:DRAW2.DWG (CR)
```

Option 5, `Copy file`, lets you copy a file with the following prompts:

```
Enter name of Source file
name: B:DRAW1.DWG (CR)
Enter name of Destination
file name: C:DRAW1.DWG (CR)
```

Option 6, `Unlock file`, is used to unlock a file as follows:

```
Enter locked file(s) specifi-
cations: \DIR1\DRAW1.DWG
The file: \DIR1\DRAW1.DWG was
locked by J. Doe at 18:81 on
10/26/1991
Do you still wish to
unlock it? <Y> Y
Lock was successfully removed.
1 file unlocked. Press RETURN
to continue: (CR)
```

SHELL The `SHELL`, or `SH`, command gives access to the DOS operating system while remaining in the **Drawing Editor** by responding as follows:

```
Command: SHELL (CR)
DOS command: DIR A: (Or similar com-
mand.) (CR)
Command: (Reappears on screen.)
```

PURGE The `PURGE` command can be used when you **first** begin editing an existing drawing in the following manner:

```
Command: PURGE (CR)
Purge unused Blocks/Layers/
LTypes/SHapes/STyles?
All: ALL (CR)
```

The `ALL` response is used to eliminate any unused objects that are in the drawing file. The other options can be used to purge specific features of a drawing. You will be prompted for each `PURGE` one at a time.

ALIASES The ALIASES (shortened abbreviations) command, can be typed at the keyboard for the commands shown in Fig. 23.18 to speed up the typing process. The command will be accepted as well if it is typed in its entirety such as `MOVE` instead of `M`.

23.9
Custom-designed lines

In Section 23.6, we introduced you to AutoCAD's `LINETYPE`s. By typing `LINETYPE`, (CR), `?`, the types and names of lines available will be displayed on the screen (Fig. 23.7). These lines can be assigned to a layer by selecting the `LTYPE` option of the `LAYER` command and typing in the preferred line type, such as hidden, center, or dashed.

These lines are composed of lines, points, and spaces. You can get long dashes or short dashes by the `LTSCALE` command and the assignment of an `LTSCALE` factor to lengthen or shorten the dashes. When this command is used, **all** lines composed of dashes and spaces are changed globally on the screen. Therefore, you may find a need for a line, such as a centerline, with smaller spaces and shorter dashes for small circles.

You can custom design line types by using the `LINETYPE` command and its `CREATE` option (Fig. 23.19). In Step 1, you will be prompted for the line's name and the file it is to be stored in. In Step 2, a description can be given by typing in keyboard dashes and dots to represent the line. When asked to enter the line pattern, begin with a dash (a positive length) or a dot, represented by a zero. Spaces are represented by minus values. Spaces, dots, and dashes are separated by commas. Only one typical pattern that will be used in multiples need be shown, for example, dash, -space, short dash, -space.

Use the `LINETYPE` command and its `LOAD` option to gain access to the newly designed line style that can be assigned to a `LAYER` by the `LTYPE` option (Step 3).

ALIASES (KEYBOARD ABBREVIATIONS)

A	ARC	M	MOVE
C	CIRCLE	MS	MSPACE
CP	COPY	P	PAN
DV	DVIEW	PS	PSPACE
E	ERASE	PL	PLINE
L	LINE	R	REDRAW
LA	LAYER	Z	ZOOM

FIGURE 23.18 The `ALIASES`, or abbreviations, command can be typed at the keyboard in their shortened form to speed up typing.

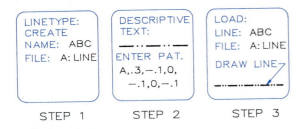

STEP 1   STEP 2   STEP 3

FIGURE 23.19 **Custom-designed** LINETYPES.

Step 1 Command: LINETYPE (CR)
?/Create/Load/Set: C (CR)
Name of line type to create: ABC (CR)
File for storage of line type
<default>: A:LINE(CR)

Step 2 Descriptive text: (**Type pattern.**) (CR)
Enter pattern (on next line): A, .3,
−.1,0,−.1,0,−.1 (CR)

Step 3 **To load the new line type:**
Command: LOAD (CR)
Name of line type to load: ABC (CR)
File to search <default>: A:LINE (CR)
(**The new line can be assigned to a layer and used.**)

23.10
Making a drawing—lines

Turn your computer on and obtain AutoCAD's **Main Menu** by the method required by your system. It might involve entering a sequence of commands, such as C:CD\ACAD (CR) and then ACAD (CR). Once you have the **Main Menu,** type 1, Begin a NEW drawing, if you are going to set your drawing parameters from scratch. If you wish to use your FORMAT file that was developed in Section 23.7 with its assigned parameters, type 2, Edit an EXIST-ING drawing. When prompted for a filename, type A:FORMAT, and the file will be loaded into the drawing editor, and the screen will appear ready for a drawing to be made.

If you wish to draw a line, you have three methods that can be used: the **Root Menu,** the keyboard, or the pull-down menu bar. If you select DRAW and then LINE from the Root Menu (Fig. 23.20), the command line on the screen will prompt you as follows:

Command: LINE (CR)
From point: (Select a point or give coordinates.)

To point: (Select second point or give coordinates.)
To point: (You can continue in this manner with a series of connected lines.)

The lines are drawn by moving the cursor about the screen and selecting endpoints by pressing the select button of the mouse (the left button). The current line will **rubber band** from its last point (Fig. 23.21), and lines can be drawn in succession until you press (CR) on the keyboard, or the right button of your mouse.

The pull-down menu bar is used by moving the cursor to the top of the screen until the menu headings appear. Select DRAW and its subcommand, LINE, and draw the line in the usual manner.

A comparison of absolute and polar coordinates are shown in Fig. 23.22. The keyboard can be used for

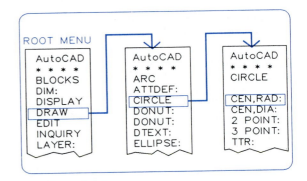

FIGURE 23.20 **You may progress from the Root Menu to levels of submenus by selecting commands with the keyboard or mouse.**

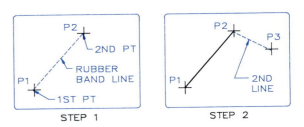

STEP 1   STEP 2

FIGURE 23.21 **Drawing a** LINE.

Step 1 Command: LINE
From point: (**Select** P1.)
To point: (**Select** P2.) (**The line is drawn.**)

Step 2 To point: (**Select** P3.) (**The line is drawn.**)
((CR) **to disengage the rubber band.**)

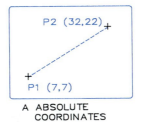
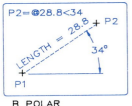

A ABSOLUTE
COORDINATES

B POLAR
COORDINATES

FIGURE 23.22 (A) Absolute coordinates can be entered at the keyboard to establish ends of a line. (B) Polar coordinates are relative to the current point on the screen and are given with a length and the angle with the horizontal measured clockwise.

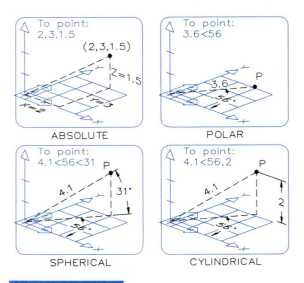

ABSOLUTE

POLAR

SPHERICAL

CYLINDRICAL

FIGURE 23.23 A point in a drawing may be located using any of the formats shown here.

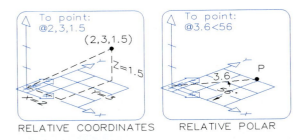

RELATIVE COORDINATES

RELATIVE POLAR

FIGURE 23.24 A point in a drawing may be located relative to the last point on the screen by either of these methods.

drawing lines in any of the following coordinates shown in Fig. 23.23: **absolute, polar, spherical,** or **cylindrical.**

ABSOLUTE COORDINATES Three-dimensional coordinates can be typed in at the keyboard in the form of 2,3,1.5, which gives the coordinates of a point from the origin of 0,0,0.

POLAR COORDINATES Two-dimensional coordinates can be typed in at the keyboard in the form of 3.6<56 to draw a line from 0,0,0 and an angle of 56 degrees with the X-axis in the X-Y plane.

SPHERICAL COORDINATES Three-dimensional coordinates can be typed in at the keyboard in the form of 4.1<56<31 to locate a point 4.1 units from 0,0,0, 56 degrees with the X-axis in the X-Y plane, and 31 degrees with the X-Y plane (Fig. 23.23).

CYLINDRICAL COORDINATES Three-dimensional coordinates can be typed in at the keyboard in the form of 4.1<56,2 to locate a point 4.1 units from the 0,0,0, 56 degrees with the X-axis in the X-Y plane, and 2 units higher in the Z-direction.

Relative coordinates can be used to locate points with respect to the current location of the cursor, as shown in Fig. 23.24.

RELATIVE COORDINATES These coordinates are X-, Y-, and Z-values from the current point on the screen of a three-dimensional point. The coordinates are preceded with the @ symbol, for example, @2,3,1.5.

RELATIVE POLAR COORDINATES These coordinates can be typed in at the keyboard to give the relative distance from the current point in the form of @3.6<56 to locate a point 3.6 units from the current point and 56 degrees with the X-axis in the X-Y plane.

LAST COORDINATES These coordinates can be found by typing @ at the keyboard to move the cursor back to the last point.

WORLD COORDINATES These coordinates are coordinates that can be used to locate points in the World Coordinate System regardless of the User Coordinate System being used by preceding the coordinates with an asterisk (*). Examples are *4,3,5; *90<44,2; and @*1,3,4.

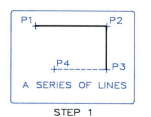

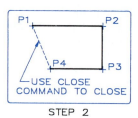

A SERIES OF LINES — STEP 1

USE CLOSE COMMAND TO CLOSE — STEP 2

FIGURE 23.25 CLOSE **command.**

Step 1 A series of lines are drawn using the LINE command from *P1* through *P3*. (The last line is a rubber-band line until the next point is selected.)

Step 2 After *P4* has been selected, select CLOSE, and the line will be drawn to the first point of the series.

You can correct errors when typing commands by one of the following methods:

Ctrl X (Deletes the line.)

Ctrl C (Cancels the current command and returns the Command: prompt.)

Backspace (Deletes one character at a time.)

The **status** line at the top of the screen shows the length of the line and angle from the last point when you are rubber-banding from point to point. When drawing a continuous series of lines, the CLOSE command can be used to draw the last line to the beginning point, as Fig. 23.25 shows.

23.11
Selection of entities

One of the most often-occurring prompts is Select objects:, which asks you to select an entity, or set of entities, that are to be ERASEd, CHANGEd, or modified in other ways. Several methods of SELECTing entities are available. You may select entities one at a time with the cursor; use a window, a crossing window, or a box; type LAST or PREVIOUS; or use the auto option.

Figure 23.26 illustrates how a single entity can be selected with the cursor. Or, when prompted, Select objects:, M (multiple) can be typed at the keyboard, and several entities can be selected one at a time.

When prompted to select, the window option (W) can be used, after which you will be prompted for the first and second diagonals of a window. Only entities totally within the window will be selected (Fig. 23.26).

The crossing option (C) shown in Fig. 23.26, selects the entities within the window and those crossing the window. When prompted to select objects, BOX can be typed at the keyboard, and a window or a crossing window can be used by selecting the second diagonal to the right or to the left, respectively (Fig. 23.26).

Selection can be made by last (L), which selects the most recently created object.

The previous option (P) allows you to recall the last selected set of objects for editing. For example, a number of objects can be selected for MOVEing, then can be MOVEd. After entering the MOVE command and typing P, the last group of entities are remembered and can be MOVEd to a new position.

When selecting a set of objects, you can remove each of them in reverse order one at a time by typing U (undo) as many times as needed.

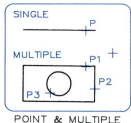

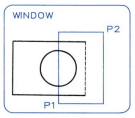

POINT & MULTIPLE — WINDOW — CROSSING — BOX

FIGURE 23.26 **Object select options. When prompted by a command to** Select objects:, **you may select them as single or multiple points. A window option** (W) **can be used to select objects that lie completely within the windows, or crossing option** (C) **can be used to select objects within or crossed by the window. The** BOX **option can be used for both the window and crossing options, determined by the sequence in which the diagonal corners are selected.**

While in the object selection process, you may remove a selected object by typing R (remove) and selecting the objects to be removed. To add others to the set, type A (add) and select the objects to be added. Once the selection is finished to your satisfaction, give a null reply (press CR) to end the Select/remove object: prompt.

By setting the selection option to SI (single) the object, or sets of objects, will be acted upon without pausing for interaction with the drafter.

23.12
Erasing and breaking lines

The ERASE command, a subcommand under EDIT, (or MODIFY), is used to remove parts of a drawing. The LAST option calls for the erasure of single entities (lines, text, circles, arcs, and blocks) one at a time working backward from the one most recently drawn. The use of a WINDOW is the second option for erasing, where entities that lie completely within the window will be erased as shown in Fig. 23.27.

```
Command: ERASE (CR)
Select  objects  or  Window  or
Last: WINDOW (or W) (CR)
First corner: (Select P1.)
Other corner: (Select P2.) 4 found
Select objects: (CR)
(The entities are erased.)
```

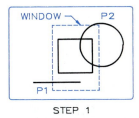
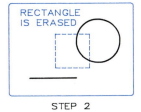

STEP 1 STEP 2

FIGURE 23.27   ERASE **command.**

Step 1 Command: ERASE **(CR)**
Select objects: WINDOW or W **(CR)**
First corner: **(Select** *P1.***)**
Other corner: **(Select** *P2.***)**

Step 2 Select objects: **(CR)**
(The rectangle within the window is erased. Entities partially within the window are not erased.)

A similar method of erasing lines is the CROSSING (C) option shown in Fig. 23.28. Any entity (line, arc, circle, or text) that is crossed by the window is removed in its entirety. The default of the ERASE command is Select Objects, which requires that you point to entities on the screen with the cursor to indicate those to erase (Fig. 23.29). By pressing (CR) the entities are erased. By using the pull-down menu, the entities are erased as they are selected.

Entities lying within the window that should not be erased can be removed by using the REMOVE command and selecting the entities with the cursor. When selected, the entities become dashed lines on the screen, indicating they have been marked for erasure. Additional lines can be marked for erasure by the ADD command in the same manner as the REMOVE command but with an opposite effect.

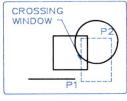
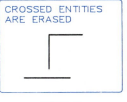

STEP 1 STEP 2

FIGURE 23.28   ERASE: **crossing option.**

Step 1 Command: ERASE **(CR)**
Select objects: C (Crossing) **(CR)**
First corner: P1 **(Select point.)**
Other corner: P2 **(Select point.)**

Step 2 Select objects: **(CR)**
(Any line crossed by the window is erased.)

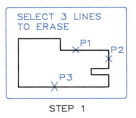
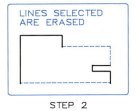

STEP 1 STEP 2

FIGURE 23.29   ERASE **command—entities.**

Step 1 Command: ERASE
Select Objects: **(Select entities, lines, with** *P1,* *P2,* **and** *P3.***)**

Step 2 Press (CR) again, and the lines are erased.
Use Command: OOPS **(CR) to recall erased entities.**

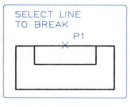

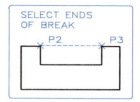

FIGURE 23.30 BREAK **command.**

Step 1 Command: <u>BREAK</u> **(CR)**
Select object: **(Select P1.) (On line to be broken)**
Enter second point or F: <u>F</u> **(CR)**

Step 2 Enter first point: **(Select P2.)**
Enter second point: **(Select P3.) (The line is broken from P2 to P3.)**

Should you mistakenly erase something, the OOPS command can be used to restore the last erasure, but only the last erasure.

The BREAK command is used to remove a part of an entity, such as a line (Fig. 23.30).

    Command: <u>BREAK</u> (CR)
    Select object: (Select point P1 on the line to be broken.)
    Enter second point or F:<u>F</u> (CR)
    Enter first point: (Select P2.)
    Enter second point: (Select P3.)
    (The line is broken from P2 to P3.)

Had the break not begun and ended at intersections of

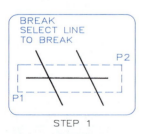

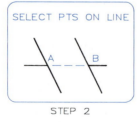

FIGURE 23.31 BREAK **command.**

Step 1 Command: <u>BREAK</u> **(CR)**
Select object: <u>WINDOW</u> or <u>W</u> **(CR)**

Step 2 Enter first point: **(Select A.)**
Enter second point: **(Select B.)**
(Line AB is removed. The OSNAP command can be used to select intersection points if needed.)

lines, you could have omitted the step in which the F response was given. But this extra step ensures that the computer understands which line is to be broken.

A portion of a line may be broken away by using the BREAK command (Fig. 23.31). By windowing the line to be broken and selecting two points at the ends of the break, that part of the line is removed.

23.13
UNDO command

The U command can be used to undo the latest entity placed on the screen. By entering U commands, one after another, the drawing can be erased back to its beginning point. Immediately after the U command, you may use the REDO command to bring back the deleted entity; OOPS will not work.

The UNDO command can be used for several different operations. Its options are as follows:

    Command: <u>UNDO</u> (CR)
    Auto/Back/Control/End/Group/
    Mark/<Number>: <u>4</u> (CR)

By entering <u>4</u>, it has the same effect as using the U command four separate times.

MARK can be used to mark a point in a drawing and then to add other experimental features that can be disposed of if desired by the BACK option. This will UNDO only that part of the drawing back to what was drawn when you used the MARK option. You will be prompted, This will undo everything. OK? <Y>. By responding <u>Y</u>, the mark will be removed also, making it possible for the next U or UNDO command to proceed backward past the mark.

The UNDO GROUP, UNDO END, and AUTO subcommands can be used to remove groups of entities at a time, but these options are meant to be used with menus.

The CONTROL subcommand has three options: ALL, NONE, and ONE. ALL turns on the full features of the UNDO command, and NONE turns off the features entirely. The ONE option limits U and UNDO commands to single operations and requires the least amount of disk space.

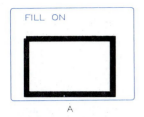

FIGURE 23.32 TRACE command. **(A) When** the TRACE command is used with FILL ON, lines **are drawn solid to the width specified. (B) When** FILL is turned off, parallel lines are drawn.

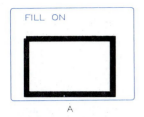

FIGURE 23.33 The pointer entities (markers) **used to locate points on a drawing can be selected and sized using the** PDMODE **and** PDSIZE SET VARs.

23.14
TRACE command

Wide lines, thicker than the point of a pen, can be drawn using the TRACE command in the following manner:

```
Command: TRACE (CR)
Width: 0.4 (CR)
From point: 2, 3 (CR)
To point: 4, 6 (CR)
To point: 6, 2 (CR)
```

With FILL on, the line will be drawn as shown in Fig. 23.32A. When FILL is off, the line will be drawn as parallel lines (Fig. 23.32B). The plotting of the line being entered will not be displayed until the endpoint of the next line is indicated in order for the program to compute the ''miter'' angles at each corner.

23.15
POINT command

The POINT command is used to locate points on a drawing. The point markers can be any of those shown in Fig. 23.33. The marker is selected by using two options under the SETVAR command: PDMODE and PDSIZE. The basic markers can be set in the following manner:

```
Command: SETVAR (CR) Variable
name or ?:PDMODE (CR)
```

```
New value for variable-name <0>:
3 (CR)
(Marker = X.)
```

The PDSIZE command is used to assign a size to the marker in the following way:

```
Command: SETVAR (CR) Variable
name or ?: PDSIZE (CR)
New value for variable-name
<0.000>: 0.1 (Size of marker.) (CR)
```

The PDSIZE variable gives the size of the marker on the drawing. If this value is a negative number, the marker will be sized as a percentage of the screen and will appear the same size regardless of the zooming that may occur. When entered as a positive number, the size of the marker will vary with each zoom.

Additional styles of markers can be obtained by adding 32, 64, and 96 to the basic values of 0 through 4 (Fig. 23.34).

When you are drawing lines and other entities, a plus sign (BLIP) is made temporarily on the screen to mark the points. When the screen is redrawn by pressing the F7 function key, they are removed since the plus signs are markers and not points on the screen. By turning the BLIPMODE command either ON or OFF, you choose whether or not to have BLIPS shown on the screen. It is recommended that beginners set the BLIPMODE to ON.

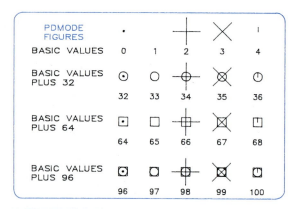

FIGURE 23.34 **The basic pointer symbols can be changed by adding 32, 64, and 96 to the basic symbols, 1 through 4.**

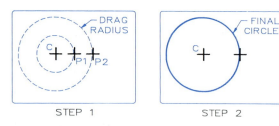

FIGURE 23.35 **Drawing a** CIRCLE.

Step 1 Command: CIRCLE (CR)
3P/2P/<Center point>: (Locate center c with cursor.)
Diameter/<Radius>: (Drag to select center P1, move to P2, and the circle will dynamically change if DRAGMODE is ON.)

Step 2 The final radius is selected with the select button, and the circle is drawn.

23.16
Drawing circles

The command for drawing circles is found under the DRAW submenu, where you may give the center and radius, the center and diameter, or three points. Used with DRAG, you can move the cursor and see the circle change until it is the size you want it to be (Fig. 23.35). DRAG can be turned ON or OFF by inserting the command, DRAGMODE, followed by ON or OFF.

23.17
Tangent options of the CIRCLE command

By using the tangent options of the CIRCLE command, a circle can be drawn tangent to a circle and a line, three lines, three circles, or two lines and a circle. Figure 23.36 shows a circle drawn tangent to a line and a circle. The TTR option is chosen, the radius is given, and points on the circle and line are selected.

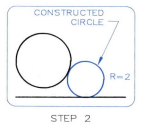

FIGURE 23.36 **Tangent options (TTR).**

Step 1 Command CIRCLE (CR)
3P/TTR/<Center Pt>: TTR
Enter Tangent spec: (Select point on circle.)
Enter second Tangent spec: (Select point on line.)

Step 2 Radius: 2 (CR)
(The circle with radius = 2 is drawn tangent to the line and circle.)

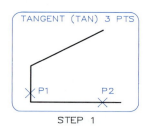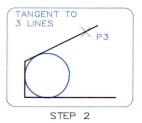

FIGURE 23.37 **Tangent to three lines.**

Step 1 Command: CIRCLE (CR).
3P/2/TTR/ <Center point>: 3P
First point: TAN (CR) to (Select P1.)
Second point: TAN (CR) to (Select P2.)

Step 2 Third point: TAN (CR) to (Select P3.)
(The circle is drawn tangent to three lines.)

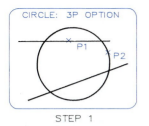
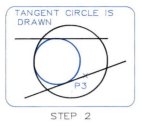

FIGURE 23.38 Tangent to lines arcs.

Step 1 Command: CIRCLE (CR)
3P/2P/TTR/<Center point>: 3P (CR)
First point: TAN (CR) (Select P1 on line.)
Second point: TAN (CR) (Select P2 on circle.)

Step 2 Third point: TAN (Select P3 on line.)
(A circle is drawn tangent to the lines and circle.)

An arc is drawn tangent to three lines in Fig. 23.37 by using the 3P option of the CIRCLE command. In this case, the radius cannot be given since it must be computed.

A circle can be drawn tangent to two lines and a circle or tangent to three circles, as Fig. 23.38 shows. These circles are also found by using the 3P option.

23.18
Drawing arcs

The ARC command is found under the DRAW menu, where arcs can be drawn using nine combinations of variables, including starting point, center, angle, ending point, length of chord, and radius. For example, the S, C, E version requires that you locate the starting point (S), the center (C), and the ending point (E) (Fig. 23.39). The arc begins at point S, but point E need not lie on the arc. Arcs are drawn in a counterclockwise direction by default.

To continue a line with an arc drawn from the last point of the line and tangent to it, respond as follows:

```
Command: ARC (CR)
Center/<Start point>: (CR)
```

An arc may now be drawn tangent to the last point of the line, which is useful for drawing runouts on a part (Fig. 23.40).

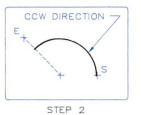

FIGURE 23.39 ARC command.

Step 1 Command: ARC (CR)
Arc Center/<Start point>: (Select S.)
Center/End/<Second point>: CENTER or C (CR)

Step 2 Angle/Length of chord/ <Endpoint>: DRAG (Select point E.) (The arc is drawn to an imaginary line from S to E in a counterclockwise direction.)

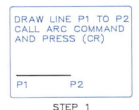
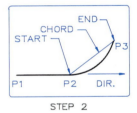

FIGURE 23.40 ARC tangent to end of line.

Step 1 Command: LINE (CR)
From point: (Select P1.)
To point: (Select P2.) (CR)

Step 2 Command: ARC (CR)
Center/<Start point>: (CR)
End point: (Select P3.)

With the only difference being the order of the commands, the same technique can be used for drawing a line from a previously drawn arc.

The DRAGMODE command can be turned on to allow the arcs to be seen before they are selected for their final positions.

23.19
FILLET command

FILLETs can be drawn to any desired radius between two nonparallel lines, whether or not they in-

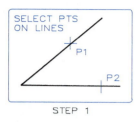

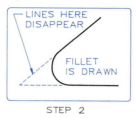

STEP 1   STEP 2

FIGURE 23.41   FILLET **command.**

Step 1 Command: FILLET (CR)
Polyline Radius/<Select two objects>:
R (CR)
Enter fillet radius <0.0000>: 1.5 (CR)
Command: (CR)

Step 2 FILLET Polyline Radius/<Select two objects>: P1 and P2 (**The fillet is drawn, and the lines are trimmed.**)

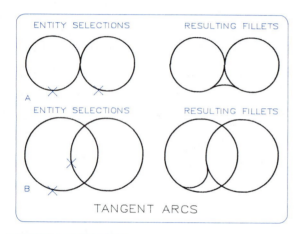

FIGURE 23.42   **Tangent arcs.**

Step 1 Command: FILLET (CR)
Polyline/Radius/<Select two objects>:
R (CR)
Enter fillet radius <0.0000>: 1.2
(**example**) (CR)

Step 2 Command: (CR)
FILLET Polyline/Radius/<Select two objects>: (**Select a point on each circle, and the fillet arc is drawn.**)

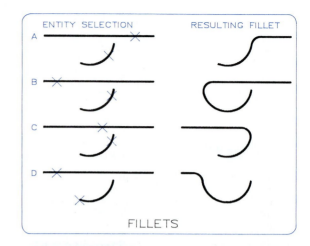

FIGURE 23.43   **When a fillet radius has been selected, points on each line or arc can be selected for filleting, as shown here. Each fillet is determined by the location of the points selected.**

tersect. The fillet is drawn, and the lines are trimmed as shown in Fig. 23.41.

By entering a fillet radius of <u>0</u>, nonintersecting lines will be automatically extended to form a perfect intersection. Once the radius is assigned, it remains in memory as the default radius for drawing additional fillets.

Arcs (fillets) can be drawn tangent to circles or arcs, as shown in Fig. 23.42. The radius must be specified and points on two circles selected. AutoCAD will draw fillet arcs that most nearly approximate the locations selected on the circles.

Figure 23.43 shows examples of fillets that connect lines and arcs. The location of the selected points on the two entities determines the position of the fillet.

23.20
CHAMFER command

The CHAMFER command is used to construct angular bevels at the intersections of lines or polylines. Select two lines, and the lines are trimmed or extended, and the CHAMFER is drawn. Press (CR) and you are ready to repeat this command using the last values if other corners are to be chamfered (Fig. 23.44). A polyline is chamfered in the same manner.

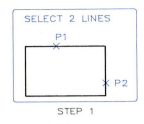

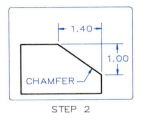

FIGURE 23.44 CHAMFER **command.**

Step 1 Command: <u>CHAMFER</u> **(CR)**
Polyline/Distance/<Select first
line>: <u>D</u> **(Used for nonpolylines)**
Enter first chamfer distance <0.4>:
<u>1.40</u> **(CR)**
Enter second chamfer distance <0.4>:
<u>1.00</u> **(CR)**

Step 2 Command: **(CR)**
CHAMFER Polyline/Distance/<Select
first line>: <u>P1</u>
Select second line: <u>P2</u>

23.21
POLYGON command

A many-sided figure composed of equal sides can be
drawn using the POLYGON command. The center is
located, the number of sides given, and inscribing or
circumscribing options entered (Fig. 23.45).

     Command: <u>POLYGON</u> (CR)
     Number of sides: <u>5</u> (CR)
     Edge/<Center of polygon>: (Lo-
     cate center.)
     Inscribed in circle\Circum-

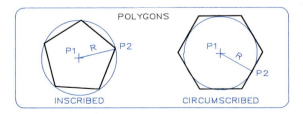

FIGURE 23.45 **The** POLYGON **command can
be used to draw polygons inscribed in or
circumscribed about a circle.**

scribed about circle (I/C): <u>I</u>
(CR)
Radius of circle: (Type length and
(CR) or select length with pointer.)

When the EDGE option is used, the next prompt asks

     First endpoint of edge: (Select point.)
     Second endpoint of edge: (Select point.)

The polygon will then be drawn in a counterclockwise
direction about the center point. A maximum of 1024
sides can be drawn with this command.

23.22
Enlarging, reducing, and panning drawings

Parts of a drawing or the entire drawing can be enlarged by the ZOOM command, a submenu under the
DISPLAY menu. A part of the drawing in Fig. 23.46
is too small to read at its present size, but by using a
ZOOM and WINDOW, you can select the area that you
want enlarged to fill the screen.

     Instead of responding with <u>WINDOW</u>, you could
have responded with <u>ALL</u>, <u>CENTER</u>, <u>DYNAMIC</u>,
<u>EXTENTS</u>, <u>LEFT</u>, <u>PREVIOUS</u>, <u>VMAX</u>, <u>SCALE</u>
<u>(X/XP)</u>, or a number to indicate the factor by which
you want the present drawing changed in size.

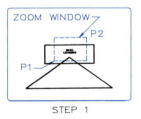

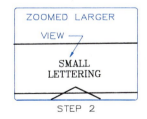

FIGURE 23.46 ZOOM **command.**

Step 1 Command: <u>ZOOM</u> **(CR)**
All/Center/Dynamic/Extents/Left/
Previous/Vmax/Window/<Scale (X-XP)>:
<u>W</u> **(CR)**
First corner: **(Select** *P1*.**)**

Step 2 Other corner: **(Select** *P2*.**)**
(The window will be enlarged to fill the screen.)

ALL The ALL response expands the drawing's LIMITS to fill the display screen.

CENTER The CENTER reply allows you to pick the center of the drawing and degree of magnification or reduction desired.

DYNAMIC The DYNAMIC option allows you to zoom and pan on the screen by selecting points with the cursor.

EXTENTS The EXTENTS response enlarges the drawing to fill the screen while disregarding its LIMITS if it does not fill them.

LEFT By responding with L (left), you can pick the lower left corner and the height of the drawing that you wish to have enlarged.

PREVIOUS The PREVIOUS reply displays the last view that was used. Views can be ZOOMed an almost infinite number of times.

VMAX By selecting V, the zoom is as far out as possible on the current viewport's virtual screen without forcing a complete regeneration.

Scale X/XP

The default is <Scale X/XP> where X represents a fraction such as 1/4 or 0.25. By typing 1/4XP or .25XP the drawing will be scaled to 1/4 inch equal to 1 inch.

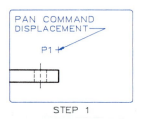
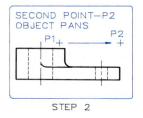

STEP 1   STEP 2

FIGURE 23.47 PAN **command.**

Step 1 Command: PAN (CR)
Displacement: **(Select** *P1.***)**

Step 2 Second point: **(Select** *P2.***)**
(The object will be moved to new position.)

PAN command

The PAN command is illustrated in Fig. 23.47, where it is used to pan the viewer's viewpoint of an object. The first point is selected as a handle to drag the drawing across the screen to the second point. The drawing is not really being relocated, just the view of the drawing is being changed.

23.23
CHANGE command

The CHANGE command permits the changes in single entities including LINES, CIRCLES, TEXT, ATTRIBUTE DEFINITIONS, and BLOCKS. Other property changes that are possible with the CHANGE command are COLOR, LAYERS, LINETYPES, and THICKNESS.

The length and direction of the line in Fig. 23.48 is modified by selecting an end of the line and its new endpoint. The size of a circle can be changed by pointing to the circle and selecting the endpoint of the new radius, and the circle will be enlarged on the screen (Fig. 23.49).

TEXT can be changed by selecting text on the screen and pressing (CR) until you are prompted for STYLE, HEIGHT, ROTATION ANGLE, and TEXT STRING (Fig. 23.50). ATTRIBUTE DEFINITIONS, including the TAG, PROMPT STRING, and DEFAULT VALUE, can be revised with the CHANGE command. BLOCKS can be moved or rotated with CHANGE command, as shown in Fig. 23.51.

Property changes of LAyer, Color, LType, and Thickness can be made with the CHANGE command by selecting the entity or entities on the screen when prompted and then typing P (for properties), as shown in Fig. 23.52. Type LA (for layer) and type the name of the new layer that the text is to be moved to.

Although multiple COLORS and LTYPES can be assigned to entities on the same layer by the CHANGE command, it is better to assign only one layer and line type to a single layer. The THICKNESS property is the length of the extrusion in Z-direction of an object drawn in three dimensions using the ELEV and THICKNESS commands. Changing the THICKNESS in a two-dimensional drawing from 0 to a nonzero value will convert the drawing into an extruded three-dimensional drawing.

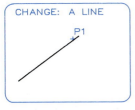

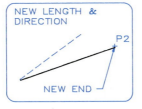

STEP 1   STEP 2

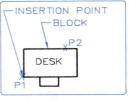

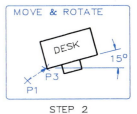

STEP 1   STEP 2

FIGURE 23.48   CHANGE **command.**

Step 1 Command: <u>CHANGE</u> (CR)
Select objects: (Select a point on line.)
Select objects: (CR)

Step 2 Properties/<Change point>:
(Select *P2*, **new point.**)

FIGURE 23.51   CHANGE **blocks command.**

Step 1 Command: <u>CHANGE</u> (CR)
Select objects: (Select *P1*.)
Select objects: (CR)
Properties/<Change point>: (CR)
Change point (or Layer or Elevation):
<u>P2</u> (Select new position.)
Enter block insertion point: **(Select new
position.)**

Step 2 New rotation angle <0>: <u>15</u> (CR)
(The block is moved and rotated 15 degrees.)

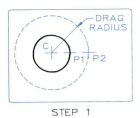

STEP 1   STEP 2

FIGURE 23.49   CHANGE **circle command.**

Step 1 Command: <u>CHANGE</u> (CR)
Select objects: (Select *P1*.) (CR)

Step 2 Properties/<Change point>:
(Select *P2* **to change radius.**)

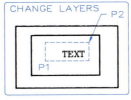

STEP 1   STEP 2

FIGURE 23.52   CHANGE **layers command.**

Step 1 Command: <u>CHANGE</u> (CR)
Select objects: <u>W</u> (CR)
First corner: (Select *P1*.) (CR)
Other corner: (Select *P2*.) (CR)

Step 2 Properties < Change point >: <u>P</u>
Change what property (Color/LAyer/
Thickness)? <u>LA</u> New layer: <u>VISIBLE</u>
(Name of existing layer) (CR)

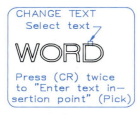

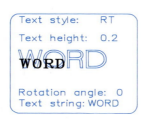

STEP 1   STEP 2

FIGURE 23.50   CHANGE **text.**

Step 1 Command: <u>CHANGE</u>
Select objects: (Select text.)
Select objects: (CR)
Properties/<Change point> (CR)
Enter text insertion point: (Select point.)

Step 2 Text Style: <u>Standard</u>
New style or RETURN for no change: <u>RT</u>
New height <.50>: <u>0.20</u>
New rotation angle <0>: (CR)
New text <WORD>: (CR)

The CHANGE command will work only if the entities being changed have extrusion directions parallel to *Z*-axis of the current User Coordinate System (UCS). The CHPROP command does not have this restriction.

23.24
CHPROP command

The CHPROP command is used to change entity properties of color, linetype, layer, and thickness regardless of their extrusion direction. This version of the CHANGE is useful in modifying entities within three-dimensional drawings.

```
Command: CHPROP
Select objects: (Select entities.)
Change what property (Color/
LAyer/LType/Thickness)?
```

Make the same choice as you would when using the CHANGE command.

23.25
POLYLINE (PLINE) command

The PLINE command is used to connect a series of lines and arcs of varying widths to form a POLY-LINE. The prompt after typing PLINE is

```
Command: PLINE: (CR)
From point: (Select starting point.)
Current line-width is 0.0000
Arc/Close/Halfwidth/Length/
Undo/Width/<Endpoint of line>:
```

The default response, End point of line, shown in brackets, will result in a line drawn from the last point on the screen at the time you executed PLINE. You will be prompted for the width of the line at its beginning and end. This command requires experimentation to become familiar with its many options.

STRAIGHT LINES The PLINE command defaults to the straight-line mode and prompts

```
From point: (Select point.)
Current line width is 0.3:
Arc/Close/Halfwidth/Length/
Undo/Width/<End point of
line>: W (For width.) (CR)
Starting width <0.000>: 0.4 (CR)
Ending width <1.000>: 0.6 (CR)
(Select endpoint.)
```

The line is drawn. The longer the line, the thinner and more tapered it becomes.

A value of zero can be entered for the thinnest lines that can be drawn by the system. CLOSE will

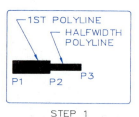
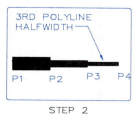

FIGURE 23.53 PLINE **command.**

Step 1 Command: PLINE (CR)
From point: P1 **(Select point.)**
Current line width is 0.0000
Arc/Close/Halfwidth/Length/Undo/
Width/<Endpoint of line>: WIDTH (CR)
Starting width <0.0000>: .12 (CR)
Ending width <.12>: (CR)
Arc/Close/ . . . /<Endpoint of line>:
P2
Arc/Close/ . . . /<Endpoint of line>:
Halfwidth (CR)
Starting half-width <0.3000>: .3 (CR)
Ending half-width <0.3000>: (CR)
Arc/Close/ . . . /<Endpoint of line>:
P3

Step 2 Arc/Close/ . . . /<Endpoint of line>: Halfwidth
Starting half-width <0.2>: .1 (CR)
Ending half-width <0.1000>: (CR)
Arc/Close/ . . . ?<Endpoint of line>:
P4

automatically close to the beginning point of the PLINE and terminate the command. LENGTH lets you continue a PLINE at its last angle by specifying the length of the segment. If the first line was an arc, this command will produce a line tangent to the arc.

The UNDO option erases the last segment of the polyline, and it can be repeated to continue erasing segments of the PLINE. The HALFWIDTH option lets you specify the width of the line from the center of a wide line, as shown in Fig. 23.53.

ARCS When you respond to the PLINE option line with an A (ARC), you can specify arc segments of a PLINE. AutoCAD will give the following command line:

```
Angle/Center/Close/Direction/
Halfwidth/Line/Radius/Second
pt/Undo/Width <End point of
arc>:
```

The default assumes an arc will be drawn tangent to the last line drawn and will pass through the next point selected.

By using the ANGLE option, you may give the Included angle: as a positive or negative value, and the next prompt will ask for Center/Radius/<End point>:. AutoCAD then draws an arc tangent to the previous line segment. If you select Center, you will be asked to give the center of the next arc segment. The next prompt asks for Angle/Length/<End point>:, where Angle refers to the included angle, and Length is the length of the arc's chord.

The CLOSE option causes the PLINE to be closed with an arc segment to the beginning point. DIRECTION allows you to override the default, which draws the next arc tangent to the last PLINE segment. AutoCAD prompts Direction from starting point:, and you can point to the desired beginning point and respond to the next prompt, Endpoint, to give the direction of the arc.

The LINE option switches the PLINE command back to the straight-line mode. The RADIUS option gives a prompt, Radius:, that allows you to specify the radius of the next arc. The following prompt, Angle/Length/<End point>:, lets you specify the included angle or the length of the arc's chord.

SECOND PT is used to select the second point and the endpoint of a three point arc. The two prompts are Second point: and Endpoint:.

23.26
PEDIT command

The PEDIT command is used to edit polylines drawn with the PLINE command. The prompts for this command are

```
Select polyline: (Select line with cur-
sor.)
Entity selected is not a
polyline
Do you want it turned into
one? Y
Close/Join/Width/Edit vertex/
Spline/Fit curve/Decurve/Undo
/exit<X>: CLOSE (Used to close PLINE.)
```

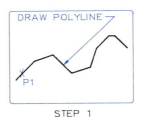 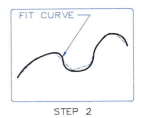

FIGURE 23.54 PEDIT **command.**

Step 1 Command: PEDIT (CR)
Select Polyline: P1 **(Select line.)**

Step 2 Close/Join/Width/Edit vertex/
Fit curve/Spline/Decurve/Undo/exit
<X>: FIT (CR) **(The curve is smoothed.)**

If the PLINE is already closed, the CLOSE command will be replaced by the OPEN option.

JOIN The JOIN option lets you respond to the prompt Select objects Window or Last: by selecting segments that are to be joined to the polyline. Once chosen, these segments become part of the polyline. Segments must have exact meeting points and must not meet with an overlapping intersection for joining to take place.

WIDTH The WIDTH option gives the prompt Enter new width for all segments:. Enter the new width from the keyboard, and the PLINE is redrawn to this new thickness.

FIT CURVE The FIT CURVE (F) option constructs a smooth curve that passes through all vertices of the PLINE with pairs of arcs that join sequential vertices (Fig. 23.54). To change the resulting curve to better suit your needs, use the EDIT VERTEX command discussed below.

SPLINE The SPLINE (S) option draws a cubic curve that passes through the first and last points, but not necessarily through the other points (Fig. 23.55). The SPLINE is useful for drawing curves representing data plotted on graphs.

DECURVE DECURVE (D) removes the arcs inserted by the FIT CURVE or SPLINE option and returns the PLINE to its straight-line form.

UNDO (U) The UNDO (U) option undoes the most recently done PEDIT editing step.

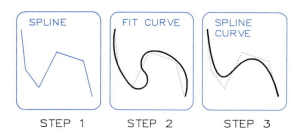

STEP 1   STEP 2   STEP 3

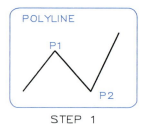

STEP 1   STEP 2

FIGURE 23.55 The SPLINE curve.

Step 1 Draw a polyline (PLINE).

Step 2 PEDIT with the FIT CURVE option.

Step 3 PEDIT with the SPLINE option for mathematical curve.

FIGURE 23.56 PEDIT—move vertex.

Step 1 Command: PEDIT (CR)
Select polyline: (Select line.)
Close/Join/Width/Edit vertex/Fit
curve/Uncurve/eXit <X>: E (Edit
vertex.) (Move X to vertex to be moved.) (CR)
Next/Previous/Break/Insert/Move /
Regen/Straighten/Tangent/
Width/eXit <N>: Move (CR)
Enter location of new vertex: (Select P1.)

Step 2 Press (CR), and the polyline will be changed to pass though the moved vertex.

EDIT VERTEX EDIT VERTEX (E) allows you to select a single vertex of the PLINE and edit it. When this option is used, the first vertex of the PLINE will be marked with an X on the screen. An arrow will be shown if you have specified a tangent direction for the vertex, and you will receive the following prompt:

```
Next/Previous/Break/Insert/
Move/Regen/Straighten/Tangent/
Width/eXit/<N>:
```

STEP 1   STEP 2

FIGURE 23.57 PEDIT—add vertex.

Step 1 Use EDIT VERTEX option and place X on line before the new vertex.
Next/Previous/Break/Insert/Move/
Regen/Straighten/Tangent/Width/eXit
<N>: Insert (CR)
Enter location of new vertex: (Select P1.)

Step 2 Press (CR), and the new vertex will be inserted, and the polyline will pass through it.

NEXT and PREVIOUS The NEXT (N) and PREVIOUS (P) options move the X marker to the next or previous vertex. To move to a vertex several vertices away, select NEXT or PREVIOUS, and press (CR) until the vertex is reached.

BREAK When the BREAK (B) option is selected, the location of the position of the X is shown, and the following prompt appears:

```
Next/Previous/Go/eXit <N>:
```

By using NEXT or PREVIOUS you can select a second point and enter GO, and the line between the two points will be erased (Fig. 23.56). Enter EXIT, and the BREAK will be canceled, and you will return to EDIT VERTEX.

INSERT The INSERT option gives the following

prompt:

```
Enter location of new vertex:
```

This lets you add a new vertex to a polyline (Fig. 23.57). A vertex can also be moved by the MOVE (M) option (Fig. 23.58).

STRAIGHTEN A polyline can be straightened by

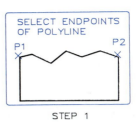

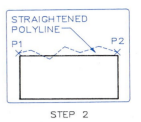

STEP 1   STEP 2   STEP 1   STEP 2

FIGURE 23.58 PEDIT—move vertex.

FIGURE 23.59 PEDIT—straighten line.

Step 1 Command: PEDIT (CR)
Select polyline: **(Select line.)**
Close/Join/Width/Edit vertex/Fit
curve/Decurve/Undo/eXit <X>: E
(Edit vertex.) (Move X to vertex to be moved.) (CR)
Next/Previous/Break/Insert/Move /
Regen/Straighten/Tangent/
Width/eXit <N>: Move (CR)
Enter location of new vertex: **(Select P1.)**

Step 2 Press (CR), and the polyline will be changed to pass though the moved vertex.

Step 1 Use EDIT VERTEX **option of** PEDIT, **and place X at the vertex at the beginning of the line to be straightened,** P1.
Next/Previous/Break/Insert/Move/
Regen/Straighten/Tangent/Width/eXit
<N>: Straighten **(CR)**

Step 2 Next/Previous/Go/eXit <N>: Next **(CR) (Move to** P2.)
Next/Previous/Go/eXit <N>: Go **(CR) (Line** P1–P2 **is straightened.)**

the STRAIGHTEN (S) option, which saves the current location of the vertex specified by an X and gives the following prompt:

Next/Previous/Go/eXit/<N>:

By moving the X to a new vertex on the line and specifying GO, the line will be straightened between the two vertices (Fig. 23.59). Enter X for eXit if you change your mind, and you will be returned to the EDIT VERTEX prompt.

The TANGENT (T) suboption lets you indicate a tangent direction at the vertex marked by the X for use in curve fitting when it is used next. The prompt is

Direction of tangent:

Enter the angle from the keyboard, or select a point with the cursor on the screen from the current point.

The WIDTH (W) suboption lets you change the beginning and ending widths of an existing line segment from the X-marked vertex. Use the NEXT and PREVIOUS options to confirm which direction the line will be drawn. To draw the changed polyline on the screen, use the REGEN (R) option.

EXIT The eXit option is used to exit from the

PEDIT command and return to the Command: prompt.

23.27
HATCH command

The HATCH command is used to cross-hatch an area that has been sectioned (Fig. 23.60). The prompts are

Command: HATCH (CR)
Pattern (? or name/U, style):
<default>: ANSI31 (CR)

By responding with ?, you will be given a list of the standard patterns in ACAD.PAT. A response of U is a user-defined pattern.

Several dialogue boxes that show the hatching patterns can be displayed on the screen by using the HATCH option under the DRAW heading on the **Menu Bar.** A pattern can be selected by the cursor.

By responding as shown below, equally spaced hatch lines will be drawn 0.2 units apart and at 45° with the horizontal.

Angle for crosshatch lines <0>:
45 (Or show with the pointer.) (CR)

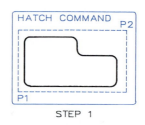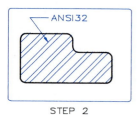

FIGURE 23.60 HATCH **command.**

Step 1 Command: <u>HATCH</u> **(CR)**
Pattern (? or name/U, style): <u>ANSI32</u> **(CR)**
Scale for pattern <default>: <u>1.00</u> **(CR)**
Angle for pattern <default>: <u>0</u> **(CR)**
Select objects: <u>W</u> **(Window object.)**

Step 2 Press **(CR)** and the hatching will be
completed. Press Ctrl C to terminate hatching if
desired.

Spacing between lines <0.1>: <u>0.2</u>
(Or show with the pointer.) (CR)
Double hatch area? <N> (CR)
Select Objects or Window or
Last:

The DRAW heading of the pull-down menu can
be selected to gain access to the HATCH option.
Drawings (icons) of the various hatch patterns are
shown in several screens from which to make pattern
selections by using the cursor arrow. Access to icons
makes it unnecessary to memorize the patterns by
their numbers.

The letters <u>N</u>, <u>O</u>, or <u>I</u> can be added to your response to the PATTERN prompt. For example, PAT-
TERN: ANSI32,N. These letters will cause the
crosshatching to be given in the style shown in Fig.

N—NORMAL   O—OUTERMOST   I—IGNORE   TEXT WINDOW

FIGURE 23.61 **Whenever you specify a
hatching pattern, enter a comma and the letter** <u>N</u>,
<u>O</u>, **or** <u>I</u> **after the pattern, and the hatching will be
applied as shown above; for example,** Pattern:
<u>ANSI31,N,</u> **to hatch the outside and alternate
layers inside the figure. Hatching automatically
leaves a window around text within the area when
windowed or selected.**

23.61 and as defined below:

> <u>N-Normal</u> (Hatches alternate areas beginning
> with the outermost.)
>
> <u>O-Outermost</u> (Hatches only the outermost
> areas.)
>
> <u>I-Ignore</u> (Hatches all inside areas ignoring
> contents.)

If a pattern is selected from the ACAD.PAT file, you
will receive the following prompt:

Scale for pattern <1.00>: <u>.75</u>
(CR)
Angle for pattern <0>: <u>0</u> (CR)
Select Objects or Window or
Last:

Both responses can be made at the keyboard or by
selecting two points on the screen for each to indicate
the angle and the scale. All hatch lines can be
ERASED as a group since they are entered as a
block. If entered with an * preceding the pattern
name, they can be edited one line at a time. The
area to be hatched is selected one line at a time or
WINDOWed as shown in Fig. 23.61.

23.28
Text and numerals

The DTEXT or TEXT command can be used for inserting text in a drawing. DTEXT displays the text,
one character at a time, as you type, enabling you to
see how long to make a line. The TEXT command
does not show the text on the screen until you have
finished typing. Therefore, you will use DTEXT almost exclusively.

The usage of the DTEXT command to begin text
at a selected left starting point is shown below.

Command: <u>DTEXT</u>
Justify/Style/<Start point>:
(Select point with cursor.)
Height <.18>: <u>.125</u>
Rotation angle <0>: (CR)
Text: <u>TYPE WORDS</u>

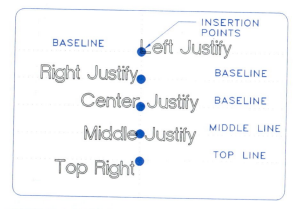

FIGURE 23.62 Text can be added to a drawing by using any of the insertion points above. For example, BC means the bottom center of a word or sentence will be located at the cursor point.

SPECIAL CHARACTERS

%%O	Start or stop Overline of text
%%U	Start or stop Underline of text
%%D	Degree symbol: 45%%D = 45°
%%P	Plus−minus: %%P0.05 = ±0.05
%%C	Diameter: %%C20 = ⌀20
%%%	Percent sign: 80%%% = 80%
%%nnn	Special character number nnn

FIGURE 23.65 These special characters beginning with %% are used while typing DTEXT and TEXT to obtain these symbols.

By selecting the JUSTIFY option of the DTEXT command, you will be prompted to select from a series of abbreviations that identify the insertion point for the string of text, as shown in Fig. 23.62. BC means bottom center, RT means right top, and so forth. Other options are shown in Fig. 23.63.

Figure 23.64 illustrates how to type multiple lines of text with DTEXT. The spacing for each succeeding line is automatically set with (CR).

Special characters can be entered from the keyboard when using DTEXT. By preceding the codes with a double percent sign, %%, the characters illustrated in Fig. 23.65 will be shown.

The QTEXT command can be used to reduce the time required to display a drawing on the screen when it is redrawn. QTEXT (quick text) can be set to ON, and the text will be drawn on the screen as a series of boxes representing the space required for the text (Fig. 23.66). When QTEXT is turned OFF, the full text will be plotted on the screen.

FIGURE 23.63 The JUSTIFY option allows you to insert text aligned with the insertion points, as shown here.

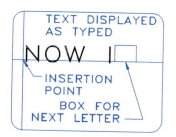

STEP 1

STEP 2

FIGURE 23.64 Inserting DTEXT.

Step 1 Command: DTEXT
Justify/Style/<Start point>: (Select point.)
Height <.18>: .125
Rotation angle <0>: (CR)
Text: NOW IS

Step 2 Press (CR) and box advances to next line.
Text: THE TIME (CR) (Box continues to space down for the next line of text.)

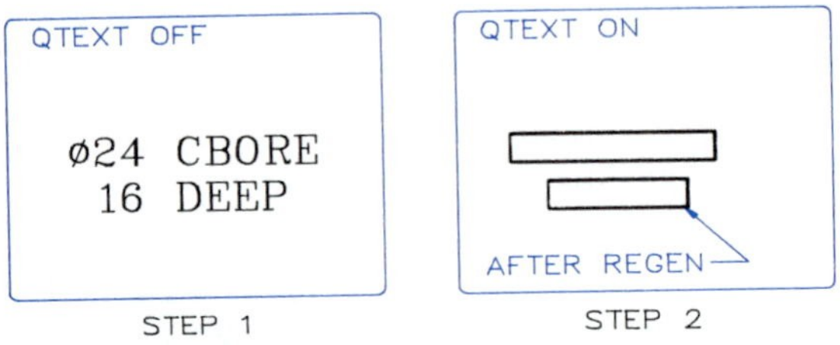

FIGURE 23.66   **QTEXT command.**

Step 1 Command: QTEXT (CR) ON/OFF <CURRENT>: ON (CR)

Step 2 Command: REGEN (CR) (Text is shown as boxes.)

23.29
The STYLE command

Many of AutoCAD's text fonts and their names are shown in Fig. 23.67. The default style is STANDARD, which uses the TXT font.

The STYLE command is used to create variations for any of the fonts in the following manner:

```
Command: STYLE (CR)
Text style name (or ?): PRETTY
(CR)
Font file <TXT>: COMPLEX (CR)
Height (0.20): 0 (CR)
Width factor <default>: 1.00
(CR)
Obliquing angle <45>: 0 (CR)
Backwards? (Y/N): N (CR)
Upside-down? <Y/N>: N (CR)
```

The text style created here is named PRETTY. The style, PRETTY, will retain its initial settings until you change them. Responding to the first prompt with ? will give a list of defined text STYLES.

To select a font for a new style, select OPTIONS from the pull-down menu, then DTEXT option, and Text Font to see examples of fonts that can be selected by cursor. If you later create a STYLE called PRETTY and select a different font, such as RO-

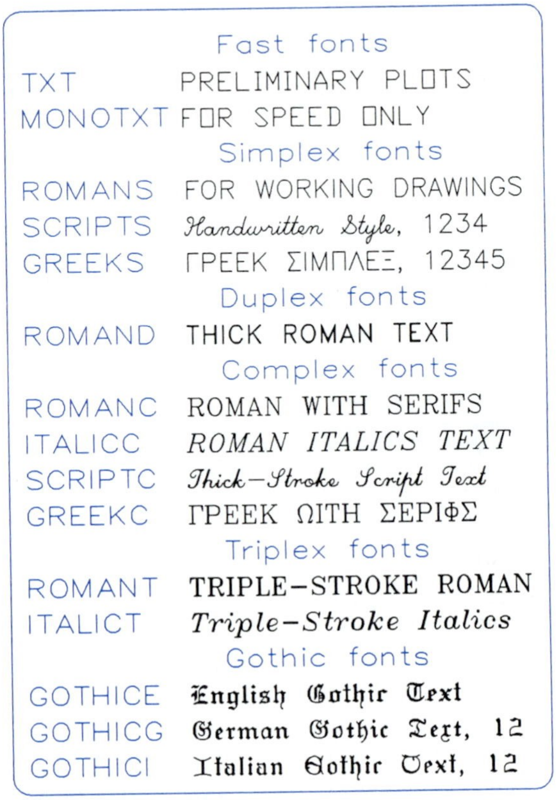

FIGURE 23.67   **Most of the fonts available in AutoCAD.**

MANS, then all the text previously drawn with the STYLE PRETTY when the COMPLEX font was assigned to it, will be redrawn on the screen with the ROMANS font.

This technique is used to change the TXT and MONOTXT fonts to a more attractive font when the drawing is ready to plot. In the meanwhile, regeneration time has been saved by using a fast font.

The VERTICAL font can be used for drawing vertical lines of text with letters stacked one under the other. A rotation angle of 270 degrees must be used when using VERTICAL.

The DDEDIT command (Fig. 23.68) can be used to edit text on the screen that is shown in a dialogue box when it has been selected with the cursor. Once corrected, the text is revised on the screen by selecting the OK button.

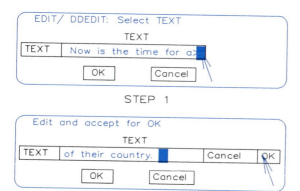

STEP 1

STEP 2

FIGURE 23.68 DDEDIT: **Editing text.**

Step 1 Command: DTEXT
<Select a TEXT or ATTDEF
object>/Undo: **(Select text.)**
(Text appears in window.)

Step 2 Text is edited with the cursor and keyboard in combination. Select OK to accept change.

STEP 1

STEP 2

FIGURE 23.69 MOVE **command.**

Step 1 Command: MOVE **(CR)**
Select objects: W **(CR)**
First corner: **(Select point.)**
Other corner: **(Select point.)** 5 found.
Select objects: **(CR)**
Base point or displacement: **(Select 1st point.)**

Step 2 Second point of displacement: **(Select 2nd point, P2.) (The object is drawn at its new position, P2, and the original drawing disappears.)**

23.30
Moving and copying drawings

A drawing can be moved to a new position by the MOVE command (Fig. 23.69), or it can be duplicated by the COPY command. When copied, the original drawing is left in its original position, and a copy of it is located where specified.

The COPY procedure is the same as MOVE except the command is COPY instead of MOVE. The COPY command has a MULTIPLE prompt that can be used for copying multiples of drawings by moving your cursor about the screen and pressing the select button.

A drawing can be moved or copied by dragging it into position by the DRAG option (Fig. 23.70). DRAG is activated after responding to Base point or displacement with DRAG. When the cursor is moved about the screen, the drawing is dynamically moved until it is set by pressing the select button of the mouse.

STEP 1

STEP 2

FIGURE 23.70 DRAG **mode.**

Step 1 Command: MOVE **(CR)**
Select objects: W **(CR) (Window object.)**
Base point or displacement: DRAG **(CR)**
Base point or displacement: **(Select P1.) (Or X,Y distance)**
Second point of displacement: **(Select new location.) (As the cursor is moved, the drawing is DRAGged about the screen.)**

Step 2 When moved to the desired location, press the Select button to draw the object in its final position. (The original drawing disappears.) If DRAGMODE is ON or AUTO, dragging will be automatic.

Besides a window, you may use the default, which is set to select the entities to be moved with the cursor, one at a time. You may also select LAST, which will move the last entity, such as a LINE, ARC, CIRCLE, TEXT, or BLOCK.

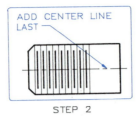

MIRROR command.

Step 1 Command: <u>MIRROR</u> **(CR)**
Select objects: <u>W</u> **(CR)**
(Window drawing to be mirrored.)
First point or mirror line: **(Select P1.)**
Second point: **(Select P2.)**

Step 2 Delete old objects? <N>: <u>N</u> **(CR)**
(The object is mirrored about the mirror line. Draw the centerline last.)

23.31
Mirroring drawings

Symmetrical objects can be drawn by drawing a portion of the figure and then mirroring the drawing about one or more axes. The schematic threads in Fig. 23.71 are drawn using the MIRROR command. If a line coincides with the MIRROR line, such as P1–P2 in Fig. 23.71, the line will be drawn twice when mirrored. Therefore, a line of this type should be drawn after the view has been mirrored.

23.32
Mirrored text (MIRRTEXT)

MIRRTEXT is a system variable that is a subcommand of the SETVAR command, which is used for mirroring a drawing that has text. By setting MIRRTEXT to 0 (off), the text will not be mirrored, but the drawing will be (Fig. 23.72). If MIRRTEXT is set to 1 (on), the text will be mirrored along with the drawing.

23.33
OSNAP
(Object snap)

By using OSNAP, you can snap to objects of a drawing on the screen rather than to the background snap grid. To activate OSNAP select the ASSIST box of the pull-down menu and the options will appear for selection (Fig. 23.73). Or, the line of stars, *****, in the Root Menu can be selected to use OSNAP. For example, Fig. 23.74 shows the steps in drawing a line from an intersection of lines to the endpoint of a line.

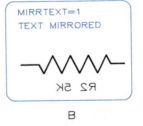

Mirrored text. (A) When the variable MIRRTEXT of the SETVAR command is given a value of 0, the text will not be mirrored. (B) When MIRRTEXT is given a value of 1, the text will be mirrored.

ASSIST	PULL–DOWN MENU
Osnap: <mode>	Snaps to:
CENter	Centers of arcs and circles
ENDpoint	End points of lines
INSert	Block insertion points
INTersection	Intersections between lines
MIDpoint	Midpoints of lines and arcs
NEArest	Nearest point on line or arc
NODe	A point
PERpendicular	Line or arc for perpendicular
QUAdrant	Nearest quad. point of circle
Quick,<mode>	Searches for nearest point
TANgent	Tangent pont on arc or circle
NONE	Turns off OSNAP

The ASSIST column under the pull-down menu gives you access to OSNAP options, letting you pick entities on the screen.

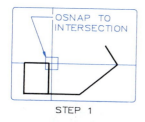
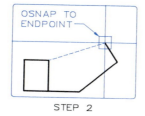

FIGURE 23.74 OSNAP **option.**

Step 1 Command: LINE (CR)
From point: (Select the stars in the Root Menu.)
(Select INTERSEC option.)
INTERSEC of: (Select intersection.)

Step 2 To point: (Select ENDPOINT option of OSNAP.)
To point: ENDPOINT of (Move cursor to endpoint of line, and press Select button. The line is drawn.)

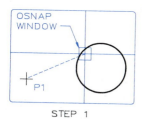
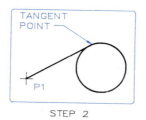

FIGURE 23.75 OSNAP **tangent option.**

Step 1 Command: LINE (CR)
From point: (Select P1.)
To point: (Select OSNAP from the menu, and select TANGENT.)

Step 2 To point: (Select point on circle.) (The line is drawn from P1 tangent to the circle on the side where the tangent point was selected.)

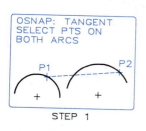
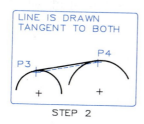

FIGURE 23.76 OSNAP **tan to 2 arcs.**

Step 1 Command: LINE (CR)
From point: (Select the stars in the Root Menu.)
(Select the TAN option.)
TANGENT to: (Select P1 on first arc.)

Step 2 To point: (Select TAN option of OSNAP.)
TANGENT to: (Select P2 on second arc and the line is drawn.)

From P1 in Fig. 23.75 a line is drawn tangent to the circle. The same procedure as shown above is used, but instead of ENDPOINT for the second point, TANGENT is selected on the side of the circle where the tangent point is to be located. The tangent point is found automatically, and the line is drawn.

By using the TANGENT option of OSNAP, a line can be drawn tangent to two arcs as shown in Fig. 23.76. The tangent line is automatically trimmed to end at its two tangent points.

The various options of OSNAP are NEAREST, ENDPOINT, MIDPOINT, CENTER, NODE, QUADRANT, INTERSECTION, INSERT, PERPENDICULAR, TANGENT, QUICK, and NONE. The NODE option snaps to a point, the QUADRANT options snaps to one of the four compass points on a circle, the INSERT option snaps to the intersection point of a BLOCK; and the NONE option turns off OSNAP for the next selection.

The QUICK option reduces searching time by selecting the first object encountered within the target area, rather than searching for the one that is closest to the center of the target.

The APERTURE command is used to assign a size to the target box that appears at the cursor when OSNAP options are in use. The APERTURE can vary from 1 to 50 pixels square.

Other OSNAP options permit you to snap to centers of arcs and circles and to insertion points of blocks, midpoints of lines, nearest points of an object, points (nodes), perpendiculars to lines, tangent to arcs and circles, and quadrant points of a circle.

One or more OSNAP settings can be made when the same options are to be used many times. For example, if you wish to repetitively snap to ENDPOINTS and circle CENTERS, do the following:

Command: OSNAP (CR)
Object snap modes: ENDPOINT, CENTER (CR)

From this point on, your cursor will have a select target at its intersection for picking endpoints and centers of arcs and circles. Remove this OSNAP setting in the following manner:

Command: OSNAP (CR)
Object snap modes: (Select NONE, OFF, or press (CR).) (CR)

23.34
ARRAY command

The `ARRAY` command is used to draw repetitive shapes in circular and rectangular patterns. For example, a series of holes can be located on a bolt circle by drawing the first hole in its desired position and then activating the `ARRAY` command, as shown in Fig. 23.77.

A rectangular `ARRAY` is begun by drawing the first part in the lower left corner of the array. Once drawn, follow the commands of the rectangular `ARRAY` shown in Fig. 23.78. Termination of the array can be caused by `Ctrl C`.

Rectangular arrays may be drawn at angles (Fig. 23.79) if the `SNAP` mode has been rotated to some angle other than zero. The first object is then drawn with the `DRAW` commands in the lower left corner of the `ARRAY`. The `ARRAY` command is activated, and the number of rows, number of columns, and cell distances are given in response to the command prompts. The array will be drawn at the rotation angle of the `SNAP` mode.

23.35
DONUT command

A "donut" can be drawn by using the `DONUT` command in which the inside diameter, outside diameter, and center point are given (Fig. 23.80). By setting the inside diameter to `0`, the `DONUT` will draw a solid circle (Fig. 23.81).

23.36
SCALE command

The `SCALE` command permits you to make drawn objects larger or smaller. For example, the desk in Fig. 23.82 is enlarged by windowing the desk, selecting a base point, and giving it a scale factor of 1.6. The text and the drawing are enlarged in the X and Y directions.

A second option of the `SCALE` command enables you to select a length of a given figure (Fig.

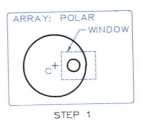 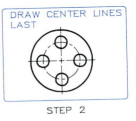

STEP 1   STEP 2

FIGURE 23.77   ARRAY—circular.

Step 1   **Begin by drawing the figure to array.**
```
Command: ARRAY (CR)
Select objects: W (Window the hole.) (CR)
Select objects: (CR)
Rectangular or Polar array (R/P): P
(CR)
Center point of array: C (Select point.)
```

Step 2 Number of items: 4
```
Angle to fill (+=ccw, -=cw)<360>: 360
(CR)
Rotate objects as they are copied?
<Y> (CR)
```
(The array is drawn.)

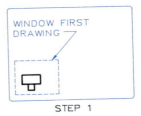 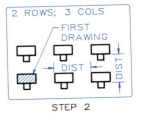

STEP 1   STEP 2

FIGURE 23.78   ARRAY—rectangular.

Step 1   **Draw desk in lower left of** `ARRAY`.
```
Command: ARRAY (CR)
Select Objects: W (Window desk.) (CR)
Rectangular or Circular array (R/P):
R
```

Step 2 Number of rows (---) <1>: 2 (CR)
```
Number of columns (|||) <1>: 3 (CR)
Unit cell distance between rows (---):
4 (CR)
Distance between columns (|||): 3.5 (CR)
(CR) (The array is drawn.)
```

23.83), specify its present length, and then specify the new length, which is a ratio of the first specified dimension. The lengths can be given by using the cursor or by typing them in at the keyboard as numeric values.

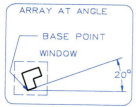

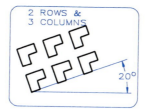

STEP 1   STEP 2

FIGURE 23.79   **Rectangular** ARRAY **at an angle.**

Step 1   The grid must be rotated using the SNAP command at the desired angle. Draw the object in the lower left corner of the array.
```
Command: ARRAY (CR)
Select objects. (Window the object.)
Rectangular or Polar array <R/P>: R
(CR)
```

Step 2   Number of rows(---) <1>: 2 (CR)
Number of columns (|||) <1>: 3 (CR)
Unit cell or distance between rows
(---): 2 (CR)
Distance between columns (|||): 3 (CR)

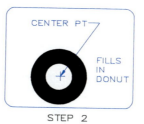

STEP 1   STEP 2

FIGURE 23.80   **The open** DONUT

Step 1   Command: DONUT (CR)
Inside diameter <0.0000>: .30 (CR)
Outside diameter <0.0000>: .60 (CR)

Step 2   Center of donut: (Select point, and donut is drawn with a center hole.)

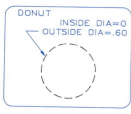

STEP 1   STEP 2

FIGURE 23.81   **The solid** DONUT.

Step 1   Command: DONUT (CR)
Inside diameter <0.0000>: 0 (CR)
Outside diameter <0.0000>: 0.6 (CR)

Step 2   Center of donut: (Select point, and donut is drawn as a solid circle.)

STEP 1   STEP 2

FIGURE 23.82   SCALE **command.**

Step 1   Command: SCALE (CR)
Select objects: (Window DESK with P1 and P2.)

Step 2   Base point: (Select base point.)
<Scale factor>/Reference: 1.6 (CR)
(The desk is drawn 60% larger.)

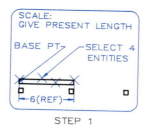

STEP 1   STEP 2

FIGURE 23.83   **Scaling—reference dimensions.**

Step 1   Command: SCALE (CR)
Select objects: (Select points on each line.)
Base point: (Select point.)
<Scale factor>/Reference: R (CR)
Reference length <1>: 6 (CR)

Step 2   New length: 12 (CR)
(The drawing is enlarged in all directions.)

23.37
STRETCH command

The STRETCH command is used to move a portion of a drawing while retaining the connections it has with other lines and entities. The window in a floor plan in Fig. 23.84 is moved to a new position by the STRETCH command while remaining attached to the line of the wall. A CROSSING window is used to select the line that will be stretched.

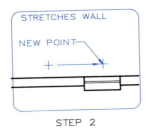

FIGURE 23.84 STRETCH **command.**

Step 1 Command: <u>STRETCH</u> **(CR)**
Select objects to stretch by
window . . .
Select objects: <u>C</u> **(Crossing window)**
First corner: **(Select P1.)**
Other corner: **(Select P2.)**
Select objects: **(CR)**
Base point: **(Select base point.)**

Step 2 New point: **(Select new point.)**
(The window symbol is repositioned.)

23.38
ROTATE command

An object drawn on the screen can be rotated about a base point by using the ROTATE command (Fig. 23.85). The object, which may comprise a number of lines, is windowed, a base point is selected, and a rotation angle is typed in at the keyboard, or the cursor is used to show the angle of rotation on the screen. Lines from several layers can be rotated if they are included within the window used to select the object.

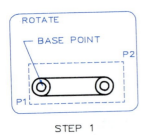

FIGURE 23.85 ROTATE **command.**

Step 1 Command: <u>ROTATE</u> **(CR)**
Select objects: <u>W</u> **(CR) (Window with P1 and P2.)**

Step 2 Base point: **(Select point.)**
<Rotation angle>/Reference: <u>45</u> **(CR)**
(Object is rotated 45° CCW.)

23.39
TRIM command

By using the TRIM command, edges of entities can be used as cutting edges to trim a line, as shown in Fig. 23.86. You are prompted to select the entities that will serve as cutting edges, and the objects between the cutting edges are selected. The cutting edges trim the lines, giving perfect intersections at the crossing points.

Using the CROSSING option of the TRIM command, a window is placed around a set of intersecting

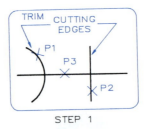 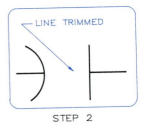

FIGURE 23.86 TRIM **edges.**

Step 1 Command: <u>TRIM</u> **(CR)**
Select cutting edge(s) . . .
Select objects: **(Select P1.)**
Select objects: **(Select P2.) (CR)**

Step 2 Select object to trim: **(Select P3.)**
(Line between cutting edges is removed.)

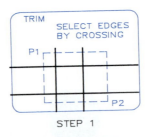 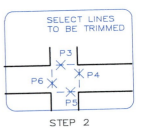

FIGURE 23.87 TRIM **by crossing.**

Step 1 Command: <u>TRIM</u> **(CR)**
Select cutting edge(s) . . .
Select objects: <u>Crossing</u> **(CR)**
First corner: **(Select P1.)**
Other corner: **(Select P2.) (CR)**

Step 2 Select object to trim: **(Select P3, P4, P5, P6.)**
(The lines are trimmed one at a time.)

lines (Fig. 23.87), and lines selected to be the cutting edges are crossed. The portions of the lines between the cutting edges are selected one at a time and are then removed or trimmed. For TRIM to work in a three-dimensional drawing, the entities selected must be parallel to the current UCS.

23.40
EXTEND command

Lines and arcs can be lengthened to intersect a selected entity by using the EXTEND command (Fig. 23.88). You are first prompted to select the boundary

entity and then the line or arc to be extended. More than one entity can be extended at a time to join the previously selected boundary entity.

A polyline can be extended with the EXTEND command to a selected boundary, as shown in Fig. 23.89. The boundary is selected, the ends of the PLINES are selected, and both PLINES are extended. The EXTEND command will not work on "closed" PLINES, such as a polygon. For EXTEND to work in a three-dimensional drawing, the entities selected must be parallel to the current UCS.

23.41
DIVIDE command

An entity can be divided into a specified number of equal segments by using the DIVIDE command (Fig. 23.90). The entity is selected by locating a point on it with the cursor. You will be prompted for the number of segments into which it is to be divided, and markers will be placed on the line at the ends of the segments. The markers will be of the type and size currently set by the PDMODE and PDSIZE variables under the SETVAR command.

The BLOCK option under the DIVIDE command allows you to select a previously saved block to mark the ends of the segments of a divided line (Fig. 23.91). The blocks can be either ALIGNED or NOT ALIGNED as shown. In this example, the BLOCKS are rectangles, but they could have been drawn in any shape.

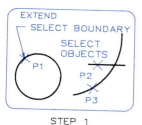 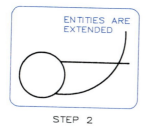

FIGURE 23.88 EXTEND **command.**

Step 1 Command: <u>EXTEND</u> (CR)
Select boundary edge(s) . . .
Select objects: **(Select** *P1*.**) (CR)**

Step 2 Select object to extend: **(Select** *P1* **and** *P2*.**)**
(The line and arc are extended to the boundary.)

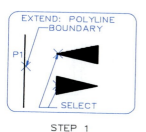 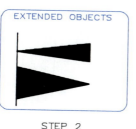

FIGURE 23.89 EXTEND **a polyline.**

Step 1: Command: <u>EXTEND</u> **(CR)**
Select boundary edge(s) . . .
Select objects: **(Select** *P1*.**) (CR)**

Step 2 Select object to extend: **(Select ends of** PLINES.**)**
(Both PLINES **are extended to the border.)**

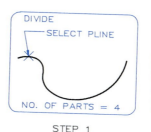 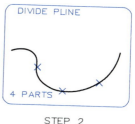

FIGURE 23.90 DIVIDE **a line.**

Step 1 Command: <u>DIVIDE</u> **(CR)**
Select object to divide: **(Select PLINE.)**

Step 2 <Number of segments>/Block: <u>4</u> **(CR)**
(PDMODE **symbols are placed along the line, dividing it.)**

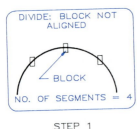
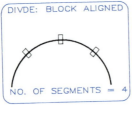

FIGURE 23.91 DIVIDE an arc.

Step 1 Command: <u>DIVIDE</u> (**CR**)
Select object to divide: (**Select arc.**)
<Number of segments>/Block: <u>B</u> (**CR**)
Block name to insert: <u>RECT</u> (**CR**)
Align block with object? <Y> <u>N</u> (**CR**)
Number of segments: <u>4</u> (**CR**)

Step 2 Align block with object? <Y> (**CR**)
(**The blocks radiate from the arc's center.**)

23.42
MEASURE command

Markers can be placed along an arc, circle, polyline, or line at a specified distance apart (Fig. 23.92) by using the MEASURE command. The segment length option asks you for the entity to be segmented, which you must select; then it asks for the segment length.

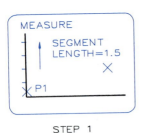
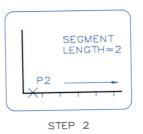

FIGURE 23.92 Measure command.

Step 1 Command: <u>MEASURE</u> (**CR**)
Select object to measure: (**Select P1.**)
<Segment length>/Block: <u>0.1</u> (**CR**)
(**The line is divided into 0.1 divisions starting at the end nearest P1.**)

Step 2 Command: <u>MEASURE</u> (**CR**)
Select object to measure: (**Select P1.**)
<Segment length>/Block: <u>0.2</u> (**CR**)
(**0.2 divisions are measured along the line.**)

Markers will be placed along the line (or entity) beginning with the end nearest to the location of the point selected. The last segment probably will not be equal to the specified segment length.

23.43
OFFSET command

An entity can be drawn parallel to and offset from another entity by the OFFSET command (Fig. 23.93). In this example, a polyline is drawn offset from a given polyline. You will be prompted for the offset dis-

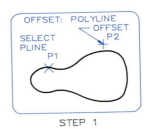
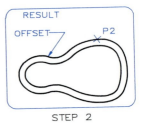

FIGURE 23.93 OFFSET command.

Step 1 Command: <u>OFFSET</u> (**CR**)
Offset distance or Through <Through>:
<u>T</u> (**CR**)
Select object to offset: (**Select P1.**)
Through point: (**Select P2.**)

Step 2 An enlarged PLINE is drawn that passes through P2.

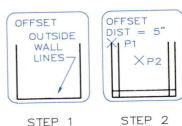
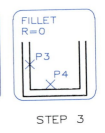

FIGURE 23.94 OFFSET—parallel lines.

Step 1 Command: <u>OFFSET</u> (**CR**)
Offset distance or Through <last>: <u>5</u>

Step 2 Select object to offset: (**Select.**)
Side to offset: (**Select.**)

Step 3 Use FILLET (R=0) to trim corners. The TRIM command can be used also.

tance or the point the offset drawing is to pass through. Next, you will be prompted for the entity (a PLINE, in this case) to be used as the pattern for the offset PLINE.

The OFFSET command is an excellent aid when drawing parallel lines, as architects would do when drawing floor plans (Fig. 23.94). The offset distance can be set to a precise value, and lines can be repetitively selected, and the side to offset is indicated with the cursor. The FILLET command, with a radius set to 0, can be used to trim the corners to perfect intersections. The TRIM command can be used as an alternate method for squaring corners. For OFFSET to work in a three-dimensional drawing, the entities selected must be parallel to the current UCS.

23.44
BLOCKS

One of the more powerful features of computer graphics is the option of building a file of drawings or symbols to be used repetitively on drawings. In AutoCAD, these drawing files are called BLOCKS.

BLOCKS, such as the SI symbol in Fig. 23.95, are drawn in the conventional manner and are blocked as shown below.

```
Command: BLOCK (CR)
BLOCK name (or ?): SI (CR)
Insertion base point: (Select point.)
Select objects: WINDOW or W (CR)
First corner: (Select point.)
Other corner: (Select point.)
Select objects: (CR)
```
(The BLOCK is filed into memory and disappears from screen.)

The BLOCK is inserted into the drawing by the steps shown in Fig. 23.95. BLOCKS are inserted as entities, which means they cannot be edited by erasing parts of them or breaking lines within them.

When an attempt is made to erase a portion of a BLOCK, the whole BLOCK is erased. However, BLOCKS can be edited if a star is inserted in front of the BLOCK name before insertion, for example, Block name (or ?): *SI (CR).

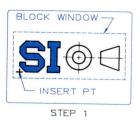

BLOCK command.

Step 1 Make a drawing that you wish to BLOCK, and respond to the BLOCK command as follows:
```
BLOCK name (or ?): SI (CR)
Insertion base point: (Select insert point.)
Select objects: WINDOW or W (CR)
```
(Window the drawing, and the object disappears into memory.)

Step 2 To Insert:
```
Command: INSERT (CR)
Block name (or ?): SI (CR)
Insertion point: (Select point with cursor.)
X-scale factor <1>/Corner/XYZ: 0.5 (CR)
Y-scale factor <default=X>: (CR)
Rotation angle <0>: (CR)
```
(Block SI is inserted at 50% size.)

BLOCKS can be used only on the current drawing file unless they are converted to WBLOCKS, Write Blocks, which become permanent. This conversion is performed as follows:

```
Command: WBLOCK (CR)
File Name: A:SI (CR) (This assigns the
```
name of the WBLOCK to the drive A.)
```
Block Name: SI (CR) (This is the name of
```
the BLOCK that is being changed to a WBLOCK.)

A library of WBLOCKS that can be inserted in different files can relieve drafters of making drawings and will greatly improve their productivity.

The EXPLODE command is used to separate a BLOCK into individual entities that can be erased one at a time. Type EXPLODE, and select any point on the BLOCK.

BLOCKS can be redefined by selecting a previously used BLOCK name and AutoCAD will ask Redefine it? <N>. Type Y for yes, and select the new drawing to be blocked. After doing this, the redefined BLOCKS will be updated on the screen automatically.

23.45
External References

External References are existing drawing files that are temporarily attached together with a current file to form a combination drawing. When the drawing session ends, the XREF is discarded, leaving only the name of the XREF and its path as part of the drawing. Therefore if the XREF is later updated, the latest version will be attached to current drawing.

For example, Fig. 23.96 shows three separate drawings, a two-view drawing of a part, and SI symbol, and a fillet and round note. To combine the three drawings together, load the PART1 file, as shown in Fig. 23.97, and use the XREF command and the AT-

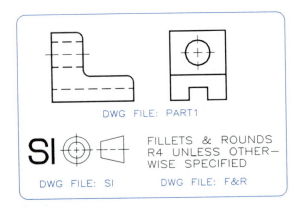

FIGURE 23.96 **These three drawing files will be used as examples to illustrate how External References (XREFs) are used to make a single drawing.**

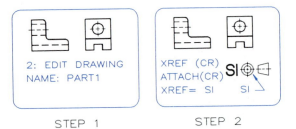

STEP 1   STEP 2

FIGURE 23.97 **A drawing with one XREF.**

Step 1 Enter 2 at the Main Menu and load the drawing, PART1.

Step 2 Command: XREF (CR)
?/Bind/Detach/Path/Reload/<Attach>: A (CR)
Xref to Attach <Default>: SI (CR)
Attach Xref SI: SI
SI is loaded.
Insertion point: (Select.)
X scale factor <1>: (CR)
Y scale factor <1>: (CR)
Rotation angle <0>: (CR) (SI symbol is inserted.)

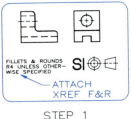

STEP 1   STEP 2

FIGURE 23.98 **A drawing with two XREFs.**

Step 1 Command: XREF
?/Bind/Detach/Path/Reload/<Attach>: A
Xref to attach: F&R
Attach Xref F&R: F&R
F&R loaded.
Insertion point: (Select.)
X scale factor <1>: (CR)
Y scale factor <1>: (CR)
Rotation angle <0>: (CR)

Step 2 Command: XREF
?/Bind/Detach/Path/Reload/<Attach>: ?
Xref(s) to list <*>: (CR) (A listing of the XREFs of the drawing is given.)

STEP 1   STEP 2

FIGURE 23.99 **BLOCKS as XREFs.**

Step 1 Attached XREFs are also listed under the BLOCK ? command.

Step 2 The blocks (A:PART1, A:SI, and A:F&R) are listed. There are one user BLOCK and two XREFs.

Layer name	State	Color	Linetype
0	ON	7	CONTIN.
VISIBLE	ON	1	CONTIN.
HIDDEN	ON	2	HIDDEN
CENTER	ON	3	CENTER
HATCH	ON	4	CONTIN.
DIMEN	ON	5	CONTIN.
CUT |VISIBLE	ON	6	PHANTOM Xdep: SI
WINDOW |VISIBLE	ON	7	CONTIN. Xdep: F&R

Combination layers

SI and F&R dependent upon CUT and WINDOW layers

FIGURE 23.100 **If the colors of layers are changed as they are being used in** XREFs, **they will retain the color of the parent drawing into which the** XREFs **are attached. The** ? **option under the** LAYER **command will give a table on the screen showing that** XREFs, SI, **and** F&R **are dependent on the** A:PART1 **file.**

TACH option to place the file, SI. Again, use the AT-TACH option of the XREF command to place the file, F&R.

By using the ? option of the XREF command, you will obtain a table on the screen that lists the names and paths of the XREFs (Fig. 23.98). By using the ? option of the BLOCK command you will obtain a listing of the current file and the XREFs that were attached to it (Fig. 23.99).

The XREFs, when inserted in the current drawing, will accept the settings of the current drawing if there is conflict in assignment of colors to layers with the same name. Figure 23.100 illustrates how AutoCAD managed a name-object definition conflict by the command, LAYER, and ?.

23.46
Transparent commands

Transparent commands can be used when another command is in progress. For example, if you are dimensioning a part and wish to use the HELP command, type 'HELP to use this command, and (CR) to complete the dimensioning command. Transparent commands are typed with apostrophes in front of them. Available transparent commands are 'GRAPHSCR, 'HELP, 'PAN, 'REDRAW, 'RE-

SUME, 'SETVAR, 'TEXTSCR, 'VIEW, and 'ZOOM.

23.47
VIEW command

A drawing can be saved as several separate named views with the VIEW command. For example, a large drawing such as a house is too large to see in its entirety. Therefore, it is desirable to show it as a series of views—one for the kitchen, one for the living room, and so forth.

The method of creating a VIEW is illustrated in Fig. 23.101, where the command prompts are given as

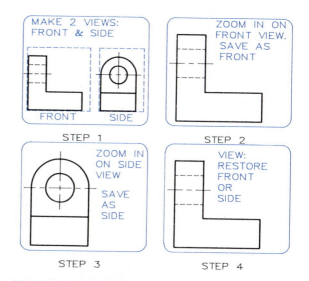

STEP 1 — MAKE 2 VIEWS: FRONT & SIDE — FRONT SIDE

STEP 2 — ZOOM IN ON FRONT VIEW. SAVE AS FRONT

STEP 3 — ZOOM IN ON SIDE VIEW SAVE AS SIDE

STEP 4 — VIEW: RESTORE FRONT OR SIDE

FIGURE 23.101 VIEW **command.**

Step 1 The two-view drawing can be saved as separate VIEWs.

Step 2 ZOOM the front view to fill the screen.
Command: VIEW (CR)
?/Delete/Restore/Save/Window: S (CR)
View name to save: FRONT (CR)

Step 3 ZOOM the side view to fill the screen.
Command: VIEW (CR)
?/Delete/Restore/Save/Window: S (CR)
View name to save: SIDE (CR)

Step 4 To display a view:
Command: VIEW (CR)
?/Delete/Restore/Save/Window: R (CR)
View name to save: FRONT (CR) **(View is displayed on screen.)**

Command: VIEW (CR) ?/Delete/
Restore/Save/Window: (Select one.)
View name: (NAME) (CR)

The exact image on the screen can be saved as it is by selecting the Save option and giving it a name when prompted. The Window option is used to select a portion of a drawing currently displayed on the screen to become the VIEW. Type Restore, give the VIEW name, and it will be displayed. The Delete subcommand removes the VIEW from the list. To review the list of VIEWS that have been made, use the ? option.

23.48
Inquiry commands

The INQUIRY commands are used to determine relationships among entities and information about the file being used. INQUIRY commands are LIST, DBLIST, STATUS, TIME, HELP, DIST, ID, and AREA.

The LIST command is illustrated in Fig. 23.102, where the circle is selected when prompted. Notice that a great deal of information about the circle is displayed on the screen.

DBLIST gives a similar listing of all the entities of a drawing. Use Ctrl S to stop and start scrolling on the screen. Ctrl C will abort the command.

UTILITY (Pull-down)

Command: LIST
Select objects: Pick circle

CIRCLE   Layer: VISIBLE
    Space: Model space
Center point, X= 2.00 Y= 1.00 Z= 0.00
    radius   1.00
circumference   6.28
    area   3.14

FIGURE 23.102 The LIST command under the UTILITY option of the pull-down menu is used to obtain information about entities on the screen, such as a circle.

DIST command is used to find the distance between two points. You will be prompted for the first and second points that can be selected with the cursor or by coordinates typed at the keyboard. At the command line, you will be given the distance, its angle, the delta X and the delta Y, where delta means the change in values from the first point.

ID is used to pick a point on the screen and obtain its X, Y, and Z coordinates.

AREA is used to find the area and perimeter of a space on the screen. You will be prompted, First point, Next point:, Next point:, and so on until you finish picking the points about the area and press (CR). An area can be calculated by having areas added or subtracted as shown in Fig. 23.103.

The STATUS command can be used by typing, Command: STATUS (CR). The data shown in Fig. 23.104 will be displayed on the screen, temporarily replacing the graphics display. This screen gives you information about the settings, layers, coordinates, and disk space.

The TIME command gives text display of information about the time spent on a drawing as shown in Fig. 23.105. The timer can be RESET and turned

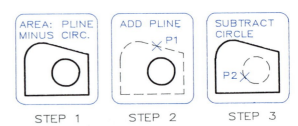

FIGURE 23.103 AREAS.

Step 1 The object drawn as a continuous PLINE and a circle can have its area determined with the AREA command.

Step 2 Command: AREA (CR)
<First point>/Entity/Add/Subtract: A (CR)
<First point>/Entity/Subtract: E (CR)
(ADD mode) Select circle or polyline: (Point to polyline, and area is given.) (CR)

Step 3 (ADD mode) Select circle or polyline: (CR)
<First point>/Entity/Subtract: S (CR)
<First point>/Entity/Add: E (CR)
(SUBTRACT mode) Select circle or polyline: (Point to circle, and its area is given along with the total area of the part.) (CR)

```
    133 entities in B:G-25-18
Limits are          X:    0.0000    3.4000   Off
                    Y:    0.0000    3.0000
Drawing uses        X:    0.1000    3.1000
                    Y:   -0.1750    3.4000  **Over
Display shows       X:    0.1000    5.3022
                    Y:   -0.1750    3.4000
Insertion base is   X:    0.0000  Y:    0.0000   Z: 0.0000
Snap resolution is  X:    0.0500  Y:    0.0500
Grid spacing is     X:    0.1000  Y:    0.1000

Current layer:      2DIMEN (Off)
Current color:      BYLAYER -- 5 (blue)
Current linetype:   BYLAYER -- CONTINUOUS
Current elevation:     0.0000 thickness:   0.0000
Axis off Fill on Grid off Ortho on Qtext off Snap on Tablet off
Object snap modes: None
Free RAM: 12250 bytes     Free disk: 764416 bytes
I/O page space: 64K bytes
```

FIGURE 23.104 The STATUS command will give this information about
your drawing.

```
TIME
Current time:           27 Apr 1992 at 19:09:49.270
Drawing created:         1  Apr 1992 at 20:11:37.950
Drawing last updated:   19 Apr 1992 at 19:24:42.360
Time in drawing editor:  0 days 00:06:18.660
Elapsed timer:           0 days 00:06:18.660
```

FIGURE 23.105 The TIME command can
be used for inspecting the time spent on a drawing
and for setting the time for an assignment.

```
File Utility Menu

  0.  Exit File Utility Menu
  1.  List Drawing files
  2.  List user specified files
  3.  Delete files
  4.  Rename files
  5.  Copy file
  6.  Unlock file

Enter selection (0 to 6) <0>:
```

FIGURE 23.106 The FILES command lets
you leave AutoCAD to make changes in existing files.

ON to measure the time for a drawing session, but the cumulative time cannot be erased without destroying the drawing file. The DISPLAY option displays the time opposite the heading, Elapsed timer:, to show how long the current session has taken after RESETting.

By typing FILES when in the Drawing Editor, the File Utility Menu (Fig. 23.106) will be displayed on the screen. You may select an option by number, use it, and return to the Drawing Editor by pressing 0 and the F1 key.

23.49
Dimensioning principles

Figure 23.107 shows the types of dimensions that can be used with AutoCAD. These options can be found in the submenu of the DIM: command. When in the LINEAR mode, you are prompted to specify if the dimension line is to be **horizontal, aligned,** or **rotated.** In Fig. 23.108, a horizontal line is dimensioned by selecting its two endpoints (*P1* and *P2*) and locating the dimension line with *P3*. The dimension of 2.40 in., which is measured by the program, is accepted by (CR) and the dimension is shown on the drawing (Step 2).

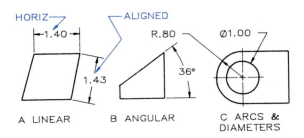

FIGURE 23.107 **The types of dimensions that appear on a drawing.**

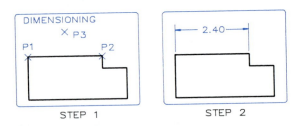

FIGURE 23.108 **Dimensioning a line.**

Step 1 Command: DIM (CR)
Dim: VERtical or HORizontal (CR)
First extension line origin or (CR) to select: (Select P1.)
Second extension line origin: (Select P2.)
Dimension line location: (Select P3.)

Step 2 Press (CR), and the measurement appears at the command line, 2.40. Press (CR) to accept this value (or give a different value), and the dimension line is drawn.

All drawings should be drawn **full size** when using computer graphics. The scaling process will take place at the time of plotting the drawings.

23.50
Dimensioning variables—introduction

Before becoming proficient with dimensioning, you must learn to control dimensioning variables, DIM VARS, shown in Fig. 23.109.

DIM VARS	DEFAULT	DEFINITION
DIMALT	OFF	Alternate units selected
DIMALTD	2	Alternate units decimal pls.
DIMALTF	25.4	Alternate units scale factor
DIMAPOST	NONE	Alternate units text suffix
DIMASO	ON	Create associative dimens.
DIMASZ	.125	Arrow size
DIMBLK	NONE	Arrow block
DIMBLK1	NONE	Separate arrow block 1
DIMBLK2	NONE	Separate arrow block 2
DIMCEN	−.05	Center mark size
DIMCLRD	BYBLOCK	Dimension line color
DIMCLRE	BYBLOCK	Extension line color
DIMCLRT	BYBLOCK	Dimension text color
DIMDLE	0	Dimension line extension
DIMDLI	.38	Dimension line increment
DIMEXE	.12	Extension line extension
DIMEXO	.06	Extension line offset
DIMGAP	.05	Dimension line gap
DIMLFAC	1	Length factor
DIMLIM	OFF	Limits dimensioning
DIMPOST	NONE	Dimension text suffix
DIMRND	0	Rounding value
DIMSAH	OFF	Separate arrow blocks
DIMSCALE	1	Dimension scale factor
DIMSE1	OFF	Suppress extension line 1
DIMSE2	OFF	Suppress extension line 2
DIMSHO	OFF	Show dragged dimensions
DIMSOXD	OFF	Suppress outside dim. lines
DIMSTYLE	UNNAMED	Dimension style
DIMTAD	OFF	Text above dimension line
DIMTFAC	1	Tolerance text scale factor
DIMTIH	ON	Text inside horizontal
DIMTIX	OFF	Text inside extension lines
DIMTM	0	Minus tolerance value
DIMTP	0	Plus tolerance value
DIMTOFL	OFF	Tolerance dimensioning
DIMTOH	ON	Text outside horizontal
DIMTOL	OFF	Tolerance dimensioning
DIMTSZ	0	Tick size
DIMTVP	0	Text vertical position
DIMTXT	.125	Text size
DIMZIN	0	Zero suppression

FIGURE 23.109 **A listing of the DIM VARS that can be selected and changed when dimensioning. Type DIM and STATUS to get this listing.**

A listing of the current DIM VARS and their assigned values (Fig. 23.109) can be displayed on the screen by using the STATUS option under the DIM: command. For an average drawing, you should reset the variables, as shown in Fig. 23.110, based on the letter height, .125″, for example. You can change

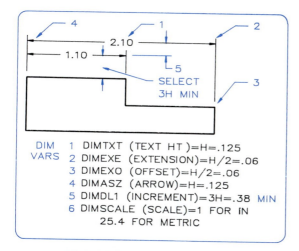

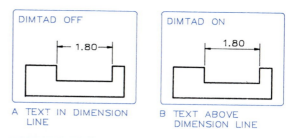

FIGURE 23.110 Dimensioning variables are based on the height of the lettering, usually about ⅛ inch.

FIGURE 23.112 (A) When DIMTAD is OFF, the text is inserted within the dimension line. (B) When DIMTAD is ON, the text is placed above the dimension line.

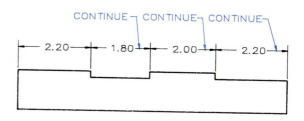

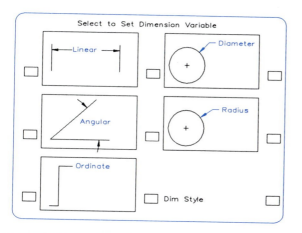

FIGURE 23.113 When dimensions are placed end to end, the option, CONTINUE, can be used to specify the Second extension line origin after the first dimension line has been drawn.

FIGURE 23.111 Dialogue boxes are available for selecting DIM VARS.

These variables will be retained in memory throughout the drawing session. They can be saved as part of your FORMAT file by the command SAVE, which will add them to this file for future use.

The UNITs command must be used to assign the number of decimals that the dimension units are to have. Two decimal places are used for inches, and none for millimeters. Architectural units are feet and inches.

Dimensions can be placed over or within the dimension lines by turning the DIMTAD command on or off (Fig. 23.112).

When a series of horizontal dimensions link together end-to-end, the CONTINUE command can be used to join successive dimension lines in a line, as shown in Fig. 23.113.

BASELINE dimensioning can be used to give a series of dimensions that originate from the same baseline. The DIMDLI variable (dimension line in-

each variable by selecting the DIM: heading in the Root Menu, and then select the DIM VARS command, which will list the names of the variables. You will need to set DIMTXT, DIMASZ, DIMEXE, DIMEXO, DIMTAD, and DIMDLI to the values shown in Fig. 23.110. Set the DIMSCALE variable to 1.

The dialogue box in Fig. 23.111 can be used to select the type of dimensioning variables that you wish to set. Other dialogue boxes will appear to aid in the further selection of these variables.

crement for continuation) automatically separates the parallel dimension lines (Fig. 23.114).

Once dimensioning variables have been set for an application, this setup can be saved as a `Dimension Style` using the dialogue box shown in Fig. 23.115. By selecting any of these options, the `Dimensioning Style` can be restored, listed, overridden, or saved (Fig. 23.116).

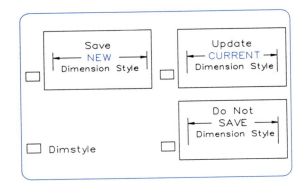

FIGURE 23.116 **This dialogue box, under the** `Entity Creation` **heading, lets you manage** `Dimension Styles`.

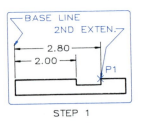 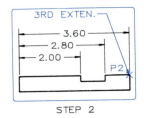

FIGURE 23.114 `BASELINE` **option**

Step 1 Command: <u>DIM</u> (CR)
Dimension the first line as shown in Fig. 23.108. When prompted for `Dim:` **for the next dimension, respond with** <u>BASELINE</u> **(CR) and you will be asked for the** `Second extension line origin:` **(Select P1.)**

Step 2 **The second dimension line will be drawn. Respond to** `Dim:` **with** <u>BASELINE</u> **(CR) again, and when asked for** `Second extension line origin:`, **select P2. Continue in this manner for any number of dimensions using the same baseline.**

23.51
Ordinate dimensions

Ordinate dimensions are used to dimension surfaces from an origin point where `X = 0` and `Y = 0`. To begin, the `UCSICON` must be set to `OR` (for `ORIGIN`) and moved to the `0,0` point on the drawing, as shown in Fig. 23.117. The *X*-ordinate is dimensioned as 2.30, which means that the vertical plane is 2.30 units along the *X*-axis from the origin.

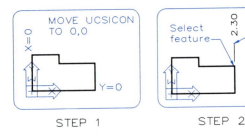

FIGURE 23.117 `ORDINATE` **dimensions:** `Xdatum`.

Step 1 **Move the** `UCSICON` **to the origin where** `X = 0` **and** `Y = 0`.

Step 2 Command: <u>ORDINATE</u>
`Select Feature:` **(Select line.)**
`Leader endpoint (Xdatum/Ydatum:` <u>X</u> **(CR)**
`Leader endpoint:` **(Select.) (CR) (The dimension is drawn.)**

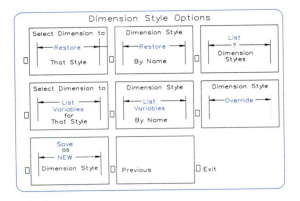

FIGURE 23.115 **This dialogue box is used for selecting and changing** `Dimensioning Styles`.

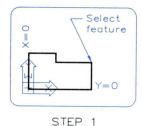

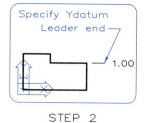

STEP 1   STEP 2

FIGURE 23.118   ORDINATE **dimensions:**
Ydatum.

Step 1 Command: DIM
DIM: ORDINATE
Select Feature: **(Select line.)**

Step 2 Leader endpoint (Xdatum/Ydatum):
Y **(CR)**
Leader endpoint: **(Select.) (CR) (The dimension is drawn.)**

A *Y*-ordinate dimension of 1.00 is found in Fig. 23.118 which means that the horizontal plane is 1.00 unit in the *Y*-direction from Y = 0.

23.52
Dimensioning circles and arcs

AutoCAD normally dimensions circles, as shown in Fig. 23.119, based on the size of the circle unless you override the standard dimensioning variables. By changing two variables, DIMTIX and DIMTOFL, circles of the same size can be dimensioned as shown in Fig. 23.120. The steps of dimensioning circles are shown in Fig. 23.121.

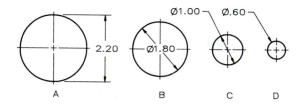

FIGURE 23.119   **Types of dimensions available for dimensioning circles with AutoCAD.**

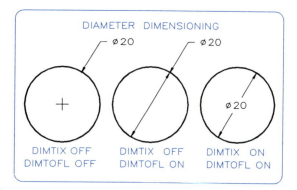

FIGURE 23.120   **Examples of dimensions with various DIM VARS settings.**

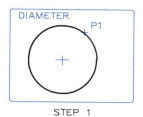

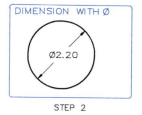

STEP 1   STEP 2

FIGURE 23.121   **Dimensioning a circle.**

Step 1   Command: Dim **(CR)**
Dim: DIAMETER **(CR)**
Select arc or circle: **(Select P1.)**

Step 2 Dimension text <2.20>: **((CR) to accept this dimension. The diametric dimension is drawn from *P1* through the center.)**

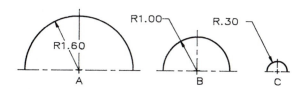

FIGURE 23.122   **Arcs will be dimensioned by one of the formats given here depending on the size of the radius.**

The DIMTIX variable forces the text to be drawn inside the extension lines regardless of the lack of space there. The DIMTOFL variable forces an extension line to be drawn between the extension lines when the text is forced to the outside.

Arcs are dimensioned with an R placed in front of the dimension of the radius (Fig. 23.122). The steps

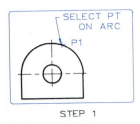

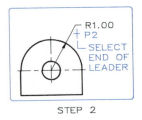

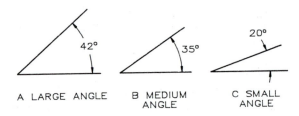

FIGURE 23.123 Dimensioning with a LEADER.

Step 1 Command: <u>DIM</u> (CR)
Dim: <u>LEADER</u> (CR)
Leader start: (Select *P1.*)

Step 2 To point: (Select *P2.*)
To point: (CR)
Dimension text <1.00>: ((CR) to accept this value or insert different value.)

FIGURE 23.125 Angles will be dimensioned in any of these three formats using AutoCAD.

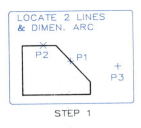

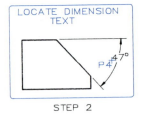

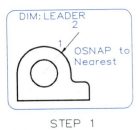

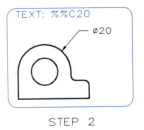

FIGURE 23.124 LEADER option.

Step 1 Dim: <u>LEADER</u>
Leader start point: (OSNAP to NEAREST.)
To point: (*P2.*)
To point: (CR)

Step 2 Dimension text <1>: <u>%%C20</u> (CR)

FIGURE 23.126 ANGULAR command.

Step 1 Command: <u>DIM</u> (CR)
Dim: <u>ANGULAR</u> (CR)
Select first line: (Select *P1.*)
Second line: (Select *P2.*)
Enter dimension line arc location: (Select *P3.*)
Dimension text <47>: (CR)

Step 2 Enter text location: (Select *P4.*) or (CR)

of dimensioning an arc with a radius and an automatic leader are shown in Fig. 23.123. The LEADER command can be used to add a dimension or a note to a drawing, but this command will not calculate the numerical value of the dimension. The value must be found first and then added to the leader end when prompted (Fig. 23.124).

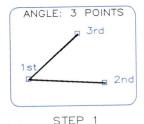

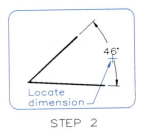

FIGURE 23.127 ANGLES: Endpoint option.

23.53

Dimensioning angles

Step 1 DIM: <u>ANGULAR</u>
Select arc, circle, line, or RETURN: (CR)
Angle Vertex: (Select *P1.*)
First angle endpoint: (Select *P2.*)
Second angle endpoint: (Select *P3.*)

Step 2 Enter dimension line arc location: (Select arc location.)
Dimension text <46>: (CR)
Enter text location: (CR) (To accept 46.)

Figure 23.125 shows variations in dimensioned angles, which occur because of inadequate space. Begin

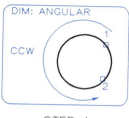
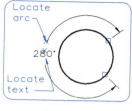

FIGURE 23.128 ANGULAR **dimensions**.

Step 1 Dim: ANGULAR
Select arc, circle, line, or RETURN:
(Select 1.)
Second line: **(Select 2.)**

Step 2 Enter dimension line arc
location: **(Select point.)**
Dimension text <280>: **(CR)**
Enter text location: **(CR)**

by selecting two lines of the angle, and then select the location for the dimension line arc. Where room permits, the dimension value will be centered in the arc between the arrows. Figure 23.126 shows the commands for dimensioning an angle.

For angles over 90 degrees as shown in Fig. 23.127, select a point on the circle and move counterclockwise to the second point. Locate the arc and the text, and the dimension is drawn. An angle can be dimensioned by beginning with its vertex and locating the endpoints of each line as shown in Fig. 23.128.

23.54
Dimensioning variables

Many of the advanced dimensioning variables shown in Fig. 23.109 are explained below.

DIMALT, if ON (1), gives dual dimensions (in two values, such as inches and millimeters). These values are taken from BIMALTF and DIMALTD settings.

DIMALTD sets the decimal places for the alternate units from 0 to 4 decimal places.

DIMALTF sets the alternate units scale factor for dual dimensions. The default is 25.4, which gives the metric conversion for measurements made in inches.

DIMPOST is used to assign a character string to follow all dimensions such as millimeters, feet, and miles. Type a period (.) when prompted for this value to disable the command.

DIMAPOST is used to assign a character string to follow all alternate dimensions (in dual dimensioning), as shown in Fig. 23.129. Type a period to disable this setting.

DIMCEN is used to mark the center of an arc or a circle. When set to 0, no center is marked; when set to larger than 0, the value is the size of the center mark; and when set to a minus value, center lines from the center marks are drawn beyond the arc.

DIMCLRD is used to set color assigned to dimension lines, arrowheads, and leaders. Any color, BYBLOCK, or BYLAYER assignments can be made.

DIMCLRE is used to set color assigned to extension lines as described above.

DIMCLRT is used to set the color assigned to dimension text as described above.

DIMDLE is used when ticks are drawn instead of arrows to make the dimension line extend past the extension line by the DIMDLE specified distance.

DIMGAP is used to set the gap in the dimension line for the placement of the text.

DIMLFAC is used to assign a global scale factor for linear dimensions before inserting the dimensions. Linear dimensions including radii, diameters, and coordinates are multiplied by the current setting of DIMLFAC.

DIMRND is used to set the rounding factor for dimensions. The number of decimal places in the rounded values is determined by the UNIT command.

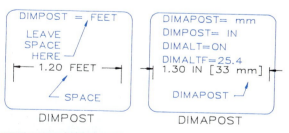

FIGURE 23.129 The DIMPOST dimensioning variable is used to add text after dimensions. The DIMAPOST adds text after alternate dimensions in dual dimensioning.

DIMSE1 is used to suppress the first extension line.

DIMSE2 is used to suppress the second extension line.

DIMSOXD is used to suppress dimension lines that would normally be drawn outside of extension lines (Fig. 23.130).

DIMTIX is used to force text inside extension lines that would normally be placed outside the extension lines (Fig. 23.130). Setting DIMTIX off forces text outside of the circle or arc. If DIMTIX is off, DIMSOXD has no effect.

DIMTOFL is turned on to force a dimension line between the extension lines even when the text is placed outside them (Fig. 23.130). When DIMTOFL is on and DIMTIX is off, dimension lines and arrowheads are drawn inside circles and arcs while the text and leader are drawn outside.

DIMTAD, if turned on (1), is used to place text above the dimension line rather than inside the line (Fig. 23.130).

DIMTVP is used with DIMTAD to position the text vertically, either above or below a dimension line. Text will be above the dimension line if the number is positive and below the dimension line if negative. The number inserted is used as the multiplier of the text height for the vertical space (DIMTVP X DIMTXT).

DIMZIN		EXAMPLES			
0	OMIT ZERO & INCHES	3/4"	9"	2'	2'-0 1/2"
1	ZERO FT & ZERO IN	0'-0 1/4"	0'-9"	2'-0"	2'-0 1/2"
2	ZERO FEET	0'-0 1/2"	0'-9"	2'	2'-0 1/2"
3	ZERO IN	3/4"	9"	2'-0"	2'-0 1/2"

FIGURE 23.131 The DIMZIN dimensioning variable controls the use of zeros when architectural units are being used. Options 0 and 1 are the most commonly used responses.

DIMTIH is turned on to position text inside of dimension lines horizontally; if turned off, the text is aligned with the dimension lines.

DIMTOH, if on, positions text outside the extension lines horizontally; if off, the text is aligned with the dimension lines.

DIMTSZ is used to specify the size of tick marks (if nonzero) instead of using arrowheads. If set to zero, arrowheads are drawn.

DIMZIN is used with architectural units of feet and inches to control the zero values that precede feet and inches as shown in Fig. 23.131.

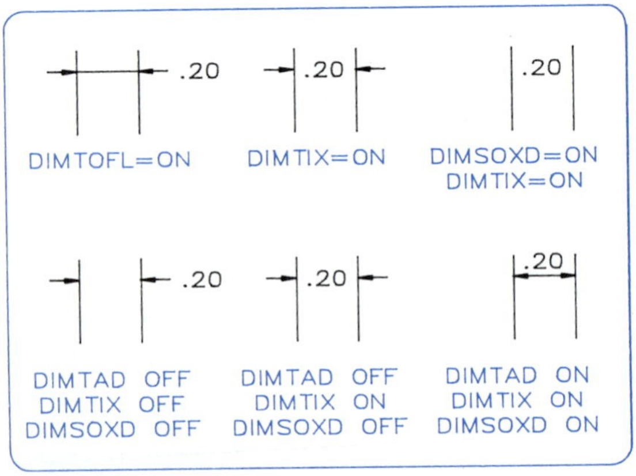

FIGURE 23.130 Examples of the application of dimensioning variables. DIMTOFL, DIMTIX, and DIMSOXD in conjunction with DIMTAD.

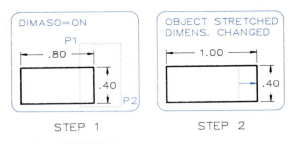

FIGURE 23.132 Associative dimensions.

Step 1 Set DIMASO to ON and apply dimensions to the part. Use the STRETCH command and WINDOW the end of the part with the C option.

Step 2 Select a new endpoint for the part and it will be lengthened and new dimensions will be calculated.

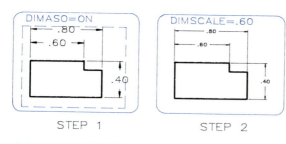

FIGURE 23.133 UPDATE **command.**

Step 1 Turn DIMASO ON, and dimension the object. Change any dimensioning variable you wish.
Dim: UPDATE **(CR)**
Select objects: **(Window drawing.)**

Step 2 If the DIMSCALE variable was changed, all dimensioning factors will be changed to their new values.

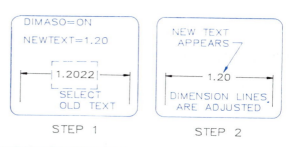

FIGURE 23.135 NEWTEXT **command.**

Step 1 Turn DIMASO ON, and dimension the object.
Dim: NEWTEXT **(CR)**
Enter new dimension text: 1.20 **(CR)**
Select objects: **(Select text to be changed.)**

Step 2 The old text will be replaced with new text.

23.55
Associative dimensioning

Associative dimensions are associated with (attached to) the objects being dimensioned so that the dimensions automatically change to show new values as the objects are lengthened or shortened with the STRETCH command. In Fig. 23.132, the box that is 0.80 wide is stretched to 1.00 wide and the associated dimension changes to show this new width.

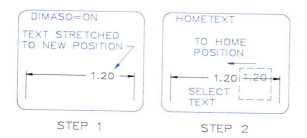

FIGURE 23.134 HOMETEXT **command.**

Step 1 Turn DIMASO ON, and dimension the object. If the object and dimensions are STRETCHed, the dimension numerals will not be centered.

Step 2 Dim: HOMETEXT **(CR)**
Select objects: **(Select text.) (The text will automatically center itself in the dimension line.)**

Associative dimensions are inserted as single entities, much as BLOCKs are inserted, so they can be erased as a unit by selecting any part of the dimension: arrowhead, extension line, or dimension line. The EXPLODE command can be used to split the parts of the dimension unit into individual entities for editing one feature at a time.

When the Dim Vars DIMASO is turned on (1), dimensions are applied as associative dimensions (Fig. 23.133); when off (0), dimensions are applied as separate entities. The DIMSHO command can be turned on (set to 1) for dynamically showing the changes in the numerical values of dimensions as they are stretched. Associative dimensions cannot be stored in a dimension style.

The following DIM: commands can be used to modify dimensions that were applied as associative dimensions:

HOMETEXT is used to reposition text to its standard position at the center of the dimension line after the dimension line has been lengthened by the STRETCH command or text changed by the TEDIT subcommand (Fig. 23.134).

NEWTEXT is used to change the dimensioning text within a dimension line, as shown in Fig. 23.135. If you give a null response (press (CR)) for new text, the actual dimension measurements will be used at the new text.

OBLIQUE is used to create linear dimensions with oblique extension lines rather than extension lines that are perpendicular to the dimension line. The

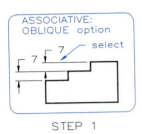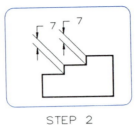

FIGURE 23.136 Associative/OBLIQUE option.

Step 1 Command: DIM
Dim: OBLIQUE
Select objects: (Select any feature of an associative dimension.) 2 selected, 2 found
Select objects: (CR)
Enter obliquing angle (RETURN for none): 45

Step 2 After pressing (CR) the extension lines are drawn at an angle and the dimension numerals are repositioned.

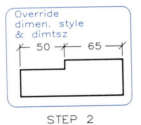

FIGURE 23.137 Associative/OVERRIDE option.

Step 1 Command: DIM
Dim: OVERRIDE
Dimension variable to override: DIMSTYLE (CR)
Current value <Standard> New value: ROMANS (CR)
Dimension variable to override: DIMTSZ (CR)
Dimension variable to override: (CR)
Select objects: (Select both dimensions.)

Step 2 After pressing (CR), the dimensioning variables are redrawn on the screen to conform to the specified changes. If any of the selected dimensions contain references to a dimension style, AutoCAD prompts: Modify dimension style "style name"? <N>: Enter Y or N.

extension lines are drawn obliquely in Fig. 23.136 by selecting the associative dimensions and specifying the desired obliquing angle.

OVERRIDE is used to override any number of dimensioning variables within an associative dimension. For example, two dimensions are selected in Fig. 23.137 and their arrowheads are changed to ticks (DIMTSZ) and the text style (DIMSTYLE) is changed.

RESTORE is used to change dimension variable settings by reading new setting from an existing dimension style:

Dim: RESTORE
Current dimension style: (Style name.)
?/Enter dimension style name or RETURN to select dimension: (Select dimensions.)

The named dimension style is restored and becomes the current style as long as no other changes are made or other styles are restored.

By responding with a ?, you will obtain a listing of the dimension styles in the current drawing.

SAVE is used to assign dimension variable settings to a dimension style:

Dim: SAVE
?/Name for new dimension style: (New dimension style name.)

TEDIT is used to change the position and orientation of dimension text of a dimension.

Dim: TEDIT
Select dimension: (Select dimension.)
Enter text location (Left/Right/Home/Angle): L
(Drag to new location.)

The HOME option moves the text back to its original position. Left and Right move the dimension to near the left and right ends of the dimension line. The Angle option rotates the text to the angle of your choice.

TROTATE is used to set the angle of rotation of dimension lines within a dimension line (Fig. 23.138):

```
Dim: TROTATE
Enter new text angle: 45
Select Objects: (Select dimensions to be
changed.)
```

This command works like the Angle option of the TEDIT command except that more than one dimension can be selected at a time.

UPDATE is used to reassign dimension variables, dimension style, current text style, and current Units setting to the current drawing. For example, the DIMSCALE in Fig. 23.133 is changed after the associative dimensions were applied, and the drawing is updated as follows:

```
DIM: UPDATE (CR)
Select objects: (Select dimension.)
```

A window can be used to select all of the dimensions of a drawing for updating.

VARIABLES is used to make a list of the variable setting of a dimension style for your inspection.

```
Dim: VAR
Current dimension style: (Style
name.)
?/Enter dimension style name or
RETURN to select dimension: (CR)
Select Dimension: (Select.)
```

The dimension style may be chosen by name as an alternative. The display of variables is identical to that of the STATUS command.

23.56

Special arrowheads

Instead of using the standard arrowhead, you can use other symbols at the ends of dimension lines. By selecting the Dim: command DIMBLK and typing DOT, dimensions will be terminated with a dot instead of an arrowhead.

To create your own arrow (a dot, in this case), draw a right-end dot with a segment of the dimension

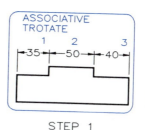
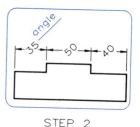

FIGURE 23.138 Associative/TROTATE option.

Step 1 Command: DIM
Dim: TROTATE
Enter new text angle: 45
Select objects: **(Select each associative dimension.)**

Step 2 When (CR) is pressed, the text is redrawn at an angle within the dimension lines.

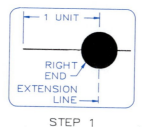
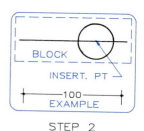

FIGURE 23.139 Custom-designed arrowhead.

Step 1 An arrowhead, a DONUT in this case, is drawn to be one unit long.

Step 2 The terminator is BLOCKed with the insertion point at the extension line. Use the DIMBLK option of the DIM VARS command to assign the BLOCK by its name. When dimensioning, the BLOCK will become the new arrowhead.

line attached and an extension of the dimension line to the right (Fig. 23.139). The total length (including the dimension line segment to where the extension line will cross) must be one drawing unit long. Make a BLOCK of the dot and dimension line, set the insertion point at the center of the dot where it will join the extension line, enter the dimensioning command, Dim:, and type DIMBLK. When prompted, type the name of the dot block which now is used at the ends of the dimension lines instead of arrowheads.

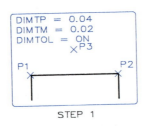

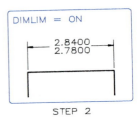

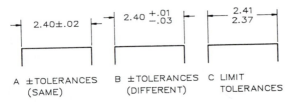

FIGURE 23.140 Special arrowhead BLOCKS can be created, saved, and named SPOT and ARR. When DIMSAH (separate arrowheads) is set to ON, and the Dim Vars DIMBLK1 is set equal to SPOT and DIMBLK2 is set equal to ARR, the first end of the dimension line will have SPOT as the arrowhead and the second end will have ARR as the arrowhead.

To disable the use of the special dot, change the DIMBLK variable to a period (.).

DIMSAH can be turned on to give separate arrowheads at each end of the dimension line (Fig. 23.140). Two right-end arrowheads are defined as DIMBLK1 and DIMBLK2 for the first and second types. When the dimension is inserted, the two types of arrowheads will be inserted. For this option to work, DIMTSZ must be set to zero, or ticks will be drawn instead.

23.57
Toleranced dimensions

Dimensions can be toleranced automatically, using any of the forms shown in Fig. 23.141, by setting the DIM VARS: DIMTOL on and assigning a plus tol-

FIGURE 23.141 Dimensions can be toleranced in any of these three formats.

FIGURE 23.142 Toleranced dimensions.

Step 1 To tolerance a dimension, set Dim Vars, DIMTOL to ON. Set DIMTP (plus tolerance) and DIMTM (minus tolerance) to the desired values. Set DIMLIM to ON to convert the tolerances into the limit form.

Step 2 Dimension the line by using P1, P2, and P3 to specify the first extension line, second extension line, and the dimension location. (The toleranced dimension is drawn.)

erance to DIMTP, and a minus tolerance to DIMTM. When DIMLIM is on, the upper and lower limits of the dimensions will be shown as in Fig. 23.142.

DIMTFAC is a scale factor setting that controls the height of the tolerance values in the text in cases where the dimension will be stacked one over the other.

23.58
Oblique pictorials

An oblique pictorial can be constructed as shown in Fig. 23.143. The front orthographic view is constructed and then COPYed behind the first view at the angle desired for the receding axis. Using OSNAP, the visible endpoints are connected, and invisible lines are erased. Circles can be drawn as true circles on the true-size front surface, but circular features should be avoided on the receding planes since their construction is complex.

23.59
Isometric pictorials

The STYLE subcommand of the SNAP command can be used to change the rectangular GRID (called STANDARD) to ISOMETRIC (I) where the dots are plotted vertically and at 30° with the horizontal

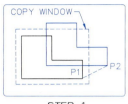
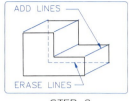

STEP 1   STEP 2

FIGURE 23.143   An oblique pictorial.

Step 1 Draw the frontal surface of the oblique, and COPY this view at *P2*, which is the desired angle and distance from *P1*.

Step 2 Connect the corner points, and erase the invisible lines to complete the oblique.

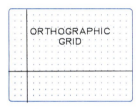
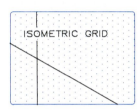

A   ORTHOGRAPHIC GRID   B   ISOMETRIC GRID

FIGURE 23.144   (A) The orthographic grid is called the STANDARD style of the SNAP mode. (B) The ISOMETRIC style of the SNAP mode.

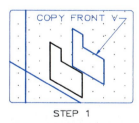
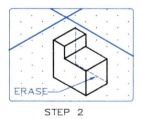

STEP 1   STEP 2

FIGURE 23.145   The isometric pictorial.

Step 1 Draw the front view of the isometric pictorial. COPY this front view at its proper location.

Step 2 Connect the corner points, and erase the invisible lines. The cursor lines can be moved into three positions using Ctrl E or ISOPLANE.

(Fig. 23.144). The lines of the cursor on the screen align with two of the isometric axes. The axes can be rotated by using Ctrl E or by activating the ISO-PLANE command from the screen menu. When OR-THO is ON, lines are forced to be drawn parallel to

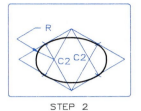
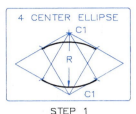

STEP 1   STEP 2

FIGURE 23.146   The four-center ellipse BLOCK.

Step 1 Construct a rhombus that is one unit along both axes. Using the four-center ellipse method, construct an ellipse using four centers and four arcs. (The rhombus should be drawn on a construction layer that can be turned off so that its lines will not show.)

Step 2 Make a WBLOCK of this ellipse using the center as the insertion point. This block can be INSERTed as a * BLOCK so that it can be broken when needed.

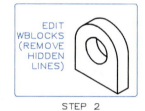
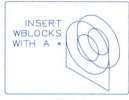

STEP 1   STEP 2

FIGURE 23.147   An isometric drawing with elliptical features.

Step 1 The UNIT BLOCK of the ellipse developed in Fig. 23.146 is inserted and sized (*X*-scale factor) to match the drawing's dimensions. The ellipse can be rotated as a BLOCK to fit any of the three isometric surfaces.

Step 2 The back side of the isometric is drawn. The hidden lines are removed with the BREAK command.

the isometric axes. Figure 23.145 shows the steps for constructing an isometric using the grid.

To depict circular features in isometric, a four-center ellipse one unit across its diameter is constructed (Fig. 23.146). This ellipse should be made into a BLOCK and inserted where needed. It can be scaled to fit different drawings by giving it an X-VALUE factor at the time of insertion, and it can be rotated to fit each isometric plane.

Figure 23.147 shows an isometric pictorial with elliptical features. The ellipse was inserted using the four-center ellipse BLOCK developed in Fig. 23.146. The size of the block is changed by entering the size factor when prompted for the X-value. It is advantageous if the BLOCK is designed to be one unit across (a UNIT BLOCK) so that its size can be scaled with ease. For example, a UNIT BLOCK will be multiplied by an X-value of 2.48 to give a BLOCK that is 2.48 in.

23.60
ELLIPSE command

The ELLIPSE command can be used to draw ellipses by several methods. In Fig. 23.148, an ellipse is drawn by selecting the endpoints of the major axis and the third point, P3, which gives the minor radius of the ellipse.

A second method (Fig. 23.149) is begun by locating the center of the ellipse, P1, then the axis endpoint, P2, and the length of the other axis, P3. The ellipse will be drawn to pass through points P2 and a point on the minor diameter specified by P3.

A third method (Fig. 23.150) requires the location of the endpoints of the axis and the specification of a rotation angle about the axis. When the rotation angle is 0°, the ellipse is a full circle; when the rotation is 90°, the ellipse is an edge.

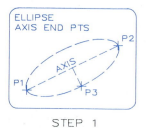
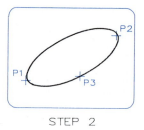

STEP 1   STEP 2

FIGURE 23.148   Ellipse—endpoints.

Step 1 Command: ELLIPSE **(CR)**
<Axis endpoint 1>/Center: **(Select P1.)**
Axis endpoint 2: **(Select P2.)**
<Other axis distance>/Rotation: **(Select P3.)**

Step 2 **The ellipse is drawn through P1 and P2 and a distance measured perpendicularly from P1–P2 by the location of P3.**

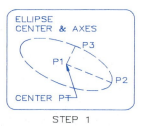
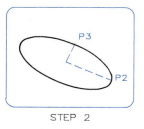

STEP 1   STEP 2

FIGURE 23.149   Ellipse—center and axes.

Step 1 Command: ELLIPSE **(CR)**
<Axis endpoint 1>/Center: C **(CR)**
Center of ellipse: **(Select P1.)**
Axis endpoint: **(Select P2.)**

Step 2 <Other axis distance>/Rotation: **(Select P3.) (The ellipse is drawn.)**

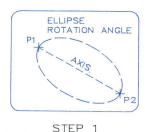
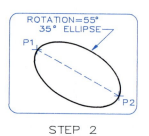

STEP 1   STEP 2

FIGURE 23.150   Ellipse—rotation angle.

Step 1 Command: ELLIPSE **(CR)**
<Axis endpoint>/Center: **(Select P1.)**
Axis endpoint 2: **(Select P2.)**

Step 2 <Other axis distance>/Rotation: R **(CR)**
Rotation around major axis: 55 **(CR)**
(The ellipse is drawn.)

When in the isometric SNAP mode, the ELLIPSE command will be

<Axis endpoint 1>
/Center/Isocircle: I (CR)

By selecting the ISOCIRCLE option, ellipses can be automatically drawn in the correct orientation on each of the three ISOPLANEs. The isometric ellipses can be selected using the previous techniques (Fig. 23.151).

Portions of ellipses drawn with these new commands can be removed by the BREAK command.

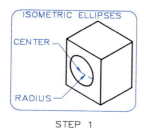

STEP 1   STEP 2

Ellipse—isometric mode.

Step 1 When in the isometric SNAP mode:
Command: <u>ELLIPSE</u> **(CR)**
<Axis endpoint 1>/Center/Isocircle: <u>I</u>
(CR)
Center of circle: **(Select center.)**
<Circle radius>/Diameter: **(Select radius.)**

Step 2 The isometric ellipse is drawn on the current
ISOPLANE.

Since ellipses are drawn with connected curves, some peculiarities can occur when the BREAK command is used.

23.61
Introduction to 3D extrusions

An elementary form of drawing three-dimensional objects is the extrusion technique that makes use of the commands ELEV, HIDE, and PLAN. Before beginning, set the UCSICON to ON so that the 3D icon will appear on the screen. The objects can be drawn with the LINE command and any of the other DRAW commands in the PLAN view, the top view, where the X- and Y-axes are true size.

The THICKNESS of the objects is extruded upward or downward by a THICKNESS of 4, in this example, as shown in Fig. 23.152. The VIEWPOINT command is used to establish a line of sight in which the object can be seen as a three-dimensional drawing. The VIEWPOINT is set at 1,-1,1 in this case, to give an isometric projection.

By using the PLAN command, the top view (plan view) will be restored. By inserting another VIEW-POINTs, the drawing reverts back to a three-dimensional drawing. All of the regular DRAW commands, such as LINE, CIRCLE, TEXT, and ARC can be

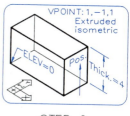

STEP 1   STEP 2

Extrusion of a box.

Step 1 Command: <u>ELEV</u>
New current elevation <0>: <u>0</u>
New current thickness <0>: <u>4</u>
Command: <u>LINE</u>
From point: **(Draw 7 × 3 rectangle as a top view.)**

Step 2 Command: <u>VPOINT</u>
Rotate/<view point><current>: <u>1,-1,1</u>
(An isometric view of the extruded box appears on the screen.)

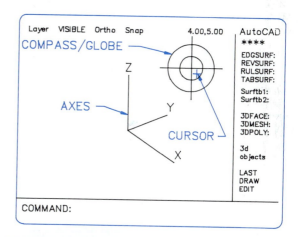

When VPOINT command is selected, and (CR) is pressed twice, a globe and a set of axes will appear on the screen for selecting a viewpoint.

used to draw features on the object in either the PLAN view or the 3D view. All features will be extruded 4 units parallel to the Z-axis until this setting of the ELEV command is changed.

You can obtain a viewpoint globe by typing <u>VPOINT</u> and pressing (CR) twice; a set of X-, Y-, and Z-axes will appear on the screen (Fig. 23.153). By moving your cursor about the globe, you can position

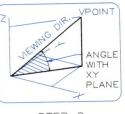

STEP 1   STEP 2

FIGURE 23.157   VPOINT **command.**

Step 1 Command: <u>VPOINT</u>
Rotate/<View Point> <current> <u>R</u>
Enter angle in X-Y plane from X axis
<current>: <u>45</u>

Step 2 Enter angle from X-Y plane
<current>: <u>30</u>

FIGURE 23.154   **By positioning the cursor within the small circle of the globe, the viewpoint will be looking down on the upper hemisphere of the globe. Any point on the small circle gives a horizontal (equatorial) view. A point between the small and large circles is a view of the lower hemisphere.**

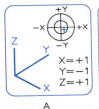

A   B   C

FIGURE 23.155   **The relationships between the points on the** VPOINT **globe and the** VPOINT **values selected from the keyboard.**

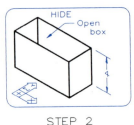

STEP 1   STEP 2

FIGURE 23.158   **Extrusion of a box:** HIDE **option.**

Step 1 (Once the box is drawn and the viewport selected, the extruded box will appear as a wire diagram.)

Step 2 Command: <u>HIDE</u> (CR)
(The vertical surface will be shown as opaque (solid) surfaces, but the top of the box will not have an opaque surface on it.)

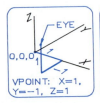

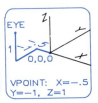

FIGURE 23.156   **Graphical examples of the selection of** VPOINT**s from the keyboard.**

the axes for the desired viewpoint and the 3D box will appear on the screen in wire-diagram form. Recycling this command lets you obtain another viewpoint of the object (Fig. 23.154).

Observation of Fig. 23.155 will help you understand the relationship of the globe to VPOINT specified numerals. Figure 23.156 also shows the graphical meaning of the numerals that are used for establishing the VPOINT. A VPOINT of 1,-1,1 means that you are looking at the origin (0,0,0) from a point

that is 1 unit in the X-direction, 1 unit in the negative Y-direction, and 1 unit in the positive Z-direction.

The VPOINT command offers the ROTATE option, as shown in Fig. 23.157. The angles with the X-axis and in the XY-plane are given. A pull-down menu under the heading VPOINT 3D, can be used to select viewpoints including the GLOBE and HIDE options.

The six principal orthographic views can be found by typing the following values when in the VPOINT

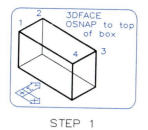

STEP 1   STEP 2

FIGURE 23.159 **Extrusion of a box:** 3DFACE.

Step 1 Command: OSNAP
Object snap modes: END
Command: 3DFACE
First point: (Select 1.)
Second point: (Select 2.)
Third point: (Select 3.)
Fourth point: (Select 4.)
Third point: (CR)

Step 2 Command: HIDE (CR)
(The top surface appears as an opaque surface.)

command:

Top views	0,0,1
Front view	0,-1,0
Right-side view	1,0,0
Left-side view	-1,0,0
Rear view	0,1,0
Bottom views	0,0-1

The HIDE command can be used to remove hidden lines, as shown in Fig. 23.158, which results in an "empty box" look. By using 3DFACE and SNAPping to the top points on the surface, a face can be applied to the top of the box to make it opaque when the HIDE command is used (Fig. 23.159).

23.62
Fundamentals of 3D drawing

Turn on the UCSICON and you are ready to experiment with three-dimensional drawing. Set the ELEV to 0 and select a THICKNESS, and you can draw the top view of an extruded part. Or, you can type in X-, Y-, and Z-coordinates of points with respect to the

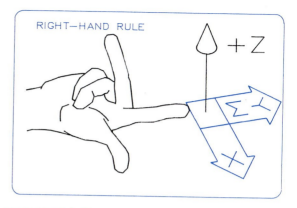

FIGURE 23.160 **By pointing the thumb of your right hand in the positive X direction, and your index finger in the positive Y direction, your middle finger will point in the positive Z direction.**

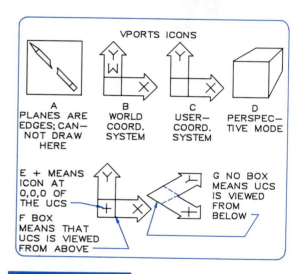

FIGURE 23.161 **The various** VPORTS ICONS that appear on the screen to show the X- and Y-axes.

0,0,0 origin and make a three-dimensional wire diagram.

To use the UCSICON, you should learn the right-hand rule (Fig. 23.160) that establishes the direction of the positive Z-axis with respect to the X and Y. Examples of directional icons are shown in Fig. 23.161. When a W appears, the icon represents the **World Coordinate System** (WCS). An icon without

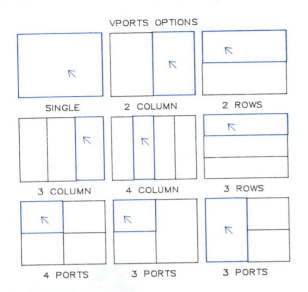

VPORTS OPTIONS

SINGLE   2 COLUMN   2 ROWS

3 COLUMN   4 COLUMN   3 ROWS

4 PORTS   3 PORTS   3 PORTS

FIGURE 23.162 **The** VPORTS OPTIONS **dialogue box that can be used for setting view ports on the screen.**

the W represents the directions of the **User Coordinate System** (UCS).

The broken-pencil icon gives a warning that a projection plane appears as an edge, making drawing impossible in that viewport. The perspective view of a box indicates that the current view of the drawing is a perspective.

A maximum of four VPORTS can be selected from a pull-down menu, VPORTS OPTION, as shown in Fig. 23.162. By using these VPORTS, you can make multiple views of a part on the screen and change them individually. VPORTS can be saved by the commands VPORTS and SAVE.

23.63
The coordinate systems

Two coordinates systems can be used in two-dimensional and three-dimensional drawings: the World Coordinate System (WCS) and the User Coordinate System (UCS).

WORLD COORDINATE SYSTEM The WCS has an origin where X, Y, and Z are 0. The normal view of the WCS is one where the X- and Y-axes are perpendicular and true length on the screen.

USER COORDINATE SYSTEM The UCS can be located within the WCS with its X- and Y-axes positioned in any direction with an origin located at any point of your choosing.

Establish a User Coordinate System in the following manner:

```
Command: UCS (CR)
Origin/ZAxis/3point/Entity/
View/X/Y/Z/Prev/Restore/Save/
Del/?/<World>: 0 (CR)
Origin point <0,0,0>: (Select desired
origin.) (CR)
```

Show the UCS icon at the UCS origin in the following manner:

```
Command: UCSICON (CR)
ON/OFF/All/Noorigin/ORigin
<ON>: OR (CR)
```

An X-Y icon is drawn at the UCS origin.

An easy way of setting the UCS to a previously drawn object is the 3point option of the UCS command as shown in Fig. 23.163. By using this option points are located at the origin, on the X-axis, and the Y-axis.

An explanation of the other options of the UCS command follows.

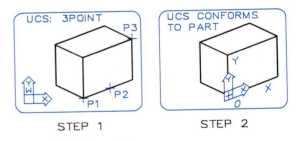

UCS: 3POINT   UCS CONFORMS TO PART

STEP 1   STEP 2

FIGURE 23.163 **The** 3Point **option.**

Step 1 Command: UCS **(Select 3 point option.)**
Origin point <0, 0, 0>: P1 **(Select origin.)**
Point on positive portion of the X axis: P2 **(On X-axis.)**
Point on positive Y portion of UCS X-Y plane: P3 **(On Y-axis.)**

Step 2 **The UCS icon will be transferred to the origin. The plus sign at the corner box indicates that it is at the origin.**

ZAXIS With the ZAXIS option, by selecting the origin and a point on the positive portion of the Z-axis, X- and Y-axes are located.

ENTITY With the ENTITY option, you can point to an entity (other than 3D Polyline or a polygon mesh) and the UCS will have the same positive Z-axis as the selected entity.

VIEW The VIEW option establishes a coordinate system with an X-Y plane that is parallel to the screen. This command allows you to apply nonpictorial text to a three-dimensional drawing.

X/Y/Z By specifying X, Y, or Z with the X/Y/Z option, you may rotate the UCS about the X-, Y-, or Z-axes.

PREVIOUS The Previous option restores the previous UCS.

RESTORE The Restore option prompts you for the name of the saved UCS and makes it the current UCS.

SAVE The Save option is used to name and save a UCS.

DELETE The Delete option removes a saved UCS from memory.

? The ? option gives a listing of the current and saved coordinate systems. If the current UCS is unnamed, it is listed as *WORLD or *NO NAME.

The UCS Dialogue box, DDUCS, can be displayed to show a list of the coordinate systems Fig. (23.164). The *WORLD* coordinate system is listed first, followed by *PREVIOUS* if other systems have been defined. By repeatedly selecting *PREVIOUS* and OK, you can step back through the systems. If the current coordinate system has not been named it will be listed as *NO NAME*.

When the Define new UCS box is selected, a dialogue box appears that lets you define a new UCS (Fig. 23.165). If you decide to not define a new UCS, select Cancel and you will be returned to the previous dialogue box.

WORLD The World option sets the UCS to the same as the WCS.

The options of the UCSICON command are explained below.

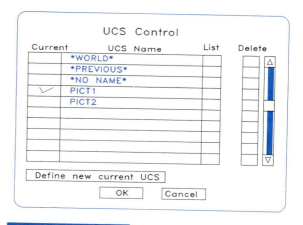

FIGURE 23.164 The UCS Control box can be selected from the SETTINGS heading of the pull-down menu and used to assign the current UCS, delete a UCS, or define a new UCS.

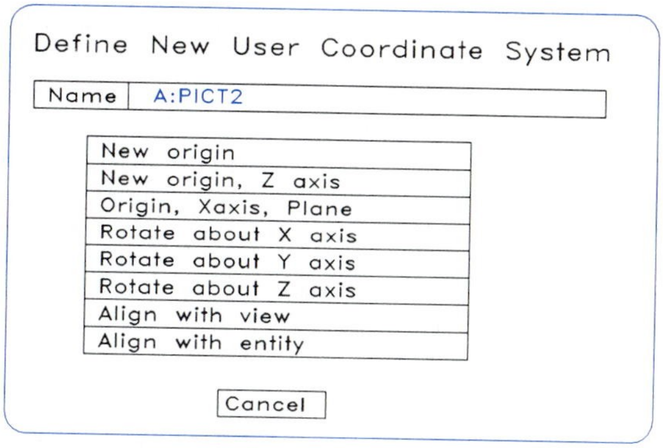

FIGURE 23.165 This dialogue box, found under the SETTINGS heading of the pull-down menu can be used for giving a new coordinate system a name, by typing the name in the box provided. Then select how you wish to define the new UCS.

ON The ON option turns the icon on.

OFF The OFF option turns the icon off.

ALL The ALL option applies the icon to all active viewports when more than one is being used.

NOORIGIN The NOORIGIN option displays the icon at the lower left corner regardless of the location of the UCS origin.

ORIGIN The ORIGIN option places the icon at the current coordinate system if space is available; otherwise, the icon is displayed at the lower left of the viewport.

Under the Settings heading of the **Menu Bar,** the UCS dialogue box entitled User Coordinate System Options can be pulled down for selecting views of a three-dimensional drawing (Fig. 23.166). For example, when a three-dimensional view of an object is on the screen, you can select the icon for the top view and you will be prompted for the origin (0,0,0). Select a point on the screen and points will appear representing the new UCS for *X-* and *Y-*axes for a top view. Use the PLAN command, and the top view will appear on the screen.

The UCSICON defines the plane of an extrusion, as shown in Fig. 23.167. By moving the UCSICON to the base of the extruded box and using UCS and X to rotate the *Y-*axis 90 degrees about the *X-*axis, an extruded circle can be drawn on the frontal plane. In Fig. 23.168, the ELEV is set to 0 and THICKNESS to −3 (using the right-hand rule), the depth of the box, and the circle is drawn and extruded.

The HIDE command shows the final view of the 3D box with all invisible lines of the wire diagram suppressed.

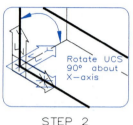

STEP 1   STEP 2

FIGURE 23.167   **Extrusion of a box: Setting** UCS.

Step 1 Command: UCSICON
ON/OFF/All/Noorigin/ORigin <current ON/OFF state>: OR
Command: UCS
Origin/ZAxis/3point/Entity/View/
X/Y/Z/Prev/Restore/Save/Del/?/
<World>: O
Origin point: (OSNAP to END at corner of box.)

Step 2 Command: UCS
Origin/ZAxis/3point/Entity/View/
X/Y/Z/ . . . <World>: X
Rotation angle about X axis <0>: 90
(The icon is placed in the plane of the front of the box.)

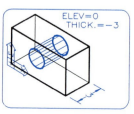
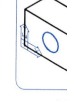

STEP 1   STEP 2

FIGURE 23.168   **Extrusion of a box: Extruded hole.**

Step 1 Command: ELEV
New current elevation <0>: 0
New current thickness <4>: −3 **(The depth of the box.)**
Command: CIRCLE 3P/2P/TTR/<Center point>: **(Select with cursor.)**
Diameter/<Radius>: .5 **(A 1" diameter cylinder is extruded 3" deep into the box.)**

Step 2 Command: HIDE **(The outline of the hole is shown, but you will be unable to see through a hole made in this manner.)**

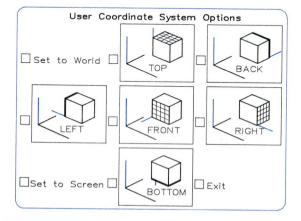

FIGURE 23.166   **When a three-dimensional view of an object is on the screen you can select an icon for the desired viewpoint. You will be prompted for origin of the selected UCS. Use the PLAN command and the designated view will appear on the screen. This works only while in the WCS.**

23.64
The DVIEW command

The DVIEW command enables you to dynamically select a viewpoint for looking at a three-dimensional drawing. In addition to obtaining axonometric drawings (parallel projections), perspectives can be drawn where parallel lines converge toward vanishing points, yielding highly realistic pictorials (Fig. 23.169).

The DVIEW commands gives the following options:

```
Command: DVIEW (CR)
CAmera/TArget?Distance/POints/
PAn/Zoom/TWist/CLip/Hide
/Off/Undo/<eXit>:
```

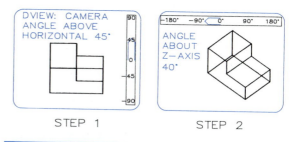

STEP 1   STEP 2

FIGURE 23.171 The DVIEW CAmera command.

Step 1 Command: DVIEW (Select CAmera option.) Select objects: (Window object) (CR) (A vertical slider bar appears at right. Select viewpoint elevation with cursor.)

Step 2 A horizontal slider bar appears at top of screen. Select desired viewpoint with cursor.

An excellent way to experiment with these commands is to use AutoCAD's DVIEWBLOCK house (Fig. 23.170) that can be displayed on the screen by entering the DVIEW command and pressing (CR) twice. The top view of the house will appear and the following commands can be experimented with, without any new construction.

With the CAmera (Camera) option your viewpoint is rotated about the object as if you were moving about it with a camera. When prompted, you specify the viewpoint at the keyboard with coordinates, or you can use slider bars. A slider bar appears at the right of the screen (Fig. 23.171) with an angle from

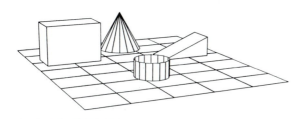

FIGURE 23.169 A perspective view is obtained when the DISTANCE option of the VPOINT command is selected.

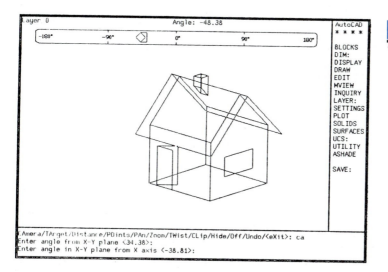

FIGURE 23.170
The DVIEWBLOCK house.

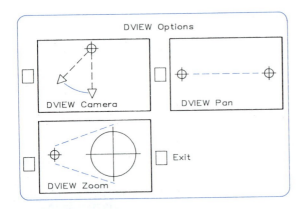

FIGURE 23.172 **Three options can be selected from the** DVIEW **options dialogue box that is pulled down from the Display heading of the Menu Bar.**

−90° to +90° for looking at the object from bottom to top.

Once the vertical viewpoint is selected, a second slider bar appears at the top of the screen that allows you to select a viewpoint from −180° to +180° about the object.

Three of the DVIEW options, including the CAmera option, can be selected from the pull-down menu under the DISPLAY heading of the menu bar (Fig. 23.172). Figure 23.173 gives an example of how the camera viewpoint can be changed for viewing a target.

TArget is used to rotate a specified target point about the camera as shown in Fig. 23.174. The prompts are identical to the CAmera option, but the target position is being rotated about the camera instead of the camera about the object.

DISTANCE places the camera with respect to the object and automatically turns on the **perspective option.** The *X-Y* icon is replaced with the perspective-box icon. You are prompted for a distance to the target, which can be entered as a number at the keyboard or specified by a slider bar from 0X to 16X. 1X is the current distance to the target. The DVIEW, ZOOM option can be used to magnify a drawing without turning the perspective option on.

POints is used to locate the target point first and the camera position second for viewing a drawing as shown in Fig. 23.175.

PAn is used to move a drawing on the screen without changing viewpoint or magnification. When in the perspective mode, the pointing device must be used; but when perspective is OFF, coordinates can be typed at the keyboard for panning coordinates.

Zoom is used to change the magnification of a drawing on the screen in the same manner as the Zoom/Center command when perspective options are not being used. If the perspective option is ON, Zoom varies the magnification of the drawing as if you were changing the lens of a camera from a wide-angle to a telescopic lens. Therefore, the view of the object is dynamically distorted.

TWist is used to rotate the three-dimensional drawing on the screen. For example, you may wish to plot a drawing with a vertical format on a sheet with a horizontal format. TWist can be used for this operation like the ROTATE command is used for two-dimensional drawings.

FIGURE 23.173 **The** CAmera **option of the** DVIEW **command can be used to obtain different views of a target point by moving the position of the camera.**

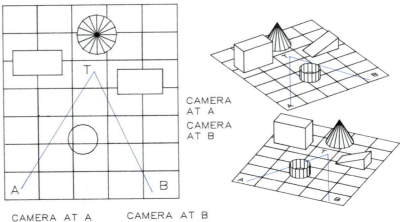

CAMERA AT A

CAMERA AT B

CAMERA AT A   CAMERA AT B

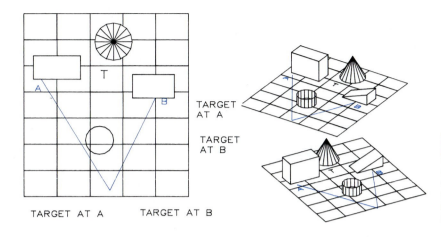

TARGET AT A

TARGET AT B

FIGURE 23.174 The TArget option of the DVIEW command can be used to obtain different views of a scene by moving the target to different locations while leaving the camera stationary.

CLip is used to place vertical planes in a drawing (perspective and parallel projections) to remove all of the drawing either in front of or in back of the plane. When the CLip option is selected, the prompts Back/Front/<Off>: appear.

By using the Back option, the clipping plane is located on the drawing for removing all that is behind it (Fig. 23.176).

By using the Front option, the clipping plane is located for removing portions of the drawing that are in front of it (Fig. 23.176).

The Off option turns clipping off. When in the perspective mode, clipping cannot be turned off, but the clipping plane is located at the camera position for front clipping.

Hide suppresses the invisible lines of a three-dimensional drawing.

OFF turns off the perspective mode that was enabled by the Distance option.

Undo reverses the previous operations performed under the DVIEW commands, one at a time.

eXit ends the DVIEW command.

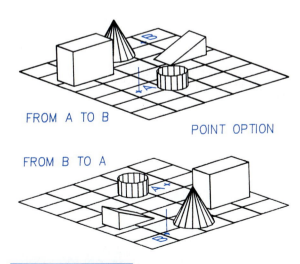

FROM A TO B

POINT OPTION

FROM B TO A

FIGURE 23.175 The POint option of the DVIEW command can be used to locate the target and camera points (with X-, Y-, Z-coordinates) to obtain the desired view. Both perspective and axonometric views can be found with this option.

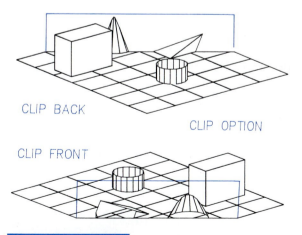

CLIP BACK

CLIP OPTION

CLIP FRONT

FIGURE 23.176 The CLip option of the DVIEW command is used to remove Back or Front portions of a drawing. The clipping plane is assumed to be parallel to the drawing screen.

When in the perspective mode, some operations cannot be performed, and a prompt will be given asking if you wish to continue with perspective off. A negative response will leave the perspective ON and a positive response will turn it OFF.

23.65
Basic 3D forms

A pull-down menu entitled 3D Construction can be found under the **Draw** heading of the **Menu Bar**. From this menu, you can select boxes, pyramids, domes, spheres, dishes, wedges, cones, and toruses (toroids). Also included in this menu are options for surface revolution, ruled surface, edge surface, tabulated surface, and meshes. Boxes may be selected for setting values for surface tabulation 1 and surface tabulation 2.

BOX Figure 23.177 illustrates how, with the BOX command, a cube or box can be drawn by selecting a corner, measuring the length and height, and selecting the angle of rotation about the Z-axis.

WEDGE Figure 23.178 shows how a wedge can be drawn in the exact manner as the box was drawn, using the WEDGE command.

PYRAMID By using the PYRAMID command, a pyramid can be drawn that extends to its apex, as il-

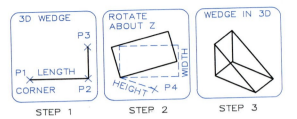

FIGURE 23.178 The WEDGE command.

Step 1 Command: <u>WEDGE</u> (CR)
Corner of wedge: (Select P1.)
Length: (Select P2.)
Width: (Select P3.)

Step 2 Height: (Select P4.)

Step 3 Rotation angle about Z axis: <u>−15</u> (CR)
(The wedge is drawn in three dimensions.)

lustrated in Fig. 23.179. An option is available for drawing a truncated pyramid with its apex removed.

CONE Figure 23.180 shows how the CONE command is used to draw a cone as a basic three-dimensional form. A truncated cone can also be drawn by following the given prompts.

SPHERE The SPHERE command is used to draw a sphere by selecting its center and its radius (Fig. 23.181).

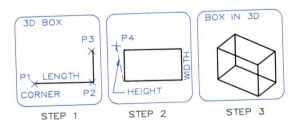

FIGURE 23.177 The BOX command.

Step 1 Command: <u>BOX</u> (CR)
Corner of box: (Select P1.)
Length: (Select P2.)

Step 2 Height: (Select P3.)

Step 3 Rotation angle about Z axis: <u>0</u> (CR) (The box is drawn in three dimensions.)

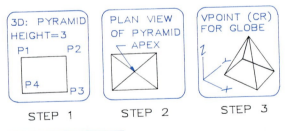

FIGURE 23.179 The PYRAMID command.

Step 1 Command: <u>PYRAMID</u> (CR)
First base point: (Select P1.)
Second base point: (Select P2.)
Third base point: (Select P3.)
Tetrahedron/<Fourth base point>: (Select P4.)

Step 2 Ridge/Top. <Apex point>: <u>.XY</u> of (need Z): <u>2</u> (CR)

Step 3 The pyramid is drawn in three dimensions.

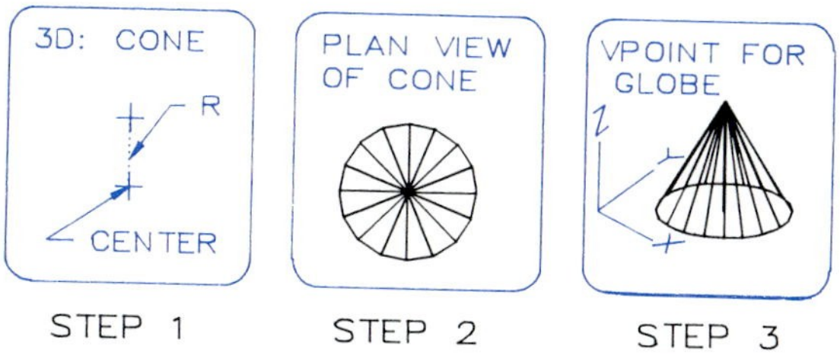

FIGURE 23.180   The CONE command.

Step 1 Command: <u>CONE</u> (**CR**)
Base center point: **(Select with pointer.)**
Diameter/<radius> of base: **(Select with pointer.)**
Diameter/<radius> of top <0>: **(CR)**
Height: **(Measure with pointer.)**
Number of segments <16>: **(CR)**

Step 2 The plan view of the cone is generated.

Step 3 Select the View point option of the VPOINT command to obtain a three-dimensional view of the cone.

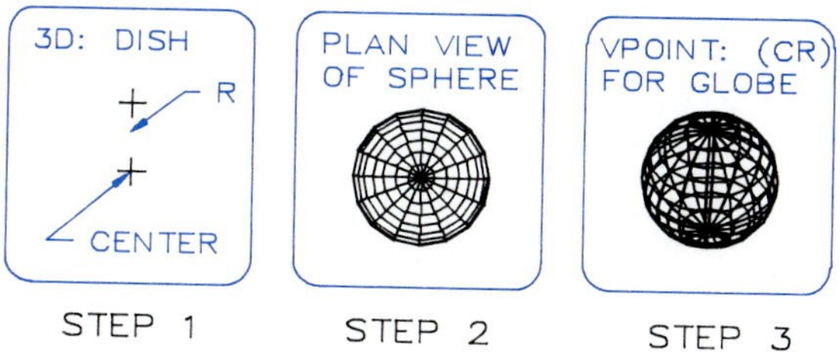

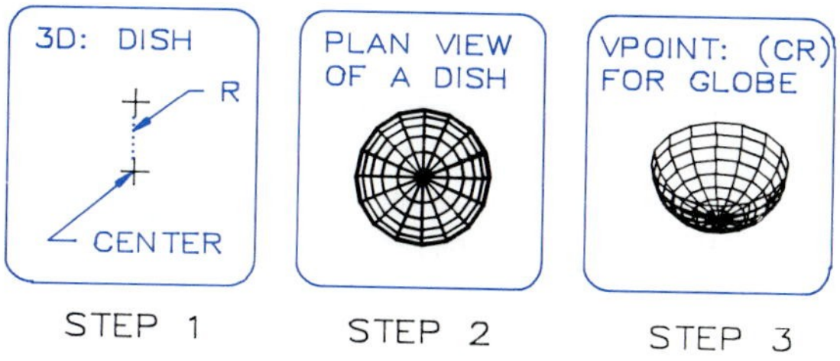

FIGURE 23.182   The DISH command.

Step 1 Command: <u>DISH</u> (**CR**)
Center of dish: **(Select with pointer.)**
Diameter/<radius>: **(Select with pointer.)**
Number of longitudinal segments <16>: **(CR)**
Number of latitudinal segments <8>: **(CR)**

Step 2 The plan view of the dish is generated.

Step 3 Select a VPOINT for a three-dimensional view of the dish.

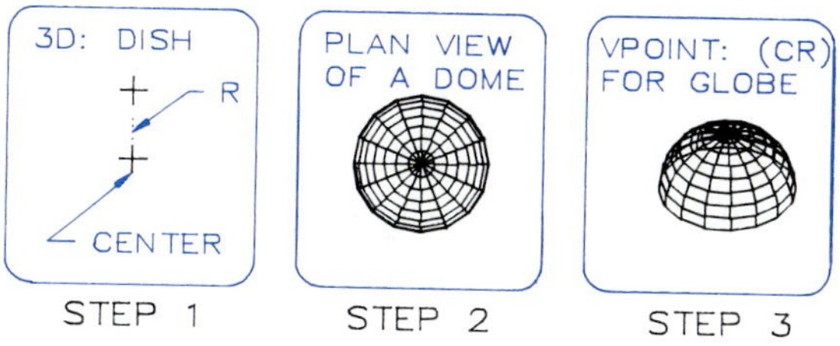

FIGURE 23.181   The SPHERE command.

Step 1 Command: <u>SPHERE</u> (**CR**)
Center of sphere: **(Select with pointer.)**
Diameter/<radius>: **(Select with pointer.)**
Number of longitudinal segments <16>: **(CR)**
Number of latitudinal segments <16>: **(CR)**

Step 2 The plan view of the sphere is generated.

Step 3 Select a VPOINT to obtain a three-dimensional view of the sphere.

FIGURE 23.183   The DOME command.

Step 1 Command: <u>DOME</u> (**CR**)
Center of dome: **(Select with pointer.)**
Diameter/<radius>: **(Select with pointer.)**
Number of longitudinal segments <16>: **(CR)**
Number of latitudinal segments <8>: **(CR)**

Step 2 The plan view of the dome is generated.

Step 3 Select a VPOINT for a three-dimensional view of the dome.

DISH   Figure 23.182 shows how a dish, the lower hemisphere of a sphere, is drawn by selecting its center and its radius using the DISH command.

DOME   Using the DOME command, Fig. 23.183 illustrates the steps of drawing a dome, the upper hemisphere of a sphere, by selecting the center and its radius.

TORUS   Figure 23.184 shows the steps of drawing a donut shape called a torus or toroid using the TORUS command.

HIDE   The HIDE command can be used to remove the hidden (invisible) lines from these three-dimensional forms to give them a more realistic three-dimensional appearance.

STEP 1   STEP 2   STEP 3

FIGURE 23.184   The TORUS command.

Step 1 Command: <u>TORUS</u> **(CR)**
Center of torus: **(Select with pointer.)**
Diameter/<radius> of torus: **(Select with pointer.)**
Diameter/<radius> of tube: **(Select with pointer.)**
Segments around tube circumference
<16>: <u>8</u> **(CR)**
Segments around torus circumference
<16>: <u>8</u> **(CR)**

Step 2 The plan view of the torus is generated.

Step 3 Select a VPOINT for a three-dimensional view of the torus.

23.66
3D polygon meshes

Three-dimensional meshes are available for showing the surfaces of an object. Also, these meshes can be edited by the PEDIT command to change the form of the meshes. The commands available for drawing meshes are: 3DMESH, RULESURF, TABSURF, REVSURF, and EDGESURF.

3DMESH is used to define a three-dimensional polygon mesh as shown in Fig. 23.185. In this example a plan view of the mesh is shown where vertices are specified in the *M*- and *N*-directions with 256 in each direction being the maximum number permitted. Once the *M*- and *N*-values of the mesh have been specified, you will be prompted for the *X*-, *Y*-, and *Z*-values for each vertex, starting from the origin. The values can be used to establish the vertices in any position, not just in a rectangular format.

When all the points have been provided the mesh will appear on the screen (Fig. 23.186). It can be viewed as a three-dimensional surface from any angle of your choosing.

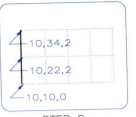

STEP 1   STEP 2

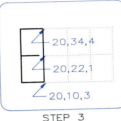

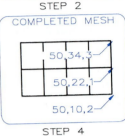

STEP 3   STEP 4

FIGURE 23.185   The 3DMESH command.

Step 1 Command: <u>3DMESH</u> **(CR)**
Mesh M size: <u>5</u>
Mesh N size: <u>3</u>

Step 2 Vertex (0,0): <u>10,10,0</u>
Vertex (0,1): <u>10,22,2</u> **(CR)**
Vertex (0,2): <u>10,34,2</u> **(CR)**

Step 3 Vertex (1,0): <u>20,10,3</u> **(CR)**
Vertex (1,1): <u>20,22,1</u> **(CR)**
Vertex (1,2): <u>20,34,4</u> **(CR)**

Step 4 (Moving to the last set.)
Vertex (4,0): <u>50,10,2</u> **(CR)**
Vertex (4,1): <u>50,22,1</u> **(CR)**
Vertex (4,2): <u>50,34,3</u> **(CR)**
(The mesh is drawn.)

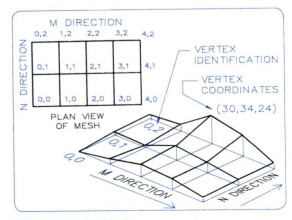

FIGURE 23.186   By selecting a VPOINT, a 3D view of the 3DMESH can be obtained.

PEDIT is a command that can be used to change a mesh:

```
Command: PEDIT (CR)
Edit vertex/Smooth surface/
Desmooth/Mclose/Nclose/Undo/
eXit<X>:
```

The Smooth option can be used to form the mesh into a smooth three-dimensional shape. Desmooth is used to discard the smoothed surface and reverts it to its previous form. The Undo command is used to undo PEDIT options that have been performed. If the mesh is closed, Mopen and Nopen will replace the commands Mclose and Nclose as options. The command eXit is used to exit from the PEDIT command.

Edit vertex enables you to edit individual vertices within the mesh:

```
Vertex (M,N): Next/Previous/
Left/Right/Up/Down/Move/REgen/
eXit<N>
```

An X appears at the origin for selecting the vertex to be changed. The Next and Previous options are used to move forward and backward from the origin toward the last vertex. The Left and Right options move the pointer in the N-direction. The Up and Down options move the pointer in the M-direction to make locating vertices faster.

Once a vertex is selected for changing, select the Move option, and you will be prompted, Enter new location. At this time, type the X-, Y-, and Z-coordinates of the new position. The revised mesh can be displayed by the REgen option.

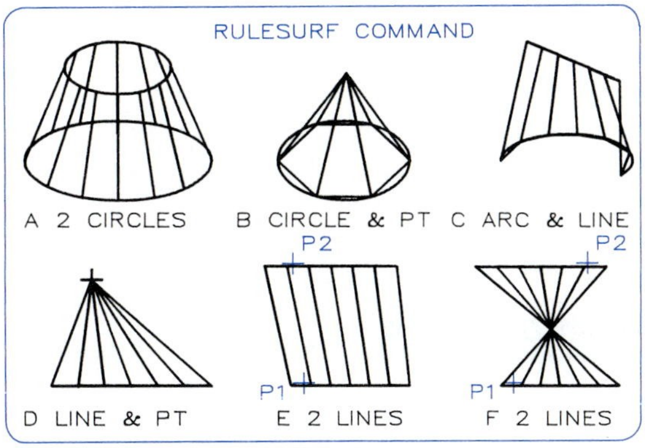

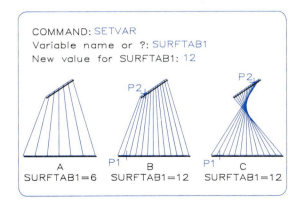

FIGURE 23.187 The RULESURF command connects entities with a series of straight lines. You can see at E and F that the positions of the selected points determine the direction of the ruled lines.

FIGURE 23.188 The SURFTAB1 system variable is used to define the density of the lines between the selected entities.

23.67
The RULESURF command

The RULESURF command is used to connect two entities with ruled lines. The entities can be curves, arcs, polylines, lines, or points (but only one of the boundaries can be a point). Examples of ruled surfaces are shown in Fig. 23.187.

The system variable, SURFTAB1, is used to assign the number of vertices that are to be placed along each entity. Figure 23.188 demonstrates the results of varying the SURFTAB1 system variable.

23.68
The TABSURF command

The TABSURF command is used to tabulate a surface from a path curve and a direction vector. In Fig. 23.189, a circle and a vector are given. By selecting the circle first, and the endpoint of the vector, a cylinder is tabulated with its elements equal in length to the vector. The SURFTAB1 system variable is used to control the density of the tabulated surface.

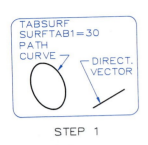
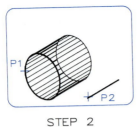

STEP 1   STEP 2

FIGURE 23.189   The TABSURF **command.**

Step 1 Command: <u>TABSURF</u>
Select path curve: **(Select circle.)**

Step 2 Select direction vector: **(Select endpoint of vector.)**
(The vector is used to tabulate the circle in accordance with the SURFTAB1 **variable.)**

23.69
The REVSURF command

A surface can be revolved about an axis with the REVSURF command as shown in Fig. 23.190. A line, polyline, arc, or circle can be revolved about an axis to form a surface of revolution. System variable SURFTAB1 controls the density of the revolved surface.

   If a circle or a closed polyline is to be revolved about an axis, as shown in Fig. 23.191, the system variable SURFTAB2 is used to control the density of the closed entity. System variable SURFTAB1 controls the density in the direction of the revolution.

23.70
The EDGESURF command

Four edges that intersect at their corners can be connected with a mesh by using the EDGESURF command as shown in Fig. 23.192. System variables SURFTAB1 and SURFTAB2 are used to control density in the *M*- and *N*-directions, respectively. The first line selected determines the *M*-direction.

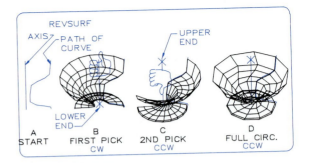

FIGURE 23.190   The REVSURF **command.**

Step 1 **An axis and a polyline defining the path curve are given.**

Step 2 Command: <u>REVSURF</u>
Select path of curve: **(Select with pointer.)**
Select axis of revolution: **(Select lower end of axis.)**
Start angle <0>: **(CR)**
Included angle (+=ccw,-=cw) <Full circle>: <u>180</u> **(Rotated counterclockwise.) (CR)**

Step 3 **By selecting a point on the upper end of the axis, the path curve is rotated clockwise.**

Step 4 **A full-circle revolution.**

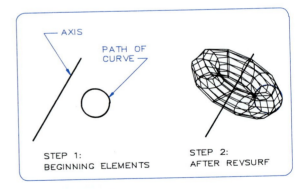

FIGURE 23.191   The REVSURF **command.**

Step 1 Command: <u>REVSURF</u> **(CR)**
Select path curve: **(Select circle.)**
Select axis of revolution: **(Select line.)**
Start angle <0>: **(CR)**

Step 2 Included angle (+=ccw, -=cw) <Full circle>: **(CR)**
(System variables SURFTAB1 **and** SURFTAB2 **are used to determine the density of the mesh.)**

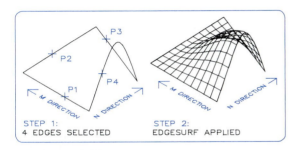

| FIGURE 23.192 | The EDGESURF command. |

FIGURE 23.192 The EDGESURF command.

Step 1 Command: EDGESURF (CR)
Select edge 1: P1 (**Determine** *M* **direction and** SURFTAB1.)
Select edge 2: P2 (**Determine** *N* **direction and** SURFTAB2.)
Select edge 3: P3
Select edge 4: P4

Step 2 The EDGESURF command is applied to mesh the surface.

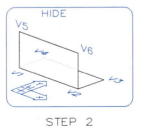

STEP 1   STEP 2

FIGURE 23.193 The Polyface command (PFACE).

Step 1 Command: PFACE
Vertex 1: (Osnap to 1.)
Vertex 2: (Osnap to 2.)
. . .
Vertex 6: (Osnap to 6.) (CR)

Step 2 Face 1,vertex 1: (**Type 1.**)
Face 1, vertex 2: (**Type 2.**)
Face 1, vertex 3: (**Type 3.**)
Face 1, vertex 4: (**Type 4.**) (CR)
Face 2, vertex 1: (**Type 1.**)
Face 2, vertex 2: (**Type 5.**)
Face 2, vertex 3: (**Type 6.**)
Face 2, vertex 4: (**Type 2.**) (CR)
Face 3, vertex 1: (CR)
Command: HIDE (**Meshed surfaces opaqued.**)

23.71
The PFACE command

The PFACE command is used to create a polyface mesh that requires less storage space than other three-dimensional meshes. To use PFACE, you define a three-dimensional polygon mesh by specifying each vertex and then each face of the mesh, as shown in Fig. 23.193.

As the vertices (either 2D or 3D) are entered, you must keep track of the vertex numbers shown in the prompts. After entering the last vertex, press (CR) and assign the vertices to each face, and press (CR) when the last assignment has been made.

Since specifying a PFACE is a tedious process, RULESURF, TABSURF, REVSURF, and EDGE-SURF commands are more convenient. The PFACE command is primarily for use with AutoLISP and ADS applications.

23.72
LINE, PLINE, and 3DPOLY commands

The command LINE will yield a two-dimensional line when only *X*- and *Y*-values are given at the keyboard, but the LINE becomes a three-dimensional line when *X*-, *Y*- and *Z*-coordinates are given. Likewise, the PLINE command is used to draw a two-dimensional polyline, which can be edited with the PEDIT command. The 3DPOLY command is used for drawing three-dimensional polylines that are input with *X*-, *Y*-, and *Z*-coordinates. 3DPOLY lines can be edited by PEDIT, but they must have a width of zero.

In Fig. 23.194, a 3DPOLY or LINE command is used to draw lines with absolute coordinates from the origin of 0,0,0. Points in three-dimensional space are located in this same manner.

Figure 23.195 shows how the ends of three-dimensional lines are located by typing relative coordinates from point to point. Each relative coordinate is written in the form, @X,Y,Z or @3,0,0, which locates an endpoint with respect to the last point.

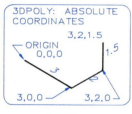
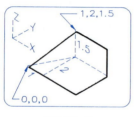

STEP 1   STEP 2

FIGURE 23.194 3DPOLY—absolute coordinates.

Step 1 Command: 3DPOLY (or LINE) (CR)
From point: 0, 0, 0 (CR)
Close/Undo/<Endpoint of line>: 3,0,0 (CR)
Close/Undo/<Endpoint of line>: 3,2,0 (CR)
Close/Undo/<Endpoint of line>:
3,2,1,5 (CR)

Step 2 Close/Undo/<Endpoint of line>:
1,2,1.5 (CR)
Close/Undo/<Endpoint of line>: C (Back to origin.) (CR)

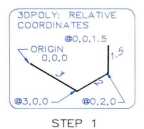
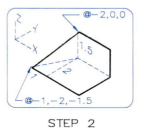

STEP 1   STEP 2

FIGURE 23.195 3DPOLY—relative coordinates.

Step 1 Command: 3DPOLY (or LINE) (CR)
From point: (Select)
Close/Undo/<Endpoint of line>:
@3,0,0 (CR)
Close/Undo/<Endpoint of line>:
@0,2,0 (CR)
Close/Undo/<Endpoint of line>:
@0,0,1.5 (CR)

Step 2 Close/Undo/<Endpoint of line>:
@-2,0,0 (CR)
Close/Undo/<Endpoint of line>:
@-1,-2,-1.5 (CR)

Three-dimensional shapes drawn with the ELEV command and THICKNESS option, as discussed in Section 23.61, can be viewed with the VPOINT command and visibility determined with the HIDE command. The SOLID command can be used to fill the top surface to make it appear opaque rather than as an open box (Fig. 23.196).

All OSNAP modes apply to LINE entities, and they also work with visible edges of polyface meshes. Three-dimensional objects can be stretched in the X and Y directions, but not in the Z direction. Height must be changed by using the ELEV command.

3DPOLY lines drawn as three-dimensional lines can be edited with the PEDIT command to form a spline curve (a helix in this case), as shown in Fig. 23.197. The accuracy of the spline curve is determined by the system variable SPLFRAME, which controls the display of the control points. If set to 0 (zero) only the spline curve is shown; if set to a 1, the curve and its defining polyline are shown.

23.73
3DFACE command

The 3DFACE command can be used to draw three-dimensional planes that are opaque when the HIDE command is applied to them.

Figure 23.198 shows how 3DFACE was used to draw sloping planes that appear as solid, opaque planes. 3DFACE can also be used to apply faces to wire-diagram drawings made with LINE by snapping the endpoints of the lines with the OSNAP option.

Figure 23.199 illustrates how four corners of a 3DFACE are selected to form an opaque plane. Four is the maximum number of points that can be selected before an area is defined, after which you are prompted for points 3 and 4, with the understanding that the previous two points will be used with the next two points that will be selected. The successive areas will be connected with splice lines creating a "patchwork" appearance that may detract from the surface.

By inserting the letter I prior to selecting a perimeter line of a 3DFACE, the splice line will be

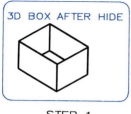
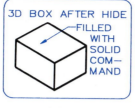

STEP 1   STEP 2

FIGURE 23.196   A box by ELEV command.

Step 1 After drawing a solid using the ELEV command and the THICKNESS option, the solids visibility can be shown by using the HIDE command.

Step 2 By using the SOLID or 3DFACE command, the top can be filled so it will appear opaque when the HIDE command is used.

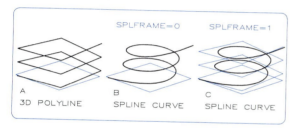

FIGURE 23.197   The PEDIT command and the SPLFRAME system variable set to zero can be used to spline a three-dimensional curve. By setting the SPLFRAME=1, both the spline curve and the defining polyline are displayed.

SOLID   3DFACE

FIGURE 23.198   The 3DFACE command can be used to draw objects with opaque planes. 3DFACE can be used to draw planes, or it can be used to apply faces to previously drawn lines by snapping to the endpoints of the lines outlining the planes.

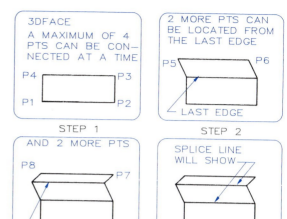

STEP 1   STEP 2   STEP 3   STEP 4

FIGURE 23.199   The 3DFACE command

Step 1 Command: 3DFACE (CR)
First point: (Select P1.)
Second point: (Select P2.)
Third point: (Select P3.).
Fourth point: (Select P4.)

Step 2 Third point: (Select P5.)
Fourth point: (Select P6.)

Step 3 Third point: (Select P7.)
Fourth point: (Select P8.)

Step 4 Third point: (CR) (Splice lines between areas will show.)

omitted, thereby eliminating the "patchwork" appearance as shown in Fig. 23.200.

When 3DFACE is used to make surfaces opaque, holes are filled in so it is impossible to see through them when the HIDE command is used. The SLOT command, written in LISP, can be used with 3DFACE to opaque the surface up to the edge of the hole as shown in Fig. 23.201.

The SLOT command is loaded by typing (LOAD "SLOT") and SLOT is typed again to use it. A slot or a hole can be drawn by responding to the prompts. The depth of the hole, or slot, is extruded upward from the point selected. By setting the system variable, SPLFRAME, to 1, it is turned on and the corner lines surrounding the hole or slot are visible.

By using the 3DFACE, with the "I" (invisible) option, the trapezoidal shapes can be made opaque.

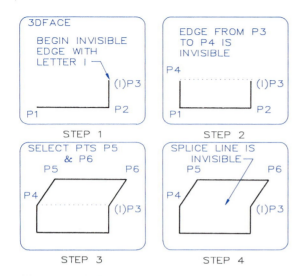

FIGURE 23.200 The 3DFACE: I option

Step 1 Command: <u>3DFACE</u> (CR)
First point: **(Select P1.)**
Second point: **(Select P2.)**
Third point: <u>I</u> **(Select P3.)**

Step 2 Fourth point: **(Select P4.)**

Step 3 Third point: **(Select P5.)**
Fourth point: **(Select P6.)**

Step 4 Third point: **(CR) (The splice line between P3 and P4 is invisible.)**

When SPLFRAME is set to 0 (off) and HIDE is applied, the surface around the hole becomes opaque, and the hole can be seen through.

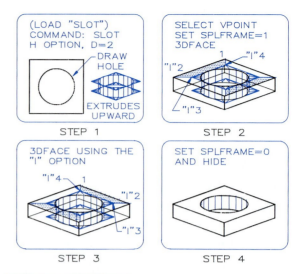

FIGURE 23.201 The SLOT command.

Step 1 Command: **(LOAD "SLOT")**
Command: <u>SLOT</u>
Hole or SLot? H/S <S>: <u>H</u>
First center point of slot: **(Select)**
Slot radius: <u>4</u> Depth: <u>2</u> **(Depth is extruded upward.)**

Step 2 Command: <u>VPOINT</u>
Rotate/<View point> <0,00,0,00,1.00>: <u>1,-1,.5</u>
(obtain 3D view of box)
Set Var: <u>SPLFRAME</u> **(Set to I and REGEN.)**
(Use 3DFACE with "I" option to connect corner point.)

Step 3 Continue using 3DFACE and the "I" option to face the other two sides.

Step 4 Set SPLFRAME=0 and apply the HIDE command. **The surface around the holes becomes opaque.**

23.74
XYZ filters

Filters are options under LINE and 3DFACE commands that can be used to attach the endpoint of a line being drawn to the coordinates of a three-dimensional point or endpoint. Figure 23.202 illustrates how filters are applied to a two-dimensional drawing. To locate the front view of a point projected from the left-side view and top view, select the .Y of the left point and the .X of the top point. The position of the front view is automatically snapped to a point located by these two coordinates.

A three-dimensional drawing can be made using filters as shown in Fig. 23.203, where two coordinates for a given point are selected using the filter option, and the third one is specified when prompted. For example, you can select the coordinates of a point in the XZ plane, and specify the distance of the point from it in the Y direction.

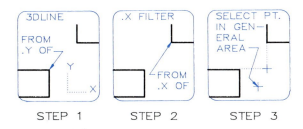

STEP 1   STEP 2   STEP 3

FIGURE 23.202   The *X-Y* filters.

Step 1 Command: LINE (CR) From point: .Y of **(Select corner.)** (need XZ)

Step 2 Close/Undo/<Endpoint of line>: .X of **(Select corner.)** (need YZ)

Step 3 Select a point in the general area of the desired position, and the point will attach itself to the *Y* coordinate of the side view and the *X* coordinate of the top view.

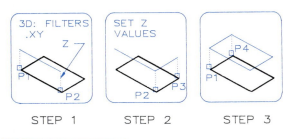

STEP 1   STEP 2   STEP 3

FIGURE 23.203   Three-dimensional filters.

Step 1 Command: 3DPOLY (or LINE) (CR) From point: .XY of **(Select *P1*.)** of (need Z) 2 (CR) Close/Undo/<Endpoint of line>: .XY of **(Select *P2*.)** (need Z) 2 (CR)

Step 2 Close/Undo/<Endpoint of line>: .XY of **(Select *P3*.)** (Need Z) 2 (CR)

Step 3 Close/Undo/<Endpoint of line>: .XY of **(Select *P4*.)** (Need Z) 2 (CR) Close/Undo/<Endpoint of line>: .XY of **(Select 1.)** (need Z) 2 (CR)

23.75
New drawing in 3D

The steps of constructing a three-dimensional object are shown in Fig. 23.204. The plan view of the object is drawn in Step 1 and moved to origin, 0,0,0. In

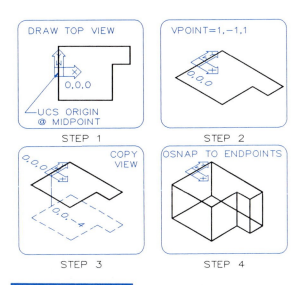

STEP 1   STEP 2

STEP 3   STEP 4

FIGURE 23.204   A drawing in three dimensions.

Step 1 Draw the top view of the part. Use the UCS command to set the origin at the midpoint of the line on the view. Use the UCSICON command to locate the icon at the UCS origin.

Step 2 Command: VPOINT (CR) Rotate/<View point> <0,0,0>: 1,-1,1 **(A three-dimensional view is displayed.)**

Step 3 Command: COPY (CR) Select objects: W **(Window 3D view.)** <Base point or displacement> /Multiple: 0,0,-4 **(CR)** Second point of displacement: **(CR)**

Step 4 Set OSNAP to END **and connect the corner points with the** LINE **command.**

Step 2, an isometric viewpoint is selected for showing a three-dimensional view of the object's surface.

In Step 3, the first view is copied and located at 0,0,-4 to establish the lower plane of the object and a height of 4 units. By using the END option of OSNAP, the corners of each plane are connected to form a wire diagram of the object.

In Fig. 23.205 (Step 1), a four-window VPORTS is selected for drawing the wire diagram in each port. In Step 2, the PLAN option of the UCS command is used to show the top view of the object. This UCS is SAVEd with that option of the UCS command.

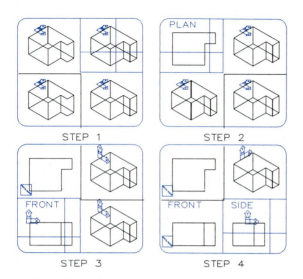

FIGURE 23.205 Multiple views of a three-dimensional object.

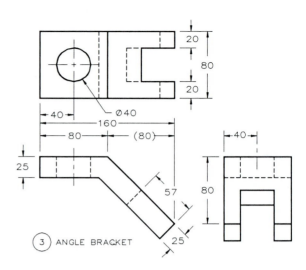

FIGURE 23.206 The angle bracket below with an inclined surface is drawn as a three-dimensional object in the following figures.

Step 1 Command: <u>VPORTS</u> **(CR)**
Save/Restore/Delete/Join/SIngle/?/2/
<3>/4: <u>4</u> **(CR)**
(The three-dimensional drawing appears in four ports.)

Step 2 (Select the upper left port.)
Command: <u>PLAN</u> **(CR)**
<Current UCS/UCS/World: **(CR)**
Command: <u>ZOOM</u>
All/Center/.... <Scale(X)>: <u>.6X</u> **(CR)**
Command: <u>UCS</u> **(CR)**
Origin/Xaxis/...<World> <u>S</u> **(CR)**
?/Name of UCS: <u>TOP</u> **(CR)**

Step 3 (Select lower left port.)
Command: <u>UCS</u> **(CR)**
Origin/Xaxis/...<World>: <u>X</u> **(CR)**
Rotation angle about X axis: <u>90</u> **(CR)**
Command: <u>PLAN</u> **(CR)**
<Current UCS>/UCS/World: **(CR)**
(Get front orthographic view.)
UCS (Command repeated.)
Origin/Xaxis...<World>: <u>S</u> **(CR)**
?/Name of UCS: <u>FRONT</u> **(CR)**

Step 4 (Select lower right port.)
Use the UCS command to rotate the UCS 90° about the Y-axis to obtain a right-side view. Obtain its orthographic view with the PLAN command and save the UCS as SIDE, as in Step 3.

In Step 3, the UCS is rotated 90 degrees about the X-axis, and the front view is found by using the PLAN command. This UCS is SAVEd, as FRONT. In Step 4, the X-axis is rotated 90° about the Y-axis, and the PLAN option is used to find the side view. Its UCS is SAVEd as SIDE.

This process has resulted in top, front, right-side views, and a pictorial view of the wire-frame drawing. Additional lines can be added in any of the four views, and the other views will be updated simultaneously.

The system variable, UCSFOLLOW, can be set to ON (1) by using the command, SETVAR. When turned on, the UCSFOLLOW variable will automatically augment the PLAN option when a UCS is RESTORED. Therefore the X- and Y-axes will automatically appear true size in the restored view. If UCSFOLLOW is set prior to selecting VPORTS, the setting will apply to all VIEWPORTS. Once VIEWPORTS have been set, each viewport can have a UCSFOLLOW variable assigned. For example, UCSFOLLOW can be active in two viewports and inactive in two others, and so forth.

Note: While working in VPORTS, the CHANGE command may not always work properly. Therefore, use the CHPROP command as an alternative method of changing layers and so forth.

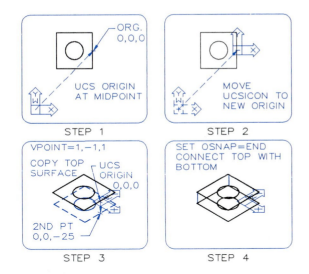

FIGURE 23.207 Three-dimensional object with an inclined surface.

Step 1 The top view of the object is drawn and the UCS origin is set at the midpoint of the line on the object.

Step 2 The UCSICON command is used to place the icon at the new origin.

Step 3 The VPOINT command is used to select a view point of 1, -1, 1 of the surface. The surface is copied to a position of 0, 0, -25 from the UCS origin.

Step 4 Set the OSNAP to the END option and connect the two planes.

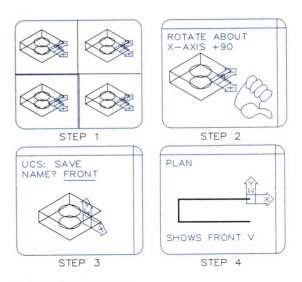

FIGURE 23.208 The 4-port construction.

Step 1 Command: VPORTS (Select 4 ports.) (The three-dimensional drawing is shown in all 4 ports.)

Step 2 (Select lower left port.) Command: UCS (CR) (Select X option.) Rotate UCS 90° about the X-axis for a front-view UCS.

Step 3 Command: UCS (Enter SAVE)?/Name of UCS: FRONT (CR)

Step 4 Command: PLAN (CR) (Display an orthographic front view. Use ZOOM to obtain desired size.)

23.76

Object with an inclined surface

The part shown in Fig. 23.206 has an inclined plane that requires an additional rotation for representation in three dimensions.

In Step 1 of Fig. 23.207, the top view of the object is drawn and the UCS origin is moved to the midpoint of one of its lines. In Step 2, the UCSICON is moved to the UCS origin. An isometric viewpoint is selected in Step 3, and the surface is copied at a location 0.25 inches below the upper surface. The corner points are connected in Step 4 with the END option of OSNAP.

In Fig. 23.208 (Step 1), a four-panel VPORTS is selected for displaying the views of the object. In Step 2, the UCS of the upper right port is rotated 90° about the X-axis for obtaining a front view. The lower left port is activated, and the UCS is SAVEd in Step 3. The PLAN command is used to obtain an orthographic front view in Step 4.

In Step 1 of Fig. 23.209, the pictorial view is RESTOREd, and the UCS is rotated 90° about the Y-axis to place the X- and Y-axes in the profile plane of the object. The lower-right panel is activated, the UCS is SAVEd, and the PLAN command is used to give an orthographic right-side view.

The top view, upper right panel is found in the manner as the previous two orthographic views were found.

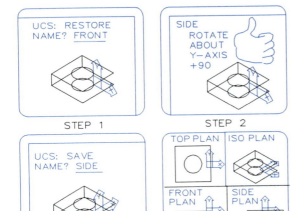

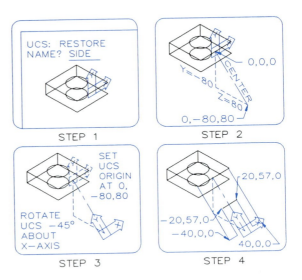

STEP 1   STEP 2

STEP 3   STEP 4

FIGURE 23.209   Top and side views.

Step 1 When desired, you may RESTORE the FRONT view by the UCS command, but it is unnecessary here.

Step 2 Select the lower right port. Command: UCS (CR) (Select Y and rotate the UCS 90° about the Y-axis.)

Step 3 Command: UCS (CR) SAVE ?/Name of UCS: SIDE (CR)

Step 4 Command: PLAN (CR) (Obtain an orthographic side view.) Use the same steps to obtain and save a top view in the upper left corner. Leave the three-dimensional view in the upper right corner.

In Step 1 of Fig. 23.210, the lower right port is activated and the UCS is RESTOREd. Draw a line from the origin to 0, -80, 80 to establish the sloping surface (Step 2). The UCS is rotated -45° about the X-axis to make it line in the plane of the inclined surface (Step 3). In Step 4, the coordinates of the notch (taken from the given view of the object) are used to construct the upper inclined surface of the part.

In Step 1 of Fig. 23.211, the OFFSET command is used to select the points of the three-dimensional pictorial of the inclined surface, the lower left port is activated, a distance of 0.25 inches assigned, and P locates the side of the offset. In Step 2, the offset surface is drawn in all views.

In Step 3, the FILLET command is applied to the line in upper right port to make them join precisely. In Step 4, the END option of the OSNAP com-

FIGURE 23.210   Drawing the inclined surface.

Step 1 Select the side-view port. Command: UCS (CR) (Select RESTORE)?/Name of UCS: SIDE (CR) Command: VIEWPORTS (CR) (SAVE this view port configuration as AUXI) Command: VPORTS (CR) (Select SIngle to obtain a full-screen image.)

Step 2 Command: UCS Origin/Zaxis/ . . . <World>: OR (Select origin.) Origin point: 0,-80,80

Step 3 Command: UCSICON (CR) (Select ORigin option to place icon at a new origin.) Command: UCS (CR) (Select X and rotate UCS - 45° about the X-axis.) Use UCS command to SAVE as INCLINED.

Step 4 Use the LINE command and dimensions from the given view of the object to specify the corners of the notch.

mand is used to connect the corner points to complete the drawing.

23.77
Model space and paper space

So far, you have been drawing in three-dimensional space where X-, Y-, and Z-coordinates for points can be inserted at the keyboard. By using the UCSICON command, the 3D icon can be turned on. It appears at the lower left of the screen, signifying that you are

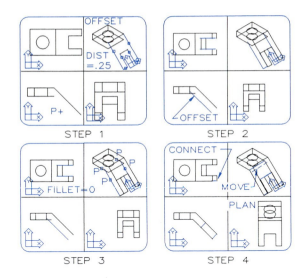

STEP 1 STEP 2

STEP 3 STEP 4

FIGURE 23.211 **Completion of the drawing.**

Step 1 Command: VPORTS (RESTORE the 4-port configuration, AUX1, to restore the four views.) Use the OFFSET command and select point on the notch in the three-dimensional view with an offset distance of 25, and locate the side of the offset in the front view port.

Step 2 The lower surface of the inclined plane is displayed.

Step 3 The FILLET command is used to trim the ends of the intersecting lines. The lines are picked in the three-dimensional view.

With OSNAP set to END, end points of lines are connected. The horizontal intersection line is MOVEd to its new position. Notice that you can use any of the views for changing or developing the drawing.

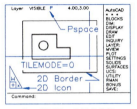

MODEL SPACE PAPER SPACE

FIGURE 23.212 **Model space and paper space. You have been drawing in** model space, **which is three-dimensional space. Turn on the** UCSICON **and the** *X*- **and** *Y*-**axes appear in the lower left corner of the screen. By setting the** TILEMODE **command to 0 (zero), you automatically enter paper space, which is two-dimensional space.**

TILEMODE=1 TILEMODE=0

FIGURE 23.213 TILEMODE=0 **and** TILEMODE=1. **When the command,** TILEMODE **is set to 1, you can only use AutoCAD's standard viewports that abut each other like tiles. Four is the maximum number you can use. By setting** TILEMODE **to 0, you automatically enter** PSPACE, **paper space.**

in a 3D mode (Fig. 23.212). However, this form of three-dimensional space makes it difficult to apply two-dimensional entities such as text, legends, and title blocks, but AutoCAD lets you bring together Mspace (model space) and Pspace (paper space) as shown in Fig. 23.212. But how do we get there?

When TILEMODE is set on 1, only a maximum of four VPORTS can be inserted on the screen at a time, and they will be "tiled," which means that they abut each other like tiles and fill the screen in several standard ways. When the TILEMODE command is set to 0, the icon in the lower left corner of the screen changes to a triangle, as shown in Fig. 23.213, a P appears in the status line, and you are in paper space (or two-dimensional space). In this mode, a two-di-

mensional border can be drawn, text can be inserted in a 2D mode, and title blocks can be drawn.

By setting TILEMODE to 0, the screen automatically goes to Pspace (two-dimensional paper space) and the screen is blank, as shown in Fig. 23.213. To open 3D ports in Pspace, you must use the MVIEW (make view) command, which asks for the diagonal corners of the port that you want as shown in Fig. 23.214. Once opened, the model viewports cannot be drawn in since you are in paper space and the cursor spans across the two-dimensional paper space of the screen.

Viewports can be copied using the COPY command (Fig. 23.215) and the same view of the drawing will appear in each of the viewports. Use the

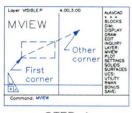

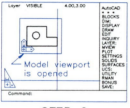

FIGURE 23.214 The MVIEW command.

Step 1 Command: MVIEW
On/OFF/Hideplot/Fit/2/3/4/Restore/
<First Point>:
Select objects: (Select first.)
Other corner: (Select other.)

Step 2 An MSPACE window is opened in paper space, and the object appears in the wndow. The cursor remains in PSPACE.

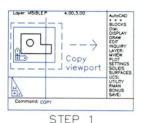

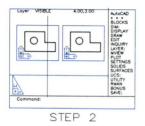

FIGURE 23.215 Copying an MSPACE viewport.

Step 1 Command: COPY
Select objects: (Window viewport and move it to a new location.)

Step 2 Both viewports are identical and contain the same view of the object. The cursor remains in PSPACE.

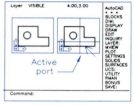

FIGURE 23.216 Using MSPACE.

Step 1 Once a model space viewport has been established, you can enter model space by the command MSPACE.

Step 2 Only one model space can be active at a time. Use the cursor to select the active viewport.

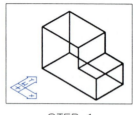

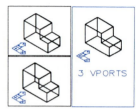

FIGURE 23.217 TILEMODE=1.

Step 1 Command: TILEMODE
New value for TILMODE <0>: 1 (Enter PSPACE.)
(Select a VPOINT to have a three-dimensional view of the object.)

Step 2 Command: VPORTS
Save/Restore/Delete/Join/SIngle/?
/2/<3>/4: 3 (Obtain three viewports.)

MSPACE command and set it to 1 (on) and you are in MSPACE (model space) as shown in Fig. 23.216 where you can draw in three but not two dimensions. You can select either viewport with the cursor to make it the active viewport where you can draw in three dimensions.

You can move between paper space (two-dimensional space) and model space (three-dimensional space) by using the PSPACE and MSPACE commands, respectively. To insert a border around a three-dimensional drawing, you must be in PSPACE so the two-dimensional border can be placed.

23.78
Drawing with tilemode=1

By leaving the TILEMODE set to 1, the default, the drawing in Fig. 23.217 is inserted and a VPOINT selected to give a three-dimensional view of the part. The VPORT command is used to give three tiled viewports on the screen, inside of which the same

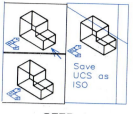

FIGURE 23.218 Tiled viewports: Saving UCS views.

Step 1 Command: UCS
Origin/ZAaxis/3point/Entity/View/X/Y
/Z/Prev/Restore/Save/Del/?/<World>:
SAVE ?/Desired UCS name: I (CR) (I for isometric, the active Vport.) (Cursor-select the top left port as the active Vport.)

Step 2 Command: PLAN (CR) (The ICON appears true size and view becomes a top view.)
Command: SAVE ?/Desired UCS name: T (CR) (T for top view.)

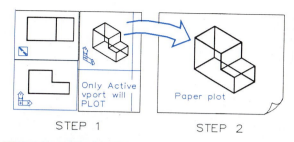

FIGURE 23.220 Tiled viewports: Plotting.

Step 1 (Cursor-select the large VPORT to make it active.)
Command: PLOT

Step 2 After specifying the plot options, only the active viewport will be plotted on the paper output.

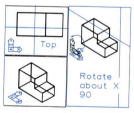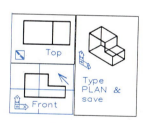

FIGURE 23.219 Tiled viewports: Rotating UCS.

Step 1 (Cursor-select the large Vport.)
Command: UCS
Origin/ZAaxis/3point/Entity/View/X/Y/Z
.../<World>: X
Rotation angle about X axis <0>: 90 (CR) (ICON moves to be parallel to front plane of part. Cursor-select front view.)

Step 2 Command: PLAN (ICON and the front view are shown true size.)
Command: UCS
Origin/ZAxis.../<World>: SAVE
?/Desired UCS name: F (F for front.)

view of the part appears, and only one viewport is active.

In Step 2 of Fig. 23.218 the USC command is used to SAVE the right view as ISO. The upper left view is selected as the activate port, the PLAN command is used to show the true top view in Step 2, and that view is SAVEd as TOP with the UCS command.

In Step 1 of Fig. 23.219, the UCS ICON is rotated 90 degrees about the X-axis, the lower-left port is selected in Step 2, and the PLAN command is used to obtain a true-size front view. The front view is SAVEd with the UCS command as FRONT.

If we were to make a paper plot at this point, only the active viewport will plot, as if it were the only drawing visible on the screen, as shown in Fig. 23.220. In order to plot all three views on single paper plot, we must use the UCS command's RESTORE option to obtain the front view and use its VIEW option too so that the front view can be BLOCKed and WBLOCKed for later insertion. Use the OOPS command to recall the lost drawing after it has been blocked (Fig. 23.221). The other two views are made into WBLOCKS in this same manner.

In Step 1 of Fig. 23.222, the three WBLOCKS of the UCS VIEWS are inserted and arranged properly on the screen. Each drawing is inserted as a two-dimensional view. When a paper plot is made in Step 2, all views plot at the same time. You can see that there are problems involved in working in three dimensions without having access to paper space.

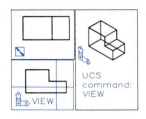
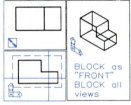

STEP 1   STEP 2

FIGURE 23.221 Tiled viewports: Blocking views.

Step 1 Command: UCS
Origin/ZAxis . . . ?/<World>: VIEW (CR)
(The ICON appears true size in the active VPORT.)

Step 2 Command: BLOCK (CR)
Block name (or ?): FRONT (CR)
Insertion base point: (Select point.)
Select objects: (Window front view.) (CR)
(View is blocked and disappears from screen in all
viewports. Type OOPS to recall. Make a WBLOCK of
the Block, FRONT. Cursor-select the other
viewports, use the VIEW option of the UCS
command, obtain the PLAN view, and block the
views as WBLOCKS.)
Command: VPORTS
Save/Restore/Delete/Join/SIngle/
?/2/<3>/4: SAVE
Name of viewport: VP1 (CR) (and QUIT.)

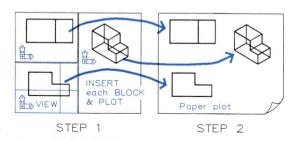

STEP 1   STEP 2

FIGURE 23.222   Tiled viewports: Inserting blocks and plotting.

Step 1 From the Main Menu, select 2, Edit a
drawing.
Name of drawing: A:FORMAT (CR)
Command: INSERT
Block named (or ?): T
Insertion point: (Cursor-select.)
Scale factor <1>: 1
Rotation angle <0>: 0 (Top view is inserted.)
Command: INSERT
Block name (or?): F (Complete the insertion
of the Block and insert Block I as well.)

Step 2 (After the UCS Views have been inserted
and aligned, type PLOT, give the plotting
specifications, and all of the views will be plotted as
a single file.

23.79
Drawing with tilemode=0

The following steps illustrate how a drawing can be
made that utilizes both paper space and model space.
In Step 1 of Fig. 21.223, TILEMODE is set to 0,
which automatically turns the screen into paper space,
and the PSPACE icon appears in the lower-left corner.

Insert an A-Size border in this space in Step 2,
and set the LIMITS large enough to contain the border, about 12 × 9.

In Step 1 of Fig. 23.224, use the MVIEW command to select diagonal corners of three-dimensional
model space. In Step 2, copy the first model space. By
typing MSPACE, one of the model spaces will become
the active viewport. Set the LIMITS in this area to
be 24,20 (large enough for the object to be drawn).

In Step 1 of Fig. 23.225, use the extrusion
method to draw a three-dimensional drawing in one
of the model spaces, which will be duplicated in the
second port. In Step 2, select a VPOINT for an isometric view of the part.

Type PSPACE to convert the screen back to paper space in Step 1 of Fig. 23.226. The cursor now
spans the screen, the two-dimensional space. Now we
must figure out the scaling of the drawing.

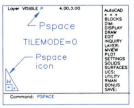 

STEP 1   STEP 2

FIGURE 23.223   Nontiled viewports: PSPACE.

Step 1 Command: TILEMODE (CR)
New value for TILEMODE <1>: 0 (CR)
(Enters PSPACE; 2D icon appears and P in status
line.)

Step 2 Command: INSERT
Block name (or?): BORDER-A
Insertion point: 0,0
X scale factor <1>/Corner/XYZ: 1
Y scale factor (default=X): 2
Rotation angle <0>: 0 (Border is inserted in
2D Pspace. Set limits to 12,9 and ZOOM/ALL.)

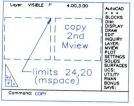

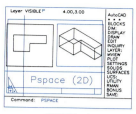

STEP 1   STEP 2   STEP 1   STEP 2

FIGURE 23.224 Nontiled viewports: MVIEW option.

Step 1 Command: MVIEW (Switching to paper space.)
ON/OFF/Hideplot/Fit/2/3/4/Restore/
<First Point>: (Cursor-select first corner.)
Other corner: (Cursor-select window.) (An empty 3D model-space port is drawn.)

Step 2: While still in paper space, make a copy of the three-dimensional model-space port and set the limits of ports to 24,20 by using the LIMITS command.

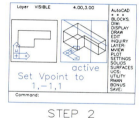

STEP 1   STEP 2

FIGURE 23.225 Nontiled Vports: **Drawing in model space.**

Step 1: Command: MSPACE (3D ICONS appear in ports and the cursor is shown in the active port.) (By using ELEV and THICKNESS, draw an extrusion of a block. It will show in both views.)

Step 2: Cursor-select the right viewport to make it active.
Command: VPOINT
Rotate/<View point> <current>: 1,-1,1 (CR) (An isometric view of the part is drawn.)

First, we set the PSPACE LIMITS to be 12,9 and inside of this two-dimensional space we want to show two views drawn in MSPACE, each with LIMITS of 24,20. These larger three-dimensional views will not fit inside the two-dimensional border unless they are scaled down, and this will require some calculations. If the two model spaces were 24 wide, their combined width would be 48. If they were scaled to half size, they would have a width of 24, still too wide. If they were scaled to 0.20 (two

FIGURE 23.226 Nontiled Vports: ZOOMing to scale.

Step 1 Command: PSPACE (CR) (The two-dimensional border reappears and the cursor spans the screen, the active area. Since the paper space had limits of 12,9, and the model space had limits of 24,20, the Mspace ports must be ZOOMed to about 0.20 of their original size for both to fit in border.)

Step 2 Command: MSPACE (CR) (One of the three-dimensional ports becomes active.)
Command: ZOOM
All/Center/.../<Scale (X/XP)>: .2XP (Active port is scaled to 2/10 size to fit within two-dimensional border. Select and ZOOM other views to same scale.)

tenths) they would be 9.6 units wide, sufficiently small enough to fit inside an A-Size border as shown in Fig. 23.226.

The X/XP (2/10P or 0.2P) option of the ZOOM command is used to size each three-dimensional port. This factor changes the width of each port in Step 2 of Fig. 23.226 from 24 units wide to 4.8 units wide, and the views are both at the same scale on the screen.

In Step 1 of Fig. 23.227, PSPACE is typed to move back to paper space where the STRETCH com-

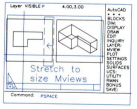

STEP 1   STEP 2

FIGURE 23.227 Nontiled Vports: **Moving ports.**

Step 1 Command: PSPACE (Switch to paper space in 2D mode. Use STRETCH command to reduce the outlines of the 3D ports.)

Step 2 Move 3D ports to their final positions within the 2D border shown in paper space.

mand is used to reduce the size of the MSPACE outlines on the screen. In Step 2, the MVIEWS are repositioned for plotting.

Plotting must be performed from PSPACE, as shown in Fig. 23.228, where both the two- and three-dimensional drawings are plotted on the same piece of paper. Notice that the outlines of the MSPACE also plot.

If you do not want the window outlines plotted, you can CHANGE them to a separate LAYER, such as one called WINDOW, and turn this layer OFF before plotting.

Dimensions can be added to the object, as shown in Fig. 23.229, when the desired dimensioning variables have been set for paper space (two-dimensional space). Set the DIMSCALE to zero and use associative dimensions for snapping to the end points of the object. A horizontal dimension is added to the top view of the object. Since DIMSCALE was set to zero, the dimensioning variables in model space will be sized to match those of paper space.

The isometric view of the object is selected as the active view port in Fig. 23.230 and the UCS icon is rotated to be parallel to the front view. The endpoints of the vertical dimension are selected and an isometric view of the vertical dimension is obtained. Notice that this dimension appears as an edge in the top view.

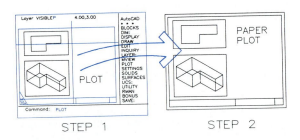

FIGURE 23.228 **Nontiled viewports: Plotting.**

Step 1 Command: PLOT (Give the plotting specifications.)

Step 2 The A-Size paper plot is plotted to show both 2D (paper space) and 3D (model space) in the same plot.

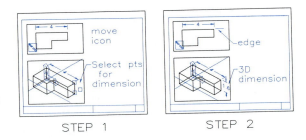

FIGURE 23.230 **Dimensioning: 3D view.**

Step 1 Move the UCSICON to the plane of the vertical dimension in the 3D view. Select the three points of the dimension.

Step 2 The dimension appears in the 3D view and in the top view. It is not readable in the top view.

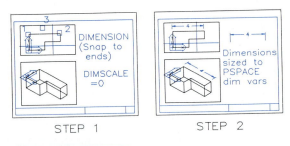

FIGURE 23.229 **Dimensioning: Model space.**

Step 1 By setting the Dim Vars DIMSCALE to 0, AutoCAD will dimension in model space viewports size the dimensioning variables to match those specified for paper space. Move the UCSICON to the plane of the dimension. Use associative dimensions and snap to the endpoints of the object.

Step 2 When the dimension is applied in one viewport, it is given in all 3D viewports. Notice the size of the dimensioning variables are the same in MSPACE as in PSPACE.

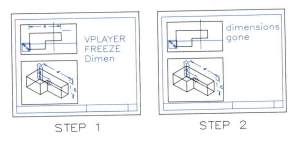

FIGURE 23.231 **Dimensioning: VPLAYER command.**

Step 1 Since the vertical dimension is not readable in the top view, use the VPLAYER command to turn off the Dimen Layer in that view.

Step 2 The dimensions are gone in the top view, but remain in the 3D view.

Since the vertical dimension (Fig. 23.231) cannot be read in the top view, select this port and use VPLAYER to FREEZE the Dimen layer on which the dimensions lie. Both the vertical and horizontal dimensions will be frozen in this view, but they will be retained in the isometric view. The resulting isometric contains width and height dimensions that are properly positioned and are easily readable. Other dimensions can be added in this same manner.

23.80
Drawing with meshes

An object can be represented by applying the previously covered commands RULESURF, TABSURF REVSURF, and EDGESURF.

By referring to Part A of Fig. 23.232, you can see where circles, arcs, and lines have been used to form a wire diagram of a part and the surface has been tabulated between the circular ends with the TABSURF command at Part B, and the ruled lines are moved to the side of the original drawing with the MOVE command and LAST option.

The RULESURF command is used in Parts D and E to rule the hole. The ruled lines are MOVEd to the same location as in the previous step. RULE-SURF is used to rule the ends of the cylinders, which are also moved (Parts H, I, and J).

In Parts K, L, and M, the upper surface of the lug is meshed using the EDGESURF command. This mesh is also moved to the right of the original drawing.

In Parts N, O, and P, the RULESURF command is used to apply mesh to the thin edges of the lug, which are moved to the right to add to the other meshed surfaces. This method can be continued to create meshes and transfer them to a removed position.

Once completed, the visibility of the object can be shown by using the HIDE command (Fig. 23.233).

23.81
Plotting a drawing

A hard copy of a drawing can be plotted with either a pen plotter or a printer plotter (a typewriter printer with graphics capability). You may initiate the com-

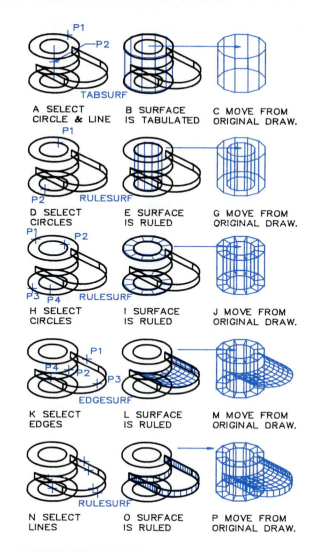

A SELECT CIRCLE & LINE   B SURFACE IS TABULATED   C MOVE FROM ORIGINAL DRAW.

D SELECT CIRCLES   E SURFACE IS RULED   G MOVE FROM ORIGINAL DRAW.

H SELECT CIRCLES   I SURFACE IS RULED   J MOVE FROM ORIGINAL DRAW.

K SELECT EDGES   L SURFACE IS RULED   M MOVE FROM ORIGINAL DRAW.

N SELECT LINES   O SURFACE IS RULED   P MOVE FROM ORIGINAL DRAW.

FIGURE 23.232   **Three-dimensional object with meshes applied.**

A, B, C A wire diagram of a three-dimensional part is given. TABSURF is used to tabulate the upper circles. The resulting lines are moved from the original drawing.

D, E, F RULESURF is used to connect the small circles. The ruled lines are moved to the same SNAP point as the tabulated lines.

H, I, J The upper and lower donut areas are connected with RULESURF. The ruled surfaces are moved to the location of the other meshes.

K, L, M The EDGESURF command is used to apply a mesh to the upper and lower surfaces of the lug, and to move the mesh to the other meshes.

N, O, P Use RULESURF to apply a mesh to the vertical edges of the lug. Transfer the mesh to the other meshes with the MOVE command.

575

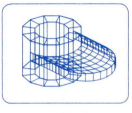
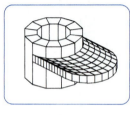

STEP 1   STEP 2

Visibility of meshed surfaces.

Step 1 The three-dimensional drawing developed in the last example is a hollow shell enclosed in meshes with no wireframe outlines.

Step 2 The HIDE command is applied to remove the hidden lines to form a more realistic three-dimensional view of the part.

mands for plotting from the **Main Menu** or the **Root Menu** when the drawing to be plotted appears on the display. The command for activating the pen plotter is PLOT, and PRPLOT is used for the printer plotter.

You must tell AutoCAD which part of the drawing to plot by responding to the following prompt:

```
What to plot Display, Extents,
Limits, View, or Window <D>: L
(CR)
```

The meanings of these options are

DISPLAY (D) plots the view currently visible on the display screen, or the last view that was displayed before SAVE, or END.

EXTENTS (E) plots all entities even though they may exceed the limits.

LIMITS (L) plots the drawing within the area defined by the limits.

VIEW (V) plots a named view that was saved by the VIEW command. You will be prompted for the VIEW name.

WINDOW (W) plots a selected part of the drawing that is enclosed in a window specified by two diagonal points. When PLOT is begun from the Drawing Editor, you can point to the corners of the window, but you must use coordinates of the window entered at the keyboard when beginning at the Main Menu.

The plotter origin for a Hewlett-Packard 7475 is located at 0.42 in., 0.35 in. from the lower left corner of an 11-×-8½-in. sheet, which gives an effective plotting area of 10.15 × 7.8 in. when you instruct the plotter to place the origin at 0,0.

After selecting the part of the drawing to be plotted, you will be given the following specifications on the display screen:

```
Sizes are in Inches
Plot origin is at (0.00, 0.00)
Plotting area is 10.50 wide by
8.00 high (A size)
Plot is NOT rotated 90 degrees
Pen width is 0.010
Area fill will NOT be adjusted
for pen width
Hidden lines will NOT be re-
moved
Scale is 1 = 1

Do you want to change
anything? <N>: N or Y (CR)
```

You can use these values by responding with N or NO, or you can change them by entering Y or YES. When YES is entered you will be given the option of changing, as shown in Fig. 23.234.

Layer Color	Pen No.	Line Type	Pen Speed	Layer Color	Pen No.
1 (Red)	1	0	9	9	1
2 (Yellow)	2	0	9	10	2
3 (Green)	2	0	9	11	2
4 (Cyan)	2	0	9	12	2
5 (Blue)	2	0	9	13	2
6 (Magenta)	1	0	9	14	1
7 (White)	2	0	9	15	2
8	2	0	9	16	2

```
Line types 0= continous
1= . . . . . . . . . .
2=___ ___ ___ ___
3=____ ____ ____ ____
4=__ . __ . __ . __ . __
5=____ _ ____ _ ____ _
6=____ __ _ ____ _ __ ____
Do you want to change any of these
parameters? <N>
```

The layer colors, pen numbers, line types, and pen speeds shown under the PLOT command.

By pressing (CR) for N O, you accept these specifications. If you enter Y for YES, you will be given a
chance to change any of these values. Do **not** change
the line types if your plotter supports multiple pen
types; leave the line type set at 0 (continuous line).
Pen speed can be set to yield the best line for the type
of pen you use. P E N N O . gives the location of the
pen in a multiple-pen plotter.

After Y has been entered to indicate that you
wish to change parameters, AutoCAD will list L a y e r
C o l o r , P e n N o . , L i n e t y p e , and P e n
S p e e d one at a time. You may accept the current
value for each by pressing (CR), or change the values
by typing the new value after each default is displayed. When completed, type S to obtain an updated
display of your changes. Type X to exit from this portion of the program when the changes are correct.

AutoCAD will now prompt you for additional
plot specifications.

```
Write the plot to a file? <N>:
N or (CR)
Size units (Inches or Millime-
ters)
<I>: I (I for inches and M for millimeters.)
(CR)
Plot origin in units<0.00,0.00>:
2,1 (CR) (Enter coordinates of the origin from
the "home" position of your plotter, using milli-
meters or inches depending on the units you have
specified.)
```

AutoCAD will list the standard plotting sizes:

```
Standard values for plotting
size

  Size      Width       Height
    A       10.50        8.00
   MAX      16.00       10.00
Enter the Size or Width,
Height (in units) <B>: A (CR)
```
(Size A sheet 11 × 8.5 inches or MAX for the
largest size the plotter will accept, perhaps a Size
C sheet.)

Instead of responding with A or MAX for sheet
size, you may type in the desired limits of the sheet,
for example, 1 0 . 1 5 , 7 . 8 . The values will be listed

as a third option, U (for User), that can be recalled
during future plots.

```
Rotate 2D plots 90 degrees
clockwise? <N>: (CR)
Pen width <0.10>: (CR) (to accept)
Adjust area fill boundaries
for pen width?<N>: (CR)
```
(By responding Y the boundaries of filled areas
will be moved inward one-half pen width for a
higher degree of accuracy.)
```
Remove hidden lines?<N> (CR)
Specify scale by entering:
Plotted units=Drawing units or
Fit or? <Fit>:
```

If your drawing was to be plotted in inches, enter
1 = 1 for a full-size drawing where a plotted inch
was equal to 1 in. on the screen. For an architectural
scale such as $\frac{1}{2}'' = 1'-0''$, the unit at the left of the
equal sign should be converted to 1. This conversion
results in an equality of $\frac{1}{2}$ in. = 12 in. or 1 in. = 24
in. If your drawing was drawn in millimeters, when
prompted, S i z e u n i t s (I n c h e s o r M i l -
l i m e t e r s) < I > : , enter M, and the drawing will
be plotted in metric units.

```
Specify scale by entering:
Plotted Millimeters=Drawing
units
or Fit or ? <F>: 1=1 (CR)
```

By responding with F, the drawing will be scaled
to fill the available space, but the scale will be a nonstandard scale. Responding with ? will give a list of
the various scaling options and their descriptions.

Plot specifications are saved by AutoCAD so that
you will not have to change plot parameters until you
wish. AutoCAD will display the following message:

```
Effective plotting area: 10.50
wide by 8.00 high
Position the paper in plotter
(Pause for this step.)
Press RETURN to continue or S
to Stop for hardware setup
Press RETURN to begin plotting
```

Some plotters have other features that can be adjusted according to manufacturer's specifications.

However, the preceding steps are typical of the ones used by most plotters and printers.

Plot scale

Figure 23.235 illustrates how the X/XP option of the ZOOM command is used to scale a drawing in model space to be half of its size in paper space. A 0.5XP is entered which gives a scale of 0.5=1, or 1=2.

A drawing in model space can be centered and scaled using the Center option of the ZOOM command, as shown in Fig. 23.236. The point that is to be positioned in the center of the screen is selected in

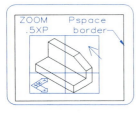
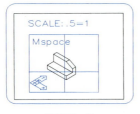

STEP 1   STEP 2

FIGURE 23.235 ZOOM X/XP.

Step 1 Command: ZOOM; X/XP **(To size a three-dimensional object in model space relative to paper space.)**

Step 2 By typing .5XP, **the size of the object in** MSPACE **is half that in** PSPACE.

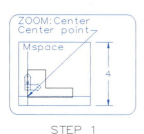

STEP 1   STEP 2

FIGURE 23.236 **Moving an object in a viewport.**

Step 1 Command: ZOOM
All/Center/Dynamic . . . <Scale (X/XP)>: C
Center points: **(Select point.)**

Step 2 Magnification or Height <default>: 8
The viewport is now 8″ high instead of 4″ high, making the object half size.

Step 1, and a height of 8 is given in Step 2 to make the dimension of 4 represent one of 8. This makes the contents of the window half size.

23.82
Attributes

Attributes are combinations of drawings saved as BLOCKS and text. A drawing can be made with attributes defined using the ATTDEF command, saved as a BLOCK, and then inserted repetitively. With each insertion, you will be prompted for attribute values that will be written on the drawing or left on file as an invisible value for listing in an attribute report.

An attribute drawing is shown in Fig. 23.237, where a title block is drawn using the DRAW command. To define the attributes, use the ATTDEF command.

Command: ATTDEF (CR)
Attribute modes—Invisible: N
Constant:N Verify:N
Enter (ICV) to change. RETURN
when done: V (CR)

By entering I, C, or V, one at a time, you can change the settings of INVISIBLE, CONSTANT, and VERIFY modes of the attributes. If the INVISIBLE mode is on, the attribute values will not be shown on the drawing to prevent clutter. If the CONSTANT mode is on, a fixed attribute value will be assigned to each BLOCK insertion. If the VERIFY mode is on, you will be able to verify the attribute values at insertion time.

The next prompts are

Attribute tag: DATE (CR) (No blanks allowed.)
Attribute prompt: DATE? (CR)
Default Attribute value: (Blank if none.) (CR or enter value.)
Attribute value: (CR)
Start point or Align/Center/Fit/
Middle/Right/Style: C (CR)
Height <0.100>: .125
Rotation angle <0>: (CR)
(The attribute tag appears on the drawing.)

 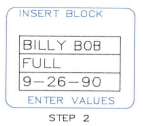

STEP 1   STEP 2

FIGURE 23.237   **Attribute definitions.**

Step 1 Command: <u>ATTDEF</u> **(CR)**
Attribute modes--Invisible:N
constant:N Verify:N Preset:N
Enter **(ICVP)** to change, RETURN when
done: <u>V</u> **(CR)**
Attribute modes--Invisible:N
Constant:N Verify:Y Preset:N
Enter: **(ICVP)** to change, RETURN when
done: **(CR)**
Attribute tag: <u>NAME</u> **(CR)**
Attribute prompt: <u>NAME?</u> **(CR)**
Default attribute value: **(Blank)**
Start point or
Align/Center/Fit/Middle/Right/Style:
(Locate tag on title block.)
Height <0.2000>: <u>.125</u> **(CR)**
Rotation angle <0>: **(CR)**

Step 2 B LOCK **the title block with a *P1* and *P2*
window.**
Command: <u>Insert</u>
Block name <or?>: <u>TITLE</u>
Insertion point: **(Select point.)**
X scale factor <1>/Corner/XYZ: **(CR)**
Y scale factor <default=X>: **(CR)**
Rotation angle <0>: **(CR)**
**(Respond to the attribute prompts to complete the
title block.)**

Multiple attributes can be added to each drawing.
Once all attributes have been added, the drawing is
windowed as a BLOCK in the usual manner.

When the BLOCK is inserted, you are prompted
for attribute values, and default values will be shown
on the screen.

      Enter attribute values
      Name?: <NONE>: <u>BILLY BOB</u> (CR)
      Scale? <full size>: (CR) (Accepts
      full size default.)
      Date: <NONE>: <u>9-26-91</u> (CR)

A listing of these attribute values will be given as you
(CR) through the list, enabling you to verify the correctness of your entries. When the end of the list is

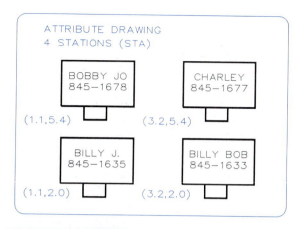

FIGURE 23.238   **A listing of the attributes on
a drawing can be made using the** ATTEXT
command in conjunction with BASIC. **More elaborate
listings can be obtained when attributes are used
with a database program.**

reached, the values will be plotted on the drawing in
the places previously assigned (Fig. 23.238).

The visibility of the attributes can be changed by
the ATTDISP command, which may be different
from the ATTDEF command specifications.

      Command: <u>ATTDISP</u> (CR)
      Normal/On/Off <current value>:
      <u>N</u> (CR)

By entering <u>N</u>, the attributes specified either hidden or
visible will appear that way on the drawing. The ON
option will display all values on the drawing, whether
hidden or visible. The OFF option does not show any
of the values on the drawing. For the values to be re-
displayed, the drawing must be REGENerated.

The ATTEDIT command is used to edit attributes one at a time or globally, where all are changed
at one time. The command is used as follows:

      Command: <u>ATTEDIT</u> (CR)
      Edit attributes one by
      one? <Y> <u>Y</u> (CR)

By responding yes (<u>Y</u>), you may edit any of the attributes currently visible on the screen. By responding
no (<u>N</u>), you may edit all attributes with global editing.
You can select the attributes to edit by responding to

the following prompts:

```
Block name specification <*>:
Attribute tag specification <*>:
Attribute  value  specification
<*>:
```

If you press (CR) after the BLOCK name specification, attributes of BLOCKS with all names will be selected for editing. If you specified the BLOCK name but entered (CR) after the tag specification, attributes for all tags will be selected. However, if you responded with NAME for the tag specification, only the attributes of the tag NAME will be selected. By entering a \, you will be able to edit null-value attributes.

You will now be prompted

```
Select ATTRIBUTES:
```

Select the attributes by pointing to them on the drawing or by windowing the group to be edited. Each attribute will be marked with an X on the drawing to indicate those eligible for editing. The next prompt is

```
Value/Position/Height/Angle/
Style/Layer/Color/Next <N>:
```

By selecting the first letter of these commands, you may change the attributes in many ways by responding to AutoCAD's prompts.

Global editing may be done by using the following sequence of commands:

```
Command: ATTEDIT
Edit  attributes  one  by  one?
<Y> N (Global option.)
Global  edit  of  Attribute  val-
ues.
Edit  only  Attributes  visible
on screen? <Y> N
```
(Screen goes to Text mode if N.)
(Drawing must be regenerated afterward.)

These next steps involve selecting the attributes in the same manner as when they were done one at a time. In this case, changes are applied uniformly to all selected blocks and attributes.

A dynamic dialogue box can be activated to edit attributes by using the command, DDATTE, as introduced in Section 23.7.

23.83
Attribute extract (ATTEXT)

The ATTEXT command is used to extract attributes from a drawing so they can be printed in tabular form. This system is excellent for preparing bills of materials, parts lists, and inventories. An elementary application is given below.

A BLOCK with attributes is inserted four times, as shown in Fig. 23.238, and the drawing is saved under the name of A:OFFICE. Next you must use a word processor to write a template file called A:DESKS.TXT. We will use the DOS editor, ED-LIN. From the drawing editor, type SH to return to DOS; type EDLIN A:DESKS.TXT, a new file. Then type I to insert the lines of the template, as shown in Fig. 23.239. End by typing Ctrl Break and enter E for end.

Return to drawing file A:OFFICE and type ATTEXT:

```
Command: ATTEXT (CR)
CDF, SDF, or DXF Attribute ex-
tract (or Entities)? <C>:C (CR)
Template file <default>:
A:DESKS (Omit .TXT.)
Extract  file  name  <A:OFFICE>:
A:OFFICE
??? entities in extract file:
```

USE EDLIN TO MAKE
TEMPLATE FILE A:DESKS.TXT

Blk name	BL:STA	C008000	8 characters
X-coord.	BL:X	N006001	6 numbers
Y-coord.	BL:Y	N006001	6 numbers
Name	NAME	C011000	11 characters
Phone	PHONE	C008000	8 characters

N=numbers   N006001 means
006=field width   000000 6 spaces
001=1 decimal pl.   0000.0

FIGURE 23.239 This template file was typed with EDLIN so that attributes could be extracted from a file similar to the one in Fig. 23.238.

```
COMMA DELIMITED FORMAT (CDF)
'STA', 1.1, 5.4, 'BOBBY JO','845-1678'
'STA', 3.2, 5.4, 'CHARLEY',845-1677'
'STA', 1.1, 2.0, 'BILLY J.','845-1635'
'STA', 3.2, 2.0, 'BILLY BOB','845-1633'

SPACE DELIMITED FORMAT (SDF)
STA     1.1   5.4   BOBBY JO    845-1678
STA     3.2   5.4   CHARLEY     845-1677
STA     1.1   2.0   BILLY J.    845-1635
STA     3.2   2.0   BILLY BOB   845-1633
```

FIGURE 23.240 **Examples of CDF and SDF attribute extractions by using the template file in Fig. 23.239.**

Now, you must type SH, go back to DOS, type ED-LIN A:OFFICE (CR), and then type LIST, and the extract file will be displayed on the screen, as shown in Fig. 23.240 as a Comma Delimited Format (CDF) table.

An alternative method is the Space Delimited Format (SDF) which uses the same template file A:DESKS.TXT and gives the table of values shown in Fig. 23.240. The Drawing Interchange File (DXF) is used to exchange data points with other software.

23.84
Grid rotation

To draw lines parallel or perpendicular to a given line, the grid on the screen can be rotated to align with existing lines. The ROTATE command is an option under the SNAP command. The prompts are as follows:

```
Command: SNAP (CR)
ON/OFF/Value/Aspect/Rotate/
Style:R (Rotate) (CR)
Base point <0, 0>: (Select a point.)
Rotation angle <0>: (Type the angle or
specify the angle on the screen.)
```

The grid will be plotted on the screen in alignment with the selected points. Figure 23.241 shows an example of a descriptive geometry problem with a true-length auxiliary view of line *AB*.

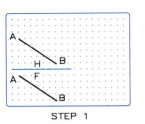
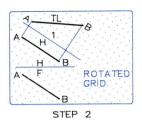

STEP 1   STEP 2

FIGURE 23.241 **Descriptive geometry problem.**

Step 1 Command: SNAP (CR)
Snap spacing or ON/OFF/Aspect/Rotate/Style <0.1250>: R (CR)
Base point <0,0>: (Select *PA*.)
Rotation angle <0>: (Select *PB*.)

Step 2 Draw H-1 reference line. OSNAP from points *A* and *B* perpendicular to H-1. Extend the projectors to locate the true-length view.

The grid is returned to its original position by selecting the SNAP command and the ROTATE option prompts.

```
Base point <2,3>: (CR)
Rotation angle <37>: 0 (CR) (Re-
aligns grid to its original position.)
```

23.85
Digitizing with the tablet

Drawings on paper can be digitized into the computer point by point when a tablet is available. Tablets vary in size from 11″ × 8.5″ to several square feet.

To calibrate a drawing and tablet for digitizing, tape the drawing to the tablet and use the following steps:

```
Command: TABLET
Option (ON/OFF/CAL/CFG): CAL (CR)
Calibrate tablet for use
Digitize first known point:
(Digitize point.)
Enter coordinates for first
point: 1,1 (CR)
Digitize second known point:
(Digitize point.)
Enter coordinates for second
point: 10,1
```

You should digitize points from left to right or from bottom to top of the drawing. Further, your limits must be large enough to contain the limits on the tablet. You may now use the AutoCAD **Root Menu** for copying the drawing. For example, activate the LINE command, select a beginning point on the tablet with your pointer, and select other points in sequence.

Use the ON or OFF commands to turn the tablet mode on or off. When turned off, your pointer can be used to access the **Root Menu** at the right of the tablet. Function key F10 on the IBM XT/AT can also be used for this purpose.

23.86
SKETCH command

The SKETCH command can be used with the tablet for tracing drawings composed of irregular lines (Fig. 23.242). Begin by attaching the drawing on the tablet and calibrate the tablet as discussed above. Enter the SKETCH command.

```
Command: SKETCH (CR)
Record increment <0.1>: 0.01
(CR)
Sketch. Pen eXit Quit Record
Erase Connect.
```

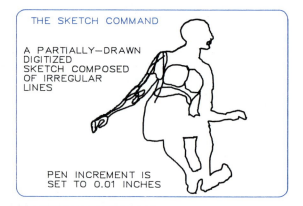

THE SKETCH COMMAND

A PARTIALLY—DRAWN DIGITIZED SKETCH COMPOSED OF IRREGULAR LINES

PEN INCREMENT IS SET TO 0.01 INCHES

FIGURE 23.242 This drawing was made using the SKETCH command at a tablet instead of a mouse. An original drawing was taped to the tablet, and the cursor was used to trace over it using increments of 0.01".

The record increment specifies the distances between the endpoints of the connecting lines that represent the irregular lines you sketch. The other command options are defined below.

Pen	Raises or lowers pen
eXit	Records lines and exits
Quit	Discards temporary lines and exits
Record	Records temporary lines
Erase	Erases selected lines
Connect	Connects current line to last endpoint
.(period)	Line from endpoint of last line to the current location of the pointer

Begin sketching by moving your pointer to the first point with the pen up, lower the pen (P), and trace over the line to be traced with the pointer. An irregular line is drawn on the screen. Erase by raising the pen (P), enter Erase (E), move the pointer backward from the current point, and select the point where you want the erasure to stop. All drawn lines are temporary until you select Record (R) or eXit (X). After recording the lines, you can begin new lines by repeating these steps. The mouse is unsatisfactory for sketching; for this type of digitizing only a tablet can be effectively used.

The SKPOLY system variable can be set by typing SETVAR, entering SKPOLY and typing 1 (on). When you enter the SKETCH command, the lines that you draw will be continuous polylines as an alternative to SKETCHed lines, which are connected but are separate segments.

23.87
Slide shows

Several commands can be used to create a slide show on the computer screen: SCRIPT, MSLIDE, RSCRIPT, DELAY, RESUME, and VSLIDE.

Enter the Drawing Editor, and make a drawing that you wish to use as a slide (Fig. 23.243). Enter the MSLIDE command and respond to the prompt with a slide name, A:FRONTSL for example. Erase the drawing and make a second drawing, enter MSLIDE, and name it A:SIDESL. You may make as many slides as you wish by following these steps.

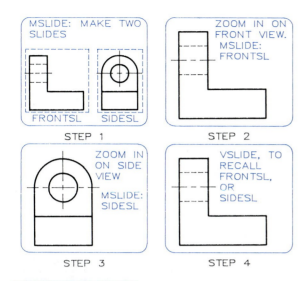

FIGURE 23.243 **Slide shows.**

Step 1 **A drawing is made that is to be saved as two separate slides.**

Step 2 ZOOM in on the front view.
Command: MSLIDE (CR) Slide file
<current>: FRONTSL (CR)

Step 3 ZOOM in on the side view.
Command: MSLIDE (CR) Slide file
<current>: SIDESL (CR)

Step 4 Command: VSLIDE (CR) Slide file
FRONTSL (CR)
(The slide is displayed on the screen.)

When you are finished, erase the last drawing from the screen.

Enter the command VSLIDE and type A:FRONTSL, and this image will be recalled and displayed on the screen. Type REDRAW to remove A:SIDESL from the screen. Recall SLIDE2 with VSLIDE in the same manner. Now that you have checked your slides, QUIT, and EXIT AutoCAD (or use the SHELL command to gain access to DOS).

To create a SCRIPT file, type SH to obtain the DOS prompt, C>, and start a script file with the EDLIN command that will be called A:DISPLAY.SCR in the following manner:

C> EDLIN A:DISPLAY.SCR
* I

Type the following script file. Notice that the disk drive, A:, is placed in front of both slide names.

1: VSLIDE A:FRONTSL (CR)
2: DELAY 2000 (CR)
3: VSLIDE A:SIDESL (CR)
4: DELAY 1000 (CR)
5: RSCRIPT (CR)
6: RESUME (CR)
7: (Ctrl Break)
*E (Saves file and ends session.)

The DELAY command is specified in milliseconds; this value will be an approximation because this speed varies with different types of computers.

To show the slides according to the SCRIPT, load AutoCAD's **Drawing Editor** by calling up any new drawing. Type SCRIPT and A:DISPLAY (the name of SCRIPT) in response to the filename prompt. The slides will be shown in sequence with the delays between each slide; then the sequence will recycle, beginning with the first slide. This sequence can be stopped by pressing two keys—Ctrl C or Backspace.

If the RSCRIPT command had been omitted from the SCRIPT file, the two slides would have been shown, and the sequence would have stopped. To repeat the show, you would then type RESUME. When you are through, type QUIT to return to the **Main Menu.**

Many more advanced slides can be developed using methods covered in AutoCAD's manual. For example, the SCRIPT file can include such variables as LIMITS, SNAP, GRID, UNITS, and TEXT, which are activated by SCRIPT each time it is called up. By using your imagination, a series of slides can be made to give animated action.

The EDLIN command is an MS-DOS line editor that can be used to write the SCRIPT file. Some of the commands used with EDLIN are as follows:

Type line number and (CR) Goes to that line for editing.

D Deletes lines. (Example: 2D (CR) to delete line 2.)

E Ends session and writes file to disk. (Example: Ctrl Break and E.)

I Inserts line into file. (Example: 4I to insert above line 4.)

L Lists lines. (Example: L (CR).)

Q   Quits editing without saving file. (Example: `Ctrl Break` and Q.)

S   Searches for a specified string. (Example: `?S` string to locate.)

`Ctrl Break`   Ends insertion mode.

`Ctrl Z`   Ends insertion mode.

23.88
SETVAR command

Many modes, sizes, limits, and variables remaining in AutoCAD's memory under the `SETVAR` command are available. These variables can be inspected and changed by means of the `SETVAR` command, unless they are read-only commands. Some variables are saved in AutoCAD's configuration file, and others are saved in drawings that are made.

To review system variables, use the `SETVAR` command.

```
Command: SETVAR (CR)
Variable name or ?: ? (CR)
```

A complete listing of the current variables will be given on the screen. To change one or more variables respond as follows:

```
Command: SETVAR (CR)
Variable name or ?: TEXTSIZE
(CR)
New value for TEXTSIZE <cur-
rent>: 0.125 (CR)
```

The default value, given in brackets, can be changed by typing a new value, `0.125`. By entering the `SETVAR` command with an apostrophe in front of it, you can use it transparently while another command is in progress.

23.89
Problems

The problems at the ends of the previous chapters can be drawn and plotted using computer-graphics techniques as covered in this chapter and previous chapters.

Additional problems are given in Figures 23.244–23.257 that can be used to develop computer graphics skills. Each can be drawn for plotting on an A-Size sheet (11 × 8.5).

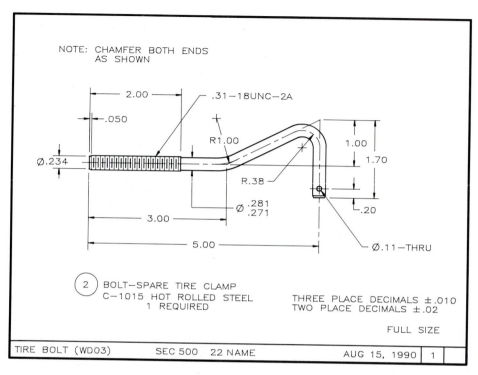

NOTE: CHAMFER BOTH ENDS
AS SHOWN

2.00

.31−18UNC−2A

.050

R1.00

1.00

1.70

Ø.234

R.38

Ø .281
.271

.20

3.00

5.00

Ø.11−THRU

② BOLT−SPARE TIRE CLAMP
C−1015 HOT ROLLED STEEL
1 REQUIRED

THREE PLACE DECIMALS ±.010
TWO PLACE DECIMALS ±.02

FULL SIZE

TIRE BOLT (WD03)   SEC 500   22 NAME   AUG 15, 1990   1

FIGURE 23.244
Problem 1. Bolt—
spare tire clamp.

FIGURE 23.245
Problem 2. Piston.

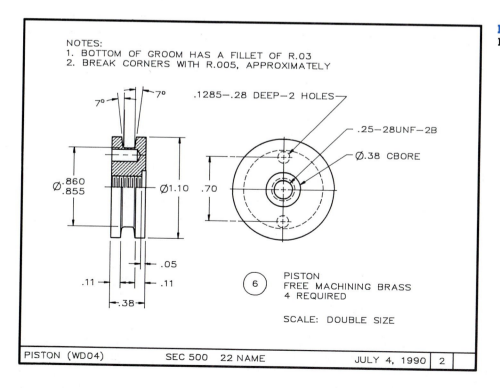

NOTES:
1. BOTTOM OF GROOM HAS A FILLET OF R.03
2. BREAK CORNERS WITH R.005, APPROXIMATELY

7°   7°

.1285−.28 DEEP−2 HOLES

.25−28UNF−2B

Ø.38 CBORE

Ø.860
.855

Ø1.10

.70

.05

.11

.11

.38

⑥ PISTON
FREE MACHINING BRASS
4 REQUIRED

SCALE: DOUBLE SIZE

PISTON (WD04)   SEC 500   22 NAME   JULY 4, 1990   2

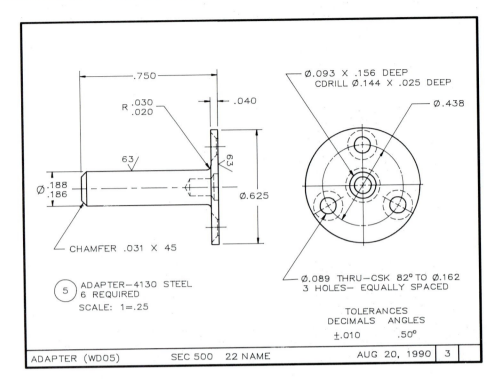

FIGURE 23.246
Problem 3. Adapter.

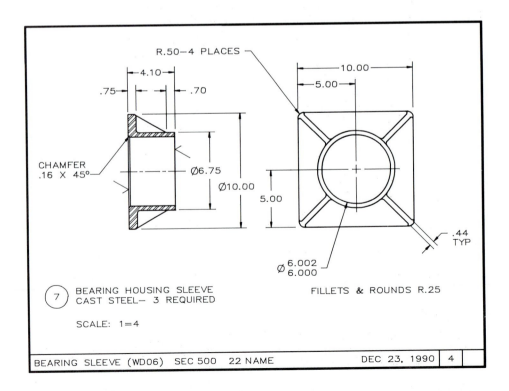

FIGURE 23.247
Problem 4. Bearing
housing sleeve.

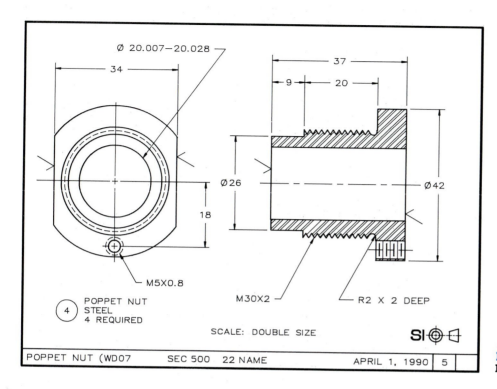

Ø 20.007–20.028

34

18

M5X0.8

POPPET NUT
STEEL
4 REQUIRED
4

37

9 20

Ø26 Ø42

M30X2 R2 X 2 DEEP

SCALE: DOUBLE SIZE

SI

POPPET NUT (WD07   SEC 500   22 NAME   APRIL 1, 1990   5

FIGURE 23.248
Problem 5. Poppet nut.

FIGURE 23.249
Problem 6. Double
bushing.

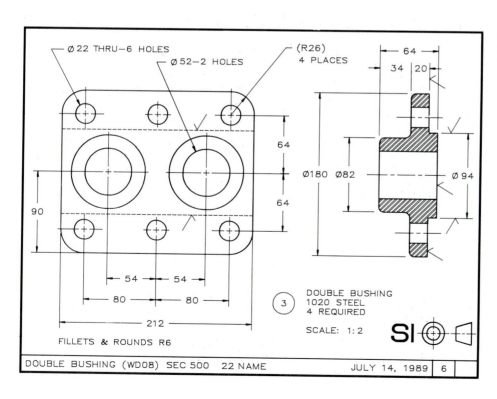

Ø 22 THRU–6 HOLES   Ø52–2 HOLES   (R26)
4 PLACES

64

90   64   64   Ø180 Ø82   Ø 94

54   54

80   80

212

FILLETS & ROUNDS R6

64

34 20

DOUBLE BUSHING
1020 STEEL
4 REQUIRED
3

SCALE: 1:2

SI

DOUBLE BUSHING (WD08) SEC 500   22 NAME   JULY 14, 1989   6

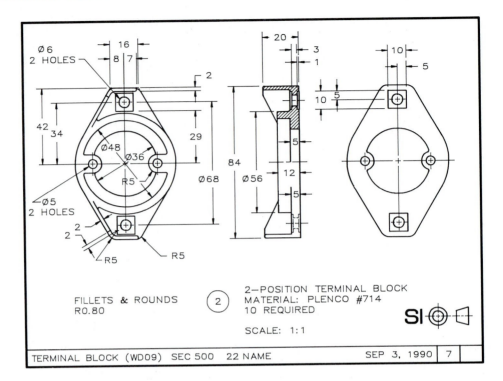

FIGURE 23.250
Problem 7.
Terminal block.

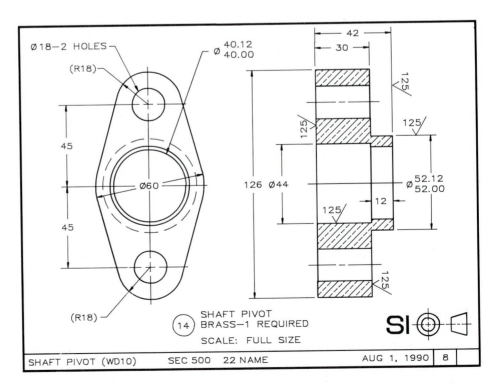

FIGURE 23.251
Problem 8. Shaft pivot.

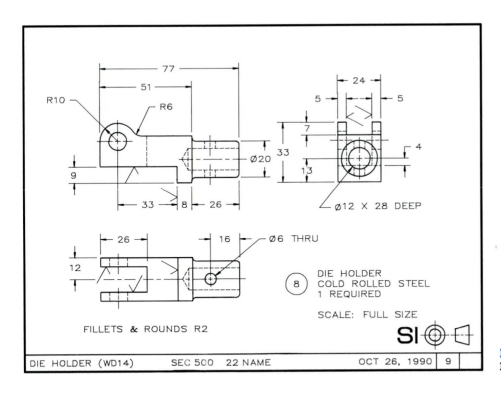

FIGURE 23.252
Problem 9. Die holder.

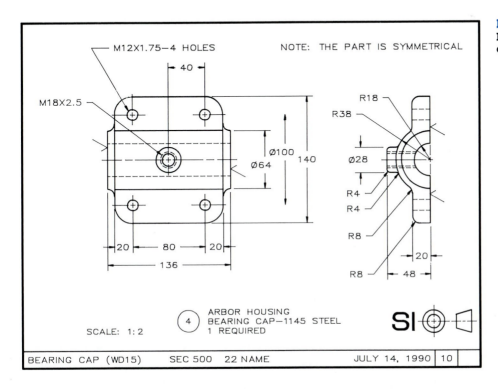

FIGURE 23.253
Problem 10. Bearing cap.

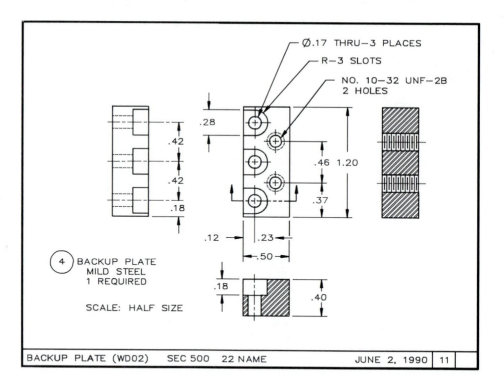

FIGURE 23.254
Problem 11.
Backup plate.

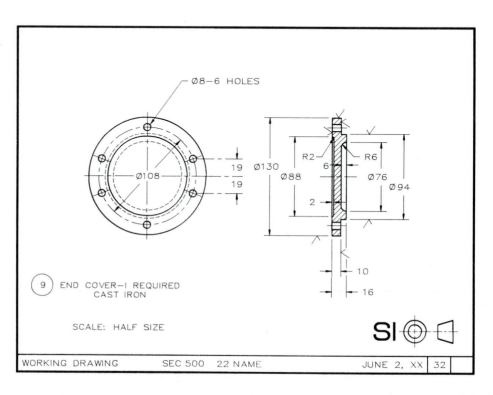

FIGURE 23.255
Problem 12. End cover.

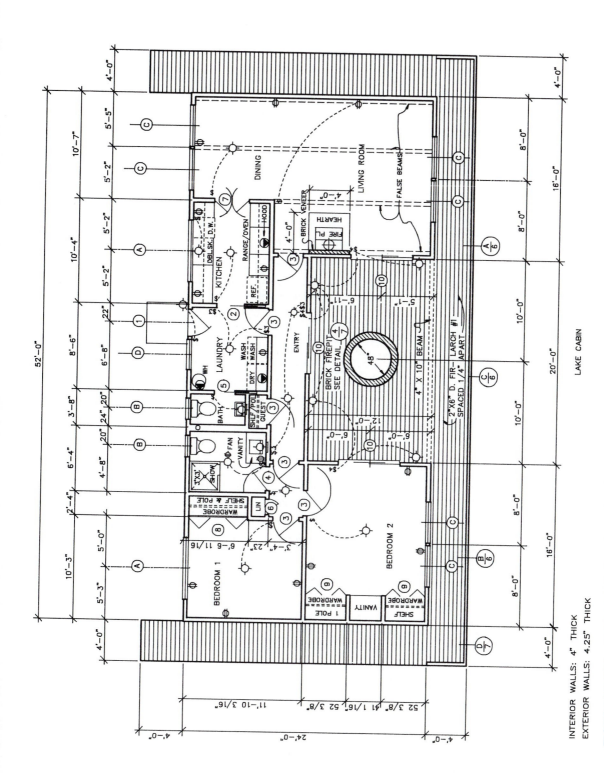

INTERIOR WALLS: 4" THICK
EXTERIOR WALLS: 4.25" THICK

LAKE CABIN

FIGURE 23.256 Architectural Plan. Draw the plan of the house on a C-Size sheet at a scale of 1/4" = 1'0". Include a door schedule, window schedule, and legend on the same sheet (Figure 23.257).

LEGEND

110V DUPLEX	12" ABOVE FLR.	CEILING, WALL AND RECESSED LIGHTS
SWITCH	48" ABOVE FLR.	220V RANGE, DRYER, WATER HEATER
3—WAY SWITCH	48" ABOVE FLR.	10" ABOVE COUNTER
4—WAY SWITCH	48" ABOVE FLR.	

WINDOW SCHEDULE

MARK	NO.	STOCK	TYPE	MFGR.	ROUGH OPENING
A	2	6030	HORIZONTAL SLIDING	ALENCO	72 3/4" X 36 1/2"
B	2	2030	HORIZONTAL SLIDING	ALENCO	24 3/4" X 36 1/2"
C	6	4040	HORIZONTAL SLIDING	ALENCO	48 3/4" X 48 1/2"
D	1	3030	HORIZONTAL SLIDING	ALENCO	36 3/4" X 36 1/2"

DOOR SCHEDULE

MARK	NO.	TYPE	MATERIAL	SIZE
1	1	1 LT. PANEL SASH DOOR	BIRCH	2'—8" X 6'—8" X 1 3/4"
2	1	FLUSH HOLLOW CORE — INTERIOR	BIRCH	2'—8" X 6'—8" X 1 3/8"
3	6	FLUSH HOLLOW CORE — INTERIOR	BIRCH	2'—6" X 6'—8" X 1 3/8"
4	1	FLUSH HOLLOW CORE — INTERIOR	BIRCH	2'—4" X 6'—8" X 1 3/8"
5	1	FLUSH HOLLOW CORE — INTERIOR	BIRCH	2'—0" X 6'—8" X 1 3/8"
6	1	FLUSH HOLLOW CORE — INTERIOR	BIRCH	1'—6" X 6'—8" X 1 3/8"
7	1 PR.	SWING. CAFE DOORS	BIRCH	1'—8" X 4'—0" X 1 1/8"
8	1	FLUSH HOLLOW CORE — FOLDING	BIRCH	6'—0" X 6'—8" X 1 3/8"
9	2	FLUSH HOLLOW CORE — INTERIOR	BIRCH	4'—0" X 6'—8" X 1 3/8"
10	3	SLIDING GLASS (6069 — 2V)	GLASS	6'—0" X 6'—8 1/2"

FIGURE 23.257
Schedules and legends of the
architectural plan.

Introduction to Solid Modeling

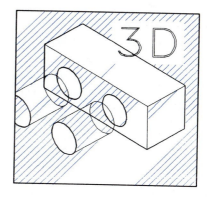

24

24.1
Introduction

An auxiliary software package, Advanced Modeling Extension (AME), is available from AutoDesk to provide solid modeling capabilities for the 386 version AutoCAD Release 11. A 386 or 486 computer with at least four megabytes of RAM is required for this software to function properly. This introduction to three-dimensional solid modeling is intended to provide a brief overview, but it should not be considered a comprehensive coverage.

Solid modeling is the technique of creating three-dimensional solids by using a combination of solid primitives such as boxes, cubes, cylinders, toruses, spheres, wedges, and cones. These primitives can be added or subtracted from each other to form the final object. By using solid primitives in combination, the possibility of creating objects that cannot be made is eliminated.

AME allows you to extrude a two-dimensional planar drawing into a three-dimensional solid. Wire diagrams can be solidified into three-dimensional objects. Solids can be modified by adding fillets or chamfers. Additional realism can be gained by using meshes and shading to render the surfaces of the objects.

Solid models can be analyzed easily to find their mathematical properties. For example, properties can be obtained such as centers of gravity, masses, surface areas, moments of inertia, products of inertia, radii of gyration, and others. This information would be difficult to calculate by hand.

This chapter will give an overview of solid modeling, which consists of (1) solid primitives, and (2) modification of solids.

24.2
Primitives: box (SOLBOX)

The most fundamental of all solids, the box, is created with the SOLBOX command, as shown in Fig. 24.1. The base of the box is always established in the plane of the current UCS.

The dimensions of the box can be created with separate widths and depths, diagonal corners of the base, or as a cube using values typed at the keyboard or cursor-selected points on the screen. The box will appear on the screen as a wire diagram.

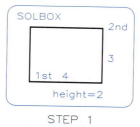
STEP 1

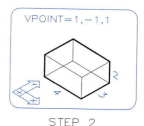
STEP 2

FIGURE 24.1   Box (SOLBOX).

Step 1 Command: <u>SOLBOX</u> **(CR)**
Corner of box: **(Select P1.)**
Cube/Length/<Other corner>: **(P2.) (CR)**
Height <1>: <u>2</u> **(CR)**

Step 2 Command: <u>VPOINT</u> **(CR)**
Enter vpoint <0,0,0>: <u>1,-1,1</u> **(CR)**
(Three-dimensional view of a solid box is obtained.)

24.3
Primitives: cone (SOLCONE)

The SOLCONE command is used to construct solid cones that have either circular or elliptical bases. Figure 24.2 illustrates how the elliptical option is used to locate the center, axis endpoints, and the height of the cone.

In Fig. 24.3, a cone with a circular base is found by specifying the center and radius or center and diameter of the base. The height is given at the keyboard or on the screen with the cursor.

24.4
Primitives: cylinder (SOLCYL)

A solid cylinder can be found with the command SOLCYL in much the same manner as the solid cone was found. Figure 24.4 illustrates the creation of a cylinder with a circular base. A cylinder with an elliptical base can also be drawn, as illustrated with the SOL-CONE command.

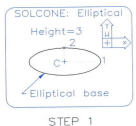 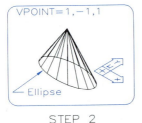

STEP 1 — STEP 2

FIGURE 24.2 **Cone with elliptical base** (SOLCONE).

Step 1 Command: SOLCONE (CR)
Elliptical/<Center point>: E (CR)
<Axis endpoint 1>/Center of base: C (CR)
Center of ellipse: (Select PC.)
Axis endpoint: (Select P1.)
Other axis distance: (Select P2.)
Height of cone: 3 (CR)

Step 2 Command: VPOINT (CR)
Enter vpoint <0,0,0>: 1,-1,1 (CR)
(**Three-dimensional view of a cone with an elliptical base is found.**)

24.5
Primitives: sphere (SOLSPHERE)

A sphere can be created with the SOLSPHERE command, as shown in Fig. 24.5. You will be prompted for the center and radius or center and diameter of the ball. The sphere is positioned so that a line connecting its north and south poles coincides with the Z-axis of the current UCS.

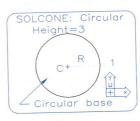 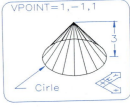

STEP 1 — STEP 2

FIGURE 24.3 **Cone with circular base** (SOLCONE).

Step 1 Command: SOLCONE (CR)
Elliptical/<Center point>: (Select PC). (CR)
Diameter/<Radius>: (Select P1.) (CR)
Height of cone: 3 (CR)

Step 2 Command: VPOINT (CR)
Enter vpoint <0,0,0>: 1,-1,1 (CR)
(**Three-dimensional view of a cone with a circular base is found.**)

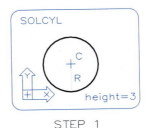 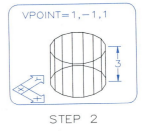

STEP 1 — STEP 2

FIGURE 24.4 **Cylinder** (SOLCYL).

Step 1 Command: SOLCYL (CR)
Elliptical/<Center point>: (Select PC.) (CR)
Height of cylinder: 3 (CR)

Step 2 Command: VPOINT (CR)
Enter vpoint <0,0,0> 1,-1,1 (CR)
(**Three-dimensional view of a solid cylinder is obtained.**)

24.6
Primitives: torus (SOLTORUS)

The torus (toroid) is a donut shape that can be created, as shown in Fig. 24.6. You will be prompted for the center of the torus, diameter or radius of the torus, and the diameter or radius of the tube. The diameter of the torus will lie in the plane of the current UCS.

24.7
Primitives: wedge (SOLWEDGE)

The wedge drawn by the SOLWEDGE command gives a rectangular base that lies in the plane of the current UCS with the upper plane of the wedge sloping downward. As shown in Fig. 24.7, you will be prompted for the length, width, and height of the wedge. Each of these dimensions can be typed in at the keyboard or located on the screen with the cursor.

24.8
Extrusions (SOLEXT)

Polylines, polygons, circles, ellipses, and three-dimensional poly entities that were drawn as two-dimensional figures can have a height added (extruded). When using AME, the sides can be tapered inward as they are extruded from the base. Polylines with crossing or intersecting segments cannot be extruded.

In Fig. 24.8, a rectangle that was drawn as a polyline is extruded upward to an assigned height.

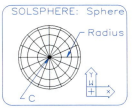
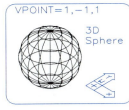

STEP 1   STEP 2

FIGURE 24.5   **Sphere** (SOLSPHERE).

Step 1 Command: SOLSPHERE **(CR)**
Center of sphere: **(Select** *PC***.)**
Diameter/<Radius> of sphere: 4 **(CR)**

Step 2 Command: VPOINT **(CR)**
Enter vpoint <0,0,0>: 1,-1,1 **(CR)**
(Three-dimensional view of a solid sphere is obtained.)

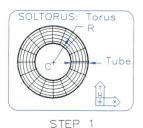
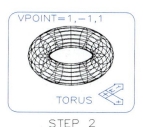

STEP 1   STEP 2

FIGURE 24.6   **Torus** (SOLTORUS).

Step 1 Command: SOLTORUS **(CR)**
Center of torus: **(Select** *PC***.)**
Diameter/<Radius> of torus: **(Select.) (CR)**
Diameter/<Radius> of tube: **(Select.) (CR)**

Step 2 Command: VPOINT **(CR)**
Enter vpoint <0,0,0>: 1,-1,1 **(CR)**
(Three-dimensional view of a solid torus is obtained.)

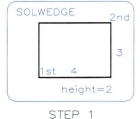
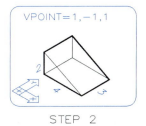

STEP 1   STEP 2

FIGURE 24.7   **Wedge** (SOLWEDGE)

Step 1 Command: SOLWEDGE **(CR)**
Corner of wedge: **(Select** *P1***.)**
Length/<Other corner>: **(Select** *P2***.) (CR)**
Height: 2 **(CR)**

Step 2 Command: VPOINT **(CR)**
Enter vpoint <0,0,0>: 1,-1,1 **(CR)**
(Three-dimensional view of a solid wedge is obtained.)

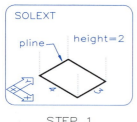

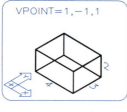

STEP 1   STEP 2

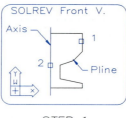

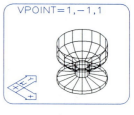

STEP 1   STEP 2

FIGURE 24.8 Extrusions (SOLEXT).

Step 1 Command: <u>SOLEXT</u> (**CR**)
Select polylines and circles for extrusion.
Select objects: (**Select rectangle.**)
Select objects: (**CR**)
Height of extrusion: <u>2</u> (**CR**)
Extrusion taper angle from Z ⟨0⟩: (**CR**)

Step 2 Command: <u>VPOINT</u> (**CR**)
Enter vpoint ⟨0,0,0⟩: <u>1,-1,1</u> (**CR**)
(**Three-dimensional view of a solid extrusion is obtained.**)

FIGURE 24.9 Solid Revolution (SOLREV).

Step 1 Command: <u>SOLREV</u> (**CR**)
Select polyline or circle for revolution.
Select objects: (**Select pline.**)
Select axis: (**Select axis.**) (**CR**)
Axis of revolution-Entity/X/Y/⟨Start point of axis⟩: <u>E</u> (**CR**)
Included angle ⟨full circle⟩: (**CR**)

Step 2 Command: <u>VPOINT</u> (**CR**)
Enter vpoint ⟨0,0,0⟩: <u>1,-1,1</u> (**CR**)
(**Three-dimensional view of a solid revolved solid is obtained.**)

24.9
Solid revolution (SOLREV)

A polyline can be revolved (swept) about an axis if the polyline has at least 3 vertices and less than 300. It is better to not revolve plines that have been Fit or Splined because of the extensive calculations required. Polylines, polygons, circles, ellipses, and three-dimensional poly objects can be revolved about an axis.

In Fig. 24.9, a polyline is revolved about an axis to make a full 360-degree revolution about the axis. The path of revolution can start and end at any point between 0 and 360 degrees.

24.10
Solidify command (SOLIDIFY)

Objects that were originally drawn as three-dimensional wire diagrams with nonzero thickness (not as solids) can be converted to three-dimensional solids by the SOLIDIFY command, as illustrated in Fig. 24.10. Polylines, polygons, circles, ellipses, traces, donuts, and AutoCAD two-dimensional solids can be solidified.

24.11
Subtracting solids (SOLSUB)

When solids intersect other solids, the solids can be subtracted from each other. In Fig. 24.11, a solid box is subtracted from a box with the SOLSUB command to give a notch in the end of the box.

24.12
Adding solids (SOLUNION)

Solids that overlap can be joined in union to form a single composite solid using the SOLUNION command. Figure 24.12 shows how a box and cylinder are joined into one solid by selecting each in succession.

24.13
Separating solids (SOLSEP)

The SOLSEP command is used to separate previously joined solids. It performs the operations of SOLSUB and SOLUNION in reverse.

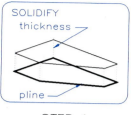

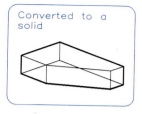

STEP 1   STEP 2

FIGURE 24.10   **Solidify a wireframe** (SOLIDIFY).

Step 1 Command: <u>SOLIDIFY</u> (CR)
Select objects: **(Select base.)**
Select objects: **(CR)**

Step 2 Command: <u>VPOINT</u> (CR)
Enter vpoint <0,0,0>: <u>1,-1,1</u> (CR)
(Three-dimensional view of a solidified wireframe is obtained.)

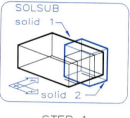

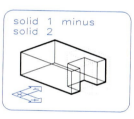

STEP 1   STEP 2

FIGURE 24.11   **Subtracting solids** (SOLSUB).

Step 1 Command: <u>SOLSUB</u> (CR)
Source objects ...
Select objects: **(Select solid 1.)**
Select objects: **(CR)**
Objects to subtract from them ...
Select objects: **(Select solid 2.)**
Select objects: **(CR)**

Step 2 Command: <u>VPOINT</u> (CR)
Enter vpoint <<u>0,0,0</u>>: <u>1,-1,1</u> (CR)
(Three-dimensional view of a solid after a subtraction.)

24.14
Chamfer (SOLCHAM)

A solid object can be chamfered by using the SOL-CHAM command and selecting the base surface, the adjoining surface, and the edges to be chamfered (Fig. 24.13). You will be prompted for the chamfer dis-

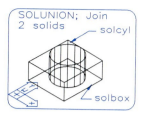

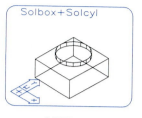

STEP 1   STEP 2

FIGURE 24.12   **Joining of solids** (SOL-UNION).

Step 1 Command: <u>SOLUNION</u> (CR)
Select objects: **(Select box.) (CR)**
Select objects: **(Select cylinder.) (CR)**
Select objects: **(CR)**

Step 2 Command: <u>VPOINT</u> (CR)
Enter vpoint <0,0,0>: <u>1,-1,1</u> (CR)
(Three-dimensional view of a solid composed of joined solids is obtained.)

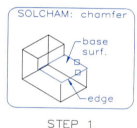

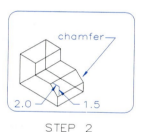

STEP 1   STEP 2

FIGURE 24.13   **Solid chamfer** (SOLCHAM).

Step 1 Command: <u>SOLCHAM</u> (CR)
Select base surface: **(Select.)**
<OK>/ Next: **(CR)**
Select edges to be chamfered (Press ENTER when done):
(Select.) (CR)
1 edge selected
Enter distance along first surface
<default>: <u>2</u>
Enter distance along second surface
<default>: <u>1.5</u>

Step 2 Command: <u>VPOINT</u> (CR)
Enter vpoint <0,0,0>: <u>1,-1,1</u> (CR)
(Three-dimensional view of a solid chamfered object is obtained.)

tances along the first surface and along the second surface. When these prompts have been satisfied, the chamfers are automatically drawn.

24.15
Fillet (SOLFILL)

The SOLFILL command lets you select the edges that are to be filleted and assign a diameter or radius of the fillet when prompted, as shown in Fig. 24.14. The SOLFILL command can be used to round the ends of cylindrical objects. It can be used for both fillets and rounds.

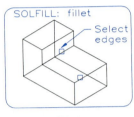
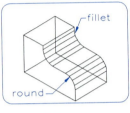

STEP 1   **STEP 2**

FIGURE 24.14 **Solid fillet** (SOLFILL).

Step 1 Command: <u>SOLFILL</u> (CR)
Select edges to be filleted (Press
ENTER when done): **(Select 3 edges.)** (CR)
3 edges selected
Diameter/<Radius> of fillet
<default>: <u>3</u>

Step 2 Command: <u>VPOINT</u> (CR)
Enter vpoint <0,0,0>: <u>1,-1,1</u> (CR)
(Three-dimensional view of a solid filleted object is obtained.)

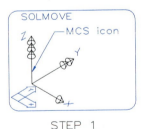
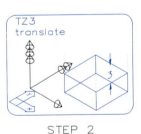

STEP 1   **STEP 2**

FIGURE 24.15 **Solid move** (SOLMOVE).

Step 1 Command: <u>SOLMOVE</u> (CR)
Select objects: **(Select box.)** (CR)
Select objects: (CR)
Redefining block SOLAXES
<Motion description>/?: <u>TZ3</u>
<Motion description>?: (CR)

Step 2 **The solid box is translated three units along the Z-axis.)**

24.16
Change solid (SOLCHP)

The SOLCHP command can be used to change the properties of a solid primitive, even if it is part of a composite solid. Changes can be made in its physical dimensions and colors. Since this is a rather involved command, it is not covered here.

24.17
Solid move (SOLMOVE)

The SOLMOVE command is used to rotate and move solids. After the solids have been selected, an icon will appear that identifies a temporary motion coordinate system (MCS) with an icon (Fig. 24.15) with one, two, and three arrows at each end. The single arrow defines the X-axis, two arrows define the Y-axis, and three arrows define the Z-axis.

When prompted <Motion description>/?:, type "?" and you will receive 15 options for moving or rotating the selected solid. The meaning of some of these options are self-explanatory, but others will require reference to HELP screens or an AutoCAD manual. Space does not permit a coverage of these prompts.

In Fig. 24.15, the solid box is translated (moved) along the Z-axis a distance of three units. The SOLMOVE option is TZ, written in the form TZ3, which means that the translation is parallel to the Z-axis.

24.18
Sections (SOLHPAT and SOLSECT)

A section can be passed through a three-dimensional solid to help with the visualization of the object. A related command, SOLHPAT (solid hatch pattern) should be set first to assign the type of pattern desired. In Fig. 24.16, SOLHPAT has been set to ANSI31 for a cast iron symbol.

The UCS icon is placed on the object to establish the plane of the section. Once set, the SOLSECT command is typed, the section automatically appears through the object, and cast iron symbols are used to hatch it. By moving the icon, other sections through the object can be found in the same manner.

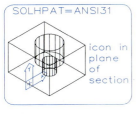
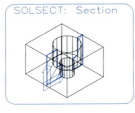

STEP 1   STEP 2

FIGURE 24.16 **Section** (SOLSECT).

Step 1 Command: SOLHPAT (CR)
Hatch pattern <default>: ANSI31
(Position UCS icon in plane of desired section.)

Step 2 Command: SOLSECT (CR)
Select objects: **(Select box.)**
Select objects: **(CR)**
(Section is drawn through the object in the plane of the icon.)

24.19
Solid inquiry commands

Inquiry commands are used to obtain information in text form about solids that have been drawn on the screen.

The SOLLIST command is used to obtain mathematical data defining edges, faces, solids, and trees. The information is sufficiently complete to provide a thorough definition of the solids.

The SOLMASSP command calculates, in text form, the mass properties of the solids that you select. You can obtain the mass (weight), volume, centroid, moments of inertia, products of inertia, radii of gyration, principal moments (in X-, Y-, and Z-directions), and the dimensions of the bounding box of the solid selected.

The SOLAREA command is used to compute the area of the selected surface on a solid.

24.20
Solid representations

Two commands, SOLMESH and SOLWIRE, can be used to represent a solid as a mesh or as a wire-frame drawing. A Set Var, SOLWDENS (solid wire density), can be set to control the mesh density.

The SOLMESH command meshes the surface of the solid and hides its wire-frame drawing. Curved surfaces are converted into straight edges.

The SOLWIRE command displays the solid as a wire-frame drawing, which is the opposite of the result of SOLMESH.

Appendix
Contents

1 Abbreviations (ANSI Z 32.13) A-3

2 Conversion Tables: Length Conversions A-4

3 Logarithms of Numbers A-5

4 Values of Trigonometric Functions A-7

5 Weights and Measures A-12

6 Decimal Equivalents and Temperature Conversion A-13

7 Weights and Specific Gravities A-14

8 Wire and Sheet Metal Gages A-16

9 American Standard Taper Pipe Threads, NPT A-17

10 American Standard 250-lb Cast Iron Flanged Fittings A-18

11 American Standard 125-lb Cast Iron Flanged Fittings A-19

12 American Standard 125-lb Cast Iron Flanges A-21

13 American National Standard 125-lb Cast Iron Screwed Fittings A-22

14 American National Standard Unified Inch Screw Threads (UN and UNR Thread Form) A-23

15 Tap Drill Sizes for American National and Unified Coarse and Fine Threads A-25

16 Length of Thread Engagement Groups A-27

17 ISO Metric Screw Thread Standard Series A-28

18 Square and Acme Threads A-31

19 American Standard Square Bolts and Nuts A-32

20 American Standard Hexagon Head Bolts and Nuts A-33

21 Fillister Head and Round Head Cap Screws A-34

22 Flat Head Cap Screws A-35

23 Machine Screws A-36

24 American Standard Machine Screws A-37

25 American Standard Machine Tapers A-37

26 American National Standard Square Head Set Screws (ANSI B18.6.2) A-38

27 American National Standard Points for Square Head Set Screw (ANSI B18.6.2) A-39

28 American National Standard Slotted Headless Set Screws (ANSI B18.6.2) A-40

29 Twist Drill Sizes A-41

30 Straight Pins A-42

31 Standard Keys and Keyways A-43

32   Woodruff Keys   A-44

33   Woodruff Keyseats   A-45

34   Taper Pins   A-46

35   Plain Washers   A-47

36   Lock Washers (ANSI B27.1)   A-49

37   Cotter Pins   A-50

38   American Standard Running and Sliding Fits   A-51

39   American Standard Clearance Locational Fits   A-53

40   American Standard Transition Locational Fits   A-55

41   American Standard Interference Locational Fits   A-56

42   American Standard Force and Shrink Fits   A-57

43   The International Tolerance Grades (ANSI B4.2)   A-58

44   Preferred Hole Basis Clearance Fits—Cylindrical Fits (ANSI B4.2)   A-59

45   Preferred Hole Basis Transition and Interference Fits—Cylindrical Fits (ANSI B4.2)   A-61

46   Preferred Shaft Basis Clearance Fits—Cylindrical Fits (ANSI B4.2)   A-63

47   Preferred Shaft Basis Transition and Interference Fits—Cylindrical Fits   A-65

48   Hole Sizes for Non-Preferred Diameters   A-67

49   Shaft Sizes for Non-Preferred Diameters   A-69

50   Grading Graph   A-71

51   Lisp Programs   A-72

APPENDIX 1

ABBREVIATIONS (ANSI Z 32.13)

Word	Abbreviation	Word	Abbreviation	Word	Abbreviation
Allowance	ALLOW	Elevation	ELEV	Not to scale	NTS
Alloy	ALY	Equal	EQ	Number	NO.
Aluminum	AL	Estimate	EST	Octagon	OCT
Amount	AMT	Exterior	EXT	On center	OC
Anneal	ANL	Fahrenheit	F	Ounce	OZ
Approximate	APPROX	Feet	(') FT	Outside diameter	OD
Area	A	Feet per minute	FPM	Parallel	PAR.
Assembly	ASSY	Feet per second	FPS	Perpendicular	PERP
Auxiliary	AUX	Fillet	FIL	Piece	PC
Average	AVG	Fillister	FIL	Plastic	PLSTC
Babbitt	BAB	Finish	FIN.	Plate	PL
Between	BET.	Finish all over	FAO	Point	PT
Between centers	BC	Flat head	FH	Polish	POL
Bevel	BEV	Foot	(') FT	Pound	LB
Bill of material	B/M	Front	FR	Pounds per square inch	PSI
Both sides	BS	Gage	GA	Pressure	PRESS.
Bottom	BOT	Gallon	GAL	Production	PROD
Brass	BRS	Galvanize	GALV	Quarter	QTR
Brazing	BRZG	Galvanized iron	GI	Radius	R
Broach	BRO	General	GEN	Ream	RM
Bronze	BRZ	Gram	G	Rectangle	RECT
Cadmium plate	CD PL	Grind	GRD	Reference	REF
Cap screw	CAP SCR	Groove	GRV	Required	REQD
Case harden	CH	Hardware	HDW	Revise	REV
Cast iron	CI	Head	HD	Revolution	REV
Cast steel	CS	Heat treat	HT TR	Revolutions per minute	RPM
Casting	CSTG	Hexagon	HEX	Right	R
Center	CTR	Horizontal	HOR	Right hand	RH
Centerline	CL	Horsepower	HP	Rough	RGH
Center to center	C to C	Hot rolled steel	HRS	Screw	SCR
Centigrade	C	Hour	HR	Section	SECT
Centigram	CG	Hundredweight	CWT	Set screw	SS
Centimeter	cm	Inch	(") IN.	Shaft	SFT
Chamfer	CHAM	Inches per second	IPS	Slotted	SLOT.
Circle	CIR	Inside diameter	ID	Socket	SOC
Clockwise	CW	Interior	INT	Spherical	SPHER
Cold rolled steel	CRS	Iron	I	Spot faced	SF
Cotter	COT	Key	K	Spring	SPG
Counterclockwise	CCW	Kip (1000 lb)	K	Square	SQ
Counterbore	CBORE	Left	L	Station	STA
Counterdrill	CDRILL	Left hand	LH	Steel	STL
Counterpunch	CPUNCH	Length	LG	Symmetrical	SYM
Countersink	CSK	Light	LT	Taper	TPR
Cubic centimeter	cc	Machine	MACH	Temperature	TEMP
Cubic feet per minute	CFM	Malleable	MALL	Tension	TENS.
Cubic foot	CU FT	Manhole	MH	Thick	THK
Cubic inch	CU IN.	Manufacture	MFR	Thousand	M
Cylinder	CYL	Material	MATL	Thousand pound	KIP
Diagonal	DIAG	Maximum	MAX	Thread	THD
Diameter	DIA	Metal	MET.	Tolerance	TOL
Distance	DIST	Meter (Instrument or		Typical	TYP
Ditto	DO	measure of length)	M	Vertical	VERT
Down	DN	Miles	MI	Volume	VOL
Dozen	DOZ	Miles per gallon	MPG	Washer	WASH.
Drafting	DFTG	Miles per hour	MPH	Weight	WT
Drawing	DWG	Millimeter	MM	Width	W
Drill	DR	Minimum	MIN	Wrought iron	WI
Each	EA	Normal	NOR	Yard	YD

APPENDIX 2
CONVERSION TABLES: LENGTH CONVERSIONS

Feet	$\times$ 12	= inches
	$\times$ 0.3048	= meters
Inches	$\times$ 2.54 $\times$ 10^8	= Angstroms
	$\times$ 25.4	= millimeters
	$\times$ 8.333 33 $\times$ 10^{-2}	= feet
Kilometers	$\times$ 3.280 839 $\times$ 10^3	= feet
	$\times$ 0.62	= miles
	$\times$ 0.539 956	= nautical miles
	$\times$ 0.621 371	= statute miles
	$\times$ 1.093 613 $\times$ 10^3	= yards
Meters	$\times$ 1 $\times$ 10^{10}	= Angstroms
	$\times$ 3.280 839 9	= feet
	$\times$ 39.370 079	= inches
	$\times$ 1.093 61	= yards
Statute miles	$\times$ 5.280 $\times$ 10^3	= feet
	$\times$ 8	= furlongs
	$\times$ 6.336 0 $\times$ 10^4	= inches
	$\times$ 1.609 34	= kilometers
	$\times$ 8.689 7 $\times$ 10^{-1}	= nautical miles
Miles	$\times$ 10^{-3}	= inches
	$\times$ 2.54 $\times$ 10^{-2}	= millimeters
	$\times$ 25.4	= micrometers
	$\times$ 0.61	= kilometers
Yards	$\times$ 3	= feet
	$\times$ 9.144 $\times$ 10^{-1}	= meters
Feet/hour	$\times$ 3.048 $\times$ 10^{-4}	= kilometers/hour
	$\times$ 1.645 788 $\times$ 10^{-4}	= knots
Feet/minute	$\times$ 0.3048	= meters/minute
	$\times$ 5.08 $\times$ 10^{-3}	= meters/second
Feet/second	$\times$ 1.097 28	= kilometers/hour
	$\times$ 18.288	= meters/minute
Kilometers/hour	$\times$ 3.280 839 $\times$ 10^3	= feet/hour
	$\times$ 54.680 66	= feet/minute
	$\times$ 0.277 777	= meters/second
	$\times$ 0.621 371	= miles/hour
Kilometers/minute	$\times$ 3.280 839 $\times$ 10^3	= feet/minute
	$\times$ 37.282 27	= miles/hour
Meters/hour	$\times$ 3.280 839	= feet/hour
	$\times$ 88	= feet/minute
	$\times$ 1.466	= feet/second
	$\times$ 1 $\times$ 10^{-3}	= kilometers/hour
	$\times$ 1.667 $\times$ 10^{-2}	= meters/minute
Feet/second^2	$\times$ 1.097 28	= kilometers/hour/second
	$\times$ 0.304 8	= meters/second^2

APPENDIX 3
LOGARITHMS OF NUMBERS

N	0	1	2	3	4	5	6	7	8	9
1.0	.0000	.0043	.0086	.0128	.0170	.0212	.0253	.0294	.0334	.0374
1.1	.0414	.0453	.0492	.0531	.0569	.0607	.0645	.0682	.0719	.0755
1.2	.0792	.0828	.0864	.0899	.0934	.0969	.1004	.1038	.1072	.1106
1.3	.1139	.1173	.1206	.1239	.1271	.1303	.1335	.1367	.1399	.1430
1.4	.1461	.1492	.1523	.1553	.1584	.1614	.1644	.1673	.1703	.1732
1.5	.1761	.1790	.1818	.1847	.1875	.1903	.1931	.1959	.1987	.2014
1.6	.2041	.2068	.2095	.2122	.2148	.2175	.2201	.2227	.2253	.2279
1.7	.2304	.2330	.2355	.2380	.2405	.2430	.2455	.2480	.2504	.2529
1.8	.2553	.2577	.2601	.2625	.2648	.2672	.2695	.2718	.2742	.2765
1.9	.2788	.2810	.2833	.2856	.2878	.2900	.2923	.2945	.2967	.2989
2.0	.3010	.3032	.3054	.3075	.3096	.3118	.3139	.3160	.3181	.3201
2.1	.3222	.3243	.3263	.3284	.3304	.3324	.3345	.3365	.3385	.3404
2.2	.3424	.3444	.3464	.3483	.3502	.3522	.3541	.3560	.3579	.3598
2.3	.3617	.3636	.3655	.3674	.3692	.3711	.3729	.3747	.3766	.3784
2.4	.3802	.3820	.3838	.3856	.3874	.3892	.3909	.3927	.3945	.3962
2.5	.3979	.3997	.4014	.4031	.4048	.4065	.4082	.4099	.4116	.4133
2.6	.4150	.4166	.4183	.4200	.4216	.4232	.4249	.4265	.4281	.4298
2.7	.4314	.4330	.4346	.4362	.4378	.4393	.4409	.4425	.4440	.4456
2.8	.4472	.4487	.4502	.4518	.4533	.4548	.4564	.4579	.4594	.4609
2.9	.4624	.4639	.4654	.4669	.4683	.4698	.4713	.4728	.4742	.4757
3.0	.4771	.4786	.4800	.4814	.4829	.4843	.4857	.4871	.4886	.4900
3.1	.4914	.4928	.4942	.4955	.4969	.4983	.4997	.5011	.5024	.5038
3.2	.5051	.5065	.5079	.5092	.5105	.5119	.5132	.5145	.5159	.5172
3.3	.5185	.5198	.5211	.5224	.5237	.5250	.5263	.5276	.5289	.5302
3.4	.5315	.5328	.5340	.5353	.5366	.5378	.5391	.5403	.5416	.5428
3.5	.5441	.5453	.5465	.5478	.5490	.5502	.5514	.5527	.5539	.5551
3.6	.5563	.5575	.5587	.5599	.5611	.5623	.5635	.5647	.5658	.5670
3.7	.5682	.5694	.5705	.5717	.5729	.5740	.5752	.5763	.5775	.5786
3.8	.5798	.5809	.5821	.5832	.5843	.5855	.5866	.5877	.5888	.5899
3.9	.5911	.5922	.5933	.5944	.5955	.5966	.5977	.5988	.5999	.6010
4.0	.6021	.6031	.6042	.6053	.6064	.6075	.6085	.6096	.6107	.6117
4.1	.6128	.6138	.6149	.6160	.6170	.6180	.6191	.6201	.6212	.6222
4.2	.6232	.6243	.6253	.6263	.6274	.6284	.6294	.6304	.6314	.6325
4.3	.6335	.6345	.6355	.6365	.6375	.6385	.6395	.6405	.6415	.6425
4.4	.6435	.6444	.6454	.6464	.6474	.6484	.6493	.6503	.6513	.6522
4.5	.6532	.6542	.6551	.6561	.6571	.6580	.6590	.6599	.6609	.6618
4.6	.6628	.6637	.6646	.6656	.6665	.6675	.6684	.6693	.6702	.6712
4.7	.6721	.6730	.6739	.6749	.6758	.6767	.6776	.6785	.6794	.6803
4.8	.6812	.6821	.6830	.6839	.6848	.6857	.6866	.6875	.6884	.6893
4.9	.6902	.6911	.6920	.6928	.6937	.6946	.6955	.6964	.6972	.6981
5.0	.6990	.6998	.7007	.7016	.7024	.7033	.7042	.7050	.7059	.7067
5.1	.7076	.7084	.7093	.7101	.7110	.7118	.7126	.7135	.7143	.7152
5.2	.7160	.7168	.7177	.7185	.7193	.7202	.7210	.7218	.7226	.7235
5.3	.7243	.7251	.7259	.7267	.7275	.7284	.7292	.7300	.7308	7316
5.4	.7324	.7332	.7340	.7348	.7356	.7364	.7372	.7380	.7388	.7396
N	0	1	2	3	4	5	6	7	8	9

APPENDIX 3
LOGARITHMS OF NUMBERS (Cont.)

N	0	1	2	3	4	5	6	7	8	9
5.5	.7404	.7412	.7419	.7427	.7435	.7443	.7451	.7459	.7466	.7474
5.6	.7482	.7490	.7497	.7505	.7513	.7520	.7528	.7536	.7543	.7551
5.7	.7559	.7566	.7574	.7582	.7589	.7597	.7604	.7612	.7619	.7627
5.8	.7634	.7642	.7649	.7657	.7664	.7672	.7679	.7686	.7694	.7701
5.9	.7709	.7716	.7723	.7731	.7738	.7745	.7752	.7760	.7767	.7774
6.0	.7782	.7789	.7796	.7803	.7810	.7818	.7825	.7832	.7839	.7846
6.1	.7853	.7860	.7868	.7875	.7882	.7889	.7896	.7903	.7910	.7917
6.2	.7924	.7931	.7938	.7945	.7952	.7959	.7966	.7973	.7980	.7987
6.3	.7993	.8000	.8007	.8014	.8021	.8028	.8035	.8041	.8048	.8055
6.4	.8062	.8069	.8075	.8082	.8089	.8096	.8102	.8109	.8116	.8122
6.5	.8129	.8136	.8142	.8149	.8156	.8162	.8169	.8176	.8182	.8189
6.6	.8195	.8202	.8209	.8215	.8222	.8228	.8235	.8241	.8248	.8254
6.7	.8261	.8267	.8274	.8280	.8287	.8293	.8299	.8306	.8312	.8319
6.8	.8325	.8331	.8338	.8344	.8351	.8357	.8363	.8370	.8376	.8382
6.9	.8388	.8395	.8401	.8407	.8414	.8420	.8426	.8432	.8439	.8445
7.0	.8451	.8457	.8463	.8470	.8476	.8482	.8488	.8494	.8500	.8506
7.1	.8513	.8519	.8525	.8531	.8537	.8543	.8549	.8555	.8561	.8567
7.2	.8573	.8579	.8585	.8591	.8597	.8603	.8609	.8615	.8621	.8627
7.3	.8633	.8639	.8645	.8651	.8657	.8663	.8669	.8675	.8681	.8686
7.4	.8692	.8698	.8704	.8710	.8716	.8722	.8727	.8733	.8739	.8745
7.5	.8751	.8756	.8762	.8768	.8774	.8779	.8785	.8791	.8797	.8802
7.6	.8808	.8814	.8820	.8825	.8831	.8837	.8842	.8848	.8854	.8859
7.7	.8865	.8871	.8876	.8882	.8887	.8893	.8899	.8904	.8910	.8915
7.8	.8921	.8927	.8932	.8938	.8943	.8949	.8954	.8960	.8965	.8971
7.9	.8976	.8982	.8987	.8993	.8998	.9004	.9009	.9015	.9020	.9025
8.0	.9031	.9036	.9042	.9047	.9053	.9058	.9063	.9069	.9074	.9079
8.1	.9085	.9090	.9096	.9101	.9106	.9112	.9117	.9122	.9128	.9133
8.2	.9138	.9143	.9149	.9154	.9159	.9165	.9170	.9175	.9180	.9186
8.3	.9191	.9196	.9201	.9206	.9212	.9217	.9222	.9227	.9232	.9238
8.4	.9243	.9248	.9253	.9258	.9263	.9269	.9274	.9279	.9284	.9289
8.5	.9294	.9299	.9304	.9309	.9315	.9320	.9325	.9330	.9335	.9340
8.6	.9345	.9350	.9355	.9360	.9365	.9370	.9375	.9380	.9385	.9390
8.7	.9395	.9400	.9405	.9410	.9415	.9420	.9425	.9430	.9435	.9440
8.8	.9445	.9450	.9455	.9460	.9465	.9469	.9474	.9479	.9484	.9489
8.9	.9494	.9499	.9504	.9509	.9513	.9518	.9523	.9528	.9533	.9538
9.0	.9542	.9547	.9552	.9557	.9562	.9566	.9571	.9576	.9581	.9586
9.1	.9590	.9595	.9600	.9605	.9609	.9614	.9619	.9624	.9628	.9633
9.2	.9638	.9643	.9647	.9652	.9657	.9661	.9666	.9671	.9675	.9680
9.3	.9685	.9689	.9694	.9699	.9703	.9708	.9713	.9717	.9722	.9727
9.4	.9731	.9736	.9741	.9745	.9750	.9754	.9759	.9763	.9768	.9773
9.5	.9777	.9782	.9786	.9791	.9795	.9800	.9805	.9809	.9814	.9818
9.6	.9823	.9827	.9832	.9836	.9841	.9845	.9850	.9854	.9859	.9863
9.7	.9868	.9872	.9877	.9881	.9886	.9890	.9894	.9899	.9903	.9908
9.8	.9912	.9917	.9921	.9926	.9930	.9934	.9939	.9943	.9948	.9952
9.9	.9956	.9961	.9965	.9969	.9974	.9978	.9983	.9987	.9991	.9996
N	0	1	2	3	4	5	6	7	8	9

APPENDIX 4
VALUES OF TRIGONOMETRIC FUNCTIONS

Degrees	Radians	Sine	Tangent	Cotangent	Cosine		
0° 00′	.0000	.0000	.0000		1.0000	1.5708	90° 00′
10′	.0029	.0029	.0029	343.77	1.0000	1.5679	50′
20′	.0058	.0058	.0058	171.89	1.0000	1.5650	40′
30′	.0087	.0087	.0087	114.59	1.0000	1.5621	30′
40′	.0116	.0116	.0116	85.940	.9999	1.5592	20′
50′	.0145	.0145	.0145	68.750	.9999	1.5563	10′
1° 00′	.0175	.0175	.0175	57.290	.9998	1.5533	89° 00′
10′	.0204	.0204	.0204	49.104	.9998	1.5504	50′
20′	.0233	.0233	.0233	42.964	.9997	1.5475	40′
30′	.0262	.0262	.0262	38.188	.9997	1.5446	30′
40′	.0291	.0291	.0291	34.368	.9996	1.5417	20′
50′	.0320	.0320	.0320	31.242	.9995	1.5388	10′
2° 00′	.0349	.0349	.0349	28.636	.9994	1.5359	88° 00′
10′	.0378	.0378	.0378	26.432	.9993	1.5330	50′
20′	.0407	.0407	.0407	24.542	.9992	1.5301	40′
30′	.0436	.0436	.0437	22.904	.9990	1.5272	30′
40′	.0465	.0465	.0466	21.470	.9989	1.5243	20′
50′	.0495	.0494	.0495	20.206	.9988	1.5213	10′
3° 00′	.0524	.0523	.0524	19.081	.9986	1.5184	87° 00′
10′	.0553	.0552	.0553	18.075	.9985	1.5155	50′
20′	.0582	.0581	.0582	17.169	.9983	1.5126	40′
30′	.0611	.0610	.0612	16.350	.9981	1.5097	30′
40′	.0640	.0640	.0641	15.605	.9980	1.5068	20′
50′	.0669	.0669	.0670	14.924	.9978	1.5039	10′
4° 00′	.0698	.0698	.0699	14.301	.9976	1.5010	86° 00′
10′	.0727	.0727	.0729	13.727	.9974	1.4981	50′
20′	.0756	.0756	.0758	13.197	.9971	1.4952	40′
30′	.0785	.0785	.0787	12.706	.9969	1.4923	30′
40′	.0814	.0814	.0816	12.251	.9967	1.4893	20′
50′	.0844	.0843	.0846	11.826	.9964	1.4864	10′
5° 00′	.0873	.0872	.0875	11.430	.9962	1.4835	85° 00′
10′	.0902	.0901	.0904	11.059	.9959	1.4806	50′
20′	.0931	.0929	.0934	10.712	.9957	1.4777	40′
30′	.0960	.0958	.0963	10.385	.9954	1.4748	30′
40′	.0989	.0987	.0992	10.078	.9951	1.4719	20′
50′	.1018	.1016	.1022	9.7882	.9948	1.4690	10′
6° 00′	.1047	.1045	.1051	9.5144	.9945	1.4661	84° 00′
10′	.1076	.1074	.1080	9.2553	.9942	1.4632	50′
20′	.1105	.1103	.1110	9.0098	.9939	1.4603	40′
30′	.1134	.1132	.1139	8.7769	.9936	1.4573	30′
40′	.1164	.1161	.1169	8.5555	.9932	1.4544	20′
50′	.1193	.1190	.1198	8.3450	.9929	1.4515	10′
7° 00′	.1222	.1219	.1228	8.1443	.9925	1.4486	83° 00′
10′	.1251	.1248	.1257	7.9530	.9922	1.4457	50′
20′	.1280	.1276	.1287	7.7704	.9918	1.4428	40′
30′	.1309	.1305	.1317	7.5958	.9914	1.4399	30′
40′	.1338	.1334	.1346	7.4287	.9911	1.4370	20′
50′	.1367	.1363	.1376	7.2687	.9907	1.4341	10′
8° 00′	.1396	.1392	.1405	7.1154	.9903	1.4312	82° 00′
10′	.1425	.1421	.1435	6.9682	.9899	1.4283	50′
20′	.1454	.1449	.1465	6.8269	.9894	1.4254	40′
30′	.1484	.1478	.1495	6.6912	.9890	1.4224	30′
40′	.1513	.1507	.1524	6.5606	.9886	1.4195	20′
50′	.1542	.1536	.1554	6.4348	.9881	1.4166	10′
9° 00′	.1571	.1564	.1584	6.3138	.9877	1.4137	81° 00′
		Cosine	Cotangent	Tangent	Sine	Radians	Degrees

APPENDIX 4
VALUES OF TRIGONOMETRIC FUNCTIONS (Cont.)

Degrees	Radians	Sine	Tangent	Cotangent	Cosine		
9° 00′	.1571	.1564	.1584	6.3138	.9877	1.4137	81° 00′
10′	.1600	.1593	.1614	6.1970	.9872	1.4108	50′
20′	.1629	.1622	.1644	6.0844	.9868	1.4079	40′
30′	.1658	.1650	.1673	5.9758	.9863	1.4050	30′
40′	.1687	.1679	.1703	5.8708	.9858	1.4021	20′
50′	.1716	.1708	.1733	5.7694	.9853	1.3992	10′
10° 00′	.1745	.1736	.1763	5.6713	.9848	1.3963	80° 00′
10′	.1774	.1765	.1793	5.5764	.9843	1.3934	50′
20′	.1804	.1794	.1823	5.4845	.9838	1.3904	40′
30′	.1833	.1822	.1853	5.3955	.9833	1.3875	30′
40′	.1862	.1851	.1883	5.3093	.9827	1.3846	20′
50′	.1891	.1880	.1914	5.2257	.9822	1.3817	10′
11° 00′	.1920	.1908	.1944	5.1446	.9816	1.3788	79° 00′
10′	.1949	.1937	.1974	5.0658	.9811	1.3759	50′
20′	.1978	.1965	.2004	4.9894	.9805	1.3730	40′
30′	.2007	.1994	.2035	4.9152	.9799	1.3701	30′
40′	.2036	.2022	.2065	4.8430	.9793	1.3672	20′
50′	.2065	.2051	.2095	4.7729	.9787	1.3643	10′
12° 00′	.2094	.2079	.2126	4.7046	.9781	1.3614	78° 00′
10′	.2123	.2108	.2156	4.6382	.9775	1.3584	50′
20′	.2153	.2136	.2186	4.5736	.9769	1.3555	40′
30′	.2182	.2164	.2217	4.5107	.9763	1.3526	30′
40′	.2211	.2193	.2247	4.4494	.9757	1.3497	20′
50′	.2240	.2221	.2278	4.3897	.9750	1.3468	10′
13° 00′	.2269	.2250	.2309	4.3315	.9744	1.3439	77° 00′
10′	.2298	.2278	.2339	4.2747	.9737	1.3410	50′
20′	.2327	.2306	.2370	4.2193	.9730	1.3381	40′
30′	.2356	.2334	.2401	4.1653	.9724	1.3352	30′
40′	.2385	.2363	.2432	4.1126	.9717	1.3323	20′
50′	.2414	.2391	.2462	4.0611	.9710	1.3294	10′
14° 00′	.2443	.2419	.2493	4.0108	.9703	1.3265	76° 00′
10′	.2473	.2447	.2524	3.9617	.9696	1.3235	50′
20′	.2502	.2476	.2555	3.9136	.9689	1.3206	40′
30′	.2531	.2504	.2586	3.8667	.9681	1.3177	30′
40′	.2560	.2532	.2617	3.8208	.9674	1.3148	20′
50′	.2589	.2560	.2648	3.7760	.9667	1.3119	10′
15° 00′	.2618	.2588	.2679	3.7321	.9659	1.3090	75° 00′
10′	.2647	.2616	.2711	3.6891	.9652	1.3061	50′
20′	.2676	.2644	.2742	3.6470	.9644	1.3032	40′
30′	.2705	.2672	.2773	3.6059	.9636	1.3003	30′
40′	.2734	.2700	.2805	3.5656	.9628	1.2974	20′
50′	.2763	.2728	.2836	3.5261	.9621	1.2945	10′
16° 00′	.2793	.2756	.2867	3.4874	.9613	1.2915	74° 00′
10′	.2822	.2784	.2899	3.4495	.9605	1.2886	50′
20′	.2851	.2812	.2931	3.4124	.9596	1.2857	40′
30′	.2880	.2840	.2962	3.3759	.9588	1.2828	30′
40′	.2909	.2868	.2994	3.3402	.9580	1.2799	20′
50′	.2938	.2896	.3026	3.3052	.9572	1.2770	10′
17° 00′	.2967	.2924	.3057	3.2709	.9563	1.2741	73° 00′
10′	.2996	.2952	.3089	3.2371	.9555	1.2712	50′
20′	.3025	.2979	.3121	3.2041	.9546	1.2683	40′
30′	.3054	.3007	.3153	3.1716	.9537	1.2654	30′
40′	.3083	.3035	.3185	3.1397	.9528	1.2625	20′
50′	.3113	.3062	.3217	3.1084	.9520	1.2595	10′
18° 00′	.3142	.3090	.3249	3.0777	.9511	1.2566	72° 00′
		Cosine	Cotangent	Tangent	Sine	Radians	Degrees

Cont.

APPENDIX 4
VALUES OF TRIGONOMETRIC FUNCTIONS (Cont.)

Degrees	Radians	Sine	Tangent	Cotangent	Cosine		
18° 00′	.3142	.3090	.3249	3.0777	.9511	1.2566	72° 00′
10′	.3171	.3118	.3281	3.0475	.9502	1.2537	50′
20′	.3200	.3145	.3314	3.0178	.9492	1.2508	40′
30′	.3229	.3173	.3346	2.9887	.9483	1.2479	30′
40′	.3258	.3201	.3378	2.9600	.9474	1.2450	20′
50′	.3287	.3228	.3411	2.9319	.9465	1.2421	10′
19° 00′	.3316	.3256	.3443	2.9042	.9455	1.2392	71° 00′
10′	.3345	.3283	.3476	2.8770	.9446	1.2363	50′
20′	.3374	.3311	.3508	2.8502	.9436	1.2334	40′
30′	.3403	.3338	.3541	2.8239	.9426	1.2305	30′
40′	.3432	.3365	.3574	2.7980	.9417	1.2275	20′
50′	.3462	.3393	.3607	2.7725	.9407	1.2246	10′
20° 00′	.3491	.3420	.3640	2.7475	.9397	1.2217	70° 00′
10′	.3520	.3448	.3673	2.7228	.9387	1.2188	50′
20′	.3549	.3475	.3706	2.6985	.9377	1.2159	40′
30′	.3578	.3502	.3739	2.6746	.9367	1.2130	30′
40′	.3607	.3529	.3772	2.6511	.9356	1.2101	20′
50′	.3636	.3557	.3805	2.6279	.9346	1.2072	10′
21° 00′	.3665	.3584	.3839	2.6051	.9336	1.2043	69° 00′
10′	.3694	.3611	.3872	2.5826	.9325	1.2014	50′
20′	.3723	.3638	.3906	2.5605	.9315	1.1985	40′
30′	.3752	.3665	.3939	2.5386	.9304	1.1956	30′
40′	.3782	.3692	.3973	2.5172	.9293	1.1926	20′
50′	.3811	.3719	.4006	2.4960	.9283	1.1897	10′
22° 00′	.3840	.3746	.4040	2.4751	.9272	1.1868	68° 00′
10′	.3869	.3773	.4074	2.4545	.9261	1.1839	50′
20′	.3898	.3800	.4108	2.4342	.9250	1.1810	40′
30′	.3927	.3827	.4142	2.4142	.9239	1.1781	30′
40′	.3956	.3854	.4176	2.3945	.9228	1.1752	20′
50′	.3985	.3881	.4210	2.3750	.9216	1.1723	10′
23° 00′	.4014	.3907	.4245	2.3559	.9205	1.1694	67° 00′
10′	.4043	.3934	.4279	2.3369	.9194	1.1665	50′
20′	.4072	.3961	.4314	2.3183	.9182	1.1636	40′
30′	.4102	.3987	.4348	2.2998	.9171	1.1606	30′
40′	.4131	.4014	.4383	2.2817	.9159	1.1577	20′
50′	.4160	.4041	.4417	2.2637	.9147	1.1548	10′
24° 00′	.4189	.4067	.4452	2.2460	.9135	1.1519	66° 00′
10′	.4218	.4094	.4487	2.2286	.9124	1.1490	50′
20′	.4247	.4120	.4522	2.2113	.9112	1.1461	40′
30′	.4276	.4147	.4557	2.1943	.9100	1.1432	30′
40′	.4305	.4173	.4592	2.1775	.9088	1.1403	20′
50′	.4334	.4200	.4628	2.1609	.9075	1.1374	10′
25° 00′	.4363	.4226	.4663	2.1445	.9063	1.1345	65° 00′
10′	.4392	.4253	.4699	2.1283	.9051	1.1316	50′
20′	.4422	.4279	.4734	2.1123	.9038	1.1286	40′
30′	.4451	.4305	.4770	2.0965	.9026	1.1257	30′
40′	.4480	.4331	.4806	2.0809	.9013	1.1228	20′
50′	.4509	.4358	.4841	2.0655	.9001	1.1199	10′
26° 00′	.4538	.4384	.4877	2.0503	.8988	1.1170	64° 00′
10′	.4567	.4410	.4913	2.0353	.8975	1.1141	50′
20′	.4596	.4436	.4950	2.0204	.8962	1.1112	40′
30′	.4625	.4462	.4986	2.0057	.8949	1.1083	30′
40′	.4654	.4488	.5022	1.9912	.8936	1.1054	20′
50′	.4683	.4514	.5059	1.9768	.8923	1.1025	10′
27° 00′	.4712	.4540	.5095	1.9626	.8910	1.0996	63° 00′
		Cosine	Cotangent	Tangent	Sine	Radians	Degrees

APPENDIX 4
VALUES OF TRIGONOMETRIC FUNCTIONS (Cont.)

Degrees	Radians	Sine	Tangent	Cotangent	Cosine		
27° 00′	.4712	.4540	.5095	1.9626	.8910	1.0996	63° 00′
10′	.4741	.4566	.5132	1.9486	.8897	1.0966	50′
20′	.4771	.4592	.5169	1.9347	.8884	1.0937	40′
30′	.4800	.4617	.5206	1.9210	.8870	1.0908	30′
40′	.4829	.4643	.5243	1.9074	.8857	1.0879	20′
50′	.4858	.4669	.5280	1.8940	.8843	1.0850	10′
28° 00′	.4887	.4695	.5317	1.8807	.8829	1.0821	62° 00′
10′	.4916	.4720	.5354	1.8676	.8816	1.0792	50′
20′	.4945	.4746	.5392	1.8546	.8802	1.0763	40′
30′	.4974	.4772	.5430	1.8418	.8788	1.0734	30′
40′	.5003	.4797	.5467	1.8291	.8774	1.0705	20′
50′	.5032	.4823	.5505	1.8165	.8760	1.0676	10′
29° 00′	.5061	.4848	.5543	1.8040	.8746	1.0647	61° 00′
10′	.5091	.4874	.5581	1.7917	.8732	1.0617	50′
20′	.5120	.4899	.5619	1.7796	.8718	1.0588	40′
30′	.5149	.4924	.5658	1.7675	.8704	1.0559	30′
40′	.5178	.4950	.5696	1.7556	.8689	1.0530	20′
50′	.5207	.4975	.5735	1.7437	.8675	1.0501	10′
30° 00′	.5236	.5000	.5774	1.7321	.8660	1.0472	60° 00′
10′	.5265	.5025	.5812	1.7205	.8646	1.0443	50′
20′	.5294	.5050	.5851	1.7090	.8631	1.0414	40′
30′	.5323	.5075	.5890	1.6977	.8616	1.0385	30′
40′	.5352	.5100	.5930	1.6864	.8601	1.0356	20′
50′	.5381	.5125	.5969	1.6753	.8587	1.0327	10′
31° 00′	.5411	.5150	.6009	1.6643	.8572	1.0297	59° 00′
10′	.5440	.5175	.6048	1.6534	.8557	1.0268	50′
20′	.5469	.5200	.6088	1.6426	.8542	1.0239	40′
30′	.5498	.5225	.6128	1.6319	.8526	1.0210	30′
40′	.5527	.5250	.6168	1.6212	.8511	1.0181	20′
50′	.5556	.5275	.6208	1.6107	.8496	1.0152	10′
32° 00′	.5585	.5299	.6249	1.6003	.8480	1.0123	58° 00′
10′	.5614	.5324	.6289	1.5900	.8465	1.0094	50′
20′	.5643	.5348	.6330	1.5798	.8450	1.0065	40′
30′	.5672	.5373	.6371	1.5697	.8434	1.0036	30′
40′	.5701	.5398	.6412	1.5597	.8418	1.0007	20′
50′	.5730	.5422	.6453	1.5497	.8403	.9977	10′
33° 00′	.5760	.5446	.6494	1.5399	.8387	.9948	57° 00′
10′	.5789	.5471	.6536	1.5301	.8371	.9919	50′
20′	.5818	.5495	.6577	1.5204	.8355	.9890	40′
30′	.5847	.5519	.6619	1.5108	.8339	.9861	30′
40′	.5876	.5544	.6661	1.5013	.8323	.9832	20′
50′	.5905	.5568	.6703	1.4919	.8307	.9803	10′
34° 00′	.5934	.5592	.6745	1.4826	.8290	.9774	56° 00′
10′	.5963	.5616	.6787	1.4733	.8274	.9745	50′
20′	.5992	.5640	.6830	1.4641	.8258	.9716	40′
30′	.6021	.5664	.6873	1.4550	.8241	.9687	30′
40′	.6050	.5688	.6916	1.4460	.8225	.9657	20′
50′	.6080	.5712	.6959	1.4370	.8208	.9628	10′
35° 00′	.6109	.5736	.7002	1.4281	.8192	.9599	55° 00′
10′	.6138	.5760	.7046	1.4193	.8175	.9570	50′
20′	.6167	.5783	.7089	1.4106	.8158	.9541	40′
30′	.6196	.5807	.7133	1.4019	.8141	.9512	30′
40′	.6225	.5831	.7177	1.3934	.8124	.9483	20′
50′	.6254	.5854	.7221	1.3848	.8107	.9454	10′
36° 00′	.6283	.5878	.7265	1.3764	.8090	.9425	54° 00′
		Cosine	Cotangent	Tangent	Sine	Radians	Degrees

Cont.

APPENDIX 4
VALUES OF TRIGONOMETRIC FUNCTIONS (Cont.)

Degrees	Radians	Sine	Tangent	Cotangent	Cosine		
36° 00′	.6283	.5878	.7265	1.3764	.8090	.9425	54° 00′
10′	.6312	.5901	.7310	1.3680	.8073	.9396	50′
20′	.6341	.5925	.7355	1.3597	.8056	.9367	40′
30′	.6370	.5948	.7400	1.3514	.8039	.9338	30′
40′	.6400	.5972	.7445	1.3432	.8021	.9308	20′
50′	.6429	.5995	.7490	1.3351	.8004	.9279	10′
37° 00′	.6458	.6018	.7536	1.3270	.7986	.9250	53° 00′
10′	.6487	.6041	.7581	1.3190	.7969	.9221	50′
20′	.6516	.6065	.7627	1.3111	.7951	.9192	40′
30′	.6545	.6088	.7673	1.3032	.7934	.9163	30′
40′	.6574	.6111	.7720	1.2954	.7916	.9134	20′
50′	.6603	.6134	.7766	1.2876	.7898	.9105	10′
38° 00′	.6632	.6157	.7813	1.2799	.7880	.9076	52° 00′
10′	.6661	.6180	.7860	1.2723	.7862	.9047	50′
20′	.6690	.6202	.7907	1.2647	.7844	.9018	40′
30′	.6720	.6225	.7954	1.2572	.7826	.8988	30′
40′	.6749	.6248	.8002	1.2497	.7808	.8959	20′
50′	.6778	.6271	.8050	1.2423	.7790	.8930	10′
39° 00′	.6807	.6293	.8098	1.2349	.7771	.8901	51° 00′
10′	.6836	.6316	.8146	1.2276	.7753	.8872	50′
20′	.6865	.6338	.8195	1.2203	.7735	.8843	40′
30′	.6894	.6361	.8243	1.2131	.7716	.8814	30′
40′	.6923	.6383	.8292	1.2059	.7698	.8785	20′
50′	.6952	.6406	.8342	1.1988	.7679	.8756	10′
40° 00′	.6981	.6428	.8391	1.1918	.7660	.8727	50° 00′
10′	.7010	.6450	.8441	1.1847	.7642	.8698	50′
20′	.7039	.6472	.8491	1.1778	.7623	.8668	40′
30′	.7069	.6494	.8541	1.1708	.7604	.8639	30′
40′	.7098	.6517	.8591	1.1640	.7585	.8610	20′
50′	.7127	.6539	.8642	1.1571	.7566	.8581	10′
41° 00′	.7156	.6561	.8693	1.1504	.7547	.8552	49° 00′
10′	.7185	.6583	.8744	1.1436	.7528	.8523	50′
20′	.7214	.6604	.8796	1.1369	.7509	.8494	40′
30′	.7243	.6626	.8847	1.1303	.7490	.8465	30′
40′	.7272	.6648	.8899	1.1237	.7470	.8436	20′
50′	.7301	.6670	.8952	1.1171	.7451	.8407	10′
42° 00′	.7330	.6691	.9004	1.1106	.7431	.8378	48° 00′
10′	.7359	.6713	.9057	1.1041	.7412	.8348	50′
20′	.7389	.6734	.9110	1.0977	.7392	.8319	40′
30′	.7418	.6756	.9163	1.0913	.7373	.8290	30′
40′	.7447	.6777	.9217	1.0850	.7353	.8261	20′
50′	.7476	.6799	.9271	1.0786	.7333	.8232	10′
43° 00′	.7505	.6820	.9325	1.0724	.7314	.8203	47° 00′
10′	.7534	.6841	.9380	1.0661	.7294	.8174	50′
20′	.7563	.6862	.9435	1.0599	.7274	.8145	40′
30′	.7592	.6884	.9490	1.0538	.7254	.8116	30′
40′	.7621	.6905	.9545	1.0477	.7234	.8087	20′
50′	.7650	.6926	.9601	1.0416	.7214	.8058	10′
44° 00′	.7679	.6947	.9657	1.0355	.7193	.8029	46° 00′
10′	.7709	.6967	.9713	1.0295	.7173	.7999	50′
20′	.7738	.6988	.9770	1.0235	.7153	.7970	40′
30′	.7767	.7009	.9827	1.0176	.7133	.7941	30′
40′	.7796	.7030	.9884	1.0117	.7112	.7912	20′
50′	.7825	.7050	.9942	1.0058	.7092	.7883	10′
45° 00′	.7854	.7071	1.0000	1.0000	.7071	.7854	45° 00′
		Cosine	Cotangent	Tangent	Sine	Radians	Degrees

APPENDIX 5
WEIGHTS AND MEASURES

UNITED STATES SYSTEM

LINEAR MEASURE

Inches	Feet	Yards	Rods	Furlongs	Miles
1.0 =	.08333 =	.02778 =	.0050505 =	.00012626 =	.00001578
12.0 =	1.0 =	.33333 =	.0606061 =	.00151515 =	.00018939
36.0 =	3.0 =	1.0 =	.1818182 =	.00454545 =	.00056818
198.0 =	16.5 =	5.5 =	1.0 =	.025 =	.003125
7920.0 =	660.0 =	220.0 =	40.0 =	1.0 =	.125
63360.0 =	5280.0 =	1760.0 =	320.0 =	8.0 =	1.0

SQUARE AND LAND MEASURE

Sq. Inches	Square Feet	Square Yards	Sq. Rods	Acres	Sq. Miles
1.0 =	.006944 =	.000772			
144.0 =	1.0 =	.111111			
1296.0 =	9.0 =	1.0 =	.03306 =	.000207	
39204.0 =	272.25 =	30.25 =	1.0 =	.00625 =	.0000098
	43560.0 =	4840.0 =	160.0 =	1.0 =	.0015625
		3097600.0 =	102400.0 =	640.0 =	1.0

AVOIRDUPOIS WEIGHTS

Grains	Drams	Ounces	Pounds	Tons
1.0 =	.03657 =	.002286 =	.000143 =	.0000000714
27.34375 =	1.0 =	.0625 =	.003906 =	.00000195
437.5 =	16.0 =	1.0 =	.0625 =	.00003125
7000.0 =	256.0 =	16.0 =	1.0 =	.0005
14000000.0 =	512000.0 =	32000.0 =	2000.0 =	1.0

DRY MEASURE

Pints	Quarts	Pecks	Cubic Feet	Bushels
1.0 =	.5 =	.0625 =	.01945 =	.01563
2.0 =	1.0 =	.125 =	.03891 =	.03125
16.0 =	8.0 =	1.0 =	.31112 =	.25
51.42627 =	25.71314 =	3.21414 =	1.0 =	.80354
64.0 =	32.0 =	4.0 =	1.2445 =	1.0

LIQUID MEASURE

Gills	Pints	Quarts	U. S. Gallons	Cubic Feet
1.0 =	.25 =	.125 =	.03125 =	.00418
4.0 =	1.0 =	.5 =	.125 =	.01671
8.0 =	2.0 =	1.0 =	.250 =	.03342
32.0 =	8.0 =	4.0 =	1.0 =	.1337
			7.48052 =	1.0

METRIC SYSTEM

UNITS

Length—Meter : Mass—Gram : Capacity—Liter
for pure water at 4°C. (39.2°F.)
1 cubic decimeter or 1 liter = 1 kilogram

$$1000 \text{ Milli} \begin{cases} meters \text{ (mm)} \\ grams \text{ (mg)} \\ liters \text{ (ml)} \end{cases} = 100 \text{ Centi} \begin{cases} meters \text{ (cm)} \\ grams \text{ (cg)} \\ liters \text{ (cl)} \end{cases} = 10 \text{ Deci} \begin{cases} meters \text{ (dm)} \\ grams \text{ (dg)} \\ liters \text{ (dl)} \end{cases} = 1 \begin{cases} meter \\ gram \\ liter \end{cases}$$

$$1000 \begin{cases} meters \\ grams \\ liters \end{cases} = 100 \text{ Deka} \begin{cases} meters \text{ (dkm)} \\ grams \text{ (dkg)} \\ liters \text{ (dkl)} \end{cases} = 10 \text{ Hecto} \begin{cases} meters \text{ (hm)} \\ grams \text{ (hg)} \\ liters \text{ (hl)} \end{cases} = 1 \text{ Kilo} \begin{cases} meter \text{ (km)} \\ gram \text{ (kg)} \\ liter \text{ (kl)} \end{cases}$$

1 Metric Ton	= 1000 Kilograms
100 Square Meters	= 1 Are
100 Ares	= 1 Hectare
100 Hectares	= 1 Square Kilometer

A-13

APPENDIX 6
DECIMAL EQUIVALENTS AND TEMPERATURE CONVERSION

DECIMAL EQUIVALENTS—INCH-MILLIMETER CONVERSION TABLE

1/2	1/4	1/8	1/16	1/32	1/64	Decimals	Millimeters
					1	.015625	.396875
				1		.031250	.793750
					3	.046875	1.190625
			1			.062500	1.587500
					5	.078125	1.984375
				3		.093750	2.381250
					7	.109375	2.778125
		1				.125000	3.175000
					9	.140625	3.571875
				5		.156250	3.968750
					11	.171875	4.365625
			3			.187500	4.762500
					13	.203125	5.159375
				7		.218750	5.556250
					15	.234375	5.953125
	1					.250000	6.350000
					17	.265625	6.746875
				9		.281250	7.143750
					19	.296875	7.540625
			5			.312500	7.937500
					21	.328125	8.334375
				11		.343750	8.731250
					23	.359375	9.128125
		3				.375000	9.525000
					25	.390625	9.921875
				13		.406250	10.318750
					27	.421875	10.715625
			7			.437500	11.112500
					29	.453125	11.509375
				15		.468750	11.906250
					31	.484375	12.303125
1						.500000	12.700000

1/2	1/4	1/8	1/16	1/32	1/64	Decimals	Millimeters
					33	.515625	13.096875
				17		.531250	13.493750
					35	.546875	13.890625
			9			.562500	14.287500
					37	.578125	14.684375
				19		.593750	15.081250
					39	.609375	15.478125
		5				.625000	15.875000
					41	.640625	16.271875
				21		.656250	16.668750
					43	.671875	17.065625
			11			.687500	17.462500
					45	.703125	17.859375
				23		.718750	18.256250
					47	.734375	18.653125
	3					.750000	19.050000
					49	.765625	19.446875
				25		.781250	19.843750
					51	.796875	20.240625
			13			.812500	20.637500
					53	.828125	21.034375
				27		.843750	21.431250
					55	.859375	21.828125
		7				.875000	22.225000
					57	.890625	22.621875
				29		.906250	23.018750
					59	.921875	23.415625
			15			.937500	23.812500
					61	.953125	24.209375
				31		.968750	24.606250
					63	.984375	25.003125
2	4	8	16	32	64	1.000000	25.400000

A-14

APPENDIX 7
WEIGHTS AND SPECIFIC GRAVITIES

Substance	Weight Lb. per Cu. Ft.	Specific Gravity	Substance	Weight Lb. per Cu. Ft.	Specific Gravity
METALS, ALLOYS, ORES			**TIMBER, U. S. SEASONED**		
			Moisture Content by Weight:		
Aluminum, cast, hammered	165	2.55-2.75	Seasoned timber 15 to 20%		
Brass, cast, rolled	534	8.4-8.7	Green timber up to 50%		
Bronze, 7.9 to 14% Sn	509	7.4-8.9	Ash, white, red	40	0.62-0.65
Bronze, aluminum	481	7.7	Cedar, white, red	22	0.32-0.38
Copper, cast, rolled	556	8.8-9.0	Chestnut	41	0.66
Copper ore, pyrites	262	4.1-4.3	Cypress	30	0.48
Gold, cast, hammered	1205	19.25-19.3	Fir, Douglas spruce	32	0.51
Iron, cast, pig	450	7.2	Fir, eastern	25	0.40
Iron, wrought	485	7.6-7.9	Elm, white	45	0.72
Iron, spiegel-eisen	468	7.5	Hemlock	29	0.42-0.52
Iron, ferro-silicon	437	6.7-7.3	Hickory	49	0.74-0.84
Iron ore, hematite	325	5.2	Locust	46	0.73
Iron ore, hematite in bank	160-180	....	Maple, hard	43	0.68
Iron ore, hematite loose	130-160	....	Maple, white	33	0.53
Iron ore, limonite	237	3.6-4.0	Oak, chestnut	54	0.86
Iron ore, magnetite	315	4.9-5.2	Oak, live	59	0.95
Iron slag	172	2.5-3.0	Oak, red, black	41	0.65
Lead	710	11.37	Oak, white	46	0.74
Lead ore, galena	465	7.3-7.6	Pine, Oregon	32	0.51
Magnesium, alloys	112	1.74-1.83	Pine, red	30	0.48
Manganese	475	7.2-8.0	Pine, white	26	0.41
Manganese ore, pyrolusite	259	3.7-4.6	Pine, yellow, long-leaf	44	0.70
Mercury	849	13.6	Pine, yellow, short-leaf	38	0.61
Monel Metal	556	8.8-9.0	Poplar	30	0.48
Nickel	565	8.9-9.2	Redwood, California	26	0.42
Platinum, cast, hammered	1330	21.1-21.5	Spruce, white, black	27	0.40-0.46
Silver, cast, hammered	656	10.4-10.6	Walnut, black	38	0.61
Steel, rolled	490	7.85			
Tin, cast, hammered	459	7.2-7.5			
Tin ore, cassiterite	418	6.4-7.0			
Zinc, cast, rolled	440	6.9-7.2			
Zinc ore, blende	253	3.9-4.2	**VARIOUS LIQUIDS**		
			Alcohol, 100%	49	0.79
			Acids, muriatic 40%	75	1.20
VARIOUS SOLIDS			Acids, nitric 91%	94	1.50
			Acids, sulphuric 87%	112	1.80
Cereals, oats bulk	32	....	Lye, soda 66%	106	1.70
Cereals, barley bulk	39	....	Oils, vegetable	58	0.91-0.94
Cereals, corn, rye bulk	48	....	Oils, mineral, lubricants	57	0.90-0.93
Cereals, wheat bulk	48	....	Water, 4°C. max. density	62.428	1.0
Hay and Straw bales	20	....	Water, 100°C.	59.830	0.9584
Cotton, Flax, Hemp	93	1.47-1.50	Water, ice	56	0.88-0.92
Fats	58	0.90-0.97	Water, snow, fresh fallen	8	.125
Flour, loose	28	0.40-0.50	Water, sea water	64	1.02-1.03
Flour, pressed	47	0.70-0.80			
Glass, common	156	2.40-2.60			
Glass, plate or crown	161	2.45-2.72	**GASES**		
Glass, crystal	184	2.90-3.00			
Leather	59	0.86-1.02	Air, 0°C. 760 mm.	.08071	1.0
Paper	58	0.70-1.15	Ammonia	.0478	0.5920
Potatoes, piled	42	....	Carbon dioxide	.1234	1.5291
Rubber, caoutchouc	59	0.92-0.96	Carbon monoxide	.0781	0.9673
Rubber goods	94	1.0-2.0	Gas, illuminating	.028-.036	0.35-0.45
Salt, granulated, piled	48	....	Gas, natural	.038-.039	0.47-0.48
Saltpeter	67	....	Hydrogen	.00559	0.0693
Starch	96	1.53	Nitrogen	.0784	0.9714
Sulphur	125	1.93-2.07	Oxygen	.0892	1.1056
Wool	82	1.32			

The specific gravities of solids and liquids refer to water at 4°C., those of gases to air at 0°C. and 760 mm. pressure. The weights per cubic foot are derived from average specific gravities, except where stated that weights are for bulk, heaped or loose material, etc.

(Courtesy of the American Institute of Steel Construction.)

APPENDIX 7
WEIGHTS AND SPECIFIC GRAVITIES (Cont.)

Substance	Weight Lb. per Cu. Ft.	Specific Gravity	Substance	Weight Lb. per Cu. Ft.	Specific Gravity
ASHLAR MASONRY			**MINERALS**		
Granite, syenite, gneiss	165	2.3-3.0	Asbestos	153	2.1-2.8
Limestone, marble	160	2.3-2.8	Barytes	281	4.50
Sandstone, bluestone	140	2.1-2.4	Basalt	184	2.7-3.2
MORTAR RUBBLE MASONRY			Bauxite	159	2.55
			Borax	109	1.7-1.8
Granite, syenite, gneiss	155	2.2-2.8	Chalk	137	1.8-2.6
Limestone, marble	150	2.2-2.6	Clay, marl	137	1.8-2.6
Sandstone, bluestone	130	2.0-2.2	Dolomite	181	2.9
DRY RUBBLE MASONRY			Feldspar, orthoclase	159	2.5-2.6
Granite, syenite, gneiss	130	1.9-2.3	Gneiss, serpentine	159	2.4-2.7
Limestone, marble	125	1.9-2.1	Granite, syenite	175	2.5-3.1
Sandstone, bluestone	110	1.8-1.9	Greenstone, trap	187	2.8-3.2
BRICK MASONRY			Gypsum, alabaster	159	2.3-2.8
Pressed brick	140	2.2-2.3	Hornblende	187	3.0
Common brick	120	1.8-2.0	Limestone, marble	165	2.5-2.8
Soft brick	100	1.5-1.7	Magnesite	187	3.0
CONCRETE MASONRY			Phosphate rock, apatite	200	3.2
Cement, stone, sand	144	2.2-2.4	Porphyry	172	2.6-2.9
Cement, slag, etc.	130	1.9-2.3	Pumice, natural	40	0.37-0.90
Cement, cinder, etc.	100	1.5-1.7	Quartz, flint	165	2.5-2.8
			Sandstone, bluestone	147	2.2-2.5
VARIOUS BUILDING MATERIALS			Shale, slate	175	2.7-2.9
			Soapstone, talc	169	2.6-2.8
Ashes, cinders	40-45	....			
Cement, portland, loose	90	....			
Cement, portland, set	183	2.7-3.2	**STONE, QUARRIED, PILED**		
Lime, gypsum, loose	53-64		Basalt, granite, gneiss	96	....
Mortar, set	103	1.4-1.9	Limestone, marble, quartz	95	....
Slags, bank slag	67-72	....	Sandstone	82	....
Slags, bank screenings	98-117	....	Shale	92	....
Slags, machine slag	96	....	Greenstone, hornblende	107	....
Slags, slag sand	49-55	....			
EARTH, ETC., EXCAVATED			**BITUMINOUS SUBSTANCES**		
Clay, dry	63	....	Asphaltum	81	1.1-1.5
Clay, damp, plastic	110	....	Coal, anthracite	97	1.4-1.7
Clay and gravel, dry	100	....	Coal, bituminous	84	1.2-1.5
Earth, dry, loose	76	....	Coal, lignite	78	1.1-1.4
Earth, dry, packed	95	....	Coal, peat, turf, dry	47	0.65-0.85
Earth, moist, loose	78	....	Coal, charcoal, pine	23	0.28-0.44
Earth, moist, packed	96	....	Coal, charcoal, oak	33	0.47-0.57
Earth, mud, flowing	108	....	Coal, coke	75	1.0-1.4
Earth, mud, packed	115	....	Graphite	131	1.9-2.3
Riprap, limestone	80-85	....	Paraffine	56	0.87-0.91
Riprap, sandstone	90	....	Petroleum	54	0.87
Riprap, shale	105	....	Petroleum, refined	50	0.79-0.82
Sand, gravel, dry, loose	90-105	....	Petroleum, benzine	46	0.73-0.75
Sand, gravel, dry, packed	100-120	....	Petroleum, gasoline	42	0.66-0.69
Sand, gravel, dry, wet	118-120	....	Pitch	69	1.07-1.15
			Tar, bituminous	75	1.20
EXCAVATIONS IN WATER					
Sand or gravel	60	....	**COAL AND COKE, PILED**		
Sand or gravel and clay	65	....	Coal, anthracite	47-58	....
Clay	80	....	Coal, bituminous, lignite	40-54	....
River mud	90	....	Coal, peat, turf	20-26	....
Soil	70	....	Coal, charcoal	10-14	....
Stone riprap	65	....	Coal, coke	23-32	....

The specific gravities of solids and liquids refer to water at 4°C., those of gases to air at 0°C. and 760 mm. pressure. The weights per cubic foot are derived from average specific gravities, except where stated that weights are for bulk, heaped or loose material, etc.

APPENDIX 8

WIRE AND SHEET METAL GAGES

WIRE AND SHEET METAL GAGES
IN DECIMALS OF AN INCH

Name of Gage	United States Standard Gage*		The United States Steel Wire Gage	American or Brown & Sharpe Wire Gage	New Birmingham Standard Sheet & Hoop Gage	British Imperial or English Legal Standard Wire Gage	Birmingham or Stubs Iron Wire Gage	Name of Gage
Principal Use	Uncoated Steel Sheets and Light Plates		Steel Wire except Music Wire	Non-Ferrous Sheets and Wire	Iron and Steel Sheets and Hoops	Wire	Strips, Bands, Hoops and Wire	Principal Use
Gage No.	Weight Oz. per Sq. Ft.	Approx. Thickness Inches	Thickness, Inches					Gage No.
7/0's			.4900		.6666	.500		7/0's
6/0's			.4615	.5800	.625	.464		6/0's
5/0's			.4305	.5165	.5883	.432	.500	5/0's
4/0's			.3938	.4600	.5416	.400	.454	4/0's
3/0's			.3625	.4096	.500	.372	.425	3/0's
2/0's			.3310	.3648	.4452	.348	.380	2/0's
0			.3065	.3249	.3964	.324	.340	0
1			.2830	.2893	.3532	.300	.300	1
2			.2625	.2576	.3147	.276	.284	2
3	160	.2391	.2437	.2294	.2804	.252	.259	3
4	150	.2242	.2253	.2043	.250	.232	.238	4
5	140	.2092	.2070	.1819	.2225	.212	.220	5
6	130	.1943	.1920	.1620	.1981	.192	.203	6
7	120	.1793	.1770	.1443	.1764	.176	.180	7
8	110	.1644	.1620	.1285	.1570	.160	.165	8
9	100	.1495	.1483	.1144	.1398	.144	.148	9
10	90	.1345	.1350	.1019	.1250	.128	.134	10
11	80	.1196	.1205	.0907	.1113	.116	.120	11
12	70	.1046	.1055	.0808	.0991	.104	.109	12
13	60	.0897	.0915	.0720	.0882	.092	.095	13
14	50	0747	.0800	.0641	.0785	.080	.083	14
15	45	.0673	.0720	.0571	.0699	.072	.072	15
16	40	.0598	.0625	.0508	.0625	.064	.065	16
17	36	.0538	.0540	.0453	.0556	.056	.058	17
18	32	.0478	.0475	.0403	.0495	.048	.049	18
19	28	.0418	.0410	.0359	.0440	.040	.042	19
20	24	.0359	.0348	.0320	.0392	.036	.035	20
21	22	.0329	.0318	.0285	.0349	.032	.032	21
22	20	.0299	.0286	.0253	.0313	.028	.028	22
23	18	.0269	.0258	.0226	.0278	.024	.025	23
24	16	.0239	.0230	.0201	.0248	.022	.022	24
25	14	.0209	.0204	.0179	.0220	.020	.020	25
26	12	.0179	.0181	.0159	.0196	.018	.018	26
27	11	.0164	.0173	.0142	.0175	.0164	.016	27
28	10	.0149	.0162	.0126	.0156	.0148	.014	28
29	9	.0135	.0150	.0113	.0139	.0136	.013	29
30	8	.0120	.0140	.0100	.0123	.0124	.012	30
31	7	.0105	.0132	.0089	.0110	.0116	.010	31
32	6.5	.0097	.0128	.0080	.0098	.0108	.009	32
33	6	.0090	.0118	.0071	.0087	.0100	.008	33
34	5.5	.0082	.0104	.0063	.0077	.0092	.007	34
35	5	.0075	.0095	.0056	.0069	.0084	.005	35
36	4.5	.0067	.0090	.0050	.0061	.0076	.004	36
37	4.25	.0064	.0085	.0045	.0054	.0068		37
38	4	.0060	.0080	.0040	.0048	.0060		38
39			.0075	.0035	.0043	.0052		39
40			.0070	.0031	.0039	.0048		40

* U. S. Standard Gage is officially a weight gage, in oz. per sq. ft. as tabulated. The Approx. Thickness shown is the "Manufacturers' Standard" of the American Iron and Steel Institute, based on steel as weighing 501.81 lbs. per cu. ft. (489.6 true weight plus 2.5 percent for average over-run in area and thickness). The A.I.S.I. standard nomenclature for flat rolled carbon steel is as follows:

Widths, Inches	Thicknesses, Inch							
	0.2500 and thicker	0.2499 to 0.2031	0.2030 to 0.1875	0.1874 to 0.0568	0.0567 to 0.0344	0.0343 to 0.0255	0.0254 to 0.0142	0.0141 and thinner
To 3½ incl..........................	Bar	Bar	Strip	Strip	Strip	Strip	Sheet	Sheet
Over 3½ to 6 incl...........	Bar	Bar	Strip	Strip	Strip	Sheet	Sheet	Sheet
" 6 to 12 "	Plate	Strip	Strip	Strip	Sheet	Sheet	Sheet	Black Plate
" 12 to 32 "	Plate	Sheet	Sheet	Sheet	Sheet	Sheet	Sheet	Sheet
" 32 to 48 "	Plate	Sheet	Sheet	Sheet	Sheet	Sheet	Sheet	—
" 48	Plate	Plate	Plate	Sheet	Sheet	Sheet	Sheet	

APPENDIX 9
AMERICAN STANDARD TAPER PIPE THREADS, NPT[1]

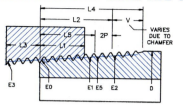

1	2	3	4	5	6	7	8	9	10	11
					Hand-Tight Engagement			Effective Thread, External		
Nominal Pipe Size	Outside Diameter of Pipe D	Threads per Inch n	Pitch of Thread p	Pitch Diameter at Beginning of External Thread E_0	Length[2] L_1 In.	Thds	Dia E_1	Length L_2 In.	Thds	Dia E_2 In.
$\frac{1}{16}$	0.3125	27	0.03704	0.27118	0.160	4.32	0.28118	0.2611	7.05	0.28750
$\frac{1}{8}$	0.405	27	0.03704	0.36351	0.180	4.86	0.37476	0.2639	7.12	0.38000
$\frac{1}{4}$	0.540	18	0.05556	0.47739	0.200	3.60	0.48989	0.4018	7.23	0.50250
$\frac{3}{8}$	0.675	18	0.05556	0.61201	0.240	4.32	0.62701	0.4078	7.34	0.63750
$\frac{1}{2}$	0.840	14	0.07143	0.75843	0.320	4.48	0.77843	0.5337	7.47	0.79179
$\frac{3}{4}$	1.050	14	0.07143	0.96768	0.339	4.75	0.98887	0.5457	7.64	1.00179
1	1.315	$11\frac{1}{2}$	0.08696	1.21363	0.400	4.60	1.23863	0.6828	7.85	1.25630
$1\frac{1}{4}$	1.660	$11\frac{1}{2}$	0.08696	1.55713	0.420	4.83	1.58338	0.7068	8.13	1.60130
$1\frac{1}{2}$	1.900	$11\frac{1}{2}$	0.08696	1.79609	0.420	4.83	1.82234	0.7235	8.32	1.84130
2	2.375	$11\frac{1}{2}$	0.08696	2.26902	0.436	5.01	2.29627	0.7565	8.70	2.31630
$2\frac{1}{2}$	2.875	8	0.12500	2.71953	0.682	5.46	2.76216	1.1375	9.10	2.79062
3	3.500	8	0.12500	3.34062	0.766	6.13	3.38850	1.2000	9.60	3.41562
$3\frac{1}{2}$	4.000	8	0.12500	3.83750	0.821	6.57	3.88881	1.2500	10.00	3.91562
4	4.500	8	0.12500	4.33438	0.844	6.75	4.38712	1.3000	10.40	4.41562
5	5.563	8	0.12500	5.39073	0.937	7.50	5.44929	1.4063	11.25	5.47862
6	6.625	8	0.12500	6.44609	0.958	7.66	6.50597	1.5125	12.10	6.54062
8	8.625	8	0.12500	8.43359	1.063	8.50	8.50003	1.7125	13.70	8.54062
10	10.750	8	0.12500	10.54531	1.210	9.68	10.62094	1.9250	15.40	10.66562
12	12.750	8	0.12500	12.53281	1.360	10.88	12.61781	2.1250	17.00	12.66562
14 OD	14.000	8	0.12500	13.77500	1.562	12.50	13.87262	2.2500	18.90	13.91562
16 OD	16.000	8	0.12500	15.76250	1.812	14.50	15.87575	2.4500	19.60	15.91562
18 OD	18.000	8	0.12500	17.75000	2.000	16.00	17.87500	2.6500	21.20	17.91562
20 OD	20.000	8	0.12500	19.73750	2.125	17.00	19.87031	2.8500	22.80	19.91562
24 OD	24.000	8	0.12500	23.71250	2.375	19.00	23.86094	3.2500	26.00	23.91562

All dimensions are given in inches.

[1] The basic dimensions of the American Standard Taper Pipe Thread are given in inches to four or five decimal places. While this implies a greater degree of precision than is ordinarily attained, these dimensions are the basis of gage dimensions and are so expressed for the purpose of eliminating errors in computations.

[2] Also length of thin ring gage and length from gaging notch to small end of plug gage.

(Courtesy of ANSI; B2.1–1960.)

APPENDIX 10
AMERICAN STANDARD 250-LB CAST IRON FLANGED FITTINGS

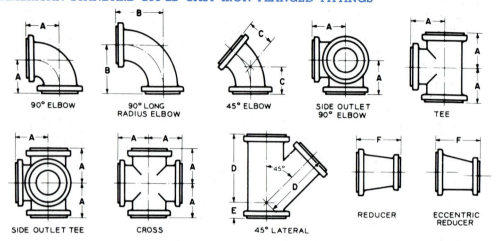

90° ELBOW   90° LONG RADIUS ELBOW   45° ELBOW   SIDE OUTLET 90° ELBOW   TEE

SIDE OUTLET TEE   CROSS   45° LATERAL   REDUCER   ECCENTRIC REDUCER

Dimensions of 250-lb Cast Iron Flanged Fittings

Nominal Pipe Size	Flanges			Fittings		Straight					
	Dia of Flange	Thickness of Flange (Min)	Dia of Raised Face	Inside Dia of Fittings (Min)	Wall Thickness	Center to Face 90 Deg Elbow Tees, Crosses and True "Y"	Center to Face 90 Deg Long Radius Elbow	Center to Face 45 Deg Elbow	Center to Face Lateral	Short Center to Face True "Y" and Lateral	Face to Face Reducer
						A	B	C	D	E	F
1	$4\frac{7}{8}$	$\frac{11}{16}$	$2\frac{11}{16}$	1	$\frac{7}{16}$	4	5	2	$6\frac{1}{2}$	2	....
$1\frac{1}{4}$	$5\frac{1}{4}$	$\frac{3}{4}$	$3\frac{1}{16}$	$1\frac{1}{4}$	$\frac{7}{16}$	$4\frac{1}{4}$	$5\frac{1}{2}$	$2\frac{1}{2}$	$7\frac{1}{4}$	$2\frac{1}{4}$	....
$1\frac{1}{2}$	$6\frac{1}{8}$	$\frac{13}{16}$	$3\frac{9}{16}$	$1\frac{1}{2}$	$\frac{7}{16}$	$4\frac{1}{2}$	6	$2\frac{3}{4}$	$8\frac{1}{2}$	$2\frac{1}{2}$	....
2	$6\frac{1}{2}$	$\frac{7}{8}$	$4\frac{3}{16}$	2	$\frac{7}{16}$	5	$6\frac{1}{2}$	3	9	$2\frac{1}{2}$	5
$2\frac{1}{2}$	$7\frac{1}{2}$	1	$4\frac{15}{16}$	$2\frac{1}{2}$	$\frac{1}{2}$	$5\frac{1}{2}$	7	$3\frac{1}{2}$	$10\frac{1}{2}$	$2\frac{1}{2}$	$5\frac{1}{2}$
3	$8\frac{1}{4}$	$1\frac{1}{8}$	$5\frac{11}{16}$	3	$\frac{9}{16}$	6	$7\frac{3}{4}$	$3\frac{1}{2}$	11	3	6
$3\frac{1}{2}$	9	$1\frac{3}{16}$	$6\frac{5}{16}$	$3\frac{1}{2}$	$\frac{9}{16}$	$6\frac{1}{2}$	$8\frac{1}{2}$	4	$12\frac{1}{2}$	3	$6\frac{1}{2}$
4	10	$1\frac{1}{4}$	$6\frac{15}{16}$	4	$\frac{5}{8}$	7	9	$4\frac{1}{2}$	$13\frac{1}{2}$	3	7
5	11	$1\frac{3}{8}$	$8\frac{5}{16}$	5	$\frac{11}{16}$	8	$10\frac{1}{4}$	5	15	$3\frac{1}{2}$	8
6	$12\frac{1}{2}$	$1\frac{7}{16}$	$9\frac{11}{16}$	6	$\frac{3}{4}$	$8\frac{1}{2}$	$11\frac{1}{2}$	$5\frac{1}{2}$	$17\frac{1}{2}$	4	9
8	15	$1\frac{5}{8}$	$11\frac{15}{16}$	8	$\frac{13}{16}$	10	14	6	$20\frac{1}{2}$	5	11
10	$17\frac{1}{2}$	$1\frac{7}{8}$	$14\frac{1}{16}$	10	$\frac{15}{16}$	$11\frac{1}{2}$	$16\frac{1}{2}$	7	24	$5\frac{1}{2}$	12
12	$20\frac{1}{2}$	2	$16\frac{7}{16}$	12	1	13	19	8	$27\frac{1}{2}$	6	14
14	23	$2\frac{1}{8}$	$18\frac{15}{16}$	$13\frac{1}{4}$	$1\frac{1}{8}$	15	$21\frac{1}{2}$	$8\frac{1}{2}$	31	$6\frac{1}{2}$	16
16	$25\frac{1}{2}$	$2\frac{1}{4}$	$21\frac{1}{16}$	$15\frac{1}{4}$	$1\frac{1}{4}$	$16\frac{1}{2}$	24	$9\frac{1}{2}$	$34\frac{1}{2}$	$7\frac{1}{2}$	18
18	28	$2\frac{3}{8}$	$23\frac{5}{16}$	17	$1\frac{3}{8}$	18	$26\frac{1}{2}$	10	$37\frac{1}{2}$	8	19
20	$30\frac{1}{2}$	$2\frac{1}{2}$	$25\frac{9}{16}$	19	$1\frac{1}{2}$	$19\frac{1}{2}$	29	$10\frac{1}{2}$	$40\frac{1}{2}$	$8\frac{1}{2}$	20
24	36	$2\frac{3}{4}$	$30\frac{5}{16}$	23	$1\frac{5}{8}$	$22\frac{1}{2}$	34	12	$47\frac{1}{2}$	10	24
30	43	3	$37\frac{3}{16}$	29	2	$27\frac{1}{2}$	$41\frac{1}{2}$	15	....	....	30

All dimensions are given in inches.
(Courtesy of ANSI; B16.1–1967.)

APPENDIX 11

AMERICAN STANDARD 125-LB CAST IRON FLANGED FITTINGS

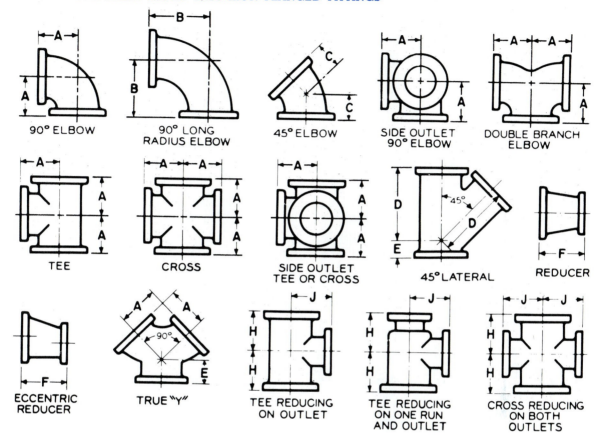

90° ELBOW

90° LONG RADIUS ELBOW

45° ELBOW

SIDE OUTLET 90° ELBOW

DOUBLE BRANCH ELBOW

TEE

CROSS

SIDE OUTLET TEE OR CROSS

45° LATERAL

REDUCER

ECCENTRIC REDUCER

TRUE "Y"

TEE REDUCING ON OUTLET

TEE REDUCING ON ONE RUN AND OUTLET

CROSS REDUCING ON BOTH OUTLETS

Cont.

APPENDIX 11

AMERICAN STANDARD 125-LB CAST IRON FLANGED FITTINGS (Cont.)

	Flanges		General		Straight Fittings						Reducing Fittings (Short Body Patterns)		
											Tees and Crosses		
Nominal Pipe Size	Dia of Flange	Thickness of Flange (Min)	Inside Dia of Flange Fittings	Wall Thickness	Center to Face 90 deg Elbow Tees, Crosses True "Y" and Double Branch Elbow A	Center to Face 90 deg Long Radius El-bow B	Center to Face 45 deg Elbow C	Center to Face Lateral D	Short Center to Face True "Y" and Lateral E	Face to Face Reducer F	Size of Outlet and Smaller	Center to Face Run H	Center to Face Outlet or Side Outlet J
1	4 1/4	7/16	1	5/16	3 1/2	5	1 3/4	5 3/4	1 3/4	. . .			
1 1/4	4 5/8	1/2	1 1/4	5/16	3 3/4	5 1/2	2	6 1/4	1 3/4	. . .			
1 1/2	5	9/16	1 1/2	5/16	4	6	2 1/4	7	2				
2	6	5/8	2	5/16	4 1/2	6 1/2	2 1/2	8	2 1/2	5			
2 1/2	7	11/16	2 1/2	5/16	5	7	3	9 1/2	2 1/2	5 1/2			
3	7 1/2	3/4	3	3/8	5 1/2	7 3/4	3	10	3	6			
3 1/2	8 1/2	13/16	3 1/2	7/16	6	8 1/2	3 1/2	11 1/2	3	6 1/2	All reducing tees and crosses, sizes 16 in. and smaller, shall have same center to face dimensions as straight size fittings, corresponding to the size of the largest opening.		
4	9	15/16	4	1/2	6 1/2	9	4	12	3	7			
5	10	15/16	5	1/2	7 1/2	10 1/4	4 1/2	13 1/2	3 1/2	8			
6	11	1	6	9/16	8	11 1/2	5	14 1/2	3 1/2	9			
8	13 1/2	1 1/8	8	5/8	9	14	5 1/2	17 1/2	4 1/2	11			
10	16	1 3/16	10	3/4	11	16 1/2	6 1/2	20 1/2	5	12			
12	19	1 1/4	12	13/16	12	19	7 1/2	24 1/2	5 1/2	14			
14	21	1 3/8	14	7/8	14	21 1/2	7 1/2	27	6	16			
16	23 1/2	1 7/16	16	1	15	24	8	30	6 1/2	18			
18	25	1 9/16	18	1 1/16	16 1/2	26 1/2	8 1/2	32	7	19	12	13	15 1/2
20	27 1/2	1 11/16	20	1 1/8	18	29	9 1/2	35	8	20	14	14	17
24	32	1 7/8	24	1 1/4	22	34	11	40 1/2	9	24	16	15	19
30	38 3/4	2 1/8	30	1 7/16	25	41 1/2	15	49	10	30	20	18	23
36	46	2 3/8	36	1 5/8	28*	49	18	. . .	. . .	36	24	20	26
42	53	2 5/8	42	1 13/16	31*	56 1/2	21	. . .	. . .	42	24	23	30
48	59 1/2	2 3/4	48	2	34*	64	24	. . .	. . .	48	30	26	34

All dimensions are given in inches.
(Courtesy of ANSI; B16.1–1967.)

APPENDIX 12

AMERICAN STANDARD 125-LB CAST IRON FLANGES*

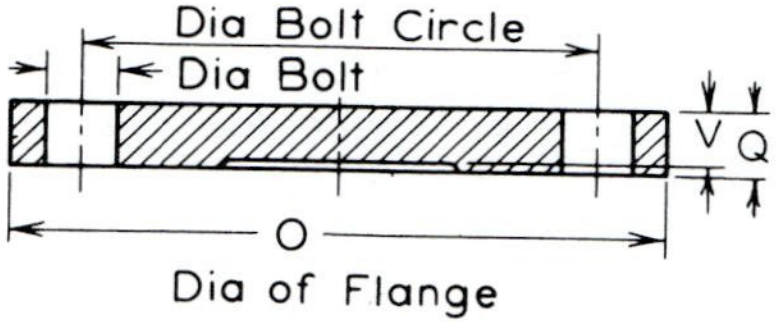

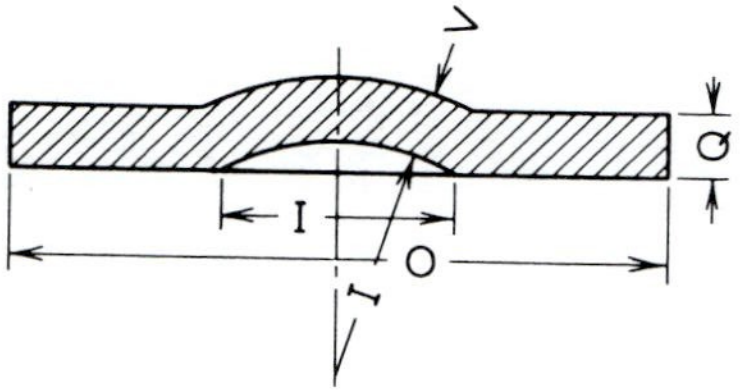

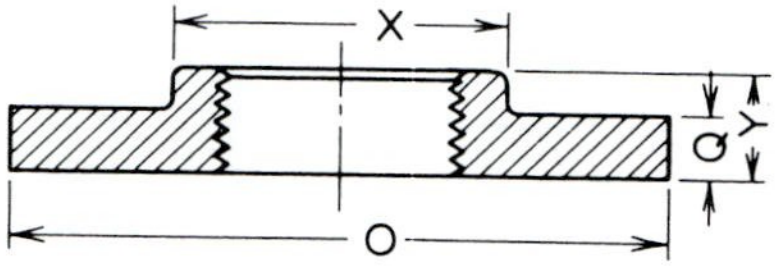

Size I	O	Q	V	X	Y	Dia. Bolt Circle	No. of Bolts	Dia. Bolts	Dia. Bolt Holes	Length of Bolts
1	4¼	7/16	—	1 15/16	0.68	3⅛	4	½	⅝	1¾
1¼	4⅝	½	—	2 5/16	0.76	3½	4	½	⅝	2
1½	5	9/16	—	2 9/16	0.87	3⅞	4	½	⅝	2
2	6	⅝	—	3 1/16	1.00	4¾	4	⅝	¾	2¼
2½	7	11/16	—	3 9/16	1.14	5½	4	⅝	¾	2½
3	7½	¾	—	4¼	1.20	6	4	⅝	¾	2½
3½	8½	13/16	—	4 13/16	1.25	7	8	⅝	¾	2¾
4	9	15/16	—	5 5/16	1.30	7½	8	⅝	¾	3
5	10	15/16	—	6 7/16	1.41	8½	8	¾	⅞	3¼
6	11	1	—	7 9/16	1.51	9½	8	¾	⅞	3¼
8	13½	1⅛	—	9 11/16	1.71	11¾	8	¾	⅞	3½
10	16	1 3/16	—	11 15/16	1.93	14¼	12	⅞	1	3¾
12	19	1¼	12/16	14 1/16	2.13	17	12	⅞	1	3¾
14 O.D.	21	1⅜	⅞	15⅜	2.25	18¾	12	1	1⅛	4¼
16 O.D.	23½	1 7/16	1	17½	2.45	21¼	16	1⅛	1⅛	4½
18 O.D.	25	1 9/16	1 1/16	19⅝	2.65	22¾	16	1⅛	1¼	4¾

All dimensions in inches.

* Extracted from American Standards, "Cast-Iron Pipe Flanges and Flanged Fittings" (ANSI B16.1), with the permission of the publisher, The American Society of Mechanical Engineers.

APPENDIX 13

AMERICAN NATIONAL STANDARD 125-LB CAST IRON SCREWED FITTINGS*

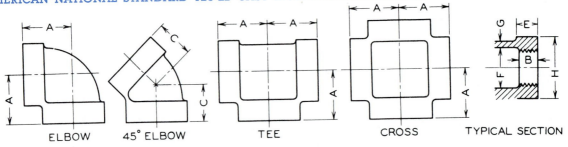

ELBOW 45° ELBOW TEE CROSS TYPICAL SECTION

Nominal Pipe Size	A	C	B Min	E Min	F Min	F Max	G Min	H Min
¼	0.81	0.73	0.32	0.38	0.540	0.584	0.110	0.93
⅜	0.95	0.80	0.36	0.44	0.675	0.719	0.120	1.12
½	1.12	0.88	0.43	0.50	0.840	0.897	0.130	1.34
¾	1.31	0.98	0.50	0.56	1.050	1.107	0.155	1.63
1	1.50	1.12	0.58	0.62	1.315	1.385	0.170	1.95
1¼	1.75	1.29	0.67	0.69	1.660	1.730	0.185	2.39
1½	1.94	1.43	0.70	0.75	1.900	1.970	0.200	2.68
2	2.25	1.68	0.75	0.84	2.375	2.445	0.220	3.28
2½	2.70	1.95	0.92	0.94	2.875	2.975	0.240	3.86
3	3.08	2.17	0.98	1.00	3.500	3.600	0.260	4.62
3½	3.42	2.39	1.03	1.06	4.000	4.100	0.280	5.20
4	3.79	2.61	1.08	1.12	4.500	4.600	0.310	5.79
5	4.50	3.05	1.18	1.18	5.563	5.663	0.380	7.05
6	5.13	3.46	1.28	1.28	6.625	6.725	0.430	8.28
8	6.56	4.28	1.47	1.47	8.625	8.725	0.550	10.63
10	8.08	5.16	1.68	1.68	10.750	10.850	0.690	13.12
12	9.50	5.97	1.88	1.88	12.750	12.850	0.800	15.47
14 O.D.	10.40	—	2.00	2.00	14.000	14.100	0.880	16.94
16 O.D.	11.82	—	2.20	2.20	16.000	16.100	1.000	19.30

All dimensions in inches.
 * Extracted from American National Standards, "Cast-Iron Screwed Fittings, 125- and 250-lb" (ANSI B16.4), with the permission of the publisher, The American Society of Mechanical Engineers.

APPENDIX 14
AMERICAN NATIONAL STANDARD UNIFIED INCH SCREW THREADS (UN AND UNR THREAD FORM)*

Sizes		Basic Major Diameter	Series with Graded Pitches			Series with Constant Pitches								Sizes
Primary	Secondary		Coarse UNC	Fine UNF	Extra Fine UNEF	4UN	6UN	8UN	12UN	16UN	20UN	28UN	32UN	
0		0.0600	—	80	—	—	—	—	—	—	—	—	—	0
	1	0.0730	64	72	—	—	—	—	—	—	—	—	—	1
2		0.0860	56	64	—	—	—	—	—	—	—	—	—	2
	3	0.0990	48	56	—	—	—	—	—	—	—	—	—	3
4		0.1120	40	48	—	—	—	—	—	—	—	—	—	4
5		0.1250	40	44	—	—	—	—	—	—	—	—	—	5
6		0.1380	32	40	—	—	—	—	—	—	—	—	UNC	6
8		0.1640	32	36	—	—	—	—	—	—	—	—	UNC	8
10		0.1900	24	32	—	—	—	—	—	—	—	—	UNF	10
	12	0.2160	24	28	32	—	—	—	—	—	—	UNF	UNEF	12
$\frac{1}{4}$		0.2500	20	28	32	—	—	—	—	—	UNC	UNF	UNEF	$\frac{1}{4}$
$\frac{5}{16}$		0.3125	18	24	32	—	—	—	—	—	20	28	UNEF	$\frac{5}{16}$
$\frac{3}{8}$		0.3750	16	24	32	—	—	—	—	UNC	20	28	UNEF	$\frac{3}{8}$
$\frac{7}{16}$		0.4375	14	20	28	—	—	—	—	16	UNF	UNEF	32	$\frac{7}{16}$
$\frac{1}{2}$		0.5000	13	20	28	—	—	—	—	16	UNF	UNEF	32	$\frac{1}{2}$
$\frac{9}{16}$		0.5625	12	18	24	—	—	—	UNC	16	20	28	32	$\frac{9}{16}$
$\frac{5}{8}$		0.6250	11	18	24	—	—	—	12	16	20	28	32	$\frac{5}{8}$
	$\frac{11}{16}$	0.6875	—	—	24	—	—	—	12	16	20	28	32	$\frac{11}{16}$
$\frac{3}{4}$		0.7500	10	16	20	—	—	—	12	UNF	UNEF	28	32	$\frac{3}{4}$
	$\frac{13}{16}$	0.8125	—	—	20	—	—	—	12	16	UNEF	28	32	$\frac{13}{16}$
$\frac{7}{8}$		0.8750	9	14	20	—	—	—	12	16	UNEF	28	32	$\frac{7}{8}$
	$\frac{15}{16}$	0.9375	—	—	20	—	—	—	12	16	UNEF	28	32	$\frac{15}{16}$
1		1.0000	8	12	20	—	—	UNC	UNF	16	UNEF	28	32	1
	$1\frac{1}{16}$	1.0625	—	—	18	—	—	8	12	16	20	28	—	$1\frac{1}{16}$
$1\frac{1}{8}$		1.1250	7	12	18	—	—	8	UNF	16	20	28	—	$1\frac{1}{8}$
	$1\frac{3}{16}$	1.1875	—	—	18	—	—	8	12	16	20	28	—	$1\frac{3}{16}$
$1\frac{1}{4}$		1.2500	7	12	18	—	—	8	UNF	16	20	28	—	$1\frac{1}{4}$
	$1\frac{5}{16}$	1.3125	—	—	18	—	—	8	12	16	20	28	—	$1\frac{5}{16}$
$1\frac{3}{8}$		1.3750	6	12	18	—	UNC	8	UNF	16	20	28	—	$1\frac{3}{8}$
	$1\frac{7}{16}$	1.4375	—	—	18	—	6	8	12	16	20	28	—	$1\frac{7}{16}$
$1\frac{1}{2}$		1.5000	6	12	18	—	UNC	8	UNF	16	20	28	—	$1\frac{1}{2}$
	$1\frac{9}{16}$	1.5625	—	—	18	—	6	8	12	16	20	—	—	$1\frac{9}{16}$
$1\frac{5}{8}$		1.6250	—	—	18	—	6	8	12	16	20	—	—	$1\frac{5}{8}$
	$1\frac{11}{16}$	1.6875	—	—	18	—	6	8	12	16	20	—	—	$1\frac{11}{16}$
$1\frac{3}{4}$		1.7500	5	—	—	—	6	8	12	16	20	—	—	$1\frac{3}{4}$
	$1\frac{13}{16}$	1.8125	—	—	—	—	6	8	12	16	20	—	—	$1\frac{13}{16}$
$1\frac{7}{8}$		1.8750	—	—	—	—	6	8	12	16	20	—	—	$1\frac{7}{8}$
	$1\frac{15}{16}$	1.9375	—	—	—	—	6	8	12	16	20	—	—	$1\frac{15}{16}$
2		2.0000	$4\frac{1}{2}$	—	—	—	6	8	12	16	20	—	—	2
	$2\frac{1}{8}$	2.1250	—	—	—	—	6	8	12	16	20	—	—	$2\frac{1}{8}$
$2\frac{1}{4}$		2.2500	$4\frac{1}{2}$	—	—	—	6	8	12	16	20	—	—	$2\frac{1}{4}$
	$2\frac{3}{8}$	2.3750	—	—	—	—	6	8	12	16	20	—	—	$2\frac{3}{8}$
$2\frac{1}{2}$		2.5000	4	—	—	UNC	6	8	12	16	20	—	—	$2\frac{1}{2}$
	$2\frac{5}{8}$	2.6250	—	—	—	4	6	8	12	16	20	—	—	$2\frac{5}{8}$
$2\frac{3}{4}$		2.7500	4	—	—	UNC	6	8	12	16	20	—	—	$2\frac{3}{4}$
	$2\frac{7}{8}$	2.8750	—	—	—	4	6	8	12	16	20	—	—	$2\frac{7}{8}$

* Series designation shown indicates the UN thread form; however, the UNR thread form may be specified by substituting UNR in place of UN in all designations for external use only.

Cont.

APPENDIX 14
AMERICAN NATIONAL STANDARD UNIFIED INCH SCREW THREADS (UN AND UNR THREAD FORM)* (Cont.)

Sizes (Primary)	Sizes (Secondary)	Basic Major Diameter	Series with Graded Pitches — Coarse UNC	Fine UNF	Extra Fine UNEF	4UN	6UN	8UN	12UN	16UN	20UN	28UN	32UN	Sizes
3		3.0000	4	—	—	UNC	6	8	12	16	20	—	—	3
	$3\frac{1}{8}$	3.1250	—	—	—	4	6	8	12	16	—	—	—	$3\frac{1}{8}$
$3\frac{1}{4}$		3.2500	4	—	—	UNC	6	8	12	16	—	—	—	$3\frac{1}{4}$
	$3\frac{3}{8}$	3.3750	—	—	—	4	6	8	12	16	—	—	—	$3\frac{3}{8}$
$3\frac{1}{2}$		3.5000	4	—	—	UNC	6	8	12	16	—	—	—	$3\frac{1}{2}$
	$3\frac{5}{8}$	3.6250	—	—	—	4	6	8	12	16	—	—	—	$3\frac{5}{8}$
$3\frac{3}{4}$		3.7500	4	—	—	UNC	6	8	12	16	—	—	—	$3\frac{3}{4}$
	$3\frac{7}{8}$	3.8750	—	—	—	4	6	8	12	16	—	—	—	$3\frac{7}{8}$
4		4.0000	4	—	—	UNC	6	8	12	16	—	—	—	4
	$4\frac{1}{8}$	4.1250	—	—	—	4	6	8	12	16	—	—	—	$4\frac{1}{8}$
$4\frac{1}{4}$		4.2500	—	—	—	4	6	8	12	16	—	—	—	$4\frac{1}{4}$
	$4\frac{3}{8}$	4.3750	—	—	—	4	6	8	12	16	—	—	—	$4\frac{3}{8}$
$4\frac{1}{2}$		4.5000	—	—	—	4	6	8	12	16	—	—	—	$4\frac{1}{2}$
	$4\frac{5}{8}$	4.6250	—	—	—	4	6	8	12	16	—	—	—	$4\frac{5}{8}$
$4\frac{3}{4}$		4.7500	—	—	—	4	6	8	12	16	—	—	—	$4\frac{3}{4}$
	$4\frac{7}{8}$	4.8750	—	—	—	4	6	8	12	16	—	—	—	$4\frac{7}{8}$
5		5.0000	—	—	—	4	6	8	12	16	—	—	—	5
	$5\frac{1}{8}$	5.1250	—	—	—	4	6	8	12	16	—	—	—	$5\frac{1}{8}$
$5\frac{1}{4}$		5.2500	—	—	—	4	6	8	12	16	—	—	—	$5\frac{1}{4}$
	$5\frac{3}{8}$	5.3750	—	—	—	4	6	8	12	16	—	—	—	$5\frac{3}{8}$
$5\frac{1}{2}$		5.5000	—	—	—	4	6	8	12	16	—	—	—	$5\frac{1}{2}$
	$5\frac{5}{8}$	5.6250	—	—	—	4	6	8	12	16	—	—	—	$5\frac{5}{8}$
$5\frac{3}{4}$		5.7500	—	—	—	4	6	8	12	16	—	—	—	$5\frac{3}{4}$
	$5\frac{7}{8}$	5.8750	—	—	—	4	6	8	12	16	—	—	—	$5\frac{7}{8}$
6		6.0000	—	—	—	4	6	8	12	16	—	—	—	6

(Courtesy of ANSI; B1.1–1974.)

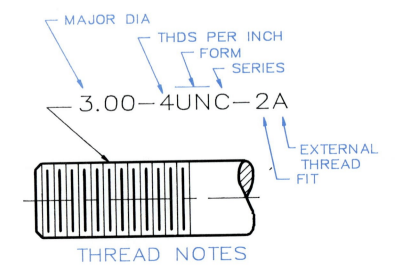

MAJOR DIA

THDS PER INCH

FORM

SERIES

3.00–4UNC–2A

EXTERNAL THREAD

FIT

THREAD NOTES

APPENDIX 15

TAP DRILL SIZES FOR AMERICAN NATIONAL AND UNIFIED COARSE AND FINE THREADS

$$p = \text{pitch} = \frac{1}{\text{No. thd per in}}$$

$$d = \text{depth} = p \times 0.650$$

$$f = \text{flat} = \frac{p}{8}$$

$$\text{pitch dia} = d - \frac{0.650}{N}$$

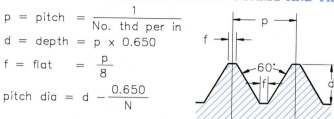

For nos. 575 and 585 screw thread micrometers

Size	Threads per inch NC UNC	Threads per inch NF UNF	Outside Diameter Inches	Pitch Diameter Inches	Root Diameter Inches	Tap Drill Approx. 75% Full Thread	Decimal Equiv. of Tap Drill
0	..	80	.0600	.0519	.0438	3/64	.0469
1	64	..	.0730	.0629	.0527	53	.0595
1	..	72	.0730	.0640	.0550	53	.0595
2	56	..	.0860	.0744	.0628	50	.0700
2	..	64	.0860	.0759	.0657	50	.0700
3	48	..	.0990	.0855	.0719	47	.0785
3	..	56	.0990	.0874	.0758	46	.0810
4	40	..	.1120	.0958	.0795	43	.0890
4	..	48	.1120	.0985	.0849	42	.0935
5	40	..	.1250	.1088	.0925	38	.1015
5	..	44	.1250	.1102	.0955	37	.1040
6	32	..	.1380	.1177	.0974	36	.1065
6	..	40	.1380	.1218	.1055	33	.1130
8	32	..	.1640	.1437	.1234	29	.1360
8	..	36	.1640	.1460	.1279	29	.1360
10	24	..	.1900	.1629	.1359	26	.1470
10	..	32	.1900	.1697	.1494	21	.1590
12	24	..	.2160	.1889	.1619	16	.1770
12	..	28	.2160	.1928	.1696	15	.1800
1/4	20	..	.2500	.2175	.1850	7	.2010
1/4	..	28	.2500	.2268	.2036	3	.2130
5/16	18	..	.3125	.2764	.2403	F	.2570
5/16	..	24	.3125	.2854	.2584	I	.2720
3/8	16	..	.3750	.3344	.2938	5/16	.3125
3/8	..	24	.3750	.3479	.3209	Q	.3320
7/16	14	..	.4375	.3911	.3447	U	.3680
7/16	..	20	.4375	.4050	.3726	25/64	.3906
1/2	13	..	.5000	.4500	.4001	27/64	.4219
1/2	..	20	.5000	.4675	.4351	29/64	.4531
9/16	12	..	.5625	.5084	.4542	31/64	.4844
9/16	..	18	.5625	.5264	.4903	33/64	.5156
5/8	11	..	.6250	.5660	.5069	17/32	.5312
5/8	..	18	.6250	.5889	.5528	37/64	.5781
3/4	10	..	.7500	.6850	.6201	21/32	.6562
3/4	..	16	.7500	.7094	.6688	11/16	.6875
7/8	9	..	.8750	.8028	.7307	49/64	.7656
7/8	..	14	.8750	.8286	.7822	13/16	.8125

Cont.

TAP DRILL SIZES FOR AMERICAN NATIONAL AND UNIFIED COARSE AND FINE THREADS (Cont.)

Size	Threads per inch NC UNC	Threads per inch NF UNF	Outside Diameter Inches	Pitch Diameter Inches	Root Diameter Inches	Tap Drill Approx. 75% Full Thread	Decimal Equiv. of Tap Drill
1	8	..	1.0000	.9188	.8376	$\frac{7}{8}$	.8750
1	..	12	1.0000	.9459	.8917	$\frac{59}{64}$	.9219
1⅛	7	..	1.1250	1.0322	.9394	$\frac{63}{64}$	.9844
1⅛	..	12	1.1250	1.0709	1.0168	$1\frac{3}{64}$	1.0469
1¼	7	..	1.2500	1.1572	1.0644	$1\frac{7}{64}$	1.1094
1¼	..	12	1.2500	1.1959	1.1418	$1\frac{11}{64}$	1.1719
1⅜	6	..	1.3750	1.2667	1.1585	$1\frac{7}{32}$	1.2187
1⅜	..	12	1.3750	1.3209	1.2668	$1\frac{19}{64}$	1.2969
1½	6	..	1.5000	1.3917	1.2835	$1\frac{11}{32}$	1.3437
1½	..	12	1.5000	1.4459	1.3918	$1\frac{27}{64}$	1.4219
1¾	5	..	1.7500	1.6201	1.4902	$1\frac{9}{16}$	1.5625
2	4½	..	2.0000	1.8557	1.7113	$1\frac{25}{32}$	1.7812
2¼	4½	..	2.2500	2.1057	1.9613	$2\frac{1}{32}$	2.0313
2½	4	..	2.5000	2.3376	2.1752	2¼	2.2500
2¾	4	..	2.7500	2.5876	2.4252	2½	2.5000
3	4	..	3.0000	3.8376	2.6752	2¾	2.7500
3¼	4	..	3.2500	3.0876	2.9252	3	3.0000
3½	4	..	3.5000	3.3376	3.1752	3¼	3.2500
3¾	4	..	3.7500	3.5876	3.4252	3½	3.5000
4	4	..	4.0000	3.3786	3.6752	3¾	3.7500

(Courtesy of the L. S. Starrett Company.)

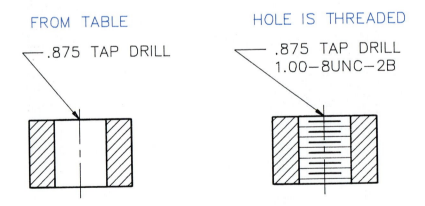

FROM TABLE — .875 TAP DRILL

HOLE IS THREADED — .875 TAP DRILL 1.00−8UNC−2B

APPENDIX 16

LENGTH OF THREAD ENGAGEMENT GROUPS

Nominal Size Diam. Over	To and Incl	Pitch P	Group S To and Incl	Group N Over	Group N To and Incl	Group L Over
1.5	2.8	0.2	0.5	0.5	1.5	1.5
		0.25	0.6	0.6	1.9	1.9
		0.35	0.8	0.8	2.6	2.6
		0.4	1	1	3	3
		0.45	1.3	1.3	3.8	3.8
2.8	5.6	0.35	1	1	3	3
		0.5	1.5	1.5	4.5	4.5
		0.6	1.7	1.7	5	5
		0.7	2	2	6	6
		0.75	2.2	2.2	6.7	6.7
		0.8	2.5	2.5	7.5	7.5
5.6	11.2	0.75	2.4	2.4	7.1	7.1
		1	3	3	9	9
		1.25	4	4	12	12
		1.5	5	5	15	15
11.2	22.4	1	3.8	3.8	11	11
		1.25	4.5	4.5	13	13
		1.5	5.6	5.6	16	16
		1.75	6	6	18	18
		2	8	8	24	24
		2.5	10	10	30	30

Nominal Size Diam. Over	To and Incl	Pitch P	Group S To and Incl	Group N Over	Group N To and Incl	Group L Over
22.4	45	1	4	4	12	12
		1.5	6.3	6.3	19	19
		2	8.5	8.5	25	25
		3	12	12	36	36
		3.5	15	15	45	45
		4	18	18	53	53
		4.5	21	21	63	63
45	90	1.5	7.5	7.5	22	22
		2	9.5	9.5	28	28
		3	15	15	45	45
		4	19	19	56	56
		5	24	24	71	71
		5.5	28	28	85	85
		6	32	32	95	95
90	180	2	12	12	36	36
		3	18	18	53	53
		4	24	24	71	71
		6	36	36	106	106
180	355	3	20	20	60	60
		4	26	26	80	80
		6	40	40	118	118

All dimensions are given in millimeters. (Courtesy of ISO Standards.)

APPENDIX 17
ISO METRIC SCREW THREAD STANDARD SERIES

Nominal Size Dia. (mm) Columnª			Series with Graded Pitches		Pitches (mm) — Series with Constant Pitches												Nominal Size Dia. (mm)
1	2	3	Coarse	Fine	6	4	3	2	1.5	1.25	1	0.75	0.5	0.35	0.25	0.2	
0.25			0.075	—	—	—	—	—	—	—	—	—	—	—	—	—	0.25
0.3			0.08	—	—	—	—	—	—	—	—	—	—	—	—	—	0.3
	0.35		0.09	—	—	—	—	—	—	—	—	—	—	—	—	—	0.35
0.4			0.1	—	—	—	—	—	—	—	—	—	—	—	—	—	0.4
	0.45		0.1	—	—	—	—	—	—	—	—	—	—	—	—	—	0.45
0.5			0.125	—	—	—	—	—	—	—	—	—	—	—	—	—	0.5
	0.55		0.125	—	—	—	—	—	—	—	—	—	—	—	—	—	0.55
0.6			0.15	—	—	—	—	—	—	—	—	—	—	—	—	—	0.6
	0.7		0.175	—	—	—	—	—	—	—	—	—	—	—	—	—	0.7
0.8			0.2	—	—	—	—	—	—	—	—	—	—	—	—	—	0.8
	0.9		0.225	—	—	—	—	—	—	—	—	—	—	—	—	—	0.9
			0.25	—	—	—	—	—	—	—	—	—	—	—	—	0.2	1
	1.1		0.25	—	—	—	—	—	—	—	—	—	—	—	—	0.2	1.1
1.2			0.25	—	—	—	—	—	—	—	—	—	—	—	—	0.2	1.2
	1.4		0.3	—	—	—	—	—	—	—	—	—	—	—	—	0.2	1.4
1.6			0.35	—	—	—	—	—	—	—	—	—	—	—	—	0.2	1.6
	1.8		0.35	—	—	—	—	—	—	—	—	—	—	—	—	0.2	1.8
2			0.4	—	—	—	—	—	—	—	—	—	—	—	0.25	—	2
	2.2		0.45	—	—	—	—	—	—	—	—	—	—	—	0.25	—	2.2
2.5			0.45	—	—	—	—	—	—	—	—	—	—	0.35	—	—	2.5
3			0.5	—	—	—	—	—	—	—	—	—	—	0.35	—	—	3
	3.5		0.6	—	—	—	—	—	—	—	—	—	—	0.35	—	—	3.5
4			0.7	—	—	—	—	—	—	—	—	—	0.5	—	—	—	4
	4.5		0.75	—	—	—	—	—	—	—	—	—	0.5	—	—	—	4.5
5			0.8	—	—	—	—	—	—	—	—	—	0.5	—	—	—	5
		5.5	—	—	—	—	—	—	—	—	—	—	0.5	—	—	—	5.5
6			1	—	—	—	—	—	—	—	—	0.75	—	—	—	—	6
		7	1	—	—	—	—	—	—	—	—	0.75	—	—	—	—	7
8			1.25	1	—	—	—	—	—	—	1	0.75	—	—	—	—	8
		9	1.25	—	—	—	—	—	—	—	1	0.75	—	—	—	—	9
10			1.5	1.25	—	—	—	—	—	1.25	1	0.75	—	—	—	—	10
		11	1.5	—	—	—	—	—	—	—	1	0.75	—	—	—	—	11
12			1.75	1.25	—	—	—	—	1.5	1.25	1	—	—	—	—	—	12
	14		2	1.5	—	—	—	—	1.5	1.25ᵇ	1	—	—	—	—	—	14
		15	—	1.5	—	—	—	—	1.5	—	1	—	—	—	—	—	15
16			2	1.5	—	—	—	—	1.5	—	1	—	—	—	—	—	16
		17	—	1.5	—	—	—	—	1.5	—	1	—	—	—	—	—	17
	18		2.5	1.5	—	—	—	2	1.5	—	1	—	—	—	—	—	18
20			2.5	1.5	—	—	—	2	1.5	—	1	—	—	—	—	—	20
	22		2.5	1.5	—	—	—	2	1.5	—	1	—	—	—	—	—	22

ª Thread diameter should be selected from columns 1, 2 or 3, with preference being in that order.
ᵇ Pitch 1.25 mm in combination with diameter 14 mm has been included for sparkplug applications.
ᶜ Diameter 35 mm has been included for bearing locknut applications.

The use of pitches shown in parentheses should be avoided wherever possible.

The pitches enclosed in the bold frame, together with the corresponding nominal diameters in columns 1 and 2, are those combinations which have been established by ISO Recommendations as a selected "coarse" and "fine" series for commercial fasteners.

APPENDIX 17

ISO METRIC SCREW THREAD STANDARD SERIES (Cont.)

Nominal Size Dia. (mm)			Pitches (mm)														Nominal Size Dia. (mm)
Column[a]			Series with Graded Pitches		Series with Constant Pitches												
1	2	3	Coarse	Fine	6	4	3	2	1.5	1.25	1	0.75	0.5	0.35	0.25	0.2	
24			3	2	—	—	—	2	1.5	—	1	—	—	—	—	—	24
		25	—	—	—	—	—	2	1.5	—	1	—	—	—	—	—	25
		26	—	—	—	—	—	—	1.5	—	1	—	—	—	—	—	26
	27		3	2	—	—	—	2	1.5	—	1	—	—	—	—	—	27
		28	—	—	—	—	—	2	1.5	—	1	—	—	—	—	—	28
30			3.5	2	—	—	(3)	2	1.5	—	1	—	—	—	—	—	30
		32	—	—	—	—	—	2	1.5	—	—	—	—	—	—	—	32
	33		3.5	2	—	—	(3)	2	1.5	—	—	—	—	—	—	—	33
		35[c]	—	—	—	—	—	—	1.5	—	—	—	—	—	—	—	35[c]
36			4	3	—	—	—	2	1.5	—	—	—	—	—	—	—	36
		38	—	—	—	—	—	—	1.5	—	—	—	—	—	—	—	38
	39		4	3	—	—	—	2	1.5	—	—	—	—	—	—	—	39
		40	—	—	—	—	3	2	1.5	—	—	—	—	—	—	—	40
42			4.5	3	—	4	3	2	1.5	—	—	—	—	—	—	—	42
	45		4.5	3	—	4	3	2	1.5	—	—	—	—	—	—	—	45
48			5	3	—	4	3	2	1.5	—	—	—	—	—	—	—	48
		50	—	—	—	—	3	2	1.5	—	—	—	—	—	—	—	50
	52		5	3	—	4	3	2	1.5	—	—	—	—	—	—	—	52
		55	—	—	—	4	3	2	1.5	—	—	—	—	—	—	—	55
56			5.5	4	—	4	3	2	1.5	—	—	—	—	—	—	—	56
		58	—	—	—	4	3	2	1.5	—	—	—	—	—	—	—	58
	60		5.5	4	—	4	3	2	1.5	—	—	—	—	—	—	—	60
		62	—	—	—	4	3	2	1.5	—	—	—	—	—	—	—	62
64			6	4	—	4	3	2	1.5	—	—	—	—	—	—	—	64
		65	—	—	—	4	3	2	1.5	—	—	—	—	—	—	—	65
	68		6	4	—	4	3	2	1.5	—	—	—	—	—	—	—	68
		70	—	—	6	4	3	2	1.5	—	—	—	—	—	—	—	70
72			—	—	6	4	3	2	1.5	—	—	—	—	—	—	—	72
		75	—	—	—	4	3	2	1.5	—	—	—	—	—	—	—	75
	76		—	—	6	4	3	2	1.5	—	—	—	—	—	—	—	76
		78	—	—	—	—	—	2	—	—	—	—	—	—	—	—	78
80			—	—	6	4	3	2	1.5	—	—	—	—	—	—	—	80
		82	—	—	—	—	—	2	—	—	—	—	—	—	—	—	82
	85		—	—	6	4	3	2	—	—	—	—	—	—	—	—	85
90			—	—	6	4	3	2	—	—	—	—	—	—	—	—	90

Cont.

METRIC THREAD NOTE

APPENDIX 17
ISO METRIC SCREW THREAD STANDARD SERIES (Cont.)

Nominal Size Dia. (mm) Column[a]			Pitches (mm) Series with Graded Pitches		Series with Constant Pitches											Nominal Size Dia. (mm)
1	2	3	Coarse	Fine	6	4	3	2	1.5	1.25	1	0.75	0.5	0.35	0.25	0.2	
	95		—	—	6	4	3	2	—	—	—	—	—	—	—	—	95
100			—	—	6	4	3	2	—	—	—	—	—	—	—	—	100
	105		—	—	6	4	3	2	—	—	—	—	—	—	—	—	105
110			—	—	6	4	3	2	—	—	—	—	—	—	—	—	110
	115		—	—	6	4	3	2	—	—	—	—	—	—	—	—	115
	120		—	—	6	4	3	2	—	—	—	—	—	—	—	—	120
125			—	—	6	4	3	2	—	—	—	—	—	—	—	—	125
	130		—	—	6	4	3	2	—	—	—	—	—	—	—	—	130
		135	—	—	6	4	3	2	—	—	—	—	—	—	—	—	135
140			—	—	6	4	3	2	—	—	—	—	—	—	—	—	140
		145	—	—	6	4	3	2	—	—	—	—	—	—	—	—	145
	150		—	—	6	4	3	2	—	—	—	—	—	—	—	—	150
		155	—	—	6	4	3	—	—	—	—	—	—	—	—	—	155
160			—	—	6	4	3	—	—	—	—	—	—	—	—	—	160
		165	—	—	6	4	3	—	—	—	—	—	—	—	—	—	165
	170		—	—	6	4	3	—	—	—	—	—	—	—	—	—	170
		175	—	—	6	4	3	—	—	—	—	—	—	—	—	—	175
180			—	—	6	4	3	—	—	—	—	—	—	—	—	—	'80
		185	—	—	6	4	3	—	—	—	—	—	—	—	—	—	185
	190		—	—	6	4	3	—	—	—	—	—	—	—	—	—	190
		195	—	—	6	4	3	—	—	—	—	—	—	—	—	—	195
200			—	—	6	4	3	—	—	—	—	—	—	—	—	—	200
		205	—	—	6	4	3	—	—	—	—	—	—	—	—	—	205
	210		—	—	6	4	3	—	—	—	—	—	—	—	—	—	210
220			—	—	6	4	3	—	—	—	—	—	—	—	—	—	220
		225	—	—	6	4	3	—	—	—	—	—	—	—	—	—	225
		230	—	—	6	4	3	—	—	—	—	—	—	—	—	—	230
		235	—	—	6	4	3	—	—	—	—	—	—	—	—	—	235
	240		—	—	6	4	3	—	—	—	—	—	—	—	—	—	240
		245	—	—	6	4	3	—	—	—	—	—	—	—	—	—	245
250			—	—	6	4	3	—	—	—	—	—	—	—	—	—	250
		255	—	—	6	4	—	—	—	—	—	—	—	—	—	—	255
	260		—	—	6	4	—	—	—	—	—	—	—	—	—	—	260
		265	—	—	6	4	—	—	—	—	—	—	—	—	—	—	265
		270	—	—	6	4	—	—	—	—	—	—	—	—	—	—	270
		275	—	—	6	4	—	—	—	—	—	—	—	—	—	—	275
280			—	—	6	4	—	—	—	—	—	—	—	—	—	—	280
		285	—	—	6	4	—	—	—	—	—	—	—	—	—	—	285
		290	—	—	6	4	—	—	—	—	—	—	—	—	—	—	290
		295	—	—	6	4	—	—	—	—	—	—	—	—	—	—	295
	300		—	—	6	4	—	—	—	—	—	—	—	—	—	—	300

[a] Thread diameter should be selected from columns 1, 2, or 3; with preference being in that order.

APPENDIX 18
SQUARE AND ACME THREADS

Size	Threads per Inch	Size	Threads per Inch
$\frac{3}{8}$	12	2	$2\frac{1}{2}$
$\frac{7}{16}$	10	$2\frac{1}{4}$	2
$\frac{1}{2}$	10	$2\frac{1}{2}$	2
$\frac{9}{16}$	8	$2\frac{3}{4}$	2
$\frac{5}{8}$	8	3	$1\frac{1}{2}$
$\frac{3}{4}$	6	$3\frac{1}{4}$	$1\frac{1}{2}$
$\frac{7}{8}$	5	$3\frac{1}{2}$	$1\frac{1}{3}$
1	5	$3\frac{3}{4}$	$1\frac{1}{3}$
$1\frac{1}{8}$	4	4	$1\frac{1}{3}$
$1\frac{1}{4}$	4	$4\frac{1}{4}$	$1\frac{1}{3}$
$1\frac{1}{2}$	3	$4\frac{1}{2}$	1
$1\frac{3}{4}$	$2\frac{1}{2}$	over $4\frac{1}{2}$	1

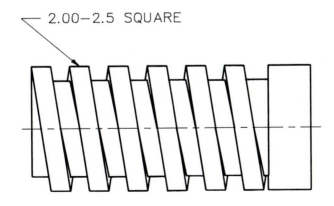

2.00−2.5 SQUARE

SQUARE THREAD NOTE

APPENDIX 19

AMERICAN STANDARD SQUARE BOLTS AND NUTS

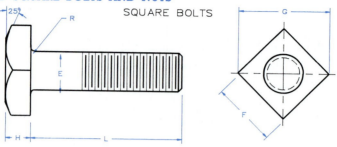

SQUARE BOLTS

Dimensions of Square Bolts

Nominal Size or Basic Product Dia		Body Dia E	Width Across Flats F			Width Across Corners G		Height H			Radius of Fillet R
		Max	Basic	Max	Min	Max	Min	Basic	Max	Min.	Max
1/4	0.2500	0.260	3/8	0.3750	0.362	0.530	0.498	11/64	0.188	0.156	0.031
5/16	0.3125	0.324	1/2	0.5000	0.484	0.707	0.665	13/64	0.220	0.186	0.031
3/8	0.3750	0.388	9/16	0.5625	0.544	0.795	0.747	1/4	0.268	0.232	0.031
7/16	0.4375	0.452	5/8	0.6250	0.603	0.884	0.828	19/64	0.316	0.278	0.031
1/2	0.5000	0.515	3/4	0.7500	0.725	1.061	0.995	21/64	0.348	0.308	0.031
5/8	0.6250	0.642	15/16	0.9375	0.906	1.326	1.244	27/64	0.444	0.400	0.062
3/4	0.7500	0.768	1 1/8	1.1250	1.088	1.591	1.494	1/2	0.524	0.476	0.062
7/8	0.8750	0.895	1 5/16	1.3125	1.269	1.856	1.742	19/32	0.620	0.568	0.062
1	1.0000	1.022	1 1/2	1.5000	1.450	2.121	1.991	21/32	0.684	0.628	0.093
1 1/8	1.1250	1.149	1 11/16	1.6875	1.631	2.386	2.239	3/4	0.780	0.720	0.093
1 1/4	1.2500	1.277	1 7/8	1.8750	1.812	2.652	2.489	27/32	0.876	0.812	0.093
1 3/8	1.3750	1.404	2 1/16	2.0625	1.994	2.917	2.738	29/32	0.940	0.872	0.093
1 1/2	1.5000	1.531	2 1/4	2.2500	2.175	3.182	2.986	1	1.036	0.964	0.093

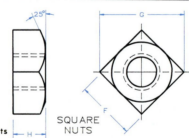

SQUARE NUTS

Dimensions of Square Nuts

Nominal Size or Basic Major Dia of Thread		Width Across Flats F			Width Across Corners G		Thickness H		
		Basic	Max	Min	Max	Min	Basic	Max	Min
1/4	0.2500	7/16	0.4375	0.425	0.619	0.584	7/32	0.235	0.203
5/16	0.3125	9/16	0.5625	0.547	0.795	0.751	17/64	0.283	0.249
3/8	0.3750	5/8	0.6250	0.606	0.884	0.832	21/64	0.346	0.310
7/16	0.4375	3/4	0.7500	0.728	1.061	1.000	3/8	0.394	0.356
1/2	0.5000	13/16	0.8125	0.788	1.149	1.082	7/16	0.458	0.418
5/8	0.6250	1	1.0000	0.969	1.414	1.330	35/64	0.569	0.525
3/4	0.7500	1 1/8	1.1250	1.088	1.591	1.494	21/32	0.680	0.632
7/8	0.8750	1 5/16	1.3125	1.269	1.856	1.742	49/64	0.792	0.740
1	1.0000	1 1/2	1.5000	1.450	2.121	1.991	7/8	0.903	0.847
1 1/8	1.1250	1 11/16	1.6875	1.631	2.386	2.239	1	1.030	0.970
1 1/4	1.2500	1 7/8	1.8750	1.812	2.652	2.489	1 3/32	1.126	1.062
1 3/8	1.3750	2 1/16	2.0625	1.994	2.917	2.738	1 13/64	1.237	1.169
1 1/2	1.5000	2 1/4	2.2500	2.175	3.182	2.986	1 5/16	1.348	1.276

(Courtesy of ANSI; B18.2.1–1965 and ANSI; B18.2.2–1965.)

APPENDIX 20

AMERICAN STANDARD HEXAGON HEAD BOLTS AND NUTS

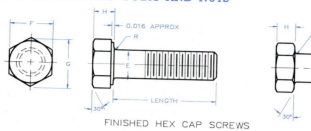

FINISHED HEX CAP SCREWS

Dimensions of Hex Cap Screws (Finished Hex Bolts)

Nominal Size or Basic Product Dia		Body Dia E		Width Across Flats F			Width Across Corners G		Height H			Radius of Fillet R	
		Max	Min	Basic	Max	Min	Max	Min	Basic	Max	Min	Max	Min
1/4	0.2500	0.2500	0.2450	7/16	0.4375	0.428	0.505	0.488	5/32	0.163	0.150	0.025	0.015
5/16	0.3125	0.3125	0.3065	1/2	0.5000	0.489	0.577	0.557	13/64	0.211	0.195	0.025	0.015
3/8	0.3750	0.3750	0.3690	9/16	0.5625	0.551	0.650	0.628	15/64	0.243	0.226	0.025	0.015
7/16	0.4375	0.4375	0.4305	5/8	0.6250	0.612	0.722	0.698	9/32	0.291	0.272	0.025	0.015
1/2	0.5000	0.5000	0.4930	3/4	0.7500	0.736	0.866	0.840	5/16	0.323	0.302	0.025	0.015
9/16	0.5625	0.5625	0.5545	13/16	0.8125	0.798	0.938	0.910	23/64	0.371	0.348	0.045	0.020
5/8	0.6250	0.6250	0.6170	15/16	0.9375	0.922	1.083	1.051	25/64	0.403	0.378	0.045	0.020
3/4	0.7500	0.7500	0.7410	1 1/8	1.1250	1.100	1.299	1.254	15/32	0.483	0.455	0.045	0.020
7/8	0.8750	0.8750	0.8660	1 5/16	1.3125	1.285	1.516	1.465	35/64	0.563	0.531	0.065	0.040
1	1.0000	1.0000	0.9900	1 1/2	1.5000	1.469	1.732	1.675	39/64	0.627	0.591	0.095	0.060
1 1/8	1.1250	1.1250	1.1140	1 11/16	1.6875	1.631	1.949	1.859	11/16	0.718	0.658	0.095	0.060
1 1/4	1.2500	1.2500	1.2390	1 7/8	1.8750	1.812	2.165	2.066	25/32	0.813	0.749	0.095	0.060
1 3/8	1.3750	1.3750	1.3630	2 1/16	2.0625	1.994	2.382	2.273	27/32	0.878	0.810	0.095	0.060
1 1/2	1.5000	1.5000	1.4880	2 1/4	2.2500	2.175	2.598	2.480	15/16	0.974	0.902	0.095	0.060
1 3/4	1.7500	1.7500	1.7380	2 5/8	2.6250	2.538	3.031	2.893	1 3/32	1.134	1.054	0.095	0.060
2	2.0000	2.0000	1.9880	3	3.0000	2.900	3.464	3.306	1 7/32	1.263	1.175	0.095	0.060
2 1/4	2.2500	2.2500	2.2380	3 3/8	3.3750	3.262	3.897	3.719	1 3/8	1.423	1.327	0.095	0.060
2 1/2	2.5000	2.5000	2.4880	3 3/4	3.7500	3.625	4.330	4.133	1 17/32	1.583	1.479	0.095	0.060
2 3/4	2.7500	2.7500	2.7380	4 1/8	4.1250	3.988	4.763	4.546	1 11/16	1.744	1.632	0.095	0.060
3	3.0000	3.0000	2.9880	4 1/2	4.5000	4.350	5.196	4.959	1 7/8	1.935	1.815	0.095	0.060

HEAVY HEX NUTS AND HEX JAM NUTS

REGULAR HEX NUTS AND HEX JAM NUTS

Dimensions of Hex Nuts and Hex Jam Nuts

Nominal Size or Basic Major Dia of Thread		Width Across Flats F			Width Across Corners G		Thickness Hex Nuts H			Thickness Hex Jam Nuts H		
		Basic	Max	Min	Max	Min	Basic	Max	Min	Basic	Max	Min
1/4	0.2500	7/16	0.4375	0.428	0.505	0.488	7/32	0.226	0.212	5/32	0.163	0.150
5/16	0.3125	1/2	0.5000	0.489	0.577	0.557	17/64	0.273	0.258	3/16	0.195	0.180
3/8	0.3750	9/16	0.5625	0.551	0.650	0.628	21/64	0.337	0.320	7/32	0.227	0.210
7/16	0.4375	11/16	0.6875	0.675	0.794	0.768	3/8	0.385	0.365	1/4	0.260	0.240
1/2	0.5000	3/4	0.7500	0.736	0.866	0.840	7/16	0.448	0.427	5/16	0.323	0.302
9/16	0.5625	7/8	0.8750	0.861	1.010	0.982	31/64	0.496	0.473	5/16	0.324	0.301
5/8	0.6250	15/16	0.9375	0.922	1.083	1.051	35/64	0.559	0.535	3/8	0.387	0.363
3/4	0.7500	1 1/8	1.1250	1.088	1.299	1.240	41/64	0.665	0.617	27/64	0.446	0.398
7/8	0.8750	1 5/16	1.3125	1.269	1.516	1.447	3/4	0.776	0.724	31/64	0.510	0.458
1	1.0000	1 1/2	1.5000	1.450	1.732	1.653	55/64	0.887	0.831	35/64	0.575	0.519
1 1/8	1.1250	1 11/16	1.6875	1.631	1.949	1.859	31/32	0.999	0.939	39/64	0.639	0.579
1 1/4	1.2500	1 7/8	1.8750	1.812	2.165	2.066	1 1/16	1.094	1.030	23/32	0.751	0.687
1 3/8	1.3750	2 1/16	2.0625	1.994	2.382	2.273	1 11/64	1.206	1.138	25/32	0.815	0.747
1 1/2	1.5000	2 1/4	2.2500	2.175	2.598	2.480	1 9/32	1.317	1.245	27/32	0.880	0.808

(Courtesy of ANSI; B18.2.1–1965 and ANSI; B18.2.2–1965.)

APPENDIX 21

FILLISTER HEAD AND ROUND HEAD CAP SCREWS

Fillister Head Cap Screws

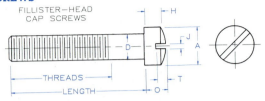

All dimensions are given in inches.

Nominal Size	D Body Diameter		A Head Diameter		H Height of Head		O Total Height of Head		J Width of Slot		T Depth of Slot	
	Max	Min	Max	Min	Max	Min	Max	Min	Max	Min	Max	Min
1/4	0.250	0.245	0.375	0.363	0.172	0.157	0.216	0.194	0.075	0.064	0.097	0.077
5/16	0.3125	0.307	0.437	0.424	0.203	0.186	0.253	0.230	0.084	0.072	0.115	0.090
3/8	0.375	0.369	0.562	0.547	0.250	0.229	0.314	0.284	0.094	0.081	0.142	0.112
7/16	0.4375	0.431	0.625	0.608	0.297	0.274	0.368	0.336	0.094	0.081	0.168	0.133
1/2	0.500	0.493	0.750	0.731	0.328	0.301	0.413	0.376	0.106	0.091	0.193	0.153
9/16	0.5625	0.555	0.812	0.792	0.375	0.346	0.467	0.427	0.118	0.102	0.213	0.168
5/8	0.625	0.617	0.875	0.853	0.422	0.391	0.521	0.478	0.133	0.116	0.239	0.189
3/4	0.750	0.742	1.000	0.976	0.500	0.466	0.612	0.566	0.149	0.131	0.283	0.223
7/8	0.875	0.866	1.125	1.098	0.594	0.556	0.720	0.668	0.167	0.147	0.334	0.264
1	1.000	0.990	1.312	1.282	0.656	0.612	0.803	0.743	0.188	0.166	0.371	0.291

All dimensions are given in inches.

The radius of the fillet at the base of the head:
For sizes 1/4 to 3/8 in. incl. is 0.016 min and 0.031 max,
7/16 to 9/16 in. incl. is 0.016 min and 0.047 max,
5/8 to 1 in. incl. is 0.031 min and 0.062 max.

Round Head Cap Screws

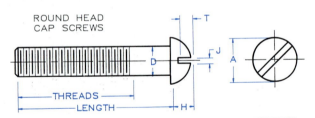

Nominal Size	D Body Diameter		A Head Diameter		H Height of Head		J Width of Slot		T Depth of Slot	
	Max	Min	Max	Min	Max	Min	Max	Min	Max	Min
1/4	0.250	0.245	0.437	0.418	0.191	0.175	0.075	0.064	0.117	0.097
5/16	0.3125	0.307	0.562	0.540	0.245	0.226	0.084	0.072	0.151	0.126
3/8	0.375	0.369	0.625	0.603	0.273	0.252	0.094	0.081	0.168	0.138
7/16	0.4375	0.431	0.750	0.725	0.328	0.302	0.094	0.081	0.202	0.167
1/2	0.500	0.493	0.812	0.786	0.354	0.327	0.106	0.091	0.218	0.178
9/16	0.5625	0.555	0.937	0.909	0.409	0.378	0.118	0.102	0.252	0.207
5/8	0.625	0.617	1.000	0.970	0.437	0.405	0.133	0.116	0.270	0.220
3/4	0.750	0.742	1.250	1.215	0.546	0.507	0.149	0.131	0.338	0.278

All dimensions are given in inches.

Radius of the fillet at the base of the head:
For sizes 1/4 to 3/8 in. incl. is 0.016 min and 0.031 max,
7/16 to 9/16 in. incl..is 0.016 min and 0.047 max,
5/8 to 1 in..incl. is 0.031 min and 0.062 max.

(Courtesy of ANSI; B18.6.2–1956.)

APPENDIX 22
FLAT HEAD CAP SCREWS

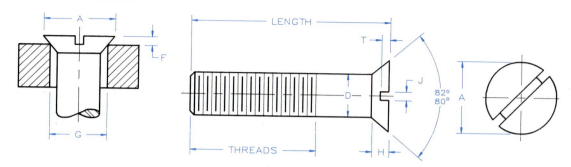

FLAT HEAD CAP SCREWS

Nominal Size	D Body Diameter		A Head Diameter			G Gaging Diameter	H Height of Head	J Width of Slot		T Depth of Slot		F Protrusion Above Gaging Diameter	
	Max	Min	Max	Min	Absolute Min with Flat		Average	Max	Min	Max	Min	Max	Min
1/4	0.250	0.245	0.500	0.477	0.452	0.4245	0.140	0.075	0.064	0.068	0.045	0.0452	0.0307
5/16	0.3125	0.307	0.625	0.598	0.567	0.5376	0.177	0.084	0.072	0.086	0.057	0.0523	0.0354
3/8	0.375	0.369	0.750	0.720	0.682	0.6507	0.210	0.094	0.081	0.103	0.068	0.0594	0.0401
7/16	0.4375	0.431	0.8125	0.780	0.736	0.7229	0.210	0.094	0.081	0.103	0.068	0.0649	0.0448
1/2	0.500	0.493	0.875	0.841	0.791	0.7560	0.210	0.106	0.091	0.103	0.068	0.0705	0.0495
9/16	0.5625	0.555	1.000	0.962	0.906	0.8691	0.244	0.118	0.102	0.120	0.080	0.0775	0.0542
5/8	0.625	0.617	1.125	1.083	1.020	0.9822	0.281	0.133	0.116	0.137	0.091	0.0846	0.0588
3/4	0.750	0.742	1.375	1.326	1.251	1.2085	0.352	0.149	0.131	0.171	0.115	0.0987	0.0682
7/8	0.875	0.866	1.625	1.568	1.480	1.4347	0.423	0.167	0.147	0.206	0.138	0.1128	0.0776
1	1.000	0.990	1.875	1.811	1.711	1.6610	0.494	0.188	0.166	0.240	0.162	0.1270	0.0870
1 1/8	1.125	1.114	2.062	1.992	1.880	1.8262	0.529	0.196	0.178	0.257	0.173	0.1401	0.0964
1 1/4	1.250	1.239	2.312	2.235	2.110	2.0525	0.600	0.211	0.193	0.291	0.197	0.1542	0.1056
1 3/8	1.375	1.363	2.562	2.477	2.340	2.2787	0.665	0.226	0.208	0.326	0.220	0.1684	0.1151
1 1/2	1.500	1.488	2.812	2.720	2.570	2.5050	0.742	0.258	0.240	0.360	0.244	0.1825	0.1245

All dimensions are given in inches.

The maximum and minimum head diameters, A, are extended to the theoretical sharp corners.

The radius of the fillet at the base of the head shall not exceed 0.4 Max. D.

*Edge of head may be flat as shown or slightly rounded.

(Courtesy of ANSI; B18.6.2–1956.)

APPENDIX 23

MACHINE SCREWS

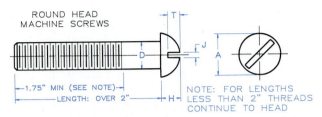

Dimensions of Slotted Round Head Machine Screws

Nominal Size	D — Diameter of Screw — Basic	A — Head Diameter		H — Head Height		J — Width of Slot		T — Depth of Slot	
		Max	Min	Max	Min	Max	Min	Max	Min
0	0.0600	0.113	0.099	0.053	0.043	0.023	0.016	0.039	0.029
1	0.0730	0.138	0.122	0.061	0.051	0.026	0.019	0.044	0.033
2	0.0860	0.162	0.146	0.069	0.059	0.031	0.023	0.048	0.037
3	0.0990	0.187	0.169	0.078	0.067	0.035	0.027	0.053	0.040
4	0.1120	0.211	0.193	0.086	0.075	0.039	0.031	0.058	0.044
5	0.1250	0.236	0.217	0.095	0.083	0.043	0.035	0.063	0.047
6	0.1380	0.260	0.240	0.103	0.091	0.048	0.039	0.068	0.051
8	0.1640	0.309	0.287	0.120	0.107	0.054	0.045	0.077	0.058
10	0.1900	0.359	0.334	0.137	0.123	0.060	0.050	0.087	0.065
12	0.2160	0.408	0.382	0.153	0.139	0.067	0.056	0.096	0.073
1/4	0.2500	0.472	0.443	0.175	0.160	0.075	0.064	0.109	0.082
5/16	0.3125	0.590	0.557	0.216	0.198	0.084	0.072	0.132	0.099
3/8	0.3750	0.708	0.670	0.256	0.237	0.094	0.081	0.155	0.117
7/16	0.4375	0.750	0.707	0.328	0.307	0.094	0.081	0.196	0.148
1/2	0.5000	0.813	0.766	0.355	0.332	0.106	0.091	0.211	0.159
9/16	0.5625	0.938	0.887	0.410	0.385	0.118	0.102	0.242	0.183
5/8	0.6250	1.000	0.944	0.438	0.411	0.133	0.116	0.258	0.195
3/4	0.7500	1.250	1.185	0.547	0.516	0.149	0.131	0.320	0.242

All dimensions are given in inches.

(Courtesy of ANSI; B18.6.3–1962.)

APPENDIX 24
AMERICAN STANDARD MACHINE SCREWS

(The proportions of the screws can be found by multiplying the major diameter, D, by the factors given below.)

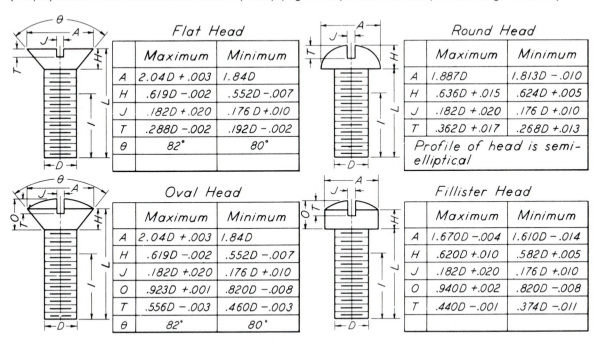

Flat Head

	Maximum	Minimum
A	2.04D + .003	1.84D
H	.619D − .002	.552D − .007
J	.182D + .020	.176 D + .010
T	.288D − .002	.192D − .002
θ	82°	80°

Round Head

	Maximum	Minimum
A	1.887D	1.813D − .010
H	.636D + .015	.624D + .005
J	.182D + .020	.176 D + .010
T	.362D + .017	.268D + .013

Profile of head is semi-elliptical

Oval Head

	Maximum	Minimum
A	2.04D + .003	1.84D
H	.619D − .002	.552D − .007
J	.182D + .020	.176 D + .010
O	.923D + .001	.820D − .008
T	.556D − .003	.460D − .003
θ	82°	80°

Fillister Head

	Maximum	Minimum
A	1.670D − .004	1.610D − .014
H	.620D + .010	.582D + .005
J	.182D + .020	.176 D + .010
O	.940D + .002	.820D − .008
T	.440D − .001	.374D − .011

APPENDIX 25
AMERICAN STANDARD MACHINE TAPERS*

No. of Taper	Taper per Foot (Basic)	Origin of Series	No. of Taper	Taper per Foot (Basic)	Origin of Series	No. of Taper	Taper per Foot (Basic)	Origin of Series	No. of Taper	Taper per Foot (Basic)	Origin of Series
0.239	0.50200	Brown & Sharpe	*	0.62326	Morse	250	0.750	¾ in. per ft.	600	0.750	¾ in. per ft.
.299	.50200	Brown & Sharpe	4½	.62400	Morse	300	.750	¾ in. per ft.	800	0.750	¾ in. per ft.
.375	.50200	Brown & Sharpe	5	.63151	Morse	350	.750	¾ in. per ft.	1000	0.750	¾ in. per ft.
1	.59858	Morse	6	.62565	Morse	400	.750	¾ in. per ft.	1200	0.750	¾ in. per ft.
2	.59941	Morse	7	.62400	Morse	450	.750	¾ in. per ft.			
3	.60235	Morse	200	.750	¾ in. per ft.	500	.750	¾ in. per ft.			

All dimensions in inches.
* Extracted from American Standards, ''Machine Tapers, Self-Holding and Steep Taper Series'' (ASA B5,10-1960), with the permission of the publisher, The American Society of Mechanical Engineers.

APPENDIX 26

AMERICAN NATIONAL STANDARD SQUARE HEAD SET SCREWS (ANSI B18.6.2)

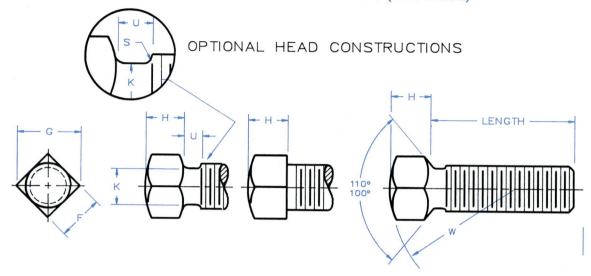

OPTIONAL HEAD CONSTRUCTIONS

Nominal Size[1] or Basic Screw Diameter		F Width Across Flats		G Width Across Corners		H Head Height		K Neck Relief Diameter		S Neck Relief Fillet Radius	U Neck Relief Width	W Head Radius
		Max	Min	Max	Min	Max	Min	Max	Min	Max	Min	Min
10	0.1900	0.188	0.180	0.265	0.247	0.148	0.134	0.145	0.140	0.027	0.083	0.48
1/4	0.2500	0.250	0.241	0.354	0.331	0.196	0.178	0.185	0.170	0.032	0.100	0.62
5/16	0.3125	0.312	0.302	0.442	0.415	0.245	0.224	0.240	0.225	0.036	0.111	0.78
3/8	0.3750	0.375	0.362	0.530	0.497	0.293	0.270	0.294	0.279	0.041	0.125	0.94
7/16	0.4375	0.438	0.423	0.619	0.581	0.341	0.315	0.345	0.330	0.046	0.143	1.09
1/2	0.5000	0.500	0.484	0.707	0.665	0.389	0.361	0.400	0.385	0.050	0.154	1.25
9/16	0.5625	0.562	0.545	0.795	0.748	0.437	0.407	0.454	0.439	0.054	0.167	1.41
5/8	0.6250	0.625	0.606	0.884	0.833	0.485	0.452	0.507	0.492	0.059	0.182	1.56
3/4	0.7500	0.750	0.729	1.060	1.001	0.582	0.544	0.620	0.605	0.065	0.200	1.88
7/8	0.8750	0.875	0.852	1.237	1.170	0.678	0.635	0.731	0.716	0.072	0.222	2.19
1	1.0000	1.000	0.974	1.414	1.337	0.774	0.726	0.838	0.823	0.081	0.250	2.50
1 1/8	1.1250	1.125	1.096	1.591	1.505	0.870	0.817	0.939	0.914	0.092	0.283	2.81
1 1/4	1.2500	1.250	1.219	1.768	1.674	0.966	0.908	1.064	1.039	0.092	0.283	3.12
1 3/8	1.3750	1.375	1.342	1.945	1.843	1.063	1.000	1.159	1.134	0.109	0.333	3.44
1 1/2	1.5000	1.500	1.464	2.121	2.010	1.159	1.091	1.284	1.259	0.109	0.333	3.75

[1] Where specifying nominal size in decimals, zeros preceding decimal and in the fourth decimal place shall be omitted.

APPENDIX 27

AMERICAN NATIONAL STANDARD POINTS FOR SQUARE HEAD SET SCREWS (ANSI B18.6.2)

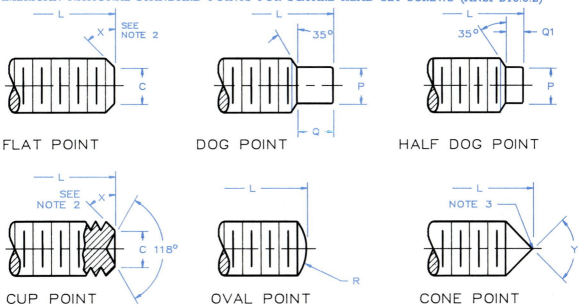

Nominal Size[1] or Basic Screw Diameter		C Cup and Flat Point Diameters		P Dog and Half Dog Point Diameters		Q Point Length				R Oval Point Radius +0.031 −0.000	Y Cone Point Angle 90° ±2° For These Nominal Lengths or Longer; 118° ±2° For Shorter Screws
						Dog		Half Dog			
		Max	Min	Max	Min	Max	Min	Max	Min		
10	0.1900	0.102	0.088	0.127	0.120	0.095	0.085	0.050	0.040	0.142	1/4
1/4	0.2500	0.132	0.118	0.156	0.149	0.130	0.120	0.068	0.058	0.188	5/16
5/16	0.3125	0.172	0.156	0.203	0.195	0.161	0.151	0.083	0.073	0.234	3/8
3/8	0.3750	0.212	0.194	0.250	0.241	0.193	0.183	0.099	0.089	0.281	7/16
7/16	0.4375	0.252	0.232	0.297	0.287	0.224	0.214	0.114	0.104	0.328	1/2
1/2	0.5000	0.291	0.270	0.344	0.334	0.255	0.245	0.130	0.120	0.375	9/16
9/16	0.5625	0.332	0.309	0.391	0.379	0.287	0.275	0.146	0.134	0.422	5/8
5/8	0.6250	0.371	0.347	0.469	0.456	0.321	0.305	0.164	0.148	0.469	3/4
3/4	0.7500	0.450	0.425	0.562	0.549	0.383	0.367	0.196	0.180	0.562	7/8
7/8	0.8750	0.530	0.502	0.656	0.642	0.446	0.430	0.227	0.211	0.656	1
1	1.0000	0.609	0.579	0.750	0.734	0.510	0.490	0.260	0.240	0.750	1 1/8
1 1/8	1.1250	0.689	0.655	0.844	0.826	0.572	0.552	0.291	0.271	0.844	1 1/4
1 1/4	1.2500	0.767	0.733	0.938	0.920	0.635	0.615	0.323	0.303	0.938	1 1/2
1 3/8	1.3750	0.848	0.808	1.031	1.011	0.698	0.678	0.354	0.334	1.031	1 5/8
1 1/2	1.5000	0.926	0.886	1.125	1.105	0.760	0.740	0.385	0.365	1.125	1 3/4

[1] Where specifying nominal size in decimals, zeros preceding decimal and in the fourth decimal place shall be omitted.

[2] Point angle X shall be 45° plus 5°, minus 0°, for screws of nominal lengths equal to or longer than those listed in Column Y, and 30° minimum for screws of shorter nominal lengths.

[3] The extent of rounding or flat at apex of cone point shall not exceed an amount equivalent to 10 per cent of the basic screw diameter.

APPENDIX 28
AMERICAN NATIONAL STANDARD SLOTTED HEADLESS SET SCREWS (ANSI B18.6.2)

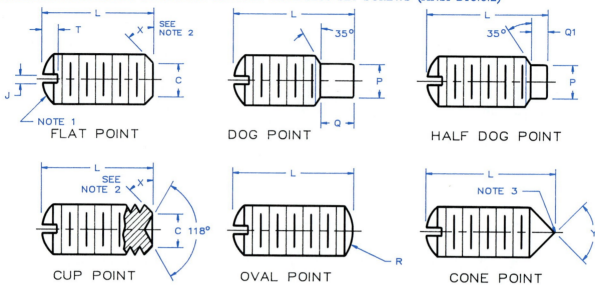

FLAT POINT　　　DOG POINT　　　HALF DOG POINT

CUP POINT　　　OVAL POINT　　　CONE POINT

Nominal Size[1] or Basic Screw Diameter		I[2] Crown Radius	J Slot Width		T Slot Depth		C Cup and Flat Point Diameters		P Dog Point Diameters		Q Point Length Dog		Q_1 Point Length Half Dog		R[2] Oval Point Radius	Y Cone Point Angle 90° ±2° For These Nominal Lengths or Longer; 118° ±2° For Shorter Screws
		Basic	Max	Min	Max	Min	Max	Min	Max	Min	Max	Min	Max	Min	Basic	
0	0.0600	0.060	0.014	0.010	0.020	0.016	0.033	0.027	0.040	0.037	0.032	0.028	0.017	0.013	0.045	5/64
1	0.0730	0.073	0.016	0.012	0.020	0.016	0.040	0.033	0.049	0.045	0.040	0.036	0.021	0.017	0.055	3/32
2	0.0860	0.086	0.018	0.014	0.025	0.019	0.047	0.039	0.057	0.053	0.046	0.042	0.024	0.020	0.064	7/64
3	0.0990	0.099	0.020	0.016	0.028	0.022	0.054	0.045	0.066	0.062	0.052	0.048	0.027	0.023	0.074	1/8
4	0.1120	0.112	0.024	0.018	0.031	0.025	0.061	0.051	0.075	0.070	0.058	0.054	0.030	0.026	0.084	5/32
5	0.1250	0.125	0.026	0.020	0.036	0.026	0.067	0.057	0.083	0.078	0.063	0.057	0.033	0.027	0.094	3/16
6	0.1380	0.138	0.028	0.022	0.040	0.030	0.074	0.064	0.092	0.087	0.073	0.067	0.038	0.032	0.104	3/16
8	0.1640	0.164	0.032	0.026	0.046	0.036	0.087	0.076	0.109	0.103	0.083	0.077	0.043	0.037	0.123	1/4
10	0.1900	0.190	0.035	0.029	0.053	0.043	0.102	0.088	0.127	0.120	0.095	0.085	0.050	0.040	0.142	1/4
12	0.2160	0.216	0.042	0.035	0.061	0.051	0.115	0.101	0.144	0.137	0.115	0.105	0.060	0.050	0.162	5/16
1/4	0.2500	0.250	0.049	0.041	0.068	0.058	0.132	0.118	0.156	0.149	0.130	0.120	0.068	0.058	0.188	5/16
5/16	0.3125	0.312	0.055	0.047	0.083	0.073	0.172	0.156	0.203	0.195	0.161	0.151	0.083	0.073	0.234	3/8
3/8	0.3750	0.375	0.068	0.060	0.099	0.089	0.212	0.194	0.250	0.241	0.193	0.183	0.099	0.089	0.281	7/16
7/16	0.4375	0.438	0.076	0.068	0.114	0.104	0.252	0.232	0.297	0.287	0.224	0.214	0.114	0.104	0.328	1/2
1/2	0.5000	0.500	0.086	0.076	0.130	0.120	0.291	0.270	0.344	0.334	0.255	0.245	0.130	0.120	0.375	9/16
9/16	0.5625	0.562	0.096	0.086	0.146	0.136	0.332	0.309	0.391	0.379	0.287	0.275	0.146	0.134	0.422	5/8
5/8	0.6250	0.625	0.107	0.097	0.161	0.151	0.371	0.347	0.469	0.456	0.321	0.305	0.164	0.148	0.469	3/4
3/4	0.7500	0.750	0.134	0.124	0.193	0.183	0.450	0.425	0.562	0.549	0.383	0.367	0.196	0.180	0.562	7/8

[1] Where specifying nominal size in decimals, zeros preceding decimal and in the fourth decimal place shall be omitted.

[2] Tolerance on radius for nominal sizes up to and including 5 (0.125 in.) shall be plus 0.015 in. and minus 0.000, and for larger sizes, plus 0.031 in. and minus 0.000. Slotted ends on screws may be flat at option of manufacturer.

[3] Point angle X shall be 45° plus 5°, minus 0°, for screws of nominal lengths equal to or longer than those listed in Column Y, and 30° minimum for screws of shorter nominal lengths.

[4] The extent of rounding or flat at apex of cone point shall not exceed an amount equivalent to 10 per cent of the basic screw diameter.

APPENDIX 29

TWIST DRILL SIZES

Number Size Drills

Size	Drill Diameter		Size	Drill Diameter		Size	Drill Diameter		Size	Drill Diameter	
	Inches	mm		Inches	mm		Inches	mm		Inches	mm
1	0.2280	5.7912	21	0.1590	4.0386	41	0.0960	2.4384	61	0.0390	0.9906
2	0.2210	5.6134	22	0.1570	3.9878	42	0.0935	2.3622	62	0.0380	0.9652
3	0.2130	5.4102	23	0.1540	3.9116	43	0.0890	2.2606	63	0.0370	0.9398
4	0.2090	5.3086	24	0.1520	3.8608	44	0.0860	2.1844	64	0.0360	0.9144
5	0.2055	5.2197	25	0.1495	3.7973	45	0.0820	2.0828	65	0.0350	0.8890
6	0.2040	5.1816	26	0.1470	3.7338	46	0.0810	2.0574	66	0.0330	0.8382
7	0.2010	5.1054	27	0.1440	3.6576	47	0.0785	1.9812	67	0.0320	0.8128
8	0.1990	5.0800	28	0.1405	3.5560	48	0.0760	1.9304	68	0.0310	0.7874
9	0.1960	4.9784	29	0.1360	3.4544	49	0.0730	1.8542	69	0.0292	0.7417
10	0.1935	4.9149	30	0.1285	3.2639	50	0.0700	1.7780	70	0.0280	0.7112
11	0.1910	4.8514	31	0.1200	3.0480	51	0.0670	1.7018	71	0.0260	0.6604
12	0.1890	4.8006	32	0.1160	2.9464	52	0.0635	1.6129	72	0.0250	0.6350
13	0.1850	4.6990	33	0.1130	2.8702	53	0.0595	1.5113	73	0.0240	0.6096
14	0.1820	4.6228	34	0.1110	2.8194	54	0.0550	1.3970	74	0.0225	0.5715
15	0.1800	4.5720	35	0.1100	2.7940	55	0.0520	1.3208	75	0.0210	0.5334
16	0.1770	4.4958	36	0.1065	2.7051	56	0.0465	1.1684	76	0.0200	0.5080
17	0.1730	4.3942	37	0.1040	2.6416	57	0.0430	1.0922	77	0.0180	0.4572
18	0.1695	4.3053	38	0.1015	2.5781	58	0.0420	1.0668	78	0.0160	0.4064
19	0.1660	4:2164	39	0.0995	2.5273	59	0.0410	1.0414	79	0.0145	0.3638
20	0.1610	4.0894	40	0.0980	2.4892	60	0.0400	1.0160	80	0.0135	0.3429

Metric Drill Sizes   Preferred sizes are in color type. Decimal-inch equivalents are for reference only.

Drill Diameter		Drill Diameter		Drill Diameter		Drill Diameter		Drill Diameter		Drill Diameter		Drill Diameter	
mm	in.	mm	in.	mm	in.	mm	in.	mm	in.	mm	in.	mm	in.
.40	.0157	1.03	.0406	2.20	.0866	5.00	.1969	10.00	.3937	21.50	.8465	48.00	1.8898
.42	.0165	1.05	.0413	2.30	.0906	5.20	.2047	10.30	.4055	22.00	.8661	50.00	1.9685
.45	.0177	1.08	.0425	2.40	.0945	5.30	.2087	10.50	.4134	23.00	.9055	51.50	2.0276
.48	.0189	1.10	.0433	2.50	.0984	5.40	.2126	10.80	.4252	24.00	.9449	53.00	2.0866
.50	.0197	1.15	.0453	2.60	.1024	5.60	.2205	11.00	.4331	25.00	.9843	54.00	2.1260
.52	.0205	1.20	.0472	2.70	.1063	5.80	.2283	11.50	.4528	26.00	1.0236	56.00	2.2047
.55	.0217	1.25	.0492	2.80	.1102	6.00	.2362	12.00	.4724	27.00	1.0630	58.00	2.2835
.58	.0228	1.30	.0512	2.90	.1142	6.20	.2441	12.50	.4921	28.00	1.1024	60.00	2.3622
.60	.0236	1.35	.0531	3.00	.1181	6.30	.2480	13.00	.5118	29.00	1.1417		
.62	.0244	1.40	.0551	3.10	.1220	6.50	.2559	13.50	.5315	30.00	1.1811		
.65	.0256	1.45	.0571	3.20	.1260	6.70	.2638	14.00	.5512	31.00	1.2205		
.68	.0268	1.50	.0591	3.30	.1299	6.80	.2677	14.50	.5709	32.00	1.2598		
.70	.0276	1.55	.0610	3.40	.1339	6.90	.2717	15.00	.5906	33.00	1.2992		
.72	.0283	1.60	.0630	3.50	.1378	7.10	.2795	15.50	.6102	34.00	1.3386		
.75	.0295	1.65	.0650	3.60	.1417	7.30	.2874	16.00	.6299	35.00	1.3780		
.78	.0307	1.70	.0669	3.70	.1457	7.50	.2953	16.50	.6496	36.00	1.4173		
.80	.0315	1.75	.0689	3.80	.1496	7.80	.3071	17.00	.6693	37.00	1.4567		
.82	.0323	1.80	.0709	3.90	.1535	8.00	.3150	17.50	.6890	38.00	1.4961		
.85	.0335	1.85	.0728	4.00	.1575	8.20	.3228	18.00	.7087	39.00	1.5354		
.88	.0346	1.90	.0748	4.10	.1614	8.50	.3346	18.50	.7283	40.00	1.5748		
.90	.0354	1.95	.0768	4.20	.1654	8.80	.3465	19.00	.7480	41.00	1.6142		
.92	.0362	2.00	.0787	4.40	.1732	9.00	.3543	19.50	.7677	42.00	1.6535		
.95	.0374	2.05	.0807	4.50	.1772	9.20	.3622	20.00	.7874	43.50	1.7126		
.98	.0386	2.10	.0827	4.60	.1811	9.50	.3740	20.50	.8071	45.00	1.7717		
1.00	.0394	2.15	.0846	4.80	.1890	9.80	.3858	21.00	.8268	46.50	1.8307		

APPENDIX 29
TWIST DRILL SIZES (Cont.)
Letter Size Drills

Size	Drill Diameter		Size	Drill Diameter		Size	Drill Diameter		Size	Drill Diameter	
	Inches	mm		Inches	mm		Inches	mm		Inches	mm
A	0.234	5.944	H	0.266	6.756	O	0.316	8.026	V	0.377	9.576
B	0.238	6.045	I	0.272	6.909	P	0.323	8.204	W	0.386	9.804
C	0.242	6.147	J	0.277	7.036	Q	0.332	8.433	X	0.397	10.084
D	0.246	6.248	K	0.281	7.137	R	0.339	8.611	Y	0.404	10.262
E	0.250	6.350	L	0.290	7.366	S	0.348	8.839	Z	0.413	10.490
F	0.257	6.528	M	0.295	7.493	T	0.358	9.093			
G	0.261	6.629	N	0.302	7.601	U	0.368	9.347			

(Courtesy of General Motors Corporation.)

APPENDIX 30
STRAIGHT PINS

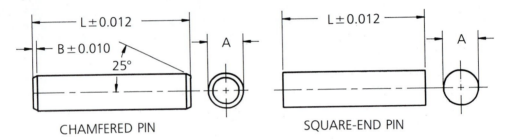

CHAMFERED PIN   SQUARE-END PIN

Nominal Diameter	Diameter A		Chamfer B
	Max	Min	
0.062	0.0625	0.0605	0.015
0.094	0.0937	0.0917	0.015
0.109	0.1094	0.1074	0.015
0.125	0.1250	0.1230	0.015
0.156	0.1562	0.1542	0.015
0.188	0.1875	0.1855	0.015
0.219	0.2187	0.2167	0.015
0.250	0.2500	0.2480	0.015
0.312	0.3125	0.3095	0.030
0.375	0.3750	0.3720	0.030
0.438	0.4375	0.4345	0.030
0.500	0.500	0.4970	0.030

All dimensions are given in inches.

These pins must be straight and free from burrs or any other defects that will affect their serviceability.

(Courtesy of ANSI; B5.20–1958.)

APPENDIX 31

STANDARD KEYS AND KEYWAYS

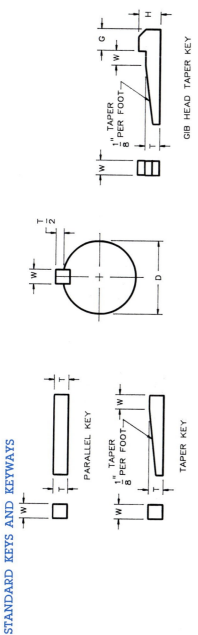

PARALLEL KEY

TAPER KEY — 1/8" TAPER PER FOOT

GIB HEAD TAPER KEY — 1/8" TAPER PER FOOT

Sprocket Bore (= Shaft Diam.) Inches D	Keyway Dimensions — Inches For Square Key Width W	Keyway Depth T/2	For Flat Key Width W	Flat Depth T/2	Key Dimensions Square Width W × Height T	Key Flat Width W × Height T	Tolerance on W and T (−)	Gib Head Square Key H	Square Key G	Gib Head Flat Key H	Flat Key G	Key Tol. W (−)	Key Tol. T (+)
1/2 — 9/16	1/8	1/16	1/8	3/64	1/8 × 1/8	1/8 × 3/32	0.002	1/4	7/32	3/16	1/8	0.002	0.002
5/8 — 7/8	3/16	3/32	3/16	1/16	3/16 × 3/16	3/16 × 1/8	0.002	5/16	9/32	1/4	3/16	0.002	0.002
13/16 — 1 1/4	1/4	1/8	1/4	3/32	1/4 × 1/4	1/4 × 3/16	0.002	7/16	11/32	5/16	1/4	0.002	0.002
1 3/16 — 1 3/8	5/16	5/32	5/16	1/8	5/16 × 5/16	5/16 × 1/4	0.002	9/16	13/32	3/8	3/16	0.002	0.002
1 7/16 — 1 3/4	3/8	3/16	3/8	1/8	3/8 × 3/8	3/8 × 1/4	0.002	11/16	15/32	7/16	3/8	0.002	0.002
1 13/16 — 2 1/4	1/2	1/4	1/2	3/16	1/2 × 1/2	1/2 × 3/8	0.0025	7/8	19/32	5/8	1/2	0.0025	0.0025
2 5/16 — 2 3/4	5/8	5/16	5/8	7/32	5/8 × 5/8	5/8 × 7/16	0.0025	1 1/16	23/32	3/4	5/8	0.0025	0.0025
2 7/8 — 3 1/4	3/4	3/8	3/4	1/4	3/4 × 3/4	3/4 × 1/2	0.0025	1 1/4	7/8	7/8	3/4	0.0025	0.0025
3 3/8 — 3 3/4	7/8	7/16	7/8	5/16	7/8 × 7/8	7/8 × 5/8	0.003	1 1/2	1	1 1/16	7/8	0.003	0.003
3 7/8 — 4 1/2	1	1/2	1	3/8	1 × 1	1 × 3/4	0.003	1 3/4	1 3/16	1 1/4	1	0.003	0.003
4 3/4 — 5 1/2	1 1/4	5/8	1 1/4	7/16	1 1/4 × 1 1/4	1 1/4 × 7/8	0.003	2	1 7/16	1 1/2	1 1/4	0.003	0.003
5 3/4 — 7 3/8	1 1/2	3/4	1 1/2	1/2	1 1/2 × 1 1/2	1 1/2 × 1	0.003	2 1/2	1 3/4	1 3/4	1 1/2	0.003	0.003
7 1/2 — 9 7/8	1 3/4	7/8	. .	. .	1 3/4 × 1 3/4	. .	0.004	3	2	. .	. .	0.004	0.004
10 — 12 1/2	2	1	. .	. .	2 × 2	. .	0.004	3 1/2	2 3/8	. .	. .	0.004	0.004

Standard Keyway Tolerances: Straight Keyway — Width (W) + .005 Depth (T/2) + .010
                                                                    − .000   − .000

Taper Keyway — Width (W) + .005 Depth (T/2) + .000
                                    − .000   − .010

APPENDIX 32
WOODRUFF KEYS

USA STANDARD

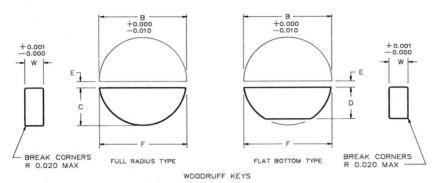

FULL RADIUS TYPE   FLAT BOTTOM TYPE

BREAK CORNERS R 0.020 MAX

WOODRUFF KEYS

Key No.	Nominal Key Size W × B	Actual Length F +0.000-0.010	Height of Key				Distance Below Center E
			C		D		
			Max	Min	Max	Min	
202	1/16 × 1/4	0.248	0.109	0.104	0.109	0.104	1/64
202.5	1/16 × 5/16	0.311	0.140	0.135	0.140	0.135	1/64
302.5	3/32 × 5/16	0.311	0.140	0.135	0.140	0.135	1/64
203	1/16 × 3/8	0.374	0.172	0.167	0.172	0.167	1/64
303	3/32 × 3/8	0.374	0.172	0.167	0.172	0.167	1/64
403	1/8 × 3/8	0.374	0.172	0.167	0.172	0.167	1/64
204	1/16 × 1/2	0.491	0.203	0.198	0.194	0.188	3/64
304	3/32 × 1/2	0.491	0.203	0.198	0.194	0.188	3/64
404	1/8 × 1/2	0.491	0.203	0.198	0.194	0.188	3/64
305	3/32 × 5/8	0.612	0.250	0.245	0.240	0.234	1/16
405	1/8 × 5/8	0.612	0.250	0.245	0.240	0.234	1/16
505	5/32 × 5/8	0.612	0.250	0.245	0.240	0.234	1/16
605	3/16 × 5/8	0.612	0.250	0.245	0.240	0.234	1/16
406	1/8 × 3/4	0.740	0.313	0.308	0.303	0.297	1/16
506	5/32 × 3/4	0.740	0.313	0.308	0.303	0.297	1/16
606	3/16 × 3/4	0.740	0.313	0.308	0.303	0.297	1/16
806	1/4 × 3/4	0.740	0.313	0.308	0.303	0.297	1/16
507	5/32 × 7/8	0.866	0.375	0.370	0.365	0.359	1/16
607	3/16 × 7/8	0.866	0.375	0.370	0.365	0.359	1/16
707	7/32 × 7/8	0.866	0.375	0.370	0.365	0.359	1/16
807	1/4 × 7/8	0.866	0.375	0.370	0.365	0.359	1/16
608	3/16 × 1	0.992	0.438	0.433	0.428	0.422	1/16
708	7/32 × 1	0.992	0.438	0.433	0.428	0.422	1/16
808	1/4 × 1	0.992	0.438	0.433	0.428	0.422	1/16
1008	5/16 × 1	0.992	0.438	0.433	0.428	0.422	1/16
1208	3/8 × 1	0.992	0.438	0.433	0.428	0.422	1/16
609	3/16 × 1 1/8	1.114	0.484	0.479	0.475	0.469	5/64
709	7/32 × 1 1/8	1.114	0.484	0.479	0.475	0.469	5/64
809	1/4 × 1 1/8	1.114	0.484	0.479	0.475	0.469	5/64
1009	5/16 × 1 1/8	1.114	0.484	0.479	0.475	0.469	5/64

(Courtesy of ANSI; B17.2-1967.)

APPENDIX 33
WOODRUFF KEYSEATS

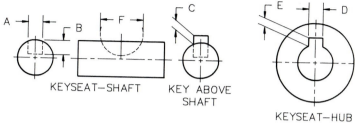

KEYSEAT—SHAFT KEY ABOVE SHAFT KEYSEAT—HUB

Keyseat Dimensions

Key Number	Nominal Size Key	Keyseat – Shaft					Key Above Shaft	Keyseat – Hub	
		Width A*		Depth B	Diameter F		Height C	Width D	Depth E
		Min	Max	+0.005 −0.000	Min	Max	+0.005 −0.005	+0.002 −0.000	+0.005 −0.000
202	1/16 × 1/4	0.0615	0.0630	0.0728	0.250	0.268	0.0312	0.0635	0.0372
202.5	1/16 × 5/16	0.0615	0.0630	0.1038	0.312	0.330	0.0312	0.0635	0.0372
302.5	3/32 × 5/16	0.0928	0.0943	0.0882	0.312	0.330	0.0469	0.0948	0.0529
203	1/16 × 3/8	0.0615	0.0630	0.1358	0.375	0.393	0.0312	0.0635	0.0372
303	3/32 × 3/8	0.0928	0.0943	0.1202	0.375	0.393	0.0469	0.0948	0.0529
403	1/8 × 3/8	0.1240	0.1255	0.1045	0.375	0.393	0.0625	0.1260	0.0685
204	1/16 × 1/2	0.0615	0.0630	0.1668	0.500	0.518	0.0312	0.0635	0.0372
304	3/32 × 1/2	0.0928	0.0943	0.1511	0.500	0.518	0.0469	0.0948	0.0529
404	1/8 × 1/2	0.1240	0.1255	0.1355	0.500	0.518	0.0625	0.1260	0.0685
305	3/32 × 5/8	0.0928	0.0943	0.1981	0.625	0.643	0.0469	0.0948	0.0529
405	1/8 × 5/8	0.1240	0.1255	0.1825	0.625	0.643	0.0625	0.1260	0.0685
505	5/32 × 5/8	0.1553	0.1568	0.1669	0.625	0.643	0.0781	0.1573	0.0841
605	3/16 × 5/8	0.1863	0.1880	0.1513	0.625	0.643	0.0937	0.1885	0.0997
406	1/8 × 3/4	0.1240	0.1255	0.2455	0.750	0.768	0.0625	0.1260	0.0685
506	5/32 × 3/4	0.1553	0.1568	0.2299	0.750	0.768	0.0781	0.1573	0.0841
606	3/16 × 3/4	0.1863	0.1880	0.2143	0.750	0.768	0.0937	0.1885	0.0997
806	1/4 × 3/4	0.2487	0.2505	0.1830	0.750	0.768	0.1250	0.2510	0.1310
507	5/32 × 7/8	0.1553	0.1568	0.2919	0.875	0.895	0.0781	0.1573	0.0841
607	3/16 × 7/8	0.1863	0.1880	0.2763	0.875	0.895	0.0937	0.1885	0.0997
707	7/32 × 7/8	0.2175	0.2193	0.2607	0.875	0.895	0.1093	0.2198	0.1153
807	1/4 × 7/8	0.2487	0.2505	0.2450	0.875	0.895	0.1250	0.2510	0.1310
608	3/16 × 1	0.1863	0.1880	0.3393	1.000	1.020	0.0937	0.1885	0.0997
708	7/32 × 1	0.2175	0.2193	0.3237	1.000	1.020	0.1093	0.2198	0.1153
808	1/4 × 1	0.2487	0.2505	0.3080	1.000	1.020	0.1250	0.2510	0.1310
1008	5/16 × 1	0.3111	0.3130	0.2768	1.000	1.020	0.1562	0.3135	0.1622
1208	3/8 × 1	0.3735	0.3755	0.2455	1.000	1.020	0.1875	0.3760	0.1935
609	3/16 × 1 1/8	0.1863	0.1880	0.3853	1.125	1.145	0.0937	0.1885	0.0997
709	7/32 × 1 1/8	0.2175	0.2193	0.3697	1.125	1.145	0.1093	0.2198	0.1153
809	1/4 × 1 1/8	0.2487	0.2505	0.3540	1.125	1.145	0.1250	0.2510	0.1310
1009	5/16 × 1 1/8	0.3111	0.3130	0.3228	1.125	1.145	0.1562	0.3135	0.1622

(Courtesy of ANSI; B17.2–1967.)

APPENDIX 34
TAPER PINS

Number	7/0	6/0	5/0	4/0	3/0	2/0	0	1	2	3	4	5	6	7	8	9	10
Size (large end)	0.0625	0.0780	0.0940	0.1090	0.1250	0.1410	0.1560	0.1720	0.1930	0.2190	0.2500	0.2890	0.3410	0.4090	0.4920	0.5910	0.7060
Length, L																	
0.375	X	X															
0.500	X	X	X	X	X	X	X										
0.625	X	X	X	X	X	X	X										
0.750			X	X	X	X	X	X									
0.875			X	X	X	X	X	X	X								
1.000			X	X	X	X	X	X	X	X							
1.250						X	X	X	X	X	X						
1.500							X	X	X	X	X	X					
1.750							X	X	X	X	X	X	X				
2.000								X	X	X	X	X	X	X			
2.250									X	X	X	X	X	X	X		
2.500									X	X	X	X	X	X	X		
2.750										X	X	X	X	X	X	X	
3.000										X	X	X	X	X	X	X	
3.250												X	X	X	X	X	
3.500													X	X	X	X	X
3.750													X	X	X	X	X
4.000													X	X	X	X	X
4.250														X	X	X	X
4.500															X	X	X
4.750															X	X	X
5.000															X	X	X
5.250																X	X
5.500																X	X
5.750																X	X
6.000																X	X

All dimensions are given in inches.

Standard reamers are available for pins given above the line.

Pins Nos. 11 (size 0.8600), 12 (size 1.032), 13 (size 1.241), and 14 (1.523) are special sizes—hence their lengths are special.

To find small diameter of pin, multiply the length by 0.02083 and subtract the result from the large diameter.

(Courtesy of ANSI; B5.20–1958.)

APPENDIX 35
PLAIN WASHERS

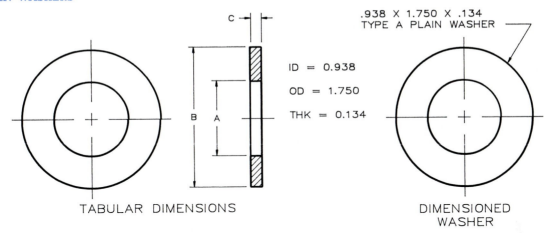

ID = 0.938

OD = 1.750

THK = 0.134

.938 X 1.750 X .134
TYPE A PLAIN WASHER

TABULAR DIMENSIONS

DIMENSIONED WASHER

Dimensions of Preferred Sizes of Type A Plain Washers[a]

When specifying washers on drawings or in notes, give the inside diameter, outside diameter, and the thickness. *Example:* 0.938 × 1.750 × 0.134 TYPE A PLAIN WASHER.

Nominal Washer Size[b]			Inside Diameter A			Outside Diameter B			Thickness C		
				Tolerance			Tolerance				
			Basic	Plus	Minus	Basic	Plus	Minus	Basic	Max	Min
—	—		0.078	0.000	0.005	0.188	0.000	0.005	0.020	0.025	0.016
—	—		0.094	0.000	0.005	0.250	0.000	0.005	0.020	0.025	0.016
—	—		0.125	0.008	0.005	0.312	0.008	0.005	0.032	0.040	0.025
No. 6	0.138		0.156	0.008	0.005	0.375	0.015	0.005	0.049	0.065	0.036
No. 8	0.164		0.188	0.008	0.005	0.438	0.015	0.005	0.049	0.065	0.036
No. 10	0.190		0.219	0.008	0.005	0.500	0.015	0.005	0.049	0.065	0.036
$\frac{3}{16}$	0.188		0.250	0.015	0.005	0.562	0.015	0.005	0.049	0.065	0.036
No. 12	0.216		0.250	0.015	0.005	0.562	0.015	0.005	0.065	0.080	0.051
$\frac{1}{4}$	0.250	N	0.281	0.015	0.005	0.625	0.015	0.005	0.065	0.080	0.051
$\frac{1}{4}$	0.250	W	0.312	0.015	0.005	0.734[c]	0.015	0.007	0.065	0.080	0.051
$\frac{5}{16}$	0.312	N	0.344	0.015	0.005	0.688	0.015	0.007	0.065	0.080	0.051
$\frac{5}{16}$	0.312	W	0.375	0.015	0.005	0.875	0.030	0.007	0.083	0.104	0.064
$\frac{3}{8}$	0.375	N	0.406	0.015	0.005	0.812	0.015	0.007	0.065	0.080	0.051
$\frac{3}{8}$	0.375	W	0.438	0.015	0.005	1.000	0.030	0.007	0.083	0.104	0.064
$\frac{7}{16}$	0.438	N	0.469	0.015	0.005	0.922	0.015	0.007	0.065	0.080	0.051
$\frac{7}{16}$	0.438	W	0.500	0.015	0.005	1.250	0.030	0.007	0.083	0.104	0.064
$\frac{1}{2}$	0.500	N	0.531	0.015	0.005	1.062	0.030	0.007	0.095	0.121	0.074
$\frac{1}{2}$	0.500	W	0.562	0.015	0.005	1.375	0.030	0.007	0.109	0.132	0.086

[a] Preferred sizes are for the most part from series previously designated "Standard Plate" and "SAE." Where common sizes existed in the two series, the SAE size is designated "N" (narrow) and the Standard Plate "W" (wide). These sizes as well as all other sizes of Type A Plain Washers are to be ordered by ID, OD, and thickness dimensions.

[b] Nominal washer sizes are intended for use with comparable nominal screw or bolt sizes.

[c] The 0.734 in., 1.156 in., and 1.469 in. outside diameters avoid washers which could be used in coin-operated devices.

Cont.

APPENDIX 35
PLAIN WASHERS (Cont.)

Nominal Washer Size[b]			Inside Diameter A			Outside Diameter B			Thickness C		
			Basic	Tolerance		Basic	Tolerance		Basic	Max	Min
				Plus	Minus		Plus	Minus			
$\frac{9}{16}$	0.562	N	0.594	0.015	0.005	1.156[c]	0.030	0.007	0.095	0.121	0.074
$\frac{9}{16}$	0.562	W	0.625	0.015	0.005	1.469[c]	0.030	0.007	0.109	0.132	0.086
$\frac{5}{8}$	0.625	N	0.656	0.030	0.007	1.312	0.030	0.007	0.095	0.121	0.074
$\frac{5}{8}$	0.625	W	0.688	0.030	0.007	1.750	0.030	0.007	0.134	0.160	0.108
$\frac{3}{4}$	0.750	N	0.812	0.030	0.007	1.469	0.030	0.007	0.134	0.160	0.108
$\frac{3}{4}$	0.750	W	0.812	0.030	0.007	2.000	0.030	0.007	0.148	0.177	0.122
$\frac{7}{8}$	0.875	N	0.938	0.030	0.007	1.750	0.030	0.007	0.134	0.160	0.108
$\frac{7}{8}$	0.875	W	0.938	0.030	0.007	2.250	0.030	0.007	0.165	0.192	0.136
1	1.000	N	1.062	0.030	0.007	2.000	0.030	0.007	0.134	0.160	0.108
1	1.000	W	1.062	0.030	0.007	2.500	0.030	0.007	0.165	0.192	0.136
$1\frac{1}{8}$	1.125	N	1.250	0.030	0.007	2.250	0.030	0.007	0.134	0.160	0.108
$1\frac{1}{8}$	1.125	W	1.250	0.030	0.007	2.750	0.030	0.007	0.165	0.192	0.136
$1\frac{1}{4}$	1.250	N	1.375	0.030	0.007	2.500	0.030	0.007	0.165	0.192	0.136
$1\frac{1}{4}$	1.250	W	1.375	0.030	0.007	3.000	0.030	0.007	0.165	0.192	0.136
$1\frac{3}{8}$	1.375	N	1.500	0.030	0.007	2.750	0.030	0.007	0.165	0.192	0.136
$1\frac{3}{8}$	1.375	W	1.500	0.045	0.010	3.250	0.045	0.010	0.180	0.213	0.153
$1\frac{1}{2}$	1.500	N	1.625	0.030	0.007	3.000	0.030	0.007	0.165	0.192	0.136
$1\frac{1}{2}$	1.500	W	1.625	0.045	0.010	3.500	0.045	0.010	0.180	0.213	0.153
$1\frac{5}{8}$	1.625		1.750	0.045	0.010	3.750	0.045	0.010	0.180	0.213	0.153
$1\frac{3}{4}$	1.750		1.875	0.045	0.010	4.000	0.045	0.010	0.180	0.213	0.153
$1\frac{7}{8}$	1.875		2.000	0.045	0.010	4.250	0.045	0.010	0.180	0.213	0.153
2	2.000		2.125	0.045	0.010	4.500	0.045	0.010	0.180	0.213	0.153
$2\frac{1}{4}$	2.250		2.375	0.045	0.010	4.750	0.045	0.010	0.220	0.248	0.193
$2\frac{1}{2}$	2.500		2.625	0.045	0.010	5.000	0.045	0.010	0.238	0.280	0.210
$2\frac{3}{4}$	2.750		2.875	0.065	0.010	5.250	0.065	0.010	0.259	0.310	0.228
3	3.000		3.125	0.065	0.010	5.500	0.065	0.010	0.284	0.327	0.249

(Courtesy of ANSI; B27.2–1965.)

APPENDIX 36
LOCK WASHERS (ANSI B27.1)

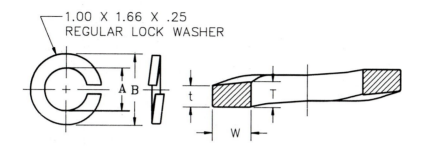

1.00 X 1.66 X .25
REGULAR LOCK WASHER

Dimensions of Regular* Helical Spring Lock Washers

Nominal Washer Size		Inside Diameter A		Outside Diameter B	Washer Section	
					Width W	Thickness $\frac{T+t}{2}$
		Min	Max	Max**	Min	Min
No. 2	0.086	0.088	0.094	0.172	0.035	0.020
No. 3	0.099	0.101	0.107	0.195	0.040	0.025
No. 4	0.112	0.115	0.121	0.209	0.040	0.025
No. 5	0.125	0.128	0.134	0.236	0.047	0.031
No. 6	0.138	0.141	0.148	0.250	0.047	0.031
No. 8	0.164	0.168	0.175	0.293	0.055	0.040
No. 10	0.190	0.194	0.202	0.334	0.062	0.047
No. 12	0.216	0.221	0.229	0.377	0.070	0.056
$\frac{1}{4}$	0.250	0.255	0.263	0.489	0.109	0.062
$\frac{5}{16}$	0.312	0.318	0.328	0.586	0.125	0.078
$\frac{3}{8}$	0.375	0.382	0.393	0.683	0.141	0.094
$\frac{7}{16}$	0.438	0.446	0.459	0.779	0.156	0.109
$\frac{1}{2}$	0.500	0.509	0.523	0.873	0.171	0.125
$\frac{9}{16}$	0.562	0.572	0.587	0.971	0.188	0.141
$\frac{5}{8}$	0.625	0.636	0.653	1.079	0.203	0.156
$\frac{11}{16}$	0.688	0.700	0.718	1.176	0.219	0.172
$\frac{3}{4}$	0.750	0.763	0.783	1.271	0.234	0.188
$\frac{13}{16}$	0.812	0.826	0.847	1.367	0.250	0.203
$\frac{7}{8}$	0.875	0.890	0.912	1.464	0.266	0.219
$\frac{15}{16}$	0.938	0.954	0.978	1.560	0.281	0.234
1	1.000	1.017	1.042	1.661	0.297	0.250
$1\frac{1}{16}$	1.062	1.080	1.107	1.756	0.312	0.266
$1\frac{1}{8}$	1.125	1.144	1.172	1.853	0.328	0.281
$1\frac{3}{16}$	1.188	1.208	1.237	1.950	0.344	0.297
$1\frac{1}{4}$	1.250	1.271	1.302	2.045	0.359	0.312
$1\frac{5}{16}$	1.312	1.334	1.366	2.141	0.375	0.328
$1\frac{3}{8}$	1.375	1.398	1.432	2.239	0.391	0.344
$1\frac{7}{16}$	1.438	1.462	1.497	2.334	0.406	0.359
$1\frac{1}{2}$	1.500	1.525	1.561	2.430	0.422	0.375

*Formerly designated Medium Helical Spring Lock Washers.

**The maximum outside diameters specified allow for the commercial tolerances on cold drawn wire.

APPENDIX 37
COTTER PINS

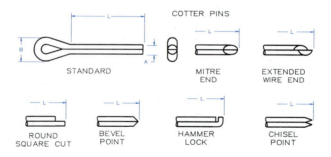

	Diameter A		Outside Eye Diameter B	Hole Sizes Recom-
Nominal Diameter	Max	Min	Min	mended
0.031	0.032	0.028	$\frac{1}{16}$	$\frac{3}{64}$
0.047	0.048	0.044	$\frac{3}{32}$	$\frac{1}{16}$
0.062	0.060	0.056	$\frac{1}{8}$	$\frac{5}{64}$
0.078	0.076	0.072	$\frac{5}{32}$	$\frac{3}{32}$
0.094	0.090	0.086	$\frac{3}{16}$	$\frac{7}{64}$
0.109	0.104	0.100	$\frac{7}{32}$	$\frac{1}{8}$
0.125	0.120	0.116	$\frac{1}{4}$	$\frac{9}{64}$
0.141	0.134	0.130	$\frac{9}{32}$	$\frac{5}{32}$
0.156	0.150	0.146	$\frac{5}{16}$	$\frac{11}{64}$
0.188	0.176	0.172	$\frac{3}{8}$	$\frac{13}{64}$
0.219	0.207	0.202	$\frac{7}{16}$	$\frac{15}{64}$
0.250	0.225	0.220	$\frac{1}{2}$	$\frac{17}{64}$
0.312	0.280	0.275	$\frac{5}{8}$	$\frac{5}{16}$
0.375	0.335	0.329	$\frac{3}{4}$	$\frac{3}{8}$
0.438	0.406	0.400	$\frac{7}{8}$	$\frac{7}{16}$
0.500	0.473	0.467	1	$\frac{1}{2}$
0.625	0.598	0.590	$1\frac{1}{4}$	$\frac{5}{8}$
0.750	0.723	0.715	$1\frac{1}{2}$	$\frac{3}{4}$

All dimensions are given in inches.

A certain amount of leeway is permitted in the design of the head; however, the outside diameters given should be adhered to.

Prongs are to be parallel; ends shall not be open.

Points may be blunt, bevel, extended prong, mitre, etc., and purchaser may specify type required.

Lengths shall be measured as shown on the above illustration (L-dimension).

Cotter pins shall be free from burrs or any defects that will affect their serviceability.

(Courtesy of ANSI; B5.20–1958.)

APPENDIX 38

AMERICAN STANDARD RUNNING AND SLIDING FITS

Limits are in thousandths of an inch.

Limits for hole and shaft are applied algebraically to the basic size to obtain the limits of size for the parts.

Data in bold face are in accordance with ABC agreements.

Symbols H5, g5, etc., are Hole and Shaft designations used in ABC System.

Nominal Size Range Inches Over — To	Class RC 1 Limits of Clearance	Hole H5	Shaft g4	Class RC 2 Limits of Clearance	Hole H6	Shaft g5	Class RC 3 Limits of Clearance	Hole H7	Shaft f6	Class RC 4 Limits of Clearance	Hole H8	Shaft f7
0 — 0.12	0.1 / 0.45	+ 0.2 / 0	− 0.1 / − 0.25	0.1 / 0.55	+ 0.25 / 0	− 0.1 / − 0.3	0.3 / 0.95	+ 0.4 / 0	− 0.3 / − 0.55	0.3 / 1.3	+ 0.6 / 0	− 0.3 / − 0.7
0.12 — 0.24	0.15 / 0.5	+ 0.2 / 0	− 0.15 / − 0.3	0.15 / 0.65	+ 0.3 / 0	− 0.15 / − 0.35	0.4 / 1.12	+ 0.5 / 0	− 0.4 / − 0.7	0.4 / 1.6	+ 0.7 / 0	− 0.4 / − 0.9
0.24 — 0.40	0.2 / 0.6	0.25 / 0	− 0.2 / − 0.35	0.2 / 0.85	+ 0.4 / 0	− 0.2 / − 0.45	0.5 / 1.5	+ 0.6 / 0	− 0.5 / − 0.9	0.5 / 2.0	+ 0.9 / 0	− 0.5 / − 1.1
0.40 — 0.71	0.25 / 0.75	+ 0.3 / 0	− 0.25 / − 0.45	0.25 / 0.95	+ 0.4 / 0	− 0.25 / − 0.55	0.6 / 1.7	+ 0.7 / 0	− 0.6 / − 1.0	0.6 / 2.3	+ 1.0 / 0	− 0.6 / − 1.3
0.71 — 1.19	0.3 / 0.95	+ 0.4 / 0	− 0.3 / − 0.55	0.3 / 1.2	+ 0.5 / 0	− 0.3 / − 0.7	0.8 / 2.1	+ 0.8 / 0	− 0.8 / − 1.3	0.8 / 2.8	+ 1.2 / 0	− 0.8 / − 1.6
1.19 — 1.97	0.4 / 1.1	+ 0.4 / 0	− 0.4 / − 0.7	0.4 / 1.4	+ 0.6 / 0	− 0.4 / − 0.8	1.0 / 2.6	+ 1.0 / 0	− 1.0 / − 1.6	1.0 / 3.6	+ 1.6 / 0	− 1.0 / − 2.0
1.97 — 3.15	0.4 / 1.2	+ 0.5 / 0	− 0.4 / − 0.7	0.4 / 1.6	+ 0.7 / 0	− 0.4 / − 0.9	1.2 / 3.1	+ 1.2 / 0	− 1.2 / − 1.9	1.2 / 4.2	+ 1.8 / 0	− 1.2 / − 2.4
3.15 — 4.73	0.5 / 1.5	+ 0.6 / 0	− 0.5 / − 0.9	0.5 / 2.0	+ 0.9 / 0	− 0.5 / − 1.1	1.4 / 3.7	+ 1.4 / 0	− 1.4 / − 2.3	1.4 / 5.0	+ 2.2 / 0	− 1.4 / − 2.8
4.73 — 7.09	0.6 / 1.8	+ 0.7 / 0	− 0.6 / − 1.1	0.6 / 2.3	+ 1.0 / 0	− 0.6 / − 1.3	1.6 / 4.2	+ 1.6 / 0	− 1.6 / − 2.6	1.6 / 5.7	+ 2.5 / 0	− 1.6 / − 3.2
7.09 — 9.85	0.6 / 2.0	+ 0.8 / 0	− 0.6 / − 1.2	0.6 / 2.6	+ 1.2 / 0	− 0.6 / − 1.4	2.0 / 5.0	+ 1.8 / 0	− 2.0 / − 3.2	2.0 / 6.6	+ 2.8 / 0	− 2.0 / − 3.8
9.85 — 12.41	0.8 / 2.3	+ 0.9 / 0	− 0.8 / − 1.4	0.8 / 2.9	+ 1.2 / 0	− 0.8 / − 1.7	2.5 / 5.7	+ 2.0 / 0	− 2.5 / − 3.7	2.5 / 7.5	+ 3.0 / 0	− 2.5 / − 4.5
12.41 — 15.75	1.0 / 2.7	+ 1.0 / 0	− 1.0 / − 1.7	1.0 / 3.4	+ 1.4 / 0	− 1.0 / − 2.0	3.0 / 6.6	+ / 0	− 3.0 / − 4.4	3.0 / 8.7	+ 3.5 / 0	− 3.0 / − 5.2
15.75 — 19.69	1.2 / 3.0	+ 1.0 / 0	− 1.2 / − 2.0	1.2 / 3.8	+ 1.6 / 0	− 1.2 / − 2.2	4.0 / 8.1	+ 1.6 / 0	− 4.0 / − 5.6	4.0 / 10.5	+ 4.0 / 0	− 4.0 / − 6.5
19.69 — 30.09	1.6 / 3.7	+ 1.2 / 0	− 1.6 / − 2.5	1.6 / 4.8	+ 2.0 / 0	− 1.6 / − 2.8	5.0 / 10.0	+ 3.0 / 0	− 5.0 / − 7.0	5.0 / 13.0	+ 5.0 / 0	− 5.0 / − 8.0
30.09 — 41.49	2.0 / 4.6	+ 1.6 / 0	− 2.0 / − 3.0	2.0 / 6.1	+ 2.5 / 0	− 2.0 / − 3.6	6.0 / 12.5	+ 4.0 / 0	− 6.0 / − 8.5	6.0 / 16.0	+ 6.0 / 0	− 6.0 / −10.0
41.49 — 56.19	2.5 / 5.7	+ 2.0 / 0	− 2.5 / − 3.7	2.5 / 7.5	+ 3.0 / 0	− 2.5 / − 4.5	8.0 / 16.0	+ 5.0 / 0	− 8.0 / −11.0	8.0 / 21.0	+ 8.0 / 0	− 8.0 / −13.0
56.19 — 76.39	3.0 / 7.1	+ 2.5 / 0	− 3.0 / − 4.6	3.0 / 9.5	+ 4.0 / 0	− 3.0 / − 5.5	10.0 / 20.0	+ 6.0 / 0	−10.0 / −14.0	10.0 / 26.0	+10.0 / 0	−10.0 / −16.0
76.39 — 100.9	4.0 / 9.0	+ 3.0 / 0	− 4.0 / − 6.0	4.0 / 12.0	+ 5.0 / 0	− 4.0 / − 7.0	12.0 / 25.0	+ 8.0 / 0	−12.0 / −17.0	12.0 / 32.0	+12.0 / 0	−12.0 / −20.0
100.9 — 131.9	5.0 / 11.5	+ 4.0 / 0	− 5.0 / − 7.5	5.0 / 15.0	+ 6.0 / 0	− 5.0 / − 9.0	16.0 / 32.0	+10.0 / 0	−16.0 / −22.0	16.0 / 36.0	+16.0 / 0	−16.0 / −26.0
131.9 — 171.9	6.0 / 14.0	+ 5.0 / 0	− 6.0 / − 9.0	6.0 / 19.0	+ 8.0 / 0	− 6.0 / −11.0	18.0 / 38.0	+ 8.0 / 0	−18.0 / −26.0	18.0 / 50.0	+20.0 / 0	−18.0 / −30.0
171.9 — 200	8.0 / 18.0	+ 6.0 / 0	− 8.0 / −12.0	8.0 / 22.0	+10.0 / 0	− 8.0 / −12.0	22.0 / 48.0	+16.0 / 0	−22.0 / −32.0	22.0 / 63.0	+25.0 / 0	−22.0 / −38.0

(Courtesy of USASI; B4.1–1955.)

Cont.

AMERICAN STANDARD RUNNING AND SLIDING FITS (Cont.)

Class RC 5			Class RC 6			Class RC 7			Class RC 8			Class RC 9			Nominal Size Range Inches	
Limits of Clearance	Hole H8	Shaft e7	Limits of Clearance	Hole H9	Shaft e8	Limits of Clearance	Hole H9	Shaft d8	Limits of Clearance	Hole H10	Shaft c9	Limits of Clearance	Hole H11	Shaft	Over	To
0.6 / 1.6	+0.6 / −0	−0.6 / −1.0	0.6 / 2.2	+1.0 / −0	−0.6 / −1.2	1.0 / 2.6	+1.0 / 0	−1.0 / −1.6	2.5 / 5.1	+1.6 / 0	−2.5 / −3.5	4.0 / 8.1	+2.5 / 0	−4.0 / −5.6	0	0.12
0.8 / 2.0	+0.7 / −0	−0.8 / −1.3	0.8 / 2.7	+1.2 / −0	−0.8 / −1.5	1.2 / 3.1	+1.2 / 0	−1.2 / −1.9	2.8 / 5.8	+1.8 / 0	−2.8 / −4.0	4.5 / 9.0	+3.0 / 0	−4.5 / −6.0	0.12	0.24
1.0 / 2.5	+0.9 / −0	−1.0 / −1.6	1.0 / 3.3	+1.4 / −0	−1.0 / −1.9	1.6 / 3.9	+1.4 / 0	−1.6 / −2.5	3.0 / 6.6	+2.2 / 0	−3.0 / −4.4	5.0 / 10.7	+3.5 / 0	−5.0 / −7.2	0.24	0.40
1.2 / 2.9	+1.0 / −0	−1.2 / −1.9	1.2 / 3.8	+1.6 / −0	−1.2 / −2.2	2.0 / 4.6	+1.6 / 0	−2.0 / −3.0	3.5 / 7.9	+2.8 / 0	−3.5 / −5.1	6.0 / 12.8	+4.0 / −0	−6.0 / −8.8	0.40	0.71
1.6 / 3.6	+1.2 / −0	−1.6 / −2.4	1.6 / 4.8	+2.0 / −0	−1.6 / −2.8	2.5 / 5.7	+2.0 / 0	−2.5 / −3.7	4.5 / 10.0	+3.5 / 0	−4.5 / −6.5	7.0 / 15.5	+5.0 / 0	−7.0 / −10.5	0.71	1.19
2.0 / 4.6	+1.6 / −0	−2.0 / −3.0	2.0 / 6.1	+2.5 / −0	−2.0 / −3.6	3.0 / 7.1	+2.5 / 0	−3.0 / −4.6	5.0 / 11.5	+4.0 / 0	−5.0 / −7.5	8.0 / 18.0	+6.0 / 0	−8.0 / −12.0	1.19	1.97
2.5 / 5.5	+1.8 / −0	−2.5 / −3.7	2.5 / 7.3	+3.0 / −0	−2.5 / −4.3	4.0 / 8.8	+3.0 / 0	−4.0 / −5.8	6.0 / 13.5	+4.5 / 0	−6.0 / −9.0	9.0 / 20.5	+7.0 / 0	−9.0 / −13.5	1.97	3.15
3.0 / 6.6	+2.2 / −0	−3.0 / −4.4	3.0 / 8.7	+3.5 / −0	−3.0 / −5.2	5.0 / 10.7	+3.5 / 0	−5.0 / −7.2	7.0 / 15.5	+5.0 / 0	−7.0 / −10.5	10.0 / 24.0	+9.0 / 0	−10.0 / −15.0	3.15	4.73
3.5 / 7.6	+2.5 / −0	−3.5 / −5.1	3.5 / 10.0	+4.0 / −0	−3.5 / −6.0	6.0 / 12.5	+4.0 / 0	−6.0 / −8.5	8.0 / 18.0	+6.0 / 0	−8.0 / −12.0	12.0 / 28.0	+10.0 / 0	−12.0 / −18.0	4.73	7.09
4.0 / 8.6	+2.8 / −0	−4.0 / −5.8	4.0 / 11.3	+4.5 / 0	−4.0 / −6.8	7.0 / 14.3	+4.5 / 0	−7.0 / −9.8	10.0 / 21.5	+7.0 / 0	−10.0 / −14.5	15.0 / 34.0	+12.0 / 0	−15.0 / −22.0	7.09	9.85
5.0 / 10.0	+3.0 / 0	−5.0 / −7.0	5.0 / 13.0	+5.0 / 0	−5.0 / −8.0	8.0 / 16.0	+5.0 / 0	−8.0 / −11.0	12.0 / 25.0	+8.0 / 0	−12.0 / −17.0	18.0 / 38.0	+12.0 / 0	−18.0 / −26.0	9.85	12.41
6.0 / 11.7	+3.5 / 0	−6.0 / −8.2	6.0 / 15.5	+6.0 / 0	−6.0 / −9.5	10.0 / 19.5	+6.0 / 0	−10.0 / −13.5	14.0 / 29.0	+9.0 / 0	−14.0 / −20.0	22.0 / 45.0	+14.0 / 0	−22.0 / −31.0	12.41	15.75
8.0 / 14.5	+4.0 / 0	−8.0 / −10.5	8.0 / 18.0	+6.0 / 0	−8.0 / −12.0	12.0 / 22.0	+6.0 / 0	−12.0 / −16.0	16.0 / 32.0	+10.0 / 0	−16.0 / −22.0	25.0 / 51.0	+16.0 / 0	−25.0 / −35.0	15.75	19.69
10.0 / 18.0	+5.0 / 0	−10.0 / −13.0	10.0 / 23.0	+8.0 / 0	−10.0 / −15.0	16.0 / 29.0	+8.0 / 0	−16.0 / −21.0	20.0 / 40.0	+12.0 / 0	−20.0 / −28.0	30.0 / 62.0	+20.0 / 0	−30.0 / −42.0	19.69	30.09
12.0 / 22.0	+6.0 / 0	−12.0 / −16.0	12.0 / 28.0	+10.0 / 0	−12.0 / −18.0	20.0 / 36.0	+10.0 / 0	−20.0 / −26.0	25.0 / 51.0	+16.0 / 0	−25.0 / −35.0	40.0 / 81.0	+25.0 / 0	−40.0 / −56.0	30.09	41.49
16.0 / 29.0	+8.0 / 0	−16.0 / −21.0	16.0 / 36.0	+12.0 / 0	−16.0 / −24.0	25.0 / 45.0	+12.0 / 0	−25.0 / −33.0	30.0 / 62.0	+20.0 / 0	−30.0 / −42.0	50.0 / 100	+30.0 / 0	−50.0 / −70.0	41.49	56.19
20.0 / 36.0	+10.0 / 0	−20.0 / −26.0	20.0 / 46.0	+16.0 / 0	−20.0 / −30.0	30.0 / 56.0	+16.0 / 0	−30.0 / −40.0	40.0 / 81.0	+25.0 / 0	−40.0 / −56.0	60.0 / 125	+40.0 / 0	−60.0 / −85.0	56.19	76.39
25.0 / 45.0	+12.0 / 0	−25.0 / −33.0	25.0 / 57.0	+20.0 / 0	−25.0 / −37.0	40.0 / 72.0	+20.0 / 0	−40.0 / −52.0	50.0 / 100	+30.0 / 0	−50.0 / −70.0	80.0 / 160	+50.0 / 0	−80.0 / −110	76.39	100.9
30.0 / 56.0	+16.0 / 0	−30.0 / −40.0	30.0 / 71.0	+25.0 / 0	−30.0 / −46.0	50.0 / 91.0	+25.0 / 0	−50.0 / −66.0	60.0 / 125	+40.0 / 0	−60.0 / −85.0	100 / 200	+60.0 / 0	−100 / −140	100.9	131.9
35.0 / 67.0	+20.0 / 0	−35.0 / −47.0	35.0 / 85.0	+30.0 / 0	−35.0 / −55.0	60.0 / 110	+30.0 / 0	−60.0 / −80.0	80.0 / 160	+50.0 / 0	−80.0 / −110	130 / 260	+80.0 / 0	−130 / −180	131.9	171.9
45.0 / 86.0	+25.0 / 0	−45.0 / −61.0	45.0 / 110	+40.0 / 0	−45.0 / −70.0	80.0 / 145	+40.0 / 0	−80.0 / −105.0	100 / 200	+60.0 / 0	−100 / −140	150 / 310	+100 / 0	−150 / −210	171.9	200

(Courtesy of ANSI; B4.1–1955.)

APPENDIX 39
AMERICAN STANDARD CLEARANCE LOCATIONAL FITS

Limits are in thousandths of an inch.

Limits for hole and shaft are applied algebraically to the basic size to obtain the limits of size for the parts.

Data in bold face are in accordance with ABC agreements.

Symbols H9,f8, etc., are Hole and Shaft designations used in ABC System.

Nominal Size Range Inches (Over)	(To)	Class LC 1 Clearance	Hole H6	Shaft h5	Class LC 2 Clearance	Hole H7	Shaft h6	Class LC 3 Clearance	Hole H8	Shaft h7	Class LC 4 Clearance	Hole H10	Shaft h9	Class LC 5 Clearance	Hole H7	Shaft g6
0	0.12	0 / 0.45	+0.25 / −0	+0 / −0.2	0 / 0.65	+0.4 / −0	+0 / −0.25	0 / 1	+0.6 / −0	+0 / −0.4	0 / 2.6	+1.6 / −0	+0 / −1.0	0.1 / 0.75	+0.4 / −0	−0.1 / −0.35
0.12	0.24	0 / 0.5	+0.3 / −0	+0 / −0.2	0 / 0.8	+0.5 / −0	+0 / −0.3	0 / 1.2	+0.7 / −0	+0 / −0.5	0 / 3.0	+1.8 / −0	+0 / −1.2	0.15 / 0.95	+0.5 / −0	−0.15 / −0.45
0.24	0.40	0 / 0.65	+0.4 / −0	+0 / −0.25	0 / 1.0	+0.6 / −0	+0 / −0.4	0 / 1.5	+0.9 / −0	+0 / −0.6	0 / 3.6	+2.2 / −0	+0 / −1.4	0.2 / 1.2	+0.6 / −0	−0.2 / −0.6
0.40	0.71	0 / 0.7	+0.4 / −0	+0 / −0.3	0 / 1.1	+0.7 / −0	+0 / −0.4	0 / 1.7	+1.0 / −0	+0 / −0.7	0 / 4.4	+2.8 / −0	+0 / −1.6	0.25 / 1.35	+0.7 / −0	−0.25 / −0.65
0.71	1.19	0 / 0.9	+0.5 / −0	+0 / −0.4	0 / 1.3	+0.8 / −0	+0 / −0.5	0 / 2	+1.2 / −0	+0 / −0.8	0 / 5.5	+3.5 / −0	+0 / −2.0	0.3 / 1.6	+0.8 / −0	−0.3 / −0.8
1.19	1.97	0 / 1.0	+0.6 / −0	+0 / −0.4	0 / 1.6	+1.0 / −0	+0 / −0.6	0 / 2.6	+1.6 / −0	+0 / −1	0 / 6.5	+4.0 / −0	+0 / −2.5	0.4 / 2.0	+1.0 / −0	−0.4 / −1.0
1.97	3.15	0 / 1.2	+0.7 / −0	+0 / −0.5	0 / 1.9	+1.2 / −0	+0 / −0.7	0 / 3	+1.8 / −0	+0 / −1.2	0 / 7.5	+4.5 / −0	+0 / −3	0.4 / 2.3	+1.2 / −0	−0.4 / −1.1
3.15	4.73	0 / 1.5	+0.9 / −0	+0 / −0.6	0 / 2.3	+1.4 / −0	+0 / −0.9	0 / 3.6	+2.2 / −0	+0 / −1.4	0 / 8.5	+5.0 / −0	+0 / −3.5	0.5 / 2.8	+1.4 / −0	−0.5 / −1.4
4.73	7.09	0 / 1.7	+1.0 / −0	+0 / −0.7	0 / 2.6	+1.6 / −0	+0 / −1.0	0 / 4.1	+2.5 / −0	+0 / −1.6	0 / 10	+6.0 / −0	+0 / −4	0.6 / 3.2	+1.6 / −0	−0.6 / −1.6
7.09	9.85	0 / 2.0	+1.2 / −0	+0 / −0.8	0 / 3.0	+1.8 / −0	+0 / −1.2	0 / 4.6	+2.8 / −0	+0 / −1.8	0 / 11.5	+7.0 / −0	+0 / −4.5	0.6 / 3.6	+1.8 / −0	−0.6 / −1.8
9.85	12.41	0 / 2.1	+1.2 / −0	+0 / −0.9	0 / 3.2	+2.0 / −0	+0 / −1.2	0 / 5	+3.0 / −0	+0 / −2.0	0 / 13	+8.0 / −0	+0 / −5	0.7 / 3.9	+2.0 / −0	−0.7 / −1.9
12.41	15.75	0 / 2.4	+1.4 / −0	+0 / −1.0	0 / 3.6	+2.2 / −0	+0 / −1.4	0 / 5.7	+3.5 / −0	+0 / −2.2	0 / 15	+9.0 / −0	+0 / −6	0.7 / 4.3	+2.2 / −0	−0.7 / −2.1
15.75	19.69	0 / 2.6	+1.6 / −0	+0 / −1.0	0 / 4.1	+2.5 / −0	+0 / −1.6	0 / 6.5	+4 / −0	+0 / −2.5	0 / 16	+10.0 / −0	+0 / −6	0.8 / 4.9	+2.5 / −0	−0.8 / −2.4
19.69	30.09	0 / 3.2	+2.0 / −0	+0 / −1.2	0 / 5.0	+3 / −0	+0 / −2	0 / 8	+5 / −0	+0 / −3	0 / 20	+12.0 / −0	+0 / −8	0.9 / 5.9	+3.0 / −0	−0.9 / −2.9
30.09	41.49	0 / 4.1	+2.5 / −0	+0 / −1.6	0 / 6.5	+4 / −0	+0 / −2.5	0 / 10	+6 / −0	+0 / −4	0 / 26	+16.0 / −0	+0 / −10	1.0 / 7.5	+4.0 / −0	−1.0 / −3.5
41.49	56.19	0 / 5.0	+3.0 / −0	+0 / −2.0	0 / 8.0	+5 / −0	+0 / −3	0 / 13	+8 / −0	+0 / −5	0 / 32	+20.0 / −0	+0 / −12	1.2 / 9.2	+5.0 / −0	−1.2 / −4.2
56.19	76.39	0 / 6.5	+4.0 / −0	+0 / −2.5	0 / 10	+6 / −0	+0 / −4	0 / 16	+10 / −0	+0 / −6	0 / 41	+25.0 / −0	+0 / −16	1.2 / 11.2	+6.0 / −0	−1.2 / −5.2
76.39	100.9	0 / 8.0	+5.0 / −0	+0 / −3.0	0 / 13	+8 / −0	+0 / −5	0 / 20	+12 / −0	+0 / −8	0 / 50	+30.0 / −0	+0 / −20	1.4 / 14.4	+8.0 / −0	−1.4 / −6.4
100.9	131.9	0 / 10.0	+6.0 / −0	+0 / −4.0	0 / 16	+10 / −0	+0 / −6	0 / 26	+16 / −0	+0 / −10	0 / 65	+40.0 / −0	+0 / −25	1.6 / 17.6	+10.0 / −0	−1.6 / −7.6
131.9	171.9	0 / 13.0	+8.0 / −0	+0 / −5.0	0 / 20	+12 / −0	+0 / −8	0 / 32	+20 / −0	+0 / −12	0 / 8	+50.0 / −0	+0 / −30	1.8 / 21.8	+12.0 / −0	−1.8 / −9.8
171.9	200	0 / 16.0	+10.0 / −0	+0 / −6.0	0 / 26	+16 / −0	+0 / −10	0 / 41	+25 / −0	+0 / −16	0 / 100	+60.0 / −0	+0 / −40	1.8 / 27.8	+16.0 / −0	−1.8 / −11.8

(Courtesy of USASI; B4.1–1955.)

Cont.

APPENDIX 39

AMERICAN STANDARD CLEARANCE LOCATIONAL FITS (Cont.)

Values in thousandths of an inch.

Class LC 6			Class LC 7			Class LC 8			Class LC 9			Class LC 10			Class LC 11			Nominal Size Range Inches	
Limits of Clearance	Hole H9	Shaft f8	Limits of Clearance	Hole H10	Shaft e9	Limits of Clearance	Hole H10	Shaft d9	Limits of Clearance	Hole H11	Shaft c10	Limits of Clearance	Hole H12	Shaft	Limits of Clearance	Hole H13	Shaft	Over	To
0.3 / 1.9	+1.0 / 0	-0.3 / -0.9	0.6 / 3.2	+1.6 / 0	-0.6 / -1.6	1.0 / 3.6	+1.6 / -0	-1.0 / -2.0	2.5 / 6.6	+2.5 / -0	-2.5 / -4.1	4 / 12	+4 / -0	-4 / -8	5 / 17	+6 / -0	-5 / -11	0	0.12
0.4 / 2.3	+1.2 / 0	-0.4 / -1.1	0.8 / 3.8	+1.8 / 0	-0.8 / -2.0	1.2 / 4.2	+1.8 / -0	-1.2 / -2.4	2.8 / 7.6	+3.0 / -0	-2.8 / -4.6	4.5 / 14.5	+5 / -0	-4.5 / -9.5	6 / 20	+7 / -0	-6 / -13	0.12	0.24
0.5 / 2.8	+1.4 / 0	-0.5 / -1.4	1.0 / 4.6	+2.2 / 0	-1.0 / -2.4	1.6 / 5.2	+2.2 / -0	-1.6 / -3.0	3.0 / 8.7	+3.5 / -0	-3.0 / -5.2	5 / 17	+6 / -0	-5 / -11	7 / 25	+9 / -0	-7 / -16	0.24	0.40
0.6 / 3.2	+1.6 / 0	-0.6 / -1.6	1.2 / 5.6	+2.8 / 0	-1.2 / -2.8	2.0 / 6.4	+2.8 / -0	-2.0 / -3.6	3.5 / 10.3	+4.0 / -0	-3.5 / -6.3	6 / 20	+7 / -0	-6 / -13	8 / 28	+10 / -0	-8 / -18	0.40	0.71
0.8 / 4.0	+2.0 / 0	-0.8 / -2.0	1.6 / 7.1	+3.5 / 0	-1.6 / -3.6	2.5 / 8.0	+3.5 / -0	-2.5 / -4.5	4.5 / 13.0	+5.0 / -0	-4.5 / -8.0	7 / 23	+8 / -0	-7 / -15	10 / 34	+12 / -0	-10 / -22	0.71	1.19
1.0 / 5.1	+2.5 / 0	-1.0 / -2.6	2.0 / 8.5	+4.0 / 0	-2.0 / -4.5	3.0 / 9.5	+4.0 / -0	-3.0 / -5.5	5 / 15	+6 / -0	-5 / -9	8 / 28	+10 / -0	-8 / -18	12 / 44	+16 / -0	-12 / -28	1.19	1.97
1.2 / 6.0	+3.0 / 0	-1.2 / -3.0	2.5 / 10.0	+4.5 / 0	-2.5 / -5.5	4.0 / 11.5	+4.5 / -0	-4.0 / -7.0	6 / 17.5	+7 / -0	-6 / -10.5	10 / 34	+12 / -0	-10 / -22	14 / 50	+18 / -0	-14 / -32	1.97	3.15
1.4 / 7.1	+3.5 / 0	-1.4 / -3.6	3.0 / 11.5	+5.0 / 0	-3.0 / -6.5	5.0 / 13.5	+5.0 / -0	-5.0 / -8.5	7 / 21	+9 / -0	-7 / -12	11 / 39	+14 / -0	-11 / -25	16 / 60	+22 / -0	-16 / -38	3.15	4.73
1.6 / 8.1	+4.0 / 0	-1.6 / -4.1	3.5 / 13.5	+6.0 / 0	-3.5 / -7.5	6 / 16	+6 / -0	-6 / -10	8 / 24	+10 / -0	-8 / -14	12 / 44	+16 / -0	-12 / -28	18 / 68	+25 / -0	-18 / -43	4.73	7.09
2.0 / 9.3	+4.5 / 0	-2.0 / -4.8	4.0 / 15.5	+7.0 / 0	-4.0 / -8.5	7 / 18.5	+7 / -0	-7 / -11.5	10 / 29	+12 / -0	-10 / -17	16 / 52	+18 / -0	-16 / -34	22 / 78	+28 / -0	-22 / -50	7.09	9.85
2.2 / 10.2	+5.0 / 0	-2.2 / -5.2	4.5 / 17.5	+8.0 / 0	-4.5 / -9.5	7 / 20	+8 / -0	-7 / -12	12 / 32	+12 / -0	-12 / -20	20 / 60	+20 / -0	-20 / -40	28 / 88	+30 / -0	-28 / -58	9.85	12.41
2.5 / 12.0	+6.0 / 0	-2.5 / -6.0	5.0 / 20.0	+9.0 / 0	-5 / -11	8 / 23	+9 / -0	-8 / -14	14 / 37	+14 / -0	-14 / -23	22 / 66	+22 / -0	-22 / -44	30 / 100	+35 / -0	-30 / -65	12.41	15.75
2.8 / 12.8	+6.0 / 0	-2.8 / -6.8	5.0 / 21.0	+10.0 / 0	-5 / -11	9 / 25	+10 / -0	-9 / -15	16 / 42	+16 / -0	-16 / -26	25 / 75	+25 / -0	-25 / -50	35 / 115	+40 / -0	-35 / -75	15.75	19.69
3.0 / 16.0	+8.0 / 0	-3.0 / -8.0	6.0 / 26.0	+12.0 / -0	-6 / -14	10 / 30	+12 / -0	-10 / -18	18 / 50	+20 / -0	-18 / -30	28 / 88	+30 / -0	-28 / -58	40 / 140	+50 / -0	-40 / -90	19.69	30.09
3.5 / 19.5	+10.0 / 0	-3.5 / -9.5	7.0 / 33.0	+16.0 / -0	-7 / -17	12 / 38	+16 / -0	-12 / -22	20 / 61	+25 / -0	-20 / -36	30 / 110	+40 / -0	-30 / -70	45 / 165	+60 / -0	-45 / -105	30.09	41.49
4.0 / 24.0	+12.0 / 0	-4.0 / -12.0	8.0 / 40.0	+20.0 / -0	-8 / -20	14 / 46	+20 / -0	-14 / -26	25 / 75	+30 / -0	-25 / -45	40 / 140	+50 / -0	-40 / -90	60 / 220	+80 / -0	-60 / -140	41.49	56.19
4.5 / 30.5	+16.0 / 0	-4.5 / -14.5	9.0 / 50.0	+25.0 / -0	-9 / -25	16 / 57	+25 / -0	-16 / -32	30 / 95	+40 / -0	-30 / -55	50 / 170	+60 / -0	-50 / 110	70 / 270	+100 / -0	-70 / -170	56.19	76.39
5.0 / 37.0	+20.0 / 0	-5 / -17	10.0 / 60.0	+30.0 / -0	-10 / -30	18 / 68	+30 / -0	-18 / -38	35 / 115	+50 / -0	-35 / -65	50 / 210	+80 / -0	-50 / -130	80 / 330	+125 / -0	-80 / -205	76.39	100.9
6.0 / 47.0	+25.0 / 0	-6 / -22	12.0 / 67.0	+40.0 / -0	-12 / -27	20 / 85	+40 / -0	-20 / -45	40 / 140	+60 / -0	-40 / -80	60 / 260	+100 / -0	-60 / -160	90 / 410	+160 / -0	-90 / -250	100.9	131.9
7.0 / 57.0	+30.0 / 0	-7 / -27	14.0 / 94.0	+50.0 / -0	-14 / -44	25 / 105	+50 / -0	-25 / -55	50 / 180	+80 / -0	-50 / -100	80 / 330	+125 / -0	-80 / -205	100 / 500	+200 / -0	-100 / -300	131.9	171.9
7.0 / 72.0	+40.0 / 0	-7 / -32	14.0 / 114.0	+60.0 / -0	-14 / -54	25 / 125	+60 / -0	-25 / -65	50 / 210	+100 / -0	-50 / -110	90 / 410	+160 / -0	-90 / -250	125 / 625	+250 / -0	-125 / -375	171.9	200

(Courtesy of ANSI; B4.1–1955.)

APPENDIX 40
AMERICAN STANDARD TRANSITION LOCATIONAL FITS

Limits are in thousandths of an inch.

Limits for hole and shaft are applied algebraically to the basic size to obtain the limits of size for the mating parts.

Data in bold face are in accordance with ABC agreements.

"Fit" represents the maximum interference (minus values) and the maximum clearance (plus values).

Symbols H7, js6, etc., are Hole and Shaft designations used in ABC System.

Nominal Size Range Inches (Over – To)	Class LT 1 Fit	LT1 Hole H7	LT1 Shaft js6	Class LT 2 Fit	LT2 Hole H8	LT2 Shaft js7	Class LT 3 Fit	LT3 Hole H7	LT3 Shaft k6	Class LT 4 Fit	LT4 Hole H8	LT4 Shaft k7	Class LT 5 Fit	LT5 Hole H7	LT5 Shaft n6	Class LT 6 Fit	LT6 Hole H7	LT6 Shaft n7
0 – 0.12	−0.10 / +0.50	+0.4 / −0	+0.10 / −0.10	−0.2 / +0.8	+0.6 / −0	+0.2 / −0.2							−0.5 / +0.15	+0.4 / −0	+0.5 / +0.25	−0.65 / +0.15	+0.4 / −0	+0.65 / +0.25
0.12 – 0.24	−0.15 / +0.65	+0.5 / −0	+0.15 / −0.15	−0.25 / +0.95	+0.7 / −0	+0.25 / −0.25							−0.6 / +0.2	+0.5 / −0	+0.6 / +0.3	−0.8 / +0.2	+0.5 / −0	+0.8 / +0.3
0.24 – 0.40	−0.2 / +0.8	+0.6 / −0	+0.2 / −0.2	−0.3 / +1.2	+0.9 / −0	+0.3 / −0.3	−0.5 / +0.5	+0.6 / −0	+0.5 / +0.1	−0.7 / +0.8	+0.9 / −0	+0.7 / +0.1	−0.8 / +0.2	+0.6 / −0	+0.8 / +0.4	−1.0 / +0.2	+0.6 / −0	+1.0 / +0.4
0.40 – 0.71	−0.2 / +0.9	+0.7 / −0	+0.2 / −0.2	−0.35 / +1.35	+1.0 / −0	+0.35 / −0.35	−0.5 / +0.6	+0.7 / −0	+0.5 / +0.1	−0.8 / +0.9	+1.0 / −0	+0.8 / +0.1	−0.9 / +0.2	+0.7 / −0	+0.9 / +0.5	−1.2 / +0.2	+0.7 / −0	+1.2 / +0.5
0.71 – 1.19	−0.25 / +1.05	+0.8 / −0	+0.25 / −0.25	−0.4 / +1.6	+1.2 / −0	+0.4 / −0.4	−0.6 / +0.7	+0.8 / −0	+0.6 / +0.1	−0.9 / +1.1	+1.2 / −0	+0.9 / +0.1	−1.1 / +0.2	+0.8 / −0	+1.1 / +0.6	−1.4 / +0.2	+0.8 / −0	+1.4 / +0.6
1.19 – 1.97	−0.3 / +1.3	+1.0 / −0	+0.3 / −0.3	−0.5 / +2.1	+1.6 / −0	+0.5 / −0.5	−0.7 / +0.9	+1.0 / −0	+0.7 / +0.1	−1.1 / +1.5	+1.6 / −0	+1.1 / +0.1	−1.3 / +0.3	+1.0 / −0	+1.3 / +0.7	−1.7 / +0.3	+1.0 / −0	+1.7 / +0.7
1.97 – 3.15	−0.3 / +1.5	+1.2 / −0	+0.3 / −0.3	−0.6 / +2.4	+1.8 / −0	+0.6 / −0.6	−0.8 / +1.1	+1.2 / −0	+0.8 / +0.1	−1.3 / +1.7	+1.8 / −0	+1.3 / +0.1	−1.5 / +0.4	+1.2 / −0	+1.5 / +0.8	−2.0 / +0.4	+1.2 / −0	+2.0 / +0.8
3.15 – 4.73	−0.4 / +1.8	+1.4 / −0	+0.4 / −0.4	−0.7 / +2.9	+2.2 / −0	+0.7 / −0.7	−1.0 / +1.3	+1.4 / −0	+1.0 / +0.1	−1.5 / +2.1	+2.2 / −0	+1.5 / +0.1	−1.9 / +0.4	+1.4 / −0	+1.9 / +1.0	−2.4 / +0.4	+1.4 / −0	+2.4 / +1.0
4.73 – 7.09	−0.5 / +2.1	+1.6 / −0	+0.5 / −0.5	−0.8 / +3.3	+2.5 / −0	+0.8 / −0.8	−1.1 / +1.5	+1.6 / −0	+1.1 / +0.1	−1.7 / +2.4	+2.5 / −0	+1.7 / +0.1	−2.2 / +0.4	+1.6 / −0	+2.2 / +1.2	−2.8 / +0.4	+1.6 / −0	+2.8 / +1.2
7.09 – 9.85	−0.6 / +2.4	+1.8 / −0	+0.6 / −0.6	−0.9 / +3.7	+2.8 / −0	+0.9 / −0.9	−1.4 / +1.6	+1.8 / −0	+1.4 / +0.2	−2.0 / +2.6	+2.8 / −0	+2.0 / +0.2	−2.6 / +0.4	+1.8 / −0	+2.6 / +1.4	−3.2 / +0.4	+1.8 / −0	+3.2 / +1.4
9.85 – 12.41	−0.6 / +2.6	+2.0 / −0	+0.6 / −0.6	−1.0 / +4.0	+3.0 / −0	+1.0 / −1.0	−1.4 / +1.8	+2.0 / −0	+1.4 / +0.2	−2.2 / +2.8	+3.0 / −0	+2.2 / +0.2	−2.6 / +0.6	+2.0 / −0	+2.6 / +1.4	−3.4 / +0.6	+2.0 / −0	+3.4 / +1.4
12.41 – 15.75	−0.7 / +2.9	+2.2 / −0	+0.7 / −0.7	−1.0 / +4.5	+3.5 / −0	+1.0 / −1.0	−1.6 / +2.0	+2.2 / −0	+1.6 / +0.2	−2.4 / +3.3	+3.5 / −0	+2.4 / +0.2	−3.0 / +0.6	+2.2 / −0	+3.0 / +1.6	−3.8 / +0.6	+2.2 / −0	+3.8 / +1.6
15.75 – 19.69	−0.8 / +3.3	+2.5 / −0	+0.8 / −0.8	−1.2 / +5.2	+4.0 / −0	+1.2 / −1.2	−1.8 / +2.3	+2.5 / −0	+1.8 / +0.2	−2.7 / +3.8	+4.0 / −0	+2.7 / +0.2	−3.4 / +0.7	+2.5 / −0	+3.4 / +1.8	−4.3 / +0.7	+2.5 / −0	+4.3 / +1.8

(Courtesy of ANSI; B4.1–1955.)

APPENDIX 41
AMERICAN STANDARD INTERFERENCE LOCATIONAL FITS

Limits are in thousandths of an inch.
Limits for hole and shaft are applied algebraically to the
basic size to obtain the limits of size for the parts.
Data in bold face are in accordance with ABC agreements,
Symbols H7, p6, etc., are Hole and Shaft designations
used in ABC System.

Nominal Size Range Inches Over	To	Class LN 1 Limits of Interference	Standard Limits Hole H6	Standard Limits Shaft n5	Class LN 2 Limits of Interference	Standard Limits Hole H7	Standard Limits Shaft p6	Class LN 3 Limits of Interference	Standard Limits Hole H7	Standard Limits Shaft r6
0	0.12	0 / 0.45	+0.25 / -0	+0.45 / +0.25	0 / 0.65	+0.4 / -0	+0.65 / +0.4	0.1 / 0.75	+0.4 / -0	+0.75 / +0.5
0.12	0.24	0 / 0.5	+0.3 / -0	+0.5 / +0.3	0 / 0.8	+0.5 / -0	+0.8 / +0.5	0.1 / 0.9	+0.5 / 0	+0.9 / +0.6
0.24	0.40	0 / 0.65	+0.4 / -0	+0.65 / +0.4	0 / 1.0	+0.6 / -0	+1.0 / +0.6	0.2 / 1.2	+0.6 / -0	+1.2 / +0.8
0.40	0.71	0 / 0.8	+0.4 / -0	+0.8 / +0.4	0 / 1.1	+0.7 / -0	+1.1 / +0.7	0.3 / 1.4	+0.7 / -0	+1.4 / +1.0
0.71	1.19	0 / 1.0	+0.5 / -0	+1.0 / +0.5	0 / 1.3	+0.8 / -0	+1.3 / +0.8	0.4 / 1.7	+0.8 / -0	+1.7 / +1.2
1.19	1.97	0 / 1.1	+0.6 / -0	+1.1 / +0.6	0 / 1.6	+1.0 / -0	+1.6 / +1.0	0.4 / 2.0	+1.0 / -0	+2.0 / +1.4
1.97	3.15	0.1 / 1.3	+0.7 / -0	+1.3 / +0.7	0.2 / 2.1	+1.2 / -0	+2.1 / +1.4	0.4 / 2.3	+1.2 / -0	+2.3 / +1.6
3.15	4.73	0.1 / 1.6	+0.9 / -0	+1.6 / +1.0	0.2 / 2.5	+1.4 / -0	+2.5 / +1.6	0.6 / 2.9	+1.4 / -0	+2.9 / +2.0
4.73	7.09	0.2 / 1.9	+1.0 / -0	+1.9 / +1.2	0.2 / 2.8	+1.6 / -0	+2.8 / +1.8	0.9 / 3.5	+1.6 / -0	+3.5 / +2.5
7.09	9.85	0.2 / 2.2	+1.2 / -0	+2.2 / +1.4	0.2 / 3.2	+1.8 / -0	+3.2 / +2.0	1.2 / 4.2	+1.8 / -0	+4.2 / +3.0
9.85	12.41	0.2 / 2.3	+1.2 / -0	+2.3 / +1.4	0.2 / 3.4	+2.0 / -0	+3.4 / +2.2	1.5 / 4.7	+2.0 / -0	+4.7 / +3.5
12.41	15.75	0.2 / 2.6	+1.4 / -0	+2.6 / +1.6	0.3 / 3.9	+2.2 / -0	+3.9 / +2.5	2.3 / 5.9	+2.2 / -0	+5.9 / +4.5
15.75	19.69	0.2 / 2.8	+1.6 / -0	+2.8 / +1.8	0.3 / 4.4	+2.5 / -0	+4.4 / +2.8	2.5 / 6.6	+2.5 / -0	+6.6 / +5.0
19.69	30.09		+2.0 / -0		0.5 / 5.5	+3 / -0	+5.5 / +3.5	4 / 9	+3 / -0	+9 / +7
30.09	41.49		+2.5 / -0		0.5 / 7.0	+4 / -0	+7.0 / +4.5	5 / 11.5	+4 / -0	+11.5 / +9
41.49	56.19		+3.0 / -0		1 / 9	+5 / -0	+9 / +6	7 / 15	+5 / -0	+15 / +12
56.19	76.39		+4.0 / -0		1 / 11	+6 / -0	+11 / +7	10 / 20	+6 / -0	+20 / +16
76.39	100.9		+5.0 / -0		1 / 14	+8 / -0	+14 / +9	12 / 25	+8 / -0	+25 / +20
100.9	131.9		+6.0 / -0		2 / 18	+10 / -0	+18 / +12	15 / 31	+10 / -0	+31 / +25
131.9	171.9		+8.0 / -0		4 / 24	+12 / -0	+24 / +16	18 / 38	+12 / -0	+38 / +30
171.9	200		+10.0 / -0		4 / 30	+16 / -0	+30 / +20	24 / 50	+16 / -0	+50 / +40

(Courtesy of ANSI; B4.1–1955.)

APPENDIX 42
AMERICAN STANDARD FORCE AND SHRINK FITS

Limits are in thousandths of an inch.
Limits for hole and shaft are applied algebraically to the basic size to obtain the limits of size for the parts.
Data in bold face are in accordance with ABC agreements.
Symbols H7, s6, etc., are Hole and Shaft designations used in ABC System.

Nominal Size Range Inches Over — To	Class FN 1 Limits of Interference	Class FN 1 Hole H6	Class FN 1 Shaft	Class FN 2 Limits of Interference	Class FN 2 Hole H7	Class FN 2 Shaft s6	Class FN 3 Limits of Interference	Class FN 3 Hole H7	Class FN 3 Shaft t6	Class FN 4 Limits of Interference	Class FN 4 Hole H7	Class FN 4 Shaft u6	Class FN 5 Limits of Interference	Class FN 5 Hole H8	Class FN 5 Shaft x7
0 — 0.12	0.05 0.5	+0.25 − 0	+ 0.5 + 0.3	0.2 0.85	+0.4 − 0	+ 0.85 + 0.6				0.3 0.95	+0.4 − 0	+ 0.95 + 0.7	0.3 1.3	+ 0.6 − 0	+ 1.3 + 0.9
0.12 — 0.24	0.1 0.6	+0.3 − 0	+ 0.6 + 0.4	0.2 1.0	+0.5 − 0	+ 1.0 + 0.7				0.4 1.2	+0.5 − 0	+ 1.2 + 0.9	0.5 1.7	+ 0.7 − 0	+ 1.7 + 1.2
0.24 — 0.40	0.1 0.75	+0.4 − 0	+ 0.75 + 0.5	0.4 1.4	+0.6 − 0	+ 1.4 + 1.0				0.6 1.6	+0.6 − 0	+ 1.6 + 1.2	0.5 2.0	+ 0.9 − 0	+ 2.0 + 1.4
0.40 — 0.56	0.1 0.8	− 0.4 − 0	+ 0.8 + 0.5	0.5 1.6	+0.7 − 0	+ 1.6 + 1.2				0.7 1.8	+ 0.7 − 0	+ 1.8 + 1.4	0.6 2.3	+ 1.0 − 0	+ 2.3 + 1.6
0.56 — 0.71	0:2 0.9	+0.4 ⊢ 0	+ 0.9 + 0.6	0.5 1.6	+0.7 − 0	+ 1.6 + 1.2				0.7 1.8	+ 0.7 − 0	+ 1.8 + 1.4	0.8 2.5	+ 1.0 − 0	+ 2.5 + 1.8
0.71 — 0.95	0.2 1.1	+0.5 − 0	+ 1.1 + 0.7	0.6 1.9	+0.8 − 0	+ 1.9 + 1.4				0.8 2.1	+ 0.8 − 0	+ 2.1 + 1.6	1.0 3.0	+ 1.2 − 0	+ 3.0 + 2.2
0.95 — 1.19	0.3 1.2	+0.5 − 0	+ 1.2 + 0.8	0.6 1.9	+0.8 − 0	+ 1.9 + 1.4	0.8 2.1	+0.8 − 0	+ 2.1 + 1.6	1.0 2.3	+ 0·8 − 0	+ 2.3 + 1.8	1.3 3.3	+ 1.2 − 0	+ 3.3 + 2.5
1.19 — 1.58	0.3 1.3	+0.6 − 0	+ 1.3 + 0.9	0.8 2.4	+1.0 − 0	+ 2.4 + 1.8	1.0 2.6	+1.0 − 0	+ 2.6 + 2.0	1.5 3.1	+1.0 − 0	+ 3.1 + 2.5	1.4 4.0	+ 1.6 − 0	+ 4.0 + 3.0
1.58 — 1.97	0.4 1.4	+0.6 − 0	+ 1.4 + 1.0	0.8 2.4	+1.0 − 0	+ 2.4 + 1.8	1.2 2.8	+1.0 − 0	+ 2.8 + 2.2	1.8 3.4	+1.0 − 0	+ 3.4 + 2.8	2.4 5.0	+ 1.6 − 0	+ 5.0 + 4.0
1.97 — 2.56	0.6 1.8	+0.7 − 0	+ 1.8 + 1.3	0.8 2.7	+1.2 − 0	+ 2.7 + 2.0	1.3 3.2	+1.2 − 0	+ 3.2 + 2.5	2.3 4.2	+1.2 − 0	+ 4.2 + 3.5	3.2 6.2	+ 1.8 − 0	+ 6.2 + 5.0
2.56 — 3.15	0.7 1.9	+0.7 − 0	+ 1.9 + 1.4	1.0 2.9	+1.2 − 0	+ 2.9 + 2.2	1.8 3.7	+1.2 − 0	+ 3.7 + 3.0	2.8 4.7	+1.2 − 0	+ 4.7 + 4.0	4.2 7.2	+ 1.8 − 0	+ 7.2 + 6.0
3.15 — 3.94	0.9 2.4	+0.9 − 0	+ 2.4 + 1.8	1.4 3.7	+1.4 − 0	+ 3.7 + 2.8	2.1 4.4	+1.4 − 0	+ 4.4 + 3.5	3.6 5.9	+1.4 − 0	+ 5.9 + 5.0	4.8 8.4	+ 2.2 − 0	+ 8.4 + 7.0
3.94 — 4.73	1.1 2.6	+0.9 − 0	+ 2.6 + 2.0	1.6 3.9	+1.4 − 0	+ 3.9 + 3.0	2.6 4.9	+1.4 − 0	+ 4.9 + 4.0	4.6 6.9	+1.4 − 0	+ 6.9 + 6.0	5.8 9.4	+ 2.2 − 0	+ 9.4 + 8.0
4.73 — 5.52	1.2 2.9	+1.0 − 0	+ 2.9 + 2.2	1.9 4.5	+1.6 − 0	+ 4.5 + 3.5	3.4 6.0	+1.6 − 0	+ 6.0 + 5.0	5.4 8.0	+1.6 − 0	+ 8.0 + 7.0	7.5 11.6	+ 2.5 − 0	+11.6 +10.0
5.52 — 6.30	1.5 3.2	+1.0 − 0	+ 3.2 + 2.5	2.4 5.0	+1.6 − 0	+ 5.0 + 4.0	3.4 6.0	+1.6 − 0	+ 6.0 + 5.0	5.4 8.0	+1.6 − 0	+ 8.0 + 7.0	9.5 13.6	+ 2.5 − 0	+13.6 +12.0
6.30 — 7.09	1.8 3.5	+1.0 − 0	+ 3.5 + 2.8	2.9 5.5	+1.6 − 0	+ 5.5 + 4.5	4.4 7.0	+1.6 − 0	+ 7.0 + 6.0	6.4 9.0	+1.6 − 0	+ 9.0 + 8.0	9.5 13.6	+ 2.5 − 0	+13.6 +12.0
7.09 — 7.88	1.8 3.8	+1.2 − 0	+ 3.8 + 3.0	3.2 6.2	+1.8 − 0	+ 6.2 + 5.0	5.2 8.2	+1.8 − 0	+ 8.2 + 7.0	7.2 10.2	+1.8 − 0	+10.2 + 9.0	11.2 15.8	+ 2.8 − 0	+15.8 +14.0
7.88 — 8.86	2.3 4.3	+1.2 − 0	+ 4.3 + 3.5	3.2 6.2	+1.8 − 0	+ 6.2 + 5.0	5.2 8.2	+1.8 − 0	+ 8.2 + 7.0	8.2 11.2	+1.8 − 0	+11.2 +10.0	13.2 17.8	+ 2.8 − 0	+17.8 +16.0
8.86 — 9.85	2.3 4.3	+1.2 − 0	+ 4.3 + 3.5	4.2 7.2	+1.8 − 0	+ 7.2 + 6.0	6.2 9.2	+1.8 − 0	+ 9.2 + 8.0	10.2 13.2	+1.8 − 0	+13.2 +12.0	13.2 17.8	+ 2.8 − 0	+17.8 +16.0
9.85 — 11.03	2.8 4.9	+1.2 − 0	+ 4.9 + 4.0	4.0 7.2	+2.0 − 0	+ 7.2 + 6.0	7.0 10.2	+2.0 − 0	+10.2 + 9.0	10.0 13.2	+2.0 − 0	+13.2 +12.0	15.0 20.0	+ 3.0 − 0	+20.0 +18.0
11.03 — 12.41	2.8 4.9	+1.2 − 0	+ 4.9 + 4.0	5.0 8.2	+2.0 − 0	+ 8.2 + 7.0	7.0 10.2	+2.0 − 0	+10.2 + 9.0	12.0 15.2	+2.0 − 0	+15.2 +14.0	17.0 22.0	+ 3.0 − 0	+22.0 +20.0
12.41 — 13.98	3.1 5.5	+1.4 − 0	+ 5.5 + 4.5	5.8 9.4	+2.2 − 0	+ 9.4 + 8.0	7.8 11.4	+2.2 − 0	+11.4 +10.0	13.8 17.4	+2.2 − 0	+17.4 +16.0	18.5 24.2	+ 3.5 + 0	+24.2 +22.0
13.98 — 15.75	3.6 6.1	+1.4 − 0	+ 6.1 + 5.0	5.8 9.4	+2.2 − 0	+ 9.4 + 8.0	9.8 13.4	+2.2 − 0	+13.4 +12.0	15.8 19.4	+2.2 − 0	+19.4 +18.0	21.5 27.2	+ 3.5 − 0	+27.2 +25.0
15.75 — 17.72	4.4 7.0	+1.6 − 0	+ 7.0 + 6.0	6.5 10.6	+2.5 − 0	+10.6 + 9.0	9.5 13.6	+2.5 − 0	+13.6 +12.0	17.5 21.6	+2.5 − 0	+21.6 +20.0	24.0 30.5	+ 4.0 − 0	+30.5 +28.0
17.72 — 19.69	4.4 7.0	+1.6 − 0	+ 7.0 + 6.0	7.5 11.6	+2.5 − 0	+11.6 +10.0	11.5 15.6	+2.5 − 0	+15.6 +14.0	19.5 23.6	+2.5 − 0	+23.6 +22.0	26.0 32.5	+ 4.0 − 0	+32.5 +30.0

(Courtesy of ANSI; B4.1–1955.)

APPENDIX 43
THE INTERNATIONAL TOLERANCE GRADES (ANSI B4.2)

Dimensions are in mm.

Basic sizes Over	Basic sizes Up to and including	IT01	IT0	IT1	IT2	IT3	IT4	IT5	IT6	IT7	IT8	IT9	IT10	IT11	IT12	IT13	IT14	IT15	IT16
0	3	0.0003	0.0005	0.0008	0.0012	0.002	0.003	0.004	0.006	0.010	0.014	0.025	0.040	0.060	0.100	0.140	0.250	0.400	0.600
3	6	0.0004	0.0006	0.001	0.0015	0.0025	0.004	0.005	0.008	0.012	0.018	0.030	0.048	0.075	0.120	0.180	0.300	0.480	0.750
6	10	0.0004	0.0006	0.001	0.0015	0.0025	0.004	0.006	0.009	0.015	0.022	0.036	0.058	0.090	0.150	0.220	0.360	0.580	0.900
10	18	0.0005	0.0008	0.0012	0.002	0.003	0.005	0.008	0.011	0.018	0.027	0.043	0.070	0.110	0.180	0.270	0.430	0.700	1.100
18	30	0.0006	0.001	0.0015	0.0025	0.004	0.006	0.009	0.013	0.021	0.033	0.052	0.084	0.130	0.210	0.330	0.520	0.840	1.300
30	50	0.0006	0.001	0.0015	0.0025	0.004	0.007	0.011	0.016	0.025	0.039	0.062	0.100	0.160	0.250	0.390	0.620	1.000	1.600
50	80	0.0008	0.0012	0.002	0.003	0.005	0.008	0.013	0.019	0.030	0.046	0.074	0.120	0.190	0.300	0.460	0.740	1.200	1.900
80	120	0.001	0.0015	0.0025	0.004	0.006	0.010	0.015	0.022	0.035	0.054	0.087	0.140	0.220	0.350	0.540	0.870	1.400	2.200
120	180	0.0012	0.002	0.0035	0.005	0.008	0.012	0.018	0.025	0.040	0.063	0.100	0.160	0.250	0.400	0.630	1.000	1.600	2.500
180	250	0.002	0.003	0.0045	0.007	0.010	0.014	0.020	0.029	0.046	0.072	0.115	0.185	0.290	0.460	0.720	1.150	1.850	2.900
250	315	0.0025	0.004	0.006	0.008	0.012	0.016	0.023	0.032	0.052	0.081	0.130	0.210	0.320	0.520	0.810	1.300	2.100	3.200
315	400	0.003	0.005	0.007	0.009	0.013	0.018	0.025	0.036	0.057	0.089	0.140	0.230	0.360	0.570	0.890	1.400	2.300	3.600
400	500	0.004	0.006	0.008	0.010	0.015	0.020	0.027	0.040	0.063	0.097	0.155	0.250	0.400	0.630	0.970	1.550	2.500	4.000
500	630	0.0045	0.006	0.009	0.011	0.016	0.022	0.030	0.044	0.070	0.110	0.175	0.280	0.440	0.700	1.100	1.750	2.800	4.400
630	800	0.005	0.007	0.010	0.013	0.018	0.025	0.035	0.050	0.080	0.125	0.200	0.320	0.500	0.800	1.250	2.000	3.200	5.000
800	1000	0.0055	0.008	0.011	0.015	0.021	0.029	0.040	0.056	0.090	0.140	0.230	0.360	0.560	0.900	1.400	2.300	3.600	5.600
1000	1250	0.0065	0.009	0.013	0.018	0.024	0.034	0.046	0.066	0.105	0.165	0.260	0.420	0.660	1.050	1.650	2.600	4.200	6.600
1250	1600	0.008	0.011	0.015	0.021	0.029	0.040	0.054	0.078	0.125	0.195	0.310	0.500	0.780	1.250	1.950	3.100	5.000	7.800
1600	2000	0.009	0.013	0.018	0.025	0.035	0.048	0.065	0.092	0.150	0.230	0.370	0.600	0.920	1.500	2.300	3.700	6.000	9.200
2000	2500	0.011	0.015	0.022	0.030	0.041	0.057	0.077	0.110	0.175	0.280	0.440	0.700	1.100	1.750	2.800	4.400	7.000	11.000
2500	3150	0.013	0.018	0.026	0.036	0.050	0.069	0.093	0.135	0.210	0.330	0.540	0.860	1.350	2.100	3.300	5.400	8.600	13.500

Tolerance grades[3]

[3] IT Values for tolerance grades larger than IT16 can be calculated by using the following formulas:
IT17 = IT12 × 10; IT18 = IT13 × 10; etc.

APPENDIX 44
PREFERRED HOLE BASIS CLEARANCE FITS—CYLINDRICAL FITS (ANSI B4.2)

AMERICAN NATIONAL STANDARD
PREFERRED METRIC LIMITS AND FITS

ANSI B4.2-1978

Dimensions in mm.

BASIC SIZE		LOOSE RUNNING Hole H11	Shaft c11	Fit	FREE RUNNING Hole H9	Shaft d9	Fit	CLOSE RUNNING Hole H8	Shaft f7	Fit	SLIDING Hole H7	Shaft g6	Fit	LOCATIONAL CLEARANCE Hole H7	Shaft h6	Fit
1	MAX	1.060	0.940	0.180	1.025	0.980	0.070	1.014	0.994	0.030	1.010	0.998	0.018	1.010	1.000	0.016
	MIN	1.000	0.880	0.060	1.000	0.955	0.020	1.000	0.984	0.006	1.000	0.992	0.002	1.000	0.994	0.000
1.2	MAX	1.260	1.140	0.180	1.225	1.180	0.070	1.214	1.194	0.030	1.210	1.198	0.018	1.210	1.200	0.016
	MIN	1.200	1.080	0.060	1.200	1.155	0.020	1.200	1.184	0.006	1.200	1.192	0.002	1.200	1.194	0.000
1.6	MAX	1.660	1.540	0.180	1.625	1.580	0.070	1.614	1.594	0.030	1.610	1.598	0.018	1.610	1.600	0.016
	MIN	1.600	1.480	0.060	1.600	1.555	0.020	1.600	1.584	0.006	1.600	1.592	0.002	1.600	1.594	0.000
2	MAX	2.060	1.940	0.180	2.025	1.980	0.070	2.014	1.994	0.030	2.010	1.998	0.018	2.010	2.000	0.016
	MIN	2.000	1.880	0.060	2.000	1.955	0.020	2.000	1.984	0.006	2.000	1.992	0.002	2.000	1.994	0.000
2.5	MAX	2.560	2.440	0.180	2.525	2.480	0.070	2.514	2.49	0.030	2.510	2.498	0.018	2.510	2.500	0.016
	MIN	2.500	2.380	0.060	2.500	2.455	0.020	2.500	2.484	0.006	2.500	2.492	0.002	2.500	2.494	0.000
3	MAX	3.060	2.940	0.180	3.025	2.980	0.070	3.014	2.994	0.030	3.010	2.998	0.018	3.010	3.000	0.016
	MIN	3.000	2.880	0.060	3.000	2.955	0.020	3.000	2.984	0.006	3.000	2.992	0.002	3.000	2.994	0.000
4	MAX	4.075	3.930	0.220	4.030	3.970	0.090	4.018	3.990	0.040	4.012	3.996	0.024	4.012	4.000	0.020
	MIN	4.000	3.855	0.070	4.000	3.940	0.030	4.000	3.978	0.010	4.000	3.988	0.004	4.000	3.992	0.000
5	MAX	5.075	4.930	0.220	5.030	4.970	0.090	5.018	4.990	0.040	5.012	4.996	0.024	5.012	5.000	0.020
	MIN	5.000	4.855	0.070	5.000	4.940	0.030	5.000	4.978	0.010	5.000	4.988	0.004	5.000	4.992	0.000
6	MAX	6.075	5.930	0.220	6.030	5.970	0.090	6.018	5.990	0.040	6.012	5.996	0.024	6.012	6.000	0.020
	MIN	6.000	5.855	0.070	6.000	5.940	0.030	6.000	5.978	0.010	6.000	5.988	0.004	6.000	5.992	0.000
8	MAX	8.090	7.920	0.260	8.036	7.960	0.112	8.022	7.987	0.050	8.015	7.995	0.029	8.015	8.000	0.024
	MIN	8.000	7.830	0.080	8.000	7.924	0.040	8.000	7.972	0.013	8.000	7.986	0.005	8.000	7.991	0.000
10	MAX	10.090	9.920	0.260	10.036	9.960	0.112	10.022	9.987	0.050	10.015	9.995	0.029	10.015	10.000	0.024
	MIN	10.000	9.830	0.080	10.000	9.924	0.040	10.000	9.972	0.013	10.000	9.986	0.005	10.000	9.991	0.000
12	MAX	12.110	11.905	0.315	12.043	11.950	0.136	12.027	11.984	0.061	12.018	11.994	0.035	12.018	12.000	0.029
	MIN	12.000	11.795	0.095	12.000	11.907	0.050	12.000	11.966	0.016	12.000	11.983	0.006	12.000	11.989	0.000
16	MAX	16.110	15.905	0.315	16.043	15.950	0.136	16.027	15.984	0.061	16.018	15.994	0.035	16.018	16.000	0.029
	MIN	16.000	15.795	0.095	16.000	15.907	0.050	16.000	15.966	0.016	16.000	15.983	0.006	16.000	15.989	0.000
20	MAX	20.130	19.890	0.370	20.052	19.935	0.169	20.033	19.980	0.074	20.021	19.993	0.041	20.021	20.000	0.034
	MIN	20.000	19.760	0.110	20.000	19.883	0.065	20.000	19.959	0.020	20.000	19.980	0.007	20.000	19.987	0.000
25	MAX	25.130	24.890	0.370	25.052	24.935	0.169	25.033	24.980	0.074	25.021	24.993	0.041	25.021	25.000	0.034
	MIN	25.000	24.760	0.110	25.000	24.883	0.065	25.000	24.959	0.020	25.000	24.980	0.007	25.000	24.987	0.000
30	MAX	30.130	29.890	0.370	30.052	29.935	0.169	30.033	29.980	0.074	30.021	29.993	0.041	30.021	30.000	0.034
	MIN	30.000	29.760	0.110	30.000	29.883	0.065	30.000	29.959	0.020	30.000	29.980	0.007	30.000	29.987	0.000

Cont.

APPENDIX 44

PREFERRED HOLE BASIS CLEARANCE FITS—CYLINDRICAL FITS (Cont.)

AMERICAN NATIONAL STANDARD
PREFERRED METRIC LIMITS AND FITS

ANSI B4.2-1978

Dimensions in mm.

BASIC SIZE		LOOSE RUNNING Hole H11	Shaft c11	Fit	FREE RUNNING Hole H9	Shaft d9	Fit	CLOSE RUNNING Hole H8	Shaft f7	Fit	SLIDING Hole H7	Shaft g6	Fit	LOCATIONAL CLEARANCE Hole H7	Shaft h6	Fit
40	MAX	40.160	39.880	0.440	40.062	39.920	0.204	40.039	39.975	0.089	40.025	39.991	0.050	40.025	40.000	0.041
	MIN	40.000	39.720	0.120	40.000	39.858	0.080	40.000	39.950	0.025	40.000	39.975	0.009	40.000	39.984	0.000
50	MAX	50.160	49.870	0.450	50.062	49.920	0.204	50.039	49.975	0.089	50.025	49.991	0.050	50.025	50.000	0.041
	MIN	50.000	49.710	0.130	50.000	49.858	0.080	50.000	49.950	0.025	50.000	49.975	0.009	50.000	49.984	0.000
60	MAX	60.190	59.860	0.520	60.074	59.900	0.248	60.046	59.970	0.106	60.030	59.990	0.059	60.030	60.000	0.049
	MIN	60.000	59.670	0.140	60.000	59.826	0.100	60.000	59.940	0.030	60.000	59.971	0.010	60.000	59.981	0.000
80	MAX	80.190	79.850	0.530	80.074	79.900	0.248	80.046	79.970	0.106	80.030	79.990	0.059	80.030	80.000	0.049
	MIN	80.000	79.660	0.150	80.000	79.826	0.100	80.000	79.940	0.030	80.000	79.971	0.010	80.000	79.981	0.000
100	MAX	100.220	99.830	0.610	100.087	99.880	0.294	100.054	99.964	0.125	100.035	99.988	0.069	100.035	100.000	0.057
	MIN	100.000	99.610	0.170	100.000	99.793	0.120	100.000	99.929	0.036	100.000	99.966	0.012	100.000	99.978	0.000
120	MAX	120.220	119.820	0.620	120.087	119.880	0.294	120.054	119.964	0.125	120.035	119.988	0.069	120.035	120.000	0.057
	MIN	120.000	119.600	0.180	120.000	119.793	0.120	120.000	119.929	0.036	120.000	119.966	0.012	120.000	119.978	0.000
160	MAX	160.250	159.790	0.710	160.100	159.855	0.345	160.063	159.957	0.146	160.040	159.986	0.079	160.040	160.000	0.065
	MIN	160.000	159.540	0.210	160.000	159.755	0.145	160.000	159.917	0.043	160.000	159.961	0.014	160.000	159.975	0.000
200	MAX	200.290	199.760	0.820	200.115	199.830	0.400	200.072	199.950	0.168	200.046	199.985	0.090	200.046	200.000	0.075
	MIN	200.000	199.470	0.240	200.000	199.715	0.170	200.000	199.904	0.050	200.000	199.956	0.015	200.000	199.971	0.000
250	MAX	250.290	249.720	0.860	250.115	249.830	0.400	250.072	249.950	0.168	250.046	249.985	0.090	250.046	250.000	0.075
	MIN	250.000	249.430	0.280	250.000	249.715	0.170	250.000	249.904	0.050	250.000	249.956	0.015	250.000	249.971	0.000
300	MAX	300.320	299.670	0.970	300.130	299.810	0.450	300.081	299.944	0.189	300.052	299.983	0.101	300.052	300.000	0.084
	MIN	300.000	299.350	0.330	300.000	299.680	0.190	300.000	299.892	0.056	300.000	299.951	0.017	300.000	299.968	0.000
400	MAX	400.360	399.600	1.120	400.140	399.790	0.490	400.089	399.938	0.208	400.057	399.982	0.111	400.057	400.000	0.093
	MIN	400.000	399.240	0.400	400.000	399.650	0.210	400.000	399.881	0.062	400.000	399.946	0.018	400.000	399.964	0.000
500	MAX	500.400	499.520	1.280	500.155	499.770	0.540	500.097	499.932	0.228	500.063	499.980	0.123	500.063	500.000	0.103
	MIN	500.000	499.120	0.480	500.000	499.615	0.230	500.000	499.869	0.068	500.000	499.940	0.020	500.000	499.960	0.000

APPENDIX 45
PREFERRED HOLE BASIS TRANSITION AND INTERFERENCE FITS—CYLINDRICAL FITS (ANSI B4.2)

AMERICAN NATIONAL STANDARD
PREFERRED METRIC LIMITS AND FITS

ANSI B4.2-1978

Dimensions in mm.

BASIC SIZE		LOCATIONAL TRANSN. Hole H7	Shaft k6	Fit	LOCATIONAL TRANSN. Hole H7	Shaft n6	Fit	LOCATIONAL INTERF. Hole H7	Shaft p6	Fit	MEDIUM DRIVE Hole H7	Shaft s6	Fit	FORCE Hole H7	Shaft u6	Fit
1	MAX	1.010	1.006	0.010	1.010	1.010	0.006	1.010	1.012	0.004	1.010	1.020	-0.004	1.010	1.024	-0.008
	MIN	1.000	1.000	-0.006	1.000	1.004	-0.010	1.000	1.006	-0.012	1.000	1.014	-0.020	1.000	1.018	-0.024
1.2	MAX	1.210	1.206	0.010	1.210	1.210	0.006	1.210	1.212	0.004	1.210	1.220	-0.004	1.210	1.224	-0.008
	MIN	1.200	1.200	-0.006	1.200	1.204	-0.010	1.200	1.206	-0.012	1.200	1.214	-0.020	1.200	1.218	-0.024
1.6	MAX	1.610	1.606	0.010	1.610	1.610	0.006	1.610	1.612	0.004	1.610	1.620	-0.004	1.610	1.624	-0.008
	MIN	1.600	1.600	-0.006	1.600	1.604	-0.010	1.600	1.606	-0.012	1.600	1.614	-0.020	1.600	1.618	-0.024
2	MAX	2.010	2.006	0.010	2.010	2.010	0.006	2.010	2.012	0.004	2.010	2.020	-0.004	2.010	2.024	-0.008
	MIN	2.000	2.000	-0.006	2.000	2.004	-0.010	2.000	2.006	-0.012	2.000	2.014	-0.020	2.000	2.018	-0.024
2.5	MAX	2.510	2.506	0.010	2.510	2.510	0.006	2.510	2.512	0.004	2.510	2.520	-0.004	2.510	2.524	-0.008
	MIN	2.500	2.500	-0.006	2.500	2.504	-0.010	2.500	2.506	-0.012	2.500	2.514	-0.020	2.500	2.518	-0.024
3	MAX	3.010	3.006	0.010	3.010	3.010	0.006	3.010	3.012	0.004	3.010	3.020	-0.004	3.010	3.024	-0.008
	MIN	3.000	3.000	-0.006	3.000	3.004	-0.010	3.000	3.006	-0.012	3.000	3.014	-0.020	3.000	3.018	-0.024
4	MAX	4.012	4.009	0.011	4.012	4.016	0.004	4.012	4.020	0.000	4.012	4.027	-0.007	4.012	4.031	-0.011
	MIN	4.000	4.001	-0.009	4.000	4.008	-0.016	4.000	4.012	-0.020	4.000	4.019	-0.027	4.000	4.023	-0.031
5	MAX	5.012	5.009	0.011	5.012	5.016	0.004	5.012	5.020	0.000	5.012	5.027	-0.007	5.012	5.031	-0.011
	MIN	5.000	5.001	-0.009	5.000	5.008	-0.016	5.000	5.012	-0.020	5.000	5.019	-0.027	5.000	5.023	-0.031
6	MAX	6.012	6.009	0.011	6.012	6.016	0.004	6.012	6.020	0.000	6.012	6.027	-0.007	6.012	6.031	-0.011
	MIN	6.000	6.001	-0.009	6.000	6.008	-0.016	6.000	6.012	-0.020	6.000	6.019	-0.027	6.000	6.023	-0.031
8	MAX	8.015	8.010	0.014	8.015	8.019	0.005	8.015	8.024	0.000	8.015	8.032	-0.008	8.015	8.037	-0.013
	MIN	8.000	8.001	-0.010	8.000	8.010	-0.019	8.000	8.015	-0.024	8.000	8.023	-0.032	8.000	8.028	-0.037
10	MAX	10.015	10.010	0.014	10.015	10.019	0.005	10.015	10.024	0.000	10.015	10.032	-0.008	10.015	10.037	-0.013
	MIN	10.000	10.001	-0.010	10.000	10.010	-0.019	10.000	10.015	-0.024	10.000	10.023	-0.032	10.000	10.028	-0.037
12	MAX	12.018	12.012	0.017	12.018	12.023	0.006	12.018	12.029	0.000	12.018	12.039	-0.010	12.018	12.044	-0.015
	MIN	12.000	12.001	-0.012	12.000	12.012	-0.023	12.000	12.018	-0.029	12.000	12.028	-0.039	12.000	12.033	-0.044
16	MAX	16.018	16.012	0.017	16.018	16.023	0.006	16.018	16.029	0.000	16.018	16.039	-0.010	16.018	16.044	-0.015
	MIN	16.000	16.001	-0.012	16.000	16.012	-0.023	16.000	16.018	-0.029	16.000	16.028	-0.039	16.000	16.033	-0.044
20	MAX	20.021	20.015	0.019	20.021	20.028	0.006	20.021	20.035	-0.001	20.021	20.048	-0.014	20.021	20.054	-0.020
	MIN	20.000	20.002	-0.015	20.000	20.015	-0.028	20.000	20.022	-0.035	20.000	20.035	-0.048	20.000	20.041	-0.054
25	MAX	25.021	25.015	0.019	25.021	25.028	0.006	25.021	25.035	-0.001	25.021	25.048	-0.014	25.021	25.061	-0.027
	MIN	25.000	25.002	-0.015	25.000	25.015	-0.028	25.000	25.022	-0.035	25.000	25.035	-0.048	25.000	25.048	-0.061
30	MAX	30.021	30.015	0.019	30.021	30.028	0.006	30.021	30.035	-0.001	30.021	30.048	-0.014	30.021	30.061	-0.027
	MIN	30.000	30.002	-0.015	30.000	30.015	-0.028	30.000	30.022	-0.035	30.000	30.035	-0.048	30.000	30.048	-0.061

Cont.

APPENDIX 45

PREFERRED HOLE BASIS TRANSITION AND INTERFERENCE FITS—CYLINDRICAL FITS (ANSI B4.2) (Cont.)

AMERICAN NATIONAL STANDARD
PREFERRED METRIC LIMITS AND FITS

ANSI B4.2-1978

Dimensions in mm.

BASIC SIZE		LOCATIONAL TRANSN. Hole H7	Shaft k6	Fit	LOCATIONAL TRANSN. Hole H7	Shaft n6	Fit	LOCATIONAL INTERF. Hole H7	Shaft p6	Fit	MEDIUM DRIVE Hole H7	Shaft s6	Fit	FORCE Hole H7	Shaft u6	Fit
40	MAX	40.025	40.018	0.023	40.025	40.033	0.008	40.025	40.042	-0.001	40.025	40.059	-0.018	40.025	40.076	-0.035
	MIN	40.000	40.002	-0.018	40.000	40.017	-0.033	40.000	40.026	-0.042	40.000	40.043	-0.059	40.000	40.060	-0.076
50	MAX	50.025	50.018	0.023	50.025	50.033	0.008	50.025	50.042	-0.001	50.025	50.059	-0.018	50.025	50.086	-0.045
	MIN	50.000	50.002	-0.018	50.000	50.017	-0.033	50.000	50.026	-0.042	50.000	50.043	-0.059	50.000	50.070	-0.086
60	MAX	60.030	60.021	0.028	60.030	60.039	0.010	60.030	60.051	-0.002	60.030	60.072	-0.023	60.030	60.106	-0.057
	MIN	60.000	60.002	-0.021	60.000	60.020	-0.039	60.000	60.032	-0.051	60.000	60.053	-0.072	60.000	60.087	-0.106
80	MAX	80.030	80.021	0.028	80.030	80.039	0.010	80.030	80.051	-0.002	80.030	80.078	-0.029	80.030	80.121	-0.072
	MIN	80.000	80.002	-0.021	80.000	80.020	-0.039	80.000	80.032	-0.051	80.000	80.059	-0.078	80.000	80.102	-0.121
100	MAX	100.035	100.025	0.032	100.035	100.045	0.012	100.035	100.059	-0.002	100.035	100.093	-0.036	100.035	100.146	-0.089
	MIN	100.000	100.003	-0.025	100.000	100.023	-0.045	100.000	100.037	-0.059	100.000	100.071	-0.093	100.000	100.124	-0.146
120	MAX	120.035	120.025	0.032	120.035	120.045	0.012	120.035	120.059	-0.002	120.035	120.101	-0.044	120.035	120.166	-0.109
	MIN	120.000	120.003	-0.025	120.000	120.023	-0.045	120.000	120.037	-0.059	120.000	120.079	-0.101	120.000	120.144	-0.166
160	MAX	160.040	160.028	0.037	160.040	160.052	0.013	160.040	160.068	-0.003	160.040	160.125	-0.060	160.040	160.215	-0.150
	MIN	160.000	160.003	-0.028	160.000	160.027	-0.052	160.000	160.043	-0.068	160.000	160.100	-0.125	160.000	160.190	-0.215
200	MAX	200.046	200.033	0.042	200.046	200.060	0.015	200.046	200.079	-0.004	200.046	200.151	-0.076	200.046	200.265	-0.190
	MIN	200.000	200.004	-0.033	200.000	200.031	-0.060	200.000	200.050	-0.079	200.000	200.122	-0.151	200.000	200.236	-0.265
250	MAX	250.046	250.033	0.042	250.046	250.060	0.015	250.046	250.079	-0.004	250.046	250.169	-0.094	250.046	250.313	-0.238
	MIN	250.000	250.004	-0.033	250.000	250.031	-0.060	250.000	250.050	-0.079	250.000	250.140	-0.169	250.000	250.284	-0.313
300	MAX	300.052	300.036	0.048	300.052	300.066	0.018	300.052	300.088	-0.004	300.052	300.202	-0.118	300.052	300.382	-0.298
	MIN	300.000	300.004	-0.036	300.000	300.034	-0.066	300.000	300.056	-0.088	300.000	300.170	-0.202	300.000	300.350	-0.382
400	MAX	400.057	400.040	0.053	400.057	400.073	0.020	400.057	400.098	-0.005	400.057	400.244	-0.151	400.057	400.471	-0.378
	MIN	400.000	400.004	-0.040	400.000	400.037	-0.073	400.000	400.062	-0.098	400.000	400.208	-0.244	400.000	400.435	-0.471
500	MAX	500.063	500.045	0.058	500.063	500.080	0.023	500.063	500.108	-0.005	500.063	500.292	-0.189	500.063	500.580	-0.477
	MIN	500.000	500.005	-0.045	500.000	500.040	-0.080	500.000	500.068	-0.108	500.000	500.252	-0.292	500.000	500.540	-0.580

APPENDIX 46

PREFERRED SHAFT BASIS CLEARANCE FITS—CYLINDRICAL FITS (ANSI B4.2)

AMERICAN NATIONAL STANDARD
PREFERRED METRIC LIMITS AND FITS

ANSI B4.2-1978

Dimensions in mm.

BASIC SIZE	MAX/MIN	LOOSE RUNNING Hole C11	LOOSE RUNNING Shaft h11	LOOSE RUNNING Fit	FREE RUNNING Hole D9	FREE RUNNING Shaft h9	FREE RUNNING Fit	CLOSE RUNNING Hole F8	CLOSE RUNNING Shaft h7	CLOSE RUNNING Fit	SLIDING Hole G7	SLIDING Shaft h6	SLIDING Fit	LOCATIONAL CLEARANCE Hole H7	LOCATIONAL CLEARANCE Shaft h6	LOCATIONAL CLEARANCE Fit
1	MAX	1.120	1.000	0.180	1.045	1.000	0.070	1.020	1.000	0.030	1.012	1.000	0.018	1.010	1.000	0.016
	MIN	1.060	0.940	0.060	1.020	0.975	0.020	1.006	0.990	0.006	1.002	0.994	0.002	1.000	0.994	0.000
1.2	MAX	1.320	1.200	0.180	1.245	1.200	0.070	1.220	1.200	0.030	1.212	1.200	0.018	1.210	1.200	0.016
	MIN	1.260	1.140	0.060	1.220	1.175	0.020	1.206	1.190	0.006	1.202	1.194	0.002	1.200	1.194	0.000
1.6	MAX	1.720	1.600	0.180	1.645	1.600	0.070	1.620	1.600	0.030	1.612	1.600	0.018	1.610	1.600	0.016
	MIN	1.660	1.540	0.060	1.620	1.575	0.020	1.606	1.590	0.006	1.602	1.594	0.002	1.600	1.594	0.000
2	MAX	2.120	2.000	0.180	2.045	2.000	0.070	2.020	2.000	0.030	2.012	2.000	0.018	2.010	2.000	0.016
	MIN	2.060	1.940	0.060	2.020	1.975	0.020	2.006	1.990	0.006	2.002	1.994	0.002	2.000	1.994	0.000
2.5	MAX	2.620	2.500	0.180	2.545	2.500	0.070	2.520	2.500	0.030	2.512	2.500	0.018	2.510	2.500	0.016
	MIN	2.560	2.440	0.060	2.520	2.475	0.020	2.506	2.490	0.006	2.502	2.494	0.002	2.500	2.494	0.000
3	MAX	3.120	3.000	0.180	3.045	3.000	0.070	3.020	3.000	0.030	3.012	3.000	0.018	3.010	3.000	0.016
	MIN	3.060	2.940	0.060	3.020	2.975	0.020	3.006	2.990	0.006	3.002	2.994	0.002	3.000	2.994	0.000
4	MAX	4.145	4.000	0.220	4.060	4.000	0.090	4.028	4.000	0.040	4.016	4.000	0.024	4.012	4.000	0.020
	MIN	4.070	3.925	0.070	4.030	3.970	0.030	4.010	3.988	0.010	4.004	3.992	0.004	4.000	3.992	0.000
5	MAX	5.145	5.000	0.220	5.060	5.000	0.090	5.028	5.000	0.040	5.016	5.000	0.024	5.012	5.000	0.020
	MIN	5.070	4.925	0.070	5.030	4.970	0.030	5.010	4.988	0.010	5.004	4.992	0.004	5.000	4.992	0.000
6	MAX	6.145	6.000	0.220	6.060	6.000	0.090	6.028	6.000	0.040	6.016	6.000	0.024	6.012	6.000	0.020
	MIN	6.070	5.925	0.070	6.030	5.970	0.030	6.010	5.988	0.010	6.004	5.992	0.004	6.000	5.992	0.000
8	MAX	8.170	8.000	0.260	8.076	8.000	0.112	8.035	8.000	0.050	8.020	8.000	0.029	8.015	8.000	0.024
	MIN	8.080	7.910	0.080	8.040	7.964	0.040	8.013	7.985	0.013	8.005	7.991	0.005	8.000	7.991	0.000
10	MAX	10.170	10.000	0.260	10.076	10.000	0.112	10.035	10.000	0.050	10.020	10.000	0.029	10.015	10.000	0.024
	MIN	10.080	9.910	0.080	10.040	9.964	0.040	10.013	9.985	0.013	10.005	9.991	0.005	10.000	9.991	0.000
12	MAX	12.205	12.000	0.315	12.093	12.000	0.136	12.043	12.000	0.061	12.024	12.000	0.035	12.018	12.000	0.029
	MIN	12.095	11.890	0.095	12.050	11.957	0.050	12.016	11.982	0.016	12.006	11.989	0.006	12.000	11.989	0.000
16	MAX	16.205	16.000	0.315	16.093	16.000	0.136	16.043	16.000	0.061	16.024	16.000	0.035	16.018	16.000	0.029
	MIN	16.095	15.890	0.095	16.050	15.957	0.050	16.016	15.982	0.016	16.006	15.989	0.006	16.000	15.989	0.000
20	MAX	20.240	20.000	0.370	20.117	20.000	0.169	20.053	20.000	0.074	20.028	20.000	0.041	20.021	20.000	0.034
	MIN	20.110	19.870	0.110	20.065	19.948	0.065	20.020	19.979	0.020	20.007	19.987	0.007	20.000	19.987	0.000
25	MAX	25.240	25.000	0.370	25.117	25.000	0.169	25.053	25.000	0.074	25.028	25.000	0.041	25.021	25.000	0.034
	MIN	25.110	24.870	0.110	25.065	24.948	0.065	25.020	24.979	0.020	25.007	24.987	0.007	25.000	24.987	0.000
30	MAX	30.240	30.000	0.370	30.117	30.000	0.169	30.053	30.000	0.074	30.028	30.000	0.041	30.021	30.000	0.034
	MIN	30.110	29.870	0.110	30.065	29.948	0.065	30.020	29.979	0.020	30.007	29.987	0.007	30.000	29.987	0.000

Cont.

APPENDIX 46

PREFERRED SHAFT BASIS CLEARANCE FITS—CYLINDRICAL FITS (Cont.)

AMERICAN NATIONAL STANDARD
PREFERRED METRIC LIMITS AND FITS

ANSI B4.2-1978

Dimensions in mm.

BASIC SIZE		LOOSE RUNNING Hole C11	Shaft h11	Fit	FREE RUNNING Hole D9	Shaft h9	Fit	CLOSE RUNNING Hole F8	Shaft h7	Fit	SLIDING Hole G7	Shaft h6	Fit	LOCATIONAL CLEARANCE Hole H7	Shaft h6	Fit
40	MAX	40.280	40.000	0.440	40.142	40.000	0.204	40.064	40.000	0.089	40.034	40.000	0.050	40.025	40.000	0.041
	MIN	40.120	39.840	0.120	40.080	39.938	0.080	40.025	39.975	0.025	40.009	39.984	0.009	40.000	39.984	0.000
50	MAX	50.290	50.000	0.450	50.142	50.000	0.204	50.064	50.000	0.089	50.034	50.000	0.050	50.025	50.000	0.041
	MIN	50.130	49.840	0.130	50.080	49.938	0.080	50.025	49.975	0.025	50.009	49.984	0.009	50.000	49.984	0.000
60	MAX	60.330	60.000	0.520	60.174	60.000	0.248	60.076	60.000	0.106	60.040	60.000	0.059	60.030	60.000	0.049
	MIN	60.140	59.810	0.140	60.100	59.926	0.100	60.030	59.970	0.030	60.010	59.981	0.010	60.000	59.981	0.000
80	MAX	80.340	80.000	0.530	80.174	80.000	0.248	80.076	80.000	0.106	80.040	80.000	0.059	80.030	80.000	0.049
	MIN	80.150	79.810	0.150	80.100	79.926	0.100	80.030	79.970	0.030	80.010	79.981	0.010	80.000	79.981	0.000
100	MAX	100.390	100.000	0.610	100.207	100.000	0.294	100.090	100.000	0.125	100.047	100.000	0.069	100.035	100.000	0.057
	MIN	100.170	99.780	0.170	100.120	99.913	0.120	100.036	99.965	0.036	100.012	99.978	0.012	100.000	99.978	0.000
120	MAX	120.400	120.000	0.620	120.207	120.000	0.294	120.090	120.000	0.125	120.047	120.000	0.069	120.035	120.000	0.057
	MIN	120.180	119.780	0.180	120.120	119.913	0.120	120.036	119.965	0.036	120.012	119.978	0.012	120.000	119.978	0.000
160	MAX	160.460	160.000	0.710	160.245	160.000	0.345	160.106	160.000	0.146	160.054	160.000	0.079	160.040	160.000	0.065
	MIN	160.210	159.750	0.210	160.145	159.900	0.145	160.043	159.960	0.043	160.014	159.975	0.014	160.000	159.975	0.000
200	MAX	200.530	200.000	0.820	200.285	200.000	0.400	200.122	200.000	0.168	200.061	200.000	0.090	200.046	200.000	0.075
	MIN	200.240	199.710	0.240	200.170	199.885	0.170	200.050	199.954	0.050	200.015	199.971	0.015	200.000	199.971	0.000
250	MAX	250.570	250.000	0.860	250.285	250.000	0.400	250.122	250.000	0.168	250.061	250.000	0.090	250.046	250.000	0.075
	MIN	250.280	249.710	0.280	250.170	249.885	0.170	250.050	249.954	0.050	250.015	249.971	0.015	250.000	249.971	0.000
300	MAX	300.650	300.000	0.970	300.320	300.000	0.450	300.137	300.000	0.189	300.069	300.000	0.101	300.052	300.000	0.084
	MIN	300.330	299.680	0.330	300.190	299.870	0.190	300.056	299.948	0.056	300.017	299.968	0.017	300.000	299.968	0.000
400	MAX	400.760	400.000	1.120	400.350	400.000	0.490	400.151	400.000	0.208	400.075	400.000	0.111	400.057	400.000	0.093
	MIN	400.400	399.640	0.400	400.210	399.860	0.210	400.062	399.943	0.062	400.018	399.964	0.018	400.000	399.964	0.000
500	MAX	500.880	500.000	1.280	500.385	500.000	0.540	500.165	500.000	0.228	500.083	500.000	0.123	500.063	500.000	0.103
	MIN	500.480	499.600	0.480	500.230	499.845	0.230	500.068	499.937	0.068	500.020	499.960	0.020	500.000	499.960	0.000

APPENDIX 47

PREFERRED SHAFT BASIS TRANSITION AND INTERFERENCE FITS—CYLINDRICAL FITS

AMERICAN NATIONAL STANDARD
PREFERRED METRIC LIMITS AND FITS

ANSI B4.2-1978

Dimensions in mm.

BASIC SIZE		LOCATIONAL TRANSN. Hole K7	LOCATIONAL TRANSN. Shaft h6	LOCATIONAL TRANSN. Fit	LOCATIONAL TRANSN. Hole N7	LOCATIONAL TRANSN. Shaft h6	LOCATIONAL TRANSN. Fit	LOCATIONAL INTERF. Hole P7	LOCATIONAL INTERF. Shaft h6	LOCATIONAL INTERF. Fit	MEDIUM DRIVE Hole S7	MEDIUM DRIVE Shaft h6	MEDIUM DRIVE Fit	FORCE Hole U7	FORCE Shaft h6	FORCE Fit
1	MAX	1.000	1.000	0.006	0.996	1.000	0.002	0.994	1.000	0.000	0.986	1.000	-0.008	0.982	1.000	-0.012
	MIN	0.990	0.994	-0.010	0.986	0.994	-0.014	0.984	0.994	-0.016	0.976	0.994	-0.024	0.972	0.994	-0.028
1.2	MAX	1.200	1.200	0.006	1.196	1.200	0.002	1.194	1.200	0.000	1.186	1.200	-0.008	1.182	1.200	-0.012
	MIN	1.190	1.194	-0.010	1.186	1.194	-0.014	1.184	1.194	-0.016	1.176	1.194	-0.024	1.172	1.194	-0.028
1.6	MAX	1.600	1.600	0.006	1.596	1.600	0.002	1.594	1.600	0.000	1.586	1.600	-0.008	1.582	1.600	-0.012
	MIN	1.590	1.594	-0.010	1.586	1.594	-0.014	1.584	1.594	-0.016	1.576	1.594	-0.024	1.572	1.594	-0.028
2	MAX	2.000	2.000	0.006	1.996	2.000	0.002	1.994	2.000	0.000	1.986	2.000	-0.008	1.982	2.000	-0.012
	MIN	1.990	1.994	-0.010	1.986	1.994	-0.014	1.984	1.994	-0.016	1.976	1.994	-0.024	1.972	1.994	-0.028
2.5	MAX	2.500	2.500	0.006	2.496	2.500	0.002	2.494	2.500	0.000	2.486	2.500	-0.008	2.482	2.500	-0.012
	MIN	2.490	2.494	-0.010	2.486	2.494	-0.014	2.484	2.494	-0.016	2.476	2.494	-0.024	2.472	2.494	-0.028
3	MAX	3.000	3.000	0.006	2.996	3.000	0.002	2.994	3.000	0.000	2.986	3.000	-0.008	2.982	3.000	-0.012
	MIN	2.990	2.994	-0.010	2.986	2.994	-0.014	2.984	2.994	-0.016	2.976	2.994	-0.024	2.972	2.994	-0.028
4	MAX	4.003	4.000	0.011	3.996	4.000	0.004	3.992	4.000	0.000	3.985	4.000	-0.007	3.981	4.000	-0.011
	MIN	3.991	3.992	-0.009	3.984	3.992	-0.016	3.980	3.992	-0.020	3.973	3.992	-0.027	3.969	3.992	-0.031
5	MAX	5.003	5.000	0.011	4.996	5.000	0.004	4.992	5.000	0.000	4.985	5.000	-0.007	4.981	5.000	-0.011
	MIN	4.991	4.992	-0.009	4.984	4.992	-0.016	4.980	4.992	-0.020	4.973	4.992	-0.027	4.969	4.992	-0.031
6	MAX	6.003	6.000	0.011	5.996	6.000	0.004	5.992	6.000	0.000	5.985	6.000	-0.007	5.981	6.000	-0.011
	MIN	5.991	5.992	-0.009	5.984	5.992	-0.016	5.980	5.992	-0.020	5.973	5.992	-0.027	5.969	5.992	-0.031
8	MAX	8.005	8.000	0.014	7.996	8.000	0.005	7.991	8.000	0.000	7.983	8.000	-0.008	7.978	8.000	-0.013
	MIN	7.990	7.991	-0.010	7.981	7.991	-0.019	7.976	7.991	-0.024	7.968	7.991	-0.032	7.963	7.991	-0.037
10	MAX	10.005	10.000	0.014	9.996	10.000	0.005	9.991	10.000	0.000	9.983	10.000	-0.008	9.978	10.000	-0.013
	MIN	9.990	9.991	-0.010	9.981	9.991	-0.019	9.976	9.991	-0.024	9.968	9.991	-0.032	9.963	9.991	-0.037
12	MAX	12.006	12.000	0.017	11.995	12.000	0.006	11.989	12.000	0.000	11.979	12.000	-0.010	11.974	12.000	-0.015
	MIN	11.988	11.989	-0.012	11.977	11.989	-0.023	11.971	11.989	-0.029	11.961	11.989	-0.039	11.956	11.989	-0.044
16	MAX	16.006	16.000	0.017	15.995	16.000	0.006	15.989	16.000	0.000	15.979	16.000	-0.010	15.974	16.000	-0.015
	MIN	15.988	15.989	-0.012	15.977	15.989	-0.023	15.971	15.989	-0.029	15.961	15.989	-0.039	15.956	15.989	-0.044
20	MAX	20.006	20.000	0.019	19.993	20.000	0.006	19.986	20.000	-0.001	19.973	20.000	-0.014	19.967	20.000	-0.020
	MIN	19.985	19.987	-0.015	19.972	19.987	-0.028	19.965	19.987	-0.035	19.952	19.987	-0.048	19.946	19.987	-0.054
25	MAX	25.006	25.000	0.019	24.993	25.000	0.006	24.986	25.000	-0.001	24.973	25.000	-0.014	24.960	25.000	-0.027
	MIN	24.985	24.987	-0.015	24.972	24.987	-0.028	24.965	24.987	-0.035	24.952	24.987	-0.048	24.939	24.987	-0.061
30	MAX	30.006	30.000	0.019	29.993	30.000	0.006	29.986	30.000	-0.001	29.973	30.000	-0.014	29.960	30.000	-0.027
	MIN	29.985	29.987	-0.015	29.972	29.987	-0.028	29.965	29.987	-0.035	29.952	29.987	-0.048	29.939	29.987	-0.061

Cont.

APPENDIX 47

PREFERRED SHAFT BASIS TRANSITION AND INTERFERENCE FITS—CYLINDRICAL FITS (Cont.)

AMERICAN NATIONAL STANDARD
PREFERRED METRIC LIMITS AND FITS

ANSI B4.2-1978

Dimensions in mm.

BASIC SIZE		LOCATIONAL TRANSN. Hole K7	Shaft h6	Fit	LOCATIONAL TRANSN. Hole N7	Shaft h6	Fit	LOCATIONAL INTERF. Hole P7	Shaft h6	Fit	MEDIUM DRIVE Hole S7	Shaft h6	Fit	FORCE Hole U7	Shaft h6	Fit
40	MAX	40.007	40.000	0.023	39.992	40.000	0.008	39.983	40.000	-0.001	39.966	40.000	-0.018	39.949	40.000	-0.035
	MIN	39.982	39.984	-0.018	39.967	39.984	-0.033	39.958	39.984	-0.042	39.941	39.984	-0.059	39.924	39.984	-0.076
50	MAX	50.007	50.000	0.023	49.992	50.000	0.008	49.983	50.000	-0.001	49.966	50.000	-0.018	49.939	50.000	-0.045
	MIN	49.982	49.984	-0.018	49.967	49.984	-0.033	49.958	49.984	-0.042	49.941	49.984	-0.059	49.914	49.984	-0.086
60	MAX	60.009	60.000	0.028	59.991	60.000	0.010	59.979	60.000	-0.002	59.958	60.000	-0.023	59.924	60.000	-0.057
	MIN	59.979	59.981	-0.021	59.961	59.981	-0.039	59.949	59.981	-0.051	59.928	59.981	-0.072	59.894	59.981	-0.106
80	MAX	80.009	80.000	0.028	79.991	80.000	0.010	79.979	80.000	-0.002	79.952	80.000	-0.029	79.909	80.000	-0.072
	MIN	79.979	79.981	-0.021	79.961	79.981	-0.039	79.949	79.981	-0.051	79.922	79.981	-0.078	79.879	79.981	-0.121
100	MAX	100.010	100.000	0.032	99.990	100.000	0.012	99.976	100.000	-0.002	99.942	100.000	-0.036	99.889	100.000	-0.089
	MIN	99.975	99.978	-0.025	99.955	99.978	-0.045	99.941	99.978	-0.059	99.907	99.978	-0.093	99.854	99.978	-0.146
120	MAX	120.010	120.000	0.032	119.990	120.000	0.012	119.976	120.000	-0.002	119.934	120.000	-0.044	119.869	120.000	-0.109
	MIN	119.975	119.978	-0.025	119.955	119.978	-0.045	119.941	119.978	-0.059	119.899	119.978	-0.101	119.834	119.978	-0.166
160	MAX	160.012	160.000	0.037	159.988	160.000	0.013	159.972	160.000	-0.003	159.915	160.000	-0.060	159.825	160.000	-0.150
	MIN	159.972	159.975	-0.028	159.948	159.975	-0.052	159.932	159.975	-0.068	159.875	159.975	-0.125	159.785	159.975	-0.215
200	MAX	200.013	200.000	0.042	199.986	200.000	0.015	199.967	200.000	-0.004	199.895	200.000	-0.076	199.781	200.000	-0.190
	MIN	199.967	199.971	-0.033	199.940	199.971	-0.060	199.921	199.971	-0.079	199.849	199.971	-0.151	199.735	199.971	-0.265
250	MAX	250.013	250.000	0.042	249.986	250.000	0.015	249.967	250.000	-0.004	249.877	250.000	-0.094	249.733	250.000	-0.238
	MIN	249.967	249.971	-0.033	249.940	249.971	-0.060	249.921	249.971	-0.079	249.831	249.971	-0.169	249.687	249.971	-0.313
300	MAX	300.016	300.000	0.048	299.986	300.000	0.018	299.964	300.000	-0.004	299.850	300.000	-0.118	299.670	300.000	-0.298
	MIN	299.964	299.968	-0.036	299.934	299.968	-0.066	299.912	299.968	-0.088	299.798	299.968	-0.202	299.618	299.968	-0.382
400	MAX	400.017	400.000	0.053	399.984	400.000	0.020	399.959	400.000	-0.005	399.813	400.000	-0.151	399.586	400.000	-0.378
	MIN	399.960	399.964	-0.040	399.927	399.964	-0.073	399.902	399.964	-0.098	399.756	399.964	-0.244	399.529	399.964	-0.471
500	MAX	500.018	500.000	0.058	499.983	500.000	0.023	499.955	500.000	-0.005	499.771	500.000	-0.189	499.483	500.000	-0.477
	MIN	499.955	499.960	-0.045	499.920	499.960	-0.080	499.892	499.960	-0.108	499.708	499.960	-0.292	499.420	499.960	-0.580

APPENDIX 48
HOLE SIZES FOR NON-PREFERRED DIAMETERS

Basic Size		C11	D9	F8	G7	H7	H8	H9	H11	K7	N7	P7	S7	U7
OVER 0 TO 3		+0.120 +0.060	+0.045 +0.020	+0.020 +0.006	+0.012 +0.002	+0.010 0.000	+0.014 0.000	+0.025 0.000	+0.060 0.000	0.000 -0.010	-0.004 -0.014	-0.006 -0.016	-0.014 -0.024	-0.018 -0.028
OVER 3 TO 6		+0.145 +0.070	+0.060 +0.030	+0.028 +0.010	+0.016 +0.004	+0.012 0.000	+0.018 0.000	+0.030 0.000	+0.075 0.000	+0.003 -0.009	-0.004 -0.016	-0.008 -0.020	-0.015 -0.027	-0.019 -0.031
OVER 6 TO 10		+0.170 +0.080	+0.076 +0.040	+0.035 +0.013	+0.020 +0.005	+0.015 0.000	+0.022 0.000	+0.036 0.000	+0.090 0.000	+0.005 -0.010	-0.004 -0.019	-0.009 -0.024	-0.017 -0.032	-0.022 -0.037
OVER 10 TO 14		+0.205 +0.095	+0.093 +0.050	+0.043 +0.016	+0.024 +0.006	+0.018 0.000	+0.027 0.000	+0.043 0.000	+0.110 0.000	+0.006 -0.012	-0.005 -0.023	-0.011 -0.029	-0.021 -0.039	-0.026 -0.044
OVER 14 TO 18		+0.205 +0.095	+0.093 +0.050	+0.043 +0.016	+0.024 +0.006	+0.018 0.000	+0.027 0.000	+0.043 0.000	+0.110 0.000	+0.006 -0.012	-0.005 -0.023	-0.011 -0.029	-0.021 -0.039	-0.026 -0.044
OVER 18 TO 24		+0.240 +0.110	+0.117 +0.065	+0.053 +0.020	+0.028 +0.007	+0.021 0.000	+0.033 0.000	+0.052 0.000	+0.130 0.000	+0.006 -0.015	-0.007 -0.028	-0.014 -0.035	-0.027 -0.048	-0.033 -0.054
OVER 24 TO 30		+0.240 +0.110	+0.117 +0.065	+0.053 +0.020	+0.028 +0.007	+0.021 0.000	+0.033 0.000	+0.052 0.000	+0.130 0.000	+0.006 -0.015	-0.007 -0.028	-0.014 -0.035	-0.027 -0.048	-0.040 -0.061
OVER 30 TO 40		+0.280 +0.120	+0.142 +0.080	+0.064 +0.025	+0.034 +0.009	+0.025 0.000	+0.039 0.000	+0.062 0.000	+0.160 0.000	+0.007 -0.018	-0.008 -0.033	-0.017 -0.042	-0.034 -0.059	-0.051 -0.076
OVER 40 TO 50		+0.290 +0.130	+0.142 +0.080	+0.064 +0.025	+0.034 +0.009	+0.025 0.000	+0.039 0.000	+0.062 0.000	+0.160 0.000	+0.007 -0.018	-0.008 -0.033	-0.017 -0.042	-0.034 -0.059	-0.061 -0.086
OVER 50 TO 65		+0.330 +0.140	+0.174 +0.100	+0.076 +0.030	+0.040 +0.010	+0.030 0.000	+0.046 0.000	+0.074 0.000	+0.190 0.000	+0.009 -0.021	-0.009 -0.039	-0.021 -0.051	-0.042 -0.072	-0.076 -0.106
OVER 65 TO 80		+0.340 +0.150	+0.174 +0.100	+0.076 +0.030	+0.040 +0.010	+0.030 0.000	+0.046 0.000	+0.074 0.000	+0.190 0.000	+0.009 -0.021	-0.009 -0.039	-0.021 -0.051	-0.048 -0.078	-0.091 -0.121
OVER 80 TO 100		+0.390 +0.170	+0.207 +0.120	+0.090 +0.036	+0.047 +0.012	+0.035 0.000	+0.054 0.000	+0.087 0.000	+0.220 0.000	+0.010 -0.025	-0.010 -0.045	-0.024 -0.059	-0.058 -0.093	-0.111 -0.146

Cont.

APPENDIX 48

HOLE SIZES FOR NON-PREFERRED DIAMETERS (Cont.)

Basic Size	c11	d9	f8	g7	h7	h8	h9	h11	k7	n7	p7	s7	u7
OVER 100 TO 120	+0.400 / +0.180	+0.207 / +0.120	+0.090 / +0.036	+0.047 / +0.012	+0.035 / 0.000	+0.054 / 0.000	+0.087 / 0.000	+0.220 / 0.000	+0.010 / -0.025	-0.010 / -0.045	-0.024 / -0.059	-0.066 / -0.101	-0.131 / -0.166
OVER 120 TO 140	+0.450 / +0.200	+0.245 / +0.145	+0.106 / +0.043	+0.054 / +0.014	+0.040 / 0.000	+0.063 / 0.000	+0.100 / 0.000	+0.250 / 0.000	+0.012 / -0.028	-0.012 / -0.052	-0.028 / -0.068	-0.077 / -0.117	-0.155 / -0.195
OVER 140 TO 160	+0.460 / +0.210	+0.245 / +0.145	+0.106 / +0.043	+0.054 / +0.014	+0.040 / 0.000	+0.063 / 0.000	+0.100 / 0.000	+0.250 / 0.000	+0.012 / -0.028	-0.012 / -0.052	-0.028 / -0.068	-0.085 / -0.125	-0.175 / -0.215
OVER 160 TO 180	+0.480 / +0.230	+0.245 / +0.145	+0.106 / +0.043	+0.054 / +0.014	+0.040 / 0.000	+0.063 / 0.000	+0.100 / 0.000	+0.250 / 0.000	+0.012 / -0.028	-0.012 / -0.052	-0.028 / -0.068	-0.093 / -0.133	-0.195 / -0.235
OVER 180 TO 200	+0.530 / +0.240	+0.285 / +0.170	+0.122 / +0.050	+0.061 / +0.015	+0.046 / 0.000	+0.072 / 0.000	+0.115 / 0.000	+0.290 / 0.000	+0.013 / -0.033	-0.014 / -0.060	-0.033 / -0.079	-0.105 / -0.151	-0.219 / -0.265
OVER 200 TO 225	+0.550 / +0.260	+0.285 / +0.170	+0.122 / +0.050	+0.061 / +0.015	+0.046 / 0.000	+0.072 / 0.000	+0.115 / 0.000	+0.290 / 0.000	+0.013 / -0.033	-0.014 / -0.060	-0.033 / -0.079	-0.113 / -0.159	-0.241 / -0.287
OVER 225 TO 250	+0.570 / +0.280	+0.285 / +0.170	+0.122 / +0.050	+0.061 / +0.015	+0.046 / 0.000	+0.072 / 0.000	+0.115 / 0.000	+0.290 / 0.000	+0.013 / -0.033	-0.014 / -0.060	-0.033 / -0.079	-0.123 / -0.169	-0.267 / -0.313
OVER 250 TO 280	+0.620 / +0.300	+0.320 / +0.190	+0.137 / +0.056	+0.069 / +0.017	+0.052 / 0.000	+0.081 / 0.000	+0.130 / 0.000	+0.320 / 0.000	+0.016 / -0.036	-0.014 / -0.066	-0.036 / -0.088	-0.138 / -0.190	-0.295 / -0.347
OVER 280 TO 315	+0.650 / +0.330	+0.320 / +0.190	+0.137 / +0.056	+0.069 / 0.017	+0.052 / 0.000	+0.081 / 0.000	+0.130 / 0.000	+0.320 / 0.000	+0.016 / -0.036	-0.014 / -0.066	-0.036 / -0.088	-0.150 / -0.202	-0.330 / -0.382
OVER 315 TO 355	+0.720 / +0.360	+0.350 / +0.210	+0.151 / +0.062	+0.075 / +0.018	+0.057 / 0.000	+0.089 / 0.000	+0.140 / 0.000	+0.360 / 0.000	+0.017 / -0.040	-0.016 / -0.073	-0.041 / -0.098	-0.169 / -0.226	-0.369 / -0.426
OVER 355 TO 400	+0.760 / +0.400	+0.350 / +0.210	+0.151 / +0.062	+0.075 / +0.018	+0.057 / 0.000	+0.089 / 0.000	+0.140 / 0.000	+0.360 / 0.000	+0.017 / -0.040	-0.016 / -0.073	-0.041 / -0.098	-0.187 / -0.244	-0.414 / -0.471
OVER 400 TO 450	+0.840 / +0.440	+0.385 / +0.230	+0.165 / +0.068	+0.083 / +0.020	+0.063 / 0.000	+0.097 / 0.000	+0.155 / 0.000	+0.400 / 0.000	+0.018 / -0.045	-0.017 / -0.080	-0.045 / -0.108	-0.209 / -0.272	-0.467 / -0.530
OVER 450 TO 500	+0.880 / +0.480	+0.385 / +0.230	+0.165 / +0.068	+0.083 / +0.020	+0.063 / 0.000	+0.097 / 0.000	+0.155 / 0.000	+0.400 / 0.000	+0.018 / -0.045	-0.017 / -0.080	-0.045 / -0.108	-0.229 / -0.292	-0.517 / -0.580

APPENDIX 49
SHAFT SIZES FOR NON-PREFERRED DIAMETERS

Basic Size	c11	d9	f7	g6	h6	h7	h9	h11	k6	n6	p6	s6	u6
OVER 0 TO 3	−0.060 / −0.120	−0.020 / −0.045	−0.006 / −0.016	−0.002 / −0.008	0.000 / −0.006	0.000 / −0.010	0.000 / −0.025	0.000 / −0.060	+0.006 / 0.000	+0.010 / +0.004	+0.012 / +0.006	+0.020 / +0.014	+0.024 / +0.018
OVER 3 TO 6	−0.070 / −0.145	−0.030 / −0.060	−0.010 / −0.022	−0.004 / −0.012	0.000 / −0.008	0.000 / −0.012	0.000 / −0.030	0.000 / −0.075	+0.009 / +0.001	+0.016 / +0.008	+0.020 / +0.012	+0.027 / +0.019	+0.031 / +0.023
OVER 6 TO 10	−0.080 / −0.170	−0.040 / −0.076	−0.013 / −0.028	−0.005 / −0.014	0.000 / −0.009	0.000 / −0.015	0.000 / −0.036	0.000 / −0.090	+0.010 / +0.001	+0.019 / +0.010	+0.024 / +0.015	+0.032 / +0.023	+0.037 / +0.028
OVER 10 TO 14	−0.095 / −0.205	−0.050 / −0.093	−0.016 / −0.034	−0.006 / −0.017	0.000 / −0.011	0.000 / −0.018	0.000 / −0.043	0.000 / −0.110	+0.012 / +0.001	+0.023 / +0.012	+0.029 / +0.018	+0.039 / +0.028	+0.044 / +0.033
OVER 14 TO 18	−0.095 / −0.205	−0.050 / −0.093	−0.016 / −0.034	−0.006 / −0.017	0.000 / −0.011	0.000 / −0.018	0.000 / −0.043	0.000 / −0.110	+0.012 / +0.001	+0.023 / +0.012	+0.029 / +0.018	+0.039 / +0.028	+0.044 / +0.033
OVER 18 TO 24	−0.110 / −0.240	−0.065 / −0.117	−0.020 / −0.041	−0.007 / −0.020	0.000 / −0.013	0.000 / −0.021	0.000 / −0.052	0.000 / −0.130	+0.015 / +0.002	+0.028 / +0.015	+0.035 / +0.022	+0.048 / +0.035	+0.054 / +0.041
OVER 24 TO 30	−0.110 / −0.240	−0.065 / −0.117	−0.020 / −0.041	−0.007 / −0.020	0.000 / −0.013	0.000 / −0.021	0.000 / −0.052	0.000 / −0.130	+0.015 / +0.002	+0.028 / +0.015	+0.035 / +0.022	+0.048 / +0.035	+0.061 / +0.048
OVER 30 TO 40	−0.120 / −0.280	−0.080 / −0.142	−0.025 / −0.050	−0.009 / −0.025	0.000 / −0.016	0.000 / −0.025	0.000 / −0.062	0.000 / −0.160	+0.018 / +0.002	+0.033 / +0.017	+0.042 / +0.026	+0.059 / +0.043	+0.076 / +0.060
OVER 40 TO 50	−0.130 / −0.290	−0.080 / −0.142	−0.025 / −0.050	−0.009 / −0.025	0.000 / −0.016	0.000 / −0.025	0.000 / −0.062	0.000 / −0.160	+0.018 / +0.002	+0.033 / +0.017	+0.042 / +0.026	+0.059 / +0.043	+0.086 / +0.070
OVER 50 TO 65	−0.140 / −0.330	−0.100 / −0.174	−0.030 / −0.060	−0.010 / −0.029	0.000 / −0.019	0.000 / −0.030	0.000 / −0.074	0.000 / −0.190	+0.021 / +0.002	+0.039 / +0.020	+0.051 / −0.032	+0.072 / +0.053	+0.106 / +0.087
OVER 65 TO 80	−0.150 / −0.340	−0.100 / −0.174	−0.030 / −0.060	−0.010 / −0.029	0.000 / −0.019	0.000 / −0.030	0.000 / −0.074	0.000 / −0.190	+0.021 / +0.002	+0.039 / +0.020	+0.051 / +0.032	+0.078 / +0.059	+0.121 / +0.102
OVER 80 TO 100	−0.170 / −0.390	−0.120 / −0.207	−0.036 / −0.071	−0.012 / −0.034	0.000 / −0.022	0.000 / −0.035	0.000 / −0.087	0.000 / −0.220	+0.025 / +0.003	+0.045 / +0.023	+0.059 / +0.037	+0.093 / +0.071	+0.146 / +0.124

Cont.

APPENDIX 49
SHAFT SIZES FOR NON-PREFERRED DIAMETERS (Cont.)

Basic Size	c11	d9	f7	g6	h6	h7	h9	h11	k6	n6	p6	s6	u6
OVER 100 TO 120	-0.180 / -0.400	-0.120 / -0.207	-0.036 / -0.071	-0.012 / -0.034	0.000 / -0.022	0.000 / -0.035	0.000 / -0.087	0.000 / -0.220	+0.025 / +0.003	+0.045 / +0.023	+0.059 / +0.037	+0.101 / +0.079	+0.166 / +0.144
OVER 120 TO 140	-0.200 / -0.450	-0.145 / -0.245	-0.043 / -0.083	-0.014 / -0.039	0.000 / -0.025	0.000 / -0.040	0.000 / -0.100	0.000 / -0.250	+0.028 / +0.003	+0.052 / +0.027	+0.068 / +0.043	+0.117 / +0.092	+0.195 / +0.170
OVER 140 TO 160	-0.210 / -0.460	-0.145 / -0.245	-0.043 / -0.083	-0.014 / -0.039	0.000 / -0.025	0.000 / -0.040	0.000 / -0.100	0.000 / -0.250	+0.028 / +0.003	+0.052 / +0.027	+0.068 / +0.043	+0.125 / +0.100	+0.215 / +0.190
OVER 160 TO 180	-0.230 / -0.480	-0.145 / -0.245	-0.043 / -0.083	-0.014 / -0.039	0.000 / -0.025	0.000 / -0.040	0.000 / -0.100	0.000 / -0.250	+0.028 / +0.003	+0.052 / +0.027	+0.068 / +0.043	+0.133 / +0.108	+0.235 / +0.210
OVER 180 TO 200	-0.240 / -0.530	-0.170 / -0.285	-0.050 / -0.096	-0.015 / -0.044	0.000 / -0.029	0.000 / -0.046	0.000 / -0.115	0.000 / -0.290	+0.033 / +0.004	+0.060 / +0.031	+0.079 / +0.050	+0.151 / +0.122	+0.265 / +0.236
OVER 200 TO 225	-0.260 / -0.550	-0.170 / -0.285	-0.050 / -0.096	-0.015 / -0.044	0.000 / -0.029	0.000 / -0.046	0.000 / -0.115	0.000 / -0.290	+0.033 / +0.004	+0.060 / +0.031	+0.079 / +0.050	+0.159 / +0.130	+0.287 / +0.258
OVER 225 TO 250	-0.280 / -0.570	-0.170 / -0.285	-0.050 / -0.096	-0.015 / -0.044	0.000 / -0.029	0.000 / -0.046	0.000 / -0.115	0.000 / -0.290	+0.033 / +0.004	+0.060 / +0.031	+0.079 / +0.050	+0.169 / +0.140	+0.313 / +0.284
OVER 250 TO 280	-0.300 / -0.620	-0.190 / -0.320	-0.056 / -0.108	-0.017 / -0.049	0.000 / -0.032	0.000 / -0.052	0.000 / -0.130	0.000 / -0.320	+0.036 / +0.004	+0.066 / +0.034	+0.088 / +0.056	+0.190 / +0.158	+0.347 / +0.315
OVER 280 TO 315	-0.330 / -0.650	-0.190 / -0.320	-0.056 / -0.108	-0.017 / -0.049	0.000 / -0.032	0.000 / -0.052	0.000 / -0.130	0.000 / -0.320	+0.036 / +0.004	+0.066 / +0.034	+0.088 / +0.056	+0.202 / +0.170	+0.382 / +0.350
OVER 315 TO 355	-0.360 / -0.720	-0.210 / -0.350	-0.062 / -0.119	-0.018 / -0.054	0.000 / -0.036	0.000 / -0.057	0.000 / -0.140	0.000 / -0.360	+0.040 / +0.004	+0.073 / +0.037	+0.098 / +0.062	+0.226 / +0.190	+0.426 / +0.390
OVER 355 TO 400	-0.400 / -0.760	-0.210 / -0.350	-0.062 / -0.119	-0.018 / -0.054	0.000 / -0.036	0.000 / -0.057	0.000 / -0.140	0.000 / -0.360	+0.040 / +0.004	+0.073 / +0.037	+0.098 / +0.062	+0.244 / +0.208	+0.471 / +0.435
OVER 400 TO 450	-0.440 / -0.840	-0.230 / -0.385	-0.068 / -0.131	-0.020 / -0.060	0.000 / -0.040	0.000 / 0.063	0.000 / -0.155	0.000 / 0.400	+0.045 / +0.005	+0.080 / +0.040	+0.108 / +0.068	+0.272 / +0.232	+0.530 / +0.490
OVER 450 TO 500	-0.480 / -0.880	-0.230 / -0.385	-0.068 / -0.131	-0.020 / -0.060	0.000 / -0.040	0.000 / -0.063	0.000 / -0.155	0.000 / -0.400	+0.045 / +0.005	+0.080 / +0.040	+0.108 / +0.068	+0.292 / +0.252	+0.580 / +0.540

A-70

APPENDIX 50
GRADING GRAPH

This graph can be used to determine the individual grades of members of a team and to compute grade averages for those who do extra assignments.

The percent participation of each team member should be determined by the team as a whole (see Chapter 2 problems).

Example: written or oral report grades

Overall team grade: 82

Team members N = 5	Contribution C = %	F = CN	Grade (graph)
J. Doe	20%	100	82.0
H. Brown	16%	80	76.4
L. Smith	24%	120	86.0
R. Black	20%	100	82.0
T. Jones	20%	100	82.0
	100%		

Example: quiz or problem sheet grades

Number assigned: 30
Number extra: 6

Total 36

Average grade for total (36): 82

$$F = \frac{\text{No. completed} \times 100}{\text{No. assigned}} = \frac{36 \times 100}{30} = 120$$

Final grade (from graph): 86.0

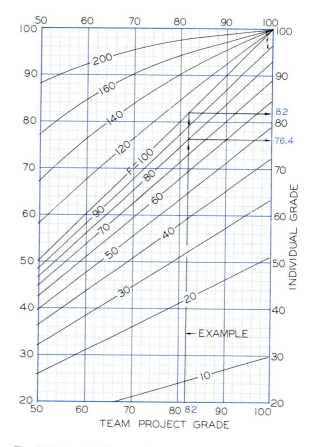

Fig. A50.–1. Grading graph.

The following programs were written by Professor Leendert Kersten of the University of Nebraska at Lincoln and they have been reproduced here with his permission. These programs were introduced in Chapter 20, in which the principles of descriptive geometry are covered. These very valuable programs can be duplicated and added as supplements to your AutoCAD software.

```
(defun C:PARALLEL ( )
    (setvar "aperture" 5)
    (setq sp (getpoint "\nSelect START point of parallel line:"))
    (setq ep (getpoint "\nSelect END point of parallel line:"))
    (setvar "osmode" 1)
    (setq sl (getpoint "\nSelect 1st point on line for parallelism:"))
    (setq el (getpoint "\nSelect 2nd point on line for parallelism:"))
    (setvar "osmode" 0)
    (setq pa (angle sl el))
    (setq la (angle sp ep))
    (setq ll (distance sp ep))
    (setq m -1)
    (setq d 0)
    (if (> pa d) (setq m 1))
    (if (> la d) (setq d 1))
    (if (/= m d) (setq pa (+ pa 3.14159)))
    (setq ep (polar sp pa ll))
    (setvar "cmdecho" 0)
    (command "line" sp ep"")
    (restore)
)

(defun C:PERPLINE ( )
    (setvar "aperture" 5) (setvar "cmdecho" 0)
    (setq sp (getpoint "\nSelect START point of perpendicular line:"))
    (setvar "osmode" 128)
    (setq cc (getpoint sp "\nSelect ANY point on line to which perp'lr:"))
    (setq beta (angle sp cc)) (setvar "osmode" 0)
(setq ep
(getpoint "\nSelect END point of desired perpendicular (for length only): "))
    (setq length (distance sp ep))
    (setq ep (polar sp beta length))
    (command "line" sp ep "")
    (restore)
)
```

```
 (setvar "aperture" 5) (setvar "cmdecho" 0)
 (setvar "osmode" 1)
 (setq aa (getpoint "\nSelect start of transfer distance:"))
 (setvar "osmode" 128)
 (setq bb (getpoint aa "\nSelect the reference plane:"))
 (setq length (distance aa bb))
 (setvar "osmode" 1)
 (setq cc (getpoint "\nSelect point to be projected:"))
 (setvar "osmode" 128)
 (setq dd (getpoint cc "\nSelect other reference plane:"))
(setvar "osmode" 0)
 (setq alpha (angle cc dd))
 (setq ep (polar dd alpha length))
 (COMMAND "CIRCLE" EP 0.05)
 (restore)
)

(defun RESTORE ( )
 (setvar "aperture" 10)
 (setvar "cmdecho" 1)
 (setvar "osmode" 0)
)
(defun command: ( )
(defun *error* (st)
 (setvar "osmode" 0)
 (princ))
 (quit))
)

(defun C:COPYDIST ( )
 (setvar "aperture"5)
 (setvar "cmdecho" 0)
 (setvar "osmode" 1)
 (setq p1 (getpoint "\nSelect start point of line distance to be copied:"))
 (setq p2 (getpoint "\nEnd point:"))
 (setvar "osmode" 0)
 (setq dist (distance p1 p2))
 (setq p1 (getpoint "\nStart point of new distance location:"))
 (setq ang (getangle p1 "\nWhich direction?:"))
 (setq p2 (polar p1 ang dist))
 (setvar "osmode" 0)
 (command "circle" p2 0.05)
 (restore)
)
)
```

```
(defun C:BISECT ( )
  (setvar "aperture" 5)
  (setvar "osmode" 32)
  (setq sp (getpoint "\nSelect Corner of angle:"))
  (setvar "osmode" 2)
  (setq aa (getpoint "\nSelect first side (remember CCW):"))
  (setq alpha (angle sp aa))
  (setq bb (getpoint "\nSelect other side:"))
  (setvar "osmode" 0)
  (setq beta (angle sp bb))
  (setq m (/ (+ alpha beta) 2))
  (if (> alpha beta) (setq ang (+ pi m)) (setq ang m))
  (setq ep
    (getpoint "\nSelect endpoint of bisecting line (for length only): "))
  (setq length (distance sp ep))
  (setq ep (polar sp ang length))
  (setvar "cmdecho" 0)
  (command "line" sp ep "")
  (restore)
)
```

Index ————————————

A

Abbreviations, A-3
ABS plastic, 244
Abscissa, 476
Absolute coordinates, 503
Acetylene gas, 313
Acme thread, 190
  drawing of, 198
Acrylic, 244
Actual size, 280
Addendum, 221, 225
Advanced modeling extension, 593
Advanced modeling extrusion, 492
AISI. *See* American Iron and Steel Institute
ALIASES command, 501
Aligned dimensions, 252
Allowance, 279–280
Alphabet of lines, 60, 108, 124
Aluminum, 238
AME. *See* Advanced Modeling Extension
American Iron and Steel Institute, 236
American Society for Testing Materials, 236
American Welding Society, 320
Ames lettering instrument, 49
Analysis, 17
Analytical geometry, 413
Angle
  acute, 77
  bisection by computer, 82
  complementary, 77
  between line and plane, 426
  obtuse, 77
  between planes, 417, 421
  right, 77
  supplementary, 77
ANGULAR option, 538
Angular distance to a line, 425
Angularity, 302

Annealing, 239
ANSI pipe symbols, 211
APERTURE command, 523
Arc
  by computer, 509
  construction by computer, 84
ARC command, 509
Arc welding, 312
Architects' scale, 64
Arcs through three points, 84
AREA command, 532
ARRAY command, 524
Arrowheads, special, 543
A-size, 329
Assembly
  exploded, 334
  orthographic, 334
  outline, 335
Assembly drawings, 333
  isometric, 334
Associative dimensioning, 259, 541
ASTM. *See* American Society for Testing Materials
ATTDEF command, 578
ATTDISP command, 579
ATTEDIT command, 579
ATTEXT command, 580
ATTRIBUTES, 330, 578
AutoCAD
  computer graphics, 491
  Release 11, 491
AutoCAD 3–D, 396
Auxiliaries
  of curved shapes, 159
  from front view, 156
  reference-line method, 155
  reference-plane method, 159
  from side view, 158
  from top view, 153, 155
Auxiliary plane, 152

Auxiliary sections, 160, 184
Auxiliary views, 152
    by computer, 157
    partial, 160
AXIS command, 496
Axonometric pictorials, 390
    by computer, 157
Axonometric projection, 370
AWS. *See* American Welding Society
Azimuth bearings, 415

B

Baker
    Favorite Seat, 29
    Manufacturing Co., 25
    Tree Stand, 25
Bar graphs, 473, 475
BASELINE dimensioning, 535
Basic dimensions, 297
Basic hole system, 281
Basic shaft system, 281
Basic size, 280–281
Bevel gear
    calculations, 225
    terms, 224
Bevel gears, drawing, 226
BISECT command, 405–406
Bisecting angles, 81–82
BLIPMODE command, 497
Blips, 507
BLOCK command, 398, 529
Blueprinting, 363–364
Bolt circle, 267
Bolt heads, 202
    hexagon, 203
Bolts
    carriage, 208
    countersunk, 208
    method of drawing, 137
    and nuts, 206
    step, 208
    stove, 208
Boring, 245, 271, 290
Boss, 138
Bottoming tap, 210
Box, by computer, 556
BOX option, 504
Braddock-Rowe lettering triangle, 49, 253
Brazing, 319
BREAK command, 506
Break corner, 274
Break, polyline, 546
Break-even graphs, 480
Breaks, conventional, 181
Brittleness, 239
Broaching, 247, 290

Broken-out sections, 181
Brush, drafting, 64
B-size, 329
Buffing, 248
Buttress threads, 190

C

Cabinet obliques, 371
CAD. *See* Computer-aided design
CAD/CAM. *See* Computer-aided design/manufacturing
CADD. *See* Computer-aided design, drafting
Cadkey, 47
Cam followers, 231
Cam motions, 230
Cam with offset follower, 234
CAMERA option, 553
Cams, 220
Cap screw, 202
Carbon-arc spot weld, 318
Case hardening, 240
Castings, 240, 336
    die, 241
    investment, 140
    permanent-mold, 240
    sand, 240
Cathode-ray tube, 45
Cavalier obliques, 371
CDF format, 580
Center of vision, 391
Centerlines, 129
    by computer, 129
Centimeter, 68
Central processing unit, 44
Chain dimensions, 288
CHAMFER command, 510
Chamfers, 271
CHANGE command, 330, 512
Chart, organizational, 485
Checking a drawing, 332
Chemical engineering, 6
CHROP command, 513, 566
CIM. *See* Computer-integrated manufacturing
Circle of centers, 267
CIRCLE command, 397, 508
Circle
    by computer, 508
    drawing of, 508
    elements of, 78
Class of fit, 191
Clearance fit, 280, 282
Clearance locational fits, A-53–A-54
CLIP option, 555
Coarse thread series, 191
Cold rolling, 243
Command line, 498

Compass, 70
  beam, 71
  bearing, 415, 455
  bow, 73
  large, 74
  point sharpening, 71
Compass globe, 547
Computer graphics
  advantages of, 41
  applications of, 42
  introduction, 41
Computer hardware, 43
Computer lines, 125
Computer-aided design, 41
  drafting, 41
  software, 47
Computer-aided design/manufacturing, 43
Computer-integrated manufacturing, 43
Concentricity, 299
Cone, 79–80
  by computer, 557
  frustum of, 79–80
  truncated, 79
Conic sections, 93
Conical tapers, 289
Conjungate diameters, 95
Construction drawings, 12
Contact area, 304
CONTINUE option, 535
Contour lines, 458
Contour maps, 458
Control-C, 499
Conventional breaks, 181
Conventional practices, 136
Conventional revolutions, 136
Coordinate system (3D), 550
Coordinates, screen, 497
Cope, 240
Copper, 238
COPY command, 521
COPYDIST command, 405–406
Cores, 240
Cotter pins, A-50
Counterboring, 256, 271
Counterdrilling, 269
Countersinking, 256, 269
CPU. *See* Central processing unit
Craftsman, 4
Crest, 190
CROSSING option, 505, 525
CRT. *See* Cathode-ray tube
Curve plotting, 135
Custom-designed lines, 501
Cut and fill, 462
Cutting planes, 170
Cylinders, 80
  measuring, 260

Cylindrical coordinates, 503
Cylindricity, 300

D

Dam design, 463
Datum planes
  primary, 295
  secondary, 296
  tertiary, 296
Datum targets, 296
Datums, cylindrical, 294
DBLIST command, 532
DBPro, 399
DDATTE, 499
DDboxes, 499
DDEDIT command, 499, 520
DDEMODES, 499
DDLMODES, 499
DDRMODES, 499
DDUCS, 499
Decimeter, 68
Decision, 17
Decision chart, 26
DECURVE option, 515
Dedendum, 221, 225
Dekameter, 68–69
DELAY command, 583
Descriptive geometry, 404, 455
  labeling programs, 404, 447
  origin of, 2
Design, 15
  drafters, 13
  drawings, 12
  problems, 29, 30, 35
  process, 15
Designer, 4
Detail
  drawings, 322
  paper, 58
  of threads, 196–197
Developed views, 137
Development
  of cones, 444
  of cylinders, 441
  of oblique cylinders, 442
  of oblique prisms, 440
  of prisms, 439
  of pyramids, 443
  of transition pieces, 445
Developments, 427, 438
Deviation, 282
Diametral pitch, 229
Diamond knurl, 271
Diamond turning, 290
Diazo printing, 363
Die casting, 290

Digitizer, 45
Digitizing with a tablet, 581
Dihedral angles, 417, 421
DIM command, 251, 536
DIM VARS command, 253–254, 256, 279, 534
DIMALT, 539
DIMALTD, 251, 539
DIMALTF, 251, 539
DIMAPOST, 539
DIMASO, 259, 541
DIMASZ, 256, 535
DIMBLK, 543
DIMBLK1, 544
DIMBLK2, 544
DIMCEN, 539
DIMCLRD, 539
DIMCLRE, 539
DIMCLRT, 539
DIMDLE, 539
DIMDLI, 535
Dimension lines, 249
Dimensioning, 249
    angles, 259, 538
    arcs, 263, 537
    circles, 537
    by computer, 256, 258–259, 534
    cones, 262
    curved surfaces, 264
    dialogue boxes, 535
    cylinders, 259
    notes, 274
    principles, 533
    prisms, 257
    pyramids, 262
    rules, 255
    slots, 268
    spheres, 262
    styles, 536
    symbols, 256
    symmetrical objects, 265
    terminology, 249
    in 3D, 574
    variables, 534, 539
    variables by computer, 254
Dimensions
    location, 266
    ordinate, 536
    placement of, 253
Dimetric projection, 390
DIMEXE, 535
DIMEXO, 535
DIMGAP, 539
DIMLFAC, 539
DIMLIM, 278, 544
DIMPOST, 539
DIMRND, 539
DIMSAH, 544

DIMSCALE, 254, 257, 535
DIMSE1, 540
DIMSE2, 540
DIMSOXD, 540
DIMTAD, 256, 535, 540
DIMTEXT, 535
DIMTFAC, 544
DIMTIH, 540
DIMTIX, 537, 540
DIMTOFL, 537, 540
DIMTOH, 540
DIMTOL, 278, 544
DIMTM, 278, 544
DIMTP, 278, 544
DIMTSZ, 256, 540, 544
DIMTVP, 540
DIMZIN, 540
Dip angle, 464
Direction of slope, 419
Directrix, 79, 95
Dish, computer, 557
Disk operating system, 494
DIST command, 532
DISTANCE option, 554
Distance
    to ore vein, 466
    point to a line, 423
DIVIDE command, 527
Dividers, 70–71
    bow, 72
    proportional, 71–72
DIVIEW command, 554
DIVIEWBLOCK house, 553
Division of lines, 83
Dodecahedron, 79
Dog points, 207
Dome, by computer, 557
DONUT command, 477, 525
DOS. See Disk operating system
DOT arrowhead, 544
Drafter's log, 333
Drafters, 12–13
Drafting instrument sets, 70
Drafting machine, 59
Drag, 240
DRAG command, 521
DRAGMODE command, 497, 508
Drawing
    enlarging by computer, 511
    instruments, 57
    sheet sizes, 329
    techniques of, 60
Drawing editor, 500
Drill press, 246
Drilling, 245, 269, 290
DTEXT command, 518
Dual dimensioning, 251

Ductility, 239
DVIEW comand, 553
DXF format, 580
Dynamic dialogue boxes, 499
DYNAMIC option, 512
DynaPerspective, 47

E

Edge joint, 314
Edge views of planes, 416
EDGESURF command, 558, 560, 575
Editing text, 520
Edison, Thomas A., 4
EDLIN command, 583
EDLIN options, 583
Einstein, Albert, 1
Elastacity, 239
ELEV command, 396, 547, 552, 562
Ellipse, 93
   by computer, 94
   isometric, 546
   templates, 94, 381
ELLIPSE command, 395–396, 546
Elliptical features, 163
END command, 493
Engineer, 3
   highway, 7
   sanitary, 7
Engineering
   aerospace, 5
   agricultural, 5
   civil, 7
   civil, applications of, 455
   designers, 13
   electrical, 8
   fields, 5
   graphics, 1
   mechanical, 9
   metallurgical, 10
   mining, 10
   nuclear, 10
   petroleum, 11
Engineers' scale, 65
English system of meaurement, 66, 250, 331
Enlargement of drawings, 83
Entities, selection of, 504
ENTITY CREATION option, 498
Equilateral triangles, 77
ERASE command, 395–396, 505
Eraser, electric, 64
Erasing shield, 63
EXIT command, 494
EXPLODE command, 541
EXTEND command, 527
Extension lines, 249
External references, 530

External thread, 189
Extra-fine threads, 191
Extrusions, by computer, 547

F

Facing, 244
Fasteners, 189
Feature control symbols, 292, 294
Ferrous metals, 236
Fiberglass, 244
FILEDIA variable, 493
FILES command, 500, 533
FILL command, 507
FILLET command, 89, 509
Fillets, 263
Fillets and rounds, 138, 140
   in isometric, 389
Fillister and round head cap screws,
      A-34
Fillister heads, 207
Film
   drafting, 59
   Mylar, 59
Filters, by computer, 564
Fine threads, 191
Finish marks, 138, 265
Finished surfaces, 265
First angle projection, 69, 142, 252
Fit, 280
Flame hardening, 240
Flanged fittings, 489
   pipe, A-18, A-19–A-20
Flash welding, 313
Flasks, 240
Flat head cap screws, A-35
Flat heads, 207
Flatness, 300
Flaws, 304
Flow welding, 312
FN. *See* Force fits
Folding a drawing, 365
Fonts for text, 520
Force fits, 287, A-57
Forge welding, 312
Forging, 241
   dies, 242
   drawing, 242
Format file, 497
Four-center ellipse, 380
   computer, 545
FREEZE layers, 495
French curves, 63
Frontal
   lines, 408
   plane, 411
Full indicator movement, 302

Full sections, 170, 173
Function keys, 497
Fundamental deviation, 282

G

Gas welding, 312
Gauging holes, 298
Gear ratios, 222
Gears, 220
  bevel, 220
  spur, 220
  worm, 220, 227
General obliques, 371
Geometric
  construction, 77
  solids, 79
  tolerances, 291
  tolerancing, rules of, 293
  tolerancing, symbols of, 292
Gib head key, A-43
Glass, 244
Gothic lettering, 49
Grading graph, A-71
Gramercy Guild, 70
Graphs, 472
  break-even, 480
  broken-line, 476
  circle, 474
  composite, 480
  logarithmic coordinate, 481
  optimization, 480
  percentage, 484
  pie, 473
  polar, 484
  semilogarthmic, 482
  smooth-line, 478
  straight line, 478
  two-scale, 479
  types of, 472
Gravity motion, 231
GRID command, 395, 496
Grid rotation by computer, 581
Grinding, 248, 290
Groove welds, 316
Ground line, 391

H

Half dog point, 207
Half sections, 176
Halftones, 365
Hardness, 239
Harmonic motion, 231–232
HATCH command, 517–518
Heat treatment, 239

Hectometer, 68
Helix, 96, 98
  conical, 98
HELP command, 500, 532
Hewlett-Packard, 46
  plotter, 576
Hexagon-socket head, 206
Hexagon, 78, 81
  bolt and nut tables, A-33
Hexahedron, 79
HIDE command, 396, 547, 557, 562
Highway plats, 13
Hole basis, 282
Hole basis fits, A-59–A-60, A-61–A-62
Holes, cylindrical, 260
HOME option, 542
HOMETEXT command, 541
Honing, 248, 290
Horizontal lines, 407
Horizontal plane, 411
Hunting seat
  analysis of, 22
  decision, 25–26
  implementation of, 27
  problem identification of, 18
Hyperbola, 93, 96–97

I

Icosahedron, 79
ID command, 532
Ideograph, welding, 315
Imperial system of unit. *See* English system of measurement
Implementation, 18
  Hunting seat, 27
India ink pens, 73
Induction welding, 312
Ingots, 239
Ink drawing, 72
Ink pens, 53
Inking compass, 73
INQUIRY commands, 532
Intersecting lines, 409
INSERT command, 139, 330
Interference fit, 280, 282
Interference location fits, A-56
International Standards Organization, 67
International tolerance grade, 282, 290, A-58
Intersections, 427
  cones and prisms, 435
  cylinders, 137, 434
  cylinders and prisms, 433
  line and plane, 428
  of lines, 129
  plane and cylinder, 430
  planes, 418, 428

planes and cones, 435
prisms, 430
pyramids and prisms, 437
Investment casting, 241
Involutes, 98, 222
Iron
    ductile, 236
    gray, 236
    malleable, 237
    white, 236
Irregular curves, 63
ISO. *See* Metric thread specifications
ISOCIRCLE mode, 395
ISOMETRIC command, 395
Isometric
    angles in, 378
    assemblies, 389
    circles, 115
    circles in, 379
    curves in, 384
    cylinders in, 382
    dimensioned, 388
    drawings, 377
    ellipse templates, 383
    grid by computer, 545
    lines, 377
    measuring angles, 383
    nonisometric planes, 386
    nuts and bolts, 387
    pictorials, 370, 376, 544
    projection, 377
    sketching, 115
    thread representation, 386
ISOPLANE command, 396
Isosceles triangles, 77
IT grades. *See* International tolerance grades

J

Joy stick, 46
JUSTIFY option, 519

K

Key, Pratt & Whitney, 212
Key, Woodruff, 212
Keyboard, computer, 45
Keys, 212
    gib-head, 212
    square, 212
    tables, A-43, A-44
Keyseats, 271
Kilometer, 68
Knurling, 271–272
Koh-I-Noor Rapidograph, Inc., 53, 73

L

LANDCADD Inc., 43
Lap joint, 314
Lapping, 248, 290
LAST option, 504
Lathe, 244
Lay, 304
LAYER command, 494
Layers
    color of, 495
    drawing of, 494
    line types, 495
    On and OFF, 495
LC fits. *See* Locational fits
Lead, 190
Leader, 262
LEADER option, 538
Least material condition, 292
Lettering, 48
    by computer, 54, 519
    guideline templates, 49
    guidelines, 49
    mechanical, 53
    single-stroke Gothic, 49
    tools of, 48,
Letters and numerals, spacing of, 52
Letters
    inclined, 51
    vertical, 50
Light pen, 45
LIMITS command, 496
LIMITS, in PSPACE, 573
LINE command, 502
Line fit, 280
Linear coordinate graphs, 475
Lines, breaking of, 506
Lines, erasing of, 505
Lines and planes, 131
LINETYPE command, 125, 495, 501
LIST command, 532
LMC. *See* Least material condition
Location dimensions, 266
Location of holes, 266
Locational fits, 287, A-53–A-54, A-55, A-56
Lock washers, 210, 274
Logarithms of numbers, A-5–A-6
LTSCALE command, 125, 497, 501
Lugs, 183
    in section, 184

M

Machine screw, 202
    tables, A-36, A-37
Machining operations, 244

Magnesium, 239
Main menu, 492, 502
Mainframes, 44
Malleability, 239
Map chart, 485
MARK option, 506
Market considerations, 19
Materials
  and processes, 236
  properties of, 239
Mating parts, 279
Maximum material condition, 292, 298
MEASURE command, 528
Measuring points, 392
Mechanical engineers' scale, 66
MegaCADD, 47
  examples of, 400
MegaCADD Inc., 43, 399
Megameter, 68
Menu bar, 517
Menus, 498
Meshes, drawing with, 575
Metallurgy, 236
Meter, 68
Metric
  conversion table, A-13
  conversions, 68, 251
  measurements, 250
  prefixes, 68
  scales, 68, 331
  symbols, 69
  system of units, 67
  thread notes, 193–194
  thread specifications, 193, 195
  threads, 193
  tolerancing, 281
  tolerancing, examples of, 285
Microcomputers, 44
Microfilming, 363–364
Micrometer, 68
Midpoint of a line, 82
Millimeter, 68, 250
Milling, 290
Milling machine, 247
Minicomputers, 44
MIRROR command, 522
MIRRTEXT variable, 522
MMC. *See* Maximum material condition
Model space, 568
Modify layer, 531
Monge, Gaspard, 2
Mouse, 46
MOVE command, 257, 521
MSLIDE command, 582
MSPACE command, 569–570
Multiple threads, 191

Multiview
  drawing with instruments, 122
  projection, 106
  sketching, 59
MVIEW command, 569
Mylar film, 59

N

Napolean, 2
Necks, 272, 274
NEWTEXT, 541
NODE, 552
Nominal size, 280
Non-preferred diameters, tables, A-67–A-68,
  A-69–A-70
NOORIGIN, 551
Normalizing, 239
North arrow, 456
Numerals, 51
  inclined, 52
Nuts
  and bolts, 206
  drawing of, 205
Nylon, 244

O

Oakes, Larry, 24
OBLIQUES command, 541
Obliques
  angles in, 372
  circles in, 374
  curves in, 375
  cylinders in, 373
  dimensioned, 376
  drawing, 371–372
  lines, 415
  pictorials, 370, 544
  projections, 370
  sketching, 376
Octagons, 78
  construction of, 81
Octahedron, 79
OFFSET command, 528
Offset sections, 178
Ogee curves, 92
One-view drawings, 134
Optimization graphs, 480
OPTIONS heading, 498
Ordinate, 476
ORDINATE command, 536
ORTHO command, 497
Orthographic projection, 106, 122, 135
  lines, 407

planes, 410
point numbering, 129
points, 406
principles of, 123
Orthographic views, selection of, 127
OSNAP command, 157, 262, 522
OSNAP mode, 497–498
Outcrop, 466
Output devices, 46
Oval
heads, 207
points, 207
Overlay drafting, 366
OVERRIDE command, 542

P

Pan command, 512
Paper and film sizes, 58
Parabola, 93, 96
PARALLEL command, 157, 405, 412
Parallel lines, construction of, 84
Parallelepiped, 79
Parallelism, 301
Part balloons, 331
Parts not sectioned, 174
Paste-on photos, 366
Patterns, 438
PDMODE variable, 507
PDSIZE option, 507
PEDIT command, 515, 559, 562
Pencils
drafting, 48
grades, for sketching, 108
lead grades of, 57
pointing, 58
sharpening of, 58
types of, 57
Pentagons, 78, 81
Pentagonal prism, 79
Percent grade, 414
Percentage graphs, 484
Perpendicularity, 302
PERPLINE command, 405
Perspective pictorials, 370
Perspective projection, 370
Perspectives, 390
by computer, 398, 554
one-point, 390
three-point, 390
two-point, 393
two-points, 390
PFACE command, 561
Photo drafting, 368
Photographic slides, 473
Photostating, 363, 365

Pictorials, 370
Pie graphs, 473
Piercing points, 417
Pin drafting, 366
Pinion, 220, 224
Pins, 210
method of drawing, 137
Pipe drawings, screwed, A-22
Pipe, threads, 211
Pitch, 190
diameter, 189, 220
Pixels, 45
PLAN, 397, 547
Plan profiles, 461
Planer, 247
Planing and shaping, 290
Plastics, 244
Plate cams, 230
construction of, 231
PLINE command, 398, 514
PLOT command, 493
Plot parameters, 576
Plot plans, 455
Plot scaling, 578
Plotter, 46
flatbed, 46
roll-fed, 46
Plotting
with AutoCAD, 492
a drawing by computer, 575
options, 576
Plug tap, 210
POINT command, 507
Point numbering, orthographic projection, 129
Point on a line, 409
Point view of a line, 412, 420
Polar arrays, 524
Polar coordinates, 503
Polar graphs, 484
Polishing, 248
Polyester film, 59
Polyethlene, 244
Polygon, 78
by computer, 81
construction of, 80
regular, 80
POLYGON command, 477, 511
Polyhedra, 79
POLYLINE command, 514
Polypropylene, 244
Polyvinal chloride, 244
Position tolerancing, 297
Powder metal-sintered, 290
Preferred fits
and non-preferred sizes, 287
and sizes, 283

Preliminary ideas, 16
  Hunting seat, 20
Pressure angle, 222
PREVIOUS option, 504
Pricing analysis, 19
Principal lines, 407
Principal projection planes, 124
Principal views, 124
Printer
  dot-matrix, 47
  impact, 47
  ink-jet, 47
  laser, 47
Prisms, 79
  truncated, 79
Problem
  identification, 15
  layouts, 74
Product design problems, 35
Professional societies, 14
Profile, 300, 458–460
  lines, 408, 411
Protractor, 62
PSPACE command, 569, 570, 573
Pull-down menus, 498
Punching, 290
PURGE command, 501
Pyramids, 79
  by computer, 556

Q

QTEXT command, 519
Quadrilaterals, 78
Quenching, 239
QUIT command, 494

R

Rapidograph lettering pens, 53
Raster-scanned screen, 45
Ratio graphs, 482
RC. *See* Running and sliding fits
Reaming, 245, 271
Rectifying arcs, 92
Reduction of drawings, 83
Refinement, 16
  Hunting seat, 21
Regardless of feature size, 292, 295
Relative coordinates, 503
Relative polar coordinates, 503
REMOVE option, 505
Removed sections, 179
RENAME command, 500
Reproduction methods, 363
Resistance welding, 312

RESTORE command, 542
RESUME command, 583
Revolution of figures, 83
Revolved sections, 178
REVSURF command, 558, 560
RFS. *See* Regardless of feature size
Rhomboid, 78
Rhombus, 78
Ribs, 176
Rivets, 212
  symbols for, 213
Robotic manufacturing, 44
Rolling, 242
Root menu, 498
ROTATE command, 155, 526
Rotating grid on screen, 545
Roughness, 303
  height, 303
  width, 303
  width cutoff, 304
Round head, 206
Roundness, 300
Rounds, 263
  and fillets, 138
RSCRIPT command, 582
RULESURF command, 558–559, 575
Ruling pens, 70, 72
Running and sliding fits, 287
  tables, A-51–A-52
Runouts, 140, 302
  circular, 302
  total, 302

S

SAE. *See* Society of Automotive Engineers
SAVE command, 493
  dimensioning style, 542
SCALE command, 524
Scale
  conversion, 69
  indication, 456
  specification, 331
  use of, 63
Scalene triangles, 77
Schematic threads, 199
Schematics, 485
Scientist, 2
Screen parameters, 496
Screen resolution, 45
Screw heads, 203
Screws, wood, 209
SCRIPT command, 583
Scroll bar, 493
SDF format, 580
Seam welds, 318
Secondary auxiliary views, 161

Section, ribs in, 175
Sectioning symbols, 171
  by computer, 172–173
Sections, 170
  auxiliary, 184
  broken-out, 181
  half, 176
  in assembly, 173
  offset, 178
  partial views, 176
  parts not section lined, 174
  removed, 179
  revolved, 178
Sector, 78
Selective assembly, 281
Set screws, 202, 207
  slotted headless table, A-40
  square head, A-38, A-39
SETVAR command, 507, 582, 584
Shaft basis, 282
  fits, A-63–A-64, A-65–A-66
Shape description, 106
Shaper, 247
Sheet sizes, 329
SHELL command, 501
Sherman, Keith, 24
Shrink fits, A-57
SI. *See* Systeme International d'Unites
SI symbol, 69, 142, 252
Silicone, 244
Single limits, 281
Six-view drawings, 107
  instruments, 125
SKETCH command, 582
Sketching, 106
  circular features, 112
  isometrics, 115
  techniques, 107
Skewed lines, 424
SKPOLY variable, 582
Slide shows, 582
Slope, 256
  angle, 414
  of a line, 414
  of a plane, 419
  ratio, 414
SLOT command, 563
SNAP command, 155, 395, 496, 581
SNAP option of STYLE, 545
Society of Automotive Engineers,
  236
SOLBOX command, 593
SOLCHAM command, 597
SOLCHP command, 598
SOLCONE command, 594
SOLCYL command, 594
Soldering, 319

SOLFILL command, 598
SOLHPAT command, 598
Solid inquiry commands, 599
Solid modeling, 593
SOLIDIFY command, 596
SOLMASSP command, 599
SOLMESH command, 599
SOLMOVE command, 598
SOLOLIST command, 599
SOLREV command, 596
SOLSEP command, 596
SOLSPHERE command, 594
SOLSUB command, 596
SOLTEXT command, 595
SOLTORUS command, 595
SOLUNION command, 596
SOLWEDGE command, 595
SOLWENS command, 599
SOLWIRE command, 599
Specifications, 322
Spheres, 79–80
  by computer, 557
Spherical coordinates, 503
Spherical radius, 256
Spider gear, 228
Spirals, 96
SPLFRAME variable, 563
SPLINE option, 515
Spline, flexible, 64
Spokes in section, 184
Spot weld, 318
Spotfacing, 256, 270
Springs, 214
  compression, 214
  drawing, 215
  extension, 214
  in schematic, 215
  torsion, 214
Sprue, 240
Spur gears, 223
  drawing, 224
Square bolt and nuts tables, A-32
Square bolts, 203
Square threads, 190
  tables, A-31
Stamping, 243
Station numbers, 458
Station point, 391
STATUS command, 532
Status line, 497
Steel, 237
  alloys, 238
  applications, 238
Stick-on drawings, 367
Straight pins, A-42
Straightness, 300
STRETCH command, 258, 525

Strike and dip, 464
Stud bolt, 202
STYLE command, 55, 257, 520
Stylist, 4
Successive auxiliary views, 420
Surface
  control symbols, 305, 307
  finishing, 248
  grinding, 290
  roughness heights, 306
  texture, 303
Surveying angle, by computer, 457
Symmetry, 299
Systeme International d'Unites, 67
System variable, 507

T

TABLET command, 581
Tablet digitizing, 581
TABSURF command, 558–559, 575
Tangency
  arc and a line, 89
  from a point to a line, 85
  indication of, 85
  line to an arc, 85
  point to an arc, 86
  points, 85
  to two lines, 86
  two arcs, 87
Tangent options, 508
Tap drill sizes, A-25–A-26
Taper, 256
Taper pins, A-46
Taper pipe threads, A-17
Taper tap, 210
Tapers, 273
  conical, 288
  table of values, A-37
Tapping holes, 210
TARGET option, 554
Technical illustration, 13, 387
Technician, 3, 12
Technological team, 2
Technologist, 3, 12
TEDIT option, 542
Tee joint, 314
Tempering, 239
Templates
  circle, 71
  drafting, 74
Terminals, 44
Tertiary datum plane, 294
Tetrahedron, 79
TEXT command, 55, 279, 398, 518
TEXTSIZE, 584
THAW layers, 495

Thermit welding, 312
THICKNESS command, 547, 562
Third angle projection, 69, 142, 252
Three-datum plane concept, 293
3D drawing, AutoCAD, 565
3D extrusions, 547
3DFACE command, 549, 562–563
3D forms, 556
3DLINE command, 561
3DMESH, 558
3DPOLY command, 562
Three-view drawing
  by computer, 133
  instruments, 126
  layout of, 130
Three-view sketch, 108
Thread
  allowance, 195
  class, 190
  detailed representation of, 196
  drawing of, 197
  drawing small, 201
  length of engagement, 195
  metric sizes, 194–195
  notes, 192
  pipe, 211
  schematic, 199
  series, 190–191
  simplified representation, 200
  single, 191
  specifications, English, 190
  square, 190, 197
  terminology, 189
Thread tables
  English, A-23–A-24
  how to use, 192
  length of, A-27
  metric, A-28–A-30
Threaded fasteners, 189
Threading, 246
Threading holes, 210
Thumb screws, 209
TILEMODE command, 569
Tiled viewports, 569
TIME command, 532
Title blocks, 330
  by computer, 330
Title strips, problems, 75
Tolerance, 277, 280, 282, 331
  angular, 291
  bilateral, 277
  by computer, 544
  form, 300
  general, 290
  limit form, 277
  limits of, 280
  location, 297

notes, 289, 331
orientation, 301
plus and minus, 278
profile, 300
runout, 302
symbols, 283
unilateral, 277
zone, 282
Tolerancing
metric, 281
symbols, 292
terminology of, 280
Toriods, by computer, 557
Toroses, by computer, 557
Toughness, 239
TRACE command, 507
Tracing
cloth, 58
paper, 58
TRANSFER command, 157, 405–406
Transition fit, 280, 282
Transition locational fits, A-55
Transition pieces, 445
Transparent commands, 531
Trapezium, 78
Trapezoid, 78
Triangles, 93
construction of, 80
drafting, 61
types of, 77
Trigonometric tables, A-7–A-11
TRIM command, 477, 526
Trimetric projection, 390
True-length
diagram, 413
lines by analytical geometry, 413
view of a line, 411
True-position tolerancing, 298
True-size view of a plane, 422
T-square, 59
TTR options, 508
Turning, 244, 290
T&W Systems, 60
TWIST option, 554
Twist drill tables, A-41–A-42
Two-view drawings, 133
TYP note, 263

U

UCS. *See* User Coordinate System
UCS control box, 551
UCSFOLLOW variable, 566
UCSICON, 549–550, 552
UNC threads. *See* Unified Nation Coarse threads
Undercuts, 272
Undercutting, 246

UNDO command, 506
Unidirectional dimensions, 252
Unified National Coarse threads, 191
Unified National Rolled threads, 190–191
Uniform acceleration, 233
UNITS command, 256, 496
Units of measurement, 250
UNR threads. *See* Unified Nation Rolled threads
UPDATE command, 541, 543
dimensions, 543
User Coordinate system, 550
Utility commands, 500

V

Vector-refreshed tubes, 45
Vellum, drafting, 58
VersaCAD, 47
VIEW command, 531
Viewpoint globe, 547
Visibility
of lines, 409
line and a plane, 409
VMAX option, 512
VPLAYER command, 574
VPOINT command, 396–397, 547–548
VPORTS command, 550
VSLIDE command, 583

W

Washer tables, A-49
Washer, helical-spring lock, 210
Washers, 210, 274
lock, 210, A-49
plain, 210, A-47–A-48
Waviness, 304
height, 304
WBLOCK command, 529, 545
WCS. *See* World Coordinate System
Webs, 176
Wedge, by computer, 556
Weights and measures, A-12
Weights and specific gravities, A-14–A-15
Weld joints, 314
Welding, 312
process symbols, 318
standards, 319
symbols, 314
Welds
built-up, 319
groove, 316
seam, 318
surface contoured, 317
types of, 315
Whiteprinter, 364

Whitworth threads, 190
Window option, 505
Wing screws, 208
Wire and sheet metal gages, A-16
Woodruff Key, 271
  table, A-44
Woodruff keyseats, A-45
Working drawings, 322
  by computer, 329
  castings, 336
  dual dimensions, 327
  forged parts, 336
  freehand, 334
  layout of, 328
  metric, 325
  sheet metal, 337

World Coordinate System, 549–550
World coordinates, 503
Worm gears, 227
  calculations, 229
  drawing, 229

X

Xerography, 363, 365
Xerox conference copier, 16
XREF command, 530
XYZ filters, 564

Z

ZOOM command, 511

The following supplements are compatible with the following textbooks from Addison-Wesley Publishing Co.:

Drafting Technology,

Engineering Design Graphics,

Design Drafting,

Graphics for Engineers,

Geometry for Engineers

DRAFTING TECHNOLOGY PROBLEMS is a problem book designed to cover basic graphics, descriptive geometry, and specialty drafting areas. It is designed to accompany *Drafting Technology,* that is available from Addison-Wesley Publ. Co., Reading, Mass. 01867. 131 pages.

BASIC DRAFTING (With Computer Graphics) is a problem book that covers the basics for a one-semester high school drafting course. 67 pages.

CREATIVE DRAFTING (With Computer Graphics) is a problem book that covers mechanical drawing and architectural drafting for the high school. 106 pages.

DRAFTING & DESIGN (With Computer Graphics) is a problem book for mechanical drawing for a high school or college course. 106 pages.

DRAFTING FUNDAMENTALS 1 is a problem book for a one-year high school course in mechanical drawing. 94 pages.

DRAFTING FUNDAMENTALS 2 is a second version of **DRAFTING FUNDAMENTALS 1** for the same level and content. 94 pages.

TECHNICAL ILLUSTRATION is a problem book for a course in pictorial drawing for the college or high school. 70 pages.

ARCHITECTURAL DRAFTING is a problem book for a first course in architectural drafting. 71 pages.

GRAPHICS FOR ENGINEERS 1 (With Computer Graphics) is a problem book for a first course in engineering graphics for the college student. 100 pages.

GRAPHICS FOR ENGINEERS 2 (With Computer Graphics) is a problem book for a first course in engineering graphics for the college student. 100 pages.

GRAPHICS FOR ENGINEERS 3 (With Computer Graphics) is a problem book for a first course in engineering graphics for the college student. 108 pages.

GEOMETRY FOR ENGINEERS 1 is a problem book for a college-level descriptive geometry course. 100 pages.

GEOMETRY FOR ENGINEERS 2 is a problem book for a college-level descriptive geometry course. 100 pages.

GEOMETRY FOR ENGINEERS 3 is a problem book for a college-level descriptive geometry course. 100 pages.

GRAPHICS & GEOMETRY 1 (With Computer Graphics) is a problem book for a college course in graphics and descriptive geometry. 119 pages.

GRAPHICS & GEOMETRY 2 (With Computer Graphics) is a problem book for a college course in graphics and descriptive geometry. 121 pages.

GRAPHICS & GEOMETRY 3 (With Computer Graphics) is a problem book for a college course in graphics and descriptive geometry. 138 pages.

DESIGN GRAPHICS 1 (With Computer Graphics) is a problem book for a college course in graphics and descriptive geometry. 140 pages.

Creative Publishing Co.
BOX 9292   COLLEGE STATION, TEXAS   77840
PHONE 409-775-6047